Cinchona Alkaloids in Synthesis and Catalysis

Edited by
Choong Eui Song

Further Reading

Crabtree, R. H. (ed.)

Handbook of Green Chemistry - Green Catalysis

2009
Hardcover
ISBN: 978-3-527-31577-2

Kollár, L (ed.)

Modern Carbonylation Methods

2008
Hardcover
ISBN: 978-3-527-31896-4

Börner, A. (ed.)

Phosphorus Ligands in Asymmetric Catalysis

Synthesis and Applications

2008
Hardcover
ISBN: 978-3-527-31746-2

Yamamoto, H., Ishihara, K. (eds.)

Acid Catalysis in Modern Organic Synthesis

2008
Hardcover
ISBN: 978-3-527-31724-0

Maruoka, K. (ed.)

Asymmetric Phase Transfer Catalysis

2008
Hardcover
ISBN: 978-3-527-31842-1

Stepnicka, P. (ed.)

Ferrocenes

Ligands, Materials and Biomolecules

Hardcover
ISBN: 978-0-470-03585-6

Christmann, M., Bräse, S. (eds.)

Asymmetric Synthesis - The Essentials

2008
Softcover
ISBN: 978-3-527-32093-6

Hudlicky, T., Reed, J. W.

The Way of Synthesis

Evolution of Design and Methods for Natural Products

2007
Hardcover
ISBN: 978-3-527-32077-6

Hiersemann, M., Nubbemeyer, U. (eds.)

The Claisen Rearrangement

Methods and Applications

2007
Hardcover
ISBN: 978-3-527-30825-5

Cornils, B., Herrmann, W. A., Muhler, M., Wong, C.-H. (eds.)

Catalysis from A to Z

A Concise Encyclopedia

2007
Hardcover
ISBN: 978-3-527-31438-6

Cinchona Alkaloids in Synthesis and Catalysis

Ligands, Immobilization and Organocatalysis

Edited by
Choong Eui Song

WILEY-VCH

WILEY-VCH Verlag GmbH & Co. KGaA

The Editor

Prof. Choong Eui Song
Sungkyunkwan University
Dept. of Chemistry
Cheoncheon-dong 300, Jangan-gu
Suwon 440-746
Republik Korea

All books published by **Wiley-VCH** are carefully produced. Nevertheless, authors, editors, and publisher do not warrant the information contained in these books, including this book, to be free of errors. Readers are advised to keep in mind that statements, data, illustrations, procedural details or other items may inadvertently be inaccurate.

Library of Congress Card No.: applied for

British Library Cataloguing-in-Publication Data
A catalogue record for this book is available from the British Library.

Bibliographic information published by the Deutsche Nationalbibliothek
The Deutsche Nationalbibliothek lists this publication in the Deutsche Nationalbibliografie; detailed bibliographic data are available on the Internet at http://dnb.d-nb.de.

© 2009 WILEY-VCH Verlag GmbH & Co. KGaA, Weinheim

All rights reserved (including those of translation into other languages). No part of this book may be reproduced in any form – by photoprinting, microfilm, or any other means – nor transmitted or translated into a machine language without written permission from the publishers. Registered names, trademarks, etc. used in this book, even when not specifically marked as such, are not to be considered unprotected by law.

Cover Grafik-Design Schulz, Fußgönheim
Typesetting Thomson Digital, Noida, India
Printing betz-druck GmbH, Darmstadt
Binding Litges & Dopf GmbH, Heppenheim

Printed in the Federal Republic of Germany
Printed on acid-free paper

ISBN: 978-3-527-32416-3

To my wife Teresa

Contents

Preface *XV*
Biography *XVII*
List of Contributors *XIX*

1 **An Overview of Cinchona Alkaloids in Chemistry** *1*
Choong Eui Song
1.1 Brief History *1*
1.2 Active Sites in Cinchona Alkaloids and Their Derivatives *3*
1.3 Structural Information on Cinchona Alkaloids *4*
1.4 How This Book Is Organized *8*
References *9*

Part One **Cinchona Alkaloid Derivatives as Chirality Inducers in Metal-Catalyzed Reactions** *11*

2 **Cinchona Alkaloids as Chirality Transmitters in Metal-Catalyzed Asymmetric Reductions** *13*
Hans-Ulrich Blaser
2.1 Introduction *13*
2.2 Homogeneous Systems for Ketone Reductions *14*
2.3 Heterogeneous Pt and Pd Catalysts Modified with Cinchona Alkaloids *15*
2.3.1 Background *15*
2.3.2 Catalysts *16*
2.3.3 Modifiers and Solvents *16*
2.3.4 Substrate Scope for Pt Catalysts *17*
2.3.4.1 α-Keto Acid Derivatives *17*
2.3.4.2 α,γ-Diketo Esters *18*
2.3.4.3 Fluorinated Ketones *18*
2.3.4.4 α-Keto Acetals *20*
2.3.4.5 α-Keto Ethers *20*
2.3.4.6 Miscellaneous Ketones *21*

Cinchona Alkaloids in Synthesis and Catalysis, Ligands, Immobilization and Organocatalysis
Edited by Choong Eui Song
Copyright © 2009 WILEY-VCH Verlag GmbH & Co. KGaA, Weinheim
ISBN: 978-3-527-32416-3

2.3.5	Substrate Scope for Pd Catalysts 21
2.4	Industrial Applications 22
2.5	Conclusions 25
	References 26

3	**Cinchona Alkaloids as Chiral Ligands in Asymmetric Oxidations** 29
	David J. Ager
3.1	Introduction 29
3.2	Asymmetric Dihydroxylation of Alkenes 30
3.2.1	Early Reactions 30
3.2.2	Bisalkaloid Ligands 33
3.2.3	Mechanism 35
3.2.4	Variations 36
3.2.5	Substrates and Selectivity 38
3.2.5.1	Simple Alkenes 38
3.2.5.2	Functionalized Alkenes 38
3.2.5.3	Polyenes 43
3.2.5.3.1	Nonconjugate Olefins 43
3.2.5.3.2	Conjugated Polyenes 43
3.2.5.4	Double Asymmetric Induction 44
3.2.5.5	Resolutions 50
3.2.6	Some Reactions of 1,2-Diols 51
3.2.6.1	Cyclic Sulfates and Sulfites 54
3.3	Aminohydroxylation 56
3.4	Sulfur Oxidations 61
3.5	Summary 61
	References 62

4	**Cinchona Alkaloids and their Derivatives as Chirality Inducers in Metal-Promoted Enantioselective Carbon–Carbon and Carbon–Heteroatom Bond Forming Reactions** 73
	Ravindra R. Deshmukh, Do Hyun Ryu, and Choong Eui Song
4.1	Introduction 73
4.2	Nucleophilic Addition to Carbonyl or Imine Compounds 74
4.2.1	Organozinc Addition 74
4.2.1.1	Dialkylzinc Addition to Aldehydes 74
4.2.1.2	Dialkylzinc Addition to Imines 75
4.2.1.3	Addition of Alkynylzincs to Carbonyls 77
4.2.2	Asymmetric Reformatsky Reaction 78
4.2.3	Indium-Mediated Addition 79
4.2.4	Asymmetric Cyanation 81
4.2.4.1	Cyanohydrin Synthesis 81
4.2.4.2	Strecker Synthesis 84
4.2.5	Reactions of Chiral Ammonium Ketene Enolates as Nucleophiles with Different Electrophiles 86

4.2.5.1	Lewis Acid Assisted Nucleophilic Addition of Ketenes (or Sulfenes) to Aldehydes: β-Lactone and β-Sultone Synthesis *86*	
4.2.5.2	Lewis Acid Assisted Nucleophilic Addition of Ketenes to Imines: β-Lactam Synthesis *90*	
4.2.5.3	Applications of Chiral Ketene Enolates to Formal [4 + 2] type Cyclization *92*	
4.2.6	Aza-Henry Reaction *92*	
4.2.7	Enantioselective Hydrophosphonylation *93*	
4.3	Miscellaneous Reactions *94*	
4.3.1	Claisen Rearrangements *94*	
4.3.2	Pd-Catalyzed Asymmetric Allylic Substitutions *95*	
4.3.3	Pauson–Khand Reaction *97*	
4.3.4	Asymmetric Dimerization of Butadiene *98*	
4.3.5	Enantiotopic Differentiation Reaction of Mesocyclic Anhydrides *98*	
4.4	Cinchona-Based Chiral Ligands in C–F Bond Forming Reactions *99*	
4.5	Conclusions *100*	
	References *101*	

Part Two Cinchona Alkaloid Derivatives as Chiral Organocatalysts *105*

5	**Cinchona-Based Organocatalysts for Asymmetric Oxidations and Reductions** *107*	
	Ueon Sang Shin, Je Eun Lee, Jung Woon Yang, and Choong Eui Song	
5.1	Introduction *107*	
5.2	Cinchona-Based Organocatalysts in Asymmetric Oxidations *108*	
5.2.1	Epoxidation of Enones and α,β-Unsaturated Sulfones Using Cinchona-Based Chiral Phase-Transfer Catalysts *108*	
5.2.1.1	Epoxidation of Acyclic Enones *108*	
5.2.1.2	Epoxidation of Cyclic Enones *113*	
5.2.1.3	Synthetic Applications of the Asymmetric Epoxidation of Enones Using Chiral PTCs *115*	
5.2.1.4	Epoxidation of α,β-Unsaturated Sulfones *117*	
5.2.2	Organocatalytic Asymmetric Epoxidation of Enones via Iminium Catalysis *118*	
5.2.3	Aziridination of Enones Using Cinchona-Based Chiral Phase-Transfer Catalyst *120*	
5.3	Cinchona-Based Organocatalysts in Asymmetric Reductions *125*	
5.4	Conclusions *127*	
	References *128*	
6	**Cinchona-Catalyzed Nucleophilic α-Substitution of Carbonyl Derivatives** *131*	
	Hyeung-geun Park and Byeong-Seon Jeong	
6.1	Introduction *131*	

6.2	Organocatalytic Nucleophilic α-Substitution of Carbonyl Derivatives *131*
6.3	Cinchona Alkaloids in Asymmetric Organocatalysis *133*
6.4	The Pioneer Works for Phase-Transfer Catalytic α-Substitution *134*
6.5	α-Substitution of α-Amino Acid Derivatives via PTC *135*
6.5.1	Monoalkylation of Benzophenone Imines of Glycine Esters *135*
6.5.2	Alkylation of α-Monosubstituted α-Amino Acid Derivatives *148*
6.6	α-Substitution of Other Carbonyl Derivatives via PTC *150*
6.6.1	α-Substitution of Monocarbonyl Compounds *150*
6.6.2	α-Substitution of β-Keto Carbonyl Compounds *153*
6.7	α-Heteroatom Substitution via PTC *156*
6.7.1	α-Hydroxylation of Carbonyl Derivatives *156*
6.7.2	α-Fluorination of Carbonyl Derivatives *157*
6.8	Nucleophilic α-Substitution of Carbonyl Derivatives via Non-PTC *157*
6.8.1	α-Arylation of Carbonyl Derivatives *158*
6.8.2	α-Hydroxylation of Carbonyl Derivatives *159*
6.8.3	α-Halogenation of Carbonyl Derivatives *160*
6.8.4	α-Amination of Carbonyl Derivatives *162*
6.8.5	α-Sulfenylation of Carbonyl Derivatives *165*
6.9	Conclusions *165*
	References *166*
7	**Cinchona-Mediated Enantioselective Protonations** *171*
	Jacques Rouden
7.1	Introduction *171*
7.2	Preformed Enolates and Equivalents *172*
7.3	Nucleophilic Addition on Ketenes *175*
7.4	Michael Additions *178*
7.5	Enantioselective Decarboxylative Protonation *184*
7.5.1	Copper-Catalyzed EDP *184*
7.5.2	Palladium-Catalyzed EDP *185*
7.5.3	Organocatalyzed EDP *188*
7.6	Proton Migration *192*
7.7	Summary of Cinchona-Mediated Enantioselective Protonations *194*
	References *194*
8	**Cinchona-Catalyzed Nucleophilic 1,2-Addition to C=O and C=N Bonds** *197*
	Hyeong Bin Jang, Ji Woong Lee, and Choong Eui Song
8.1	Introduction *197*
8.2	Aldol and Nitroaldol (Henry) Reactions *198*
8.2.1	Aldol Reactions *198*
8.2.1.1	Mukaiyama-Type Aldol Reactions *198*
8.2.1.2	Direct Aldol Reactions *200*

8.2.2	Henry Reactions	206
8.3	Mannich and Nitro-Mannich Reactions	209
8.3.1	Mannich Reactions	209
8.3.2	Nitro-Mannich (Aza-Henry) Reactions	215
8.4	Aldol- and Mannich-Related Reactions	218
8.4.1	Darzens Reactions	218
8.4.2	Morita–Baylis–Hillman Reactions and Aza-Morita–Baylis–Hillman Reactions	221
8.4.2.1	Morita–Baylis–Hillman Reactions	221
8.4.2.2	Aza-Morita–Baylis–Hillman Reactions	225
8.4.3	Nucleophilic Addition of Ammonium Ketene Enolate to C=O or C=N Bonds	228
8.5	Cyanation Reactions	229
8.5.1	Cyanohydrin Synthesis	229
8.5.2	Strecker Synthesis	232
8.6	Trifluoromethylation	234
8.7	Friedel–Crafts Type Alkylation	237
8.8	Hydrophosphonylation	240
8.9	Conclusions	244
	References	244
9	**Cinchona-Catalyzed Nucleophilic Conjugate Addition to Electron-Deficient C=C Double Bonds**	**249**
	Ji Woong Lee, Hyeong Bin Jang, and Choong Eui Song	
9.1	Introduction	249
9.2	Conjugate Reaction of α,β-Unsaturated Ketones, Amides, and Nitriles	249
9.2.1	Natural Cinchona Alkaloids as Catalysts	249
9.2.2	PTC-Catalyzed Enantioselective Michael Addition Reactions	252
9.2.3	Non-PTC-Catalyzed Enantioselective Michael Addition Reactions	261
9.3	Conjugate Addition of Nitroalkenes	274
9.4	Conjugate Addition of Vinyl Sulfones and Vinyl Phosphates	284
9.4.1	Vinyl Phosphate	286
9.5	Cyclopropanation and Other Related Reactions	288
9.5.1	Cyclopropanation	288
9.5.2	Epoxidation and Aziridination	292
9.6	Conclusions	293
	References	293
10	**Cinchona-Catalyzed Cycloaddition Reactions**	**297**
	Yan-Kai Liu and Ying-Chun Chen	
10.1	Introduction	297
10.2	Asymmetric Cycloadditions Catalyzed by Quinuclidine Tertiary Amine	297

10.3	Asymmetric Cycloadditions Catalyzed by Bifunctional Cinchona Alkaloids *308*
10.4	Asymmetric Cycloaddition Reactions Catalyzed by Cinchona-Based Primary Amines *312*
10.5	Asymmetric Cycloaddition Catalyzed by Cinchona-Based Phase-Transfer Catalysts *320*
10.6	Conclusion *323*
	References *323*
11	**Cinchona-Based Organocatalysts for Desymmetrization of *meso*-Compounds and (Dynamic) Kinetic Resolution of Racemic Compounds** *325*
	Ji Woong Lee, Hyeong Bin Jang, Je Eun Lee, and Choong Eui Song
11.1	Introduction *325*
11.2	Desymmetrization of *meso*-Compounds *326*
11.2.1	Desymmetrization of *meso*-Cyclic Anhydrides *326*
11.2.1.1	Applications *336*
11.2.2	Desymmetrization of *meso*-Diols *336*
11.2.3	Desymmetrization of *meso*-Endoperoxides *341*
11.2.4	Desymmetrization of *meso*-Phospholenes via Alkene Isomerization *344*
11.2.5	Desymmetrization of *meso*-Epoxy Phospholenes to Allyl Alcohols via Rearrangement *345*
11.2.6	Desymmetrization of Prochiral Ketones by Means of Horner–Wadsworth–Emmons Reaction *346*
11.3	(Dynamic) Kinetic Resolution of Racemic Compounds *346*
11.3.1	(Dynamic) Kinetic Resolution of Racemic Cyclic Anhydrides *346*
11.3.2	(Dynamic) Kinetic Resolution of Racemic *N*-Cyclic Anhydrides *348*
11.3.3	Dynamic Kinetic Resolution of Racemic Azlactones *350*
11.3.4	Catalytic Sulfinyl Transfer Reaction via Dynamic Kinetic Resolution of Sulfinyl Chlorides *351*
11.4	Conclusions *354*
	References *355*

Part Three Organic Chemistry of Cinchona Alkaloids *359*

12	**Organic Chemistry of *Cinchona* Alkaloids** *361*
	Hans Martin Rudolf Hoffmann and Jens Frackenpohl
12.1	Introduction *361*
12.2	Preparation of Quincorine and Quincoridine: Discovery of a Novel Cleavage Reaction of Cinchona Alkaloids *364*
12.3	Transformations of the Quinoline Moiety *366*
12.4	Basic Transformations of the Vinyl Side Chain *368*
12.4.1	Alkyne Cinchona Alkaloids, Their Derivatives, and Basic Transformations *368*

12.4.1.1	The Ethynyl Group is Anything but a Spectator Substituent	371
12.4.2	Fluorination of the Vinyl Side Chain	374
12.4.3	Oxidation and Oxidative Cleavage of the Vinyl Group	375
12.4.3.1	Oxidative Functionalization of Quincorine and Quincoridine	380
12.4.4	Degradation of the Vinyl Side Chain: Synthesis of Cinchona Alkaloid Ketones	381
12.5	Selected Novel Transformations of the Quinuclidine Moiety of Cinchona Alkaloids	382
12.5.1	Cage Helicity and Further Consequences: Nucleophilic Attack on Quinuclidin-3-ones	382
12.5.2	Transformations at Carbon C6 and Formation of Bridgehead Bicyclic Lactams	389
12.5.3	Functionalization of Other Quinuclidine Carbons, for Example, C5 and C7	393
12.6	Nucleophilic Substitution at Carbon C9	394
12.6.1	Unusual Steric Course of Solvolysis of C9-Activated Alkaloids: Access to C9-*epi*-Configured Stereoisomers	394
12.6.2	Replacing the C9-Hydroxy Group by Alternative Substituents	396
12.6.3	Transformations of QCI and QCD at Carbon C9: Access to a Novel Class of Small Molecule Ligands	399
12.7	Novel Rearrangements of the Azabicyclic Moiety	403
12.7.1	First Cinchona Rearrangement	403
12.7.2	Second Cinchona Rearrangement	407
12.8	General Experimental Hints and Obstacles	409
12.9	Conclusions	412
	References	415

Part Four Cinchona Alkaloid and Their Derivatives in Analytics *419*

13 Resolution of Racemates and Enantioselective Analytics by Cinchona Alkaloids and Their Derivatives *421*
Karol Kacprzak and Jacek Gawronski

13.1	Introduction	421
13.2	Resolution of Racemates by Crystallization and Extraction of Diastereoisomers	423
13.2.1	Resolution of Racemates by Crystallization	423
13.2.2	Resolution of Racemates by Enantioselective Extraction	430
13.3	Enantioselective Chromatography and Related Techniques	433
13.3.1	Early Attempts and Current Status	434
13.3.2	Cinchona 9-*O*-Carbamates as CSPs in HPLC	436
13.3.2.1	Applications	437
13.3.2.2	Mechanistic Studies on Chiral Discrimination	443
13.3.3	Other Cinchona-Based Selectors: Toward "Receptor-Like" CSPs	447
13.3.4	Cinchona-Based Chiral Modifiers and Phases in Capillary Electrophoresis and Capillary Electrochromatography	450

13.3.5	Other Chromatographic Techniques	*452*
13.4	Cinchona Alkaloids as Chiral Solvating (Shift) Agents in NMR Spectroscopy	*453*
13.5	Cinchona-Based Sensors, Receptors, and Materials for Separation and Analytics	*455*
	References	*464*

Appendix: Tabular Survey of Selected Cinchona-Promoted Asymmetric Reactions *471*
Ji Woong Lee and Choong Eui Song

Index *507*

Preface

Since the pioneering works of H. Wynberg in the late 1970s and early 1980s, cinchona alkaloids have been intensively applied as either standalone catalysts or chiral ligands in catalytic asymmetric reactions and are now regarded as one of the most privileged chirality inducers. Indeed, today, nearly all classes of organic reactions can be effectively carried out with the use of cinchona alkaloids in a highly stereoselective fashion. Some of them are even used in large-scale processes, for example, the heterogeneous hydrogenation of α-ketoesters catalyzed by cinchona alkaloid-modified platinum, the Sharpless asymmetric dihydroxylation of olefins, and the asymmetric alkylation of indanones using cinchona alkaloid-derived chiral phase-transfer catalysts, and so on. Of these reactions, the osmium-catalyzed asymmetric dihydroxylation of olefins using cinchona alkaloid derivatives as chiral ligands has had the greatest impact on modern asymmetric catalysis. In 2001, the Nobel Prize in Chemistry was awarded to Professor Sharpless "for his pioneering work on chirally catalyzed oxidation reactions".

In spite of the huge amount of attention that has been given to this research area and the immense success that has been obtained, surprisingly, no book on this topic has been published to date. So far, on this topic, only a few review articles on the use of cinchona alkaloids in asymmetric synthesis (Pracejus (up to 1967), Morrison and Mosher (up to 1970), Wynberg (up to 1986), Song (up to 1999), Gawronski (up to 2000), Deng (up to 2004), and Lectka (up to 2008), and so on) have appeared. However, these reviews are either dated or deal only with a specific reaction class. Thus, when I was invited to do so by Wiley-VCH Verlag GmbH, I felt that it was indeed an honor to be asked to serve as editor of this new and first handbook on cinchona alkaloids. I accepted the invitation with full confidence that this new and timely book would be warmly welcomed by many researchers.

This multiauthor handbook will cover the whole spectrum of cinchona alkaloid chemistry ranging from the fundamentals to industrial applications. This book is organized in four units, namely, the use of cinchona alkaloids as chirality inducers in metal-promoted reactions (Chapters 2–4), the use of cinchona alkaloids as chiral organocatalysts (Chapters 5–11), the organic chemistry of cinchona alkaloids themselves (Chapter 12), and the use of cinchona alkaloids as chiral discriminating agents in modern analysis (Chapter 13), reflecting the comprehensive current state of the art on cinchona alkaloid chemistry. All of the chapters are written by organic chemists at the forefront of research in this field and are able to provide an insider's view.

In addition, a collection of carefully selected representative catalytic examples, organized by reaction type, is given in the Appendix. These tables will offer a great deal of information and be invaluable to anyone who wants to get the information of the current state of the art on this topic within a short time.

I hope this book will be of interest to all those involved in this field, from graduate students to independent organic researchers, both academic and industrial. Especially, this book should be a must to read for anyone working in the field of asymmetric synthesis.

Last but definitely not the least, I am very grateful to all of my colleagues for their excellent contributions to this book, despite their busy time schedules. Grateful acknowledgments are offered to the Wiley-VCH editorial staff, in particular to Dr Elke Maase, who gave me the good fortune of being the editor of this first handbook and who offered a great deal of help at the beginning of this project. I also thank Dr Stefanie Volk for her professional work during the production process.

Finally, I can hardly wait to see the spectacular achievements that will doubtless be made in this rapidly evolving research field in the near future and that will require this manual to be rapidly updated.

Suwon, February 2009 *Choong Eui Song*

Biography

Choong Eui Song has been a full professor at the Sungkyunkwan University since 2004. He received his B.S. in 1980 from Chungang University and obtained a diploma (1985) and a Ph.D. (1988) at RWTH Aachen in Germany. After completing his Ph.D., he worked as Principal Research Scientist at the Korea Institute of Science and Technology (KIST). In 2001, he was appointed as head of the National Research Laboratory for Green Chirotechnology in Korea. In 2004, he moved to his current position and in 2006 he was appointed as a director at the Research Institute of Advanced Nanomaterials and Institute of Basic Sciences at the Sungkyunkwan University. His research interests focus on asymmetric catalysis, ionic liquid chemistry, and nanochemistry. He received the Scientist of the Month Award from the Ministry of Science and Technology of Korea in 2001.

List of Contributors

David J. Ager
DSM Pharmaceutical Chemicals
PMB 150, 9650 Strickland Road,
Suite 103
Raleigh, NC 27615
USA

Hans-Ulrich Blaser
Solvias AG
P.O. Box
CH-4002 Basel
Switzerland

Ying-Chun Chen
Sichuan University
West China School of Pharmacy
Department of Medicinal Chemistry
Chengdu 610041
China

Ravindra R. Deshmukh
Sungkyunkwan University
Department of Chemistry
300 Cheoncheon, Jangan, Suwon
Gyeonggi 440-746
Korea

Jens Frackenpohl
Bayer CropScience AG
Gebäude G 836, Industriepark Höchst
65926 Frankfurt am Main
Germany

Jacek Gawronski
Adam Mickiewicz University
Department of Chemistry
Grunwaldzka 6
60-780 Poznan
Poland

Hans Martin Rudolf Hoffmann
University of Hannover
Department of Organic Chemistry
Schneiderberg 1B
30167 Hannover
Germany

Hyeong Bin Jang
Sungkyunkwan University
Department of Chemistry
300 Cheoncheon, Jangan, Suwon
Gyeonggi 440-746
Korea

Byeong-Seon Jeong
Yeungnam University
College of Pharmacy
Gyeongsan 712-749
South Korea

Cinchona Alkaloids in Synthesis and Catalysis, Ligands, Immobilization and Organocatalysis
Edited by Choong Eui Song
Copyright © 2009 WILEY-VCH Verlag GmbH & Co. KGaA, Weinheim
ISBN: 978-3-527-32416-3

Karol Kacprzak
Adam Mickiewicz University
Department of Chemistry
Grunwaldzka 6
60-780 Poznan
Poland

Je Eun Lee
Sungkyunkwan University
Department of Chemistry
300 Cheoncheon, Jangan, Suwon
Gyeonggi 440-746
Korea

Ji Woong Lee
Sungkyunkwan University
Department of Chemistry
300 Cheoncheon, Jangan, Suwon
Gyeonggi 440-746
Korea

Yan-Kai Liu
Sichuan University
West China School of Pharmacy
Department of Medicinal Chemistry
Chengdu 610041
China

Hyeung-geun Park
Seoul National University
College of Pharmacy
Seoul 151-742
South Korea

Jacques Rouden
Université de Caen-Basse Normandie
Laboratoire de Chimie Moléculaire et
Thio-organique, ENSICAEN, CNRS
6 Boulevard du Maréchal Juin
14050 Caen
France

Do Hyun Ryu
Sungkyunkwan University
Department of Chemistry
300 Cheoncheon, Jangan, Suwon
Gyeonggi 440-746
Korea

Ueon Sang Shin
Sungkyunkwan University
Department of Chemistry
300 Cheoncheon, Jangan, Suwon
Gyeonggi 440-746
Korea

Choong Eui Song
Sungkyunkwan University
Department of Chemistry
300 Cheoncheon, Jangan, Suwon
Gyeonggi 440-746
Korea

Jung Woon Yang
Sungkyunkwan University
Department of Chemistry
300 Cheoncheon, Jangan, Suwon
Gyeonggi 440-746
Korea

1
An Overview of Cinchona Alkaloids in Chemistry
Choong Eui Song

1.1
Brief History

Cinchona alkaloids (Figure 1.1), isolated from the bark of several species of cinchona trees, are the organic molecules with the most colorful biography [1]. Their history dates back to the early seventeenth century when they were first introduced into the European market after the discovery of the antimalarial property of cinchona bark and the subsequent isolation of its active compound, quinine, by Pierre-Joseph Pelletier and Joseph Bienaimé Caventou in 1820. Since then, cinchona alkaloids (especially, quinine) have played a pivotal medicinal role in human society for over 300 years. Approximately 700 metric tons of cinchona alkaloids is now extracted from the bark of *Cinchona ledgeriana* annually. Nearly half of this is used in the food and beverages industry as a bitter additive, and much of the remaining quinine and quinidine is used as an important antimalarial drug and muscle relaxant compound and as a cardiac depressant (antiarrhythmic), respectively.

The role of cinchona alkaloids in organic chemistry was firmly established with the discovery of their potential as resolving agents by Pasteur in 1853, which ushered in an era of racemate resolutions by the crystallization of diastereomeric salts [3]. Today, there are countless examples in which cinchona alkaloids are used as chiral resolving agents [4]. Besides the classical resolution process, significant progress has also been made in the past two decades in the field of cinchona-based enantioseparation, as well as in their use as enantioselective analytical tools (Chapter 13). The considerable effort made to accomplish the stereoselective synthesis of quinine over the past 150 years, which was initially triggered by the supply problem caused by political vagaries of the producing countries, has also undoubtedly laid the foundation for much of modern organic chemistry [5]. However, possibly the most interesting application of cinchona alkaloids in chemistry resides in their ability to promote enantioselective transformations in both homogeneous and heterogeneous catalyses (Chapters 2–11). The first asymmetric reaction carried out using a cinchona base was published by Bredig and Fiske [6] as early as in 1912. These two German chemists reported

Figure 1.1 Quinine and other cinchona alkaloids are extracted from the bark of the cinchona tree [2], which is mainly cultivated in Africa, Latin America, and Indonesia. Approximately 700 metric tons of cinchona alkaloids is harvested annually.

that the addition of HCN to benzaldehyde is accelerated by the pseudoenantiomeric alkaloids, quinine and quinidine, and that the resulting cyanohydrins are optically active and are of opposite chirality. However, the optical yields achieved were in the range of <10% ee. After about four decades, Pracejus was first to obtain useful levels of enantioselectivity (74% ee) by using O-acetylquinine as a catalyst (1 mol%) in the addition of methanol to phenylmethylketene, affording (−)-α-phenyl methylpropionate [7]. Two decades later (in the late 1970s and early 1980s) after Pracejus' seminal study, Wynberg and coworkers began a new era in asymmetric catalysis driven by cinchona alkaloids [8]. Their extensive studies on the use of cinchona alkaloids as chiral Lewis base/nucleophilic catalysts demonstrated that this class of alkaloids could serve as highly versatile catalysts for a broad spectrum of enantioselective transformations (e.g., conjugate additions and the addition of ketenes to carbonyl compounds, resulting in β-lactones). Since their pioneering studies, the popularity of cinchona derivatives in asymmetric catalysis has increased considerably. During the late 1980s and early 1990s, quite successful examples in terms of the catalytic activity and enantioselectivity have been reported, where the asymmetry was induced by cinchona alkaloids. In particular, Sharpless and coworkers developed the osmium-catalyzed asymmetric dihydroxylation (AD) of olefins [9], which is one of the reactions that has had the greatest impact on synthetic chemistry and for which Sharpless was awarded the Nobel Prize in chemistry in 2001. Furthermore, since 2000, explosively expanding interest in chiral organocatalysis as a new stream of catalysis [10] has sparked a second renaissance in the use of cinchona alkaloids as organocatalysts. Thus, nowadays, cinchona alkaloids and their derivatives are classified as the most "privileged organic chirality inducers," efficiently catalyzing nearly all classes of organic reactions in a highly stereoselective fashion (Chapters 5–11).

1.2 Active Sites in Cinchona Alkaloids and Their Derivatives

As mentioned in the previous section, nowadays, readily available and inexpensive cinchona alkaloids with pseudoenantiomeric forms, such as quinine and quinidine or cinchonine and cinchonidine, are among the most privileged chirality inducers in the area of asymmetric catalysis. The key feature responsible for their successful utility in catalysis is that they possess diverse chiral skeletons and are easily tunable for diverse types of reactions (Figure 1.2). The presence of the 1,2-aminoalcohol subunit containing the highly basic and bulky quinuclidine, which complements the proximal Lewis acidic hydroxyl function, is primarily responsible for their catalytic activity.

The presence of the quinuclidine base functionality makes them effective ligands for a variety of metal-catalyzed processes (Chapters 2–4). The most representative example is the osmium-catalyzed asymmetric dihydroxylation of olefins [9]. The metal binding properties of the quinuclidine nitrogen also allow to use cinchona alkaloids as metal surface modifiers, for example, in the highly enantioselective heterogeneous asymmetric hydrogenation of α-keto esters (Chapter 2). Both

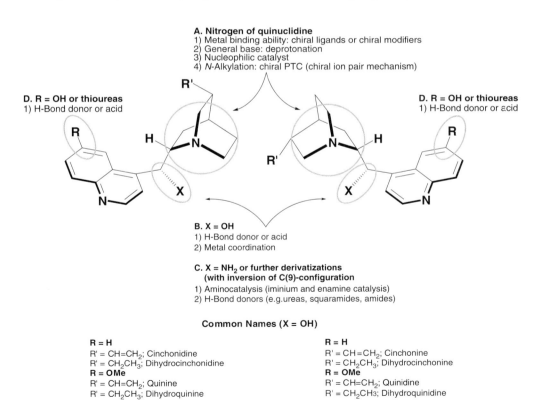

Figure 1.2 Active sites in cinchona alkaloids and their derivatives.

reactions are classified as ligand-accelerated catalyses (LAC) [11]. In addition to its utility for metal binding, the quinuclidine nitrogen can be used as a chiral base or a chiral nucleophilic catalyst promoting the vast majority of organocatalytic reactions (Chapters 5–11). Finally, the related quaternized ammonium salts of cinchona alkaloids have proved to catalyze numerous reactions under phase-transfer conditions, where asymmetric inductions occur through a chiral ion pairing mechanism between the cationic ammonium species and an anionic nucleophile [12].

The secondary 9-hydroxy group can serve as an acid site or hydrogen bond donor [13]. The derivatization of the OH group into ureas, amides, and so on, with either the retention or inversion of the configuration, provides a more powerful acidic site or hydrogen bond donor. The 6′-methoxy group of quinine and quinidine can also be readily derivatized to the free OH group or thiourea moiety, which can serve as an effective H-bond donor. Moreover, the substitution of 9-OH into the free amino group with the inversion of the configuration enables enantioselective aminocatalysis, which includes reactions of the so-called generalized enamine cycle [14] and charge accelerated reactions via the formation of iminium intermediates [15]. Representative examples of modified cinchona alkaloids are depicted in Figure 1.3.

However, in general, these active sites in cinchona alkaloids and their derivatives act in catalysis not independently but cooperatively; that is, they activate the reacting molecules simultaneously. Furthermore, in many cases, the catalysis is also supported by a π–π interaction with the aromatic quinoline ring or by its steric hindrance.

1.3
Structural Information on Cinchona Alkaloids

Cinchona alkaloids have characteristic structural features for their diverse conformations and self-association phenomena. Therefore, knowledge of their "real structure" in solution can provide original information on the chiral inducing and discriminating ability of these alkaloids.

Conformational investigations of this class of alkaloids, based on computational and spectroscopic methods, have been undertaken with the aim of providing information that would help understand these chiral induction and discrimination processes. Dijkstra *et al.* were the first to investigate in 1989 the conformational behavior by means of NMR spectroscopic and molecular mechanics (MM) calculations and identify that the C8–C9 and C4′–C9 bonds are the most important in determining the overall conformation, resulting in four low-energy conformers (*syn*-closed, *syn*-open, *anti*-closed, and *anti*-open conformers) (Figure 1.4) [16]. MM calculations showed that the parent alkaloids preferentially adopt an *anti*-open conformation in nonpolar solvents [16]. More sophisticated *ab initio* calculations conducted later also revealed that *anti*-open is the most stable conformer in apolar solvents [17, 18]. In polar solvents, two other conformers, *syn*-closed and *anti*-closed, are strongly stabilized compared to the *anti*-open conformer, due to the greater support provided by their large dipole moments [18]. For example, in polar solvents, the fraction of cinchonidine adopting a closed conformation is more than 50% at

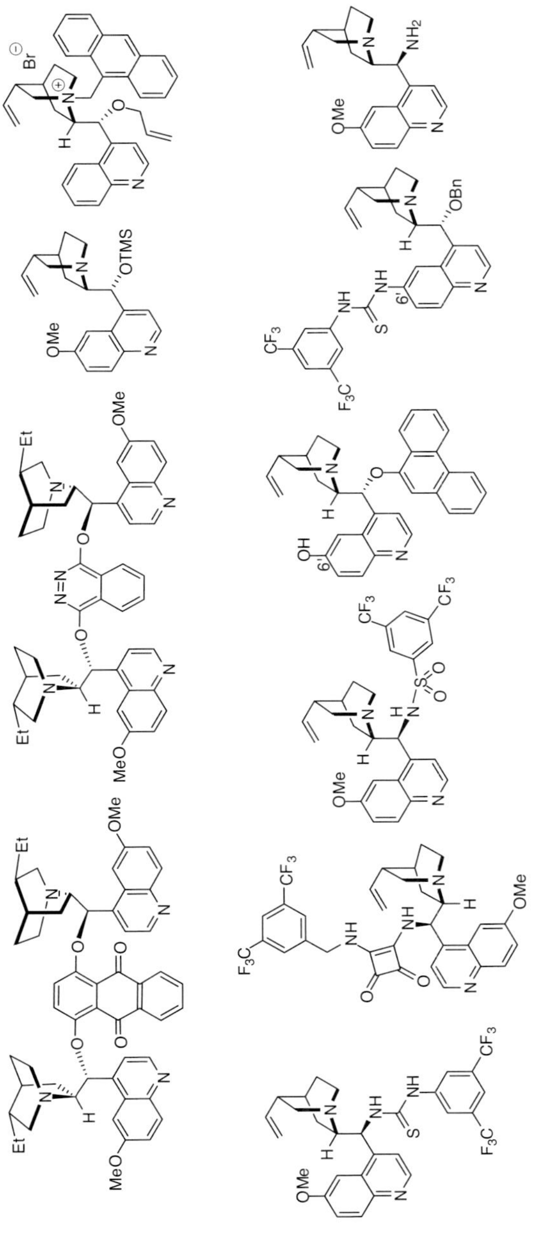

Figure 1.3 Representative examples of cinchona alkaloid derivatives.

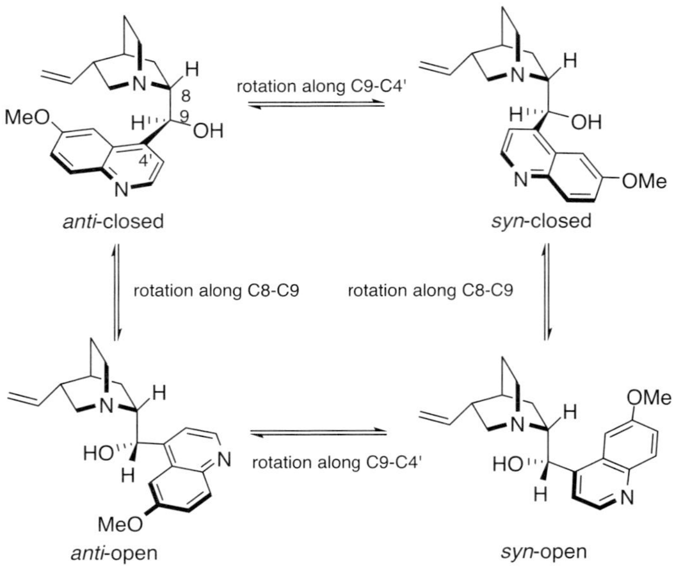

Figure 1.4 The four conformers of quinidine showing the lowest energy.

room temperature. However, upon protonation, the *anti*-open conformation is observed exclusively [18]. The protonation of cinchona alkaloids appears to hinder their rotation around the C4′–C9 and C9–C8 bonds and favor only a narrow range of the conformational space of the molecule [19].

The pivotal role of the conformational behavior of a cinchona alkaloid (e.g., cinchonidine) in its enantioselectivity was nicely illustrated in the platinum-catalyzed enantioselective hydrogenation of ketopantolactone in different solvents [18]. The achieved enantiomeric excess shows the same solvent dependence as the fraction of *anti*-open conformer in solution, suggesting that this conformer plays a crucial role in the enantiodifferentiation. As a more dramatic example, the solvent affects the absolute chirality of the product in the 1,3-hydron transfer reaction catalyzed by dihydroquinidine [20]. An NMR study revealed that the changes in the ratio between the two conformers of dihydroquinidine can explain the observed reversal of the sense of the enantioselectivity for this reaction when the solvent is changed from *o*-dichlorobenzene (open/closed ∼60 : 40) to DMSO (open/closed ∼20 : 80).

Undoubtedly, the modification of the structure of the cinchona alkaloid also has a significant effect on its conformational behavior in solution; esters [17] and 9-*O*-carbamoyl derivatives [21] exist as a mixture of two major *anti*-closed and *anti*-open conformers, while C9 methyl ethers prefer an *anti*-closed arrangement in noncoordinating solvents [17]. Here again, protonation provides the *anti*-open conformation as the sole stable form [16b]. In addition to the solvent polarity, many other factors such as intermolecular interactions are also responsible for the complex conformational behavior of cinchona alkaloids in solution.

Another characteristic structural feature of cinchona alkaloids is their multifunctional character and, thus, autoassociation phenomena are possible that could result in the strong dependency of their efficiency on the concentration and temperature [22, 23].

In 1969, Uskokovic and coworkers observed a concentration-dependent intermolecular interaction between dihydroquinine molecules [24]. The ^{1}H NMR spectra of (−)-dihydroquinine and racemic dihydroquinine are clearly different under concentrated conditions; that is, this molecule can generate its own nonequivalence. However, at high dilution (0.01 M), this interaction disappeared. This indicates the existence of a self-association process mediated by intermolecular hydrogen bonding [24]. Moreover, the measurement of the molecular weight of quinine using osmometry conducted by Hiemstra and Wynberg revealed the presence of particles larger than monomeric quinine at 37 °C for a 16 mM solution in toluene. For concentrations less than 4 mM, on the other hand, quinine was almost completely monomeric [22].

In 1992, the coexistence of the monomer and dimers of quinine in chloroform solution was established by Salvadori and coworkers by investigating the temperature and concentration dependence of the NMR spectral parameters (chemical shift, NOE effect, relaxation time, etc.). The mole fraction of dimers was about 40% in 0.6 M solution [25]. The structure of the dimer determined from NOESY and the relaxation rate data is given in Figure 1.5. As shown in Figure 1.5, the dimer of quinine was shown to be a π–π complex with nearly parallel quinoline rings. A study carried out on the self-aggregation of chloroquine [26] gave similar results. Recently, the changes in the ^{13}C chemical shifts at various concentrations were also used to study the self-association of quinine. T-shaped dimers formed by the quinoline rings were proposed [27].

Quite recently, we also observed that quinine-based thiourea derivatives showed dramatic concentration and temperature effects on the enantioselectivity in the alcoholytic desymmetrization of meso-cyclic anhydrides, which can also be attributed to the self-association of the catalyst [23]. Of course, the possibility that the variation in

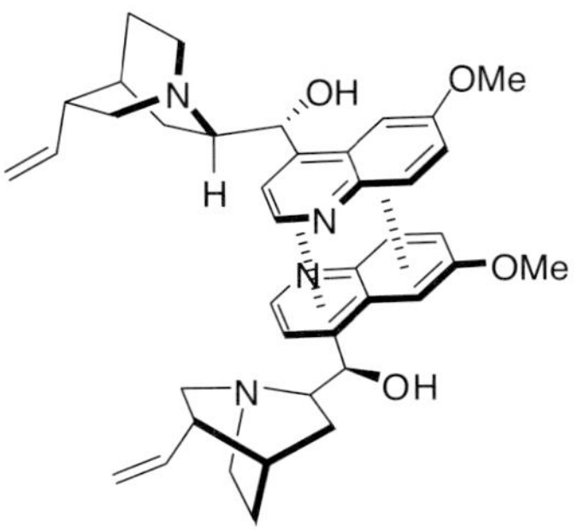

Figure 1.5 Conformation of quinine dimer from NMR results.

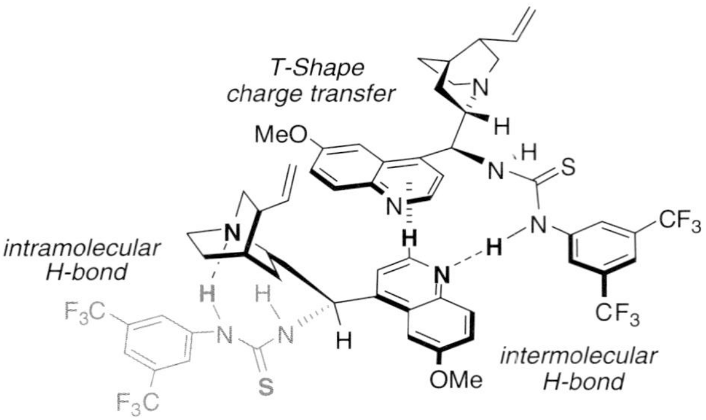

Figure 1.6 Dimeric structure of 9-epiquinine thiourea.

the enantioselectivity is caused by a concentration-dependent change in the conformational composition cannot be ruled out. However, according to the results obtained by Salvadori [25], at least in the case of the parent alkaloids, the conformation is similar in the dimer and monomer. Quite recently, Soós and coworkers proved by means of NMR NOESY experiments and computational studies that the quinine thiourea catalyst exists in a dimeric form under concentrated conditions due to H-bond and T-type π–π interactions (Figure 1.6) [28].

As discussed, the solvent dipole moments, concentration, and temperature play a significant role in determining the structure of cinchona alkaloids and their derivatives in solution. In order to delineate the intimate details of the mechanism of action of cinchona alkaloids and their derivatives, a thorough understanding of their real structure in solution is needed. Furthermore, such detailed information on the real structure in solution would make it possible to develop new and more powerful chiral catalysts and discriminators.

1.4
How This Book Is Organized

The goal of this handbook is to provide up-to-date information on the whole spectrum of cinchona alkaloid chemistry. The authors have attempted to provide those who want to learn about the current state of the art on this topic and are willing to contribute actively to the extraordinary developments taking place in this field with an insider's view of this subject. This book is organized in four units, namely, the use of cinchona alkaloids as chirality inducers in metal-promoted reactions (Chapters 2–4), as chiral organocatalysts (Chapters 5–11), the organic chemistry of cinchona alkaloids themselves (Chapter 12), and their use as chiral discriminating agents in modern analysis (Chapter 13). In addition, a collection of carefully selected representative catalytic examples organized according to the reaction type is given in the form of an appendix.

References

1 Garfield, S. (2000) *Mauve*, Faber & Faber, London, p. 224.
2 From http://www.quinine-buchler.com/cinchona.htm.
3 (a) Pasteur, L. (1853) *Acad. Sci.*, **37**, 162; (b) Pasteur, L. (1853) *Liebigs Ann. Chem.*, **88**, 209.
4 (a) Newman, P. (1981) *Optical Resolution Procedures for Chemical Compounds: Acids*, vol. 2, Optical Resolution Information Center, Manhattan College, Riverdale, NY, pp. 7–22; (b) Jacques, J., Collet, A., and Wilen, S.H. (1981) *Enantiomers, Racemates and Resolutions*, John Wiley & Sons, Inc., New York, pp. 254, 257; (c) Sheldon, R.A. (1993) *Chirotechnology*, Marcel Dekker, New York, Chapter 6; (d) Kozma, D. (2002) *CRC Handbook of Optical Resolutions via Diastereomeric Salts*, CRC Press, Boca Raton, FL.
5 Excellent reviews about quinine total synthesis with historical background: (a) Kaufmann, T.S. and Rúveda, E.A. (2005) *Angew. Chem. Int. Ed.*, **44**, 854; (b) Weinreb, S.M. (2001) *Nature*, **411**, 429; (c) Nicolaou, K.C. and Sorensen, E.J. (1996) in *Classics in Total Synthesis: Targets, Strategies, Methods*, Wiley-VCH Verlag GmbH, Weinheim, Chapter 15.
6 (a) Bredig, G. and Fiske, P.S. (1912) *Biochem. Z.*, **46**, 7; (b) Bredig, G. and Minaeff, M. (1932) *Biochem. Z.*, **249**, 241.
7 (a) Pracejus, H. (1960) *Justus Liebigs Ann. Chem.*, **634**, 9; (b) Pracejus, H. and Mätje, H. (1964) *J. Prakt. Chem.*, **24**, 195.
8 Wynberg, H. (1986) *Top. Stereochem.*, **16**, 87.
9 Selected reviews: (a) Kolb, H.C., VanNieuwenhze, M.S., and Sharpless, K.B. (1994) *Chem. Rev.*, **94**, 2483; (b) Kolb, H.C. and Sharpless, K.B. (1998) in *Transition Metals for Organic Synthesis*, vol. 2 (eds M. Beller and C. Bolm), Wiley-VCH Verlag GmbH, Weinheim, p. 219;(c) Johnson, R.A. and Sharpless, K.B. (2000) in *Catalytic Asymmetric Synthesis*, 2nd edn (ed. I. S Ojima), John Wiley & Sons, Inc., New York, p. 357; (d) Beller, M. and Sharpless, K.B. (2002) in *Applied Homogeneous Catalysis with Organometallic Compounds*, 2nd edn, vol. 3 (eds B.H. Cornils and A. Wolfgang), Wiley-VCH Verlag GmbH, Weinheim, p. 1149.
10 (a) Berkessel, A. and Gröger, H. (2005) *Asymmetric Organocatalysis*, Wiley-VCH Verlag GmbH, Weinheim; (b) Dalko, P.I.(ed.) (2007) *Enantioselective Organocatalysis*, Wiley-VCH Verlag GmbH, Weinheim; (c) List, B.(ed.) (2007) *Chem. Rev.*, **107**, 5413 (special issue on organocatalysis); (d) Houk, K.N. and List, B.(eds) (2004) *Acc. Chem. Res.*, **37**, 487 (special issue on asymmetric organocatalysis).
11 Berrisford, D.J., Bolm, C., and Sharpless, K.B. (1995) *Angew. Chem. Int. Ed.*, **34**, 1059.
12 For recent reviews on asymmetric phase-transfer catalysis, see (a) Maruoka, K.(ed.) (2008) *Asymmetric Phase Transfer Catalysis*, Wiley-VCH Verlag GmbH, Weinheim; (b) Hashimoto, T. and Maruoka, K. (2007) *Chem. Rev.*, **107**, 5656.
13 For a recent review on organocatalysis via hydrogen bonding, see Doyle, A.G. and Jacobsen, E.N. (2007) *Chem. Rev.*, **107**, 5713.
14 For a recent review on enamine catalysis, see Mukherjee, S., Yang, J.W., Hoffmann, S., and List, B. (2007) *Chem. Rev.*, **107**, 5471.
15 For a recent review on iminium catalysis, see Erkkilä, A., Majander, I., and Pihko, P.M. (2007) *Chem. Rev.*, **107**, 5416.
16 (a) Dijkstra, G.D.H., Kellogg, R.M., and Wynberg, H. (1989) *Recl. Trav. Chim. Pays-Bas*, **108**, 195; (b) Dijkstra, G.D.H., Kellogg, R.M., Wynberg, H., Svendsen, J.S., Marko, I., and Sharpless, K.B. (1989) *J. Am. Chem. Soc.*, **111**, 8069.
17 Dijkstra, G.D.H., Kellogg, R.M., and Wynberg, H. (1990) *J. Org. Chem.*, **55**, 6121.
18 Bürgi, T. and Baiker, A. (1998) *J. Am. Chem. Soc.*, **120**, 12920.

19 Olsen, R.A., Borchardt, D., Mink, L., Agarwal, A., Mueller, L.J., and Zaera, F. (2006) *J. Am. Chem. Soc.*, **128**, 15594.
20 Aune, M., Gogoll, A., and Matsson, O. (1995) *J. Org. Chem.*, **60**, 1356.
21 Maier, N.M., Schefzick, S., Lombardo, G.M., Feliz, M., Rissanen, K., Lindner, W., and Lipkowitz, K.B. (2002) *J. Am. Chem. Soc.*, **124**, 8611.
22 Hiemstra, H. and Wynberg, H. (1981) *J. Am. Chem. Soc.*, **103**, 417.
23 Rho, H.S., Oh, S.H., Lee, J.W., Lee, J.Y., Chin, J., and Song, C.E. (2008) *Chem. Commun.*, 1208.
24 Williams, T., Pitcher, R.G., Bommer, P., Gutzwiller, J., and Uskokovic, M. (1969) *J. Am. Chem. Soc.*, **91**, 1871.
25 Uccello-Barretta, G., Bari, L.D., and Salvadori, P. (1992) *Magn. Reson. Chem.*, **30**, 1054.
26 Marchettini, N., Valensin, G., and Gaggelli, E. (1990) *Biophys. Chem.*, **36**, 65.
27 Casabianca, L.B. and de Dios, A.C. (2004) *J. Phys. Chem. A*, **108**, 8505.
28 Tárkányi, G., Király, P., Varga, S., Vakulya, B., and Soós, T. (2008) *Chem. Eur. J.*, **14**, 6078.

Part One
Cinchona Alkaloid Derivatives as Chirality Inducers in Metal-Catalyzed Reactions

2
Cinchona Alkaloids as Chirality Transmitters in Metal-Catalyzed Asymmetric Reductions

Hans-Ulrich Blaser

2.1
Introduction

The enantioselective reduction of prochiral C=O, C=N, and C=C moieties to the corresponding saturated chiral products is one of the most important stereoselective transformations both on the laboratory [1] and on the industrial scale [2]. It is therefore somewhat surprising that only very few effective cinchona-based reduction catalysts are known despite the fact that the use of cinchona alkaloids as chiral auxiliaries or chirality transmitters in organic synthesis has a long and impressive history [3]. The first report appeared in 1965, when Cervinka [4] described the reduction of aryl alkyl ketones with stoichiometric amounts of LiAlH$_4$ complexed with the parent cinchona alkaloids, cinchonidine (Cd), quinine (Qn), cinchonine (Cn), and quinidine (Qd), depicted in Figure 2.1 with ee values up to 48%. The most important breakthrough in the field was achieved by Orito and coworkers in 1979 [5] who found that the heterogeneous Pt catalysts modified with the parent cinchona alkaloids gave excellent results for the hydrogenation of α-keto esters with up to 82% ee. In 1981, Julia *et al.* [6] described a ketone reduction using NaBH$_4$ in the presence of an *N*-alkylated Qn or Cd derivative under phase-transfer conditions but with only 29% ee. The first reduction of a C=C bond was described by Perez *et al.* [7] in 1985 with a heterogeneous Pd catalyst modified with Cd that reduced substituted cinnamic acids with 30% ee. The first homogeneous catalytic transfer hydrogenation was reported only recently by He *et al.* [8] who obtained up to 97% ee for acetophenone derivatives using Rh and Ir complexes of epicinchona alkaloids where the 9-OH was replaced with an NH$_2$ group.

In this chapter, we do not attempt to give a comprehensive overview of the field, but we would rather concentrate on results where both enantioselectivity and catalyst activity are relevant to preparative application. In the first section, results obtained with cinchona-mediated homogeneous systems for the reduction of ketones are briefly reviewed. Then, heterogeneous cinchona-modified Pt catalysts applied to the hydrogenation of α-functionalized ketones and cinchona-modified Pd catalysts for the hydrogenation of activated C=C bonds are discussed from a synthetic point

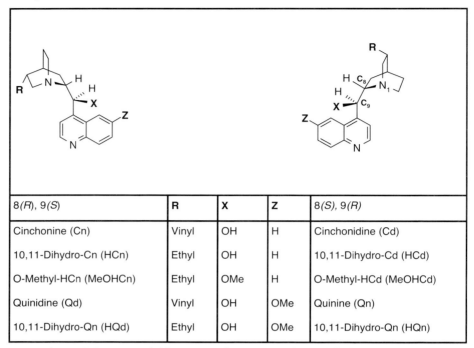

Figure 2.1 Structure and abbreviations of important cinchona alkaloids.

8(R), 9(S)	R	X	Z	8(S), 9(R)
Cinchonine (Cn)	Vinyl	OH	H	Cinchonidine (Cd)
10,11-Dihydro-Cn (HCn)	Ethyl	OH	H	10,11-Dihydro-Cd (HCd)
O-Methyl-HCn (MeOHCn)	Ethyl	OMe	H	O-Methyl-HCd (MeOHCd)
Quinidine (Qd)	Vinyl	OH	OMe	Quinine (Qn)
10,11-Dihydro-Qn (HQd)	Ethyl	OH	OMe	10,11-Dihydro-Qn (HQn)

of view. Finally, the application of cinchona alkaloid-based catalytic system to industrial problems is described in some detail.

2.2
Homogeneous Systems for Ketone Reductions

Only very few publications have appeared describing cinchona-based homogeneous reduction systems, all of them restricted to the reduction of aromatic ketones. As already described, adding stoichiometric amounts of cinchona alkaloids to metal hydrides was the first successful application of a cinchona auxiliary for an enantioselective reduction [4]. The results were later confirmed, but ee values never exceeded 50% [9]. Variations of this procedure were tried, but in all cases enantioselectivities were too low for synthetic applications. The $NaBH_4$ reduction under phase-transfer conditions in the presence of an N-alkylated cinchona derivative as pioneered by Julia et al. [6] was later taken up by Hofstetter et al. [10], but ee values never exceeded 30%. Cha and colleagues [11] described the $NaBH_4$ reduction of a variety of substituted acetophenone derivatives catalyzed by Pd complexes of copolymers of methacrylate and Qn derivatives, but again enantioselectivity did not exceed 48%. Takeuchi and Ohgo [12] described the reduction of benzil with SmI_2 in the presence of excess Qn with ≤56% ee.

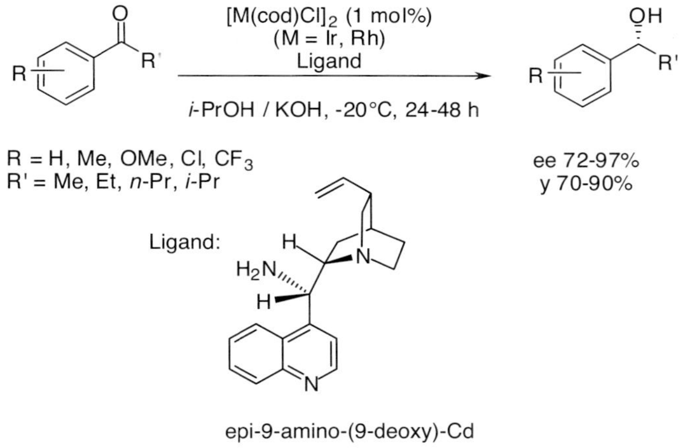

Figure 2.2 Homogeneous Rh- and Ir-catalyzed transfer hydrogenation.

Reaction conditions: [M(cod)Cl]$_2$ (1 mol%) (M = Ir, Rh), Ligand, i-PrOH / KOH, -20°C, 24-48 h

R = H, Me, OMe, Cl, CF$_3$
R' = Me, Et, n-Pr, i-Pr

ee 72-97%
y 70-90%

Ligand: epi-9-amino-(9-deoxy)-Cd

Somewhat better results were obtained for the hydrosilylation of ketones, first described by Vannoorenberghe and Buono [13] using a Rh phosphinite Qn or Qd complex (Figure 2.1, X = O−PPh$_2$) with good yields and up to 58% ee. Enantioselectivities up to 78% were achieved by Drew et al. [14] using N-aralkylated Qn and Qd fluorides as organocatalysts.

The only homogeneous catalysts with some preparative potential are Rh and Ir complexes of *epi*-9-amino-(9-deoxy) derivatives of various cinchona alkaloids as reported only recently by He et al. [8] with up to 97% ee for acetophenone derivatives. As depicted in Figure 2.2, the reactions are carried out at rather low temperature with 1 mol% of catalyst and long reaction times. Yields for a variety of aryl alkyl ketones are generally good to high and enantioselectivities range from 70 up to 97%. Cd and Qn derivatives lead to (R)-alcohols whereas Cn and Qd give the (R)-enantiomers. As is typical of transfer hydrogenation reactions using *i*-PrOH/KOH as reducing agent, the reactions have to be carried out at rather high dilution (0.05 M in ketone). In general, Ir complexes are superior giving 5–25% better yields and 2–10% higher ee values than the corresponding Rh complexes. Interestingly, the intact Rh and Ir complexes can be extracted with HCl after completion of the reduction and reused up to five times without much effect on ee and rate [8b].

2.3
Heterogeneous Pt and Pd Catalysts Modified with Cinchona Alkaloids

2.3.1
Background

As mentioned in Section 2.1, the use of cinchona alkaloids as modifiers of the classical heterogeneous hydrogenation catalyst platinum on a support was pioneered by Orito. In the late 1970s and early 1980s, he published a series of groundbreaking papers on their application to the enantioselective hydrogenation of α-keto esters [5, 15]. Since then, many research groups have worked on this topic, thereby significantly expanding the scope of the catalytic system to other metals (especially, Pd, Rh, and Ir) and to other substrate classes. At the same time, the question of the mode of action of the cinchona modifier was addressed, and today there is some basic understanding how these catalytic systems work. Several recent reviews give a comprehensive overview on all these topics [16]. Here, we do not intend to discuss all this information but focus on selected aspects relevant to synthetic applications.

2.3.2
Catalysts

Pt is the metal of choice for the hydrogenation of functionalized ketones. Recently, it was shown that Rh is effective for aromatic ketones with up to 80% ee [17]. Different supports are suitable, but Al_2O_3, SiO_2, TiO_2, and zeolites give the best performance. Results obtained with colloids show that the support plays only an indirect role. In general, high metal dispersions seem to be detrimental to high enantioselectivities; however, Pt colloids with 1–2 nm particle size also give very high ee values. Two commercial 5% Pt/Al_2O_3 catalysts have shown superior performance: E 4759 from Engelhard and JMC 94 from Johnson Matthey. While both catalysts have dispersions around 0.2–0.3, E 4759 has rather small pores and a low pore volume while JMC 94 is a wide-pore catalyst with a large pore volume. E 4759 from Engelhard has emerged as "standard" catalysts for many groups working with the Pt-cinchona system. For most catalysts, a reductive pretreatment in flowing hydrogen at 400 °C just before catalyst use significantly increases enantioselectivity and reaction rate. Although this is not much of a problem on a small scale, it complicates large-scale applications. Cat*A*Sium® F214, which does not need pretreatment for optimal performance, was developed recently by Degussa in collaboration with Solvias [18]. The reason for the better performance of the Cat*A*Sium F214 catalyst was attributed to a lower reducible residue level and a narrower Pt crystal size distribution.

Pd is preferred for the enantioselective hydrogenation of C=C bonds, but with few exceptions, enantioselectivities are rather low. Preferred supports for Pd catalysts are carbon, TiO_2, Al_2O_3, or SiO_2. Several reports show that the Pd dispersion, the nature and texture of the support, as well as the catalyst preparation and pretreatment can have significant effects on the enantioselectivity of the Pd catalysts.

2.3.3
Modifiers and Solvents

Apart from the parent cinchona derivatives described by Orito (see Figure 2.1), a large number of other cinchona derivatives and cinchona mimics have been prepared and tested for the hydrogenation of various activated ketones [19]. From these studies, the following conclusions can be drawn:

- Best results are usually obtained with cinchonidine or slightly altered derivatives such as HCd or 9-methoxy-HCd (MeOHCd) to give the (*R*)-alcohol and with cinchonine derivatives to give the (S)-enantiomer, albeit with somewhat lower enantioselectivity. The most effective cinchona modifiers are commercially available or can be easily prepared from these.

- Three structural elements in the cinchona molecule were identified to affect rate and ee of the enantioselective hydrogenation of α-keto acid derivatives: (i) an extended aromatic moiety, (ii) the substitution pattern of the quinuclidine (the absolute configuration at C_8 controls the sense of induction; N-alkylation yields racemate), and (iii) the substituents at C_9 (OH or MeO is optimal; larger groups reduce the enantioselectivity).

- The choice of the solvent had a significant effect on enantioselectivity and rate. MeOHCd/acetic acid and HCd/toluene are often the most effective modifier/solvent combinations.

2.3.4
Substrate Scope for Pt Catalysts

A pictorial overview of the substrate scope is presented in Figure 2.3. More detailed results for synthetically useful reactions are presented in the next sections. Besides α-keto acid derivatives, α-keto acetals, α-keto ethers, and some trifluoromethyl ketones have been shown to give high ee values with Pt-based catalysts. Pd-based catalysts give moderate enantioselectivities for most α,β-unsaturated acids and up to 94% ee for selected pyrones. Nevertheless, for the synthetic chemist, the substrate scope is still relatively narrow, and it is not expected that new important substrate classes will be found soon. On the other hand, the chemoselectivity of this system has not yet been exploited to its full value, and this might hold a potential for synthetically useful applications in the future.

2.3.4.1 α-Keto Acid Derivatives

α-Keto esters such as methyl and ethyl pyruvate and phenyl glyoxylic acid esters are the substrates giving the highest ee values (Table 2.1) and activities. The following strategies were applied for improving the ee values: addition of trifluoroacetic acid [20] and amines [21], use of special colloidal catalysts [22], use of ultrasound [23], and slow addition of modifier [24] and solvent mixtures [25, 26]. Not surprisingly, the reaction conditions have to be optimized for every substrate. An analysis of the results shows that AcOH in combination with MeOHCd or HCd gave the best results for most aliphatic keto ester derivatives. For aromatic and conjugated systems, the combination of HCd and toluene was usually optimal. In most cases, the question of activity and productivity of the catalytic systems was not addressed.

Szöri et al. [29] investigated a series of α-keto ester RCOCOOR′ and found that bulky R, and to a lower extent R′ groups, caused a significant decrease in ee values and rate. Similar results were reported for various R′ with Pt colloids [22]. For the corresponding α-keto acids, much less work has been carried out; a preliminary study revealed somewhat lower ee values and different optimal solvent systems than for the corresponding esters [25]. Keto amides related to **5** also showed significantly lower ee values [28].

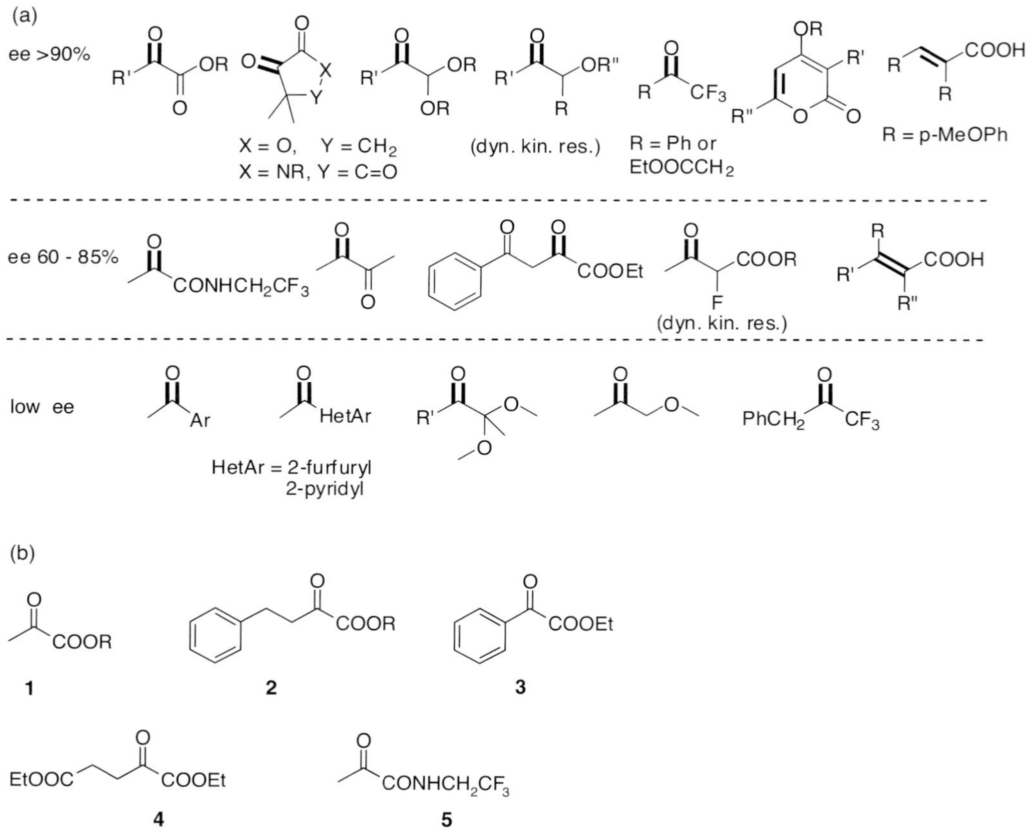

Figure 2.3 (a) Structures of "good," "medium," and "bad" substrates for cinchona-modified catalysts. (b) Structures of suitable α-keto acid derivatives.

2.3.4.2 α,γ-Diketo Esters

The hydrogenation of 2,4-diketo acid derivatives to the corresponding 2-hydroxy compounds with cinchona-modified Pt catalysts as depicted in Figure 2.4 can be carried out with chemoselectivities more than 99% and enantioselectivities up to 87% (R) and 68% (S), respectively [30a]. Enrichment to more than 98% ee was possible for several substrates by recrystallization, giving rise to an efficient technical synthesis of (R)-2-hydroxy-4-phenyl butyric acid ethyl ester [30b], a building block for several ACE (angiotensin-converting enzyme) inhibitors, as well as some enantiomerically enriched α-hydroxy and α-amino acid esters (see below) [30c].

2.3.4.3 Fluorinated Ketones

The trifluoromethyl group was identified to have similar activating properties as an ester group (Figure 2.5), and several papers by the Baiker's group elaborate

Table 2.1 Best ee values for various α-keto acid derivatives (for substrate structures, see Figure 2.3b).

Substrate	ee (%)	Catalyst, modifier, solvent, reaction conditions, remarks	Reference
1, R = Me	98	Pt colloids, Cd, AcOH, 40 bar, 25 °C	[22]
1, R = Et	97	5% Pt/Al$_2$O$_3$ (E 4759), MeOHCd, AcOH, 10 bar, 25 °C, ultrasound	[23a]
2, R = Et	96	5% Pt/Al$_2$O$_3$ (E 4759), MeOHCd, AcOH, 10 bar, 25 °C, ultrasound	[23a]
2, R = Et	94	1% Pt/Al$_2$O$_3$ (Aldrich), HCd, AcOH, 5.8 bar, 17 °C, dosing of modifier	[24]
2, R = H	85	5% Pt/Al$_2$O$_3$ (JMC 94), MeOHCd, EtOH/H$_2$O 9:1, 100 bar, 20–30 °C	[25]
3	98	5% Pt/Al$_2$O$_3$ (E 4759), HCd, AcOH/toluene, 25 bar, 0 °C	[26]
4	96	5% Pt/Al$_2$O$_3$ (JMC 94), MeOHCd, AcOH, 20 bar, 20 °C	[27]
5	60	5% Pt/Al$_2$O$_3$ (E 4759), Cd, AcOH, 60 bar, rt	[28]

on this topic [31]. Me, Et, and i-Pr esters of 4,4,4-trifluoroacetoacetate gave ee values of 90–96% in AcOH or trifluoroacetic acid/THF mixtures, and MeOHCd was often significantly more efficient than HCd. Various trifluoroacetophenone derivatives with additional CF$_3$ or N(Et)$_2$ substitutents on the aromatic ring gave ee values between 36 and 81%, but ee values and TOFs were highest without any substituent. Other CF$_3$-substituted ketones were tested, but except 2-trifluoroacetylpyrrole (63% ee) none of them gave ee values significantly above 20% [23b,31e]. Ethyl 2-fluoroacetoacetate and 2-fluorocyclohexanone can also be hydrogenated with a cinchona-modified Pt catalyst, the only examples where a single fluorine atom sufficiently activates the keto group. Owing to dynamic kinetic resolution, enantioselectivities up to 82 and 59%, respectively, and diastereoselectivities of 85–98% at high conversions were obtained [32].

α,γ-Diketo Esters

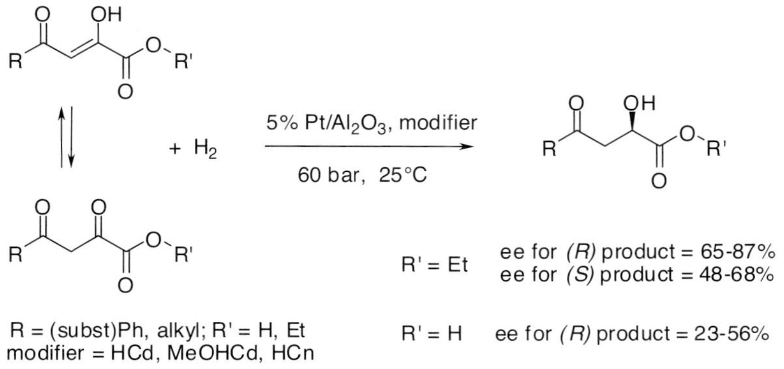

Figure 2.4 Hydrogenation of α,γ-diketo acid derivatives.

Fluorinated Ketones

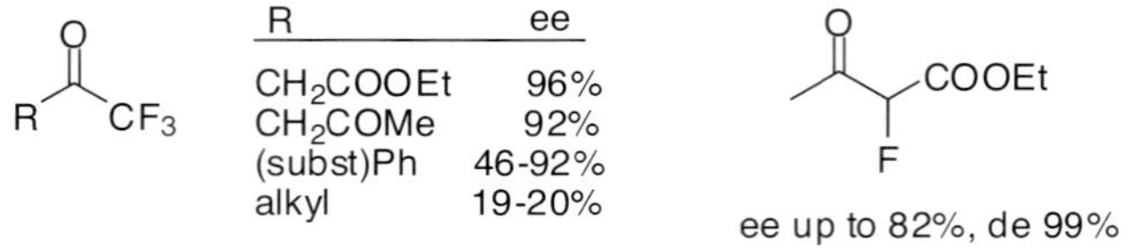

R	ee
CH$_2$COOEt	96%
CH$_2$COMe	92%
(subst)Ph	46-92%
alkyl	19-20%

ee up to 82%, de 99%

Conditions: 5% Pt/Al$_2$O$_3$ (E 4759), Cd or MeOHCd, 10-30 bar, 20-25°C

Figure 2.5 Hydrogenation of fluorinated ketones.

2.3.4.4 α-Keto Acetals

As simultaneously shown by two different groups, α-keto acetals can be hydrogenated using cinchona-modified Pt catalysts with high rates and ee values up to 97% (Figure 2.6) [33]. The highest ee and rate values were obtained with methyl glyoxal acetals. In these cases, the addition of the modifier led to a rate acceleration on the order of 10, comparable to that observed for ethyl pyruvate [33a]. Other aliphatic and aromatic α-keto acetals with relatively low bulkiness also gave high ee values, but with significantly lower rates. Significantly lower enantioselectivities and very much lower rates were observed for keto acetals with more bulky R and especially with larger R′. Aromatic and aliphatic ethers as well as esters and amides are tolerated as functional groups in the R residue and do not seem to affect the enantioselectivity very much. α-Keto ketals are hydrogenated very slowly and with negligible induction. Enantiomerically enriched α-hydroxy acetals are interesting synthons and can be transformed to a variety of chiral building blocks such as 1,2-diols, α-hydroxy acids, or 1,2-amino alcohols.

2.3.4.5 α-Keto Ethers

Substituted aliphatic and aromatic α-keto ethers (see Figure 2.7) are also amenable to enantioselective hydrogenation catalyzed by cinchona-modified Pt catalysts [34]. However, as opposed to the achiral ketones discussed above, kinetic resolution is observed for these chiral substrates. At conversions of 20–42%, ee values of 91–98% were obtained when starting with a racemic substrate. While the very high initial ee values were impressive, it was also clear that this method with yields of less than 50% and gradually decreasing ee values is of little preparative value. The obvious solution was to attempt dynamic kinetic resolution in the presence of a base. Indeed, with OH$^-$-activated Amberlites dynamic kinetic resolution was observed. Both (R,S)-2-methoxy cyclohexanol and (R,S)-2-methoxy-1,2-diphenyl ethanol can be obtained

α-Keto Acetals

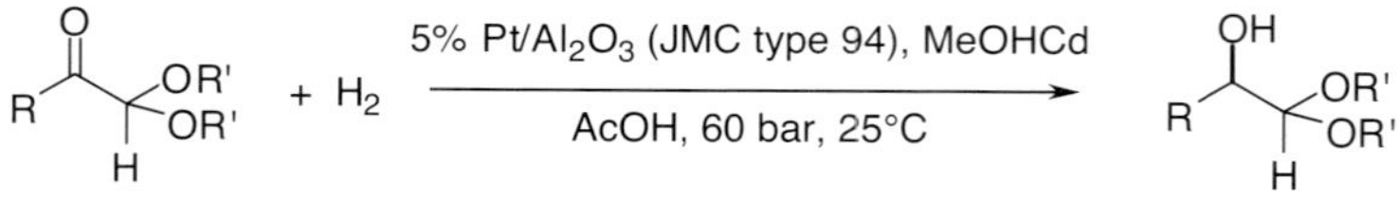

R = Ph, alkyl; R′ = Me, Et, n-Bu, -(CH$_2$)$_3$-

ee 50-97%

Figure 2.6 Hydrogenation of α-keto acetals.

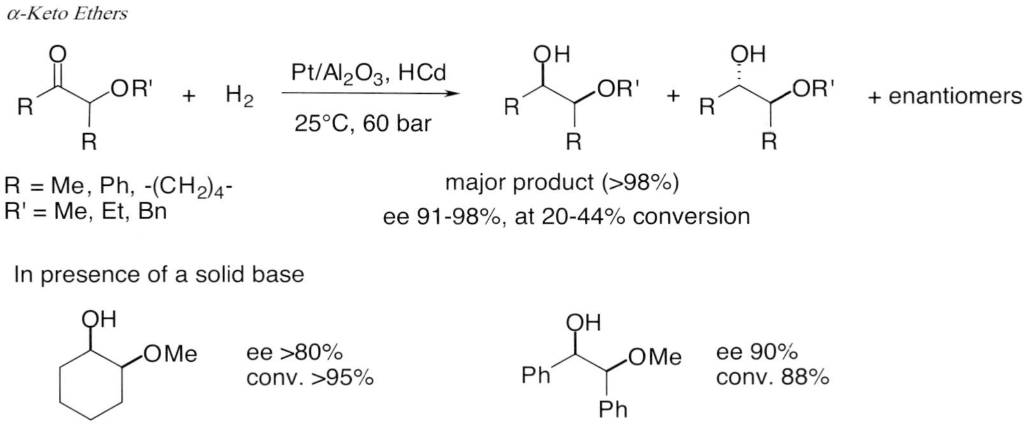
α-Keto Ethers
Pt/Al2O3, HCd
25°C, 60 bar
+ enantiomers
R = Me, Ph, -(CH2)4-
R' = Me, Et, Bn
major product (>98%)
ee 91-98%, at 20-44% conversion
In presence of a solid base
ee >80%
conv. >95%
ee 90%
conv. 88%

Figure 2.7 (Dynamic) kinetic resolution of α-keto ethers.

with satisfactory enantioselectivities at high conversion. However, this transformation has low scope and little synthetic potential.

2.3.4.6 Miscellaneous Ketones

Ketopantolactone [35] and a cyclic imidoketone [36] were hydrogenated with ee values ≥90% under optimized reaction conditions. The hydrogenation of 1,2-butanedione [37, 38] and 1-phenyl-1,2-propanedione [16b, 39] gave initial ee values of 65 and 50% for the keto alcohol, respectively (see Figure 2.8). The ee values of the keto alcohol increased slowly to more than 90% during the reaction due to a preferential hydrogenation of the minor enantiomer. Hydroxy and methoxy acetophenone are hydrogenated with ee values up to 81% [40]. Interestingly, Rh/Al_2O_3 in the presence of Cd gives about the same enantioselectivity [17].

2.3.5 Substrate Scope for Pd Catalysts

Even though the first report on Pd/cinchona-catalyzed hydrogenation of a C=C bond was published in 1985 [7], systematic investigations started only in the mid-1990s

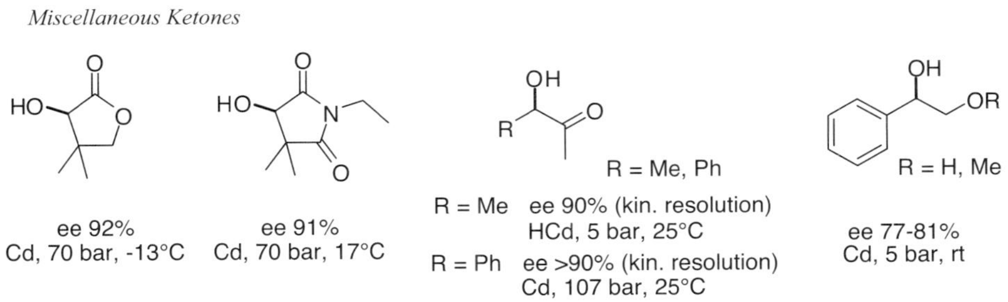
Miscellaneous Ketones
ee 92%
Cd, 70 bar, -13°C
ee 91%
Cd, 70 bar, 17°C
R = Me, Ph
R = Me ee 90% (kin. resolution)
HCd, 5 bar, 25°C
R = Ph ee >90% (kin. resolution)
Cd, 107 bar, 25°C
R = H, Me
ee 77-81%
Cd, 5 bar, rt

Figure 2.8 Products obtained from various activated ketones.

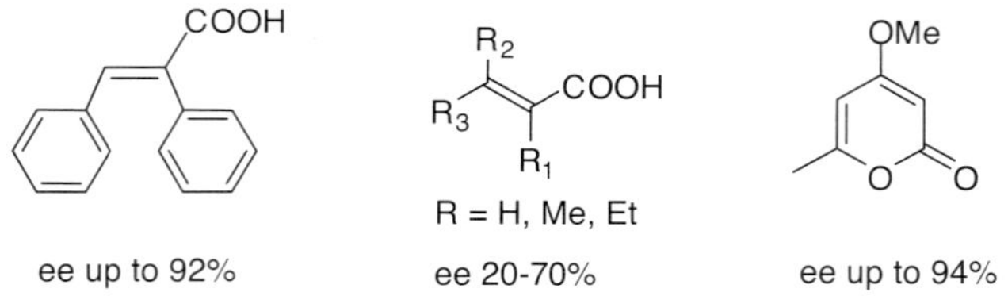

Figure 2.9 Hydrogenation of activated C=C bonds.

ee up to 92%   ee 20–70%   ee up to 94%

and focused on α,β-unsaturated acids and ketones. We will restrict this section to a short summary of important results (see Figure 2.9) because despite considerable efforts, ee values are usually low to moderate and cannot compete with the enantioselectivities observed for homogeneous catalysts (for an extensive review, see Ref. [16f]). For α,β-unsaturated acids (ee values up to 92% [41]) and hydroxymethylpyrone derivatives (ee values up to 94% [42]) cinchonidine is the preferred modifier. Best enantioselectivities have been obtained for α,β-unsaturated acids with aryl and alkyl substituents whereas amido groups lead to low ee values. The nature of the solvent and the presence of water, acids, or bases (especially benzylamine [43]) can have strong effects on activity and enantioselectivity of these catalysts. With the possible exception of the hydrogenation of hydroxymethylpyrone derivatives, no application with real synthetic potential has been reported yet.

2.4
Industrial Applications

The first technical application of a cinchona-modified Pt catalyst was reported by Ciba-Geigy in 1986 for the synthesis of methyl (R)-2-hydroxy-4-phenyl butyrate (R)-HPB ester), an intermediate for the ACE inhibitor benazepril depicted in Figure 2.10 [44a].

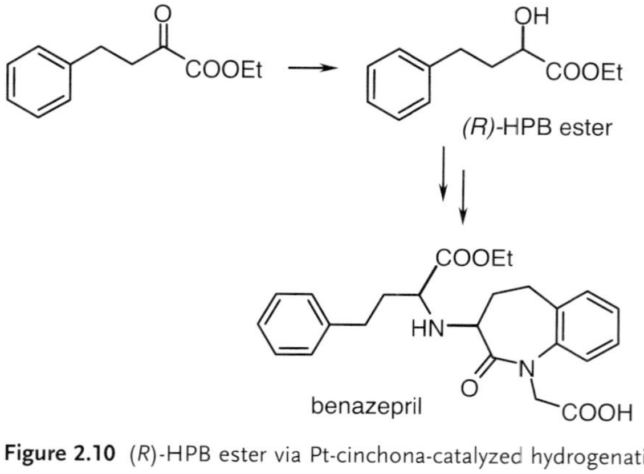

Figure 2.10 (R)-HPB ester via Pt-cinchona-catalyzed hydrogenation of an α-keto ester.

The development of a viable process for the HPB ester took more than a year. In the course of process development, more than 200 hydrogenation reactions were carried out. The most important results of this development work can be summarized as follows:

- **Catalyst**: 5% Pt/Al$_2$O$_3$ catalysts gave the best overall performance, and the E 4759 from Engelhard was the final choice.
- **Modifier**: About 20 modifiers were tested; HCd (in toluene) and MeOHCd (an AcOH) gave best results and were chosen for further development.
- **Solvent**: It was found that acetic acid was far superior to all classical solvents, allowing up to 92% ee for the HPB ester and 95% for ethyl pyruvate (then a new world record!) [45]. For technical reasons, toluene was chosen as solvent for the production process.
- **Reaction conditions**: Best results (full conversion after 3–5 h, high yield, 80% ee) were obtained at 70 bar and room temperature with 0.5% w/w 5% Pt/Al$_2$O$_3$ (pretreated in H$_2$ at 400 °C) and 0.03% w/w modifier.
- **Substrate quality**: The enantioselective hydrogenation of α-keto esters proved to be exceptionally sensitive to the origin of the substrate [44b].

After about 2 years, the production process was developed, patented, and scaled up, and in 1987 several hundred kilograms were manufactured in a 500 l autoclave.

A few years later, a new process for the (R)-HPB ester was reported by Solvias in collaboration with Ciba SC (see Figure 2.11). Claisen condensation of cheap acetophenone and diethyl oxalate was carried out, followed by chemo- and enantio-selective hydrogenation of the resulting diketo ester and hydrogenolysis to the HPB ester [30]. Even though the 2,4-dioxo ester was a new substrate type, it took only a few months to develop, scale up, and implement the new process. The following aspects were the key to success: (i) the low price of the diketo ester prepared via Claisen

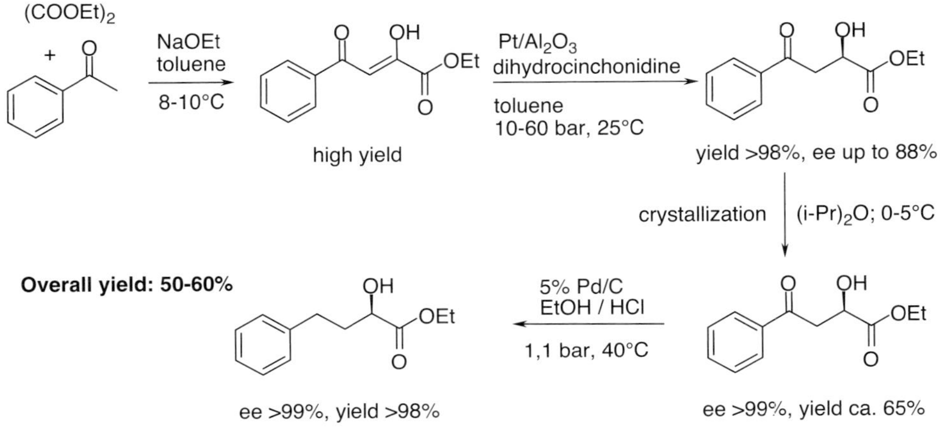

Figure 2.11 Second-generation synthesis of (R)-HPB ester.

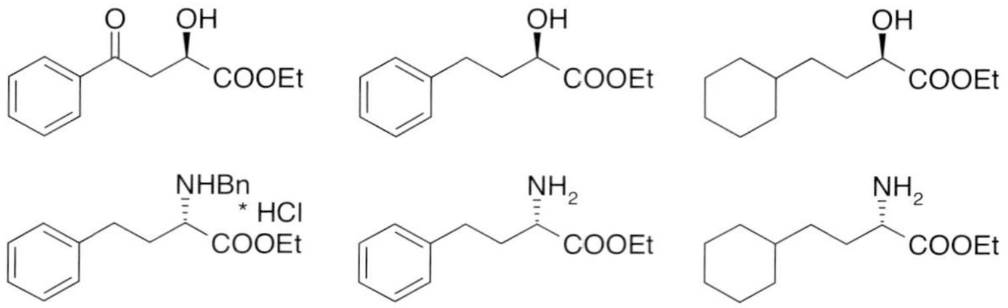

Figure 2.12 HPB ester-related compounds available from Fluka.

condensation of acetophenone and diethyl oxalate; (ii) the high chemoselectivity in the Pt-cinchona hydrogenation; and (iii) the possibility to enrich the hydroxy ketone intermediate with ee values ranging from as low as 70 to >99% in one crystallization step. The removal of the second keto group via Pd-catalyzed hydrogenolysis did not lead to any racemization. Derived from the keto hydroxy intermediate, a whole range of chiral building blocks is now available in laboratory quantities from Fluka both in the (R)- and in the (S)-form (see Figure 2.12) [30c].

For a comparison of the various processes developed for industrial manufacturing of (R)-HPB ester, see Ref. [46].

Bench-scale processes were developed for the enantioselective hydrogenation of p-chlorophenylglyoxylic acid derivatives using both homogeneous and heterogeneous catalysts (Figure 2.13) [47]. For the production of kilogram amounts of (S)-p-chloromandelic acid, a Ru/MeO-BIPHEP catalyst achieved 90–93% ee (s/c 4000 and TOF up to $210\,h^{-1}$). A modified Pt catalyst achieved 93% ee for (R)-methyl p-chloromandelate and 86% ee for (S)-methyl p-chloromandelate using HCd and isocinchonine as modifier, respectively. For the HCd-Pt system, a scale-up from 100 mg to 15 g presented no problems, indicating that the Pt-cinchona system

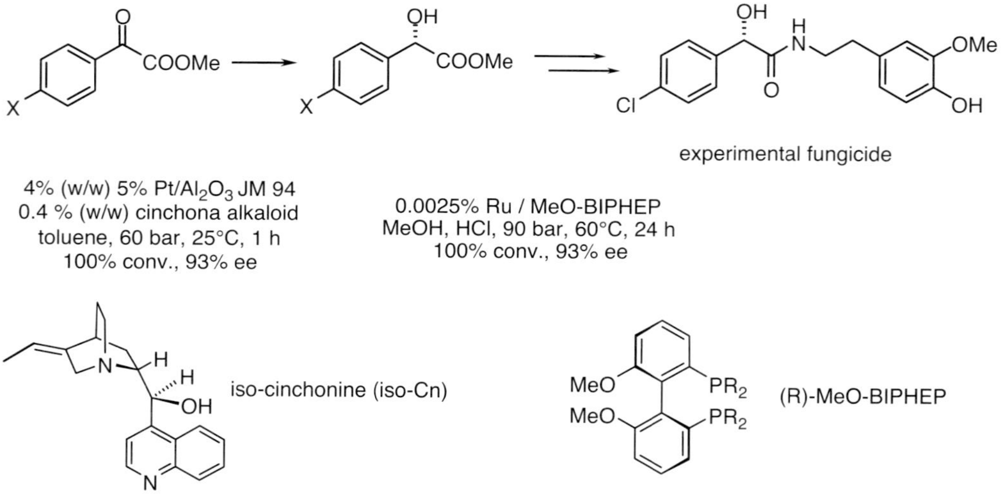

Figure 2.13 Hydrogenation of p-chlorophenylglyoxylic acid derivatives.

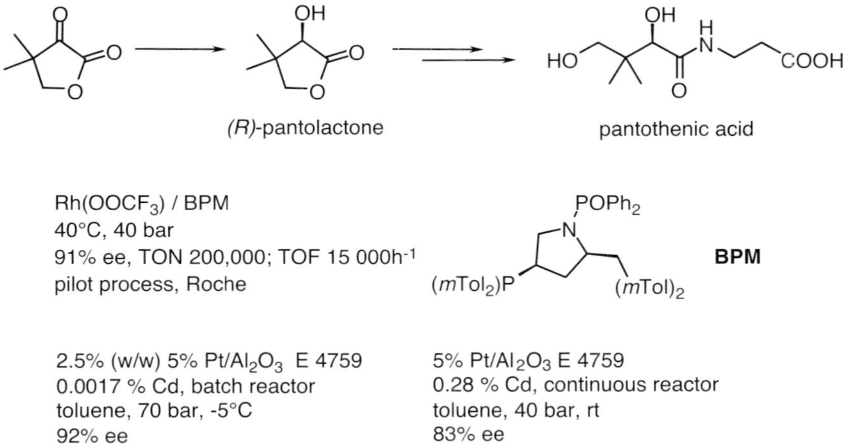

(R)-pantolactone
HO
OH
COOH
pantothenic acid
Rh(OOCF3) / BPM
40°C, 40 bar
91% ee, TON 200,000; TOF 15 000h-1
pilot process, Roche
POPh2
(mTol2)P
(mTol)2
BPM
2.5% (w/w) 5% Pt/Al2O3 E 4759
0.0017 % Cd, batch reactor
toluene, 70 bar, -5°C
92% ee
5% Pt/Al2O3 E 4759
0.28 % Cd, continuous reactor
toluene, 40 bar, rt
83% ee

Figure 2.14 Hydrogenation of ketopantolactone.

might be a viable alternative to the homogeneous catalyst for the production of the (*R*)-enantiomer.

The hydrogenation of an α-ketolactone depicted in Figure 2.14 is the key step for an enantioselective synthesis of pantothenic acid, which is produced by Roche via racemate separation of pantoic acid. A homogeneous pilot process was developed by Roche [48] and (*R*)-pantolactone was produced in multihundred kilogram quantities. A Rh/BPM catalyst proved to be highly active with satisfactory selectivity of 91% ee. As already described by Niwa *et al.* [49], α-keto lactones are also suitable substrates for Pt/cinchona catalysts. This was confirmed by the Baiker group who developed a continuous bench-scale process for the hydrogenation of ketopantolactone [50a]. The reaction was carried out in a fixed-bed reactor using the standard E 4759 catalyst at 40 bar. Cinchonidine was added continuously with the feed, otherwise ee values dropped very fast. Productivity was 94 mmol/(g_{cat} h) and up to 83% ee was obtained. Compared to the heterogeneous batch process with ee values up to 92% at much lower Cd concentrations [50b] or to the homogeneous process, the continuous variant is probably not competitive. At the moment, pantothenic acid is still produced using the established racemate separation route.

2.5 Conclusions

The scope of cinchona-based chiral auxiliary or chirality transmitters for enantioselective reductions is at the moment restricted to heterogeneous Pt and Pd catalysts and primarily to the reduction of α-functionalized ketones and to a lesser degree of activated C=C bonds. Up to now, very few homogeneous catalysts have been described, and with the exception of a transfer hydrogenation system, none shows any promise.

For the hydrogenation of α-keto esters and α-keto acetals, the performance of the heterogeneous cinchona-modified Pt catalysts is equal to and in some cases superior to the best homogenous catalysts. Indeed, several industrial applications have been described that underline this statement. For most other substrates, the performance of the cinchona-modified Pd or Pt catalysts is not (yet) on a level where the application to "real-world" substrates has been demonstrated.

References

1 Jacobsen, E.N., Hayashi, T., and Pfaltz, A. (eds) (1999) *Comprehensive Asymmetric Catalysis*, Springer, Berlin.
2 Blaser, H.U. and Schmidt, E. (eds) (2003) *Large Scale Asymmetric Catalysis*, Wiley-VCH Verlag GmbH, Weinheim.
3 For a recent overview, see Kacprzak, K. and Gawronski, J. (2001) *Synthesis*, 961.
4 Cervinka, O. (1965) *Collect. Czech. Chem. Commun.*, **30**, 1684.
5 (a) Orito, Y., Imai, S., Niwa, S., and Nguyen, G.-H. (1979) *J. Synth. Org. Chem. Jpn.*, **37**, 173; (b) Orito, Y., Imai, S., and Niwa, S. (1979) *J. Chem. Soc. Jpn.*, 670, 1118, 1980.
6 Julia, S., Gienbreda, A., Guixer, J., Masana, J., and Tomas, A. (1981) *J. Chem. Soc., Perkin Trans. I*, 574.
7 Perez, J.R.G., Malthete, J., and Jacques, J. (1985) *C. R. Acad. Sci. Paris Ser. II*, 169.
8 (a) He, W., Zhang, B.-L., Jiang, R., Liu, P., Sun, X.-L., and Zhang, S.-Y. (2006) *Tetrahedron Lett.*, **47**, 5367; (b) He, W., Liu, P., Zhang, B.-L., Sun, X.-L., and Zhang, S.-Y. (2006) *Appl. Organomet. Chem.*, **20**, 328.
9 Cervinka, O., Fabryova, A., and Sablukova, I. (1986) *Collect. Czech. Chem. Commun.*, **51**, 401.
10 Hofstetter, C., Wilkinson, P.S., and Pochapsky, T.C. (1999) *J. Org. Chem.*, **64**, 8794.
11 Cha, R.-T., Li, Q.-L., Chen, N.-L., and Wang, Y.-P. (2007) *J. Appl. Polym. Sci.*, **103**, 148.
12 Takeuchi, S. and Ohgo, Y. (1988) *Chem. Lett.*, 403.
13 Vannoorenberghe, Y. and Buono, G. (1988) *Tetrahedron Lett.*, **29**, 3235.
14 Drew, M.D., Lawrence, N.J., Watson, W., and Bowles, S.A. (1997) *Tetrahedron Lett.*, **38**, 5857.
15 (a) Orito, Y., Imai, S., and Niwa, S. (1980) *J. Chem. Soc. Jpn.*, 670; (b) Orito, Y., Imai, S., and Niwa, S. (1982) *J. Chem. Soc. Jpn.*, 137.
16 (a) Studer, M., Blaser, H.U., and Exner, C. (2003) *Adv. Synth. Catal.*, **345**, 45; (b) Murzin, D., Maki-Arvela, P., Toukoniitty, E., and Salmi, T. (2005) *Catal. Rev. Sci. Eng.*, **47**, 175; (c) Baiker, A. (2005) *Catal. Today*, **100**, 159; (d) Blaser, H.U. and Studer, M. (2005) in *Handbook of Chiral Chemicals*, 2nd edn (ed. D.J. Ager), CRC Press, Boca Raton, FL, p. 345; (e) Bartók, M. (2006) *Curr. Org. Chem.*, **10**, 1533; (f) Mallat, T., Diezi, S., and Baiker, A. (2008) in *Handbook of Heterogeneous Catalysis*, 2nd edn (eds G. Ertl, H. Knözinger, F. Schüth, and J. Weitkamp), Wiley-VCH Verlag GmbH, Weinheim, p. 3603; (g) Mallat, T., Orglmeister, E., and Baiker, A. (2007) *Chem. Rev.*, **107**, 4863.
17 Sonderegger, O.J., Ho, G.M.-W., Bürgi, T., and Baiker, A. (2005) *J. Catal.*, **230**, 499.
18 Ostgard, D.J., Hartung, R., Krauter, J.G.E., Seebald, S., Kukula, P., Nettekoven, U., Studer, M., and Blaser, H.U. (2005) *Chem. Ind. (Catal. Org. React.)*, **104**, 553.
19 For recent reviews on the effect of modifier structure, see (a) Pfaltz, A. and Heinz, T. (1997) *Top. Catal.*, **4**, 229; (b) Blaser, H.U., Jalett, H.P., Lottenbach, W., and Studer, M. (2000) *J. Am. Chem. Soc.*, **122**, 12675.

20 Török, B., Balázsik, K., Felföldi, K., and Bartók, M. (2000) *Stud. Surf. Sci. Catal.*, **130**, 3381.
21 Margitfalvi, J.L., Tálas, E., and Hegedüs, M. (1999) *Chem. Commun.*, 645.
22 Zuo, X., Liu, H., Guo, D., and Yang, X. (1999) *Tetrahedron*, **55**, 7787.
23 (a) Török, B., Balázsik, K., Török, M., Szöllösi, G., and Bartók, M. (2000) *Ultrason. Sonochem.*, **7**, 151; (b) Balázsik, K., Török, B., Felföldi, K., and Bartók, M. (1999) *Ultrason. Sonochem.*, **5**, 149; (c) Török, B., Felföldi, K., Szakonyi, G., and Bartók, M. (1997) *Ultrason. Sonochem.*, **4**, 301; (d) Török, B., Felföldi, K., Szakonyi, G., Balázsik, K., and Bartók, M. (1998) *Catal. Lett.*, **52**, 81.
24 LeBlond, C., Wang, J., Liu, J., Andrews, A.T., and Sun, Y.-K. (1999) *J. Am. Chem. Soc.*, **121**, 4920; Wang, J., Andrews, A.T., and Sun, Y.-K. (2000) *Top. Catal.*, **13**, 169.
25 Blaser, H.U. and Jalett, H.P. (1993) *Stud. Surf. Sci. Catal.*, **78**, 139.
26 Sutyinszki, M., Szöri, K., Felföldi, K., and Bartók, M. (2002) *Catal. Commun.*, **3**, 125.
27 Balázsik, K., Szöri, K., Felföldi, K., Török, B., and Bartók, M. (2000) *Chem. Commun.*, 555.
28 Wang, G.Z., Mallat, T., and Baiker, A. (1997) *Tetrahedron: Asymmetry*, **8**, 2133.
29 Szöri, K., Török, B., Felföldi, K., and Bartók, M. (2001) *Chem. Ind. (Dekker)*, **82**, 489.
30 (a) Studer, M., Burkhardt, S., Indolese, A.F., and Blaser, H.U. (2000) *Chem. Commun.*, 1327; (b) Herold, P., Indolese, A.F., Studer, M., Jalett, H.P., and Blaser, H.U. (2000) *Tetrahedron*, **56**, 6497; (c) Blaser, H.U., Burkhardt, S., Kirner, H.-J., Mössner, T., and Studer, M. (2003) *Synthesis*, 1679.
31 (a) von Arx, M., Bürgi, T., Mallat, T., and Baiker, A. (2002) *Chem. Eur. J.*, **8**, 1430; (b) von Arx, M., Mallat, T., and Baiker, A. (2002) *Catal. Lett.*, **78**, 267 and references cited therein; (c) Bodmer, M., Mallat, T., and Baiker, A. (1998) *Chem. Ind. (Dekker)*, **75**, 75; (d) von Arx, M., Mallat, T., and Baiker, A. (2001) *J. Catal.*, **202**, 169; (e) von Arx, M., Mallat, T., and Baiker, A. (2001) *Spec. Pub. Royal Chem. Soc.*, **266**, 247; (f) von Arx, M., Mallat, T., and Baiker, A. (2001) *Tetrahedron: Asymmetry*, **12**, 3089.
32 (a) Szöri, K., Szöllösi, G., and Bartók, M. (2006) *Adv. Synth. Catal.*, **348**, 515; (b) Szöri, K., Szöllösi, G., and Bartók, M. (2006) *J. Catal.*, **244**, 255.
33 (a) Studer, M., Burkhardt, S., and Blaser, H.U. (1999) *Chem. Commun.*, 1727; (b) Török, B., Felföldi, K., Balázsik, K., and Bartók, M. (1999) *Chem. Commun.*, 1725.
34 Studer, M., Blaser, H.U., and Burkhardt, S. (2002) *Adv. Synth. Catal.*, **344**, 511.
35 Wandeler, R., Künzle, N., Schneider, M.S., Mallat, T., and Baiker, A. (2001) *Chem. Commun.*, 673.
36 Künzle, N., Szabó, A., Schürch, M., Wang, G., Mallat, T., and Baiker, A. (1998) *Chem. Commun.*, 1377.
37 Studer, M., Blaser, H.U., and Okafor, V. (1998) *Chem. Commun.*, 1053.
38 Slipszenko, J.A., Griffith, S.P., Johnston, P., Simons, K.E., Vermeer, W.A., and Wells, P.B. (1998) *J. Catal.*, **179**, 267.
39 (a) Toukoniitty, E., Mäki-Arvela, P., Kuzma, M., Villela, A., Neyestanaki, A.K., Salmi, T., Sjöholm, R., Leino, R., Laine, E., and Murzin, D.Y. (2001) *J. Catal.*, **204**, 281; (b) Toukoniitty, E., Mäki-Arvela, P., Wärnå, J., and Salmi, T. (2001) *Catal. Today*, **66**, 411; (c) Toukoniitty, E., Mäki-Arvela, P., Sjöholm, R., Leino, R., Salmi, T., and Murzin, D.Y. (2002) *React. Kinet. Catal. Lett.*, **75**, 21.
40 Sonderegger, O.J., Ho, G.M.-W., Bürgi, T., Baiker, A. (2005) *J. Mol. Catal. A: Chem.*, **229**, 19.
41 (a) Nitta, Y., Watanabe, J., Okuyama, T., and Sugimura, T. (2005) *J. Catal.*, **236**, 164; (b) Sugimura, T., Watanabe, J., Uchida, T., Nitta, Y., Okuyama, T., and Tadashi, (2006) *Catal. Lett*, **12**, 27 and references cited therein.
42 (a) Huck, W.R., Mallat, T., and Baiker, A. (2000) *J. Catal.*, **193**, 1; (b) Huck, W.R.,

Bürgi, T., Mallat, T., and Baiker, A. (2001) *J. 2Catal.*, **200**, 171; 2002, **205**, 213; (c) Huck, W.R., Mallat, T., and Baiker, A. (2002) *Catal. Lett.*, **80**, 87.
43 Kun, I., Török, B., Felföldi, K., and Bartók, M. (2000) *J. Mol. Catal. A: Gen.*, **203**, 71; Szöllösi, G., Fülöp, F., and Bartók, M. (2007) *Appl. Catal. A: Gen.*, **331**, 39 and references cited therein.
44 (a) Sedelmeier, G.H., Blaser, H.U., and Jalett, H.P. (1986) EP 206993 assigned to Ciba-Geigy AG; (b) Blaser, H.U., Jalett, H.P., and Spindler, F. (1996) *J. Mol. Catal. A: Chem.*, **107**, 85.
45 Blaser, H.U., Jalett, H.P., and Wiehl, J. (1991) *J. Mol. Catal.*, **68**, 215.
46 Blaser, H.U., Eissen, M., Fauquex, P.F., Hungerbühler, K., Schmidt, E., Sedelmeier, G., and Studer, M. (2003) in *Large Scale Asymmetric Catalysis* (eds H.U. Blaser and E. Schmidt), Wiley-VCH Verlag GmbH, Weinheim, p. 91.
47 Cederbaum, F., Lamberth, C., Malan, C., Naud, F., Spindler, F., Studer, M., and Blaser, H.U. (2004) *Adv. Synth. Catal.*, **346**, 842.
48 Schmid, R. and Scalone, M. (1999) in *Comprehensive Asymmetric Catalysis* (eds E.N. Jacobsen, H. Yamamoto, and A. Pfaltz), Springer, Berlin, p. 1439.
49 Niwa, S. Imamura, J., and Otsuka, K. (1987) JP 62158268 (CAN 108, 128815).
50 (a) Künzle, N., Hess, R., Mallat, T., and Baiker, A. (1999) *J. Catal.*, **186**, 239; (b) Schürch, M., Künzle, N., Mallat, T., and Baiker, A. (1998) *J. Catal.*, **176**, 569.

3
Cinchona Alkaloids as Chiral Ligands in Asymmetric Oxidations
David J. Ager

3.1
Introduction

For oxidation reactions, the cinchona alkaloids have been mainly employed to control the osmium-catalyzed conversion of an alkene to give a 1,2-diol or vicinal functionalized alcohol. As these are important asymmetric reactions, they have been the subject of a number of reviews [1–18]. This chapter discusses the uses of these alkaloids as chiral ligands in asymmetric oxidation reactions. Oxidation reactions where an alkaloid is used in a phase-transfer sense are discussed in Chapter 5.

The use of osmium tetroxide for the conversion of an alkene to a 1,2-diol is a well-established reaction [19–22]. The formation of an intermediate cyclic ester accounts for the *cis*-stereochemistry [21, 23–32] as reaction occurs on the least hindered face of the alkene [21, 30, 33–38]. This steric effect is amplified in cyclic substrates [39, 40]. The reaction conditions have to be carefully controlled to avoid oxidative cleavage of the diol product [28].

There is a marked rate acceleration in the presence of a tertiary amine or pyridine [19, 41]. This finding provided the background for the asymmetric dihydroxylation (AD) and, later, the asymmetric aminohydroxylation (AA) reactions as it is this ligand acceleration effect (LAE) that ensures the reaction pathway involving the ligand.

The use of a cooxidant can reduce the amount of osmium required for a complete reaction of an alkene from stoichiometric to catalytic; some examples of oxidants that can achieve this are peroxides [20, 22, 35, 37, 42–44] including hydrogen peroxide [20, 42], chlorates [45], periodate [46, 47], hypochlorite [48], *N*-methyl-morpholine-*N*-oxide (NMMO) [22, 34, 35, 37, 49], potassium ferricyanide [50, 51], and even air [52, 53].

Stereoselection for the OsO_4 reactions itself can also be observed when a "directing" group is adjacent to the alkene, such as in an allyl alcohol. An empirical rule has been devised, where the reagent approaches from the face opposite to the preexisting oxygen functionality. Although the hydroxy group may be protected, the presence of an acyl group reduces stereoselectivity. *cis*-Alkenes afford better selectivity than their *trans*-counterparts [54–58], while cyclic substrates can offer stereoselection through the inherent stereofacial requirements of the substrate [39, 59] or from chiral auxiliaries [60].

The use of chiral ligands has allowed the asymmetric oxidation of alkenes with osmium reagents in catalytic amounts [61–66]. The cinchona alkaloids have played, and continue to play, a significant role in this useful synthetic methodology.

There are still some issues associated with the use of osmium, especially osmium tetroxide at scale. The presence of a heavy metal, even in trace amounts, in a pharmaceutical product can lead to expensive procedures for its removal. In addition, the exposure limits and relatively high vapor pressure of osmium tetroxide give rise to handling and containment issues. At scale, the handling and disposal of waste containing heavy metals can be costly. For laboratory uses, some of these issues have been addressed by the necessary mixtures of reagents being sold with the osmium contained within this mixture as a nonvolatile salt (see below). The metal itself can be immobilized and reused, and this has been advocated to alleviate some of the handling and safety problems [5, 67–72].

3.2
Asymmetric Dihydroxylation of Alkenes

3.2.1
Early Reactions

Early work on the oxidation of an alkene by osmium tetroxide in the presence of a chiral ligand, such as dihydroquinine acetate (**1**, R = Ac) or dihydroquinidine acetate (**2**, R = Ac), led to 1,2-diol formation with some enantiomeric excess [73]. The asymmetric induction was greatly improved by the use of cinchona alkaloid esters (**1** and **2**, R = p-ClC$_6$H$_4$) together with a catalytic amount of osmium tetroxide [2, 74–76]. The alkaloid esters act as pseudoenantiomeric ligands (Scheme 3.1) [77–81]. The use of NMMO as the stoichiometric oxidant gave better catalytic turnover and yields [73, 78]. However, the use of NMMO gave lower enantiomeric excesses compared to the stoichiometric reaction [78, 82]. This gave rise to the hypothesis of two catalytic cycles (Scheme 3.2) [2]. The procedure was modified to overcome the low ee problem by using a slow addition of the alkene; ironically, this modification results in a faster reaction. The use of low alkene concentrations effectively removes the second, low enantioselective cycle [77, 82].

The use of potassium ferricyanide in place of NMMO as oxidant also improves the level of asymmetric induction [50, 51, 83], as the slow addition of oxidant may not be necessary because of the formation of a biphasic system where the osmium is the only oxidant in the organic layer [51, 74, 83]. Hydrolysis of the resultant osmium(VI) monoglycolate ester **3** releases the diol product and the ligand to the organic layer and the osmium(VI) to the aqueous layer, preventing its entry into the second cycle. Thus, the two-cycle system shown in Scheme 3.2 is modified to the simpler system that is shown in Scheme 3.3 [2, 5].

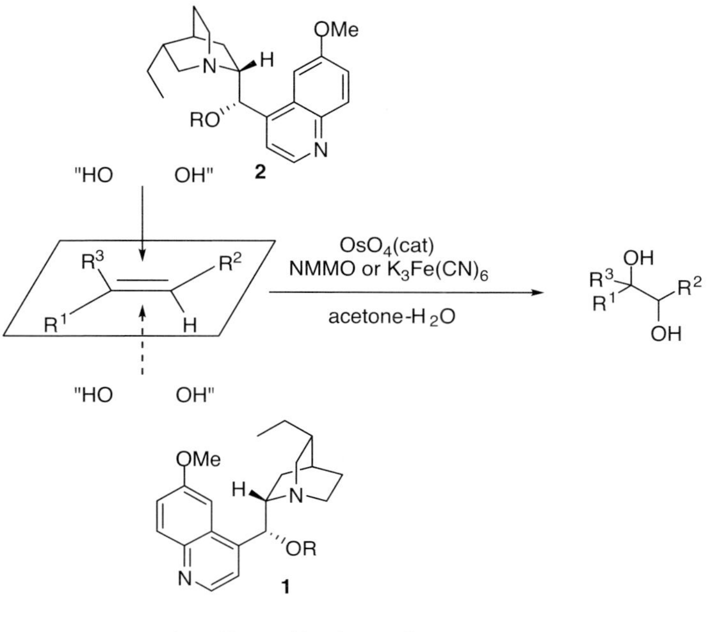

3.2 Asymmetric Dihydroxylation of Alkenes

where R = p-chlorobenzoyl

Scheme 3.1

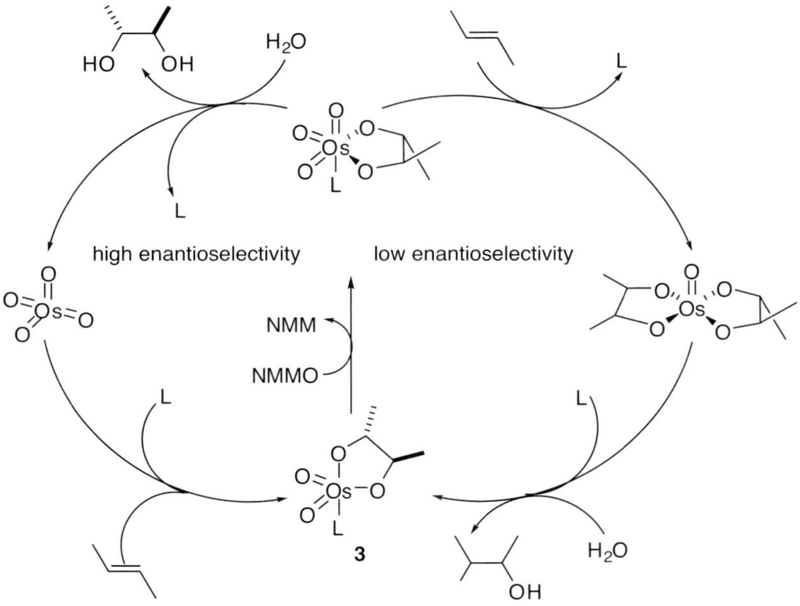

Scheme 3.2

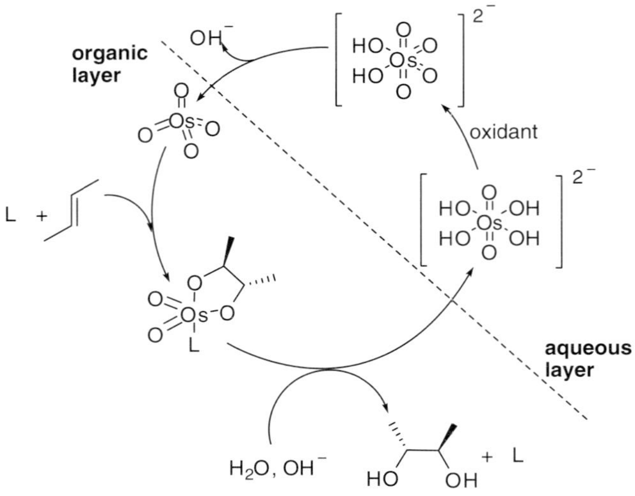

32 *3 Cinchona Alkaloids as Chiral Ligands in Asymmetric Oxidations*

Scheme 3.3

The use of a sulfonamide aids the hydrolysis of the osmate esters, which improves the catalyst turnover. A sulfonamide, such as methanesulfonamide, can routinely be added to the reaction mixture. Only in the case of terminal alkenes, the presence of this additive slows down the reaction [23, 76, 84].

Although the use of cinchona alkaloids as chiral ligands provides good asymmetric induction with a number of types of alkene, ee values were usually not high and the search for better systems continued [1, 2]. Selectivities and scope continued to expand, but the major breakthrough was the use of the bisligand systems (Scheme 3.4) [76, 84–88]. Some of the more common, current systems are summarized in Figure 3.1 [2, 5, 14].

The bisalkaloid ligands are simple to make from the dichloride of the heterocyclic spacer with KOH and K_2CO_3 as solid bases in toluene (Scheme 3.5) [2].

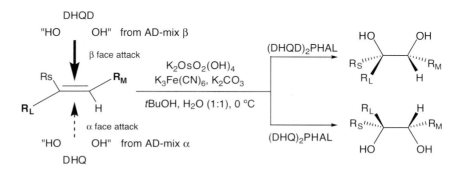

$MeSO_2NH_2$ should be added if the alkene is nonterminal as this increases the rate of ester hydrolysis.

Scheme 3.4

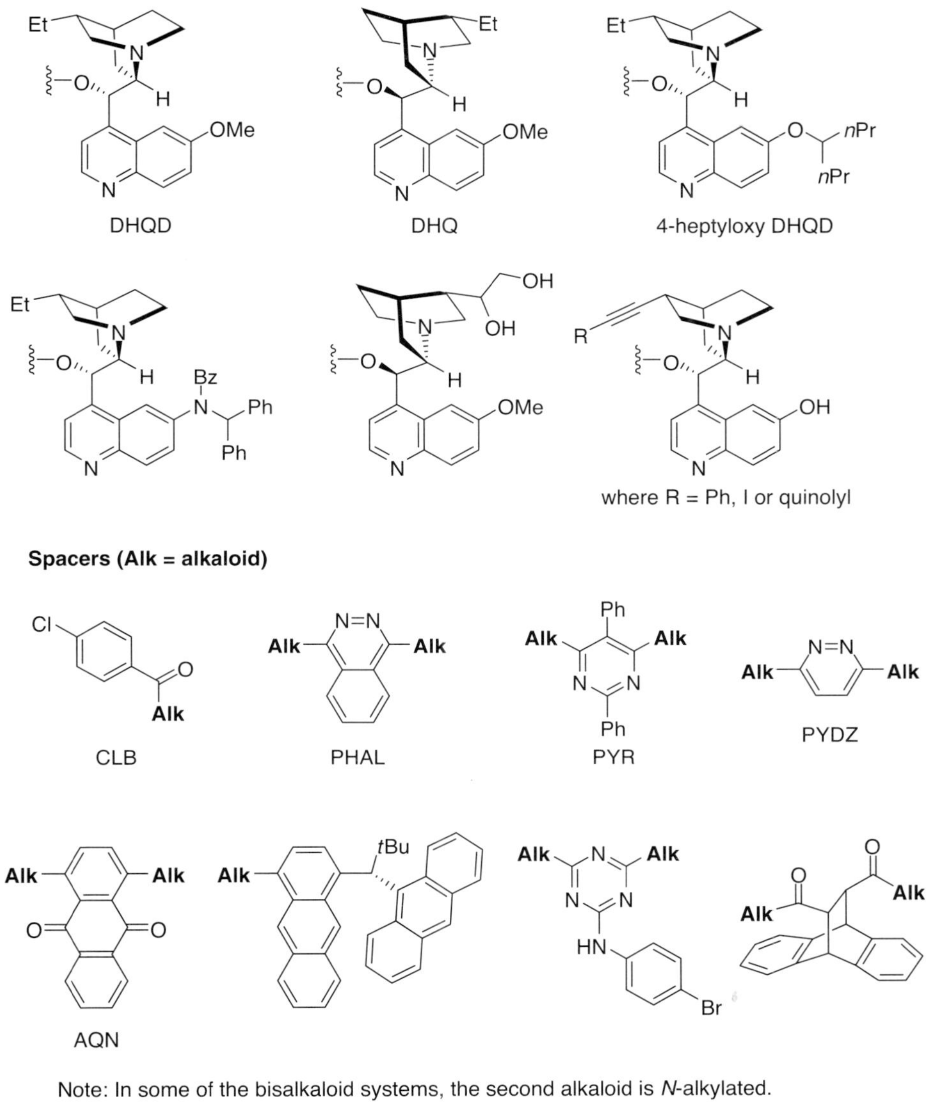

Spacers (Alk = alkaloid)

Note: In some of the bisalkaloid systems, the second alkaloid is N-alkylated.

Figure 3.1 Common cinchona alkaloid derived ligand systems for asymmetric oxidations.

3.2.2
Bisalkaloid Ligands

For many AD reactions, the bisalkaloid systems (cf. Figure 3.1), in particular the (DHQ)$_2$PHAL and (DHQD)$_2$PHAL systems, provide good asymmetric induction. Studies with mixed ligand systems showed that the (DHQD)$_2$PHAL complex was

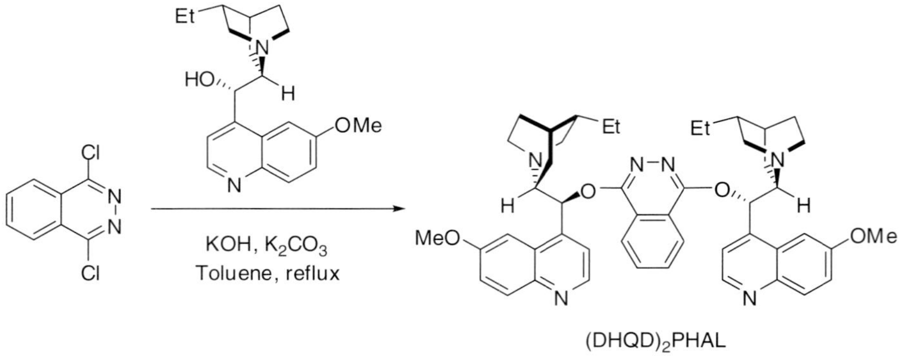

Scheme 3.5

much more reactive in the AD reaction than the complex from DHQ-CLB, indicating the high reactivity associated with these bisalkaloid ligands [89].

Although the bisligand systems provide faster reactions and better selectivities in the majority of cases, more hindered substrates, such as tetrasubstituted alkenes or alkenes with a bulky substituent can benefit from the use of a monoalkaloid ligand system [90]. For a discussion on relative rates of reaction for different substrate classes and selectivities see Section 3.2.5.1.

With the bisalkaloid ligands, potassium ferricyanide can be used as the stoichiometric oxidant [84, 91]. As with the parent achiral osmium oxidation, NMMO can also be used as the oxidant (see above) [92]. However, rather than using NMMO in stoichiometric amounts, this morpholine component can be used in catalytic amounts by the addition of the biomimetic flavin **4** to set up a triple catalytic system where hydrogen peroxide is the oxidant [93–95]. Methyltrioxorhenium can be used in place of the flavin mimic [96], as can tungsten(VI) [97] and carbon dioxide [98].

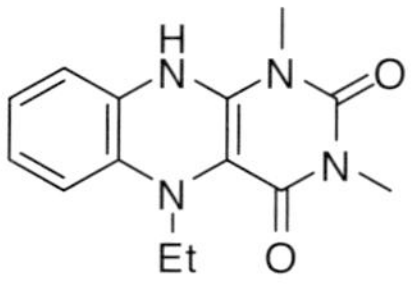

Other alternatives for the oxidant for stoichiometric oxidations include the use of a selenoxide [99], including a photochemical oxidation of catalytic selenium [100], iodine [101], sodium chlorite [102], hypochlorite [103], and electrochemical methods [101, 104]. Even air can be used as the oxidant [99, 100], but care has to be taken with regard to the choice of solvent as cleavage of the product 1,2-diol can occur, especially when the alkene has an aryl substituent [53, 105, 106].

The use of a constant pH (12) for the reaction with internal alkenes provides faster reaction rates, and the need for the hydrolysis aid, methanesulfonamide, is alleviated. For terminal alkenes, the use of a constant pH 10 affords higher enantioselectivities compared to reactions where the pH is not controlled [107].

$K_2OsO_2(OH)_4$ can be used as the nonvolatile form of osmium and this can be mixed with the oxidant, potassium ferricyanide, and the ligand to allow ease of use.

These mixtures have been called AD-mixes. The AD-mix α contains the (DHQ)$_2$PHAL ligand while AD-mix β contains (DHQD)$_2$PHAL with a ligand to metal ratio of ~2.5 : 1 [2]. These mixtures are commercially available. For larger scale applications, it is possible to recover the ligand by extraction [2].

3.2.3
Mechanism

X-Ray, NMR, kinetic analyses, and theoretical approaches have provided insight into the mechanism for both the achiral and asymmetric osmium-catalyzed oxidations of alkenes [64, 77, 81, 108–119].

This work has allowed an empirical mnemonic to be proposed (cf. Scheme 3.4), which has later been refined. The mnemonic allows for the prediction of the reaction outcome and reactivity. Initially, the steric requirements of the system indicated that a hydrogen atom (i.e., a trisubstituted alkene is the best substrate) is necessary for a good asymmetric induction (Figure 3.2a) [84].

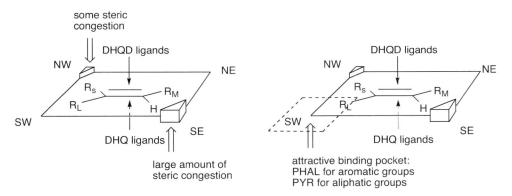

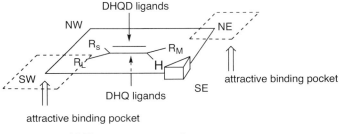

Figure 3.2 Steric requirements for the asymmetric dihydroxylation procedure as indicated by the mnemonic, which has been modified over time.

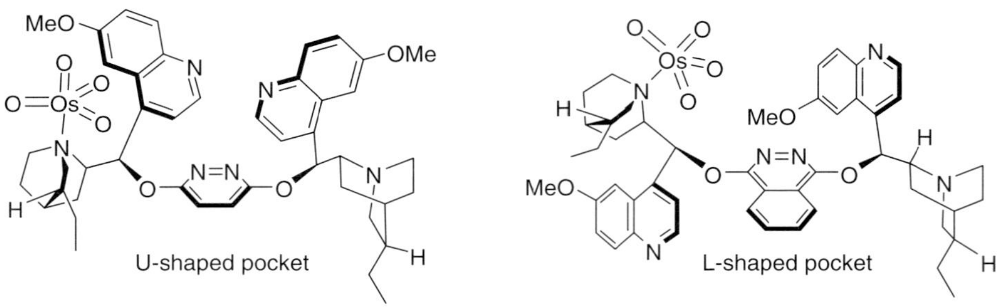

Figure 3.3 The U-shaped (Corey model) and L-shaped (Sharpless model) binding pockets.

The face selectivity is believed to arise from how the alkene fits into the binding pocket created by the aromatic groups of the ligand. Indeed, there has been controversy about the exact structure of this pocket with a U-shaped one being proposed by Corey [120, 121], and an L-shaped one being proposed by Sharpless (Figure 3.3) [114, 122]. The U-shaped pocket with the enzyme-like kinetics is now the accepted model.

The initial model (Figure 3.2a) for the transition state was based on a [2 + 2] cycloaddition reaction model and, although NMR studies support this, there is now evidence and consensus that a [3 + 2] model is followed [123–130]. Computational studies suggest that solvent can play a key role in account for the apparent reversal of reaction stereochemical outcome compared to that predicted by Sharpless's original mnemonic (Figure 3.2a). This model has been modified to include attractive forces based on kinetic and molecular modeling studies [114, 122]. Some indication of which the best ligand system is for different classes of substrates can be advanced base on the attractive forces (Figure 3.2b) [131].

A further model has been suggested and this can be refined by calculation [127]. This modified mnemonic considers the interactions for a specific substrate and the hydrogen bonding that is affected by the solvent system. These points have been combined into one model, which contains attractive and repulsive quadrants (Figure 3.2c) [132]. Note that the NW quadrant is now open. Further computational studies have shown this facial model to be in agreement with the substrate–complex interactions within the transition state and experimental results [133].

3.2.4
Variations

The presence of water in the reaction medium is required for the hydrolysis of the intermediate osmate esters. Although aqueous *tert*-butanol is the solvent system of choice, MTBE has been used in a large-scale application [2].

A number of variations of the ligand system were discussed in the preceding section. In addition to solution chemistry, the ligands can also be supported on a polymer [134–143], including polystyrene [144], mesoporous molecular sieve [145], ion exchange resin [146, 147], macroporous resins [148], silica [149, 150], and amorphous silica gel [145, 151]. The use of a titanium silicate–silica gel system also enables the use of hydrogen peroxide as the oxidant [151].

A soluble polymer-bound version of the asymmetric dihydroxylation ligand is provided by DHQD-PHAL-OPEG-OMe, a polyethylene glycol derivative [152, 153]. The use of a PEG system with a ligand at each end allowed the dihydroxylation reaction to be run in a continuous flow system with a membrane to retain the ligand. Under these conditions, metal leaching was observed [154].

The catalyst system can be reused if an ionic liquid–water or ionic liquid–water–*tert*-butanol solvent system is used [155–157]. Some care has to be taken with the choice of ligand under these conditions, as the presence of an alkene within the ligand itself results in dihydroxylation during the reaction and, as a result, the modified ligand is not extracted from the ionic liquid into the organic layer [158]. Supercritical CO_2 can be used in place of the organic solvent to extract the product [159, 160].

A method has been developed to entrap the osmium at the end of an AD reaction and allow the metal to be reused in another cycle with no loss of yield or stereoselectivity [161]. Another alternative is the use of nanofiltration to separate the product from the metal and ligand components. These expensive materials can then be used in another cycle of the reaction [162].

The use of a dihydroxylation procedure can provide high stereoselectivity compared to a direct epoxidation method. The conditions for the dihydroxylation of **5** had to be modified due to the low solubility of the substrate and the base sensitivity of the product (Scheme 3.6). The result was a more efficient process for scale-up compared to an epoxidation [163].

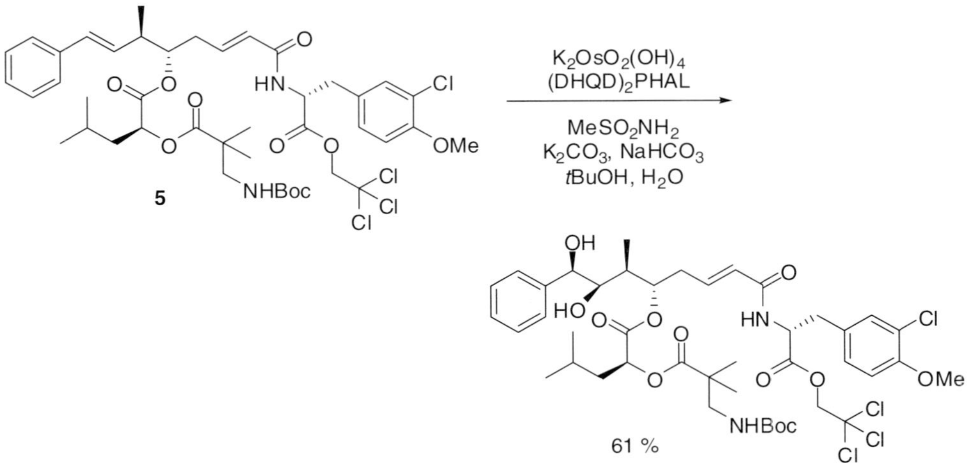

Scheme 3.6

3.2.5
Substrates and Selectivity

In the first AD reactions, the emphasis was on face selectivity and reactivity. As the reaction has to be used in more complex systems, the competition between multiple alkenes can occur within a single substrate. To simplify the discussion within this section, molecules with only one alkene are considered first, followed by polyenes. Within each of these sections, olefins are considered first followed by functionalized alkenes. Tables and figures have been used to summarize the vast number of examples in the literature; the survey is not intended to be exhaustive, just illustrative. The preferred site of reaction is indicated for many of these, but it should be remembered that oxidation occurs at more than one site and only the major product is indicated. The reader should consult the original citations for more information.

3.2.5.1 Simple Alkenes
A kinetic study was undertaken to look at ligand acceleration effects in the AD reaction with various alkene substitution patterns. The effect is most dramatic on trisubstituted alkenes and some idea of relative reactivity can be obtained (Figure 3.4) [164]. Table 3.1 summarizes the types of simple alkenes that can be used in an AD reaction.

Studies with $(DHQD)_2PYDZ$ and terminal alkenes showed that the enantioselectivity of the reaction is a function of chain length [177]. The use of allyl ethers is a way to circumvent this problem [178].

3.2.5.2 Functionalized Alkenes
Table 3.2 summarizes some of the substrates that can be used for an AD reaction. Compounds with more than one alkene unit are considered in the polyene sections below.

Some substrates do require modification of the standard reaction conditions, for example, with allyl halides as substrates, the reaction medium needs to be buffered, usually with $NaHCO_3$, to avoid epoxide formation [2, 85, 254].

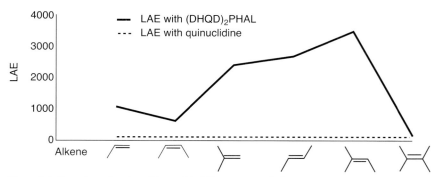

Figure 3.4 Ligand accelerating effects with different types of alkenes.

Table 3.1 Alkene substrates for an asymmetric dihydroxylation reaction.[a,b]

Substrate	References	Substrate	References
R⌐⌐	[2]	Ar⌐⌐	[2, 165–170]
R¹⌐⌐R²	[2]	Ar⌐⌐R	[2, 167, 171]
Ar¹⌐⌐Ar²	[2, 172, 173]	R¹⌐⌐R²	[2]
Ar¹⌐⌐R	[2]	R²⌐⌐R¹	[2]
R⌐⌐Ar	[2, 174]	Ar²⌐⌐Ar¹	[2, 175]
R²⌐⌐R¹⌐⌐R³	[2, 176]	Ar⌐⌐R¹⌐⌐R²	[2]
R²⌐⌐R¹⌐⌐R³⌐⌐R⁴	[2]	R¹⌐⌐Ar⌐⌐R²⌐⌐R³	[2]
R¹⌐⌐Ar¹⌐⌐Ar²⌐⌐R²	[2]	—	—

[a]R denotes an alkyl group of any size but not hydrogen, Ar denotes an aryl group including substituted phenyl and aromatic heterocycles, and X denotes a halogen.
[b]When a methyl group is shown, only examples with this substituent have been used.

When the vinyl ether **6** was the substrate, AD-mix β led to significant amounts of cleavage of the diol product. A change to the pyrimidine-based ligands, in particular the trimethoxy derivative **7** gave high selectivity for the formation of the desired α-hydroxy aldehyde **8** (Scheme 3.7) [255].

With vinyl sulfones as the substrates for the dihydroxylation procedure, a mixture of products results of which the α-hydroxy aldehyde is the major component (Scheme 3.8) [199].

As already illustrated, the AD reaction still requires some investigation to obtain good selectivity other than that with simple substrates. Another example is in work toward the synthesis of Bryostatin-C where reasonable selectivity was seen with **9** as a substrate for an AD reaction with AD-mix β. However, the selectivity fell significantly with the more substituted cyclic and acyclic analogues of **9** required for the synthesis of the natural product. A survey of commercially

Table 3.2 Functionalized alkene substrates for an asymmetric dihydroxylation reaction.[a,b]

Substrate	References	Substrate	References
⟶CF$_3$	[2]	⟶SiMe$_3$	[2]
⟶CO$_2$R	[2, 179]	⟶CH(OR^1)R^2 (X)	[180]
⟶SiR$_3$	[53]	⟶SAr	[2]
⟶OAr	[2, 178, 181]	⟶CH(OR)Ar	[2, 182]
⟶OC(O)Ar	[181, 183]	⟶OC(O)NHR	[184]
⟶N(R)C(O)R	[185]	⟶CH(OR^1)(OR^2)	[2, 186, 187]
⟶CH(O$_2$R^1)R^2	[188]	⟶C(Ar)=CF$_3$	[189]
⟶C(R)=CF$_3$	[2]	⟶C(Ar)–X	[2, 182]
⟶C(R)–OAr	[181]	⟶C(R^1)–OR^2	[2, 190]
⟶C(R^1)–CONR^{2}R^3	[2, 191]	⟶C(CF$_3$)–CO$_2$R	[192]
⟶C(CF$_3$)–CONR^{1}R^2	[192]	⟶C(R^1)–CH(NHR^2)CO$_2$R^3	[193]
Ar–CH=CH–CO$_2$Et	[194]	R–CH=CH–OH	[195]
R^1–CH=CH–OR^2	[196]	R–CH=CH–CF$_3$	[197, 198]
R–CH=CH–X	[199]	R–CH=CH–SO$_2$Ar	[199]
R^1–CH=CH–CO$_2$R^2	[2, 84, 200–203]	Ar–CH=CH–CO$_2$R	[2, 84, 167, 204–208]
R^1–CH=CH–CONR^{2}R^3	[2, 209]	Ar–CH=CH–CONR^{1}R^2	[2]

Table 3.2 (Continued)

Substrate	References	Substrate	References
R¹−CH=CH−C(O)−R²	[2, 210]	Ar−CH=CH−C(O)−R	[2]
R¹−CH=CH−P(O)(OR²)₂	[211–213]	R−CH(NHR)−CH=CH−CO₂R	[214]
R¹−CH(NHR²)−CH=CH−C(O)−Ar	[215]	Ar-epoxide−CH=CH−CO₂R	[216]
R¹−CH=CH−CH₂−X	[2, 217–219]	Ar−CH=CH−CH₂−X	[2]
R¹−CH=CH−CH₂−SAr	[2, 220, 221]	Ar¹−CH=CH−CH₂−SAr²	[2, 220]
R¹−CH=CH−CH₂−SiR²₃	[222–224]	R−CH=CH−CH₂−OH	[219, 225, 226]
R¹−CH=CH−C(R²)(R³)−OH	[227]	R¹−CH=CH−CH₂−OR²	[196, 228]
R¹−CH=CH−CH₂−OAr	[181]	R¹−CH=CH−CH₂−O−C(O)−Ar	[181, 183]
R¹−CH=CH−CH₂−NR²R³	[229]	R¹−CH=CH−CH(OR²)(OR³)	[230]
R¹−CH=CH−CH₂−P(O)(OR²)₂	[231]	R¹−CH=CH−CH(OR²)(OR₃)	[232]
R¹−CH=CH−CH₂−CO₂R²	[233–237]	R¹−CH=CH−CH(R²)−CO₂R³	[238]
R¹−CH=CH−CH₂−CONR²R³	[209]	R¹−CH=CH−CH₂−CH₂−CO₂R²	[239]
R¹−C(OR²)=CH−R³	[2]	Ar¹−C(OR¹)=CH−Ar²	[2]
Ar¹−C(OR)=CH−Ar²	[2]	R¹−C(OR²)=CH−OR³	[240]
R¹−CH=C(R²)−P(O)Ar₂	[90]	R¹−CH=C(R²)−CO₂R³	[241–243]

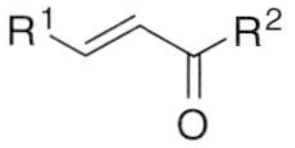

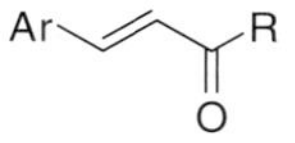

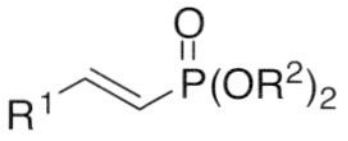

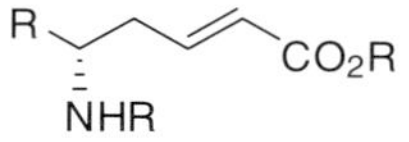

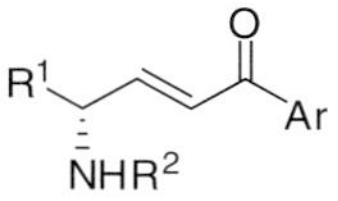

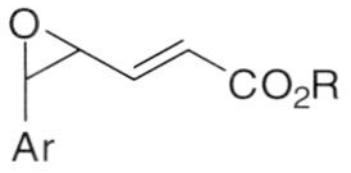

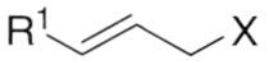

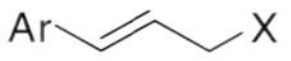

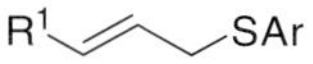

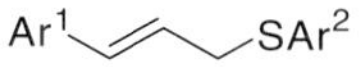

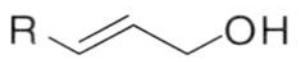

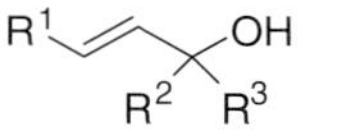

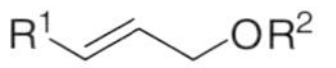

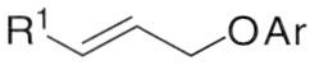

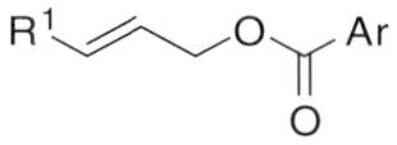

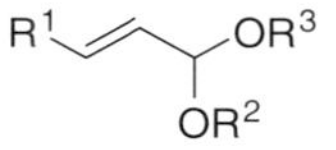

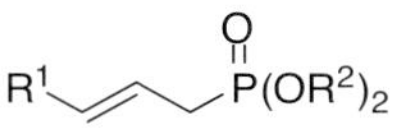

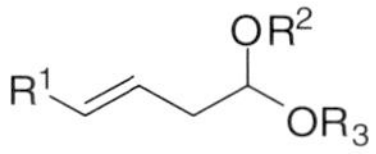

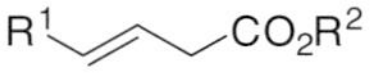

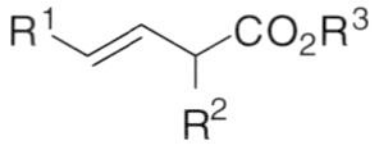

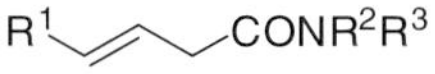

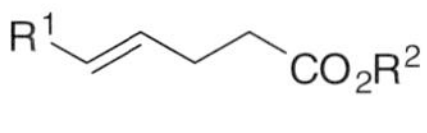

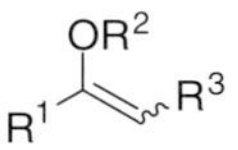

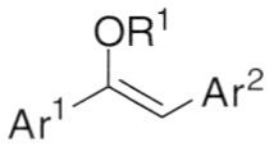

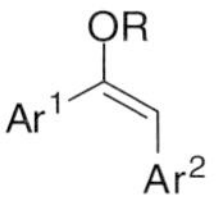

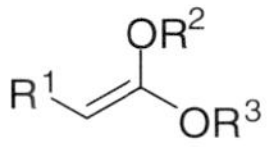

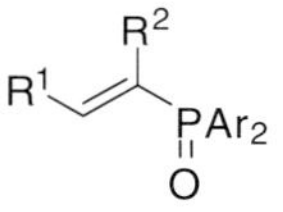

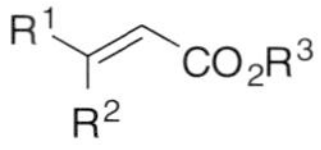

(Continued)

Table 3.2 (Continued)

Substrate	References	Substrate	References
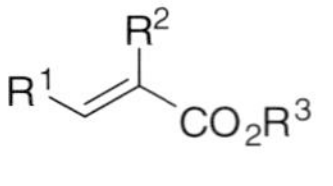 R^{1}CH=CR^2–CO$_2$R^3	[244]	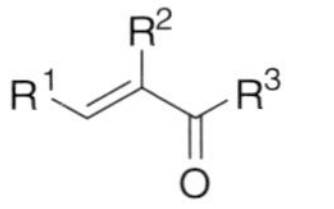 R^{1}CH=CR^2–C(=O)R^3	[2]
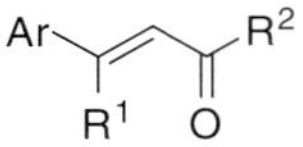 Ar–CR^1=CH–C(=O)R^2	[2]	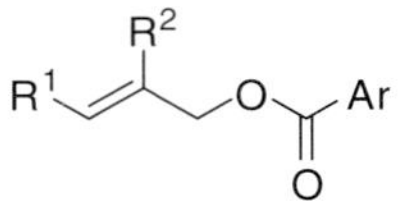 R^{1}CH=CR^2–O–C(=O)–O–Ar	[181, 183, 245, 246]
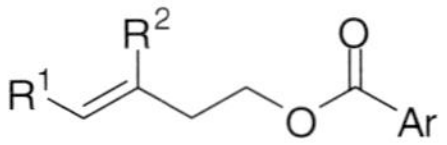 R^{1}CH=CR^2–CH$_2$–O–C(=O)Ar	[246]	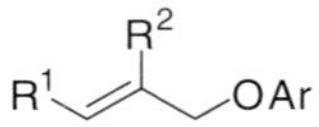 R^{1}CH=CR^2–OAr	[181]
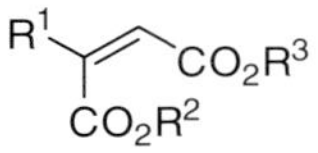 R^{1}CH=C(CO$_2$R^3)(CO$_2$R^2)	[247, 248]	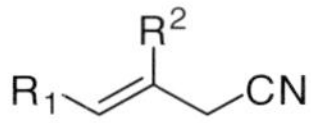 R$_1$CH=CR^2–CH$_2$CN	[249]
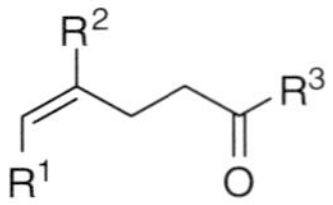 R^{1}CH=CR^2–CH$_2$–C(=O)R^3	[250]	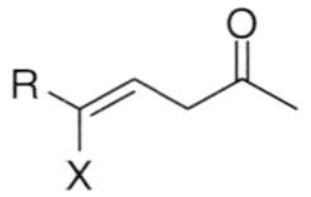 R–CX=CH–C(=O)CH$_3$	[251]
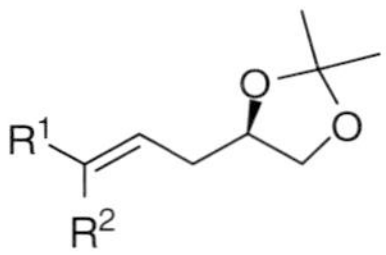 R^{1}CH=CR^2–CH$_2$–(dioxolane)	[252]	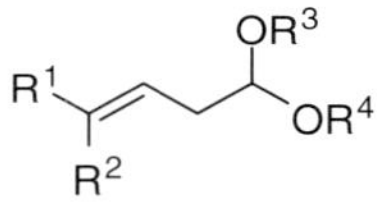 R^{1}CH=CR^2–CH$_2$–CH(OR^3)(OR^4)	[253]
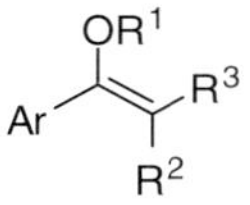 Ar–C(OR^1)=CR^2–R^3	[2]	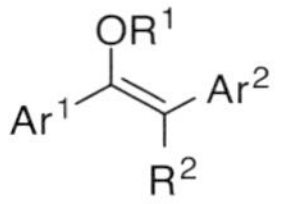 Ar^1–C(OR^1)=CR^2–Ar^2	[2]
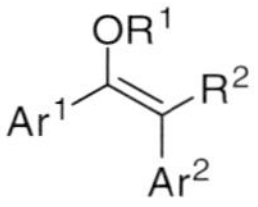 Ar^1–C(OR^1)=C(Ar^2)R^2	[2]	—	—

^{a}R denotes an alkyl group of any size but not hydrogen, Ar denotes an aryl group including substituted phenyl and aromatic heterocycles, and X denotes a halogen.
^{b}When a methyl group is shown, only examples with this substituent have been used.

available ligands and the new alkyne compounds **10** did not alleviate the problem [256].

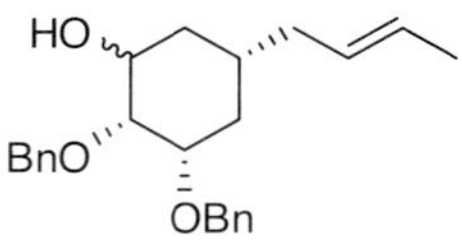

9

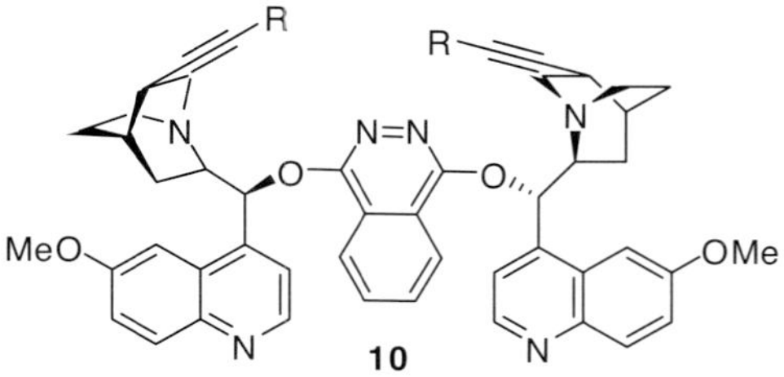

10

where R = Ph, I, or quinolyl

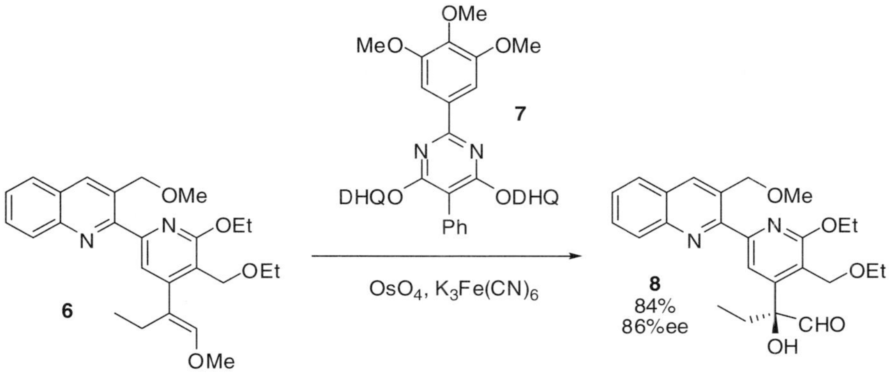

3.2 Asymmetric Dihydroxylation of Alkenes

Scheme 3.7

3.2.5.3 Polyenes

3.2.5.3.1 Nonconjugate Olefins The presence of the chiral ligands within the reagent does influence selectivity with substrates containing more than one olefin. The reactivity of the chiral agents is low with tetrasubstituted alkenes, but otherwise it is usually the more electron-rich alkene that reacts faster (Figure 3.5) [6, 257].

Table 3.3 contains examples of nonconjugated alkenes with the major site of reaction indicated. This table also includes the examples of functionalized nonconjugated alkenes.

The reactivity of unfunctionalized, nonconjugated polyenes in an AD reaction can be generalized. The substitution pattern and the steric environments determine the regioslectivity, and it will follow the reactivity outlined in Figure 3.5, which, in turn, follows that shown in Figure 3.4. If the disubstituted double bond is part of a "normal" ring system, then the reaction at a terminal double bond is favored. For polycyclic alkenes in "normal" size rings with only Z-alkenes, then poor enantioselectivity is usually observed. For medium and large rings, with at least one E-alkene, high enantioselectivity can be observed as for acyclic systems [259].

3.2.5.3.2 Conjugated Polyenes Table 3.4 contains examples of conjugated alkenes with the major site of reaction indicated. This table also included some examples of functionalized polyenes.

For conjugated trienes, the central double bond is not reactive; while a *trans*-alkene is more reactive than a terminal one, a mixture usually results with the major product arising from reaction of the *trans*-alkene. If the end of the triene unit is a *cis*-alkene rather than a *trans*-alkene, then only reaction at the terminal alkene is observed.

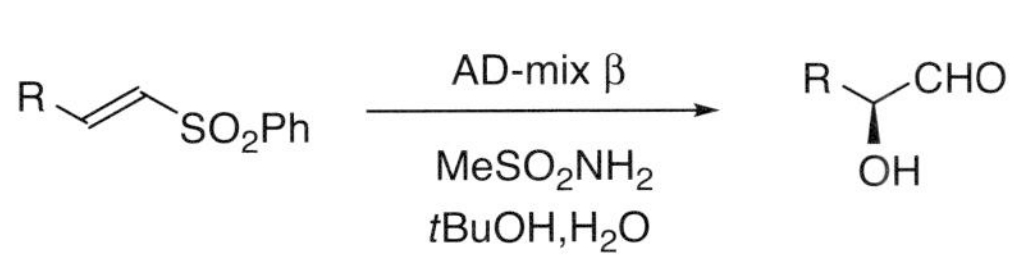

Scheme 3.8

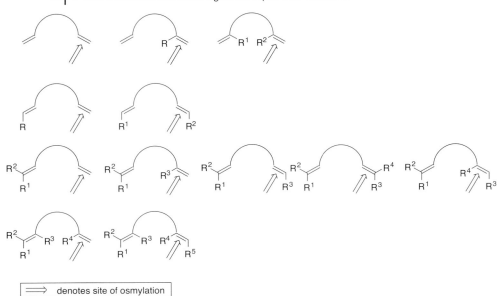

Figure 3.5 Major site of reaction for nonconjugated polyenes in AD reactions.

For the oxidation of conjugated dienes with an aryl substituent at one of the systems and both alkenes being *trans*, the major product arises from reaction of the alkene not attached to the aryl unit, unless that moiety is naphthyl [259].

For conjugated dienes, Sharpless found that a biphasic system gave rise to formation of the diol [258], but if a monophasic system was employed, the tetraol was formed [304]. The catalytic system provides the asymmetric induction. These findings were exploited in a two-step oxidation of the diene **11**, where the diol is formed first and a second oxidation is subsequently undertaken to give higher overall yield and better stereochemical control (Scheme 3.9) [305].

For polyene systems, a number of factors control subsequent oxidations including the facial bias of the substrate and the steric environment around the unsaturation. Reaction of the triene **12** with AD mix α gave reaction at the double bond farthest from the ester group (Scheme 3.10). Although face selectivity was seen in a second reaction, there was no regiochemical control. A similar loss of regiochemical control was seen with the analogous nonchiral reaction. Removal of the ketal protection did not solve this regiochemical problem [257, 306].

By contrast, the dienoates **13** provide reasonable stereocontrol although the second oxidation uses the facial bias of an osmium oxidation of an allyl alcohol (see Sections 3.1 and 3.2.5.4) as diastereocontrol problems occurred when a chiral ligand was present (Scheme 3.11) [307].

3.2.5.4 Double Asymmetric Induction

As shown in Schemes 3.10 and 3.11, although the alkaloid system is the dominant factor in the control of the stereochemical outcome of the AD reaction, other stereogenic centers within the substrate can also have an influence.

Table 3.3 Nonconjugated polyalkene substrates for the asymmetric dihydroxylation reaction.[a,b]

Substrate	References	Substrate	References
(pentadiene)	[2, 258]	(isopropyl methylcyclohexadiene)	[259]
(methylenecyclohexane with isopropylidene)	[259]	(dialkyl benzene with R$_1$, R^2)	[260–263]
(methylenehexadiene)	[2, 258]	(methylheptadiene with R^2)	[2, 258]
(hexadiene)	[264]	(diene with R, R^1, R^2)	[265]
(heptadiene)	[266]	(methylheptadiene with R^1, R^2)	[179]
(Ar-substituted methylheptadiene with R^2)	[267, 268]	(allyl amine with SiR$^1{_3}$, R^1, N)	[181]
(dioxolane diene with R^1)	[2, 259]	(dialkyl benzene with R^2, R^2)	[260, 261]
		(Continued)	

Table 3.3 (Continued)

Substrate	References	Substrate	References
(allylic alcohol, prenyl-type with OH)	[269]	(allylic acetate, prenyl-type with OAc)	[269–271]
(allylic acetate, prenyl-type)	[270, 272]	(diene with R¹ and OR²)	[273, 274]
(geranyl acetate-type)	[270, 275–277]	(geranyl ester with CO₂R)	[278–280]
(cyclic diol, HO/OH)	[2, 259]	(cyclic diether, RO/OR)	[2, 259]
(HO, vinyl substituted)	[87]	—	—

[a] R denotes an alkyl group of any size but not hydrogen, Ar denotes an aryl group including substituted phenyl and aromatic heterocycles, and X denotes a halogen.
[b] When a methyl group is shown, only examples with this substituent have been used.

3.2 Asymmetric Dihydroxylation of Alkenes

Table 3.4 Conjugated polyalkene substrates for the asymmetric dihydroxylation reaction.[a,b]

Substrate	References	Substrate	References
Ar-CH=C=CH₂ (allene)	[281]	R-CH=CH-CH=CH₂	[2, 258]
R-CH₂-CH(R)-CH=CH-CH=CH₂	[282]	CH₃-CH=CH-CH=CH-CH₃	[258]
R-CH=CH-CH=CH-R	[2, 258]	Ar-CH=CH-CH=CH-Ar	[2, 258]
CH=CH-CH=CH-CH₂-R	[2]	R-CH=CH-CH=CH-Ar	[259]
R¹-CH=CR²-CH=CH-R³	[227, 259, 283–286]	R₂C=CH-CH=CR₂	[2, 258]
α-terpinene (p-isopropyl methylcyclohexadiene)	[258, 259]	CH₂=C=C-C≡C-R	[287, 288]
CH₂=C=CR-C≡C-R	[287, 288]	R-CH=C=CH-C≡C-R	[287, 288]
R-CH=CH-C≡C-R	[287, 288]	R-C(CH₃)=CH-C≡C-R	[289]
R-CH=CH-CH=CH-CH₂-O-C(O)-Ar	[2]	R¹-CH=CH-CH=CH-CO₂R²	[290]
R¹-CH=CH-CH=CH-CO₂R₂	[2, 291–295]	R¹-CH=CH-CH=C(X)-CO₂R₂	[296]
R-CH=CH-CH=CH-CHO	[2]	R-CH=CH-CH=CH-C(O)-Ar	[188]
R¹-CH=CH-CH=CH-CONR²R³	[297]	R¹-CH=CH-C(R²)=CH-CO₂R³	[298]
R-C≡C-CH=CH-CH₂-CO₂R	[299, 300]	HC≡C-C(CH₃)=CH-CH₂-OR	[301]
HC≡C-C(CH₃)=CH-CH₂-OR	[301]	R-CH=CH-CH=CH₂	[259]

(*Continued*)

Table 3.4 (Continued)

Substrate	References	Substrate	References
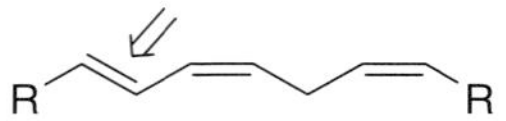 R⌇═╲╱═╲R (triene)	[284]	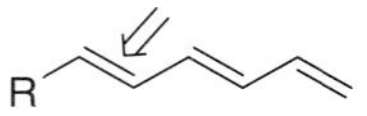 R⌇═╲═╱═╲ (triene)	[259]
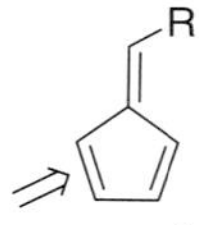 R-CH=cyclopentadienyl	[302, 303]	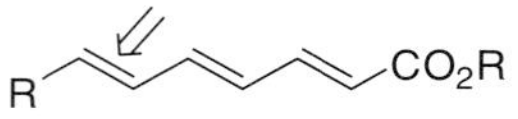 R╲═╱═╲═╱CO$_2$R	[2]
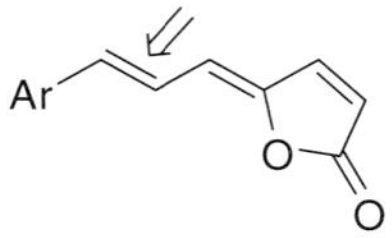 Ar╲═╱═(furanone)	[194]	—	—

[a] R denotes an alkyl group of any size but not hydrogen, Ar denotes an aryl group including substituted phenyl and aromatic heterocycles, and X denotes a halogen.
[b] When a methyl group is shown, only examples with this substituent have been used.

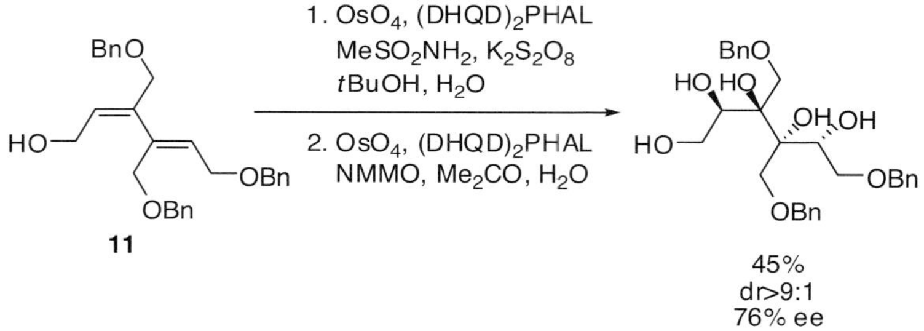

Scheme 3.9

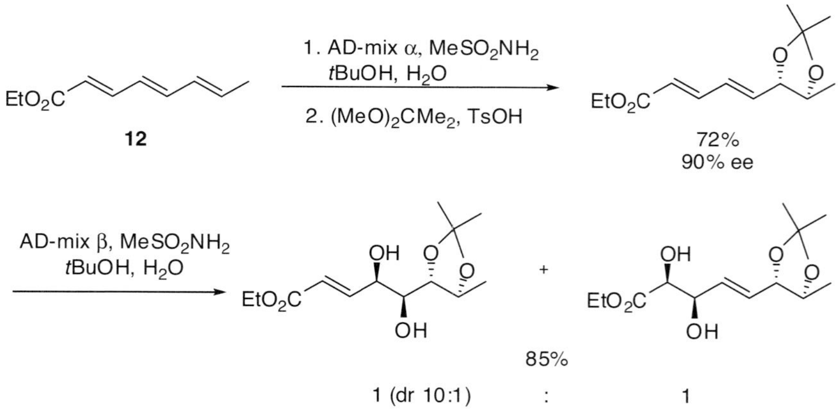

Compound **11**
1. OsO$_4$, (DHQD)$_2$PHAL, MeSO$_2$NH$_2$, K$_2$S$_2$O$_8$, tBuOH, H$_2$O
2. OsO$_4$, (DHQD)$_2$PHAL, NMMO, Me$_2$CO, H$_2$O

→ product, 45%, dr>9:1, 76% ee

Scheme 3.10

Compound **12** (EtO$_2$C-pentadienyl)

1. AD-mix α, MeSO$_2$NH$_2$, tBuOH, H$_2$O
2. (MeO)$_2$CMe$_2$, TsOH

→ product, 72%, 90% ee

AD-mix β, MeSO$_2$NH$_2$, tBuOH, H$_2$O

→ two products, 85%, 1 (dr 10:1) : 1

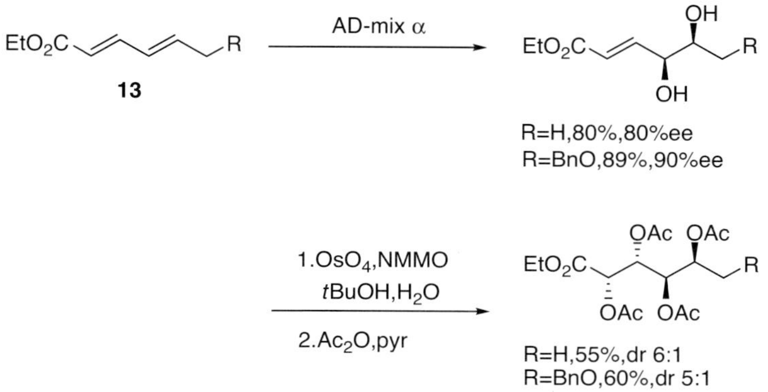

3.2 Asymmetric Dihydroxylation of Alkenes | 49

Scheme 3.11

EtO$_2$C—CH=CH—CH=CH—CH$_2$R (**13**)

AD-mix α → EtO$_2$C—CH=CH—CH(OH)—CH(OH)—CH$_2$R
R=H, 80%, 80%ee
R=BnO, 89%, 90%ee

1. OsO$_4$, NMMO, tBuOH, H$_2$O
2. Ac$_2$O, pyr

→ EtO$_2$C—CH(OAc)—CH(OAc)—CH(OAc)—CH(OAc)—CH$_2$R
R=H, 55%, dr 6:1
R=BnO, 60%, dr 5:1

Oxidation of one unit of a conjugated diene, followed by protection and then reaction of the second alkene can lead to good degrees of stereoinduction for the matched series (Scheme 3.12) [308].

In the case of the AD reaction with γ-amino-α,β-unsaturated esters **14**, it was found that the nature of the protecting group on nitrogen could be used to control the double induction phenomenon (Scheme 3.13) [309]. This work was performed before the final version of the mnemonic was developed; the changes in the electronic nature of the groups on nitrogen account for the change in selectivity.

A further double asymmetric induction example is given by the alkene **15**, derived from α-methylserine (Scheme 3.14). Reaction with the AD-mix β was the mismatched system while AD-mix α gave good stereoselection for the *syn*-diol [310].

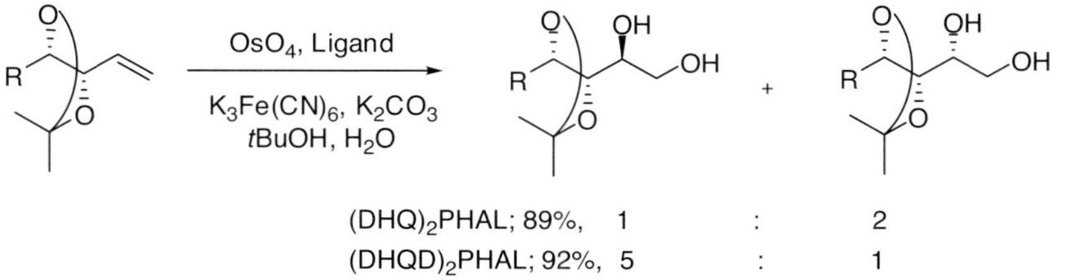

Scheme 3.12

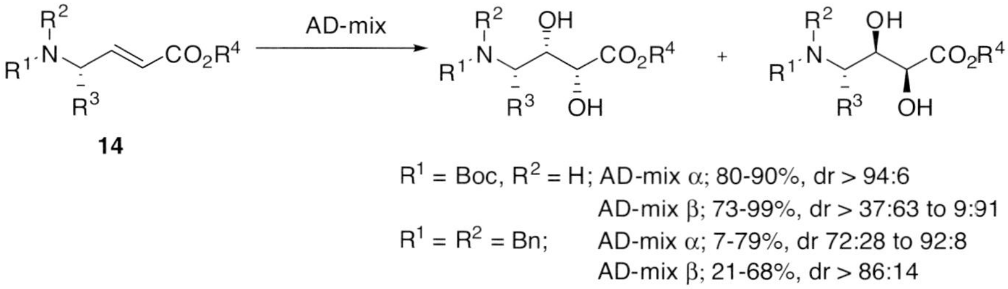

OsO$_4$, Ligand, K$_3$Fe(CN)$_6$, K$_2$CO$_3$, tBuOH, H$_2$O

(DHQ)$_2$PHAL; 89%, 1 : 2
(DHQD)$_2$PHAL; 92%, 5 : 1

Scheme 3.13

R^1—N(R^3)—CH=CH—CO$_2$R^4 (**14**)

AD-mix →

R^1—N(R^3)—CH(OH)—CH(OH)—CO$_2$R^4 + R^1—N(R^3)—CH(OH)—CH(OH)—CO$_2$R^4

R^1 = Boc, R^2 = H; AD-mix α; 80-90%, dr > 94:6
AD-mix β; 73-99%, dr > 37:63 to 9:91

R^1 = R^2 = Bn; AD-mix α; 7-79%, dr 72:28 to 92:8
AD-mix β; 21-68%, dr > 86:14

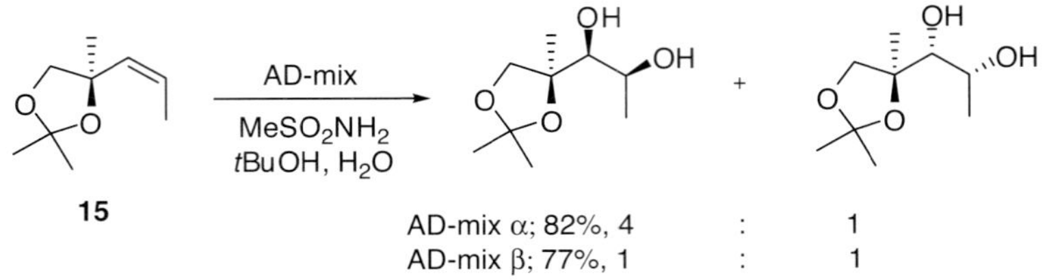

Scheme 3.14

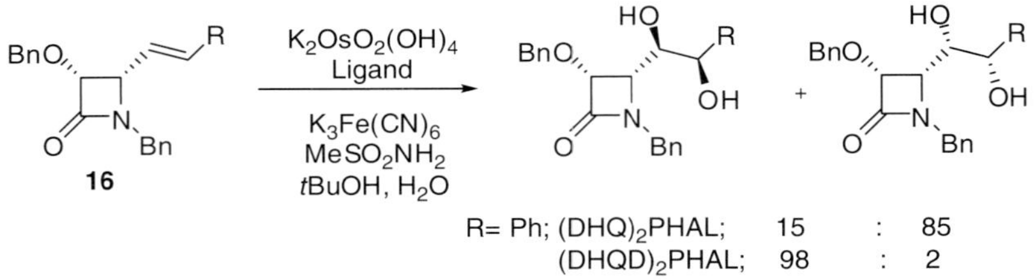

Scheme 3.15

Use of the dihydroxylation procedure with the β-lactam-substituted alkenes **16** showed a matched–mismatched phenomenon (Scheme 3.15). Use of an achiral reagent showed little facial bias. With an alkyl or ester group rather than aryl as the other alkene substituent (R^1), the selectivity diminished [311].

3.2.5.5 Resolutions

Kinetic resolutions can be accomplished by the dihydroxylation approach [312, 313]. To achieve this, there has to be a difference in the rate of reaction between the facial modes of reaction. Although high ee values could be observed with the alkene **17**, diastereoselectivity was low (∼2 : 1) and kinetic resolution was ineffective [314, 315].

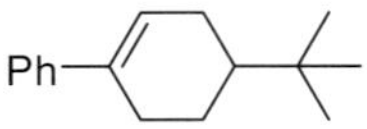

Asymmetric dihydroxylation of the alkene **18** could be used to access either enantiomer by a kinetic resolution (Scheme 3.16). In this approach, the desired alkene is the unreacted material [316].

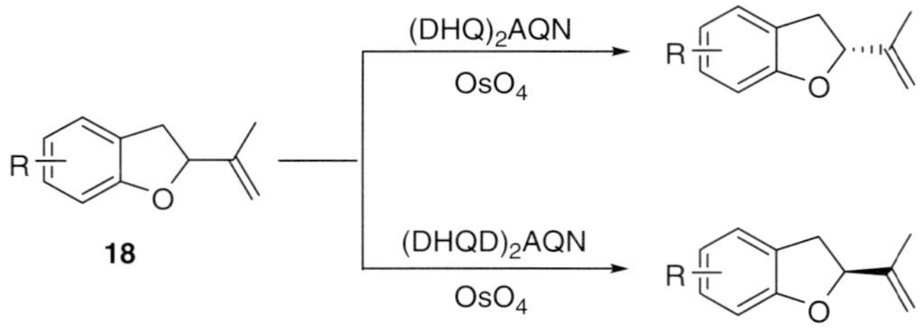

Scheme 3.16

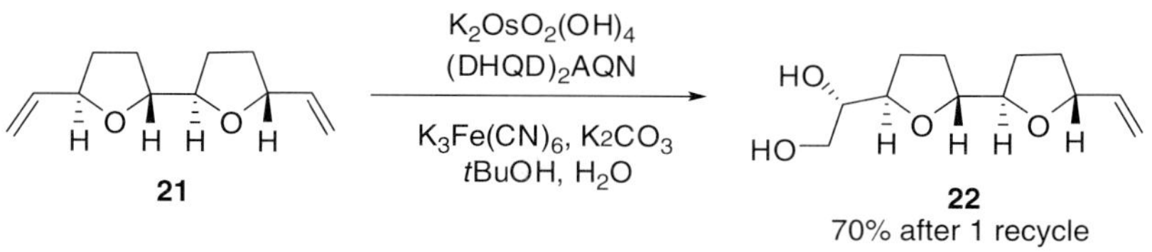

3.2 Asymmetric Dihydroxylation of Alkenes

Scheme 3.17

Kinetic resolutions can also be performed on atropisomers [317], as for the amides **19** [318], and has been used as the first nonenzymatic method for planar–chiral ferrocenes **20** [319].

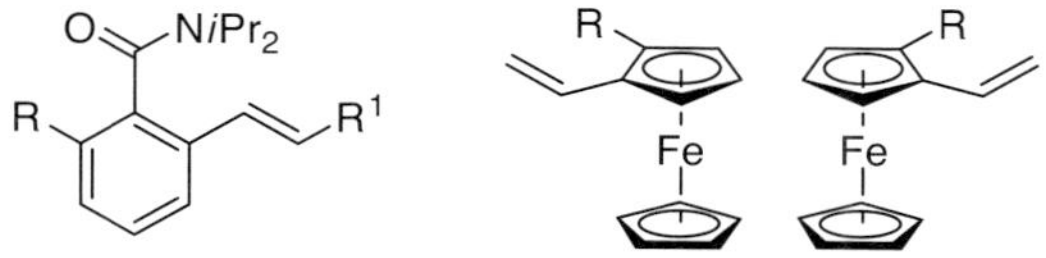

where R¹= H or CO₂Et

Related to a resolution, a desymmetrization reaction was used in an approach to Uvaricin. Use of AD-mix β gave only moderate diastereoselection with **21**, but the use of (DHQD)₂AQN gave **22** after two cycles; the first reaction was run to 50% completion and the recycled diene **21** was then subjected to the reaction conditions for a second time (Scheme 3.17) [320].

3.2.6
Some Reactions of 1,2-Diols

The 1,2-diols formed by the asymmetric oxidation can be used as substrates in a wide variety of transformations. Conversion of the hydroxy groups to p-toluenesulfonates then allows nucleophilic displacement by azide at both centers with the inversion of configuration (Scheme 3.18) [321].

Monosubstitution by a nitrogen nucleophile can also be achieved with acetonitrile as the reagent (Scheme 3.19). The mechanism is postulated to proceed through an oxonium intermediate, and the presence of the aryl group may control the regiochemistry [322].

The diol **23**, prepared by a Sharpless dihydroxylation, was converted to the epoxide **24** through the mesylate (Scheme 3.20) [323].

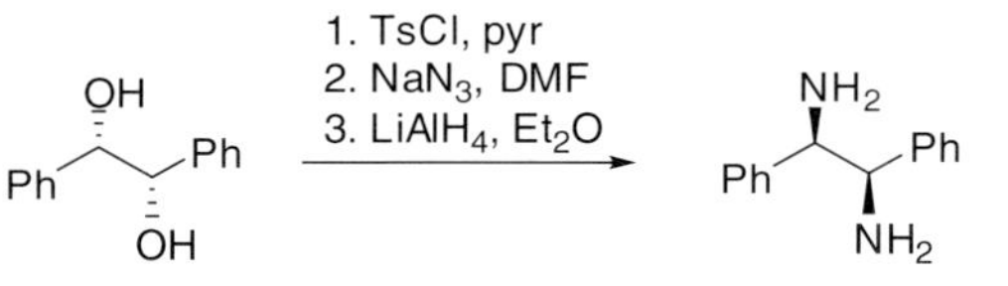

Scheme 3.18

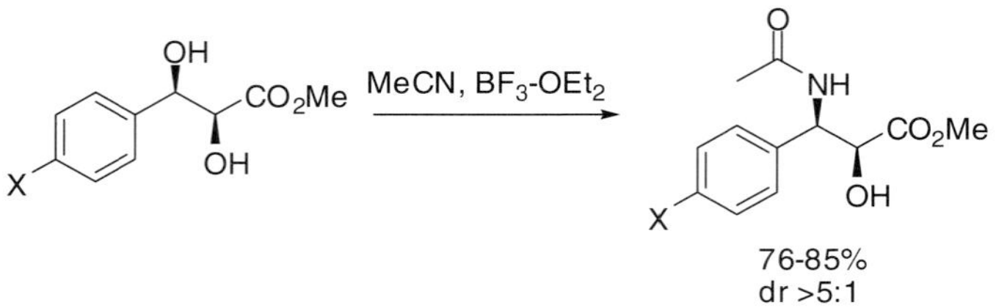

Scheme 3.19

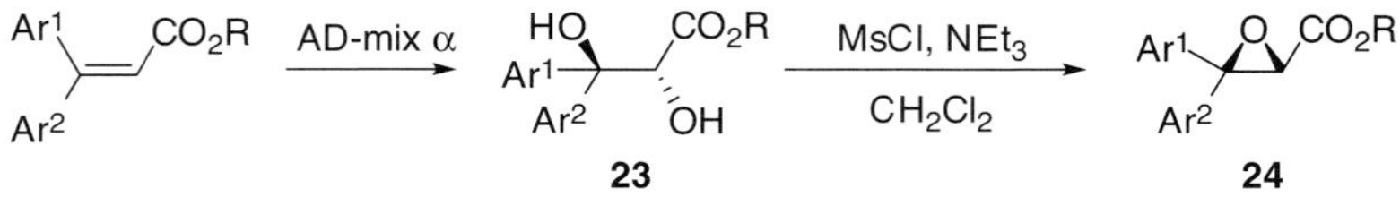

Scheme 3.20

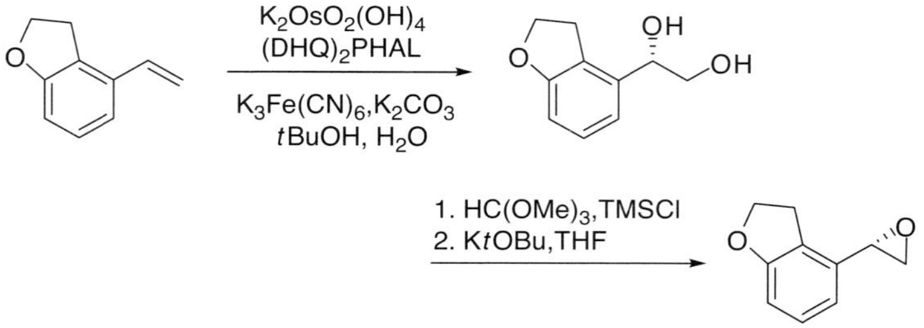

Scheme 3.21

Another study compared a Jacobsen asymmetric epoxidation to epoxide formation through an asymmetric dihydroxylation. The latter process (Scheme 3.21) was found to be advantageous as the overall yield was higher and better stereochemical control could be achieved [324].

This conversion of a 1,2-diol to an epoxide has been used as an approach to 2-arylpropanoic acids, as per the members of the nonsteroidal anti-inflammatory drugs (NSAIDs) family of drugs [325]. However, the sequence can be shortened by a selective hydrogenolysis, as illustrated for naproxen (**25**) (Scheme 3.22) [326].

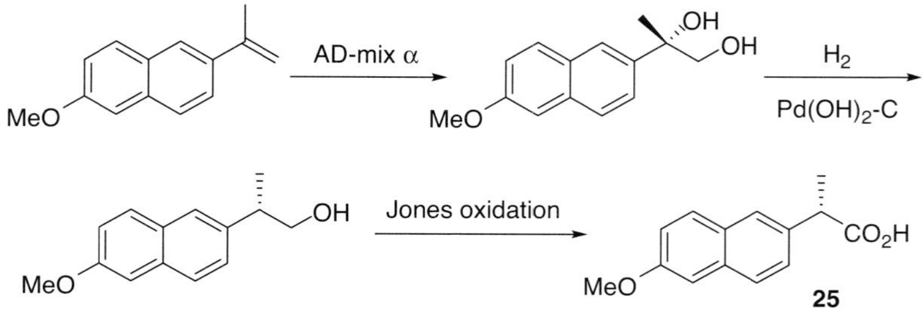

Scheme 3.22

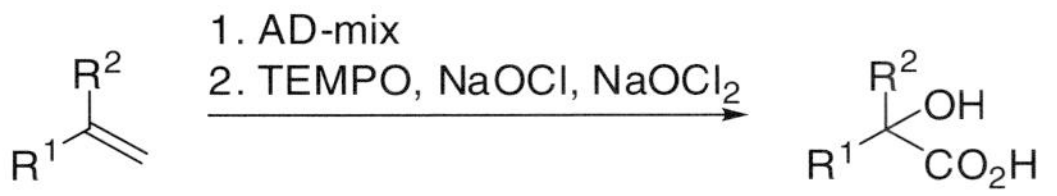

3.2 Asymmetric Dihydroxylation of Alkenes | 53

Scheme 3.23

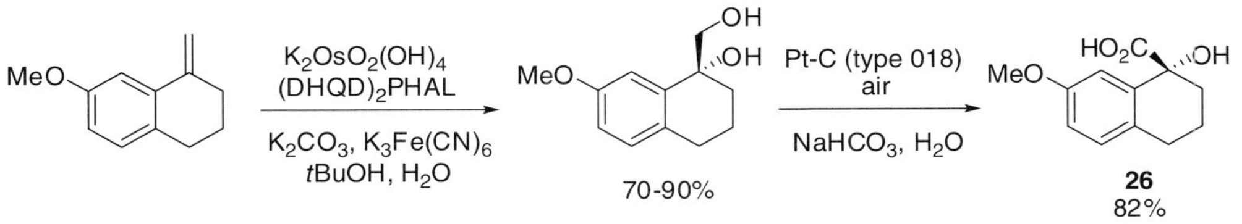

Scheme 3.24

Another useful variation on the AD reaction is to oxidize the primary hydroxy group of the diol while leaving the tertiary hydroxy untouched [327]. This reaction sequence can also be performed as a "one-pot" reaction (Scheme 3.23) [328].

A variation of this approach has been used to prepare the α-hydroxy carboxylic acid **26** with the oxidation reaction being performed at 0.5 mole scale (Scheme 3.24) [329].

A method to prepare 1,2-amino alcohols through the dihydroxylation of an enol ether has been reported by Merck (Scheme 3.25) [330]. The ee was reported to diminish if shorter chain alcohols were used although the TBDMS enol ether was used successfully in a structurally related system [331]. Indeed, early work by Sharpless with acyclic methyl and *tert*-butyldimethylsilyl ethers showed high enantioselectivities in AD reactions (cf. Table 3.2) [332].

The use of an enol ether substrate for an AD reaction has been the key step in a number of approaches to (+)-camptothecin (**27**) (e.g., Scheme 3.26) [333–337].

An AD reaction with the cyclic silyl enol ether **28** provides rapid access to the natural product (+)-pantolactone (**29**) (Scheme 3.27) [338]. In effect, these reactions are asymmetric α-hydroxylations of carbonyl compounds.

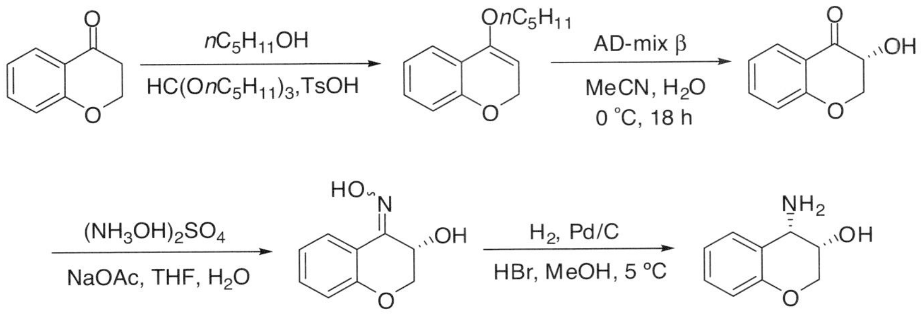

Scheme 3.25

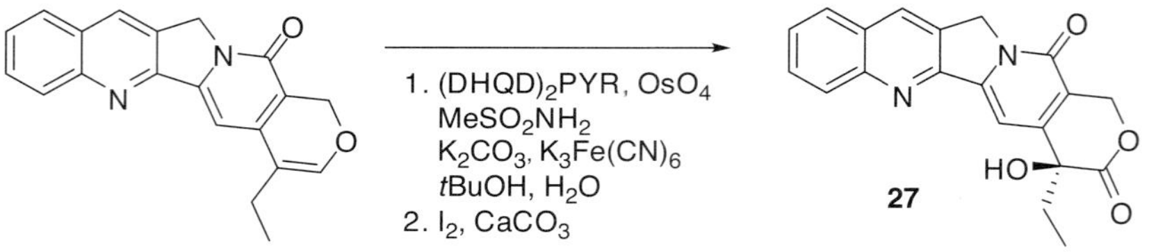

Scheme 3.26

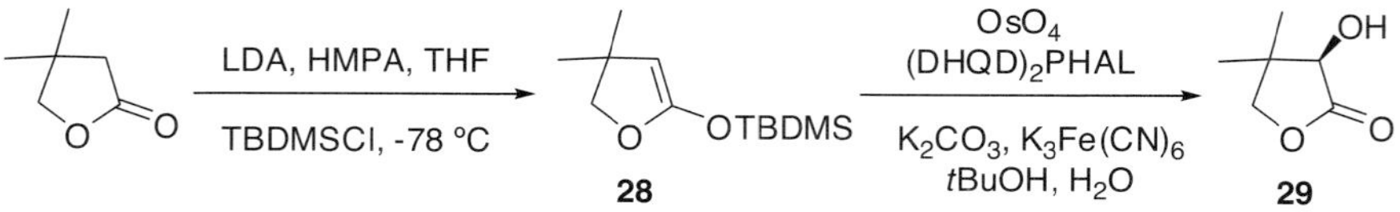

Scheme 3.27

The synthesis of the (+)-compactin lactone **30**, an important component of the statins, illustrates the power of the dihydroxylation methodology when coupled with regioselective sulfite ring opening (Scheme 3.28) [339].

Reaction of bromide with the diol derived from a cinnamate provides access to β-aryl α-hydroxy esters (Scheme 3.29) [340]. The regiochemical control is not complete.

3.2.6.1 Cyclic Sulfates and Sulfites

Cyclic sulfates provide a useful alternative to epoxides. These cyclic compounds are prepared by reaction of the diol with thionyl chloride, followed by ruthenium-catalyzed oxidation of the sulfur (Scheme 3.30) [341–343]. The cyclic sulfates can also be accessed by direct reaction of the diol with sulfuryl chloride (Scheme 3.30) [344].

The cyclic sulfates undergo ring opening with a wide variety of nucleophiles, such as hydride, azide, fluoride, benzoate, amines, and Grignard reagents. In some

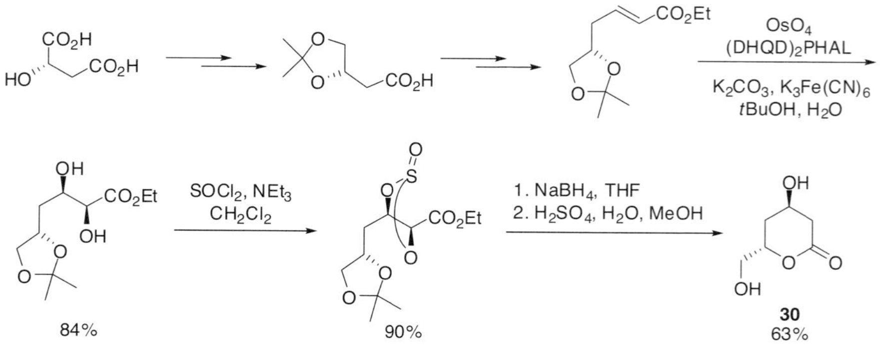

Scheme 3.28

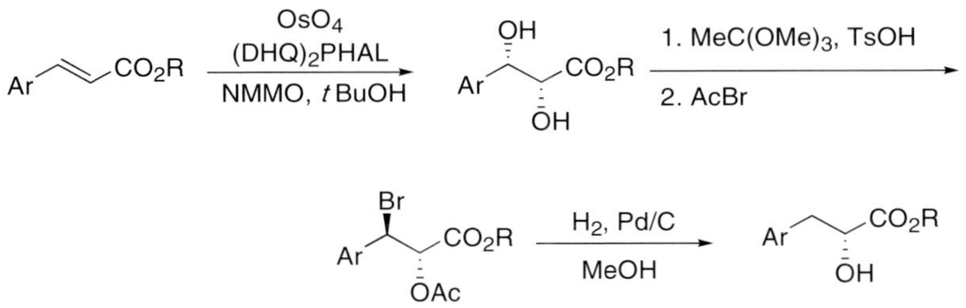

3.2 Asymmetric Dihydroxylation of Alkenes

Scheme 3.29

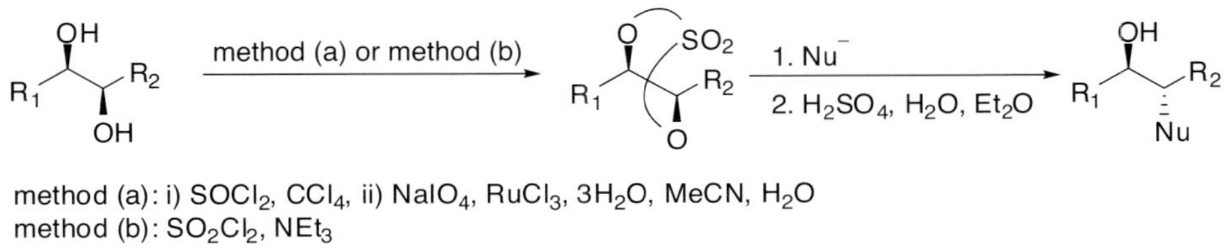

Scheme 3.30

method (a): i) $SOCl_2$, CCl_4, ii) $NaIO_4$, $RuCl_3$, $3H_2O$, MeCN, H_2O
method (b): SO_2Cl_2, NEt_3

instances, the use of a nucleophilic substitution by nitrogen after an AD reaction is more efficient than an asymmetric aminohydroxylation reaction.

A synthesis of (R)-reticuline included a comparison of epoxide and cyclic sulfate chemistry (Scheme 3.31) [345].

The analogous sulfites, obtained by the reaction of 1,2-diol with thionyl chloride (cf. Schemes 3.28, 3.32 and 3.33), also undergo facile ring opening with concurrent

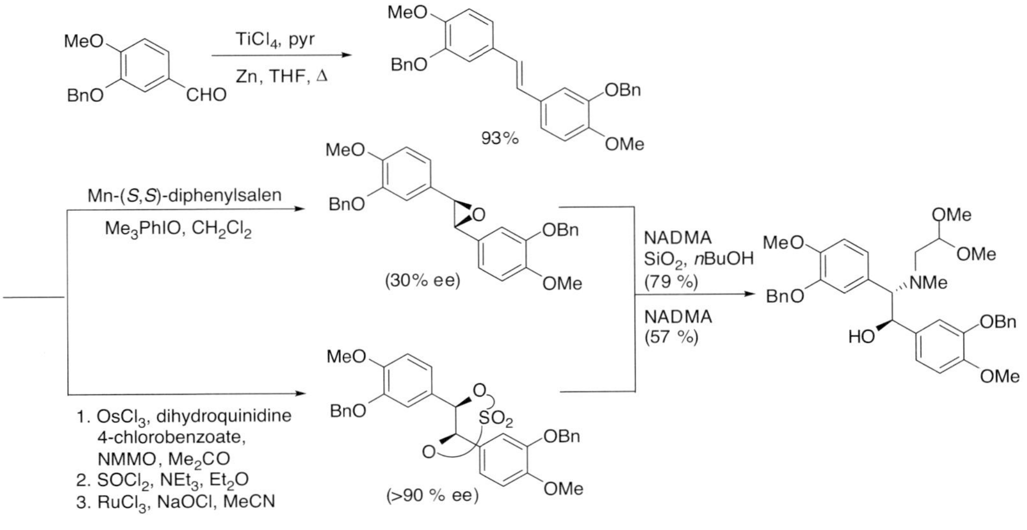

Scheme 3.31

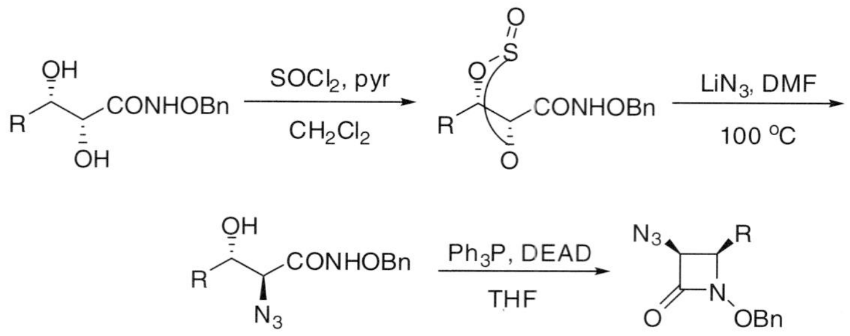

Scheme 3.32

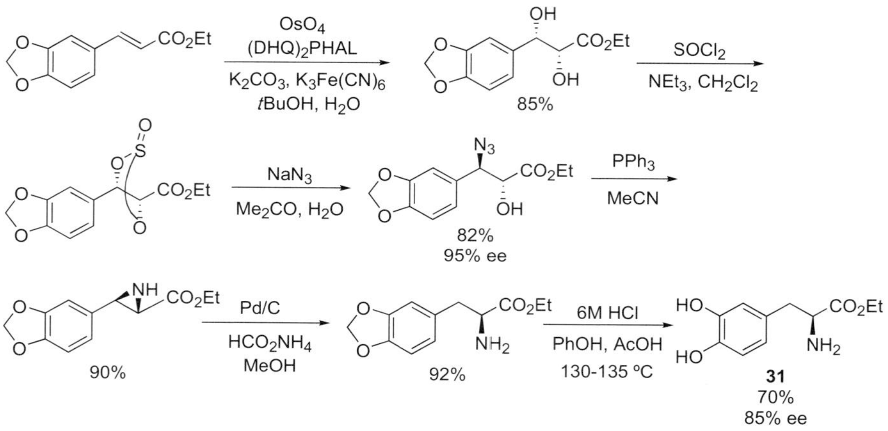

Scheme 3.33

inversion at the reaction center when treated with a nucleophile such as azide [346]. However, sulfates are less prone to side reactions.

For cyclic sulfites and sulfates, reaction is favored adjacent to the aryl group while for alkyl enoic acid derivatives, nucleophilic substitution is favored next to the ester group [347]. This ring-opening selectivity was exploited in a synthesis of β-lactams (Scheme 3.32) [348].

The change in regiocontrol exerted by an aryl group on the reaction is illustrated in a synthesis of L-DOPA (**31**) (Scheme 3.33) [349].

3.3
Aminohydroxylation

A variation on the Sharpless dihydroxylation methodology allows for the preparation of amino alcohols. This reaction is known as the AA reaction, and it has been

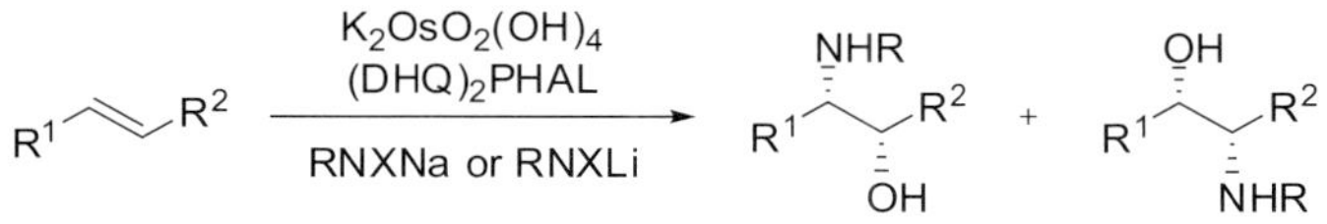

3.3 Aminohydroxylation

Scheme 3.34

reviewed in Refs [3, 4, 10, 18]. As the groups are introduced in a *syn* manner, this compliments an epoxide opening or a similar reaction that employs substitution at one center. As with the AD reaction, the stereochemical outcome of the AA reaction can be predicted by the same mnemonic (cf. Figure 3.2). The general reaction is shown in Scheme 3.34, and the main differences are the nature of the nitrogen nucleophile [3]. The reaction conditions are very similar for each nucleophile; the major ones are tosyl (R = Ts) [350], methanesulfonyl (R = Ms) [351], benzyloxycarbonyl (R = Cbz) [352], *tert*-butoxycarbonyl (R = Boc) [353–355], 2-trimethylsilylethoxycarbonyl (R = Teoc) [356], and acetyl (R = Ac) [357].

Higher selectivity is observed when one of the alkene substituents is electron withdrawing, as in an ester group (Scheme 3.35) [350, 351, 358]. The electronic properties of the aryl group can also influence the reaction outcome [359].

However, there is some variation in the selectivity when different nitrogen nucleophiles are used (Schemes 3.36 and 3.37) [3].

The use of dibromoisocyanuric acid (DBI) with a primary amide allows for the preparation of an N-bromoamide, which can be used in the aminohydroxylation reaction in the presence of a base, such as lithium hydroxide [357, 360].

With styrenes, the use of *tert*-butyl carbamates was found to give better regiocontrol [354]. The role of the solvent can also be important (Scheme 3.38) [3, 353, 357]. In addition, the regiochemical control can change between the use of (DHQ)$_2$PHAL

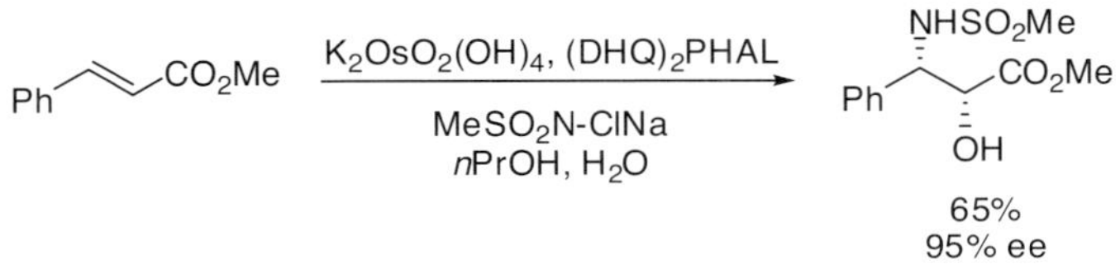

Scheme 3.35

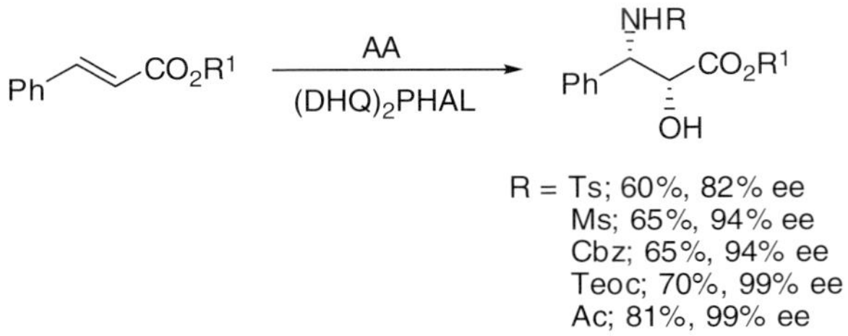

Scheme 3.36

R = Ts; 60%, 82% ee
Ms; 65%, 94% ee
Cbz; 65%, 94% ee
Teoc; 70%, 99% ee
Ac; 81%, 99% ee

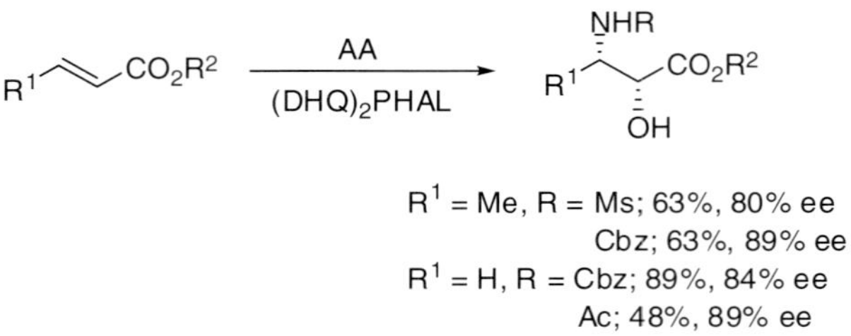

3 Cinchona Alkaloids as Chiral Ligands in Asymmetric Oxidations

Scheme 3.37

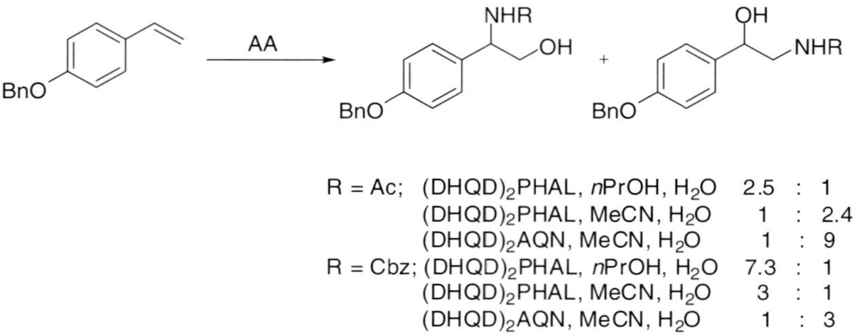

R^1 = Me, R = Ms; 63%, 80% ee
  Cbz; 63%, 89% ee
R^1 = H, R = Cbz; 89%, 84% ee
  Ac; 48%, 89% ee

R = Ac; $(DHQD)_2PHAL$, nPrOH, H_2O 2.5 : 1
  $(DHQD)_2PHAL$, MeCN, H_2O 1 : 2.4
  $(DHQD)_2AQN$, MeCN, H_2O 1 : 9
R = Cbz; $(DHQD)_2PHAL$, nPrOH, H_2O 7.3 : 1
  $(DHQD)_2PHAL$, MeCN, H_2O 3 : 1
  $(DHQD)_2AQN$, MeCN, H_2O 1 : 3

Scheme 3.38

and $(DHQD)_2PHAL$ even though the ee values are essentially the same, although of opposite configuration [354, 355].

With styrenes, the typical reaction conditions provide the α-amino β-hydroxy α-arylethanes as the major product. This regioselectivity can be changed by the use of a pH-controlled method when a carbamate is the nitrogen nucleophile [361].

The use of a carbamate as the nitrogen source provides for the direct conversion of an alkene to an oxazolidinone [356, 362]. However, fresh *tert*-butyl hypochlorite was needed as the stoichiometric oxidant. The need for larger scale applications of the reaction led to the finding that 1,3-dichloro-5,5-dimethylhydantoin or the sodium salt of dichloroisocyanuric acid could be used as alternatives to the hypochlorite (Scheme 3.39) [363].

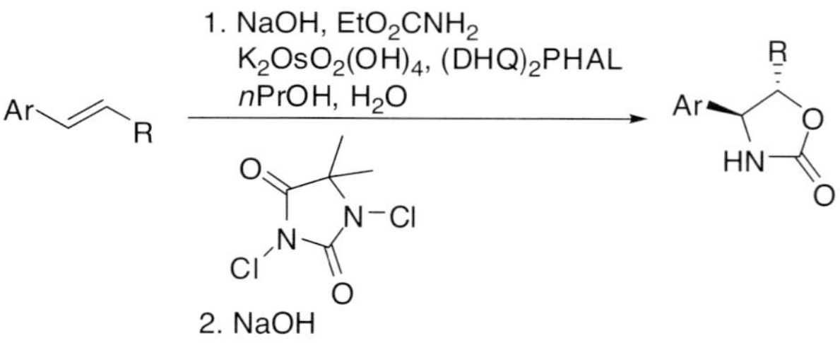

Scheme 3.39

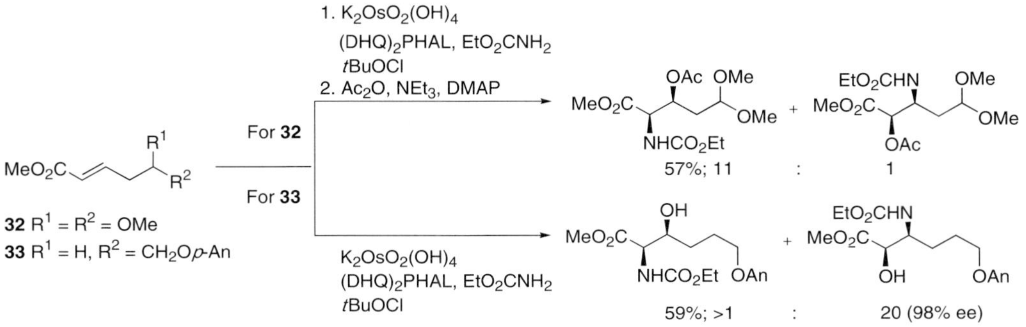

3.3 Aminohydroxylation

Scheme 3.40

Reagents for Scheme 3.40:
1. K$_2$OsO$_2$(OH)$_4$, (DHQ)$_2$PHAL, EtO$_2$CNH$_2$, tBuOCl
2. Ac$_2$O, NEt$_3$, DMAP

32 R^1 = R^2 = OMe
33 R^1 = H, R^2 = CH$_2$O*p*-An

For 32: 57%; 11 : 1 (major with NHCO$_2$Et adjacent to OMe/OAc)
For 33: K$_2$OsO$_2$(OH)$_4$, (DHQ)$_2$PHAL, EtO$_2$CNH$_2$, tBuOCl; 59%; >1 : 20 (98% ee)

Compared to the dihydroxylation procedure, the aminohydroxylation analogue requires optimization of the ligand, the oxidant, and the nitrogen nucleophile. Regioselectivity is another problem that is not encountered in the AD reaction itself. For the unsaturated esters **32**, the ligand system (DHQ)$_2$AQN provided the best regiochemical control with good enantioselectivity (Scheme 3.40). However, the simple change to the aryl ether, as in **33**, rather than the acetal led to the other regioisomeric product with an excellent ee [364].

The AA reaction can be performed with the nitrogen nucleophile being delivered in an intramolecular manner (Scheme 3.41). Of course, this approach circumvents the regiochemical problem associated with the site of attack of the nucleophile. In this example, the product oxazolidinone undergoes equilibrium under the reaction conditions to give the two products, but the nitrogen is at the 4-position of the ring system in both [365].

The 4-nitrophenyl ether group can be used to provide good regiochemical selectivity. A simple change to the ligand backbone changed the major product regioisomer (Scheme 3.42) [366].

The nature of the alkene substitutent can play a key role in the success of the AA methodology. Although the furyl and thiophenyl derivatives of the ester **34** gave good regio- and enantioselectivities, the pyrrol analogues did not undergo reaction (Scheme 3.43) [367, 368].

For applications, the AA reaction can be used to prepare α-arylglycinols from styrenes [353, 355]. The approach has also been used to access the side chain of paclitaxel from cinnamate (Scheme 3.44, cf. Scheme 3.35) [369].

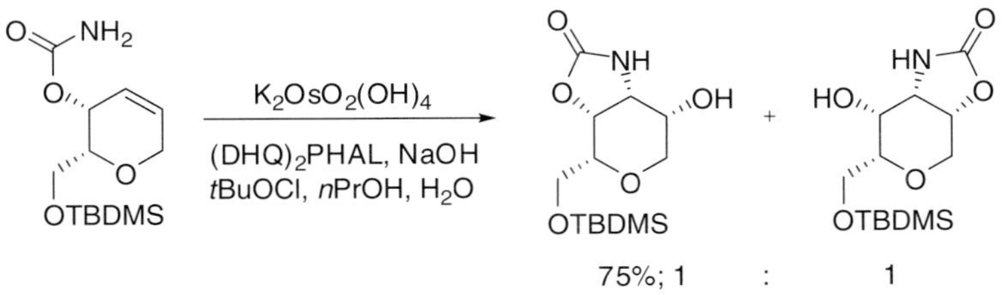

Scheme 3.41

K$_2$OsO$_2$(OH)$_4$, (DHQ)$_2$PHAL, NaOH, tBuOCl, nPrOH, H$_2$O

75%; 1 : 1

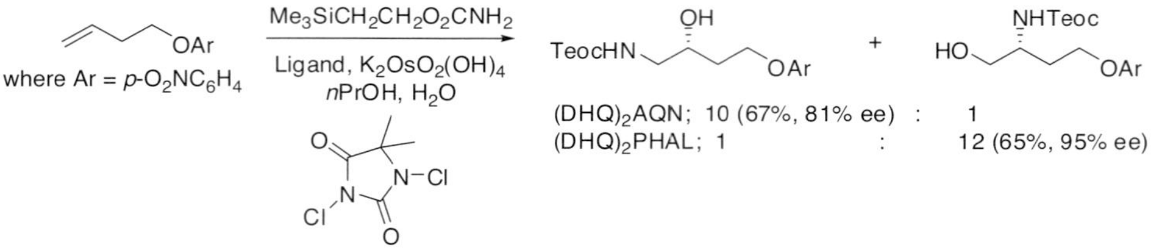

Scheme 3.42

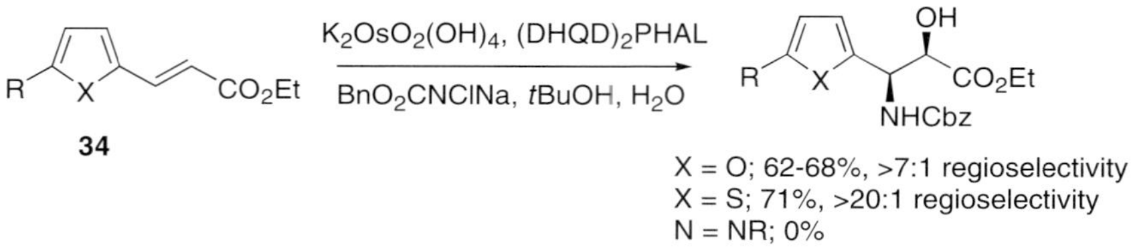

Scheme 3.43

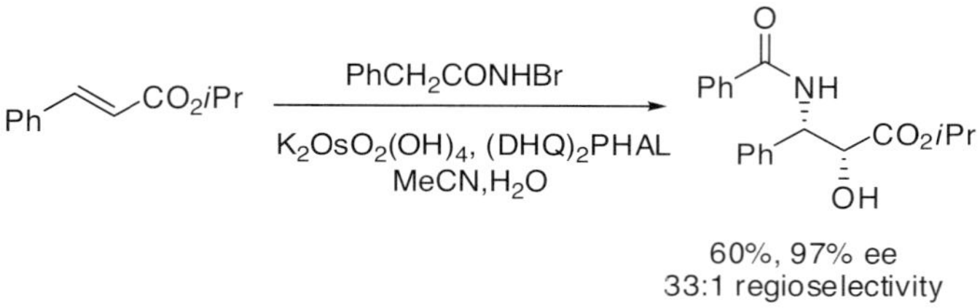

X = O; 62–68%, >7:1 regioselectivity
X = S; 71%, >20:1 regioselectivity
N = NR; 0%

Scheme 3.44

60%, 97% ee
33:1 regioselectivity

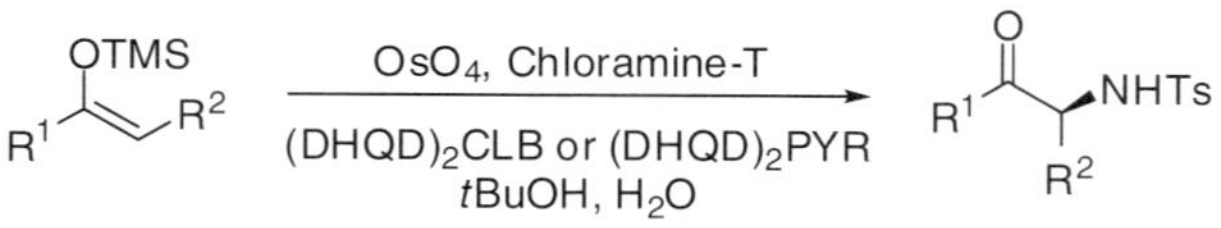

Scheme 3.45

Silyl enol ethers can be used as substrates for AA reactions, and they provide α-amino ketones (Scheme 3.45) [370, 371].

The AA reaction has added the concern of the regiochemical outcome of the reaction. Often, both the regioisomers are formed. In some cases, changes in the reaction conditions as with alkyl-substituted butenoates, the regiochemical outcome can favor the α-amino product by using an aryl ester; 4-halophenyl esters give the best enantioselectivities with good regiochemical control, with 4-bromophenyl providing the best compromise [372].

In other cases, it can be better to perform an AD reaction followed by nucleophilic substitution with a nitrogen nucleophile on the derived cyclic sulfate of the diol (cf. Section 3.2.6.1) [373, 374].

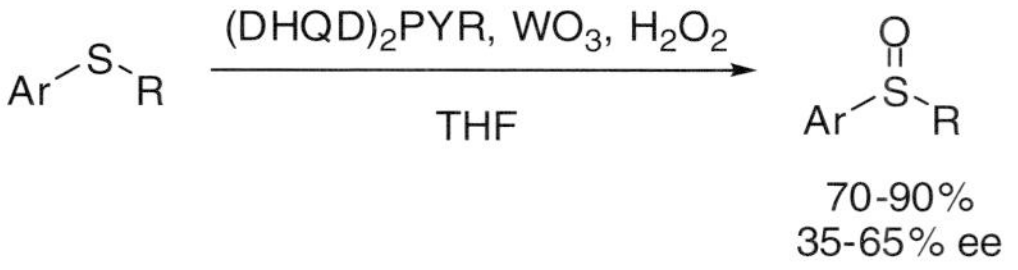

Scheme 3.46

Ar–S–R → (DHQD)₂PYR, WO₃, H₂O₂, THF → Ar–S(=O)–R

70-90%
35-65% ee

For the AA reaction, immobilization of the ligands and osmium catalyst can offer some advantages, as for the AD reaction (see Section 3.2.4).

3.4
Sulfur Oxidations

The biscinchona alkaloid ligands can also be used for the asymmetric oxidation of sulfides to sulfoxides with hydrogen peroxide as the oxidant in the presence of tungsten(VI) oxide, or layered double hydroxide (LDH) supported OsO_4 as catalyst (Scheme 3.46) [375, 376]. The approach can also be used to affect a kinetic resolution of racemic sulfoxides by oxidation of one enantiomer to the sulfone (Scheme 3.47) [376].

3.5
Summary

The cinchona alkaloids have opened up the field of asymmetric oxidations of alkenes without the need for a functional group within the substrate to form a complex with the metal. Current methodology is limited to osmium-based oxidations. The power of the asymmetric dihydroxylation reaction is exemplified by the thousands (literally) of examples for the use of this reaction to establish stereogenic centers in target molecule synthesis. The usefulness of the AD reaction is augmented by the bountiful chemistry of cyclic sulfates and sulfites derived from the resultant 1,2-diols.

The asymmetric aminohydroxylation reaction provides ready access to 1,2-amino alcohol derivatives and is a useful alternative not only to the AD methodology but also to peroxide-based approaches. The power of both the approaches in synthesis is helped by the development of the mnemonic device, which allows the stereochemical outcome of the reactions to be predicted with a large degree of certainty.

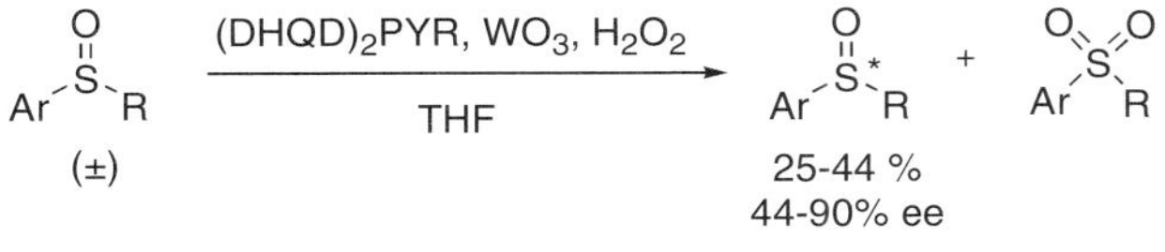

Scheme 3.47

Ar–S(=O)–R (±) → (DHQD)₂PYR, WO₃, H₂O₂, THF → Ar–S*(=O)–R + Ar–S(=O)₂–R

25-44%
44-90% ee

References

1 Lohray, B.B. (1992) *Tetrahedron: Asymmetry*, **3**, 1317.
2 Kolb, H.C., VanNieuwenhze, M.S. and Sharpless, K.B. (1994) *Chem. Rev.*, **94**, 2483.
3 O'Brien, P. (1999) *Angew. Chem. Int. Ed.*, **38**, 326.
4 Kacprzak, K. and Gawroński, J. (2001) *Synthesis*, 961.
5 Zaitsev, A.B. and Adolfsson, H. (2006) *Synthesis*, 1725.
6 Français, A., Bedel, O., and Haudrechy, A. (2008) *Tetrahedron*, **64**, 2495.
7 Kolb, H.C. and Sharpless, K.B. (1998) in *Transition Metals for Organic Synthesis*, vol. 2 (eds M. Beller and C. Bolm), Wiley-VCH Verlag GmbH, Weinheim, p. 219.
8 Markó, I. and Svendsen, J.S. (1999) in *Comprehensive Asymmetric Catalysis*, vol. 2 (eds E.N. Jacobsen, A. Pfaltz, and H. Yamamoto), Springer, Berlin, p. 713.
9 Johnson, R.A. and Sharpless, K.B. (2000) in *Catalytic Asymmetric Synthesis*, 2nd edn (ed. I. Ojima), Wiley, New York, p. 357.
10 Bolm, C., Hildebrand, J.P., and Muñiz, K. (2000) in *Catalytic Asymmetric Reactions*, 2nd edn (ed. I. Ojima), John Wiley & Sons, New York, p. 399.
11 Beller, M. and Sharpless, K.B. (2002) in *Applied Homogeneous Catalysis with Organometallic Compounds*, 2nd edn, vol. 3 (ed. B.H. Cornils and A. Wolfgang), Wiley-VCH Verlag GmbH, Weinheim, p. 1149.
12 Muñiz, K. (2004) in *Transition Metals in Organic Synthesis*, vol. 2 (eds M. Beller and C. Bolm), Wiley-VCH Verlag GmbH, Weinheim, p. 298.
13 Drudis-Solé, G., Ujaque, G., Maseras, F. and Lledós, A. (2005) *Topics Organometal. Chem.*, **12**, 79.
14 Kolb, H.C. and Sharpless, K.B. (2004) in *Transition Metals in Organic Synthesis*, vol. 2 (eds M. Beller and C. Bolm), Wiley-VCH Verlag GmbH, Weinheim, p. 275.
15 Mehrman, S.J., Abdel-Magid, A.F., Maryanoff, C.A., and Medaer, B.P. (2004) *Topics Organometal. Chem.*, **6**, 153.
16 Sundermeier, U., Döbler, C., and Beller, M. (2004) in *Modern Oxidation Methods* (ed. J.-E. Bäckvall), Wiley-VCH Verlag GmbH, Weinheim, p. 1.
17 Sharpless, K.B. (1998) in *Transition Metal for Organic Synthesis*, vol. 2 (eds M. Beller and C. Bolm), Wiley-VCH Verlag GmbH, Weinheim, p. 219.
18 Sharpless, K.B. (1998) in *Transition Metal for Organic Synthesis*, vol. 2 (eds M. Beller and C. Bolm), Wiley-VCH Verlag GmbH, Weinheim, p. 243.
19 Criegee, R. (1936) *Justus Liebigs Ann. Chem.*, **522**, 75.
20 Milas, N.A. and Sussman, S. (1936) *J. Am. Chem. Soc.*, **58**, 1302.
21 Schröder, M. (1980) *Chem. Rev.*, **80**, 187.
22 Akashi, K., Palermo, R.E., and Sharpless, K.B. (1978) *J. Org. Chem.*, **43**, 2063.
23 Göbel, T. and Sharpless, K.B. (1993) *Angew. Chem. Int. Ed.*, **32**, 1329.
24 Lee, D.G. and Brown, K.C. (1982) *J. Am. Chem. Soc.*, **104**, 5076.
25 Lee, D.G. and Brownridge, J.R. (1973) *J. Am. Chem. Soc.*, **95**, 3033.
26 Lee, D.G. and Brownridge, J.R. (1974) *J. Am. Chem. Soc.*, **96**, 5517.
27 Ogino, T. (1980) *Tetrahedron Lett.*, **21**, 177.
28 Ogino, T. and Mochizuki, K. (1979) *Chem. Lett.*, 443.
29 Simandi, L.I. and Jaky, M. (1976) *J. Am. Chem. Soc.*, **98**, 1995.
30 Sivik, M.R., Gallucci, J.C., and Paquette, L.A. (1990) *J. Org. Chem.*, **55**, 391.
31 Stewart, R. (1964) *Oxidation Mechanisms*, Benjamin, New York.
32 Wiberg, K.B., Deutsch, C.J., and Rocek, J. (1973) *J. Am. Chem. Soc.*, **95**, 3034.
33 Corey, E.J., Pan, B.-C., Hua, D.H., and Deardorff, D.R. (1982) *J. Am. Chem. Soc.*, **104**, 6816.
34 Ray, R. and Matteson, D.S. (1980) *Tetrahedron Lett.*, **21**, 449.

35 VanRheenan, V., Kelly, R.C., and Cha, D.Y. (1976) *Tetrahedron Lett.*, **17**, 1973.
36 Cainelli, G., Contento, M., Manescalchi, F., and Plessi, L. (1989) *Synthesis*, 45.
37 Sharpless, K.B. and Akashi, K. (1976) *J. Am. Chem. Soc.*, **98**, 1986.
38 Paaren, H., Schnoes, H.K., and DeLuca, H.F. (1983) *J. Org. Chem.*, **48**, 3819.
39 Ager, D.J. and East, M.B. (1993) *Tetrahedron*, **49**, 5683.
40 Daniels, R. and Fischer, J.L. (1963) *J. Org. Chem.*, **28**, 320.
41 Criegee, R., Marchand, B., and Wannwins, H. (1942) *Justus Liebigs Ann. Chem.*, **550**, 99.
42 Milas, N.A. and Sussman, S. (1937) *J. Am. Chem. Soc.*, **59**, 2345.
43 Byers, A. and Hickinbottom, J. (1948) *J. Chem. Soc.*, 1328.
44 Milas, N.A., Trepagnier, J.H., Nolan, J.T., Jr., and Iliopulos, M.I. (1959) *J. Am. Chem. Soc.*, **81**, 4730.
45 Hofmann, K.A. (1912) *Chem. Ber.*, **45**, 3329.
46 McMurry, J.E., Andrus, A., Ksander, G.M., Musser, J.H., and Johnsonn, M.A. (1979) *J. Am. Chem. Soc.*, **101**, 1330.
47 Wiesner, K. and Santroch, J. (1966) *Tetrahedron Lett.*, **7**, 5939.
48 Foglia, T.A., Barr, P.A., Malloy, A.J., and Costanzo, M.J. (1977) *J. Am. Oil Chem. Soc.*, **54**, 870.
49 Corey, E.J., Danheiser, R.L., Chandrasekaran, S., Siret, P., Keck, G.E., and Grass, J.L. (1978) *J. Am. Chem. Soc.*, **100**, 8031.
50 Singh, M.P., Singh, H.S., Arya, A.K., Singh, A.K., and Sisodia, A.K. (1975) *Indian J. Chem.*, **13**, 112.
51 Minato, M., Yamamoto, K., and Tsuji, J. (1990) *J. Org. Chem.*, **55**, 766.
52 Cairns, J.F. and Roberts, H.L. (1968) *J. Chem. Soc. C*, 640.
53 Döbler, C., Mehltretter, G.M., Sundermeier, U., and Beller, M. (2000) *J. Am. Chem. Soc.*, **122**, 10289.
54 Cha, J.K., Christ, W.J., and Kishi, Y. (1983) *Tetrahedron Lett.*, **24**, 3943.
55 Christ, W.J., Cha, J.K., and Kishi, Y. (1983) *Tetrahedron Lett.*, **24**, 3947.
56 Evans, D.A. and Kaldor, S.W. (1990) *J. Org. Chem.*, **55**, 1698.
57 Cha, J.K., Christ, W.J., and Kishi, Y. (1984) *Tetrahedron*, **40**, 2247.
58 Vedejs, E. and Dent, W.H. (1989) *J. Am. Chem. Soc.*, **111**, 6861.
59 Ager, D.J. and East, M.B. (1992) *Tetrahedron*, **48**, 2803.
60 Johnson, C.R. and Barbachyn, M.R. (1984) *J. Am. Chem. Soc.*, **106**, 2459.
61 Hirama, M., Oishi, T., and Ito, S. (1989) *J. Chem. Soc., Chem. Commun.*, 665.
62 Tokles, M. and Snyder, J.K. (1986) *Tetrahedron Lett.*, **27**, 3951.
63 Yamada, T. and Narasaka, K. (1986) *Chem. Lett.*, 131.
64 Corey, E.J. and Lotto, G.I. (1990) *Tetrahedron Lett.*, **31**, 2665.
65 Oishi, T. and Hirama, M. (1989) *J. Org. Chem.*, **54**, 5834.
66 Tomioka, K., Nakajima, M., and Koga, K. (1990) *Tetrahedron Lett.*, **31**, 1741.
67 Kobayashi, S., Ishida, T., and Akiyama, R. (2001) *Org. Lett.*, **3**, 2649.
68 Choudary, B.M., Chowdari, N.S., Jyothi, K., and Kantam, M.L. (2003) *J. Mol. Catal. A: Chem.*, **196**, 151.
69 Choudary, B.M., Chowdari, N.S., Madhi, S., and Kantam, M.L. (2003) *J. Org. Chem.*, **68**, 1736.
70 Kobayashi, S., Endo, M., and Nagayama, S. (1999) *J. Am. Chem. Soc.*, **121**, 11229.
71 Choudary, B.M., Roy, M., Roy, S., Kantam, M.L., Sreedhar, B., and Kumar, K.V. (2006) *Adv. Synth. Catal.*, **348**, 1734.
72 Reddy, S.M., Srinivasulu, M., Reddy, Y.V., Narasimhulu, M., and Venkateswarlu, Y. (2006) *Tetrahedron Lett.*, **47**, 5285.
73 Hentges, S.G. and Sharpless, K.B. (1980) *J. Am. Chem. Soc.*, **102**, 4263.
74 Ogino, Y., Chen, H., Manoury, E., Shibata, T., Beller, M., Lübben, D., and Sharpless, K.B. (1991) *Tetrahedron Lett.*, **32**, 5761.

75 Sharpless, K.B., Amberg, W., Beller, M., Chen, H., Hartung, J., Kawanami, Y., Lübben, D., Manoury, E., Ogino, Y., Shibata, T., and Ukita, T. (1991) *J. Org. Chem.*, **56**, 4585.

76 Sharpless, K.B. (2002) *Angew. Chem. Int. Ed.*, **41**, 2024.

77 Jacobsen, E.N., Marko, I., France, M.B., Svendsen, J.S., and Sharpless, K.B. (1989) *J. Am. Chem. Soc.*, **111**, 737.

78 Jacobsen, E.N., Markó, I., Mungall, W.S., Schröder, G., and Sharpless, K.B. (1988) *J. Am. Chem. Soc.*, **110**, 1968.

79 Lohray, B.B., Kalantar, T.H., Kim, B.M., Park, C.Y., Shibata, T., Wai, J.S.M., and Sharpless, K.B. (1989) *Tetrahedron Lett.*, **30**, 2041.

80 Shibata, T., Gilheany, D.G., Blackburn, B.K., and Sharpless, K.B. (1990) *Tetrahedron Lett.*, **31**, 3817.

81 Jorgensen, K.A. (1990) *Tetrahedron Lett.*, **31**, 6417.

82 Wai, J.S.M., Marko, I., Svendsen, J.S., Finn, M.G., Jacobsen, E.N., and Sharpless, K.B. (1989) *J. Am. Chem. Soc.*, **111**, 1123.

83 Kwong, H.-L., Sorato, C., Ogino, Y., Chen, H., and Sharpless, K.B. (1990) *Tetrahedron Lett.*, **31**, 2999.

84 Sharpless, K.B., Amberg, W., Bennani, Y.L., Crispino, G.A., Hartung, J., Jeong, K.-S., Kwong, H.-L., Morikawa, K., Wang, Z.-M., Xu, D., and Zhang, X.-L. (1992) *J. Org. Chem.*, **57**, 2768.

85 Arrington, M.P., Bennani, Y.L., Göbel, T., Walsh, P., Zhao, S.-H., and Sharpless, K.B. (1993) *Tetrahedron Lett.*, **34**, 7375.

86 Morikawa, K., Park, J., Andersson, P.G., Hashiyama, T., and Sharpless, K.B. (1993) *J. Am. Chem. Soc.*, **115**, 8463.

87 Vidari, G., Giori, A., Dapiaggi, A., and Lanfranchi, G. (1993) *Tetrahedron Lett.*, **34**, 6925.

88 Wang, L. and Sharpless, K.B. (1992) *J. Am. Chem. Soc.*, **114**, 7568.

89 Zhang, S.Y., Girard, C., and Kagan, H.B. (1995) *Tetrahedron: Asymmetry*, **6**, 2637.

90 Nelson, A., O'Brien, P., and Warren, S. (1995) *Tetrahedron Lett.*, **36**, 2685.

91 Ogino, Y., Chen, H., Kwong, H.-L., and Sharpless, K.B. (1991) *Tetrahedron Lett.*, **32**, 3965.

92 Ahrgren, L. and Sutin, L. (1997) *Org. Proc. Res. Dev.*, **1**, 425.

93 Bergstad, K., Jonsson, S.Y., and Bäckvall, J.-E. (1999) *J. Am. Chem. Soc.*, **121**, 10424.

94 Jonsson, S.Y., Adolfsson, H., and Bäckvall, J.-E. (2001) *Org. Lett.*, **3**, 3463.

95 Jonsson, S.Y., Färnegårdh, K., and Bäckvall, J.-E. (2001) *J. Am. Chem. Soc.*, **123**, 1365.

96 Jonsson, S.Y., Adolfsson, H., and Bäckvall, J.-E. (2003) *Chem. Eur. J.*, **9**, 2783.

97 Choudary, B.M., Jyothi, K., Roy, M., Kantam, M.L., and Sreedhar, B. (2004) *Adv. Synth. Catal.*, **345**, 1471.

98 Balagam, B., Mitra, R., and Richardson, D.E. (2008) *Tetrahedron Lett.*, **49**, 1071.

99 Krief, A. and Castillo-Colaux, C. (2001) *Synlett*, 501.

100 Krief, A. and Colaux-Castillo, C. (1999) *Tetrahedron Lett.*, **40**, 4189.

101 Torii, S., Liu, P., Bhuvaneswari, N., Amatore, C., and Jutand, A. (1996) *J. Org. Chem.*, **61**, 3055.

102 Junttila, M.H. and Hormi, O.E.O. (2004) *J. Org. Chem.*, **69**, 4816.

103 Mehltretter, G.M., Bhor, S., Klawonn, M., Döbler, C., Sundermeier, U., Eckert, M., Militzer, H.-M., and Beller, M. (2003) *Synthesis*, 295.

104 Torii, S., Liu, P., and Tanaka, H. (1995) *Chem. Lett.*, 319.

105 Döbler, C., Mehltretter, G.M., and Beller, M. (1999) *Angew. Chem. Int. Ed.*, **38**, 3026.

106 Döbler, C., Mehltretter, G.M., Sundermeier, U., and Beller, M. (2001) *J. Organometal. Chem.*, **621**, 70.

107 Mehltretter, G.M., Döbler, C., Sundermeier, U., and Beller, M. (2000) *Tetrahedron Lett.*, **41**, 8083.

108 Amberg, W., Bennani, Y.L., Chadha, R.K., Crispino, G.A., Davis, W.D., Hartung, J., Jeong, K.-S., Ogino, Y., Shibata, T., and Sharpless, K.B. (1993) *J. Org. Chem.*, **58**, 844.

References

109 Corey, E.J., Noe, M.C., and Sarshar, S. (1993) *J. Am. Chem. Soc.*, **115**, 3828.

110 Dijkstra, G.D.H., Kellogg, R.M., Wynberg, H., Svendsen, J.S., Marko, I., and Sharpless, K.B. (1989) *J. Am. Chem. Soc.*, **111**, 8069.

111 Kolb, H.C., Andersson, P.G., Bennani, Y.L., Crispino, G.A., Jeong, K.-S., Kwong, H.-L., and Sharpless, K.B. (1993) *J. Am. Chem. Soc.*, **115**, 12226.

112 Pearlstein, R.M., Blackburn, B.K., Davis, W.M., and Sharpless, K.B. (1990) *Angew. Chem. Int. Ed.*, **29**, 639.

113 Svendsen, J.S., Marko, I., Jacobsen, E., Rao, C.P., Bott, S., and Sharpless, K.B. (1989) *J. Org. Chem.*, **54**, 2263.

114 Kolb, H.C., Andersson, P.G., and Sharpless, K.B. (1994) *J. Am. Chem. Soc.*, **116**, 1278.

115 Lohray, B.B., Bhushan, V., and Nandanan, E. (1994) *Tetrahedron Lett.*, **35**, 4209.

116 Corey, E.J. and Noe, M.C. (1996) *J. Am. Chem. Soc.*, **118**, 319.

117 Norrby, P.-O., Becker, H., and Sharpless, K.B. (1996) *J. Am. Chem. Soc.*, **118**, 35.

118 Corey, E.J. and Noe, M.C. (1996) *J. Am. Chem. Soc.*, **118**, 11038.

119 Norrby, P.-O., Kolb, H.C., and Sharpless, K.B. (1994) *Organometallics*, **13**, 344.

120 Corey, E.J., Noe, M.C., and Grogan, M.J. (1994) *Tetrahedron Lett.*, **35**, 6427.

121 Corey, E.J., Noe, M.C., and Sarshar, S. (1994) *Tetrahedron Lett.*, **35**, 2861.

122 Norrby, P.-O., Kolb, H.C., and Sharpless, K.B. (1994) *J. Am. Chem. Soc.*, **116**, 8470.

123 Deubel, D.V. and Frenking, G. (2003) *Acc. Chem. Res.*, **36**, 645.

124 Becker, H., Ho, P.T., Kolb, H.C., Loren, S., Norrby, P.-O., and Sharpless, K.B. (1994) *Tetrahedron Lett.*, **35**, 7315.

125 Ujaque, G., Maseras, F., and Lledós, A. (1999) *J. Am. Chem. Soc.*, **121**, 1317.

126 Norrby, P.-O., Rasmussen, T., Haller, J., Strassner, T., and Houk, K.N. (1999) *J. Am. Chem. Soc.*, **121**, 10186.

127 Moitessier, N., Henry, C., Len, C., and Chapleur, Y. (2002) *J. Org. Chem.*, **67**, 7275.

128 Periasamy, M., Kumar, S.S., and Kumar, N.S. (2008) *Tetrahedron Lett.*, **49**, 4416.

129 DelMonte, A.J., Haller, J., Houk, K.N., Sharpless, K.B., Singleton, D.A., Strassner, T., and Thomas, A.A. (1997) *J. Am. Chem. Soc.*, **119**, 9907.

130 Torrent, M., Deng, L., Duran, M., Sola, M., and Ziegler, T. (1997) *Organometallics*, **16**, 13.

131 Vanhessche, K.P.M. and Sharpless, K.B. (1996) *J. Org. Chem.*, **61**, 7978.

132 Fristrup, P., Tanner, D., and Norrby, P.-O. (2003) *Chirality*, **15**, 360.

133 Fristrup, P., Jensen, G.H., Andersen, M.L.N., Tanner, D., and Norrby, P.-O. (2006) *J. Organometal. Chem.*, **691**, 2182.

134 Kim, B.M. and Sharpless, K.B. (1990) *Tetrahedron Lett.*, **31**, 3003.

135 Han, H. and Janda, K.D. (1996) *J. Am. Chem. Soc.*, **118**, 7632.

136 Dickerson, T.J., Reed, N.N., and Janda, K.D. (2002) *Chem. Rev.*, **102**, 3325.

137 Nandanan, E., Sudalai, A., and Ravindranathan, T. (1997) *Tetrahedron Lett.*, **38**, 2577.

138 Bolm, C. and Gerlach, A. (1998) *Eur. J. Org. Chem.*, 21.

139 Riedl, R., Tappe, R., and Berkessel, A. (1998) *J. Am. Chem. Soc.*, **120**, 8994.

140 Song, C.E., Yang, J.W., Ha, H.J., and Lee, S.-G. (1996) *Tetrahedron: Asymmetry*, 7, 645.

141 Cha, R.-t., Wang, S.-y., and Cheng, S.-l. (2008) *J. Appl. Poly. Sci.*, **108**, 845.

142 Achkar, J., Hunt, J.R., Beingessner, R.L., and Fenniri, H. (2008) *Tetrahedron: Asymmetry*, **19**, 1049.

143 Han, H. and Janda, K.D. (1997) *Angew. Chem. Int. Ed.*, **36**, 1731.

144 Lohray, B.B., Thomas, A., Chittari, P., Ahuja, J.R., and Dhal, P.K. (1992) *Tetrahedron Lett.*, **33**, 5453.

145 Motorina, I. and Crudden, C.M. (2001) *Org. Lett.*, **3**, 2325.

146 Yang, J.W., Han, H., Roh, E.J., Lee, S.-G., and Song, C.E. (2002) *Org. Lett.*, **4**, 4685.

147 Choudary, B.M., Chowdari, N.S., Jyothi, K., and Kantam, M.L. (2002) *J. Am. Chem. Soc.*, **124**, 5341.

148 Jo, C.H., Han, S.-H., Yang, J.W., Roh, E.J., Shin, U.-S., and Song, C.E. (2003) *Chem. Commun.*, 1312.

149 Lee, H.M., Kim, S.-W., Hyeon, T., and Kim, B.M. (2001) *Tetrahedron: Asymmetry*, **12**, 1537.

150 Song, C.E., Yang, J.W., and Ha, H.-J. (1997) *Tetrahedron: Asymmetry*, **8**, 841.

151 Choudary, B.M., Chowdari, N.S., Jyothi, K., Madhu, S., and Kantam, M.L. (2002) *Adv. Synth. Catal.*, **344**, 503.

152 Kuang, Y.-Q., Zhang, S.-Y., and Wei, L.-L. (2001) *Tetrahedron Lett.*, **42**, 5925.

153 Han, H. and Janda, K.D. (1997) *Tetrahedron Lett.*, **38**, 1527.

154 Wöltinger, J., Henniges, H., Krimmer, H.-P., Bommarius, A.S., and Druz, K. (2001) *Tetrahedron: Asymmetry*, **12**, 2095.

155 Branco, L.C. and Afonso, C.A.M. (2002) *Chem. Commun.*, 3036.

156 Song, C.E., Jung, D.-u., Roh, E.J., Lee, S.-G., and Chi, D.Y. (2002) *Chem. Commun.*, 3038.

157 Branco, L.C. and Afonso, C.A.M. (2004) *J. Org. Chem.*, **69**, 4381.

158 Liu, Q., Zhang, Z., van Rantwijk, F., and Sheldon, R.A. (2004) *J. Mol. Catal. A: Chem.*, **224**, 213.

159 Branco, L.C., Serbanovic, A., da Ponte, M.N., and Afonso, C.A.M. (2005) *Chem. Commun.*, 107.

160 Serbanovic, A., Branco, L.C., da Ponte, M.N., and Afonso, C.A.M. (2005) *J. Organometal. Chem.*, **690**, 3600.

161 Lee, D., Lee, H., Kim, S., Yeom, C.-E., and Kim, B.M. (2006) *Adv. Synth. Catal.*, **348**, 1021.

162 Ferreira, F.C., Branco, L.C., Verma, K.K., Crespo, J.G., and Afonso, C.A.M. (2007) *Tetrahedron: Asymmetry*, **18**, 1637.

163 Liang, J., Moher, E.D., Moore, R.E., and Hoard, D.W. (2000) *J. Org. Chem.*, **65**, 3143.

164 Andersson, P.G. and Sharpless, K.B. (1993) *J. Am. Chem. Soc.*, **115**, 7047.

165 Philippo, C.M.G., Mougenot, P., Braun, A., Defosse, G., Auboussier, S., and Bovy, P.R. (2000) *Synthesis*, 127.

166 Lawrence, N.J. and Bushell, S.M. (2001) *Tetrahedron Lett.*, **42**, 7671.

167 Jary, W.G. and Baumgartner, J. (1998) *Tetrahedron: Asymmetry*, **9**, 2081.

168 Balachari, D. and O'Doherty, G.A. (2000) *Org. Lett.*, **2**, 863.

169 Taniguchi, T., Nakamura, K., and Ogasawara, K. (1996) *Synlett*, 971.

170 Harris, J.M., Keranen, M.D., and O'Doherty, G.A. (1999) *J. Org. Chem.*, **64**, 2982.

171 Nascimento, I.R., Lopes, L.M.X., Davin, L.B., and Lewis, N.G. (2000) *Tetrahedron*, **56**, 9181.

172 Wyatt, P., Warren, S., McPartlin, M., and Woodroffe, T. (2001) *J. Chem. Soc., Perkin Trans. 1*, 279.

173 Donnoli, M.I., Scafato, P., Superchi, S., and Rosini, C. (2001) *Chirality*, **13**, 258.

174 Moreno, R.M., Bueno, A., and Moyano, A. (2006) *J. Org. Chem.*, **71**, 2528.

175 Wang, X., Zak, M., Maddess, M., O'Shea, P., Tillyer, R., Grabowski, E.J.J., and Reider, P.J. (2000) *Tetrahedron Lett.*, **41**, 4865.

176 Zhou, X.-D., Cai, F., and Zhou, W.-S. (2001) *Tetrahedron Lett.*, **42**, 2537.

177 Drudis-Solé, G., Ujaque, G., Maseras, F., and Lledós, A. (2005) *Chem. Eur. J.*, **11**, 1017.

178 Wang, Z.-M., Zhang, X.-L., and Sharpless, K.B. (1993) *Tetrahedron Lett.*, **34**, 2267.

179 Becker, H. and Sharpless, K.B. (1996) *Angew. Chem. Int. Ed.*, **35**, 448.

180 Paterson, I., Anderson, E.A., Dalby, S.M., and Loiseiur, O. (2005) *Org. Lett.*, **7**, 4121.

181 Corey, E.J., Guzman-Perez, A., and Noe, M.C. (1995) *J. Am. Chem. Soc.*, **117**, 10805.

182 Wang, Z.-M. and Sharpless, K.B. (1993) *Synlett*, 603.

183 Corey, E.J., Guzman-Perez, A., and Noe, M.C. (1994) *J. Am. Chem. Soc.*, **116**, 12109.

184 Kawashima, E., Naito, Y.-k., and Ishido, Y. (2000) *Tetrahedron Lett.*, **41**, 3903.

185 Kulig, K., Holzgrabe, U., and Malawska, B. (2001) *Tetrahedron: Asymmetry*, **12**, 2533.

186 Oi, R. and Sharpless, K.B. (1992) *Tetrahedron Lett.*, **33**, 2095.
187 Henderson, I., Sharpless, K.B., and Wong, C.-H. (1994) *J. Am. Chem. Soc.*, **116**, 558.
188 Li, M. and O'Doherty, G.A. (2004) *Tetrahedron Lett.*, **45**, 6407.
189 Bennani, Y., Vanhessche, K.P.M., and Sharpless, K.B. (1994) *Tetrahedron: Asymmetry*, **5**, 1473.
190 Hale, K.J., Manaviazar, S., and Peak, S.A. (1994) *Tetrahedron Lett.*, **35**, 425.
191 Avenoza, A., Cativiela, C., Peregrina, J.M., Sucunza, D., and Zurbano, M.M. (2001) *Tetrahedron: Asymmetry*, **12**, 1383.
192 Avenoza, A., Busto, J.H., Jiménez-Osés, G., and Peregrina, J.M. (2005) *J. Org. Chem.*, **70**, 5721.
193 Kiyota, H., Takai, T., Shimasaki, Y., Saitoh, M., Nakayama, O., Takada, T., and Kuwahara, S. (2007) *Synthesis*, 2471.
194 Xu, D. and Sharpless, K.B. (1994) *Tetrahedron Lett.*, **35**, 4685.
195 VanNieuwenhze, M.S. and Sharpless, K.B. (1994) *Tetrahedron: Asymmetry*, **35**, 843.
196 Ko, S.Y., Malik, M., and Dickinson, A.F. (1994) *J. Org. Chem.*, **59**, 2570.
197 Wang, B.-L., Yu, F., Qiu, X.-L., Jiang, Z.-X., and Qing, F.-L. (2006) *J. Fluorine Chem.*, **127**, 580.
198 Jiang, Z.-X., Qin, Y.-Y., and Qing, F.-L. (2003) *J. Org. Chem.*, **68**, 7544.
199 Evans, P. and Leffray, M. (2003) *Tetrahedron*, **59**, 7973.
200 Loh, T.-P. and Feng, L.-C. (2001) *Tetrahedron Lett.*, **42**, 6001.
201 Nicolaou, K.C., Li, J., and Zenke, G. (2000) *Helv. Chim. Acta*, **83**, 1977.
202 Luzzio, F.A., Thomas, E.M., and Figg, W.D. (2000) *Tetrahedron Lett.*, **41**, 7151.
203 Yadav, J.S., Rajaiah, G., and Raju, A.K. (2003) *Tetrahedron Lett.*, **44**, 5831.
204 Pearson, A.J. and Heo, J.-N. (2000) *Tetrahedron Lett.*, **41**, 5991.
205 Feng, Z.-X. and Zhou, W.-S. (2003) *Tetrahedron Lett.*, **44**, 493.
206 Catasús, M., Bueno, A., Moyano, A., Maestro, M.A., and Mahía, J. (2002) *J. Organometal. Chem.*, **642**, 212.
207 Bonini, C., Chiummiento, L., De Bonis, M., Funicello, M., Lupattelli, P., and Pandolfo, R. (2006) *Tetrahedron: Asymmetry*, **17**, 2919.
208 Lu, X., Xu, Z., and Yang, G. (2000) *Org. Proc. Res. Dev.*, **4**, 575.
209 Bennani, Y. and Sharpless, K.B. (1993) *Tetrahedron Lett.*, **34**, 2079.
210 Lee, D.-H. and Rho, M.-D. (2000) *Tetrahedron Lett.*, **41**, 2573.
211 Yokomatsu, T., Yamagishi, T., Suemune, K., Yoshida, Y., and Shibuya, S. (1998) *Tetrahedron*, **54**, 767.
212 Yokomatsu, T., Yoshida, Y., Suemune, K., Yamagishi, T., and Shibuya, S. (1995) *Tetrahedron: Asymmetry*, **6**, 365.
213 Kobayashi, Y., William, A.D., and Tokoro, Y. (2001) *J. Org. Chem.*, **66**, 7903.
214 Chandrasekhar, S., Sultana, S.S., Kiranmai, N., and Narsihmulu, C. (2007) *Tetrahedron Lett.*, **48**, 2373.
215 Reddy, J.S. and Rao, B.V. (2007) *J. Org. Chem.*, **72**, 2224.
216 Yadav, J.S., Raju, A.K., Rao, P.P., and Rajaiah, G. (2005) *Tetrahedron: Asymmetry*, **16**, 3283.
217 Vanhessche, K.P.M., Wang, Z.-M., and Sharpless, K.B. (1994) *Tetrahedron Lett.*, **35**, 3469.
218 Fernandes, R.A. (2008) *Tetrahedron: Asymmetry*, **19**, 15.
219 Zhang, Z.-B., Wang, Z.-M., Wang, Y.-X., Liu, H.-Q., Lei, G.-X., and Shi, M. (2000) *J. Chem. Soc., Perkin Trans. I*, 53.
220 Walsh, P.J., Ho, P.T., King, S.B., and Sharpless, K.B. (1994) *Tetrahedron Lett.*, **35**, 5129.
221 Sammakia, T., Hurley, T.B., Sammond, D.M., Smith, R.S., Sobolov, S.B., and Oeschger, T.R. (1996) *Tetrahedron Lett.*, **37**, 4427.
222 Okamoto, S., Tani, K., Sato, F., Sharpless, K.B., and Zargarian, D. (1993) *Tetrahedron Lett.*, **34**, 2509.
223 Soderquist, J.A., Rane, A.M., and López, C.J. (1993) *Tetrahedron Lett.*, **34**, 1893.

224 Bassindale, A.R., Taylor, P.G., and Xu, Y. (1994) *J. Chem. Soc., Perkin Trans. I*, 1061.
225 Fernandes, R.A. and Kumar, P. (2000) *Eur. J. Org. Chem.*, 3447.
226 Naidu, S.V. and Kumar, P. (2003) *Tetrahedron Lett.*, **44**, 1035.
227 Wang, Z.-M. and Sharpless, K.B. (1993) *Tetrahedron Lett.*, **34**, 8225.
228 Allevi, P., Tarocco, G., Longo, A., Anastasia, M., and Cajone, F. (1997) *Tetrahedron: Asymmetry*, **8**, 1315.
229 Walsh, P.J., Bennani, Y.L., and Sharpless, K.B. (1993) *Tetrahedron Lett.*, **34**, 5545.
230 Uenishi, J. and Ohmiya, H. (2003) *Tetrahedron*, **59**, 7011.
231 Yamagishi, T., Fujii, K., Shibuya, S., and Yokomatsu, T. (2004) *Synlett*, 2505.
232 Jung, M.E. and Gardiner, J.M. (1994) *Tetrahedron Lett.*, **35**, 6755.
233 Harcken, C. and Brückner, R. (1997) *Angew. Chem. Int. Ed.*, **36**, 2750.
234 Wang, Z.-M., Zhang, X.-L., Sharpless, K.B., Sinha, S.C., Sinha-Bagchi, A., and Keinan, E. (1992) *Tetrahedron Lett.*, **33**, 6407.
235 Sabitha, G., Yadagiri, K., and Yadav, J.S. (2007) *Tetrahedron Lett.*, **48**, 8065.
236 Martín, T. and Martín, V.S. (2000) *Tetrahedron Lett.*, **41**, 2503.
237 García, C., Soler, M.A., and Martín, V.S. (2000) *Tetrahedron Lett.*, **41**, 4127.
238 He, Y.-T., Yang, H.-N., and Yao, Z.-J. (2002) *Tetrahedron*, **58**, 8805.
239 Keinan, E., Sinha, S.C., Sinha-Bagchi, A., Wang, Z.-M., Zhang, X.-L., and Sharpless, K.B. (1992) *Tetrahedron Lett.*, **33**, 6411.
240 Kirschning, A., Dräger, G., and Jung, A. (1997) *Angew. Chem. Int. Ed.*, **36**, 253.
241 Taillier, C., Gille, B., Bellosta, V., and Cossy, J. (2005) *J. Org. Chem.*, **70**, 2097.
242 Qin, D.-G., Zha, H.-Y., and Yao, Z.-J. (2002) *J. Org. Chem.*, **67**, 1038.
243 Shen, J.-W., Qin, D.-G., Zhang, H.-W., and Yao, Z.-J. (2003) *J. Org. Chem.*, **68**, 7479.
244 Mulzer, J., Karig, G., and Pojarliev, P. (2000) *Tetrahedron Lett.*, **41**, 7635.
245 Corey, E.J., Noe, M.C., and Ting, A.Y. (1996) *Tetrahedron Lett.*, **37**, 1735.
246 Noe, M.C. and Corey, E.J. (1996) *Tetrahedron Lett.*, **37**, 1739.
247 Toshima, H., Saito, M., and Yoshihara, T. (1999) *Biosci. Biotechnol. Biochem.*, **63**, 964.
248 Toshima, H., Saito, M., and Yoshihara, T. (1999) *Biosci. Biotechnol. Biochem.*, **63**, 1934.
249 Devaux, J.-M., Goré, J., and Vatèle, J.-M. (1998) *Tetrahedron: Asymmetry*, **9**, 1619.
250 Turpin, J.A. and Weigel, L.O. (1992) *Tetrahedron Lett.*, **33**, 6563.
251 Trost, B.M. and Pinkerton, A.B. (2002) *J. Am. Chem. Soc.*, **124**, 7376.
252 Miyashita, K., Ikejiri, M., Kawasaki, H., Maemura, S., and Imanishi, T. (2003) *J. Am. Chem. Soc.*, **125**, 8238.
253 Chavan, S.P., Sharma, P., Sivappa, R., and Kalkote, U.R. (2007) *Synlett*, 79.
254 Kolb, H.C., Bennani, Y.L., and Sharpless, K.B. (1993) *Tetrahedron: Asymmetry*, **4**, 133.
255 Blagg, B.S.J. and Boger, D.L. (2002) *Tetrahedron*, **58**, 6343.
256 Seidel, M.C., Smits, R., Stark, C.B.W., Frackenpohl, J., Gaertzen, O., and Hoffmann, H.M.R. (2004) *Synthesis*, 1391.
257 Gao, D. and O'Doherty, G.A. (2005) *J. Org. Chem.*, **70**, 9932.
258 Xu, D., Crispino, G.A., and Sharpless, K.B. (1992) *J. Am. Chem. Soc.*, **114**, 7570.
259 Becker, H., Soler, M.A., and Sharpless, K.B. (1995) *Tetrahedron*, **51**, 1345.
260 Landais, Y. and Zekri, E. (2001) *Tetrahedron Lett.*, **42**, 6547.
261 Landais, Y. and Zekri, E. (2002) *Eur. J. Org. Chem.*, 4037.
262 Angelaud, R. and Landais, Y. (1996) *J. Org. Chem.*, **61**, 5202.
263 Angelaud, R., Babot, O., Charvat, T., and Landais, Y. (1999) *J. Org. Chem.*, **64**, 9613.
264 Gogoi, S., Barua, N.C., and Kalita, B. (2004) *Tetrahedron Lett.*, **45**, 5577.
265 Belley, M.L., Hill, B., Mitenko, H., Scheigetz, J., and Zamboni, R. (1996) *Synlett*, 92.

266 Neighbors, J.D., Beutler, J.A., and Wiemer, D.F. (2005) *J. Org. Chem.*, **70**, 925.
267 Caruso, T. and Spinella, A. (2002) *Tetrahedron: Asymmetry*, **13**, 2071.
268 Spinella, A., Caruso, T., and Coluccini, C. (2002) *Tetrahedron Lett.*, **43**, 1681.
269 Xu, D., Park, C.Y., and Sharpless, K.B. (1994) *Tetrahedron Lett.*, **35**, 2495.
270 Vidari, G., Dapiaggi, A., Zanoni, G., and Garlaschelli, L. (1993) *Tetrahedron Lett.*, **34**, 6485.
271 Corey, E.J., Noe, M.C., and Shieh, W.-C. (1993) *Tetrahedron Lett.*, **34**, 5995.
272 Couladouros, E.A. and Vidali, V.P. (2004) *Chem. Eur. J.*, **10**, 3822.
273 Andrus, M.B., Lepore, S.D., and Sclafani, J.A. (1997) *Tetrahedron Lett.*, **38**, 4043.
274 Andrus, M.B., Lepore, S.D., and Turner, T.M. (1997) *J. Am. Chem. Soc.*, **119**, 12159.
275 Corey, E.J., Noe, M.C., and Lin, S. (1995) *Tetrahedron Lett.*, **36**, 8741.
276 Madden, B.A. and Prestwich, G.D. (1997) *Bioorg. Med. Chem. Lett.*, **7**, 309.
277 Huang, A.X., Xiong, Z., and Corey, E.J. (1999) *J. Am. Chem. Soc.*, **121**, 9999.
278 Crispino, G.A. and Sharpless, K.B. (1993) *Synthesis*, 777.
279 Robustell, B.J., Abe, I., and Prestwich, G.D. (1998) *Tetrahedron Lett.*, **39**, 9385.
280 Robustell, B.J., Abe, I., and Prestwich, G.D. (1998) *Tetrahedron Lett.*, **39**, 957.
281 Fleming, S.A., Carroll, S.M., Hirschi, J., Liu, R., Pace, J.L., and Redd, J.T. (2004) *Tetrahedron Lett.*, **45**, 3341.
282 Dounay, A.B., Florence, G.J., Saito, A., and Forsyth, C.J. (2002) *Tetrahedron*, **58**, 1865.
283 Ariza, X., Fernández, N., Garcia, J., López, M., Montserrat, L., and Ortiz, J. (2004) *Synthesis*, 128.
284 Fernandes, R.A. and Kumar, P. (2002) *Tetrahedron*, **58**, 6685.
285 Anson, C.E., Dave, G., and Stephenson, G.R. (2000) *Tetrahedron*, **56**, 2273.
286 Guzman-Perez, A. and Corey, E.J. (1997) *Tetrahedron Lett.*, **38**, 5941.
287 Jeong, K.-S., Sjö, P., and Sharpless, K.B. (1992) *Tetrahedron Lett.*, **33**, 3833.
288 Gardiner, J.M., Giles, P.E., and Martín, M.L.M. (2002) *Tetrahedron Lett.*, **43**, 5415.
289 Brummond, K.M., Lu, J., and Petersen, J. (2000) *J. Am. Chem. Soc.*, **122**, 4915.
290 Ahmed, M.M. and O'Doherty, G.A. (2005) *J. Org. Chem.*, **70**, 10576.
291 Ferrié, L., Capdevielle, P., and Cossy, J. (2005) *Synlett*, 1933.
292 Matsushima, Y. and Kino, J. (2005) *Tetrahedron Lett.*, **46**, 8609.
293 Hunter, T.J. and O'Doherty, G.A. (2001) *Org. Lett.*, **3**, 2777.
294 Garaas, S.D., Hunter, T.J., and O'Doherty, G.A. (2002) *J. Org. Chem.*, **67**, 2002.
295 Hunter, T.J. and O'Doherty, G.A. (2001) *Org. Lett.*, **3**, 1049.
296 Braun, N.A., Bürkle, U., Klein, I., and Spitzner, D. (1997) *Tetrahedron Lett.*, **38**, 7057.
297 Ferrié, L., Reymond, S., Capdevielle, P., and Cossy, J. (2007) *Synlett*, 2891.
298 Mortensen, M.S., Osbourn, J.M., and O'Doherty, G.A. (2007) *Org. Lett.*, **9**, 3105.
299 Ghosh, A.K. and Gong, G. (2004) *J. Am. Chem. Soc.*, **126**, 3704.
300 Chavan, S.P. and Praveen, C. (2004) *Tetrahedron Lett.*, **45**, 421.
301 Alvarez, S., Alvarez, R., and de Lera, A.R. (2004) *Tetrahedron: Asymmetry*, **15**, 839.
302 Armstrong, A. and Hayter, B.R. (1997) *Tetrahedron: Asymmetry*, **8**, 1677.
303 Wang, Z.-M., Kakiuchi, K., and Sharpless, K.B. (1994) *J. Org. Chem.*, **59**, 6895.
304 Park, C.Y., Kim, B.M., and Sharpless, K.B. (1991) *Tetrahedron Lett.*, **32**, 1003.
305 Armstrong, A., Barsanti, P.A., Jones, L.H., and Ahmed, G. (2000) *J. Org. Chem.*, **65**, 7020.
306 Gao, D. and O'Doherty, G.A. (2005) *Org. Lett.*, **7**, 1069.
307 Ahmed, M.M., Berry, B.P., Hunter, T.J., Tomcik, D.J., and O'Doherty, G.A. (2005) *Org. Lett.*, **7**, 745.
308 Fernandes, R.A. and Kumar, P. (2000) *Tetrahedron Lett.*, **41**, 10309.
309 Reetz, M.T., Strack, T.J., Mutulis, F., and Goddard, R. (1996) *Tetrahedron Lett.*, **37**, 9293.

310 Avenoza, A., Busto, J.H., Corzana, F., Peregrina, J.M., Sucunza, D., and Zurbano, M.M. (2003) *Tetrahedron: Asymmetry*, **14**, 1037.

311 Palomo, C., Oiarbide, M., Landa, A., Esnai, A., and Linden, A. (2001) *J. Org. Chem.*, **66**, 4180.

312 VanNieuwenhze, M.S. and Sharpless, K.B. (1993) *J. Am. Chem. Soc.*, **115**, 7864.

313 Corey, E.J., Noe, M.C., and Guzman-Perez, A. (1995) *J. Am. Chem. Soc.*, **117**, 10817.

314 Hamon, D.P.G., Tuck, K.L., and Christie, H.S. (2001) *Tetrahedron*, **57**, 9499.

315 Christie, H.S., Hamon, D.P.G., and Tuck, K.L. (1999) *Chem. Commun.*, 1989.

316 Yamaguchi, S., Muro, S., Kobayashi, M., Miyazawa, M., and Hirai, Y. (2003) *J. Org. Chem.*, **68**, 6274.

317 Dai, W.-M., Zhang, Y., and Zhang, Y. (2004) *Tetrahedron: Asymmetry*, **15**, 525.

318 Rios, R., Jimeno, C., Carroll, P.J., and Walsh, P.J. (2002) *J. Am. Chem. Soc.*, **124**, 10272.

319 Bueno, A., Rosol, M., García, J., and Moyano, A. (2006) *Adv. Synth. Catal.*, **348**, 2590.

320 Burke, S.D. and Jiang, L. (2001) *Org. Lett.*, **3**, 1953.

321 Pini, D., Iuliano, A., Rosini, C., and Salvadori, P. (1990) *Synthesis*, 1023.

322 Voronlov, M.V., Gontcharov, A.V., Wang, Z.-M., and Kolb, H.C. (2001) *ARKIVOK*, (ix), 160.

323 Jansen, R. (2001) PCT. Appl. Pat. Patent Appl. WO 0153281.

324 Prasad, J.S., Vu, T., Totleben, M.J., Crispino, G.A., Kaesur, D.J., Swaminathan, S., Thornton, J.E., Fritz, A., and Singh, A.K. (2003) *Org. Proc. Res. Dev.*, **7**, 821.

325 Griesbach, R.C., Hamon, D.P.G., and Kennedy, R.J. (1997) *Tetrahedron: Asymmetry*, **8**, 507.

326 Ishibashi, H., Maeki, M., Yagi, J., Ohba, M., and Kanai, T. (1999) *Tetrahedron*, **55**, 6075.

327 Gupta, P., Fernandes, R.A., and Kumar, P. (2003) *Tetrahedron Lett.*, **44**, 4231.

328 Aladro, F.J., Guerra, F.M., Moreno-Dorado, F.J., Bustamante, J.M., Jorge, Z.D., and Massanet, G.M. (2000) *Tetrahedron Lett.*, **41**, 3209.

329 Ainge, D., Ennis, D., Gidlund, M., Stefinovic, M., and Vaz, L.-M. (2003) *Org. Proc. Res. Dev.*, **7**, 198.

330 Marcune, B.F., Karady, S., Reider, P.J., Miller, R.A., Biba, M., DiMichele, L., and Reamer, R.A. (2003) *J. Org. Chem.*, **68**, 8088.

331 Couché, E., Fkyerat, A., and Tabacchi, R. (2003) *Helv. Chim. Acta*, **86**, 210.

332 Hashiyama, T., Morikawa, K., and Sharpless, K.B. (1992) *J. Org. Chem.*, **57**, 5067.

333 Chavan, S.P. and Venkatraman, M.S. (2005) *ARKIVOK*, (iii), 165.

334 Zhou, H.-B., Liu, G.-S., and Yao, Z.-J. (2007) *Org. Lett.*, **9**, 2003.

335 Fang, F.G., Xie, S., and Lowery, M.W. (1994) *J. Org. Chem.*, **59**, 6142.

336 Fang, F.G., Bankston, D.D., Huie, E.M., Johnson, M.R., Kang, M.-C., LeHoullier, C.S., Lewis, G.C., Lovelace, T.C., Lowery, M.W., McDougald, D.L., Meerholz, C.A., Partridge, J.J., Sharp, M.J., and Xie, S. (1997) *Tetrahedron*, **53**, 10953.

337 Josien, H., Ko, S.-B., Bom, D., and Curran, D.P. (1998) *Chem. Eur. J.*, **4**, 67.

338 Upadhya, T.T., Gurunath, S., and Sudalai, A. (1999) *Tetrahedron: Asymmetry*, **10**, 2899.

339 Fernandes, R.A. and Kumar, P. (2002) *Eur. J. Org. Chem.*, 2921.

340 Lawrence, N.J. and Brown, S. (2002) *Tetrahedron*, **58**, 613.

341 Gao, Y. and Sharpless, K.B. (1988) *J. Am. Chem. Soc.*, **110**, 7538.

342 Denmark, S.E. (1981) *J. Org. Chem.*, **46**, 3144.

343 Lowe, G. and Salamone, S.J. (1983) *J. Chem. Soc., Chem. Commun.*, 1392.

344 Alonso, M. and Riera, A. (2005) *Tetrahedron: Asymmetry*, **16**, 3908.

345 Hirsenkorn, R. (1990) *Tetrahedron Lett.*, **31**, 7591.

346 Lohray, B.B. and Ahuja, J.R. (1991) *J. Chem. Soc., Chem. Commun.*, 95.

347 Xiong, C., Wang, W., and Hruby, V.J. (2002) *J. Org. Chem.*, **67**, 3514.
348 Kim, B.M. and Sharpless, K.B. (1990) *Tetrahedron Lett.*, **31**, 4317.
349 Sayyed, I.A. and Sudalai, A. (2004) *Tetrahedron: Asymmetry*, **15**, 3111.
350 Li, G., Chang, H.-T., and Sharpless, K.B. (1996) *Angew. Chem. Int. Ed.*, **35**, 451.
351 Rudolph, J., Sennhenn, P.C., Vlaar, C.P., and Sharpless, K.B. (1996) *Angew. Chem. Int. Ed.*, **35**, 2810.
352 Li, G., Angert, H.H., and Sharpless, K.B. (1996) *Angew. Chem. Int. Ed.*, **35**, 2813.
353 Reddy, K.L. and Sharpless, K.B. (1998) *J. Am. Chem. Soc.*, **120**, 1207.
354 O'Brien, P., Osborne, S.A., and Parker, D.D. (1998) *Tetrahedron Lett.*, **39**, 4099.
355 O'Brien, P., Osborne, S.A., and Parker, D.D. (1998) *J. Chem. Soc., Perkin Trans. 1*, 2519.
356 Reddy, K.L., Dress, K.R., and Sharpless, K.B. (1998) *Tetrahedron Lett.*, **39**, 3667.
357 Bruncko, M., Schlingloff, G., and Sharpless, K.B. (1997) *Angew. Chem. Int. Ed.*, **36**, 1483.
358 Rubin, A.E. and Sharpless, K.B. (1997) *Angew. Chem. Int. Ed.*, **36**, 2637.
359 Park, H., Cao, B., and Joullié, M.M. (2001) *J. Org. Chem.*, **66**, 7223.
360 Demko, Z.P., Bartsch, M., and Sharpless, K.B. (2000) *Org. Lett.*, **2**, 2221.
361 Nesterenko, V., Byers, J.T., and Hergenrother, P.J. (2003) *Org. Lett.*, **5**, 281.
362 Li, G., Lenington, R., Willis, S., and Kim, S.H. (1998) *J. Chem. Soc., Perkin Trans. 1*, 1753.
363 Barta, N.S., Sidler, D.R., Somerville, K.B., Weissman, S.A., Larsen, R.D., and Reider, P.J. (2000) *Org. Lett.*, **2**, 2821.
364 Davey, R.M., Brimble, M.A., and McLeod, M.D. (2000) *Tetrahedron Lett.*, **41**, 5141.
365 Cohen, J.L. and Chamberlin, A.R. (2007) *Tetrahedron Lett.*, **48**, 2533.
366 Harding, M., Bodkin, J.A., Hutton, C.A., and McLeod, M.D. (2005) *Synlett*, 2829.
367 Zhang, H., Xia, P., and Zhou, W. (2000) *Tetrahedron: Asymmetry*, **11**, 3439.
368 Bonini, C., D'Auria, M., and Fedeli, P. (2002) *Tetrahedron Lett.*, **43**, 3813.
369 Song, C.E., Oh, C.R., Roh, E.J., Lee, S.-G., and Choi, J.H. (1999) *Tetrahedron: Asymmetry*, **10**, 671.
370 Phukan, P. and Sudalai, A. (1998) *Tetrahedron: Asymmetry*, **9**, 1001.
371 Muñiz, K., Hövelmann, C.H., Villar, A., Vicente, R., Streuff, J., and Nieger, M. (2006) *J. Mol. Catal. A: Chem.*, **251**, 277.
372 Morgan, A.J., Masse, C.E., and Panek, J.S. (1999) *Org. Lett.*, **1**, 1949.
373 Avenoza, A., Cativiela, C., Corzana, F., Peregrina, J.M., Sucunza, D., and Zurbano, M.M. (2001) *Tetrahedron: Asymmetry*, **12**, 949.
374 Xiong, C., Wang, W., Cai, C., and Hruby, V.J. (2002) *J. Org. Chem.*, **67**, 1399.
375 Thakur, V.V. and Sudalai, A. (2003) *Tetrahedron: Asymmetry*, **14**, 407.
376 Kantam, M.L., Prakash, B.V., Bharathi, B., and Reddy, C.V. (2005) *J. Mol. Catal. A: Chem.*, **226**, 119.

4
Cinchona Alkaloids and their Derivatives as Chirality Inducers in Metal-Promoted Enantioselective Carbon–Carbon and Carbon–Heteroatom Bond Forming Reactions

Ravindra R. Deshmukh, Do Hyun Ryu, and Choong Eui Song

4.1
Introduction

Stereoselective carbon–carbon and carbon–heteroatom bond forming reactions are among the most fundamental reactions employed for the construction of the molecular frameworks of various biologically active molecules in synthetic organic chemistry. A number of metal-catalyzed asymmetric processes have been developed and have gained wide acceptance, and some are even used on an industrial scale [1].

The naturally occurring cinchona alkaloids (Figure 4.1) have been invaluable in the field of asymmetric catalysis. They are multifunctional and easily tunable for diverse catalytic reactions through different mechanisms, which make them privileged ligands or catalysts. Thus, today, cinchona alkaloids by themselves and their synthetic derivatives have been recognized as the most useful chirality transmitters in metal-catalyzed hydrogenation (Chapter 2) and oxidation (Chapter 3) reactions. Over the past few decades, remarkable scientific achievements have also been made in the area of metal-promoted asymmetric carbon–carbon and carbon–heteroatom bond forming catalytic reactions, in which cinchona alkaloids and their derivatives are utilized as chiral ligands or chiral cobase catalysts.

In this chapter, the current state of the art on the applications of cinchona alkaloids and their derivatives as chiral ligands or chiral cobase catalysts in metal-promoted asymmetric carbon–carbon and carbon–heteroatom bond forming catalytic reactions will be discussed.

Cinchona Alkaloids in Synthesis and Catalysis, Ligands, Immobilization and Organocatalysis
Edited by Choong Eui Song

ISBN: 978-3-527-32416-3

R = H
R' = CH=CH$_2$; Cinchonidine (**Cd**)
R' = CH$_2$CH$_3$; Dihydrocinchonidine (**HCd**)

R = OMe
R' = CH=CH$_2$; Quinine (**Qn**)
R' = CH$_2$CH$_3$; Dihydroquinine (**HQn**)

R = H
R' = CH=CH$_2$; Cinchonine (**Cn**)
R' = CH$_2$CH$_3$; Dihydrocinchonine (**HCn**)

R = OMe
R' = CH=CH$_2$; Quinidine (**Qd**)
R' = CH$_2$CH$_3$; Dihydroquinidine (**HQd**)

Figure 4.1 Structures of various naturally occurring cinchona alkaloids.

4.2
Nucleophilic Addition to Carbonyl or Imine Compounds

4.2.1
Organozinc Addition

4.2.1.1 Dialkylzinc Addition to Aldehydes

The enantioselective addition of dialkylzinc compounds to carbonyls is a reaction that has been frequently used to construct a chiral center that results in the formation of optically active alcohols [2]. Till now, a variety of chiral ligands have been utilized for the addition of organozinc compounds to carbonyls. Among these chiral ligands, the use of cinchona alkaloids has also been studied [3, 4]. However, the results have showed that, in general, cinchona alkaloids do not surpass other chiral β-amino alcohols such as 3-*exo*-dimethylaminoisoborneol (DAIB) in terms of their catalytic efficiency [5]. Only *ortho*-alkoxy substituted aromatic aldehydes give satisfactory levels of asymmetric induction (83–92% ee) [3]. For example, when diethylzinc is added to a solution of o-ethoxybenzaldehyde in toluene in the presence of catalytic amounts (2 mol%) of quinine (**Qn**), (+)-o-ethoxyphenylethylcarbinol is formed in an ee of 92% [3]. Nonalkoxy-substituted aromatic aldehydes give only moderate ee values. However, a quite remarkable effect of the reaction temperature on the enantioselectivity was observed [4]. The best ee (73%) in the addition of diethylzinc to benzaldehyde was obtained at elevated temperature (100 °C) and not, as would be expected, at a lower temperature (48% ee at −10 °C) (Table 4.1).

Quite recently, Lin and coworkers developed the highly enantioselective bifunctional cinchona-based ligand **1** for the enantioselective addition of diethylzinc to aldehydes. The catalytically active zinc complex **2** can be generated *in situ* from the ligand **1** during the addition of diethylzinc to aldehydes [6]. The reaction of diethylzinc with 2-methoxybenzaldehyde using 10 mol% of **1** was completed within 3 h at room temperature, affording the desired (S)-carbinol in 94% yield with 92% ee (Table 4.2, entry 1). The stereochemical outcome of the reactions is determined by the alkaloid, which is the only source of chirality. Replacing the quinine moiety in **1** with

Table 4.1 Influence of reaction temperature on enantioselectivity in the cinchona alkaloid catalyzed addition of diethylzinc to benzaldehyde.

PhCHO + Et$_2$Zn $\xrightarrow{\text{Qn or Qd (14 mol\%)}}$ Ph*(OH)Et

Alkaloid catalyst	Reaction time (h)	Reaction temperature (°C)	Yield (%)	ee (%)	Configuration
Qn	38	−10	90	48	R
Qn	16	rt	97	64	R
Qn	0.25	100	95	73	R
Qd	38	−10	92	45	S
Qd	16	rt	98	51	S
Qd	0.25	100	96	69	S

quinidine affords the enantiomeric (R)-carbinol in 93% ee (entry 5). However, when using 9-O-benzylquinine (**Bn-Qn, 3**) as a Lewis base or the Zn complex of the Schiff base **4** as a Lewis acid alone, the same reaction proceeds sluggishly, affording a nearly racemic product (entries 2 and 3). Moreover, the combination of the non-tethered **Bn-Qn (3)** and **4** also did not result in any asymmetric induction (entry 4). All of these results indicate that the intramolecular bifunctional activation is crucial for rate acceleration and asymmetric induction.

4.2.1.2 Dialkylzinc Addition to Imines

The addition of dialkylzinc compounds to imines in an enantioselective manner, leading to the formation of chiral amines, has also been studied extensively [7]. Beresford [8] recently reported that cinchona alkaloids promote the addition of diethylzinc to N-diphenylphosphinoylimines, affording the corresponding amines with excellent ee values (up to 96% ee). For example, the reaction of P,P-diphenyl-N-(phenylmethylene)phosphinic amide with diethylzinc using 1.0 equiv of **Cd** or **HCd** in toluene at ambient temperature furnished the (R)-enantiomer of N-(phenylpropyl)-P,P-diphenylphosphinic amide in 93% ee and 96% ee, respectively (entries 1 and 2 of Table 4.3). Under the same reaction conditions using **Cn**, the (S)-enantiomer was obtained in 91% ee and 77% yield (entry 6). Similarly to the addition of diethylzinc to aldehydes, both the yield and enantioselectivity markedly increased (from 67% ee to 93% ee) as the reaction temperature was increased from −18 °C to room temperature (entries 5 and 1, respectively). Reducing the quantity of **Cd** to 0.2 equiv gave a lower ee value and, moreover, the reaction proceeded sluggishly, which is indicative of a ligand acceleration effect (entry 3). Surprisingly, however, the use of achiral additives such as methanol was found to significantly improve the rate and enantioselectivity under these catalytic conditions. Thus, using 0.2 equiv of **Cd** and 2.0 equiv of methanol, under the same reaction conditions, the product was obtained in 70% yield and 93% ee (entry 4). This value for the ee is the same as that achieved using a stoichiometric amount of **Cd**. This result demonstrates the potential that achiral additives have for

Table 4.2 Bifunctional catalytic activity of bifunctional ligand **1**.

Entry	Ligand (10 mol%)	Yield (%)	ee (%)
1	1	94	92 (S)
2	3	35	<2
3	4	47	0
4	3 + 4	55	<2
5	ent-1	86	93 (R)

Table 4.3 Addition of diethylzinc to P,P-diphenyl-N-(phenylmethylene)phosphinic amide.

Entry	Ligand (equiv)	Temperature	Yield (%)	ee (%)	Configuration
1	Cd (1.0)	rt	76	93	R
2	HCd (1.0)	rt	70	96	R
3	Cd (0.2)	rt	52	80	R
4	Cd (0.2) + MeOH (2.0)	rt	70	93	R
5	Cd (1.0)	−18 °C	33	67	R
6	Cn (1.0)	rt	77	91	S

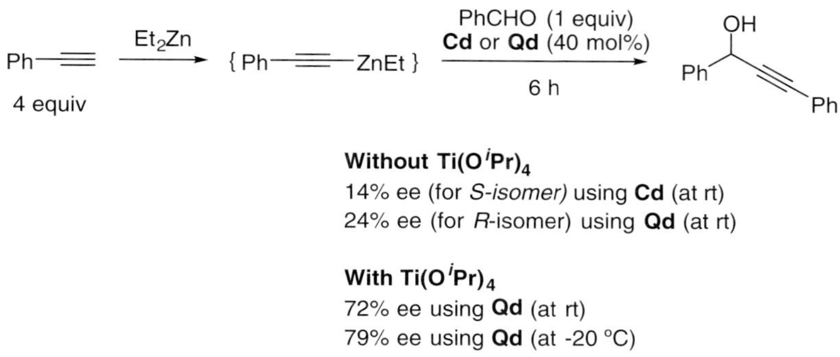

Scheme 4.1 Enantioselective phenylacetylene addition to benzaldehyde.

Without Ti(O^{i}Pr)$_4$
14% ee (for S-isomer) using **Cd** (at rt)
24% ee (for R-isomer) using **Qd** (at rt)

With Ti(O^{i}Pr)$_4$
72% ee using **Qd** (at rt)
79% ee using **Qd** (at -20 °C)

enhancing the enantioselectivity with asymmetric catalytic systems ([9], this paper has been withdrawn by the authors).

4.2.1.3 Addition of Alkynylzincs to Carbonyls

The addition of acetylenes to carbonyl compounds represents the most straightforward approach to obtain propargylic alcohols developed so far. Most of the asymmetric versions of this reaction employ chiral β-aminoalcohols as ligands and stoichiometric amounts of organozinc reagents, to generate the reactive zinc acetylenides [10]. Recently, cinchona alkaloids were also employed as chiral β-aminoalcohol ligands in the Et$_2$Zn-mediated addition of phenylacetylene to various aldehydes, affording the corresponding propargylic alcohols in up to 85% ee [11]. As shown in Scheme 4.1, in the case of benzaldehyde, the cinchona alkaloid ligands, **Cd** and **Qd**, alone gave very poor enantioselectivity (14 and 24% ee, respectively) [11, 12]. However, the ee values were dramatically improved by the addition of titanium isopropoxide (e.g., up to 72% ee using **Qd** at room temperature). At −20 °C, the enantioselectivities further increased to up to 79% ee (Scheme 4.1) [11]. The marked increase of the ee values in the presence of triethylaluminum as an additive was also observed in the asymmetric addition of alkynylzincs to inactivated aromatic ketones catalyzed by cinchona alkaloids (Scheme 4.2). The triethylaluminum-promoted

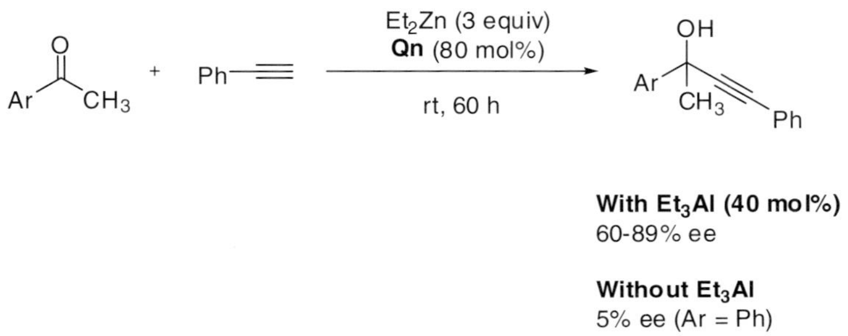

With Et$_3$Al (40 mol%)
60-89% ee

Without Et$_3$Al
5% ee (Ar = Ph)

Scheme 4.2 Asymmetric addition of phenylacetylene to various aromatic ketones using quinine as a chiral ligand.

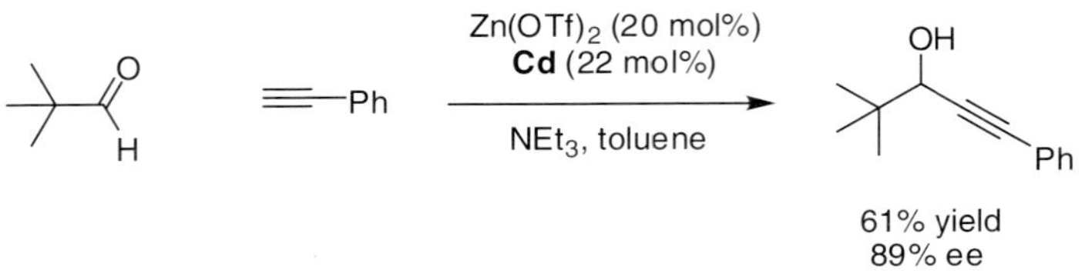

Scheme 4.3 Addition of phenylacetylene to 2-methylpropionaldehyde in the presence of cinchonidine as a chiral ligand.

addition of phenylacetylene to various ketones in the presence of **Qn** was reported to give up to 89% ee. On the other hand, very poor enantioselectivity (5% ee) was obtained in the absence of the triethylaluminum additive [13].

All of the results described above were obtained by using stoichiometric amounts of Et_2Zn to generate zinc acetylenides. A breakthrough was made by Carreira and coworkers who found that zinc triflate can be used in catalytic amounts to generate zinc acetylenides. By this protocol, the highly efficient, asymmetric alkynylation (up to 99% ee) of aliphatic aldehydes was realized with N-methylephedrine as a chiral ligand [14]. The use of cinchona alkaloids (22 mol%) as ligands in combination with a catalytic amount (20 mol%) of $Zn(OTf)_2$ also gave relatively good results for aliphatic aldehydes (e.g., 89% ee for 2,2-dimethylpropionaldehyde using **Cd**) (Scheme 4.3) [15]. Reducing the ligand loading from 22 mol% to 10 or 5 mol% resulted in a considerable drop in the yield; however, the enantioselectivity was almost unchanged.

4.2.2
Asymmetric Reformatsky Reaction

The asymmetric Reformatsky reaction is a versatile and straightforward approach for obtaining chiral β-hydroxy esters [16, 17]. However, there have been few reports so far on its efficient use for ketone substrates. A representative example of the use of this reaction for ketone substrates was reported by Soai's research group, in which chiral tertiary alcohols were obtained from aryl ketones, albeit in low to moderate enantioselectivity, using ephedra alkaloids as chiral ligands [18]. The difficulty encountered in achieving high enantioselectivity might be attributed to the poor enantioface discrimination of the ketone, due to the bulkiness of both carbonyl substituents. Recently, Yamano and coworkers [19] overcame this difficulty by the introduction of a substituent that participates in the formation of a geometrically defined complex between zinc and the ketone as a reactive intermediate. Thus, in the presence of pyridine additive, up to 97% ee was obtained using cinchona alkaloids as chiral ligands with ketones having the sp^2-nitrogen adjacent to the reactive carbonyl center. In the absence of pyridine, however, the same reaction affords only moderate ee (68% ee) (Scheme 4.4). In the case of 4-imidazolyl and 2-pyridyl ketones, an sp^2-nitrogen adjacent to the carbonyl group serves as a coordinating site with the zinc for the formation of a chelate, which plays a pivotal role in the induction of the high enantioselectivity. However, in the case of 3- and 4-pyridyl ketones, the sp^2-nitrogen

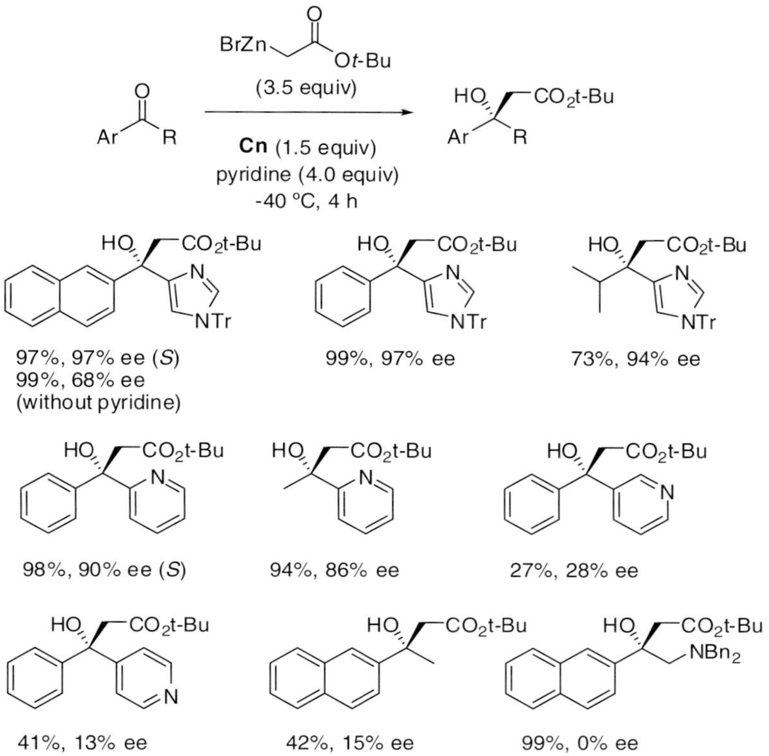

Scheme 4.4 Asymmetric Reformatsky reaction with various ketones using cinchonine as a chiral ligand.

on the pyridine ring is not able to form a chelate, which ultimately results in a decrease in the reactivity as well as the enantioselectivity. The reaction of the ketone having an sp^3-nitrogen was also found to give the product, but in the form of a racemic mixture. Although the enantioselectivity of this protocol is quite impressive, the substrate is limited to carbonyl compounds possessing an adjacent sp^2-nitrogen. The improvement of the enantioselectivity by the addition of basic additives such as pyridine is another noticeable point. The addition of basic additives that could interact with the zinc would change the chiral environment.

4.2.3
Indium-Mediated Addition

The indium-promoted allylation of carbonyl compounds in the presence of cinchona alkaloids (2 equiv), as a route to chiral allylic alcohols, was recently reported by Loh [20, 21]. The reaction is applicable to a variety of aldehydes and ketones. The enantioselectivities obtained with allyl bromides as the substrate are at the best moderate (up to 75% ee) [20]. However, the use of prenyl bromides as substrates

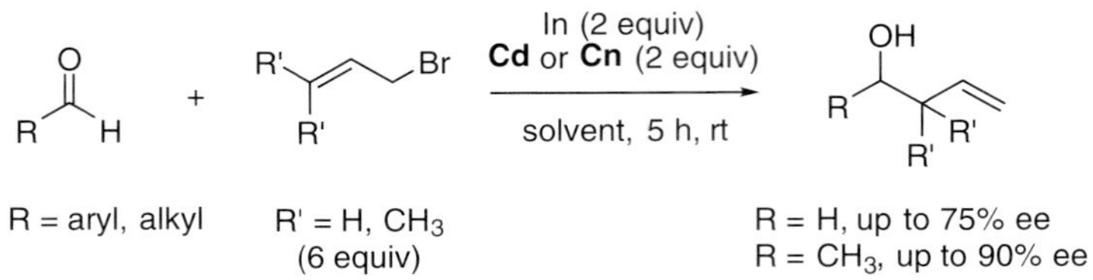
In (2 equiv)
Cd or Cn (2 equiv)
solvent, 5 h, rt
OH
R
H
R'
Br
R = aryl, alkyl
R' = H, CH3
(6 equiv)
R = H, up to 75% ee
R = CH3, up to 90% ee

Scheme 4.5 Enantioselective indium-mediated allylation reaction of aldehydes.

afforded almost quantitative yields and much higher enantioselectivities (up to 90% ee) (Scheme 4.5) [21]. The use of catalytic or substoichiometric amounts of the ligand afforded much lower enantioselectivities (19% ee) [21], indicating the importance of the formation of the 1 : 1 allylindium–ligand complex. Although the identity of the allylindium–alkaloid ligand complex is obscure, the presence of the free 9-OH group of the alkaloids seems to be important for the asymmetric induction. The use of acetylated cinchona alkaloids yielded a racemic product [22]. In the absence of the alkaloid ligand, indium could not insert into the allylic bromides at room temperature. This indicates that the ligand facilitates the formation of the allylic indium species. It is also noteworthy that the stereoselectivity increased as the reaction temperature was increased from −78 to 25 °C (Scheme 4.6) [20].

Recently, Loh's indium-mediated enantioselective allylation procedure was applied by Frejd and coworkers to the synthesis of optically active bicyclic hydroxyketones, such as (−)-*endo*-**5**, which are useful chiral building blocks in asymmetric catalysis and natural product synthesis (Scheme 4.7) [22].

Similarly to the indium-mediated allylation, Loh's research group also reported the enantioselective indium-mediated propargylation of different types of aldehydes using stoichiometric amounts of the cinchona alkaloids, **Cd** and **Cn**, as chiral ligands. Using the simple propargylic bromide as a substrate, chiral homopropargylic alcohols were obtained in moderate to good yields with up to 85% ee (Scheme 4.8) [23]. Notably, no detectable amount of allenic alcohol was observed. On the other hand, the use of the sterically demanding silicon and phenyl substituents completely eradicated the enantioselectivity. Interestingly, when 1-phenylpropargylic bromide was used as the substrate, the allenic alcohol was obtained almost exclusively. On the other hand,

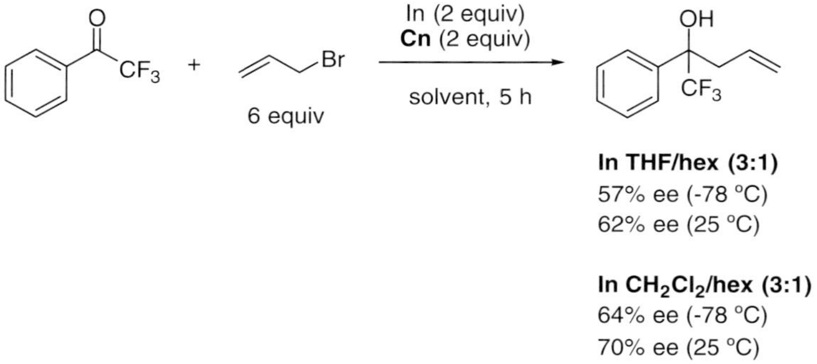
O
CF3
Br
6 equiv
In (2 equiv)
Cn (2 equiv)
solvent, 5 h
OH
CF3
In THF/hex (3:1)
57% ee (-78 °C)
62% ee (25 °C)
In CH2Cl2/hex (3:1)
64% ee (-78 °C)
70% ee (25 °C)

Scheme 4.6 Indium-mediated allylation using cinchonidine as a ligand.

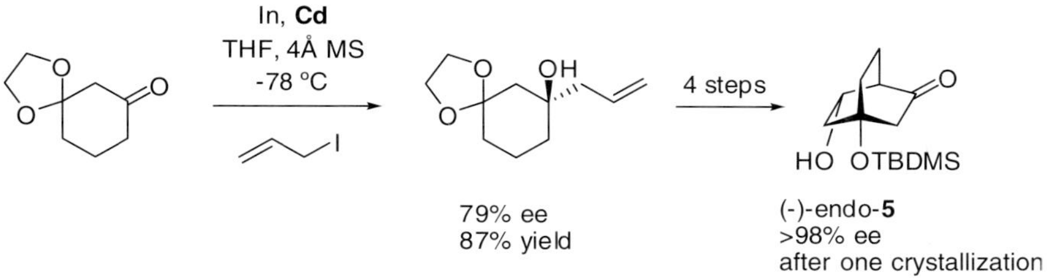

Scheme 4.7 Synthesis of optically active bicyclic hydroxyketone **5**.

when the trimethylsilylpropargyl bromide was used, the propargylic alcohol was obtained exclusively (Scheme 4.8). This is interesting because the homopropargylic indium species equilibrate in solution to give a mixture of the homopropargylic and allenylic indium species, consequently resulting in poor product selectivity (Scheme 4.9).

4.2.4
Asymmetric Cyanation

4.2.4.1 Cyanohydrin Synthesis

Optically pure cyanohydrins serve as highly versatile synthetic building blocks [24]. Much effort has, therefore, been devoted to the development of efficient catalytic systems for the enantioselective cyanation of aldehydes and ketones using HCN or trimethylsilyl cyanide (TMSCN) as a cyanide source [24]. More recently, cyanoformic esters (ROC(O)CN), acetyl cyanide ($CH_3C(O)CN$), and diethyl cyanophosphonate have also been successfully employed as cyanide sources to afford the corresponding functionalized cyanohydrins. It should be noted here that, as mentioned in Chapter 1, the cinchona alkaloid catalyzed asymmetric hydrocyanation of aldehydes discovered

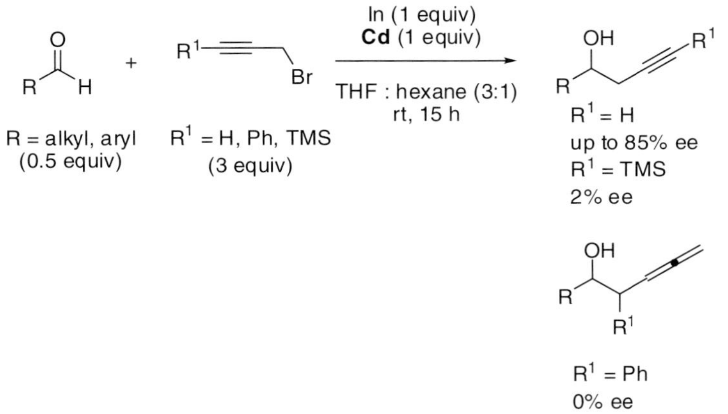

Scheme 4.8 Enantioselective indium-mediated addition of propargyl bromide to various aldehydes.

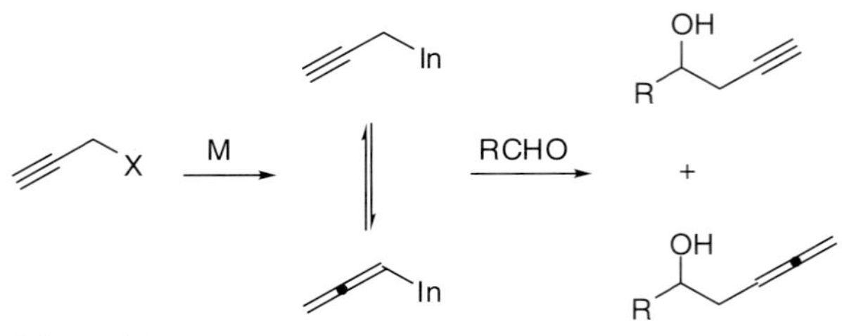

Scheme 4.9

by Bredig and Fiske in 1912 [25], and the reinvestigation of this work by Prelog [26] during the mid-1950s actually promoted the concept of asymmetric catalysis. Since then, many cinchona alkaloid derivatives have been used by themselves (e.g., (DHQD)$_2$AQN) as highly efficient catalysts for enantioselective cyanation reactions [27]. This highly interesting metal-free version will be discussed in Chapter 8 in detail. Here, we will discuss only some recent examples in which cinchona alkaloids are used as chiral ligands or cobase catalysts.

In 1991, Mukaiyama et al. reported that high enantioselectivity can be achieved using the chiral tin (II) complex **6** derived from cinchonine (**Cn**) as a Lewis acid catalyst in the reactions of aldehydes and TMSCN [28]. However, the utilization of this protocol is limited to aliphatic aldehydes, where up to 96% ee was obtained (Table 4.4). However, when this protocol was applied to an aromatic aldehyde (benzaldehyde), no reaction was observed with TMSCN under the same conditions.

Quite recently, Feng and coworkers reported that by employing a discrete bifunctional catalyst system, consisting of the heterobimetallic (S)-AlLi(binaphthoxide) ((S)-ALB) complex (**7**) combined with cinchonine), the highly enantioselective cyanation of aldehydes with cyanoformic esters (EtOCOCN), affording the corresponding cyanohydrins ethyl carbonate, could be achieved [29]. Most aromatic, α,β-unsaturated and aliphatic aldehydes could be smoothly converted to the corresponding cyanohydrin derivatives in nearly quantitative yields with up to 95% ee (Scheme 4.10). Using other bases such as DMAP and NEt$_3$ instead of **Cn** also improved the reactivity, but afforded very low asymmetric inductions (up to 16% ee). More importantly, when the (S)-ALB complex (**7**) was used alone as a catalyst, it did not show any activity. Also, when **Cn** alone was used as a catalyst, the same reaction proceeded sluggishly with very low asymmetric induction (13% ee). These results indicate the importance of the bifunctional activation mechanism in this reaction. A possible catalytic cycle proposed by the authors is illustrated in Scheme 4.11. In this mechanism, **Cn** coordinates with the lithium to form the (S)-ALB (**7**)-**Cn** complex, and the aldehyde coordinates with the Lewis acidic aluminum of the (S)-ALB (**7**)-**Cn** complex. EtOCOCN can also be activated by the quinuclidine base of **Cn** as proposed by Deng et al. [27]. Finally, the activated cyanide would react with the activated aldehyde to generate the product (Scheme 4.11).

Moberg and coworkers also achieved the highly enantioselective cyanation of aldehydes by using the dual activation concept (Table 4.5) [30]. It is known that the Lewis acidic dimeric salen–Ti complex **8** catalyzes the cyanation of benzaldehyde with

Table 4.4 Asymmetric addition reaction of TMSCN with aldehydes using Ti(II)-Lewis acid derived from cinchonidine.

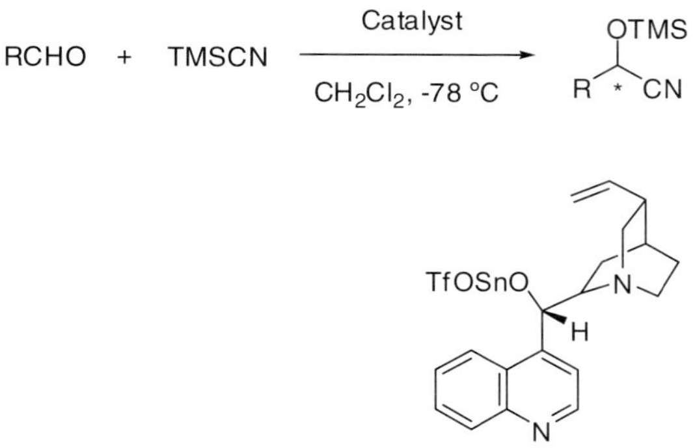

RCHO + TMSCN $\xrightarrow[\text{CH}_2\text{Cl}_2, -78\,°\text{C}]{\text{Catalyst}}$ R*CH(OTMS)(CN)

Tin (II) complex (**6**)
(unspecified)

R	Yield (%)	ee (%)
c-C$_8$H$_{11}$	89	72
n-C$_8$H$_{17}$	79	96
iPr	67	95
t-Bu	49	83
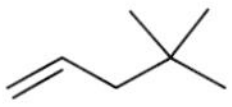 (allyl neopentyl)	27	93
Ph	0	—

(**S**)-ALB (**7**)

RCHO + NC-CO-OEt $\xrightarrow[\text{CH}_2\text{Cl}_2, -20\,°\text{C}]{\text{10 mol\% 7, 10 mol\% Cn}}$ Ph-CH(OC(O)OEt)(CN) (*S*)

up to 99%, 74-95% ee

with **7** alone; no reaction
with **Cn** alone; 13% ee

R = alkyl, aryl

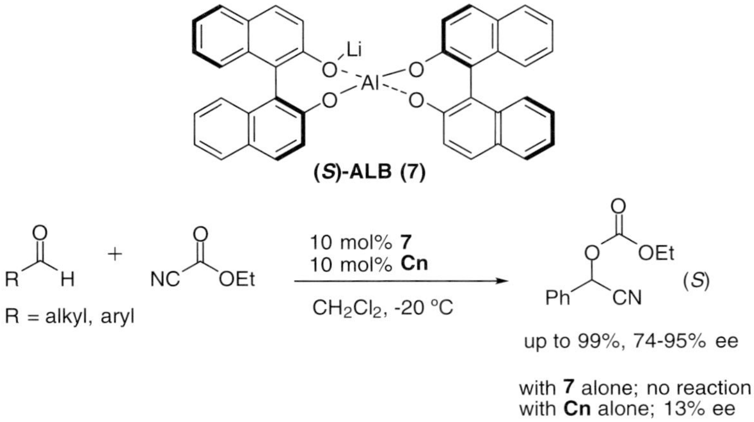

Scheme 4.10 Enantioselective cyanation of aldehydes with cyanoformic esters catalyzed by (*S*)-ALB (**7**) and cinchonine.

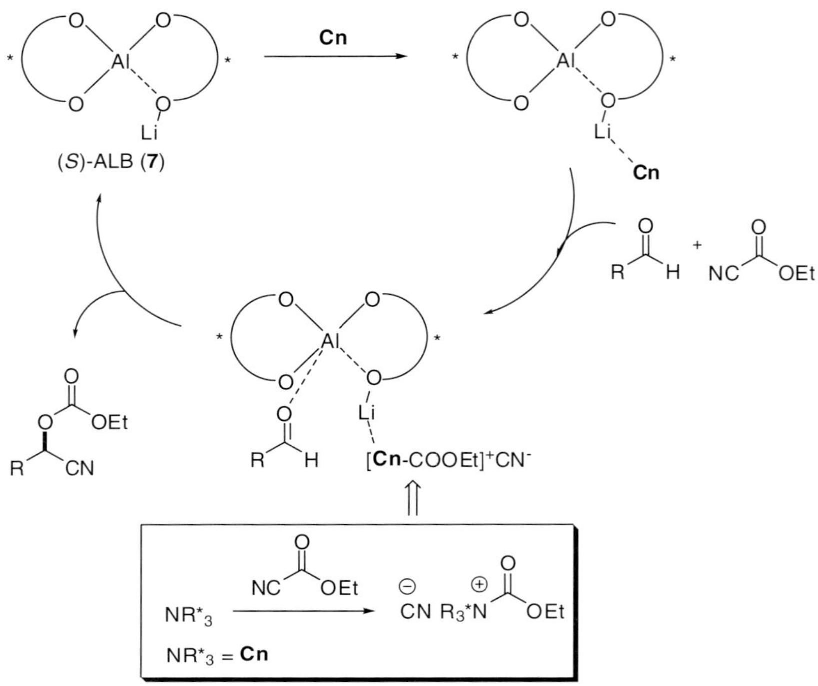

Scheme 4.11

trimethylsilyl cyanide [31]. However, no reaction occurred between benzaldehyde and acetyl cyanide at −40 °C when the Ti complex 8 alone was used as a catalyst (entry 1). However, the simultaneous activation of acetyl cyanide with the addition of a base improved the reactivity of acetyl cyanide dramatically. For example, in the presence of NEt$_3$ or DMAP, the reaction proceeded smoothly to give the product with 94% ee after 6 h at −40 °C (entries 2 and 3). The use of chiral bases such as cinchona alkaloids also allowed the reaction to proceed to completion under the same conditions, but showed only a little effect on the enantioselectivity (up to 96% ee using cinchonidine (**Cd**)) (entry 4). The same reaction with the chiral base alone as a catalyst resulted in low conversion and low enantioselectivity (entries 8 and 9). These results demonstrate that dual activation is required to achieve sufficient reactivity as well as enantioselectivity.

4.2.4.2 Strecker Synthesis

Asymmetric Strecker reactions of aldimines and ketoimines are ideal means of synthesis of chiral α-amino acids including α-quaternary amino acids [32]. Great endeavors have been devoted to the development of catalytic methods of this reaction [33].

Very recently, Feng and coworkers reported that a highly efficient catalytic asymmetric Strecker reaction with a broad substrate scope can be achieved by a

Table 4.5 Dual Lewis acid–Lewis base activation in enantioselective cyanation of aldehydes using acetyl cyanide.

Entry	Ti complex (5 mol%)	Base (10%)	Time (h)	Conversion (%)	% ee (S)
1	8	—	24	0	—
2	8	DMAP	6	57	94
3	8	NEt₃	8	96	94
4	8	Cd	9	78	96
5	8	Qn	9	80	92
6	—	DMAP	8	<1	—
7	—	NEt₃	8	13	—
8	—	Cd	8	9	40
9	—	Qn	8	2	15

catalyst system generated from the asymmetric activation of axially flexible 2,2′-biphenol (bipol) derivative with a chiral activator cinchonine through coordinative interaction with titanium [34] (Scheme 4.12).

Screening of modified bipol ligands showed that the bipol ligand bearing a sterically demanding aromatic substituent at the 3,3′-position such as **9** greatly enhanced the enantioselectivity to 94–95% ee. When the catalyst loading was lowered to below 10 mol%, the presence of 1.2 equiv of iPrOH was essential to maintain the high enantioselectivity. Using 5 mol% of cinchonine as the chiral base in combination with Ti(O*i*Pr)₄ and bipol **9** (Scheme 4.20), asymmetric cyanation of a series of imines **10** derived from aromatic, heterocyclic, α,β-unsaturated, and aliphatic aldehydes provided chiral α-alkyl or aryl α-amino nitriles **11** in up to 97% ee. In addition, this catalyst system could be extended to the cyanation of ketoimines including *ortho*-substituted diarylketoimines to generate chiral α-amino nitriles in excellent enantioselectivity (up to 99% ee).

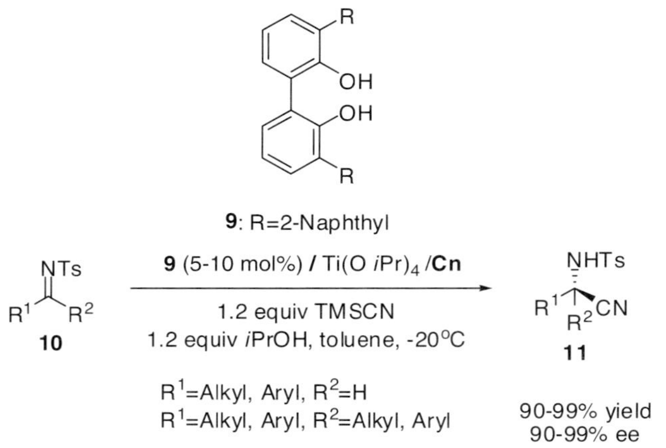

9: R=2-Naphthyl

$$\underset{10}{\overset{NTs}{R^1 \diagdown R^2}} \xrightarrow[\text{1.2 equiv } i\text{PrOH, toluene, -20°C}]{\textbf{9 (5-10 mol\%) / Ti(O}i\text{Pr})_4\textbf{/Cn}} \underset{11}{\overset{NHTs}{R^1 \diagdown R^2 CN}}$$

1.2 equiv TMSCN

R¹=Alkyl, Aryl, R²=H
R¹=Alkyl, Aryl, R²=Alkyl, Aryl

90-99% yield
90-99% ee

Scheme 4.12 Catalytic asymmetric Strecker reaction of N-Ts aldimines and ketoimines by activation of 2,2′-biphenol with cinchonine.

4.2.5
Reactions of Chiral Ammonium Ketene Enolates as Nucleophiles with Different Electrophiles

Cinchona alkaloids possess a nucleophilic quinuclidine structure and can act as versatile Lewis bases to react with ketenes generated *in situ* from acyl halides in the presence of an acid scavenger. By acting as nucleophiles, the resulting ketene enolates can react with electrophilic C=O or C=N bonds to deliver formal [2 + 2]- or [4 + 2]-cycloadducts. Although this catalytic strategy has been well established, cycloaddition based on ammonium enolates is limited to highly reactive electrophiles, probably due to the relatively limited nucleophilicity of the former. As described in the following section, a breakthrough resulting in the expansion of the substrate scope and increased yields was achieved independently by Nelson and Lectka by employing a Lewis acid cocatalyst that can activate the electrophiles without interfering with the nucleophilic enolates.

4.2.5.1 Lewis Acid Assisted Nucleophilic Addition of Ketenes (or Sulfenes) to Aldehydes: β-Lactone and β-Sultone Synthesis

In 1982, Wynberg and coworkers discovered the cinchona alkaloid catalyzed enantioselective aldol lactonization of ketenes with chloral or trichloroacetone [35], in which the zwitterionic acyl ammonium enolate provides the carbon nucleophile. This work is probably one of the most important early contributions to enantioselective organocatalysis [36]. One drawback associated with this process is the severe substrate limitations. The aldehydes should be highly reactive, presumably due to the relatively limited nucleophilicity of ammonium enolates. Nelson and coworkers first addressed the scope and reactivity problems associated with Wynberg's original protocol by combining a cinchona alkaloid derivative (O-trimethylsilylquinine (**12**) or O-trimethylsilylquinidine (**13**)) with a metal Lewis acid as a cocatalyst to

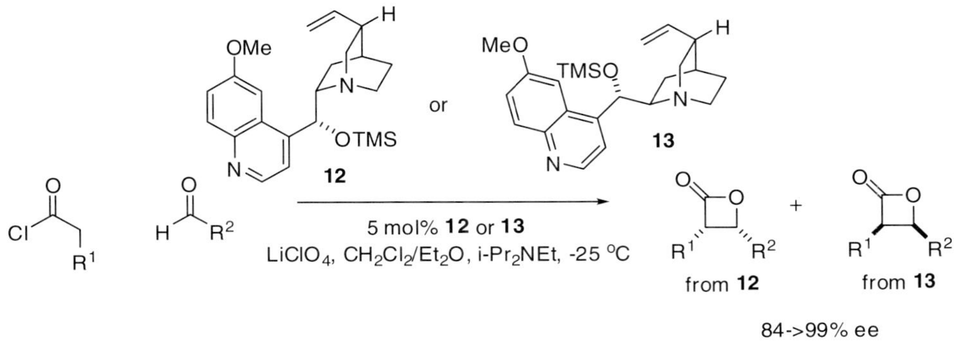

Scheme 4.13

simultaneously activate both the nucleophilic (enolate) and the electrophilic (aldehyde) reagents, respectively [37]. Among the various Lewis acids examined, LiClO$_4$ was found to be the optimal cocatalyst. As a result, a wide range of unreactive, sterically hindered alkyl and aromatic aldehydes and simple acid chlorides (precursors to ketenes) were converted to 4-substituted and cis-3,4-disubstituted-β-lactones in high yield with near perfect absolute and relative stereocontrol (up to >96% de and >99% ee) (Scheme 4.13). The authors proposed a mechanism via an Li-organized closed Zimmerman–Traxler transition state (Scheme 4.14). The utility of this methodology in synthesis activities was exemplified by the catalytic asymmetric total synthesis of the polyketide natural product (−)-pironetin (Scheme 4.15) [38].

Soon after, Calter and coworkers reported that lanthanide and pseudolanthanide triflates (e.g., Sc(OTf)$_3$, Yb(OTf)$_3$, and Er(OTf)$_3$) can also serve as efficient cocatalysts for the TMSQd **13** catalyzed addition of ketenes to unactivated aromatic aldehydes, delivering β-lactones with almost quantitative ee values [39]. It is worth to note that, while Nelson's protocol (i.e., the combination of a cinchona alkaloid with a lithium salt) provides the cis-β-lactone products, the use of lanthanide and pseudolanthanide triflates as cocatalysts, in some cases, affords the trans isomers in high diastereoselectivity. The sense of the diastereoselectivity depends on the substitution of the acid

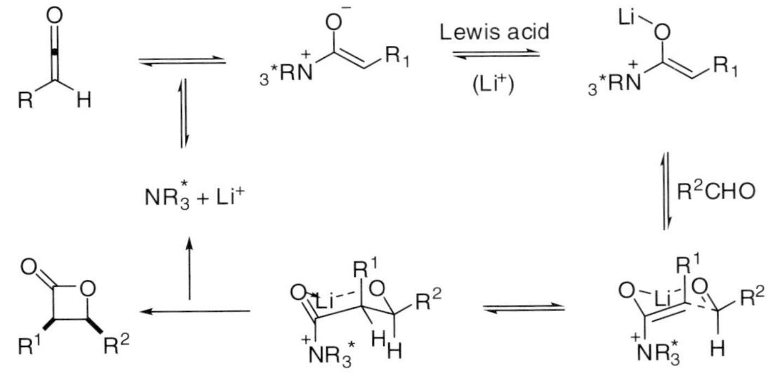

Scheme 4.14

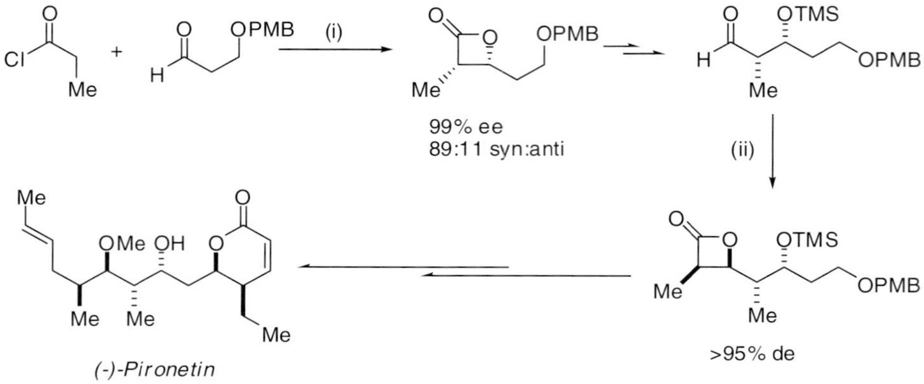

Scheme 4.15 Catalytic asymmetric total synthesis of natural product (−)-pironetin.

(i) 10 mol% TMSQd (**13**), LiClO$_4$, iPr$_2$NEt, CH$_2$Cl$_2$/Et$_2$O, -78 °C
(ii) 10 mol% TMSQn (**12**), EtCOCl, LiI, iPr$_2$NEt

chloride. The reactions of aliphatic chlorides predominantly gave the trans isomer, while those of aryloxy- and alkoxyacetyl chlorides favored the formation of the cis isomer (Scheme 4.16).

Very recently, a similar bifunctional catalyst system (metal triflates and cinchona alkaloid) was successfully applied by Peters and coworkers for the synthesis of β-sultones (Scheme 4.17) [40] and chiral α,β-unsaturated δ-lactones [41].

Quite recently, Doyle and coworkers found that dirhodium(II) complexes such as rhodium(II) acetate and Rh$_2$(4S-MEAZ)$_4$ (**14**) also act as highly active Lewis acid catalysts (1 mol%) for the reaction of trimethylsilylketene and ethyl glyoxalate, affording the β-lactone **15** (Table 4.6) [42]. However, the use of the chiral Rh-complex, Rh$_2$(4S-MEAZ)$_4$, alone afforded almost no asymmetric induction (5% ee for (S)-isomer) (entry 2). The use of quinine (10 mol%) as a cobase catalyst to activate the ketene simultaneously provided exceptional enantiocontrol (99% ee) and enhanced reactivity (entry 5).

As a logical extension to the discrete bifunctional catalytic systems described above, Lin et al. developed a catalyst containing two catalytic moieties (i.e., an alkaloid as a

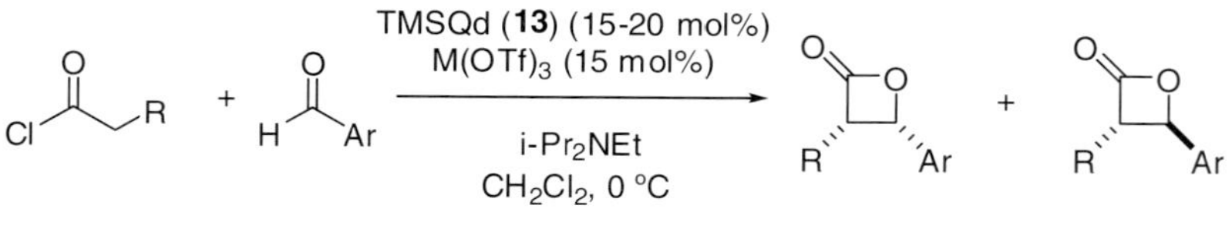

Using Er(OTf)$_3$
R = OPh, OBn; 92:8 (cis:trans), >99% ee for cis

Using Sc(OTf)$_3$
R = Me, Et; 8:92 (cis:trans), 98-99% ee for trans

Scheme 4.16

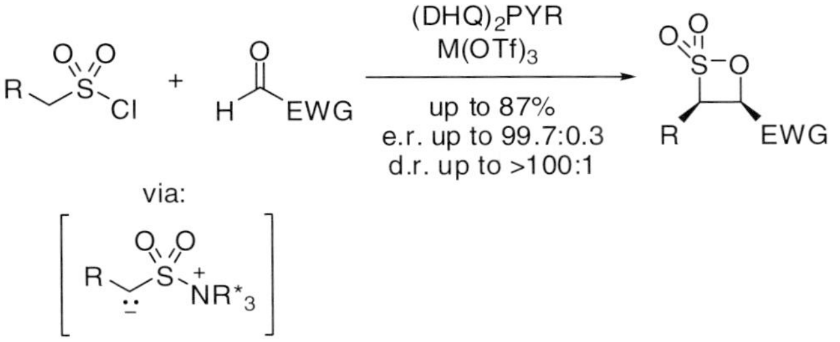

4.2 Nucleophilic Addition to Carbonyl or Imine Compounds

Scheme 4.17

(DHQ)$_2$PYR: dihydroquinine-2,5-diphenyl-4,6-pyrimidinediyl diether
M(OTf)$_3$: In(OTf)$_3$, Bi(OTf)$_3$, etc.
EWG: CCl$_3$, CO$_2$Et

nucleophilic catalyst and Co(II)salen as a Lewis acid) in a single entity [43, 44]. This bifunctional alkaloid catalyst **16** intramolecularly tethered with the Co(II)salen complex exhibited a remarkable catalytic activity and enantioselectivity for the [2 + 2] cycloaddition between ketenes and benzyloxyacetaldehyde or aromatic aldehydes. The use of only 1 mol% of the catalyst resulted in the completion of the reaction within 1 h, producing the corresponding β-lactones in good yields and excellent ee values (up to >99%) (Scheme 4.18a). An intramolecular dual activation

Table 4.6 Dual Lewis acid–Lewis base activation in the enantioselective [2 + 2] cycloaddition of TMS-ketene and ethyl glyoxalate.

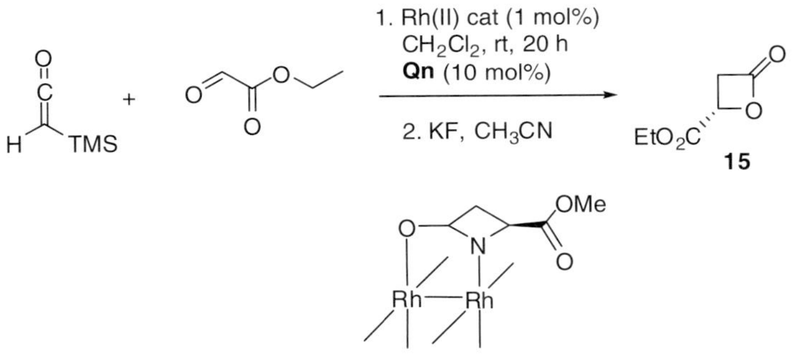

Entry	Rh-Catalyst	Qn	Yield (%)	ee (%)
1	Rh$_2$(OAc)$_4$	None	90	—
2	14	None	79	5 (S)
3	None	10 mol%	NR	—
4	Rh$_2$(OAc)$_4$	10 mol%	68	90 (R)
5	14	10 mol%	87	99 (R)

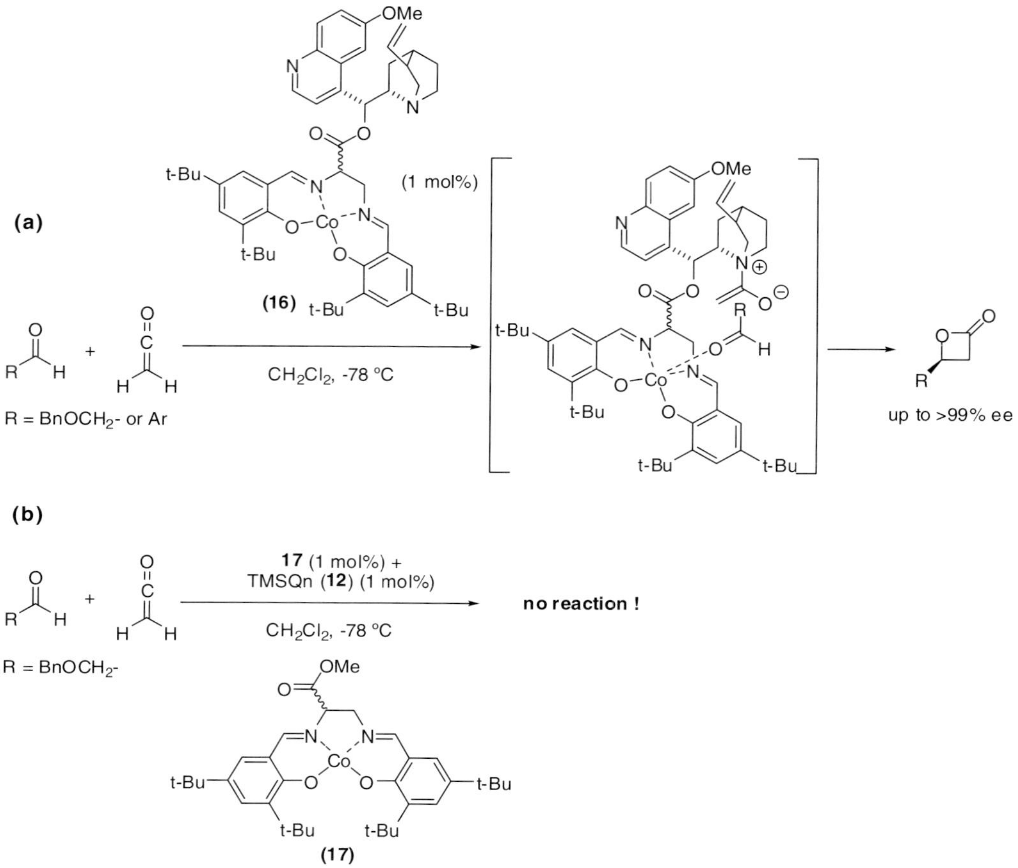

Scheme 4.18

mechanism was suggested for this remarkable catalytic activity; that is, the quinuclidine-bound enolate is presumed to attack the Co(II)-activated aldehyde intramolecularly (Scheme 4.18a). This bifunctional mechanism was proven by the control experiments; a catalytic system composed of a discrete Co(II)salen complex **17** (1 mol%), and a quinine derivative (**TMSQn** (**12**), 1 mol%) showed no catalytic activity (Scheme 4.18b). It is also notable that the configuration of the diaminopropanoic acid linker proved to be of negligible importance for chiral induction; the chiral induction is determined solely by the alkaloid.

4.2.5.2 Lewis Acid Assisted Nucleophilic Addition of Ketenes to Imines: β-Lactam Synthesis

Lectka and coworkers first demonstrated that the ammonium enolates **18** generated from acid chlorides in the presence of a cinchona alkaloid catalyst (**BQn**, **19**) and

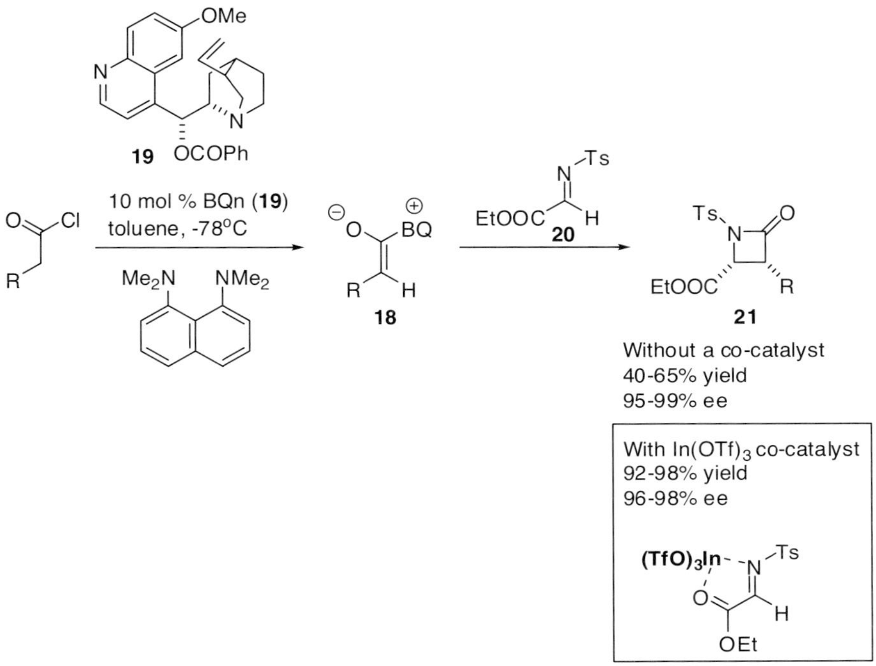

Scheme 4.19

proton sponge can react with the imine **20** to give the β-lactam product **21** [45]. The enantioselectivities and diastereoselectivities in these reactions were high enough, but the chemical yields were not of a preparatively useful level (40–65%). By employing a metal triflate as a cocatalyst to increase the electrophilicity of the imine, β-lactam products were obtained in remarkably increased yields (92–98%) and with preserved stereoselectivity (Scheme 4.19) [46]. The screening of different metals revealed that indium(III) triflate was the best Lewis acid cocatalyst for promoting this reaction. A reason behind the effect of the acidic In(III) source was attributed to the low affinity of In(III) to the alkaloid and fast "on/off" rates, consequently minimizing the "self-quenching" between the acid–base catalysts. Although the overall efficiency of the β-lactam process is greatly improved by the addition of a Lewis acid, this protocol is still limited to the use of highly reactive imines as electrophiles.

As a logical extension of the discrete bifunctional catalytic systems described above, Lectka et al. developed the homogeneous salicylate indium complex **22**, containing the chiral nucleophile (alkaloid) and the Lewis acid (In(OTf)$_3$) in a single unit, which exhibited excellent catalytic activity and stereoselectivity (90% yield, 99% ee, and 10:1 diastereomeric ratio (dr)) (Scheme 4.20) [47, 48]. Mechanistic studies revealed that the chiral nucleophiles form zwitterionic enolates that react with the metal-coordinated imines to form a ternary complex **23** in which C−C bond formation occurs.

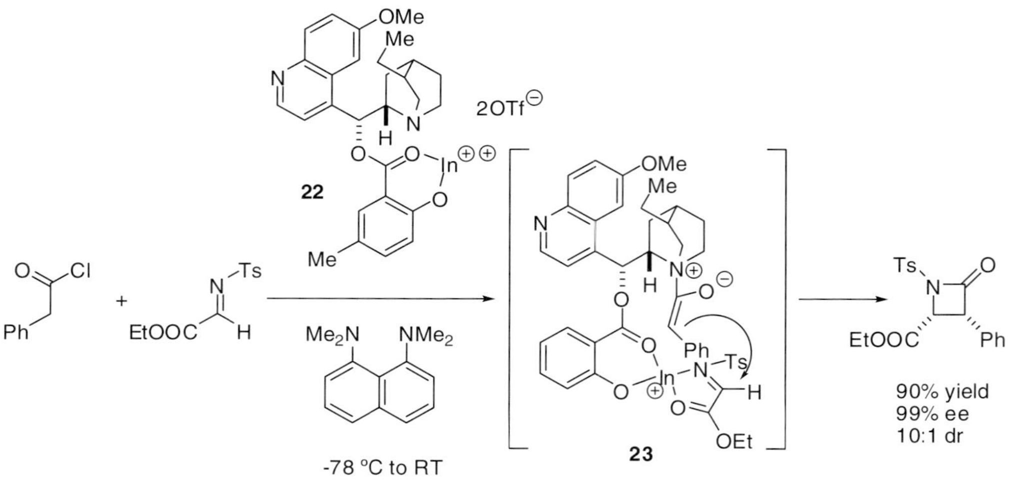

Scheme 4.20

4.2.5.3 Applications of Chiral Ketene Enolates to Formal [4 + 2] type Cyclization

Further applications of chiral ketene enolates to formal [4 + 2] type cyclization using the above-mentioned bifunctional catalyst systems have recently been discovered independently by the Lectka [48] and Nelson groups [49]. This highly interesting topic is discussed in Chapter 9 in detail.

4.2.6
Aza-Henry Reaction

Very recently, Jørgensen and coworkers reported that a highly diastereoselective and enantioselective aza-Henry reaction could also be achieved by the dual activation of a nucleophile (nitroalkanes) and electrophile (imines) with a cinchona alkaloid as a chiral base and copper(II) bisoxazoline complex as a chiral Lewis acid, respectively [50]. Using quinine as the chiral base in combination with (R)-Ph-BOX-Cu(OTf)$_2$ (24) (Scheme 4.21), the reaction of 2-nitropropanoic acid tert-butyl ester 25 with (p-methoxyphenylimino)acetic acid ethyl ester 26 afforded the aza-Henry adduct 27 in 98% ee and a dr of 14 : 1 with complete conversion. Lowering the alkaloid loading from 20 to 5 mol% did not result in any decrease in the activity or stereoselectivity (98% ee and 14 : 1 of dr). Performing the same reaction with other cinchona alkaloids (e.g., hydroquinine, quinidine, hydroquinidine, cinchonine, and hydrocinchonine) afforded only a slight decrease in the stereoselectivity. However, changing the cinchona alkaloids to achiral bases such as NEt$_3$ or Hünig base (iPr$_2$NEt) lowered the ee values of the products significantly and, moreover, no diastereoselectivity was achieved (1 : 1). In the absence of an organic base, no reaction took place.

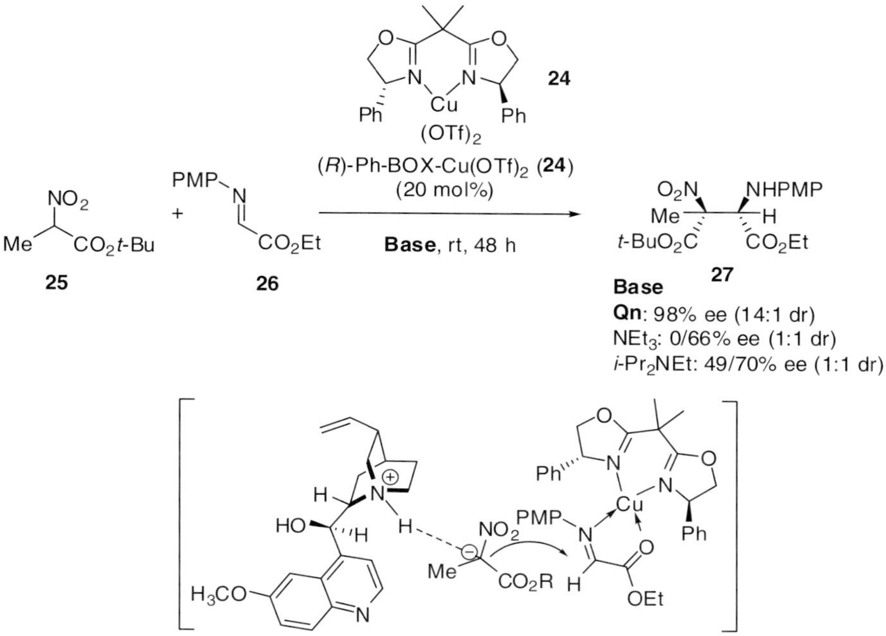

Scheme 4.21 Various combinations of cinchona alkaloids as chiral base with (R)-Ph-BOX-Cu(OTf)$_2$ for the aza-Henry reaction.

4.2.7
Enantioselective Hydrophosphonylation

For the construction of P−C bond, hydrophosphonylation of aldehydes is the most straightforward and general method, which is widely used for the synthesis of α-hydroxy phosphonic acid and related derivatives. The first highly enantioselective addition of aldehyde to dimethyl phosphate was described by Shibasaki and coworkers [51] where they used a heterobimetallic multifunctional catalyst based on 1,1-bi-2-naphthol. However, the above-described catalyst system was only applicable to aliphatic aldehydes while affording moderate ee values. Very recently, You and coworkers developed a new type of bifunctional catalysts generated from metal–organic self-assembly of substituted binols and cinchona alkaloids in association with Ti(OiPr)$_4$ [52]. A range of aliphatic and aromatic aldehydes was explored in the catalytic hydrophosphonylation with dialkyl phosphite using this type of catalysts (2.5–10 mol%). The best results (up to 99% yield and up to 99% ee) were obtained when cinchonidine was employed in combination with 3,3′-iodine substituted binol 28 in a molar ratio of 1:1 (Scheme 4.22). In the transition state model 29, proposed by authors, Ti(IV) metal center captures the aldehyde while the basic quinuclidine nitrogen of cinchona alkaloids simultaneously reacts with the phosphite.

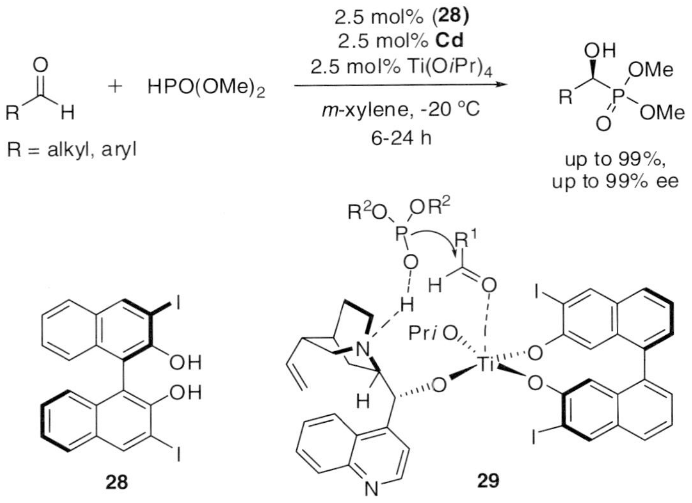

Scheme 4.22 Catalytic hydrophosphonylation of various aldehydes with dimethyl phosphite using bifunctional catalyst.

4.3
Miscellaneous Reactions

4.3.1
Claisen Rearrangements

Asymmetric sigmatropic rearrangements have received a great deal of attention during the past few decades because they are among the most powerful tools for stereoselective C−C forming reactions. The Claisen rearrangement proceeds through a concerted reaction mechanism via the formation of a chair-like transition state, which is feasible for chirality transfer [53]. The successful catalytic asymmetric ester enolate Claisen rearrangement of chelate-bridged enolates affording chiral γ,δ-unsaturated amino acids was reported by Kazmaier et al., in which they employed cinchona alkaloids as chiral ligands [54, 55]. The asymmetric induction occurs by the coordination of the chiral ligands to the chelating metal. In the presence of quinine and a second metal salt such as Al(OiPr)$_3$, the rearrangement of the N-protected glycine allylic esters **30** gave the corresponding γ,δ-unsaturated amino acids **31** with high ee values (up to 93% ee) (Scheme 4.23). Notably, in all cases, the observed yields and diastereoselectivities (98% ds) were clearly higher than those obtained in the corresponding rearrangements without the addition of the chiral ligand [54b]. However, when the reaction was carried out in the absence of lithium, the selectivity dropped dramatically. When NaHMDS were used as a base instead of LHMDS, yields of up to 18% and near 14% ee were the best that could be obtained (for $R^1 = H$ and $R^2 = Me$). In the case where KHMDS was used, no rearrangement occurred. The choice of the second metal salts is also important for the enantioselectivity. For

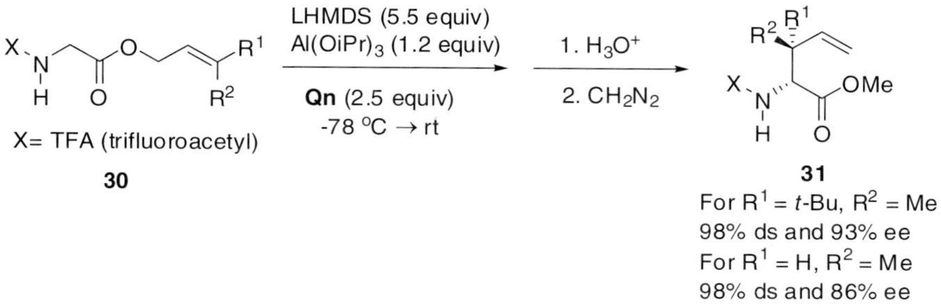

Scheme 4.23 Asymmetric ester enolate Claisen rearrangement of (N)-trifluoroacetyl-glycine allylic esters.

For $R^1 = t\text{-Bu}$, $R^2 = \text{Me}$
98% ds and 93% ee
For $R^1 = H$, $R^2 = \text{Me}$
98% ds and 86% ee

example, the use of $ZnCl_2$ instead of $Al(OiPr)_3$ gave excellent yields but very poor ee values (10% ee for $R^1 = H$ and $R^2 = Me$).

A bimetallic intermediate formed with the bidentate quinine or quinidine ligand, coordinating to the lithium enolate that undergoes Claisen rearrangement to afford the enantioenriched products, was proposed (Figure 4.2).

4.3.2
Pd-Catalyzed Asymmetric Allylic Substitutions

Pd-catalyzed asymmetric alkylation reactions are useful synthetic methods for asymmetric C—C bond formation, which operate by allowing the Pd-catalyzed substitution of a suitable leaving group in an allylic position by a nucleophile [56]. A number of chiral ligands have been developed for this kind of reaction. It has also been reported that the seleno (**32–35**) and sulfurethers (**36–39**) of cinchona alkaloids can serve as chiral ligands in Pd-catalyzed allylic substitutions [57]. In the absence of a ligand, no product could be detected in the reaction of diethyl malonate with rac-1,3-diphenyl-2-propenyl acetate, when using N,O-bis-(trimethylsilyl)acetamide (BSA)-KOAc as a base. However, in the presence of a ligand, the substitution product was obtained with moderate ee values (46–76% ee) (Scheme 4.24), but a long reaction time was required (3–8 d). The long reaction time and moderate ee values might be due to the less favorable complexation of the sterically demanding cinchona alkaloid

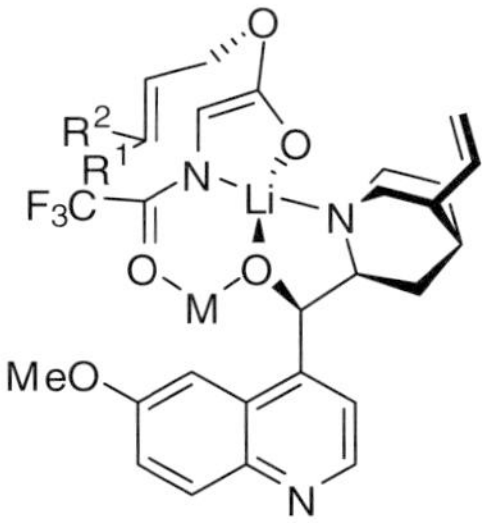

Figure 4.2 Bimetallic chelate-bridged ester enolate complex.

X = Se or S

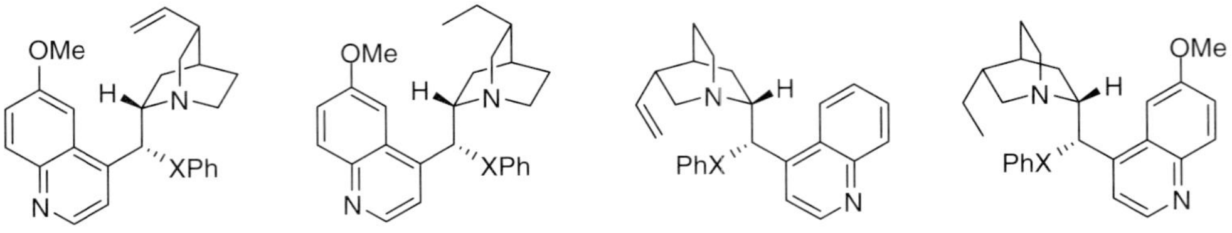

9-PhSe-*epi*-QN (**32**)
9-PhS-*epi*-QN (**36**)

9-PhSe-*epi*-DHQN (**33**)
9-PhS-*epi*-DHQN (**37**)

9-PhSe-*epi*-CN (**34**)
9-PhS-*epi*-CN (**38**)

9-PhSe-*epi*-DHQD (**35**)
9-PhS-*epi*-DHQD (**39**)

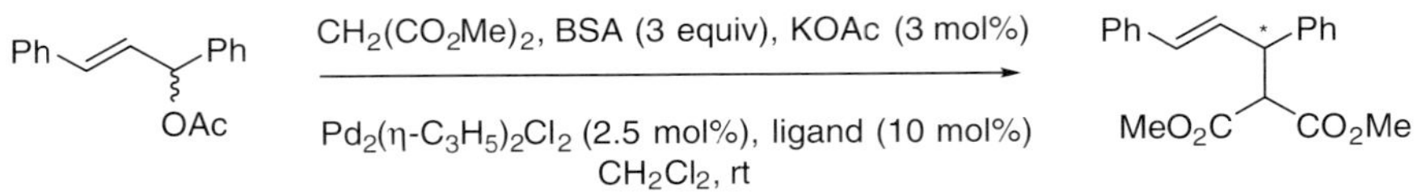

46–76% ee
16–67% yield after 3–8 days

Scheme 4.24 Cinchona alkaloid derived phenyl selenides in Pd-catalyzed alkylation of dimethyl malonate with 1,3-diphenyl-2-propenyl acetate.

ligands with palladium. The examination of molecular models and the stereochemical outcomes clearly indicated that, due to the trans effect, the nucleophilic attack is directed toward the allylic carbon, which is located in the trans position with respect to the chalcogen center, which is more π-accepting than the σ-donating sp^3-nitrogen atom in the M-shaped complex (Figure 4.3).

Very recently, Zhang and coworkers developed a new family of cinchona-based P, N-bidentate ligands **40a–d** for the palladium(II)-catalyzed asymmetric allylic alkylation reaction, which can be easily prepared starting from the enantiopure (S,S)-1,2-diphenyl-1,2-ethandiol and cinchona alkaloids in two steps [58]. 1,3-Diphenyl-3-acetoxyprop-1-ene was reacted with different nucleophiles using 2 mol% Pd and 6 mol% cinchona-based phosphite ligands **40a–d** to afford the substituted products in up to 100% yield and up to 94% ee (Scheme 4.25). Interestingly, this reaction afforded the same enantiomer regardless of which pseudoenantiomer of the cinchona alkaloid is employed. Thus, the ligands **40a–d** derived from (S,S)-1,2-

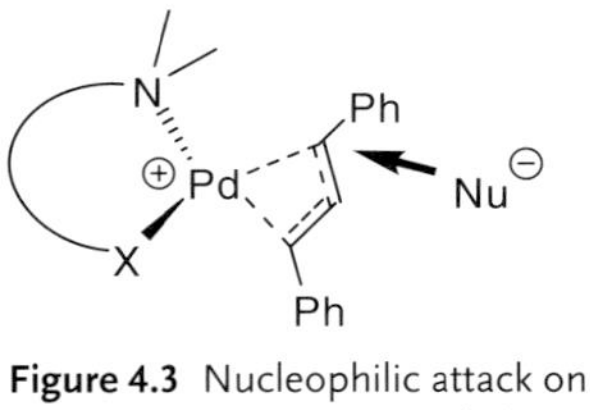

Figure 4.3 Nucleophilic attack on the allylic carbon, located *trans* to the more π-accepting chalocogen center.

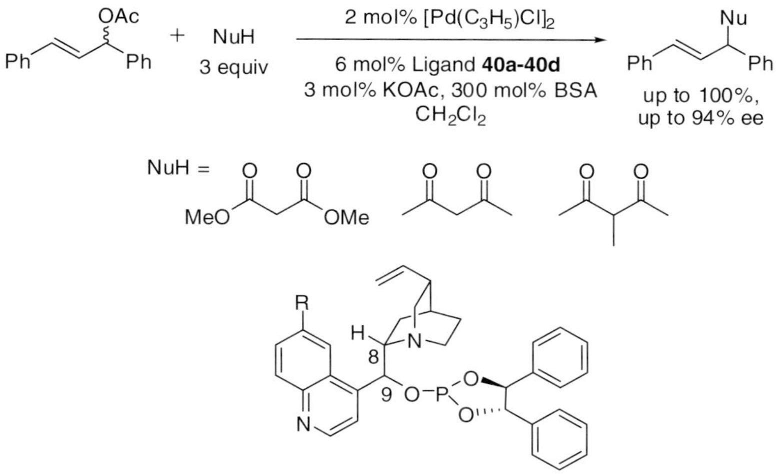

4.3 Miscellaneous Reactions

Scheme 4.25 Cinchona alkaloids derived P,N-bidentate ligands **40a–d** for the asymmetric allylic alkylation reaction.

40a R= H; Cinchonine (8R, 9S)
40b R= H; Cinchonidine (8S, 9R)
40c R= OMe; Quinine (8S, 9R)
40d R= OMe; Quinidine (8R, 9S)

diphenyl-1,2-ethanediol afforded only (R)-configured product. This result indicates that not the backbone structure of cinchona alkaloid in ligands **40a–d** but that of 1,2-diphenyl-1,2-ethanediol is responsible for the configuration of the products. The oppositely configured (S)-product was obtained when the ligands derived from the (R,R)-1,2-diphenyl-1,2-ethanediol were used.

4.3.3
Pauson–Khand Reaction

Although the cobalt-mediated Pauson–Khand reaction was discovered more than 30 years ago, few asymmetric versions of this reaction have so far been developed. Up to now, the only direct method of controlling the enantioselectivity of intermolecular cobalt-mediated Pauson–Khand reactions involves the use of alkaloid N-oxides [59–62].

Norbornene was reacted with various terminal alkynes in the presence of quinine-N-oxide (**41**) to afford the corresponding cyclopentenone derivatives (**42**) in low yields (27–68%) and ee values (2–30% ee) (Table 4.7) [62]. From the results listed in Table 4.7, hydrogen bonding between the alkyne and N-oxide plays an important role in controlling the enantioselectivity (Figure 4.4). Thus, alkynes bearing a tethered alcohol moiety result in enantioselectivities that are typically one order of magnitude higher than those obtained with nonfunctionalized alkynes. This observation therefore indicates that there is a chance of improving the enantioselectivity by optimizing the tethering structure of the alkynes.

Table 4.7 Cinchona alkaloid *N*-oxide promoted asymmetric cobalt-mediated Pauson–Khand reaction.

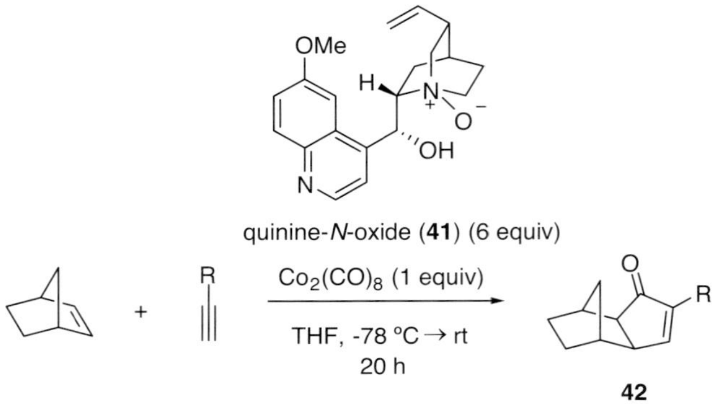
OMe
H
N
O
OH
N
quinine-N-oxide (41) (6 equiv)
R
Co2(CO)8 (1 equiv)
THF, -78 °C → rt
20 h
O
R
42

Entry	R	Yield (%)	ee (%)
1	Pr	28	2
2	*t*-Bu	39	6
3	Ph	65	7
4	Me_2COH	48	8
5	CH_2CH_2OH	68	30
6	CH_2OBn	27	7

4.3.4
Asymmetric Dimerization of Butadiene

The dimerization of butadiene catalyzed by $Ni(COD)_2$ and quinidine-DPP (**43**) as a chiral ligand gave rise to (*R*)-(+)-vinylcyclohexene with an ee of approximately 30%, accompanied by a selectivity for vinylcyclohexene versus cyclooctadiene of approximately 1 (Scheme 4.26) [63, 64].

4.3.5
Enantiotopic Differentiation Reaction of Mesocyclic Anhydrides

Several monofunctional and bifunctional cinchona alkaloid derivatives were successfully utilized as organocatalysts for the stereoselective alcoholysis of meso-anhydrides (for details, see Chapter 10) [65]. It was also reported that the zinc complex

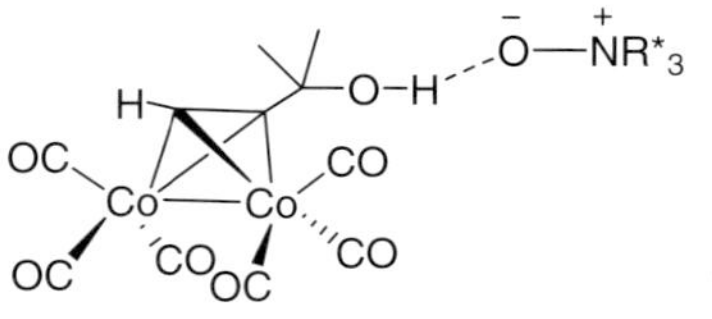
O—NR*3
O-H
H
OC
CO
Co
Co
OC
CO
OC
CO

Figure 4.4 Cobalt alkyne complex in the presence of cinchona *N*-oxide in Pauson–Khand reaction.

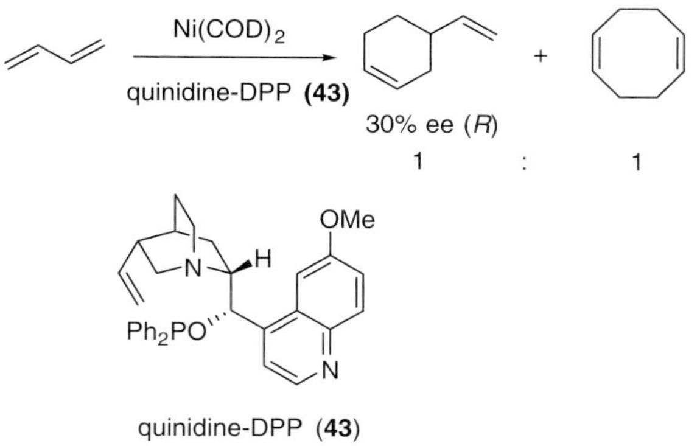

Scheme 4.26 Dimerization of butadiene catalyzed by Ni(COD)$_2$ and quinidine-DPP (**40**).

generated by the cinchona alkaloids and diethylzinc can efficiently catalyze this reaction. The zinc complex of the cinchona alkaloids provided better enantioselectivity (up to 91% ee) [66] than the unmodified cinchona alkaloids (up to 76% ee) [67] (Scheme 4.27). This improved enantioselectivity in the presence of zinc might originate from the bifunctional mechanism of action of the active catalyst. The quinuclidine group of the catalyst may be able to function as a general base catalyst and activate the nucleophile, while the Lewis acidic zinc may be able to simultaneously activate the electrophile. However, attempts to extend this methodology to other substrates revealed that the process was highly sensitive to the substrate structure.

4.4
Cinchona-Based Chiral Ligands in C−F Bond Forming Reactions

The quest for methods of synthesizing chiral fluoroorganic compounds is one of the most fascinating topics in modern organofluorine chemistry because the replacement of hydrogen with a fluorine atom often results in a significant improvement in the physicochemical properties and biological activities of the molecules.

A variety of highly enantioselective chiral N−F fluorinating reagents derived from Cinchona alkaloids were developed for the direct enantioselective fluorination of C−H acidic substrates (for details, see Chapter 6). Simultaneously, metal-catalyzed

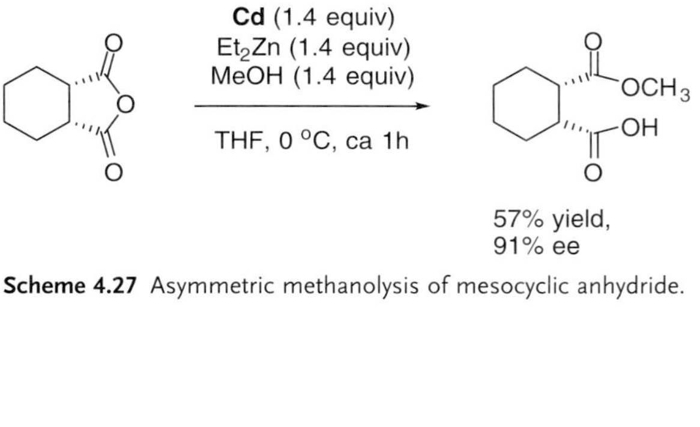

Scheme 4.27 Asymmetric methanolysis of mesocyclic anhydride.

Table 4.8 Cinchona alkaloids and their derivatives promoted the enantioselective fluorination of 2-oxo-cyclopentanecarboxylic acid tert-butyl ester.

Entry	Catalyst (mol%)	Solvent	Time (h)	Yield (%)	ee (%)
1	Qn-ZnEt$_2$(10)	THF	36	98	22(+)
2	44-Cu(OTf)$_2$(3)	THF	5	96	20(+)
3	45-Cu(OTf)$_2$(3)	THF	5	90	12(−)

enantioselective electrophilic fluorination using an achiral fluorinating reagent was also reported. However, in contrast to the successful use of chiral electrophilic N–F fluorinating reagents, the results obtained so far are rather disappointing [68, 69]. Cinchona alkaloids–oxophilic Lewis acid (Cu(OTf)$_2$ and ZnEt$_2$) complexes were examined for the enantioselective electrophilic fluorination of β-ketoesters using N-fluorobenzenesulfonamide (NFSI) as an achiral fluorinating reagent [70d]. Using Qn, 44, and 45 as ligands, 2-oxo-cyclopentanecarboxylic acid tert-butyl ester was converted smoothly into the fluorinated product in excellent yields. However, a very poor enantiomeric excess (up to 22%) was observed (Table 4.8).

4.5
Conclusions

This chapter has presented the current stage in the development of metal-promoted asymmetric C–C and C–X bond forming reactions, in which cinchona alkaloids are utilized as chirality inducers. As shown in many of the examples discussed above, cinchona alkaloids and their derivatives have great potential to serve as chiral ligands or cobase catalysts in diverse metal-promoted asymmetric C–C and C–X bond forming reactions. However, despite the scientific achievements that have been made in this field, only a few such reactions have reached the level of synthetic practicability as regards the catalyst activity and enantioselectivity. Thus, more systematic studies designed to understand the details of the asymmetric induction step should be performed that will lead to more efficient and practical catalyst systems. There is no doubt that such studies will be conducted and will provide exciting results in the near future.

Acknowledgments

This work was supported by grants KRF-2008-005-J00701 (MOEHRD), R11-2005-008-00000-0 (SRC program of MEST/KOSEF) and R31-2008-000-10029-0 (WCU program).

References

1 (a) Collins, A.N., Shedrake, G.N., and Crosby, J. (eds) (1992) *Chirality in Industry*, John Wiley & Sons, Ltd, Chichester; (b) Collins, A.N., Shedrake, G.N., and Crosby, J. (eds) (1996) *Chirality in Industry II*, John Wiley & Sons, Ltd, Chichester; (c) Blaser, H.U. and Schmidt, E.(eds) (2004) *Asymmetric Catalysis on Industrial Scale*, Wiley-VCH Verlag GmbH, Weinheim.

2 (a) Soai, K. and Shibata, T. (1999) in *Comprehensive Asymmetric Catalysis II* (eds E.N. Jacobsen, A. Pfaltz, and H. Yamamoto), Springer, Berlin, pp. 911–922; (b) Denmark, S.E. and Nicaise, O.J.-C. (1999) in *Comprehensive Asymmetric Catalysis II*, (eds E.N. Jacobsen, A. Pfaltz, and H. Yamamoto), Springer, Berlin, pp. 911–922.

3 Smaardijk, A.A. and Wynberg, H. (1987) *J. Org. Chem.*, **52**, 135–137.

4 Muchow, G., Vannoorenberghe, Y., and Buono, G. (1987) *Tetrahedron Lett.*, **28**, 6163–6166.

5 Kitamura, M., Suga, S., Kawai, K., and Noyori, R. (1986) *J. Am. Chem. Soc.*, **108**, 6071–6072.

6 Casarotto, V., Li, Z., Boucau, J., and Lin, Y.-M. (2007) *Tetrahedron Lett.*, **48**, 5561–5564.

7 Kobayashi, S. and Ishitani, H. (1999) *Chem. Rev.*, **99**, 1069–1094.

8 (a) Beresford, K.J.M. (2002) *Tetrahedron Lett.*, **43**, 7175–7177; (b) Beresford, K.J.M. (2004) *Tetrahedron Lett.*, **45**, 6041–6044.

9 Vogl, E.M., Gröger, H., and Shibasaki, M. (1999) *Angew. Chem. Int. Ed.*, **38**, 1570–1577.

10 (a) Pu, L. and Yu, H.-B. (2001) *Chem. Rev.*, **101**, 757–824; (b) Frantz, D.E., Fässler, R., Tomooka, C.S., and Carreira, E.M. (2000) *Acc. Chem. Res.*, **33**, 373–381.

11 Kamble, R.M. and Singh, V.K. (2003) *Tetrahedron Lett.*, **44**, 5347–5349.

12 Dahmen, S. (2004) *Org. Lett.*, **6**, 2113–2116.

13 Liu, L., Wang, R., Kang, Y.-F., Chen, C., Xu, Z.-Q., Zhou, Y.F., Ni, M., Cai, H.-Q., and Gong, M.Z. (2005) *J. Org. Chem.*, **70**, 1084–1086.

14 (a) Anand, N.K. and Carreira, E.M. (2001) *J. Am. Chem. Soc.*, **123**, 9687–9688; (b) El-Sayed, E., Anand, N.K., and Carreira, E.M. (2001) *Org. Lett.*, **3**, 3017–3020; (c) Fässler, R., Tomooka, C.S., Frantz, D.E., and Carreira, E.M. (2004) *Proc. Natl. Acad. Sci. USA*, **101**, 5843–5845; (d) Boyall, D., Frantz, D.E., and Carreira, E.M. (2002) *Org. Lett.*, **4**, 2605–2606; (e) Cozzi, P.G., Hilgraf, R., and Zimmermann, N. (2004) *Eur. J. Org. Chem.*, 4095–4105.

15 Ekström, J., Zaitsev, A.B., and Adolfsson, H. (2006) *Synlett*, **6**, 885–888.
16 (a) Riberio, C.M.R., Cordeiro, F.M., and de Farias, M.C. (2006) *Mini-Rev. Org. Chem.*, **3**, (1), 1–10; (b) Cozzi, P.G. (2007) *Angew. Chem. Int. Ed.*, **46**, 2568–2571.
17 Johar, P.S., Araki, S., and Butsugan, Y. (1992) *J. Chem. Soc., Perkin Trans I*, 711–713.
18 Soai, K., Oshio, A., and Saito, T. (1993) *J. Chem. Soc., Chem. Commun.*, 811–812.
19 Ojida, A., Yamano, T., Taya, N., and Tasaka, A. (2002) *Org. Lett.*, **4**, 3051–3054.
20 Loh, T.-P., Zhou, J.-R., and Li, X.-R. (1999) *Tetrahedron Lett.*, **40**, 9333–9336.
21 Loh, T.-P., Zhou, J.-R., and Yin, Z. (1999) *Org. Lett.*, **11**, 1855–1857.
22 Thornqvist, V., Manner, S., and Frejd, T. (2006) *Tetrahedron: Asymmetry*, **17**, 410–415.
23 Loh, T.-P., Lin, M.-J., and Tan, K.-L. (2003) *Tetrahedron Lett.*, **44**, 507–509.
24 (a) North, M. (1993) *Synlett*, 807–820; (b) Effenberger, F. (1994) *Angew. Chem. Int. Ed.*, **33**, 1555–1564; (c) Gregory, R.J.H. (1999) *Chem. Rev.*, **99**, 3649–3682; (d) Mori, A. and Noue, I.S. (1999) in *Comprehensive Asymmetric Catalysis II* (eds E.N. Jacobsen, A. Pfaltz, and H. Yamamoto), Springer, Berlin, pp. 983–992.
25 Bredig, G. and Fiske, P.S. (1912) *Biochem. Z*, **46**, 7.
26 Prelog, V. and Wilhelm, M. (1954) *Helv. Chim. Acta*, **37**, 1634–1660.
27 Tian, S.-K. and Deng, L. (2001) *J. Am. Chem. Soc.*, **123**, 6195–6196.
28 Kobayashi, S., Tsuchiya, Y., and Mukaiyama, T. (1991) *Chem. Lett.*, 541–544.
29 Gou, S., Wang, J., Liu, X., Wang, W., Chen, F.-X., and Feng, X. (2007) *Adv. Synth. Catal.*, **349**, 343–349.
30 Lundgren, S., Wingstrand, E., Penhoat, M., and Moberg, C. (2005) *J. Am. Chem. Soc.*, **127**, 11592–11593.
31 Belokon, Y.N., Caveda-Cepas, S., Green, B., Ikonnikov, N.S., Khrustalev, V.N., Larichev, V.S., Moscalenko, M.A., North, M., Orizu, C., Tararov, V.I., Tasinazzo, M., Timofeeva, G.I., and Yashkina, L.V. (1999) *J. Am. Chem. Soc.*, **121**, 3968–3973.
32 Spino, C. (2004) *Angew. Chem. Int. Ed.*, **43**, 1764–1766.
33 (a) Pan, S.C., Zhou, J., and List, B. (2007) *Angew. Chem. Int. Ed.*, **46**, 612–614. (b) Kato, N., Mita, T., Kanai, M., Therrien, B., Kawano, M., Yamaguchi, K., Danjo, H., Sei, Y., Sato, A., Furusho, S., and Shibasaki, M. (2006) *J. Am. Chem. Soc.*, **128**, 6768–6769.
34 Wang, J., Hu, X., Jiang, J., Gou, S., Huang, X., Liu, X., and Feng, X. (2007) *Angew. Chem. Int. Ed.*, **46**, 8468–8470.
35 (a) Wynberg, H. and Staring, E.G.J. (1982) *J. Am. Chem. Soc.*, **104**, 166–168; (b) Wynberg, H. and Staring, E.G.J. (1985) *J. Org. Chem.*, **50**, 1977–1979.
36 (a) Berkessel, A. and Gröger, H. (2005) *Asymmetric Organocatalysis*, Wiley-VCH Verlag GmbH, Weinheim, (b) Dalko, P.I. (ed.) (2007) *Enantioselective Organocatalysis*, Wiley-VCH Verlag GmbH, Weinheim; (c) List, B. (ed.) (2007) *Chem. Rev.*, **107**, 5413–5883 (special issue for organocatalysis); (d) Houk, K.N. and List, B. (eds), (2004) *Acc. Chem. Res.*, **37**, 487–631 (special issue on asymmetric organocatalysis).
37 Zhu, C., Shen, X. and Nelson, S.G. (2004) *J. Am. Chem. Soc.*, **126**, 5352–5353.
38 Shen, X., Wasmuth, A.S., Zhao, J., Zhu, C., and Nelson, S.G. (2006) *J. Am. Chem. Soc.*, **128**, 7438–7439.
39 Calter, M.A., Tretyak, O.A., and Flaschenriem, C. (2005) *Org. Lett.*, **7**, 1809–1812.
40 Koch, F.M. and Peters, R. (2007) *Angew. Chem. Int. Ed.*, **46**, 2685–2689.
41 Tiseni, P.S. and Peters, R. (2007) *Angew. Chem. Int. Ed.*, **46**, 5325–5328.
42 Forslund, R.E., Cain, J., Colyer, J., and Doyle, M.P. (2005) *Adv. Synth. Catal.*, **347**, 87–92.
43 Lin, Y.-M., Boucau, J., Li, Z., Casarotto, V., Lin, J., Nguyen, A.N., and Ehrmantraut, J. (2007) *Org. Lett.*, **9**, 567–570.
44 Lin, Y.-M., Li, Z., and Boucau, J. (2007) *Tetrahedron Lett.*, **48**, 5275–5278.
45 (a) Taggi, A.E., Hafez, A.M., Wack, H., Young, B., Drury, III, W.J., and Lectka, T.

(2000) *J. Am. Chem. Soc.*, **122**, 7831–7832; (b) Taggi, A.E., Hafez, A.M., Wack, H., Young, B., Ferraris, D., and Lectka, T. (2002) *J. Am. Chem. Soc.*, **124**, 6626–6635.
46. France, S., Wack, H., Hafez, A.M., Taggi, A.E., Witsil, D.R., and Lectka, T. (2002) *Org. Lett.*, **4**, 1603–1605.
47. France, S., Shah, M.H., Weatherwax, A., Wack, H., Roth, J.P., and Lectka, T. (2005) *J. Am. Chem. Soc.*, **127**, 1206–1215.
48. (a) Abraham, C.J., Paull, D.H., Scerba, M.T., Grebinski, J.W., and Lectka, T. (2006) *J. Am. Chem. Soc.*, **128**, 13370–13371; (b) Paull, D.H., Alden-Danforth, E., Wolfer, J., Dogo-Isonagie, C., Abraham, C.J., and Lectka, T. (2007) *J. Org. Chem.*, **72**, 5380–5382; (c) Wolfer, J., Bekele, T., Abraham, C.J., Dogo-Isonagie, C., and Lectka, T. (2006) *Angew. Chem. Int. Ed.*, **45**, 7398–7400; (d) Paull, D.H., Abraham, C.J., Scerba, M.T., Alden-Danforth, E., and Lectka, T. (2008) *Acc. Chem. Res.*, **41**, 655–663; (e) Abraham, C.J., Paull, D.H., Bekele, T., Scerba, M.T., Dudding, T., and Lectka, T. (2008) *J. Am. Chem. Soc.*, **130**, 17085–17094.
49. Xu, X., Wang, K., and Nelson, S.G. (2007) *J. Am. Chem. Soc.*, **129**, 11690–11691.
50. Knudsen, K.R. and Jørgensen, K.A. (2005) *Org. Biomol. Chem.*, **3**, 1362–1364.
51. Arai, T., Bougauchi, M., Sasai, H., and Shibasaki, M. (1996) *J. Org. Chem.*, **61**, 2926–2927.
52. Yang, F., Zhao, D., Lan, J., Xi, P., Yang, L., Xiang, S., and You, J. (2008) *Angew. Chem. Int. Ed.*, **47**, 5646–5649.
53. Frauenrath, H. (1995) in *Houben Weyl*, vol. **E21d** *Stereoselective Synthesis* (eds G. Helmchen, R.W. Hoffmann, J. Mulzer, and E. Schaumann), Thieme, Stuttgart, p. 3301.
54. (a) Kazmaier, U. and Krebs, A. (1995) *Angew. Chem. Int. Ed. Engl.*, **34**, 2012–2014; (b) Kazmaier, U. (1994) *Angew. Chem. Int. Ed. Engl.*, **33**, 998–999.
55. (a) Kazmaier, U., Mues, H., and Krebs, A. (2002) *Chem. Eur. J.*, **8**, 1850–1855; (b) Krebs, A. and Kazmaier, U. (1996) *Tetrahedron Lett.*, **37**, 7945–7946.
56. (a) Pfaltz, A. and Lautens, M. (1999) in *Comprehensive Asymmetric Catalysis II* (eds E.N. Jacobsen, A. Pfaltz, and H. Yamamoto), Springer, Berlin, pp. 833–884; (b) Trost, B.M. and Crawley, M.L. (2003) *Chem. Rev.*, **103**, 2921–2943.
57. Zielińska-Błajet, M., Siedlecka, R. and Skarżewski, J. (2007) *Tetrahedron: Asymmetry*, **18**, 131–136.
58. Wang, Q.-F., He, W., Liu, X.-Y., Chen, H., Qin, X.-Y., and Zhang, S.-Y. (2008) *Tetrahedron: Asymmetry*, **19**, 2447–2450.
59. Kerr, W.J., Kirk, G.G., and Middlemiss, D. (1995) *Synlett*, 1085–1086.
60. Carpenter, N.E. and Nicholas, K.M. (1999) *Polyhedron*, **18**, 2027–2034.
61. Derdau, V., Laschat, S., and Jones, P.G. (1998) *Heterocycles*, **48**, 1445–1453.
62. Derdau, V. and Laschat, S. (2002) *J. Organomet. Chem.*, **642**, 131–136.
63. Rubina, K.I., Goldberg, Y.S., Shymanska, M.V., and Lukevics, E. (1987) *Appl. Organomet. Chem.*, **1**, 435–439.
64. Cros, P., Peiffer, G., Dennis, D., Mortreux, A., Buono, G., and Petit, F. (1987) *New J. Chem.*, **11**, 573–579.
65. Reviews: (a) Atodiresei, I., Schiffers, I., and Bolm, C., (2007) *Chem. Rev.*, **107**, 5683–5712; (b) Tian, S.-K., Chen, Y., Hang, J., Tang, L., McDaid, P., and Deng, L. (2004) *Acc. Chem. Res.*, **37**, 621–631; Recent papers on bifunctional catalysis: (c) Rho, H.S., Oh, S.H., Lee, J.W., Lee, J.Y., Chin, J., and Song, C.E. (2008) *Chem. Commun.*, 1208–1210; (d) Oh, S.H., Rho, H.S., Lee, J.W., Lee, J.E., Youk, S.H., Chin, J., and Song, C.E. (2008) *Angew. Chem. Int. Ed.*, **47**, 7872–7875.
66. Shimizu, M., Matsukawa, K., and Fusisawa, T. (1993) *Bull. Chem. Soc. Jpn.*, **66**, 2128–2130.
67. Hiratake, J., Yamamoto, Y., and Oda, J. (1985) *J. Chem. Soc., Chem. Commun.*, 1717–1719.
68. (a) Shibata, N., Suzuki, E., Asahi, T., and Shiro, M. (2001) *J. Am. Chem. Soc.*, **123**, 7001–7009; (b) Shibata, N., Suzuki, E., and Takeuchi, Y. (2000) *J. Am. Chem. Soc.*, **122**, 10728–10729; (c) Shibata, N.,

Ishimaru, T., Nakamura, S., and Toru, T. (2007) *J. Fluorine. Chem.*, **128**, 469–483.

69 (a) Cahard, D., Audouard, C., Plaquevent, J.-C., and Roques, N. (2000) *Org. Lett.*, **2**, 3699–3701; (b) Mohar, B., Baudoux, J., Plaquevent, J.C., and Cahard, D. (2001) *Angew. Chem. Int. Ed. Engl.*, **40**, 4214–4216; (c) Mohar, B., Sterk, D., Ferron, L., and Cahard, D. (2005) *Tetrahedron Lett.*, **46**, 5029–5031; (d) Ma, J.-A. and Cahard, D. (2004) *J. Fluorine Chem.*, **125**, 1357–1361.

Part Two
Cinchona Alkaloid Derivatives as Chiral Organocatalysts

5
Cinchona-Based Organocatalysts for Asymmetric Oxidations and Reductions
Ueon Sang Shin, Je Eun Lee, Jung Woon Yang, and Choong Eui Song

5.1
Introduction

The asymmetric oxidation and reduction of sp^2 carbons are the most important approaches for controlling the absolute configuration of asymmetric sp^3 carbons that can be achieved through facial selectivity. Although transition-metal-based reactions have been well established for these oxidations and reductions, the sensitivity of the reaction conditions toward moisture and oxygen, as well as toxic metal contamination in the product, restrict their large-scale application. Thus, at present, there is much interest in chiral organocatalysts, as they tend to be less toxic and more environment friendly than traditional metal-based catalysts [1]. They are usually robust and thus tolerate moisture and oxygen, so that they usually do not demand any special reaction conditions. In this chapter, we will review the asymmetric oxidation of electron-deficient olefins and the asymmetric reduction of prochiral ketones, where the control of the stereochemical outcome is achieved by using cinchona alkaloid derivatives as chiral organocatalysts.

Two complementary methodologies have been developed for the asymmetric epoxidation of electron-poor olefins, where either cinchona-based phase-transfer catalysts (PTCs) or 9-amino-9-(deoxy)-*epi*-cinchona alkaloids are used as organocatalysts. Mechanistically, in these two methodologies, the reaction proceeds via a chiral ion pairing mechanism and iminium catalysis, respectively. The catalytic asymmetric epoxidation of electron-deficient olefins has been regarded as one of the most representative asymmetric PTC reactions. However, one drawback associated with this process is the substrate limitation. High enantioselectivities can be achieved only with acyclic enones. With most cyclic enones, cinchona-based chiral PTCs gave only poor to moderate enantioselectivities. The inherent difficulties in the enantioselective epoxidation of cyclic enones using PTCs have, quite recently, been overcome by adopting the iminium catalysis approach, in which 9-amino-*epi*-cinchona alkaloids are employed as catalysts. Using this approach, most β-substituted cyclic enones afford the corresponding epoxides with excellent ee values. In this chapter, the scope and limitation of the cinchona-catalyzed asymmetric aziridination of enones are also

Cinchona Alkaloids in Synthesis and Catalysis, Ligands, Immobilization and Organocatalysis
Edited by Choong Eui Song
Copyright © 2009 WILEY-VCH Verlag GmbH & Co. KGaA, Weinheim
ISBN: 978-3-527-32416-3

discussed. Here again, iminium catalysis using 9-amino-*epi*-cinchona alkaloids has proven to be quite successful.

On the contrary, to achieve a successful cinchona-catalyzed asymmetric oxidation chemistry, cinchona-catalyzed asymmetric reduction has been explored very little despite the importance of this reaction. Previous reports on this subject are restricted to the reduction of aromatic ketones and, moreover, the enantioselectivities achieved to date remain far from satisfactory when compared with metal catalysis.

The schemes exemplified in this chapter will demonstrate the indispensable role of cinchona alkaloids as catalysts in these important research areas.

5.2
Cinchona-Based Organocatalysts in Asymmetric Oxidations

5.2.1
Epoxidation of Enones and α,β-Unsaturated Sulfones Using Cinchona-Based Chiral Phase-Transfer Catalysts

5.2.1.1 Epoxidation of Acyclic Enones

The asymmetric epoxidation of electron-deficient olefins using a stoichiometric oxidant (e.g., hydrogen peroxide, an alkyl hydroperoxide or sodium hypochlorite) has been regarded as one of the most representative asymmetric PTC reactions. The pioneering studies by Wynberg during the 1970s established that cinchona-based chiral PTCs can be used to effect this reaction. However, only low to moderate enantioselectivities (up to 55% ee) were obtained in the epoxidation of *trans*-chalcone derivatives with basic hydrogen peroxide using N-benzylquininium chloride **1** as a phase-transfer catalyst (Scheme 5.1) [2]. Interestingly, the absolute configuration of the product highly depended on the type of oxidant used. For example, the opposite absolute configuration of the products was obtained on switching from bleach (NaOCl) to H_2O_2 [3].

An improvement of the enantioselectivity (ee values of more than 90%) was reported independently by the Arai, Lygo, Corey, and Park–Jew groups. Arai and coworkers observed that a dramatic jump in the ee values could be achieved simply

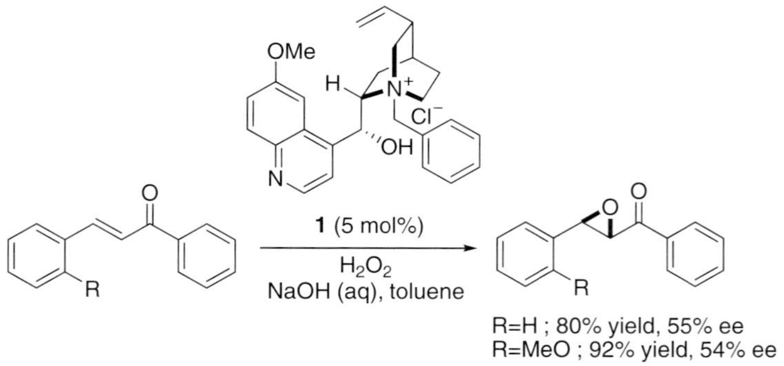

Scheme 5.1

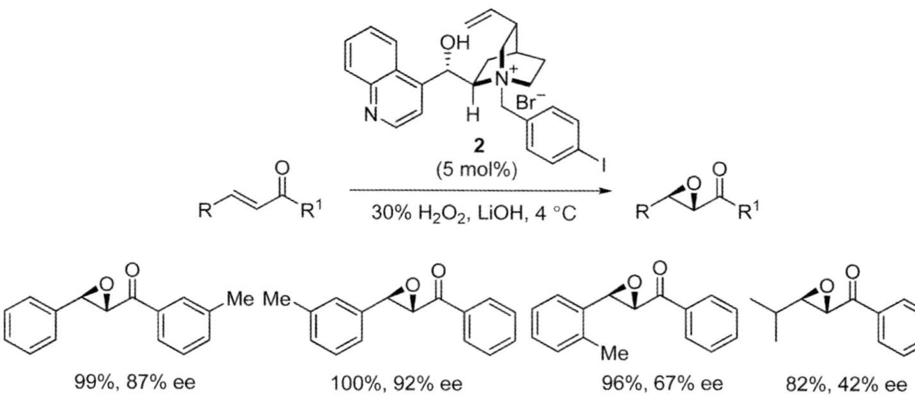

5.2 Cinchona-Based Organocatalysts in Asymmetric Oxidations

Scheme 5.2

by introducing the iodo group at the para position on the phenyl ring of the N-benzyl moiety of the Wynberg type PTC **1**. Using 5 mol% of the N-4-iodobenzyl derivative **2** as a catalyst along with aqueous H_2O_2 as an oxidant, the epoxidation of trans-chalcones proceeded smoothly under biphasic conditions (aqueous LiOH, n-Bu$_2$O, 4 °C) to afford the corresponding epoxides in quantitative yields with much higher ee values (up to 92% ee) than those obtained using Wynberg's catalyst **1** (Scheme 5.2) [4].

The Lygo and Corey groups independently reported that the cinchona-based PTCs **3** and **4** bearing an N-anthracenylmethyl function exhibited significantly increased enantioselectivity. Using O-benzylated N-anthracenylmethyl dihydrocinchondinium hydroxide **3** (10 mol%) as a catalyst with aqueous sodium hypochlorite as an oxidant, acyclic enones can be smoothly oxidized to the corresponding epoxides in excellent yields with high ee values (76–89%) (Scheme 5.3) [5]. A further

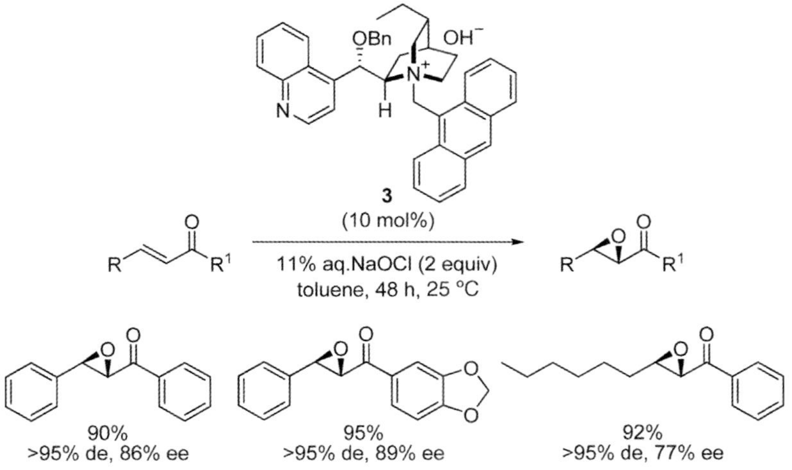

Scheme 5.3

improvement was achieved by replacing the hydroxide anion of **3** with the bromide anion. The reactions of various α,β-enones with aqueous KOCl as an oxidant were found to take place quite efficiently in the presence of 10 mol% of catalyst **4** in toluene at −40 °C. Very high enantioselectivities (ranging from 91 to 98.5% ee) of the epoxide products were obtained along with excellent yields (70–96%) in most cases (Scheme 5.4) [6]. To explain the observed sense of the stereoselectivity and rate acceleration effect of **4**, Corey proposed a plausible transition-state structure that allows for the charge-accelerated, face-selective conjugate addition of the ion-paired hypochlorite.

Another optimization study by Lygo's group revealed that only 1 mol% of catalyst **4** is sufficient to give almost the same results (Scheme 5.5) [7]. Under asymmetric PTC conditions using **4** as a catalyst, Lygo also reported that allyl alcohols with aromatic and aliphatic side chains could be converted directly to epoxyketones with moderate to good ee values [8].

Later on, Liang and coworkers successfully employed trichloroisocyanuric acid (TCCA) as a new type of stoichiometric oxidant for the asymmetric epoxidation of acyclic enones in the presence of 10 mol% of catalyst **4** (Scheme 5.6) [9]. The desired epoxy ketones were obtained in good yields (69–93%) with high enantioselectivities (73–93% ee) under nonaqueous solid–liquid conditions [9b]. In this reaction, TCCA reacts with an inorganic base (KOH) to form a hypochlorite salt, which is transferred to the organic phase by the phase-transfer catalyst and oxidizes

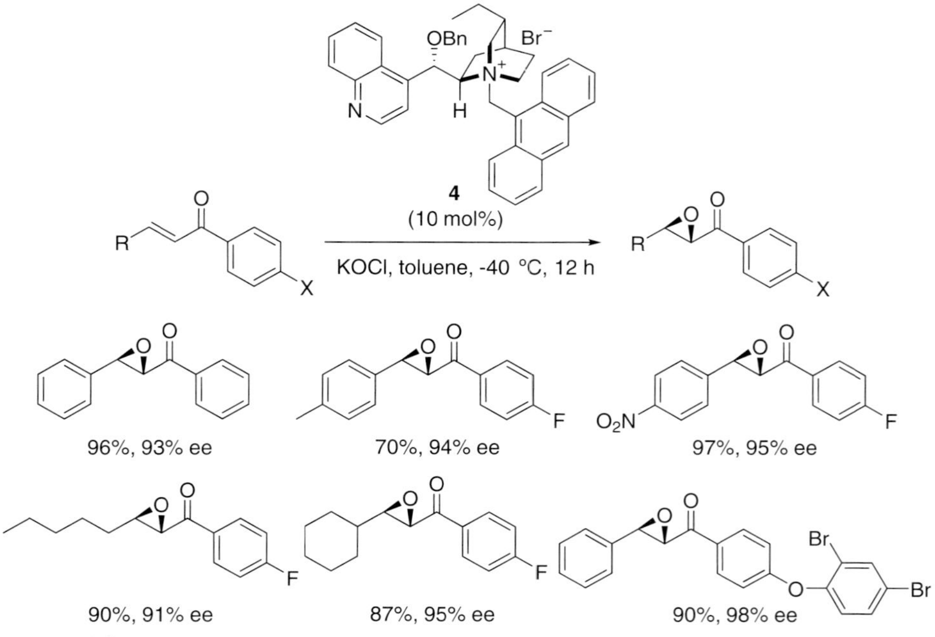

Scheme 5.4

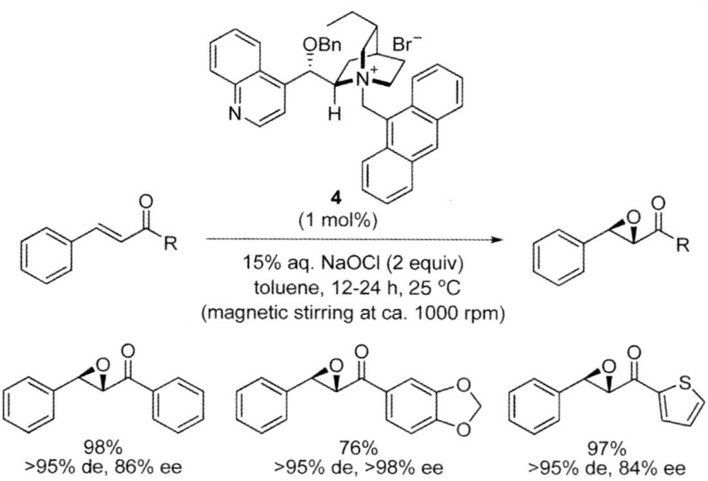

Scheme 5.5

the enones to the corresponding epoxy ketones (Scheme 5.7). Later, Liang and coworkers successfully extended their protocol to a tandem Claisen–Schmidt condensation–epoxidation sequence, providing a one-pot entry to enantioenriched α,β-epoxy ketones (Scheme 5.8) [10].

Quite recently, one of the most efficient phase-transfer-catalyzed epoxidation methods for chalcone-type enones was developed by the Park–Jew group [11]. A series of *meta*-dimeric cinchona PTCs with modified phenyl linkers were prepared. Among this series, the 2-fluoro substituted catalyst **5**, exhibited unprecedented activity and enantioselectivity for the epoxidation of various *trans*-chalcones in the

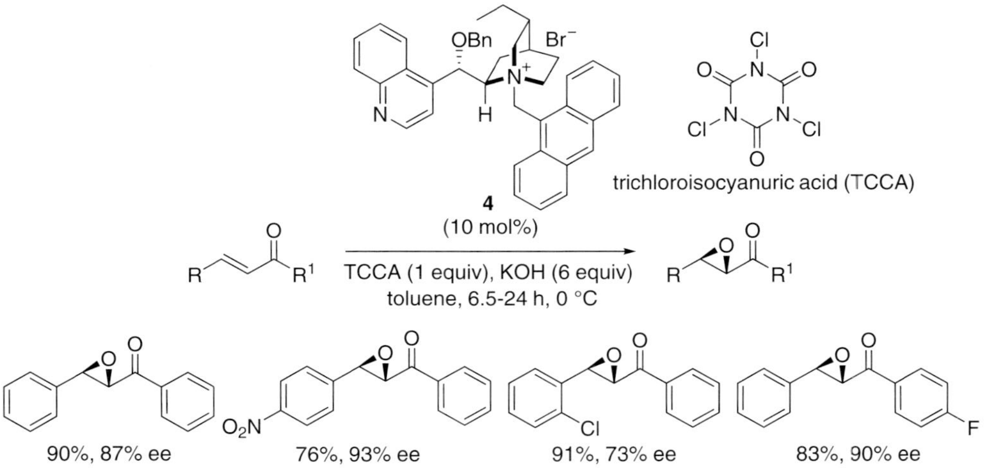

Scheme 5.6

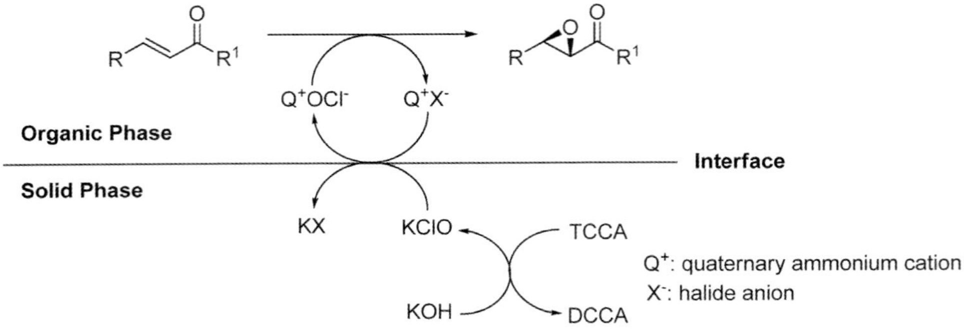

112 | *5 Cinchona-Based Organocatalysts for Asymmetric Oxidations and Reductions*

Scheme 5.7

presence of a surfactant such as Triton X-100, Tween 20, or Span 20. Using just 1 mol % of the dimeric alkaloid catalyst **5**, 1 mol% of Span 20, 30% aqueous H_2O_2 and 50% KOH in diisopropyl ether, the corresponding epoxides were obtained in nearly quantitative yields and excellent enantioselectivities (97% to >99% ee) at room temperature (Scheme 5.9).

A plausible transition-state model was proposed by the authors, in which the chalcone is located between two cinchona units in the catalyst **5**, and the β-phenyl moiety of the chalcone has a π–π stacking interaction with one of the quinoline moieties in **5**. The carbonyl oxygen atom is placed as close to one of the ammonium cations as permitted by the van der Waals forces. The other ammonium cation is ion paired with the hydrogen peroxide ion through hydrogen bonding with the oxygen of the 6′-OMe group in the quinoline moiety. As a consequence, the hydrogen peroxide anion can only approach the β-carbon of the chalcone from above in the 1,4-addition to afford the α*S*,β*R*-stereoisomer (Figure 5.1).

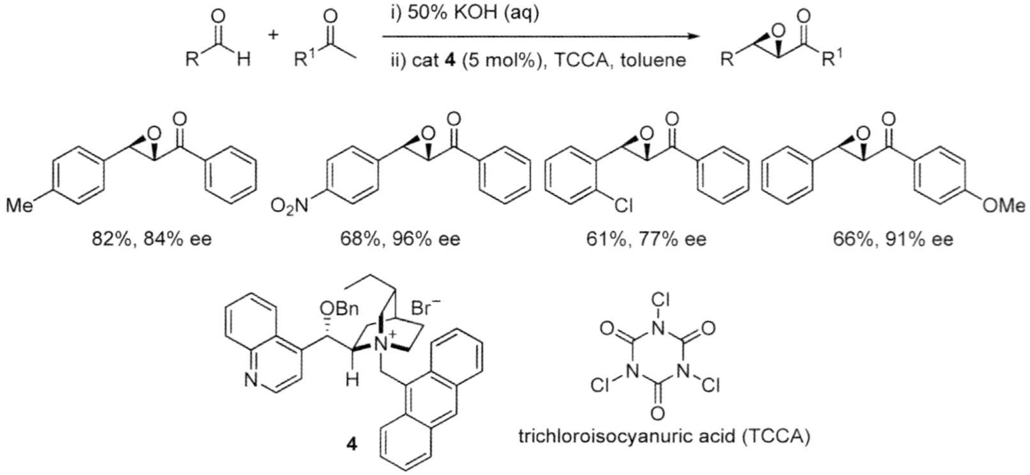

Scheme 5.8

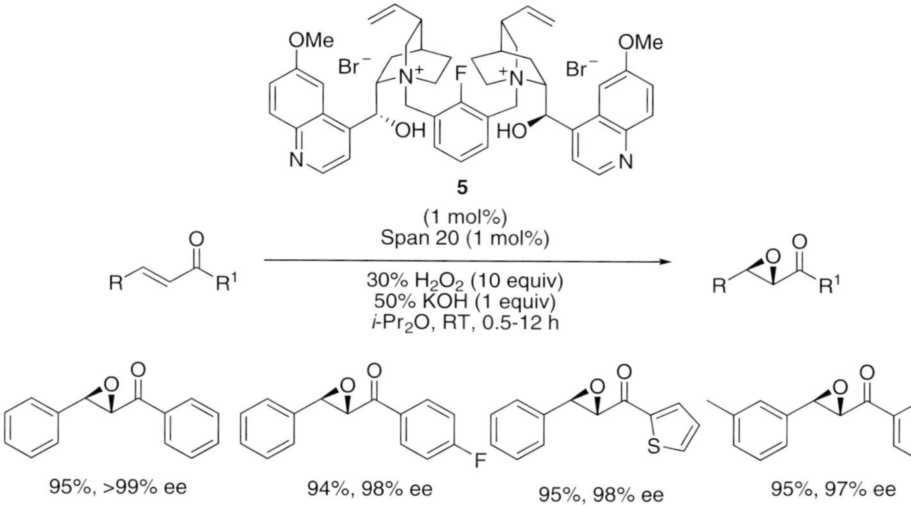

5.2 Cinchona-Based Organocatalysts in Asymmetric Oxidations

Scheme 5.9

Reaction conditions: catalyst **5** (1 mol%), Span 20 (1 mol%), 30% H_2O_2 (10 equiv), 50% KOH (1 equiv), i-Pr_2O, RT, 0.5–12 h.

Product yields and ee:
- 95%, >99% ee
- 94%, 98% ee
- 95%, 98% ee
- 95%, 97% ee

Wang and coworkers reported the asymmetric epoxidation of chalcone derivatives using polyethylene glycol (PEG) supported cinchona-based dimeric PTCs and *tert*-butyl hydroperoxide as an oxidant. However, only low to moderate ee values (33–86% ee) were obtained [12].

5.2.1.2 Epoxidation of Cyclic Enones

In 1976, using the parent quinine as a catalyst and 30% aqueous hydroperoxide as an oxidant, the asymmetric epoxidation of cyclic enones, such as naphthoquinones, was explored by Wynberg and coworkers [13]. However, the resulting enantioselectivity was minimal. Their further attempt under biphasic conditions with N-benzyl-quininium chloride **1** as the catalyst and t-BuOOH as the oxidant also resulted in very low enantioselectivities (up to 20% ee for the epoxidation of various cyclohexenones) [14]. However, the use of the dimeric form of Wynberg's catalysts **6** and **7** resulted in somewhat better (up to 63% ee with cyclohexenone) asymmetric

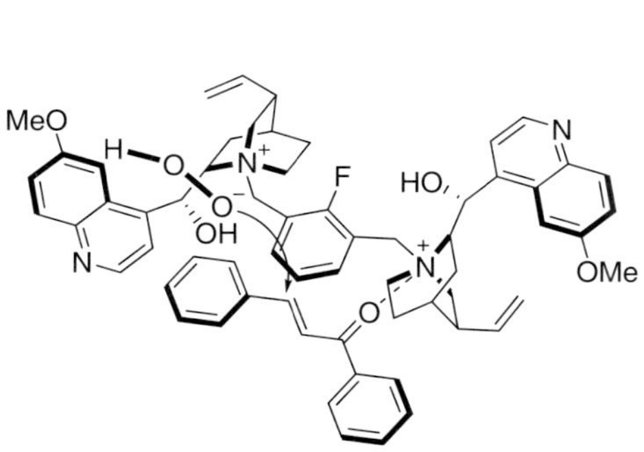

Figure 5.1 Plausible transition state for the asymmetric epoxidation catalyzed by **5**.

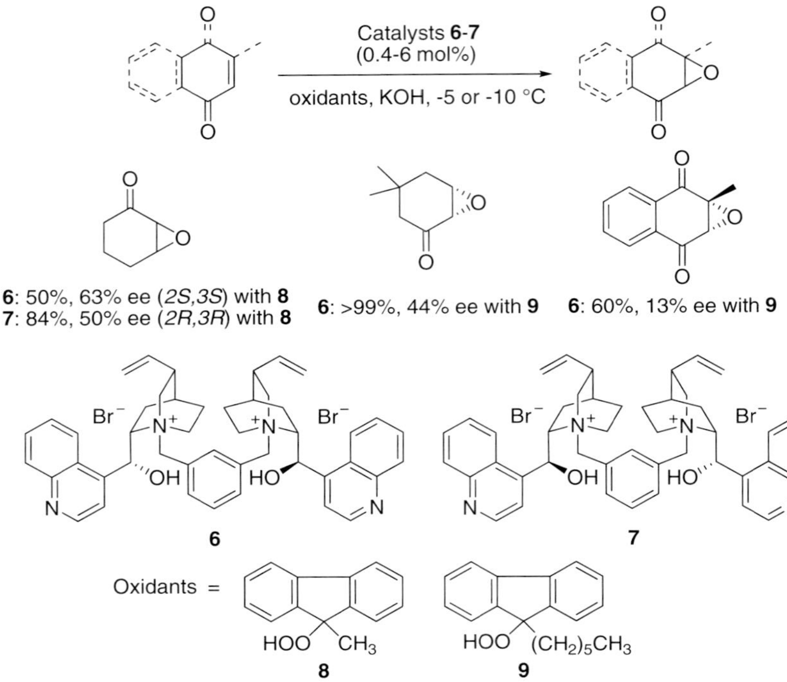

5 Cinchona-Based Organocatalysts for Asymmetric Oxidations and Reductions

Scheme 5.10

6: 50%, 63% ee (2S,3S) with 8
7: 84%, 50% ee (2R,3R) with 8

6: >99%, 44% ee with 9

6: 60%, 13% ee with 9

induction in the epoxidation of cyclic enones using the 9-alkylfluorenyl peroxides **8** or **9** as an oxidant (Scheme 5.10) [15].

Further improved ee values were obtained by using N-(α-naphthylmethyl) quinidinium chloride **10**. The enantioselectivity strongly depends on the size of the α-substituent in the quinones. The use of bulkier substituents (e.g., i-Pr and Ph) at the 2-position of naphthoquinones afforded better enantioselectivities (Scheme 5.11) [16].

Dehmlow and coworkers screened several analogues of cinchona-based PTCs bearing an N-(9-anthracenylmethyl) group [17]. Especially, in the case of 2-isopropyl naphthoquinones, the nonnatural deazacinchonidine derivative catalyst **11** showed better results compared to those obtained with the natural cinchonidine-derived analogue **12**, in terms of both the catalytic activity and the enantioselectivity (84% ee) in this reaction (Scheme 5.12).

Very recently, using structurally varied PTCs based on quinine, quinidine, dihydroquinine, and dihydroquinidine, Berkessel and coworkers conducted the asymmetric epoxidation of 2-methylnaphthoquinone (precursor of vitamin K_3) with an aqueous solution of NaOCl at $-10\,°C$ in chlorobenzene [18]. Among these new catalysts, the phase-transfer catalyst **13** bearing an extra chiral moiety at the quinuclidine nitrogen atom provided an enantioselectivity of 79% ee with good yield (86%). However, it was found that the best results were achieved with the readily

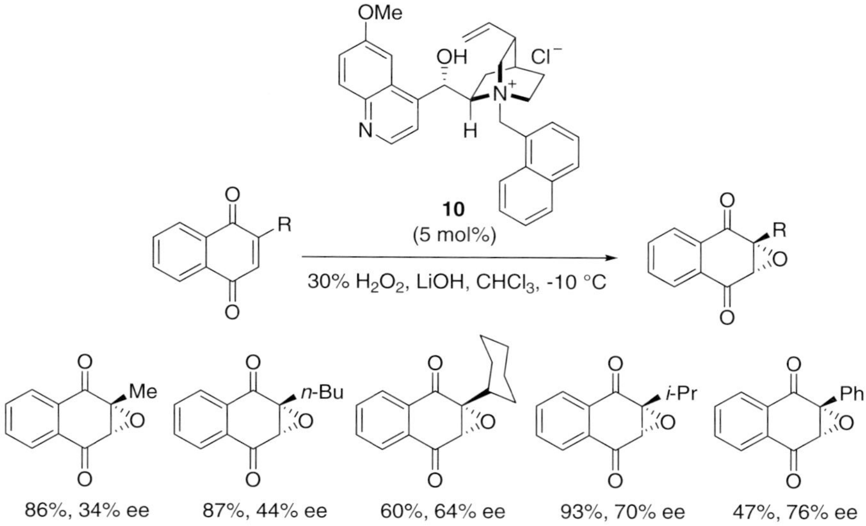

Scheme 5.11

86%, 34% ee | 87%, 44% ee | 60%, 64% ee | 93%, 70% ee | 47%, 76% ee

available ammonium salt **14** bearing a hydroxyl group at the C6′ atom of the quinoline system, which afforded a good yield (73%) and higher enantioselectivity (85% ee) for this biologically active compound (Scheme 5.13).

5.2.1.3 Synthetic Applications of the Asymmetric Epoxidation of Enones Using Chiral PTCs

In 1998, Wynberg's epoxidation procedure using chiral phase-transfer catalysts was applied by Taylor and coworkers to the total synthesis of (+)-manumycin A [19], (−)-alisamycin [20], and (+)-MT 35214 [20] (Scheme 5.14). The epoxidation

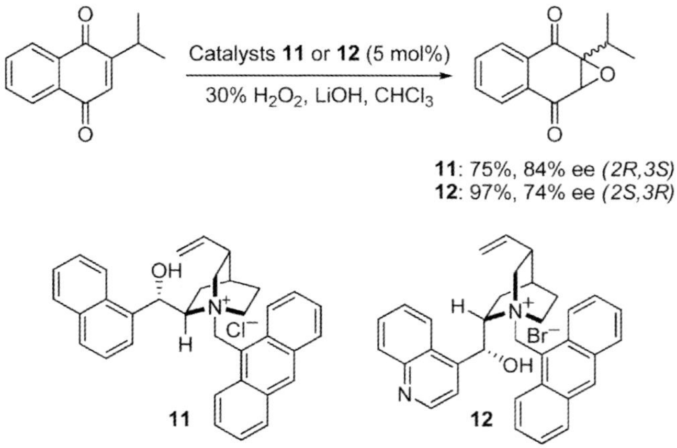

11: 75%, 84% ee (2R,3S)
12: 97%, 74% ee (2S,3R)

Scheme 5.12

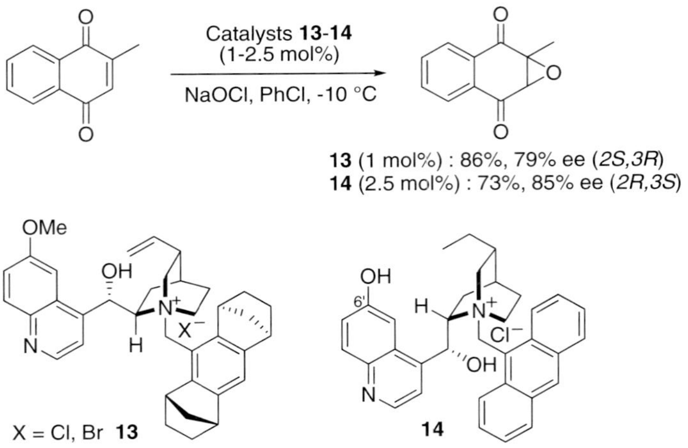
Catalysts 13-14
(1-2.5 mol%)
NaOCl, PhCl, -10 °C
13 (1 mol%) : 86%, 79% ee (2S,3R)
14 (2.5 mol%) : 73%, 85% ee (2R,3S)
OMe
OH
N
H
X⁻
6'
Cl⁻
X = Cl, Br 13
14

Scheme 5.13

of the quinone acetal **15** with *tert*-butyl hydroperoxide using *N*-benzylcinchonidinium chloride **16** as the catalyst gave the epoxide (−)-**17** in 32% yield and with 89% ee. Two recrystallizations of the reaction product from dichloromethane-hexane gave the enantiomerically pure epoxide (−)-**17** (>99.5% ee). However, surprisingly, the use of the pseudoenantiomeric *N*-benzylcinchonium chloride **1** gave (+)-**17** with only 10% ee.

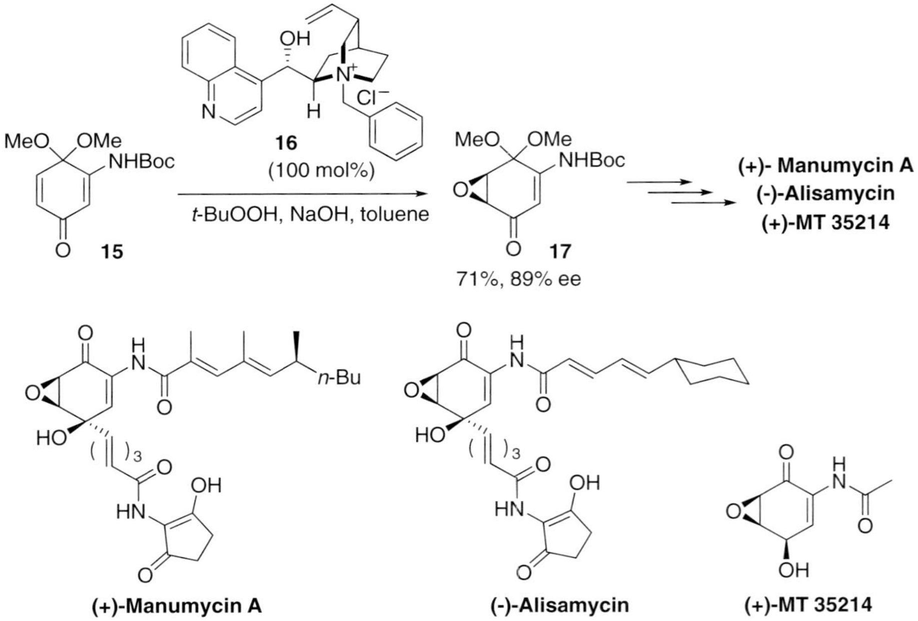
16
(100 mol%)
t-BuOOH, NaOH, toluene
MeO OMe
NHBoc
15
17
71%, 89% ee
(+)- Manumycin A
(-)-Alisamycin
(+)-MT 35214
n-Bu
HO
OH
HN
(+)-Manumycin A
(-)-Alisamycin
(+)-MT 35214

Scheme 5.14

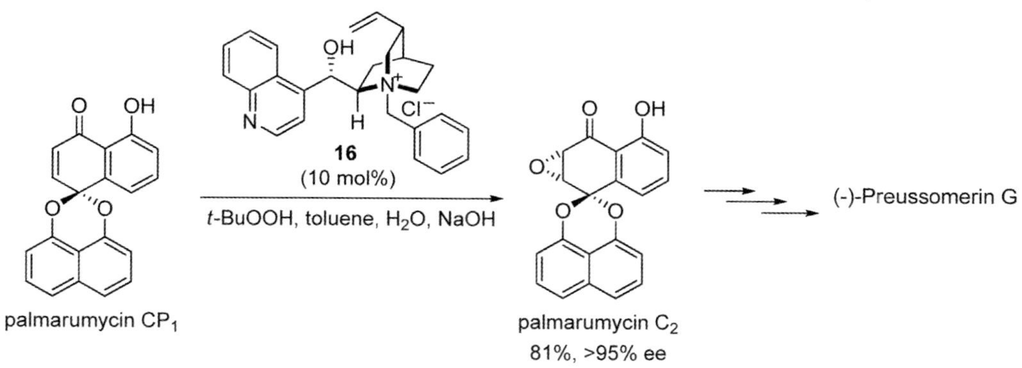

Scheme 5.15

In particular, the catalyst **16** has also been used by Barrett *et al.* for the epoxidation of the naphthoquinone palmarumycin CP$_1$ [21]. The resulting palmarumycin C$_2$ as a key intermediate in the synthesis of (−)-preussomerin G was obtained in 81% yield with >95% ee (Scheme 5.15).

Adam *et al.* also successfully applied Wynberg's condition to the asymmetric epoxidation of isoflavones [22]. Using the chiral PTC **18** and cumyl hydroperoxide (CHP) as an oxidant, the isoflavone epoxide was obtained almost quantitatively and with excellent enantioselectivity (up to 98% ee) even at a low catalyst loading (1 mol%) (Scheme 5.16).

5.2.1.4 Epoxidation of α,β-Unsaturated Sulfones

The asymmetric phase-transfer epoxidation of (*E*)-α,β-unsaturated sulfones has recently been achieved by Dorow and coworker using *N*-anthracenylmethyl cinchona alkaloid derivatives as catalysts and KOCl as an oxidant at low temperature [23]. The screening of several etheral functional groups at the C9(O) position of the catalyst moiety indicated that the steric size and the electronic factor of the ether substituent has a significant effect on both the reaction conversion and the enantioselectivity.

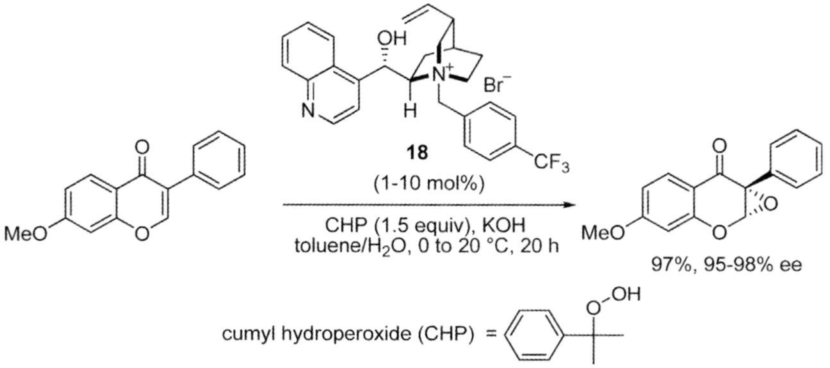

Scheme 5.16

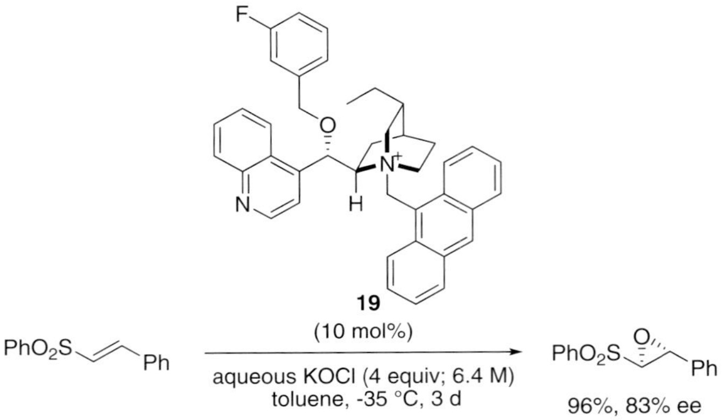

Scheme 5.17

PhO$_2$S⟶Ph + aqueous KOCl (4 equiv; 6.4 M), toluene, -35 °C, 3 d, catalyst **19** (10 mol%) → PhO$_2$S-(epoxide)-Ph, 96%, 83% ee

Catalyst **19**, which contains the 3-fluorophenyl methyl ether, was found to be the most effective for the epoxidation of (E)-α,β-unsaturated sulfones (Scheme 5.17). Of note, the (Z)-configured sulfone was converted under the same optimized conditions into the corresponding cis-α,β-epoxysulfone, but with a lower enantioselectivity (16% ee).

5.2.2
Organocatalytic Asymmetric Epoxidation of Enones via Iminium Catalysis

As described in Section 5.2.1.2, the asymmetric epoxidations of cyclic α,β-enones using cinchona-based chiral PTCs gave generally poor or moderate enantioselectivities with most cyclic enones. Thus, a method of achieving the highly enantioselective organocatalytic asymmetric epoxidation of cyclic enones has long been sought after. A breakthrough was made quite recently by the List group [24]. In the presence of the TFA salt of 9-amino-9-(deoxy)-*epi*-quinine (**20**), the epoxidation of β-substituted cyclopentenones, cyclohexenones, and cycloheptenones afforded the corresponding epoxides in good yields and excellent ee values (92% to >99%) (Scheme 5.18). The corresponding opposite enantiomers could also be obtained with the same degree of ee using a quinidine-derived catalyst. Based on the observations that the α-substituted enones were unreactive under the same reaction conditions and the corresponding nonprotonated amines, as well as quinine or quinidine itself, are much less active and enantioselective, the reaction mechanism via the iminium ion was proposed by the authors. The second basic tertiary amine site of the catalyst may provide the necessary chiral environment by directing the attack of hydrogen peroxide toward one enantioface of the double bond (Figure 5.2).

Soon afterward, Deng et al. adapted this primary iminium catalysis strategy for the epoxidation of acyclic enones [25]. Interestingly, the treatment of the various acyclic aliphatic enones **21** with the hydroperoxides **22** in the presence of the catalyst **20** provided either the epoxides **24** or the corresponding peroxides **23**

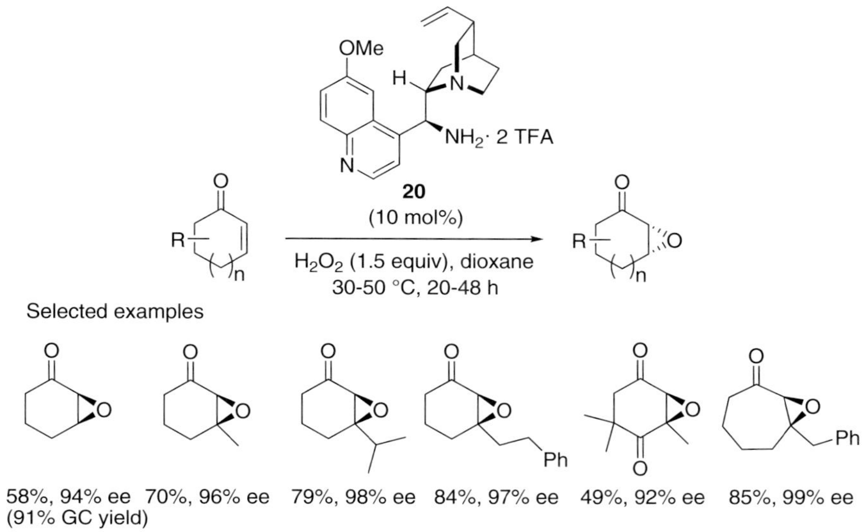

5.2 Cinchona-Based Organocatalysts in Asymmetric Oxidations

Scheme 5.18

Selected examples: 58%, 94% ee (91% GC yield); 70%, 96% ee; 79%, 98% ee; 84%, 97% ee; 49%, 92% ee; 85%, 99% ee

depending upon the reaction temperature (Scheme 5.19). At elevated temperature (55 °C) or with a prolonged reaction time at 23 °C, the reactions proceeded smoothly, affording the corresponding epoxides **24** as the predominant products in moderate to good yields with excellent enantioselectivities (96–97% ee, entries 7–12 of Table 5.1). On the contrary, at low temperature (0 °C) or with a short reaction time (4 h) at 23 °C, the peroxides **23** were obtained as the major products in high enantioselectivities (entries 1–6 of Table 5.1). The authors proposed that a cinchona-based catalyst such as **20** can not only increase the nucleophilicity of the peroxide toward the eniminium ion intermediate but also strongly influence the inclination of the resulting peroxyenamine intermediate **25** toward epoxidation or peroxidation.

Quite recently, List and coworkers observed that when hydrogen peroxide is used as an oxidant, the same reaction furnishes stable and isolable cyclic peroxyhemiketals [26]. When using the primary amine catalyst **20**-trichloroacetic acid (10 mol%), the oxidation of both linear and branched acyclic enones with aqueous hydrogen peroxide (3 equiv) at 30 °C in dioxane gave the peroxyhemiketals **26** in reasonable yields with high enantioselectivities (94–95% ee) (Scheme 5.20). Cyclic peroxyhemiketals were further transformed via a simple protocol (treatment with 1 N NaOH)

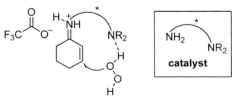

Figure 5.2 Pretransition state assembly.

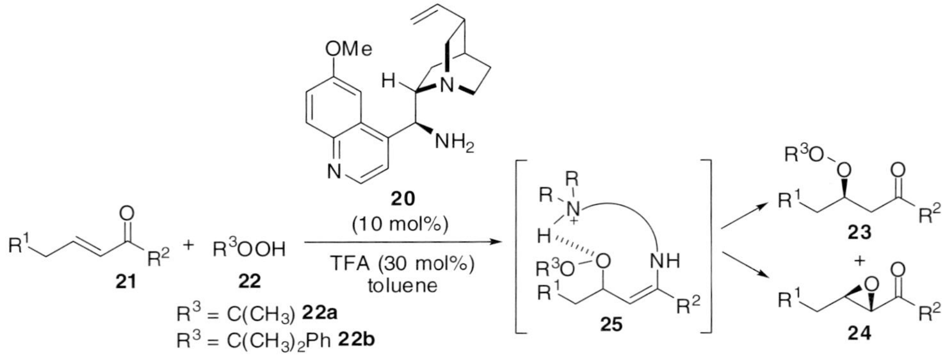

Scheme 5.19

to afford the corresponding epoxides without any loss of enantiopurity. Interestingly, regardless of the olefin configuration (E or Z) of the enone, the epoxide with the same configuration was obtained in high enantioselectivity (Scheme 5.21). These results indicate that, as would be expected, this epoxidation proceeds via a stepwise mechanism (i.e., the nucleophilic addition of hydroperoxide to give β-peroxyenamine followed by epoxide ring closure), affording predominantly the thermodynamically favorable *trans*-product **27**. Moreover, the reduction of cyclic peroxyhemiketals with P(OEt)$_3$ can provide the corresponding aldol products, namely, the β-hydroxy ketones **28** (Scheme 5.22).

5.2.3
Aziridination of Enones Using Cinchona-Based Chiral Phase-Transfer Catalyst

Aziridines are useful precursors for the preparation of biologically active species such as amino acids, β-lactams, and alkaloids [27]. Numerous stereoselective variants

Table 5.1 Asymmetric epoxidation vs. peroxidation of α,β-unsaturated ketones.

Entry	R^1	R^2	Peroxide	Temperature (°C)	Time (h)	23 : 24	Yield (%)	ee (%) of major product
1	Bn	Me	22a	23	4	92 : 8	85	91
2	n-Bu	Me	22a	23	4	93 : 7	86	93
3	H	n-Bu	22a	23	4	94 : 6	89	87
4	Bn	Me	22b	0	16	86 : 14	64	95
5	n-Bu	Me	22b	0	16	87 : 13	77	95
6	H	n-Bu	22b	0	24	85 : 15	66	96
7	Bn	Me	22b	23	72	1 : 99	88	97
8	Et	Me	22b	23	72	1 : 99	91	97
9	n-Bu	Me	22b	23	72	1 : 99	91	97
10	H	Et	22b	55	24	32 : 68	55	97
11	n-Hex	Et	22b	55	24	33 : 67	54	96
12	H	n-Bu	22b	55	24	13 : 87	71	97

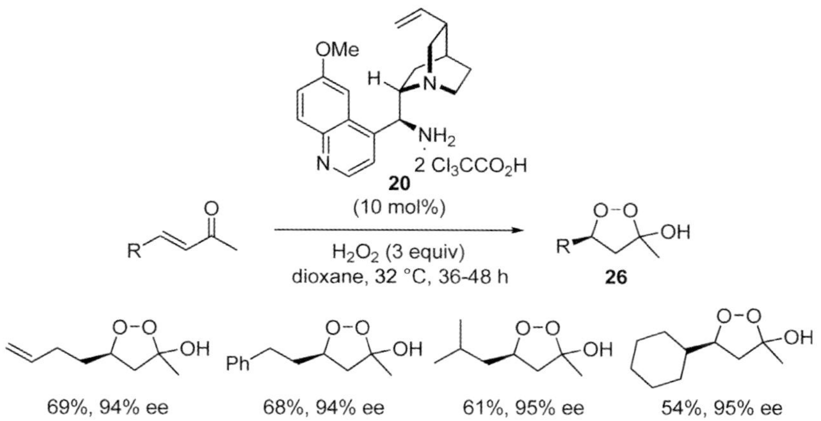

Scheme 5.20

26: R—[dioxolane with O-O]—OH

- 69%, 94% ee (allyl-CH₂)
- 68%, 94% ee (Ph-CH₂)
- 61%, 95% ee (iPr)
- 54%, 95% ee (cyclohexyl)

Conditions: catalyst **20** (10 mol%), H₂O₂ (3 equiv), dioxane, 32 °C, 36–48 h; 2 Cl₃CCO₂H.

have been achieved by means of different catalysts, but commonly they involve transition-metal complexes [28]. However, a nontoxic organocatalytic system would be needed for industrial-scale applications from an environmental point of view. In 1996, Prabhakar and coworkers demonstrated for the first time that in the presence of the cinchona-based PTC **29**, the asymmetric aziridination of acrylates can be achieved with the aryl hydroxamic acids **30** under phase-transfer conditions [29a]. However, only low to moderate ee values (0–61% ee) were obtained (Scheme 5.23). Their further optimization studies failed to improve the enantioselectivity [29b].

Significantly increased ee values were obtained by using the new chiral PTCs **31** and **32**, which can be easily prepared starting from 2-hydroxy-3-chloromethyl-5-methyl benzaldehyde and cinchonine or cinchonidine, respectively [30]. By using **31** or **32**, up to 95% ee was achieved in the reaction of electron-deficient olefins with N-acyl-N-arylhydroxylamines as nitrogen transfer reagents under biphasic conditions (toluene/aqueous NaOH) at room temperature (Scheme 5.24).

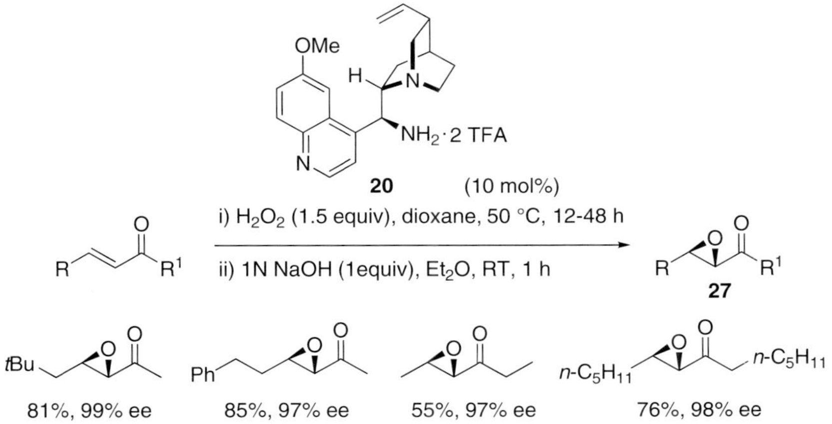

Scheme 5.21

Catalyst **20**·2 TFA (10 mol%)
i) H₂O₂ (1.5 equiv), dioxane, 50 °C, 12–48 h
ii) 1N NaOH (1 equiv), Et₂O, RT, 1 h

Product **27**: epoxy ketones R—[epoxide]—C(O)—R¹

- 81%, 99% ee (tBu, R¹ group)
- 85%, 97% ee (Ph)
- 55%, 97% ee
- 76%, 98% ee (n-C₅H₁₁, n-C₅H₁₁)

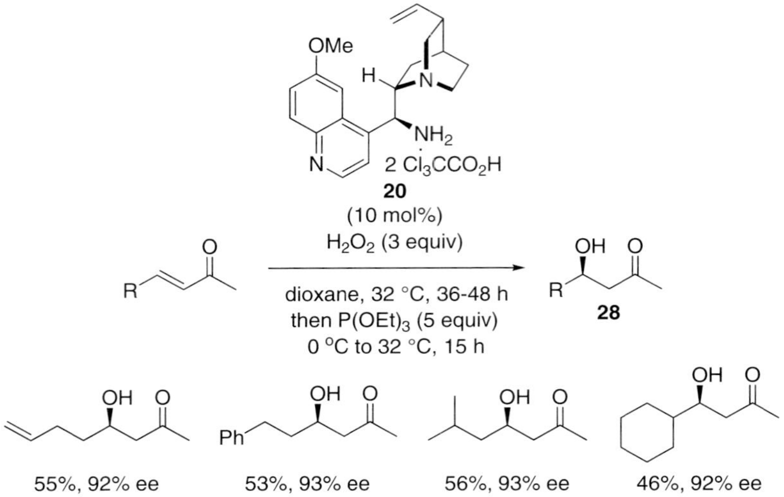

Scheme 5.22

Conditions: catalyst **20** (10 mol%), H$_2$O$_2$ (3 equiv), dioxane, 32 °C, 36–48 h; then P(OEt)$_3$ (5 equiv), 0 °C to 32 °C, 15 h.

Catalyst **20**: quinine-derived amine with OMe, NH$_2$, and 2 Cl$_3$CCO$_2$H.

Substrate: R–CH=CH–C(O)–R' → product **28** (R–CH(OH)–CH$_2$–C(O)–R')

Products:
- 55%, 92% ee
- Ph-substituted: 53%, 93% ee
- 56%, 93% ee
- cyclohexyl: 46%, 92% ee

The author proposed that this observation might arise from the fact that the incorporation of an R$_4$N$^+$ moiety in proximity to the N-acyloxy anion should facilitate their electrostatic attraction via ion pair formation. The π-bond of the olefins also has an electrostatic interaction with the R$_4$N$^+$ moiety of the catalyst (Figure 5.3).

In 2004, Fioravanti, Pellacani, and Tardella reported the asymmetric aza-Michael-initiated ring closure additions of ethyl nosyloxycarbamate to 2-(phenylsulfanyl)-2-cycloalkenones using the cinchona-based PTCs **11** or **29**, affording the corresponding aziridines with moderate ee values (Scheme 5.25) [31]. Interestingly, this reaction afforded the same enantiomer, regardless of which pseudoenantiomer of the cinchona-derived catalyst was employed. However, the absolute configuration of the products was not determined in this study.

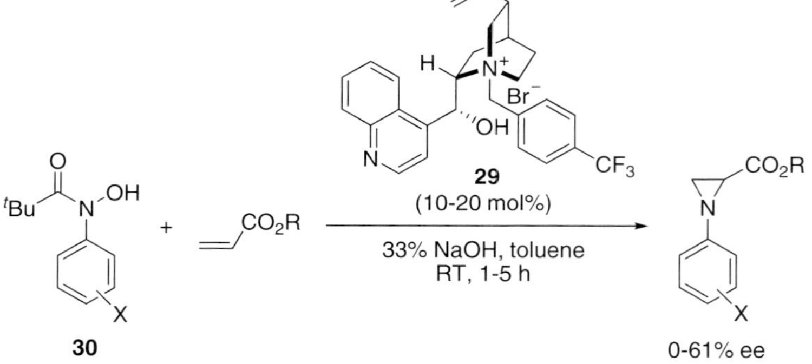

Scheme 5.23

^{t}Bu–C(O)–N(OH)–Ar(X) (**30**) + CH$_2$=CH–CO$_2$R → aziridine with CO$_2$R

Conditions: catalyst **29** (10–20 mol%), 33% NaOH, toluene, RT, 1–5 h.

Catalyst **29**: cinchona-derived ammonium bromide with OH and CF$_3$-aryl group.

0–61% ee

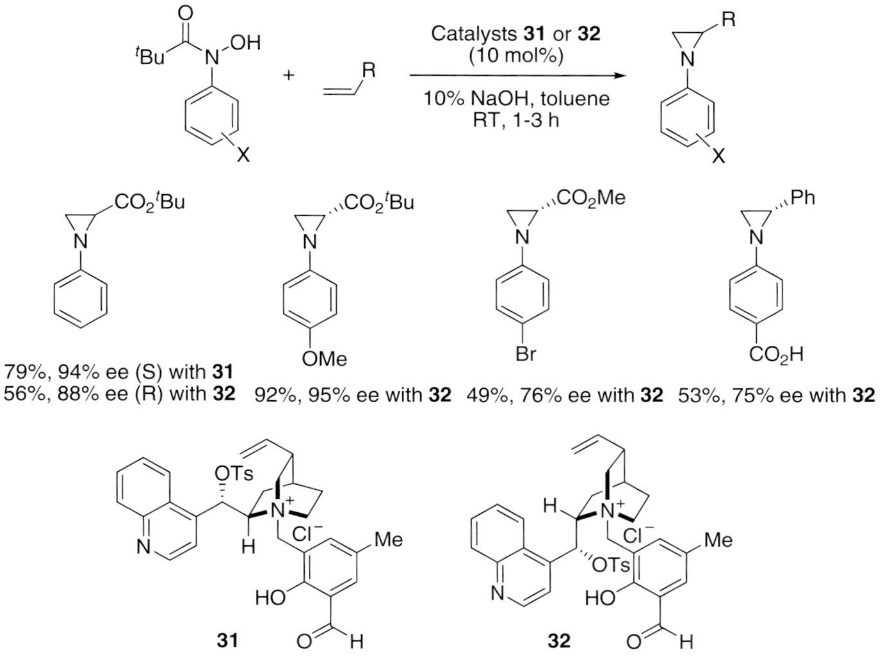

5.2 Cinchona-Based Organocatalysts in Asymmetric Oxidations

Scheme 5.24

79%, 94% ee (S) with **31**
56%, 88% ee (R) with **32**   92%, 95% ee with **32**   49%, 76% ee with **32**   53%, 75% ee with **32**

Minakata and coworkers also reported the asymmetric aziridination reactions of electron-deficient olefins using cinchona-based PTCs **33–37** and N-chloro-N-sodio carbamate as an oxidant [32]. Moderate values of the ee were obtained. Comparative experiments revealed that the electron-deficient olefin bearing both the dimethylpyrazole and the di-isopropylpyrazole groups turned out to be a better Michael acceptor than that bearing the oxazolidinone substituent. The modification of the R-substituent of the cinchona-derived anthracenylmethylated ammonium salts **35–37** did not have any effect on the enantioselectivity (Scheme 5.26).

As discussed above, the use of phase-transfer conditions has not usually provided satisfactory results for asymmetric aziridination reactions. Recently, Melchiorre and coworkers developed a highly versatile and efficient catalytic methodology for

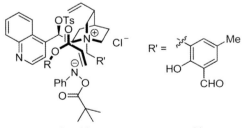

Figure 5.3 Plausible transition state assembly.

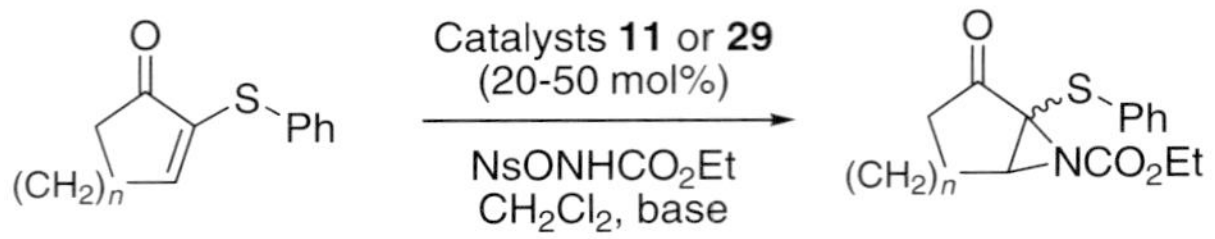

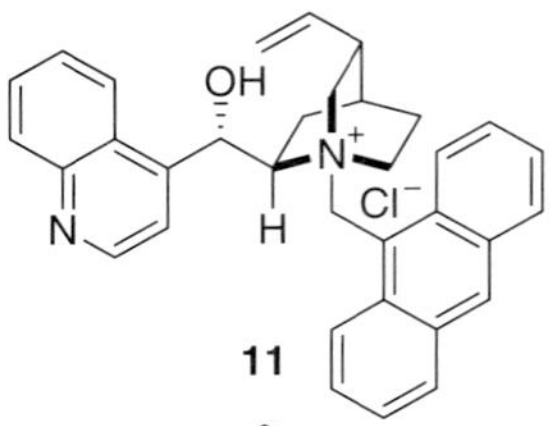

124 | 5 Cinchona-Based Organocatalysts for Asymmetric Oxidations and Reductions

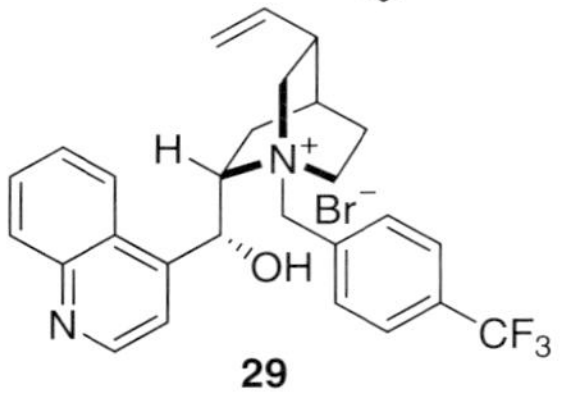

Scheme 5.25

		Cat 11		Cat 29	
base		yield (%)	ee (%)	yield (%)	ee (%)
NaHCO₃ (aq)		65	35	25	48
CaO (s)		98	17	18	25
NaHCO₃ (aq)		77	21	25	71
CaO (s)		96	61	93	75

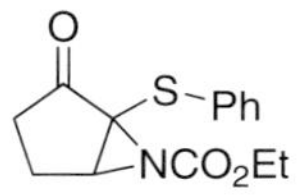

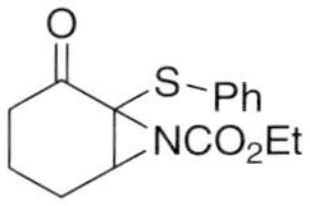

stereoselective aziridination by adopting the iminium–enamine catalysis approach. By using the cinchona-derived catalyst salt **39** generated by combining 9-amino-9-(deoxy)-*epi*-hydroquinine with D-*N*-Boc-phenylglycine, a highly remarkable degree

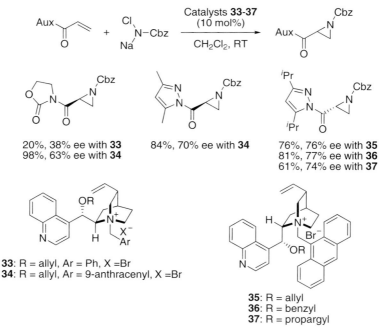

20%, 38% ee with **33**
98%, 63% ee with **34**

84%, 70% ee with **34**

76%, 76% ee with **35**
81%, 77% ee with **36**
61%, 74% ee with **37**

33: R = allyl, Ar = Ph, X = Br
34: R = allyl, Ar = 9-anthracenyl, X = Br

35: R = allyl
36: R = benzyl
37: R = propargyl

Scheme 5.26

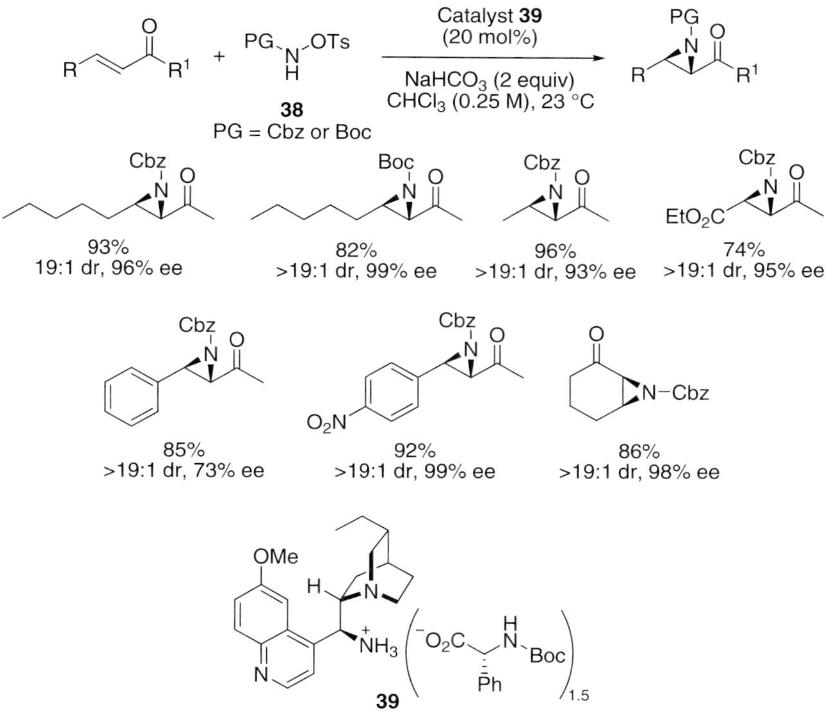

Scheme 5.27

of stereoselectivity was achieved in the aziridination reaction of various α,β-unsaturated ketones with the commercially available N-protected hydroxyl amines **38** [33]. This reaction proceeds via the conjugate addition of the O-tosylated hydroxycarbamate **38** to the iminium intermediate, which can be generated from the α,β-enone and chiral primary amine catalyst salt. Then, the resulting enamine intermediate is intramolecularly cyclized to form the corresponding N-Boc- or Cbz-protected aziridines in high yields (74–96%) and excellent stereoselectivities (>19 : 1 dr, up to 99% ee) (Scheme 5.27). Importantly, this organocatalyst system is also applicable to cyclic enones such as cyclohexenone, which are known to be difficult substrates, giving the corresponding aziridine in high yield (86%) with excellent stereoselectivity (>19 : 1 dr, 98% ee).

5.3 Cinchona-Based Organocatalysts in Asymmetric Reductions

As mentioned briefly in Chapter 2, very few publications describing cinchona alkaloid-based asymmetric reduction systems have appeared, despite the importance of this reaction, and they are restricted to the reduction of aromatic ketones.

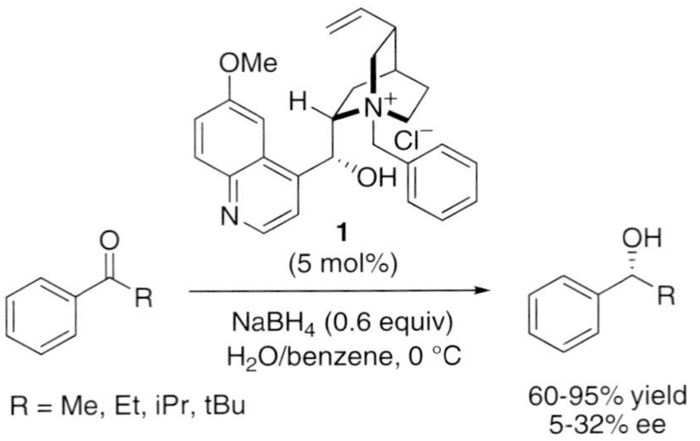

Scheme 5.28

Reagents: ketone (R = Me, Et, iPr, tBu) with catalyst **1** (5 mol%), NaBH$_4$ (0.6 equiv), H$_2$O/benzene, 0 °C, giving the alcohol in 60–95% yield, 5–32% ee.

In 1978, Colonna and coworkers first demonstrated that in the presence of cinchona-derived PTCs, alkyl aryl ketones can be reduced asymmetrically with the inexpensive and easily handled sodium borohydride under aqueous–organic biphasic conditions. However, the enantiomeric excess of the corresponding alcohols was very disappointing. The best ee obtained in their study was 32% when pivalophenone was reduced in the presence of 5 mol% of benzylquininium chloride **1** (Scheme 5.28) [34]. Variants of this procedure were tried later by several research groups, but in all cases the enantioselectivities were too low for synthetic applications [35].

Later, Lawrence and coworker obtained somewhat better results by employing silanes as hydride donors and cinchona-based ammonium fluorides as nucleophilic catalysts (Scheme 5.29) [36]. Among the N-alkylated quinine and quinidine fluorides screened in their study, the N-benzylquinidinium fluoride **40** was identified as the

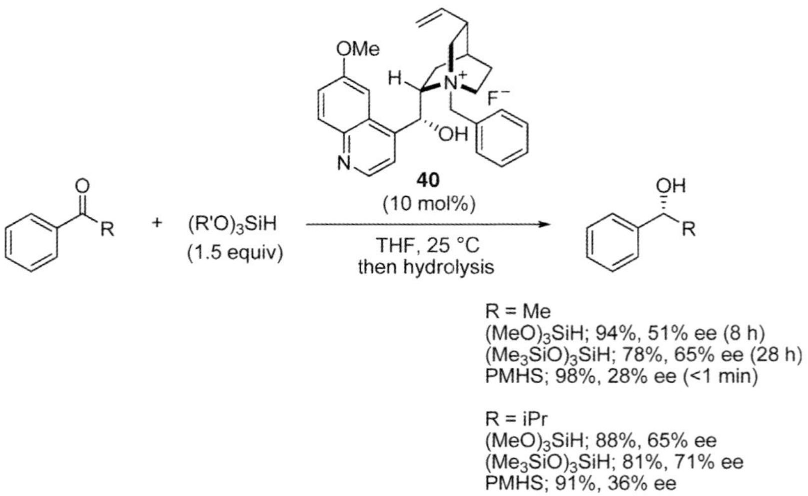

Scheme 5.29

Ketone + (R'O)$_3$SiH (1.5 equiv), catalyst **40** (10 mol%), THF, 25 °C, then hydrolysis → alcohol.

R = Me
(MeO)$_3$SiH; 94%, 51% ee (8 h)
(Me$_3$SiO)$_3$SiH; 78%, 65% ee (28 h)
PMHS; 98%, 28% ee (<1 min)

R = iPr
(MeO)$_3$SiH; 88%, 65% ee
(Me$_3$SiO)$_3$SiH; 81%, 71% ee
PMHS; 91%, 36% ee

most enantioselective catalyst. For example, acetophenone could be reduced with tris (trimethylsiloxy)silane in the presence of 10 mol% of **40**, affording the corresponding alcohol in 78% ee. The pseudoenantiomeric catalyst, N-benzylquininium fluoride, also afforded the desired product with the opposite configuration, albeit with slightly lower ee values. The enantioselectivity showed a slight dependence on the size of the silane. Tris(trimethylsiloxy)silane with its bulky substituents gave slightly better enantioselectivities than that of trimethoxysilane. However, a prolonged reaction time was required. Polymethylhydrosiloxane (PMHS), an inexpensive and readily available hydrosilane, was also studied. This reducing agent was shown to be very active (for example, acetophenone is reduced in less than 1 min) but, unfortunately, it gave too low ee values (28%).

5.4 Conclusions

This chapter described the cinchona-catalyzed asymmetric oxidation of enones and the reduction of prochiral ketones.

The optimal control of the stereochemical outcome of the epoxidation of enones with peroxide has been achieved by using cinchona-based organocatalysts. Cinchona-based PTCs have been shown to be highly efficient for the enantioselective epoxidation of acyclic enones. On the other hand, iminium catalysis using cinchona-based bifunctional primary amine catalysts has been proved to be suitable for the epoxidation of cyclic enone systems. Thus, these two complementary methodologies will be highly useful for synthetic chemists. In the case for aziridination, there are rare examples where good enantioselectivity has been achieved using chiral PTCs. Here too, a simple and highly versatile method for asymmetric aziridination has been achieved by adopting iminium catalysis.

In contrast to asymmetric oxidation chemistry, cinchona-catalyzed asymmetric reduction reactions have been explored very little, despite the importance of this reaction. Previous reports on this topic are restricted to the reduction of aromatic ketones, and the enantioselectivities achieved to date remain far from satisfactory when compared with metal catalysis. Moreover, Hantsch esters, another type of useful organic hydrides, have not yet been studied in combination with cinchona catalysts. However, as is well known, the structures of cinchona alkaloids are easily modifiable, thus permitting the easy tuning of the reaction course. The successful use of cinchona catalysts for this reaction will therefore likely be reported in the very near future.

Acknowledgments

This work was supported by grants KRF-2008-005-J00701 (MOEHRD), R11-2005-008-00000-0 (SRC program of MEST/KOSEF) and R31-2008-000-10029-0 (WCU program).

References

1. (a) Berkessel, A. and Gröger, H. (2005) *Asymmetric Organocatalysis*, Wiley-VCH Verlag GmbH, Weinheim; (b) Dalko, P.I. (ed.) (2005) *Enantioselective Organocatalysis*, Wiley-VCH Verlag GmbH, Weinheim; (c) List, B. (ed.) (2007) *Chem. Rev.*, **107**, 5413–5883 (special issue on organocatalysis); (d) Houk, K.N., and List, B. (eds) (2004) *Acc. Chem. Res.*, **37**, 487–631 (special issue on asymmetric organocatalysis).
2. (a) Helder, R., Hummelen, J.C., Laane, R.W.P.M., Wiering, J.S., and Wynberg, H. (1976) *Tetrahedron Lett.*, **17**, 1831–1834; (b) Wynberg, H. and Greijdanus, B. (1978) *J. Chem. Soc., Chem. Commun.*, 427–428.
3. Hummelen, J.C. and Wynberg, H. (1978) *Tetrahedron Lett.*, **19**, 1089–1092.
4. (a) Arai, S., Tsuge, H., and Shioiri, T. (1998) *Tetrahedron Lett.*, **39**, 7563–7566; (b) Arai, S., Tsuge, H., Oku, M., Miura, M., and Shioiri, T. (2002) *Tetrahedron*, **58**, 1623–1630.
5. (a) Lygo, B. and Wainwright, P.G. (1998) *Tetrahedron Lett.*, **39**, 1599–1602; (b) Lygo, B. and Wainwright, P.G. (1999) *Tetrahedron*, **55**, 6289–6300.
6. Corey, E.J. and Zhang, F.-Y. (1999) *Org. Lett.*, **1**, 1287–1290.
7. (a) Lygo, B. and To, D.C.M. (2001) *Tetrahedron Lett.*, **42**, 1343–1346; (b) Lygo, B., Gardiner, S.D., McLeod, M.C., and To, D.C.M. (2007) *Org. Biomol. Chem.*, **5**, 2283–2290.
8. Lygo, B. and To, D.C.M. (2002) *Chem. Commun.*, 2360–2361.
9. (a) Ye, J., Wang, Y., Liu, R., Zhang, G., Zhang, Q., Chen, J., and Liang, X. (2003) *Chem. Commun.*, 2714–2715; (b) Ye, J., Wang, Y., Chen, J., and Liang, X. (2004) *Adv. Synth. Catal.*, **346**, 691–696.
10. Wang, Y., Ye, J., and Liang, X. (2007) *Adv. Synth. Catal.*, **349**, 1033–1036.
11. Jew, S.-s., Lee, J.-H., Jeong, B.-S., Yoo, M.-S., Kim, M.-J., Lee, Y.-J., Lee, J., Choi, S.-h., Lee, K., Lah, M.S., and Park, H.-g. (2005) *Angew. Chem. Int. Ed.*, **44**, 1383–1385.
12. Lv, J., Wang, X., Liu, J., Zhang, L., and Wang, Y. (2006) *Tetrahedron: Asymmetry*, **17**, 330–335.
13. Helder, R., Hummelen, J.C., Laane, R.W.P.M., Wiering, J.S., and Wynberg, H. (1976) *Tetrahedron Lett.*, **17**, 1831–1834.
14. Wynberg, H. and Marsman, B. (1980) *J. Org. Chem.*, **45**, 158–161.
15. Baba, N., Oda, J., and Kawaguchi, M. (1986) *Agric. Biol. Chem.*, **50**, 3113–3117.
16. (a) Arai, S., Oku, M., Miura, M., and Shioiri, T. (1998) *Synlett*, 1201–1202; (b) Arai, S., Tsuge, H., Oku, M., Miura, M., and Shioiri, T. (2002) *Tetrahedron*, **58**, 1623–1630.
17. Dehmlow, E.V., Düttmann, S., Neumann, B., and Stammler, H.-G. (2002) *Eur. J. Org. Chem.*, 2087–2093.
18. Berkessel, A., Guixà, M., Schmidt, F., Neudörfl, J.M., and Lex, J. (2007) *Chem. Eur. J.*, **13**, 4483–4498.
19. Alcaraz, L., Macdonald, G., Ragot, J.P., Lewis, N., and Taylor, R.J.K. (1998) *J. Org. Chem.*, **63**, 3526–3527.
20. Macdonald, G., Alcaraz, L., Lewis, N.J., and Taylor, R.J.K. (1998) *Tetrahedron Lett.*, **39**, 5433–5436.
21. Barrett, A.G.M., Blaney, F., Campbell, A.D., Hamprecht, D., Meyer, T., White, A.J.P., Witty, D., and Williams, D.J. (2002) *J. Org. Chem.*, **67**, 2735–2750.
22. (a) Adam, W., Rao, P.B., Degen, H.-G., and Saha-Möller, C.R. (2001) *Tetrahedron: Asymmetry*, **12**, 121–125; (b) Adam, W., Rao, P.B., Degen, H.-G., Levai, A., Patonay, T., and Saha-Möller, C.R. (2002) *J. Org. Chem.*, **67**, 259–264.
23. Dorow, R.L. and Tymonko, S.A. (2006) *Tetrahedron Lett.*, **47**, 2493–2495.
24. Wang, X., Reisinger, C.M., and List, B. (2008) *J. Am. Chem. Soc.*, **130**, 6070–6071.
25. Lu, X., Liu, Y., Sun, B., Cindric, B., and Deng, L. (2008) *J. Am. Chem. Soc.*, **130**, 8134–8135.

26 Reisinger, C.M., Wang, X., and List, B. (2008) *Angew. Chem. Int. Ed.*, **47**, 8112–8115.

27 (a) Tanner, D. (1994) *Angew. Chem. Int. Ed.*, **33**, 599–619; (b) Jacobsen, E.N. (1999) *Comprehensive Asymmetric Catalysis*, vol. 2 (eds E.N. Jacobsen, A. Pfaltz, and H. Yamamoto), Springer, Berlin, pp. 607–618; (c) Yudin, A.K. (ed.) (2006) *Aziridines and Epoxides in Organic Synthesis*, Wiley-VCH Verlag GmbH, Weinheim; (d) Sweeney, J.B. (2002) *Chem. Soc. Rev.*, **31**, 247–258.

28 For reviews, see (a) Halfen, J.A., (2005) *Curr. Org. Chem.*, **9**, 657–669; (b) Müller, P. and Fruit, C. (2003) *Chem. Rev.*, **103**, 2905–2920.

29 (a) Aires-de-Sousa, J., Lobo, A.M., and Prabhakar, S. (1996) *Tetrahedron Lett.*, **37**, 3183–3186; (b) Aires-de-Sousa, J., Prabhakar, S., Lobo, A.M., Rosa, A.M., Gomes, M.J.S., Corvo, M.C., Williams, D.J., and White, A.J.P. (2001) *Tetrahedron: Asymmetry*, **12**, 3349–3365.

30 Murugan, E. and Siva, A. (2005) *Synthesis*, 2022–2028.

31 Fioravanti, S., Mascia, M.G., Pellacani, L., and Tardella, P.A. (2004) *Tetrahedron*, **60**, 8073–8077.

32 Minakata, S., Murakami, Y., Tsuruoka, R., Kitanaka, S., and Komatsu, M. (2008) *Chem. Commun.*, 6363–6365.

33 Pesciaioli, F., De Vincentiis, F., Galzerano, P., Bencivenni, G., Bartoli, G., Mazzanti, A., and Melchiorre, P. (2008) *Angew. Chem. Int. Ed.*, **47**, 8703–8706.

34 (a) Balcells, J., Colonna, S., and Fornasier, R. (1976) *Synthesis*, 266–267; (b) Colonna, S. and Fornasier, R. (1978) *J. Chem. Soc., Perkin Trans. I*, 371–373.

35 (a) Juliá, S., Ginebreda, A., Guixer, J., Masana, J., Tomás, A., and Colonna, S. (1981) *J. Chem. Soc., Perkin Trans. I*, 574–577; (b) Dehmlow, E.V., Wagner, S., and Müller, A. (1999) *Tetrahedron*, **55**, 6335–6346; (c) Hofstetter, C., Wilkinson, P.S., and Pochapsky, T.C. (1999) *J. Org. Chem.*, **64**, 8794–8800.

36 Drew, M.D. and Lawrence, N.J. (1997) *Tetrahedron Lett.*, **38**, 5857–5860.

6
Cinchona-Catalyzed Nucleophilic α-Substitution of Carbonyl Derivatives

Hyeung-geun Park and Byeong-Seon Jeong

6.1
Introduction

A stereoselective introduction of a variety of functional groups to active methylene or methine position of diverse carbonyl derivatives is one of the essential transformations in organic and medicinal chemistry. As a matter of course, tremendous efforts have been made on this issue and highly efficient methods have been developed in the past several decades. Stereoselective reaction systems for nucleophilic α-substitutions of carbonyl derivatives have mainly been focused on the development of chiral substrates, chiral reagents, or chiral auxiliaries, all of these are usually needed more than stoichiometric amount, until the advent of chiral catalysts such as transition-metal catalysts or small organic molecule catalysts. Succeeding to the great achievements in the transition-metal catalyzed organic transformation, especially in the area of the direct asymmetric α-substitution of carbonyl derivatives, asymmetric catalytic organic reactions with chiral small organic molecule catalysts, called "asymmetric organocatalysis," have rushed like a flood since the late twentieth century [1].

6.2
Organocatalytic Nucleophilic α-Substitution of Carbonyl Derivatives

Two representative organocatalytic reaction systems can be considered for nucleophilic α-substitution of carbonyl compounds, the issue of this chapter. One involves the *in situ* formation of a chiral enamine through covalent bond between organocatalyst (mainly a chiral secondary amine such as proline) and substrate (mainly an aldehyde), followed by asymmetric formation of new bond between the α-carbon of carbonyl compound and electrophile. Detachment of organocatalyst provides optically active α-substituted carbonyl compound, and the free organocatalyst then participates in another catalytic cycle (Figure 6.1a) [2].

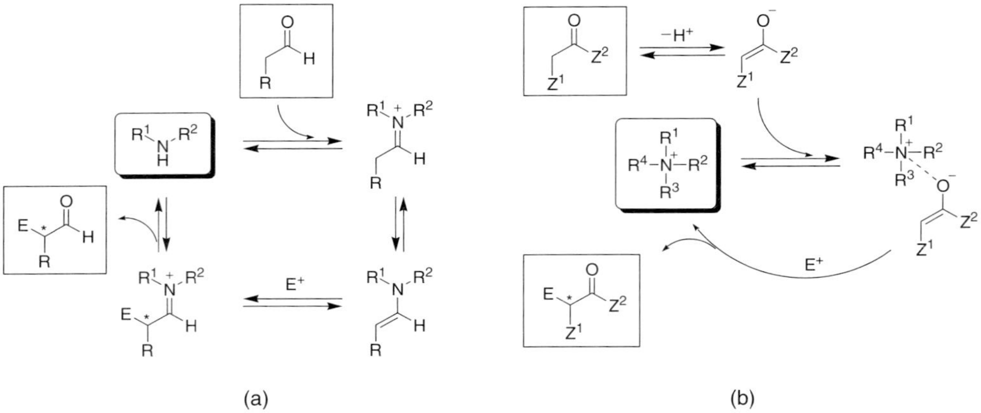

6 Cinchona-Catalyzed Nucleophilic α-Substitution of Carbonyl Derivatives

Figure 6.1 Two representative pathways for organocatalytic α-substitution of carbonyls.

The other system realizes a stereoselective introduction of electrophile to α-position of carbonyl compound by way of noncovalent interaction such as hydrogen bonding [3] or ion-pair formation [4] between the organocatalyst and the substrate. Cinchona alkaloids have shown their great talent for asymmetric α-substitution of carbonyl derivatives, which leads to the construction of stereogenic carbon centers. Cinchona alkaloids usually take part in the transformation of type (b) shown in Figure 6.1b as their quaternary ammonium salt forms. Chiral ionic pair can lead electrophile to approach exclusively toward one of the two faces of the enolate, providing optically enriched α-substituted carbonyl derivatives (Figure 6.1b). These reactions frequently proceed via phase-transfer catalysis – a reaction between chemical species located in different phases facilitated by a small quantity of a phase-transfer catalyst (PTC).

Phase-transfer catalytic reaction has been recognized as one of the most useful methods for practical synthesis because of its operational simplicity, high yielding process, mild reaction conditions, use of safe, inexpensive, and environmental friendly reagents and solvents, and ease in scaling up, which enables this method to be applied to the industrial process [4, 5]. In addition to these basic advantages, a new stereogenic center can be generated in stereoselective manner in phase-transfer catalytic reaction by employing a chiral PTC [6]. In particular, a cation type of chiral PTC called chiral onium PTC plays a critical role in creating stereoselectivity in the alkylation of enolate, the issue of this chapter. Together with chiral phosphonium salts, chiral quaternary ammonium salts occupy the central position in chiral onium PTC category.

A representative mechanistic illustration for the asymmetric substitution of active methylene (or methine) compounds including carbonyl derivatives under catalytic phase-transfer conditions is presented in Figure 6.2 [5f, 7]. This process generally proceeds via the following three main steps: In first step, the substrate (**A**) having active methylene (or methine) group reacts with inorganic base, mostly metal hydroxides, at the interface between organic phase and aqueous (or solid) phase,

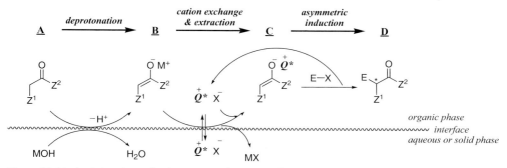

Figure 6.2 Mechanistic scheme for the α-monosubstitution of carbonyl compounds under phase-transfer conditions.

providing the corresponding metal enolate (**B**), this polar metal enolate (**B**) is still situated near the interface until cation exchange (the second step) occurs between metal cation and lipophilic phase-transfer onium cation, and the resulting lipophilic and chiral enolate–onium ion pair (**C**) is able to move into the organic phase at which it reacts with electrophile (the third step). Owing to the chirality of the onium moiety, substitution reaction can proceed in a stereoselective manner to give the chiral α-substituted product (**D**). The level of optical purity of product (**D**) is governed by the following typical factors: (i) Direct reaction of the metal enolate (**B**) with electrophile leads to racemic product, (ii) substitution of wrong onium–enolate ion pair produces the enantiomer of the desired product, which causes a decrease in stereoselectivity, (iii) deprotonation of the chiral product by inorganic base followed by reprotonation leads to racemization, and (iv) control of the geometry of the enolate (*E* versus *Z*) is obviously important for high level of stereoinduction. Besides these direct factors affecting stereoselectivity, there are some issues that are related to chemical yield such as disubstitution or decomposition of functional groups vulnerable in both substrate and product under basic reaction conditions, and so on. All of these factors should be considered to design highly stereoselective organocatalytic α-substitution reaction system, especially substrate and chiral catalyst.

6.3
Cinchona Alkaloids in Asymmetric Organocatalysis

Cinchona alkaloids have unique structural nature (Figure 6.3). They have a quite sterically hindered tertiary amine that can be derivatized to provide a variety of quaternary ammonium salts. In addition to the bridgehead tertiary amine, cinchona alkaloids have versatile functional groups such as the 9-hydroxy group, the 6'-methoxy group in quinoline, and the 10,11-vinyl group in the quinuclidine, all of which sometimes play critical roles in chirality-creating steps, either themselves or in chemically modified forms. Owing to these useful functional groups as well as their characteristic structural features, the parent natural cinchona alkaloids have

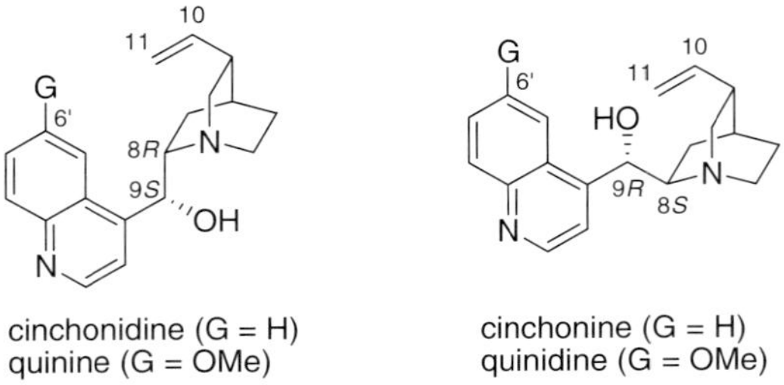

6 Cinchona-Catalyzed Nucleophilic α-Substitution of Carbonyl Derivatives

cinchonidine (G = H)
quinine (G = OMe)

cinchonine (G = H)
quinidine (G = OMe)

Figure 6.3 Representative cinchona alkaloids.

frequently been served as valuable sources in the field of catalytic asymmetric organic synthesis, especially as chiral PTCs. Moreover, the high accessibility in both pseudoenantiomeric forms at low cost is an additional major attraction for their utilization.

As mentioned above, quaternary ammonium salts derived from cinchona alkaloids have occupied the central position as efficient PTCs in various organic transformations, especially in the asymmetric α-substitution reaction of carbonyl derivatives. A cinchona alkaloidal quaternary ammonium salt, which acts as a PTC in various organic reactions, is prepared by a simple and easy chemical transformation of the bridgehead tertiary nitrogen with a variety of active halides, mainly arylmethyl halides. Other moieties of cinchona alkaloids (the 9-hydroxy, the 6′-methoxy, or the 10,11-vinyl) are occasionally modified for the enhancement of both chemical and optical yields (Figure 6.4).

6.4
The Pioneer Works for Phase-Transfer Catalytic α-Substitution

In 1984, the first successful monumental use of cinchona PTC for asymmetric α-substitution of carbonyls was reported by Dolling and coworkers of the Merck research group (Scheme 6.1) [8]. In this work, cinchoninium salt (**1**) was employed in the catalytic asymmetric methylation of 6,7-dichloro-5-methoxy-2-phenyl-1-indanone (**2**) under phase-transfer conditions. The methylated product **3**, which was finally transformed to (+)-indacrinone through three further steps, was obtained in 95% conversion with 92% enantiomeric excess (ee). Through the systematic investigation, the group reported the relationship between the chemical/optical yield and the reaction variables (e.g., amount or concentration of each chemical species, halide of

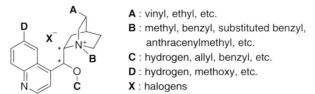

A : vinyl, ethyl, etc.
B : methyl, benzyl, substituted benzyl, anthracenylmethyl, etc.
C : hydrogen, allyl, benzyl, etc.
D : hydrogen, methoxy, etc.
X : halogens

Figure 6.4 Quaternary ammonium salts derived from cinchona alkaloids.

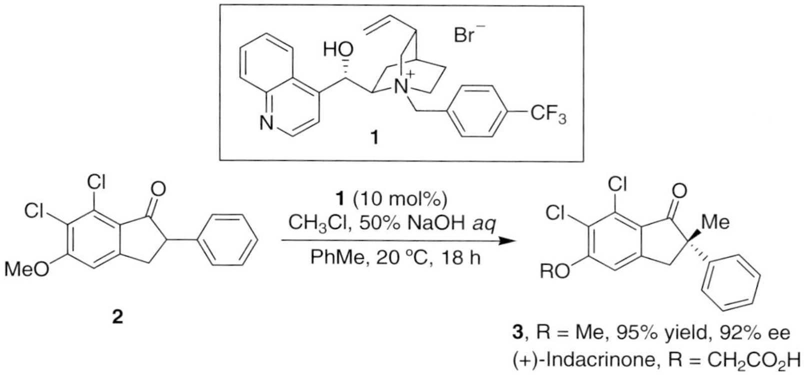

Scheme 6.1

[Scheme showing PTC **1** (cinchona alkaloid-based with quinoline, HO, N+, allyl, and CF3-benzyl groups, Br⁻ counterion)]

2 (Cl, Cl, MeO-substituted indanone with phenyl group)

→ **1** (10 mol%), CH₃Cl, 50% NaOH aq, PhMe, 20 °C, 18 h →

3, R = Me, 95% yield, 92% ee
(+)-Indacrinone, R = CH₂CO₂H

methylating reagent, solvent, reaction temperature, and stirring rate, etc.) [9]. Moreover, the group proposed the way of asymmetric induction shown in Figure 6.5. A preferred conformation of the PTC **1** may be that in which the quinoline ring, the C9–O bond, and the N-arylmethyl group lie in one plane. The enolate anion of **2** also has an almost planar structure. Both molecules fit on top of each other, and the electrostatic interaction, hydrogen-bonding, and π–π stacking interactions make tight and stable ion pair between the PTC and the substrate. Because the PTC efficiently blocks the rear side of the enolate, the alkylating agent preferentially approaches the ion pair from the front side, giving the optically active product **3**.

6.5 α-Substitution of α-Amino Acid Derivatives via PTC

6.5.1 Monoalkylation of Benzophenone Imines of Glycine Esters

The Merck group's report has undoubtedly sparked the development of efficient catalytic organic reaction systems using structurally well-defined chiral organocatalysts. Cinchona alkaloids have taken the lead in this research area, and, as a matter of course, a variety of cinchona PTCs have been newly developed and applied to diverse

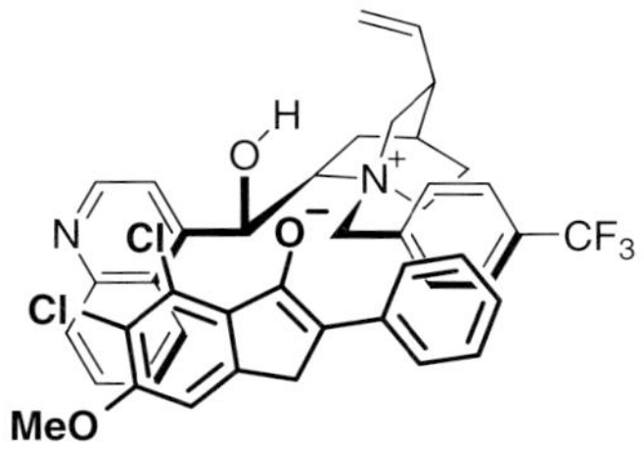

Figure 6.5 Complex ion pair between the PTC **1** and the anion of **2** proposed by the Dolling group.

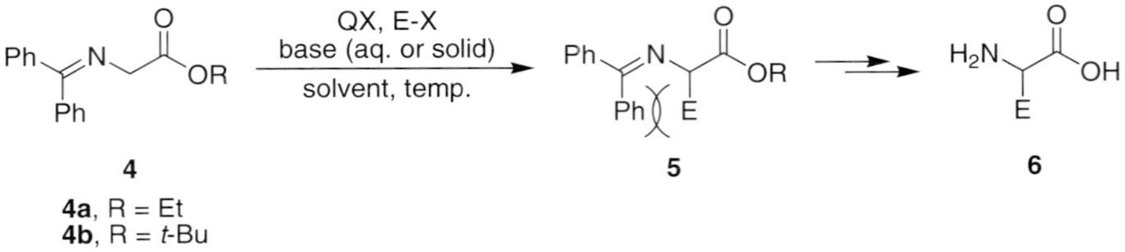

Scheme 6.2

organic transformations. In the early stage of the cinchona PTC history, the structural fine-tuning of catalyst has been mainly focused on the enantioselective alkylation of α-amino acid derivatives (Scheme 6.2). Although there have been lots of examples on the asymmetric phase-transfer catalysis, it is not too much to say that the research on the asymmetric alkylation of benzophenone imines of glycine esters **4** is the most useful and fruitful one. Most cinchona PTCs, to date, have been developed through this research, and the newly prepared cinchona PTCs have been gradually employed in other various organic catalytic transformations.

In 1978, O'Donnell and coworkers developed the benzophenone imines of glycine alkyl esters **4** as glycine anion equivalents, which have been found to be perfect to use in the phase-transfer catalysis [10]. An essential feature of this reaction system lies in the selective "mono" substitution of the starting Schiff base, the O'Donnell substrate **4**. This can be possible because of the significant difference in acidity of α-hydrogen between starting substrate **4** [pK_a(DMSO) 18.7 (R = Et)] and α-monosubstituted product **5** [pK_a(DMSO) 22.8 (R = Et, E = Me), 23.2 (R = Et, E = CH$_2$Ph)] [11]. This dramatic acidity difference makes it possible for selective formation of only monoalkylated product without concomitant production of undesired dialkylated product or racemization.

In 1989, the O'Donnell group first reported the asymmetric version of this alkylation using cinchona PTC (Scheme 6.3) [12]. The asymmetric alkylation of **4b** proceeded smoothly under mild phase-transfer conditions using N-benzylcinchoninium

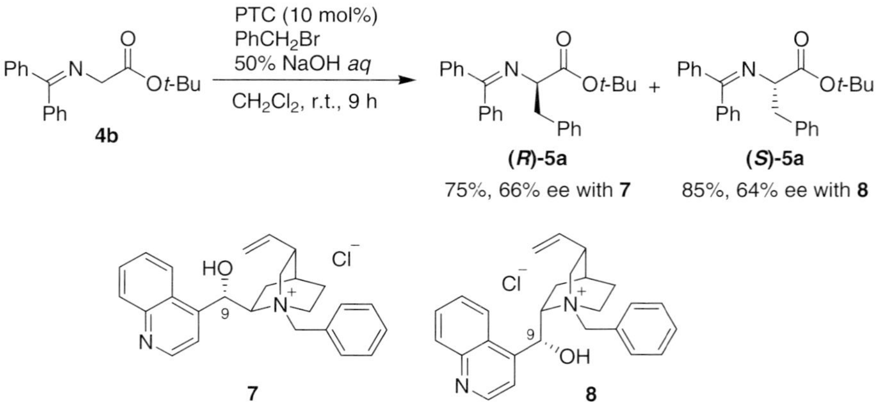

Scheme 6.3

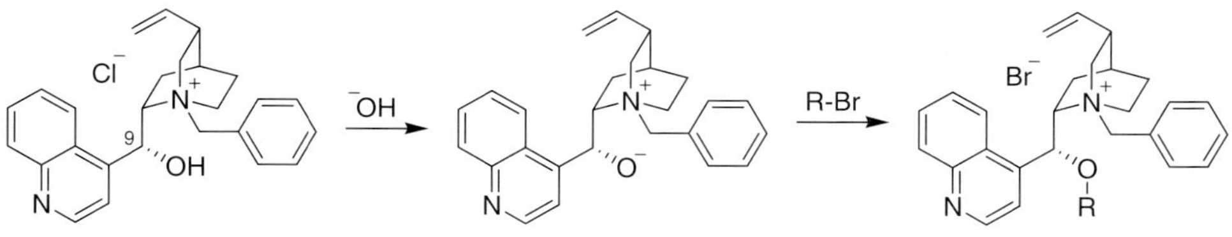

6.5 α-Substitution of α-Amino Acid Derivatives via PTC | 137

Scheme 6.4

chloride (**7**), affording the alkylated product (**R**)-**5a** in good yield with moderate enantioselectivity, while the opposite enantiomer (**S**)-**5a** was obtained in similar level of stereoselectivity by using pseudoenantiomeric N-benzylcinchonidinium chloride (**8**). The group finally obtained enantiomerically pure α-alkylated amino acids by a single recrystallization and subsequent deprotection in acidic media.

In 1994, the same group made the remarkable improvement in the enantioselectivity by the introduction of alkyl moieties to C9—O atom, in particular allyl or benzyl group (Scheme 6.4) [13]. With modified O-alkylated cinchona PTCs, the dramatic enhancement of enantiomeric excess by almost 20% was achieved (e.g., (**S**)-**5a** with 87% yield and 81% ee using **10a**). During this research, the group found that C9—OH functional group in PTC **8** is deprotonated by hydroxide base to produce Zwitterion **9**, and subsequently, it is alkylated by reaction with an alkyl halide used for α-alkylation of substrate **4b**, affording the corresponding O-alkylated PTCs **10**, the catalytically active species. It is also found that the level of enantiomeric excess when using cinchonidinium PTC is generally higher than using its pseudoenantiomeric cinchoninium one.

Based on the groundbreaking efforts of the O'Donnell group, a great improvement of the enantioselectivity with ee values to the range over 90% ee was reported by the two independent groups, that is, the Corey and the Lygo groups, in 1997 (Scheme 6.5). The key to achieve the dramatic jump in enantioselectivity was the introduction of the bulky 9-anthracenylmethyl moiety to the bridgehead nitrogen of the parent cinchona alkaloids for quaternarization. Lygo and coworker prepared the C9—OH PTCs **11**, **12** and applied them to the asymmetric phase-transfer alkylation of **4b** with a much higher enantioselectivity of approximately 90% ee [14]. Corey and coworkers used the preprepared O-allyl PTC **13**, and they adopted solid inorganic base system rather than aqueous one to enable the reaction at a low temperature of −78 °C [15]. Such a low reaction temperature makes the conformations of both the enolate and the catalyst more rigid, providing better enantioselectivity. Excellent enantioselectivities and generally high chemical yields were realized in a wide spectrum of electrophiles.

The Corey group clearly elucidated the origin of the stereoselectivity on the basis X-ray analysis of O-allyl-N-anthracenylmethylcinchonidinium p-nitrophenoxide. The group suggested a preferred three-dimensional arrangement of the ion pair from **13** and the E-enolate of **4b** (Figure 6.6). Alkylation of the enolate occurs by attack of the electrophile at Si-face (front) of the enolate for steric reasons [15].

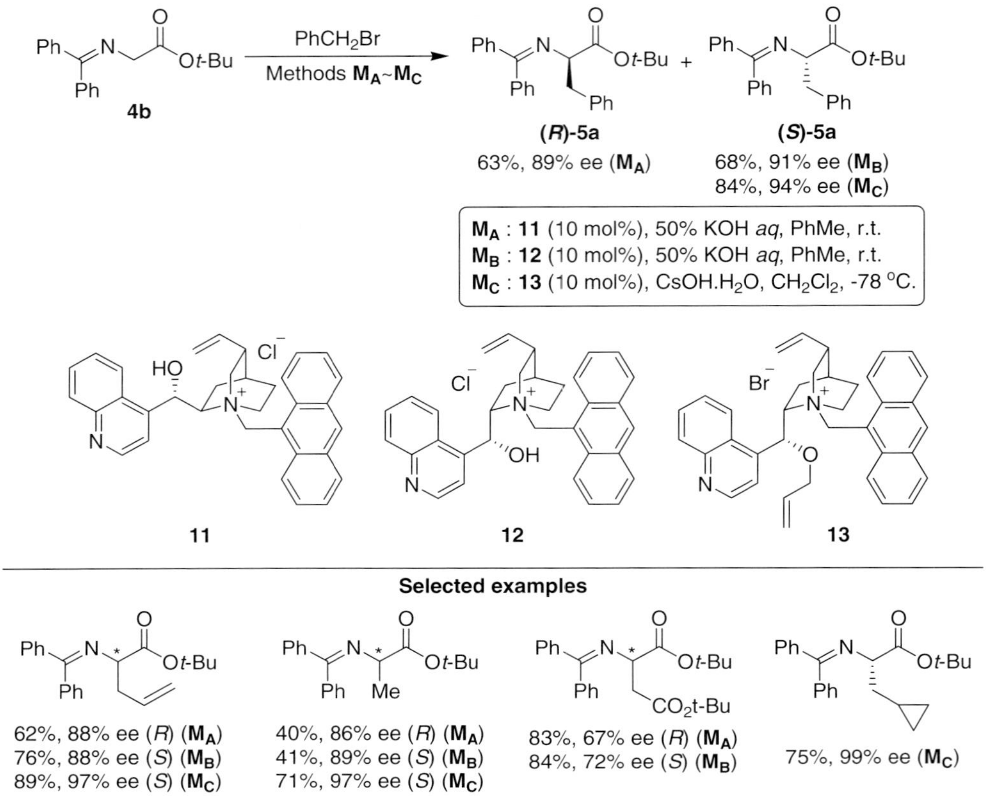

6 Cinchona-Catalyzed Nucleophilic α-Substitution of Carbonyl Derivatives

4b + PhCH₂Br →(Methods M_A~M_C) (R)-5a 63%, 89% ee (M_A) + (S)-5a 68%, 91% ee (M_B); 84%, 94% ee (M_C)

M_A : **11** (10 mol%), 50% KOH aq, PhMe, r.t.
M_B : **12** (10 mol%), 50% KOH aq, PhMe, r.t.
M_C : **13** (10 mol%), CsOH·H₂O, CH₂Cl₂, −78 °C.

11, **12**, **13**

Selected examples

62%, 88% ee (R) (M_A)
76%, 88% ee (S) (M_B)
89%, 97% ee (S) (M_C)

40%, 86% ee (R) (M_A)
41%, 89% ee (S) (M_B)
71%, 97% ee (S) (M_C)

83%, 67% ee (R) (M_A)
84%, 72% ee (S) (M_B)

75%, 99% ee (M_C)

Scheme 6.5

On the basis of the results from both the O'Donnell and the Corey groups [12b, 15], it can be generalized that a quaternary ammonium salt derived from cinchona alkaloid has an imaginary tetrahedron composed of adjacent four carbons to the bridgehead nitrogen. As shown in Figure 6.7, to be an efficient catalyst in this

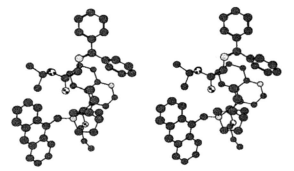

Figure 6.6 Stereopair representation of the preferred three-dimensional arrangement of the ion pair from the PTC **13** and the (E)-enolate of **4b** proposed by the Corey group.

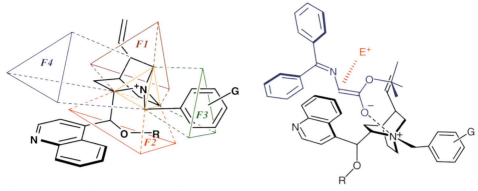

Figure 6.7 Origin of stereoselectivity of cinchona PTCs.

alkylation, the cinchona-derived PTC (a cinchonidinium PTC model is used for explanation in this figure) should be designed to provide effective steric screening that can inhibit an approach of the enolate of imine **1** to three faces (F1–F3) of this tetrahedron, while the remainder (F4) should be sufficiently open to allow close contact between the enolate anion and the ammonium cation [16]. Taking into account all of the data and rationales, an arylmethyl group for quaternarization of the bridgehead nitrogen is placed in F3 face and exercised great influence on whole conformation. In the cases of the PTCs **11–13**, the fixed orientation of the bulky 9-anthracenylmethyl substituent in **11–13** provides steric screening of F3 face and also rigidifies the cation, which provides a beneficial conformation to make a better-shaped ion pair with the enolate. Like this successful case, it plays a critical role to design a newly introducing moiety into F3 face in the development of highly efficient chiral cinchona PTCs. The well-established theory along with the supporting experimental results on the cinchona PTCs has stimulated many researchers into the development of more efficient and practical PTCs and their application to various organic reaction systems.

From 2001 to 2003, Park, Jew, and coworkers successively reported a series of polymeric cinchona PTCs, which were found to be highly effective in the asymmetric alkylation of **4b** under phase-transfer conditions [17]. The group devoted attention to the fact that the significant improvement was achieved in Sharpless asymmetric dihydroxylation when the dimeric ligands were used. The group applied this advantage of polymerization to the design of cinchona PTCs as depicted in Figure 6.8.

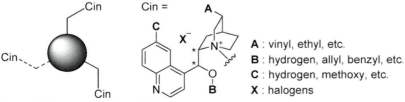

Figure 6.8 Polymeric cinchona PTCs.

A : vinyl, ethyl, etc.
B : hydrogen, allyl, benzyl, etc.
C : hydrogen, methoxy, etc.
X : halogens

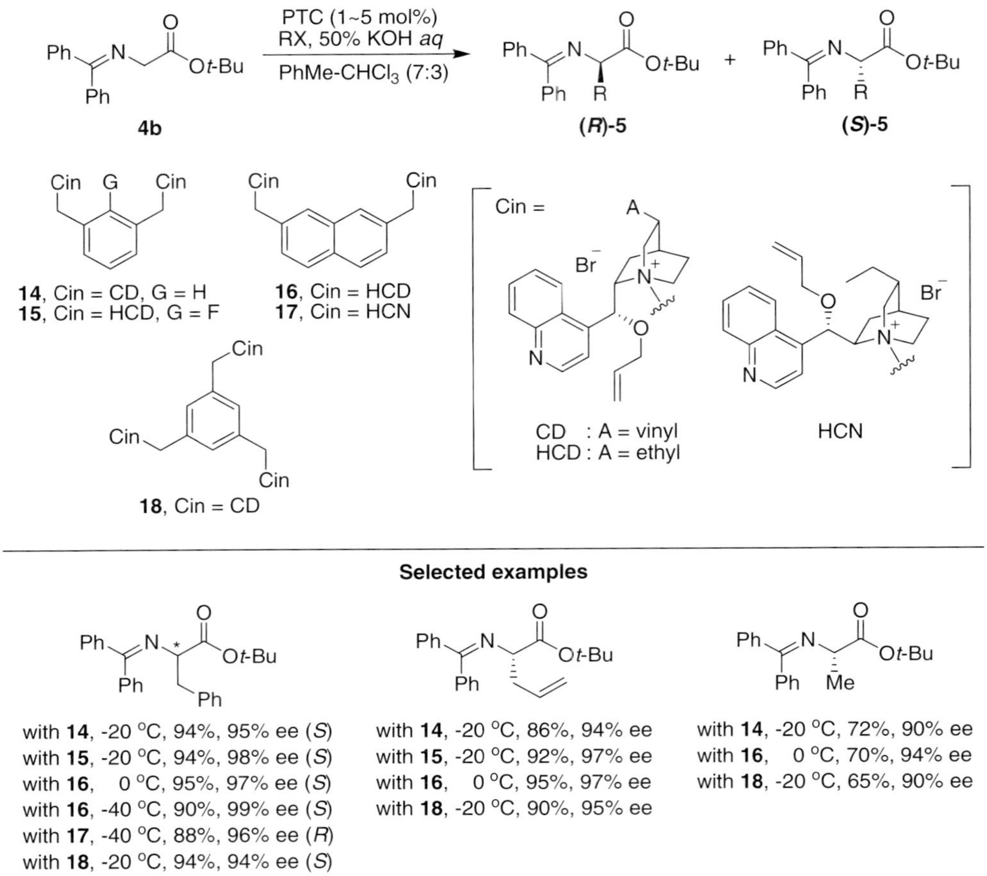

Scheme 6.6

From the systematic investigation of the Park and Jew group, several highly efficient and practical polymeric cinchona PTCs were developed (Scheme 6.6). Interestingly, polymeric catalysts with a specific direction of attachment between aromatic linkers (e.g., benzene or naphthalene) and each cinchona unit were found to be effective in the asymmetric alkylation of **4b**. The phenyl-based polymeric PTCs with the meta-relationship between cinchona units such as **14**, **15**, and **18** showed their high catalytic efficiencies. Furthermore, the 2,7-dimethylnaphthalene moiety as in **16** and **17** was ultimately found to be the ideal spacer for dimeric cinchona PTC for this asymmetric alkylation. For example, with 5 mol% of **16**, the benzylation of **4b** was completed within a short reaction time of 30 min at 0 °C, affording **(S)-5a** in 95% yield with 97% ee. Almost optically pure (>99% ee) **(S)-5a** was obtained at lower reaction temperature (−40 °C) with **16**, and moreover, even with a smaller quantity (1 mol%), its high catalytic efficiency in terms of both reactivity and enantioselectivity was well conserved.

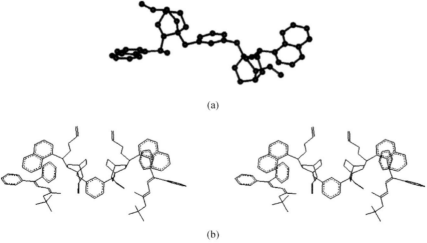

Figure 6.9 (a) The probable structure of the dimeric cinchona, C9—OH analogue of **14** (a) and (b) stereoview of plausible model of the ion pair from the PTC **14** and one (or two) *E*-enolate(s) of **4b** proposed by the Park–Jew group.

The Park and Jew group proposed the probable structure of the 1,3-phenyl-dimeric PTC, C9—OH analogue of **14**, based on the X-ray crystallographic study (Figure 6.9a) [16]. The two cinchona units are placed in a direction of antirelationship to each other, and each cinchona unit has a same conformation and is situated in an identical circumstance. Therefore, the same result will be obtained even if an enolate anion of **4b** approaches either of the two ammonium cations in a dimeric PTC. Unlike the monomeric catalyst such as **10a**, the rotations of phenyl ring in the dimeric catalyst get restricted, especially when two cinchona alkaloid units are connected through phenyl spacer in the meta-direction. The bulkiness of cinchona alkaloid unit can restrict free rotation of both N^+—CH_2 (benzylic) bond and CH_2 (benzylic)—C (phenyl) bond, which makes the whole conformation of the dimeric catalyst rigid, providing an efficient blocking of F3 face from an access of enolate to bridgehead nitrogen cation. Stereopair representation of the presumed plausible transition state, consisting of the enolate of **4b** and the PTC **14**, is also shown in Figure 6.9b. Alkylation of the enolate occurs by attack of the electrophile at *si* face of the enolate for steric reasons, leading to the enantiomeric products.

Inspired by the reports by the Park and Jew group, further structural variations of polymeric cinchona PTC were performed by several research groups. In 2002, Najera and coworkers prepared dimeric PTCs in which the 9,10-dimethylanthracenyl moiety is incorporated as a linker, expecting the advantageous effect shown in monomeric series (Scheme 6.7) [18a]. This class of dimeric PTCs gave generally good stereoinduction in the alkylation of **4b**; however, it did not come up to the level of enantioselectivity using the corresponding monomeric PTCs **11–13**. The preprepared *O*-allyl derivatives **20** and **22** did not give the better stereoselectivities than the free OH analogues **19** and **21** under the same reaction conditions, which is contrary to

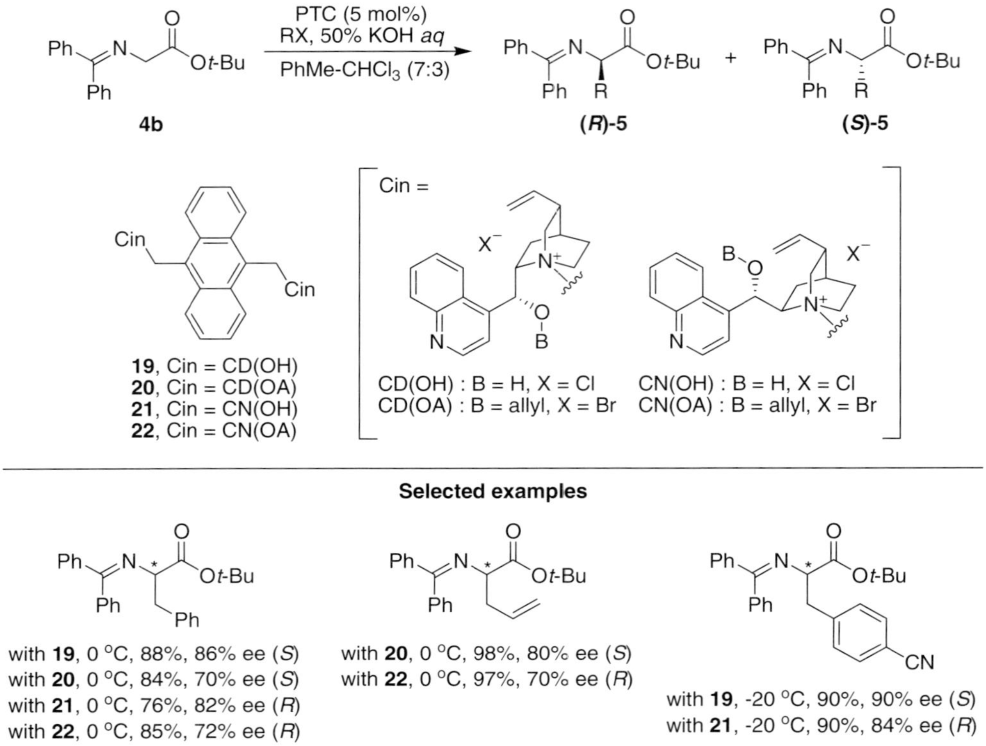
PTC (5 mol%)
RX, 50% KOH aq
PhMe-CHCl3 (7:3)
4b
(R)-5
(S)-5
Cin =
19, Cin = CD(OH)
20, Cin = CD(OA)
21, Cin = CN(OH)
22, Cin = CN(OA)
CD(OH) : B = H, X = Cl
CD(OA) : B = allyl, X = Br
CN(OH) : B = H, X = Cl
CN(OA) : B = allyl, X = Br
Selected examples
with 19, 0 °C, 88%, 86% ee (S)
with 20, 0 °C, 84%, 70% ee (S)
with 21, 0 °C, 76%, 82% ee (R)
with 22, 0 °C, 85%, 72% ee (R)
with 20, 0 °C, 98%, 80% ee (S)
with 22, 0 °C, 97%, 70% ee (R)
with 19, -20 °C, 90%, 90% ee (S)
with 21, -20 °C, 90%, 84% ee (R)

Scheme 6.7

the tendency. The group also investigated the counteranion effect by exchanging the chloride or bromide anions with tetrafluoroborate (BF_4^-) or hexafluorophosphate (PF_6^-) anions, but the effect of changing the anion was found to be minimal [18b].

In 2005 and 2006, Siva and coworkers reported several types of new dimeric and trimeric cinchona-PTCs **23–27** and their application to the asymmetric alkylation of **4b** (Scheme 6.8). While, generally, high chemical and optical yields were obtained using dimeric **23** and **24**, trimeric PTCs **25–27** showed slightly lower stereoselectivities [19].

A new paradigm was suggested in the design of cinchona PTCs by Park, Jew, and coworkers in 2002 [20]. While the successful cinchona PTCs have been developed primarily on the basis of the introduction of sterically bulky moiety contributing to make favorable conformation of the catalyst for good asymmetric induction, the group focused on the role of the electronic factor in cinchona PTCs. During this research, the group found the intriguing result that the catalysts containing *ortho*-F-substituted benzyl moieties were found to be effective in the asymmetric alkylation of **4b** (Scheme 6.9). For example, the 2′-F derivative **28a** affords the benzylated product of **4a** with the enantioselectivity of 89% ee whereas the nonsubstituted analogue **28′** gave the lower enantiomeric excess (74% ee) under the same reaction conditions, despite no significant difference in terms of steric effect between **28a**

6.5 α-Substitution of α-Amino Acid Derivatives via PTC

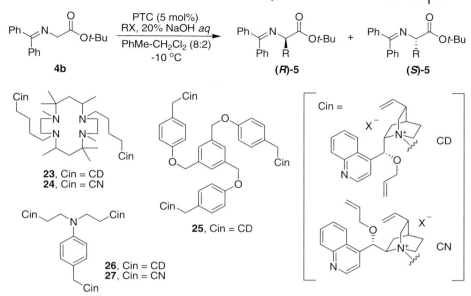

Scheme 6.8

and **28'**. Positive effect in the enantioselectivity was observed by the additional introduction of fluoro atoms on the *meta*- and/or *para*-position, but only when all substituents are positioned in the same direction as in **28b–d**.

The group proposed the origin of the stereoselectivity of the 2'-F-containing cinchona PTCs based on the X-ray crystallographic investigation of **28a** (Figure 6.10). The group noticed that **28a** exists in two different conformations in unequal amounts, four parts as F(a) and one part as F(b). This might be as a result of the unfavorable anionic repulsion between the bromide anion and F(b). Another possibility is that conformation F(a) could be favored owing to the internal hydrogen bonding involving water between F(a) and the C9–OH group. The 2'-F might be involved in internal hydrogen bonding involving water resulting in good whole conformation of the catalyst for high stereoselectivity.

The same group also reported the several evidence, which can strongly support their hypothesis of the electronic factor in the cinchona PTCs [21]. First, benzylations

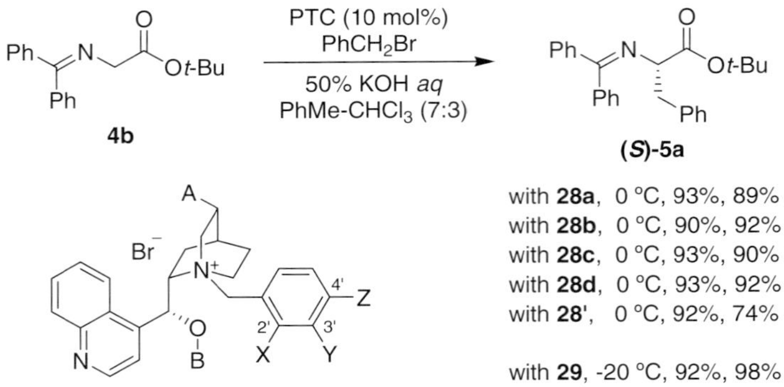

6 Cinchona-Catalyzed Nucleophilic α-Substitution of Carbonyl Derivatives

Scheme 6.9

with **28a**, 0 °C, 93%, 89% ee
with **28b**, 0 °C, 90%, 92% ee
with **28c**, 0 °C, 93%, 90% ee
with **28d**, 0 °C, 93%, 92% ee
with **28'**, 0 °C, 92%, 74% ee

with **29**, −20 °C, 92%, 98% ee

28a, A = vinyl, B = H, X = F, Y = Z = H
28b, A = vinyl, B = H, X = Y = F, Z = H
28c, A = vinyl, B = H, X = Y = F, Z = H
28d, A = vinyl, B = H, X = Y = Z = F
28', A = vinyl, B = H, X = Y = Z = H
29, A = ethyl, B = allyl, X = Y = Z = F

of **4a** using the O-allyl derivative of **28a** and **28'** under anhydrous conditions and NaH in dry toluene–CH$_2$Cl$_2$ at 0 °C, both gave the same enantioselectivity of 62% ee. Second, when two sets of PTCs, **30**, **31a** and **32**, **33a** that were designed to have similar steric effect and different electronic effect, were applied to the asymmetric benzylation, the PTCs having ability to form hydrogen bonding, **31a** and **33a** showed better enantioselectivities (Scheme 6.10). The hydro- and O-allyl derivatives, **31b** and **33b**, were found to be highly effective electronically modified cinchona PTCs along with the series of the 2'-F-containing PTCs such as **29**.

Various monomeric cinchona PTCs have been developed for the catalytic asymmetric organic reactions, especially for the alkylation of **4b**, and several noticeable PTCs are summarized in Scheme 6.11. Nájera, van Koten, and coworkers prepared the cinchonidinium salts bearing 3,5-dialkoxybenzyl group such as **34** [22]. The group found unusual reversal of enantioselectivity depending on the kind of alkaline metal of base; the use of potassium hydroxide afforded the (S)-enantiomer, whereas using sodium hydroxide under the same conditions gave the corresponding (R)-enantiomer.

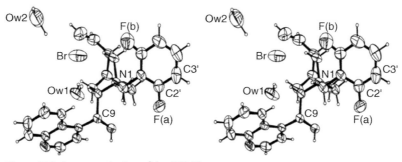

Figure 6.10 Stereoscopic view of the PTC **28a**.

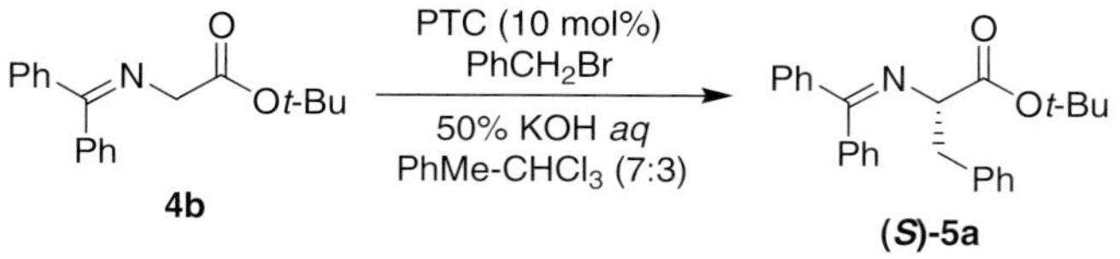

6.5 α-Substitution of α-Amino Acid Derivatives via PTC

Scheme 6.10

Reaction of **4b** (Ph₂C=N-CH₂-CO₂t-Bu) with PhCH₂Br, PTC (10 mol%), 50% KOH aq, PhMe-CHCl₃ (7:3) gives (S)-5a.

- with **30**, 0 °C, 90%, 75% ee
- with **31a**, 0 °C, 95%, 92% ee
- with **32**, 0 °C, 95%, 61% ee
- with **33a**, 0 °C, 92%, 90% ee
- with **31b**, -20 °C, 96%, 98% ee
- with **33b**, -20 °C, 95%, 98% ee

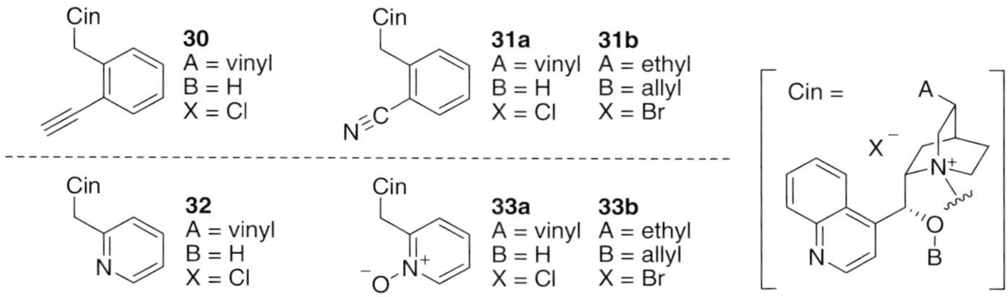

30: Cin, A = vinyl, B = H, X = Cl (benzyl with ortho alkynyl)
31a: Cin, A = vinyl, B = H, X = Cl (benzyl with CN)
31b: Cin, A = ethyl, B = allyl, X = Br
32: Cin, A = vinyl, B = H, X = Cl (pyridyl)
33a: Cin, A = vinyl, B = H, X = Cl (N-oxide pyridyl)
33b: Cin, A = ethyl, B = allyl, X = Br

Cin = cinchona-derived quaternary ammonium framework with substituents A, B and counterion X.

Scheme 6.11

Reaction of **4b** with PhCH₂Br under various conditions gives (R)-5a + (S)-5a.

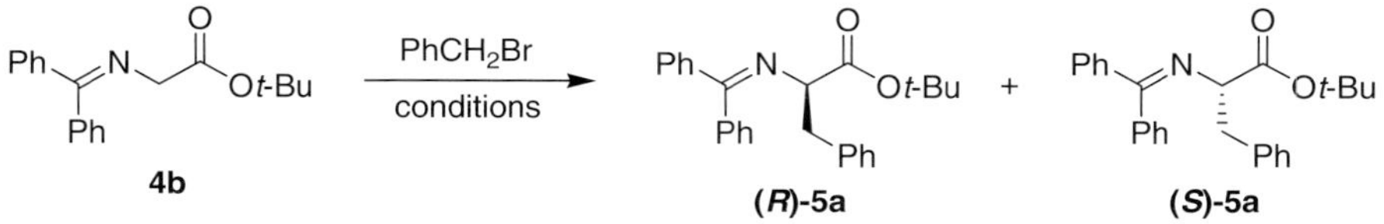

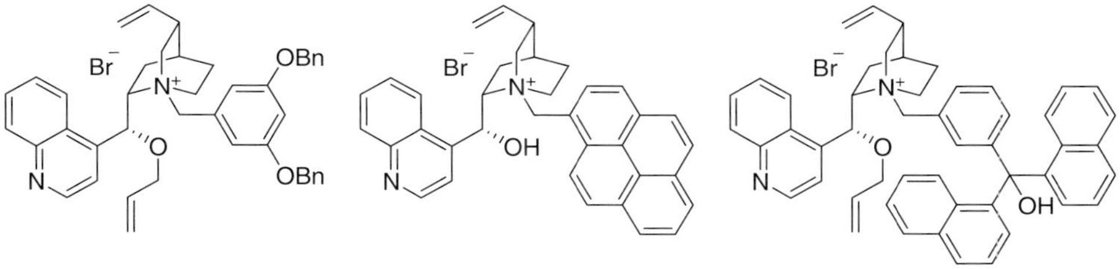

34 (10 mol%), 50% KOH aq, PhMe-CHCl₃ (7:3), -20 °C, 1 h
91%, 72% ee, **(S)-5a**
by Najera and van Koten

35 (5 mol%), 50% NaOH aq, PhMe, 25 °C, 9 h
91%, 88% ee, **(S)-5a**
by Elango

36 (10 mol%), 50% KOH aq, PhMe-CH₂Cl₂ (7:3), -20 °C, 6 h
93%, 92% ee, **(S)-5a**
by Ramachandran

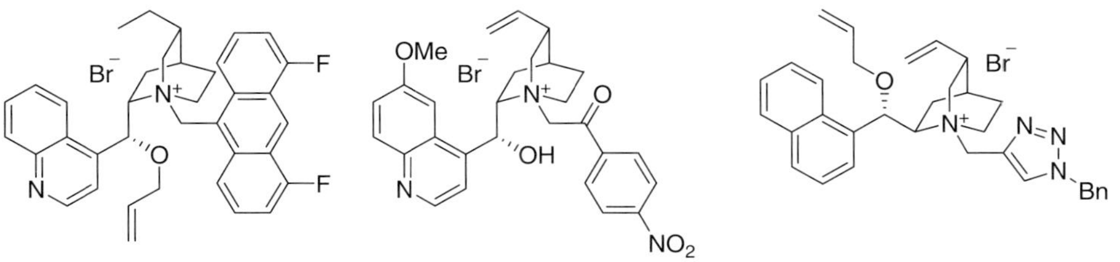

37 (10 mol%), 50% KOH aq, PhMe-THF (7:3), -20 °C, 16 h
82%, 97% ee, **(S)-5a**
by Andrus

38 (5 mol%), 10M KOH aq, PhMe-CHCl₃ (7:3), r.t., 6 h
83%, 96% ee, **(S)-5a**
by Wang

39 (10 mol%), 50% KOH aq, PhMe-CHCl₃ (7:3), 0 °C, 24 h
84%, 71% ee, **(R)-5a**
by Zhang

Elango and coworkers introduced 13-picenylmethyl (not shown in Scheme 6.11) and 1-pyrenylmethyl group in **35**, bulkier moieties than the 9-anthracenylmethyl group in the Lygo and the Corey PTCs, in the design of cinchona PTCs [23]. The Ramachandran group designed cinchona PTCs having diaryl substitution at the 3- and 4-positions of the N-benzyl group in cinchonidinium salts [24]. The *meta*-di (1-naphthyl)-substituted catalyst **36** was found to be the most effective in the asymmetric alkylation of **4b**. The Andrus group combined the good steric screening power of 9-anthracenylmethyl group and the beneficial electronic character of fluorine substitution. Several fluorinated anthracenylmethyl cinchonidinium salts were prepared including **37** [25]. Wang and coworkers prepared a series of acetophenone-based quininium salts (e.g., **35**) in which halogen (Cl, Br) or nitro group is incorporated at the *meta*- or *para*-position of the phenyl moiety of acetophenone part [26]. Among the prepared PTCs, 4-NO_2-derivative **35** gave the best result. The group proposed that the electron-withdrawing N-acetophenone moiety increases the overall electron deficiency of the positive charge, leading enhancement of the degree of the ion-pairing with the enolate, and the tighter ion pair can provide better enantioselectivity. Zhang and coworkers prepared the triazole-based cinchoninium PTC **39**, which showed moderate enantioselectivity, by Cu(I)-mediated 1,3-dipolar cycloaddition with propargyl cinchoninium salt and benzyl azide [27].

In connection with the development of cinchona PTCs, considerable efforts have been made simultaneously on the preparation of the recyclable polymer-supported cinchona PTCs for the asymmetric alkylation (Scheme 6.12). Polymer-bound catalysts present advantages such as simplified workup for product purification, easy recovery, good stability, reduced toxicity, and potential recycling. In 2000, the Nájera group prepared the PTCs supported with Merrifield resin to the bridgehead nitrogen such as **40** and applied them to the asymmetric alkylation [28]. The polymer-supported PTCs were found to be more effective in the case of alkylation of isopropyl analogue (**4b′**) than that in **4b**. The recovered catalyst showed almost identical chemical and optical yield at its second use in the alkylation. In 2001, Cahard and coworkers reported two papers on this issue. First, the group designed the similar polymer-supported PTCs such as **41** to the Nájera's PTC **40**, but the group incorporated spacers composed of several methylene units [29]. Interestingly, the same major enantiomer was always obtained irrespective of the nature of catalyst used, even though cinchonine and cinchonidine are known to behave as pseudoenantiomers. Second, the group also reported the modified Lygo–Corey analogue **42** prepared by the attachment of Merrifield resin on the C(9)−OH position [30]. With this PTC, (**S**)-**5a** was obtained in the enhanced enantioselectivity of 94% ee using CsOH·H_2O at −50 °C. In 2008, Park, Jew, and coworkers reported the Merrifield resin-supported PTCs in which hydrogen bond inducing functional groups are built-in such as **43** and **44**, expecting that the advantageous electronic effect already confirmed in their previous data can also be operative in polymer-supported version [31]. Two types of resin-bound electronically modified PTCs, O-supported PTCs such as **43** and N-supported PTCs such as **44**, were prepared and evaluated. Among them, N-oxypyridine-based series especially provided the highest stereoselectivity in each series. In addition, the recovered PTC after the first use gave almost similar results in the next several cycles.

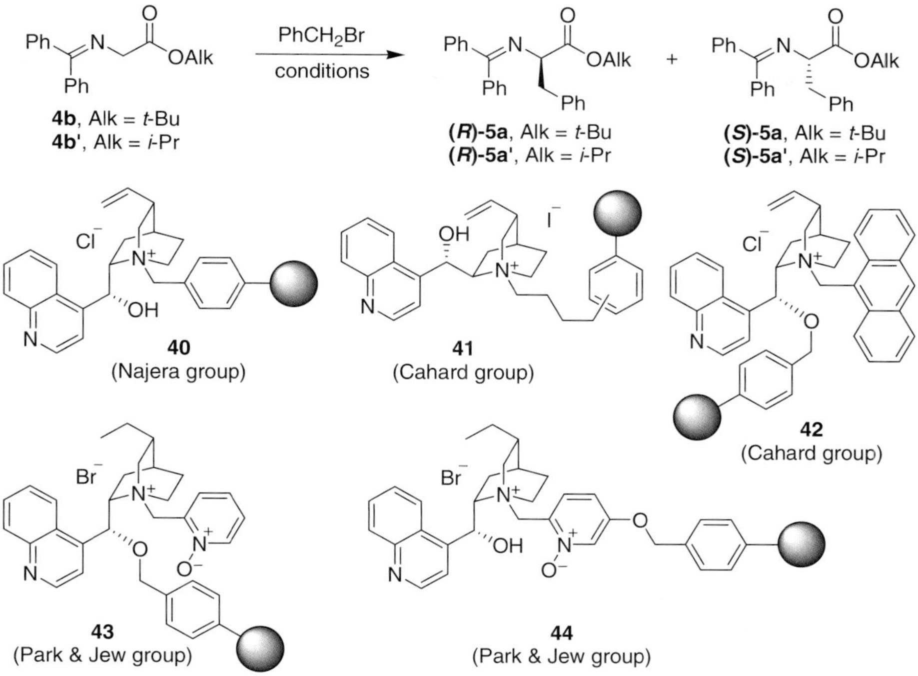

6.5 α-Substitution of α-Amino Acid Derivatives via PTC

Scheme 6.12

- **4b**, Alk = t-Bu
- **4b'**, Alk = i-Pr

PhCH$_2$Br, conditions →

- (**R**)-**5a**, Alk = t-Bu
- (**R**)-**5a'**, Alk = i-Pr

+

- (**S**)-**5a**, Alk = t-Bu
- (**S**)-**5a'**, Alk = i-Pr

40 (Najera group)
41 (Cahard group)
42 (Cahard group)
43 (Park & Jew group)
44 (Park & Jew group)

with **40** (10 mol%): 25% NaOH aq, PhMe, 0 °C, 17 h, 90%, 90% ee, (**S**)-**5a'**
with **41** (10 mol%): 50% KOH aq, PhMe, 0 °C, 15 h, 60%, 81% ee, (**R**)-**5a**
with **42** (10 mol%): CsOH-H$_2$O, PhMe, -50 °C, 30 h, 67%, 94% ee, (**S**)-**5a**
with **43** (20 mol%): 50% KOH aq, PhMe-CHCl$_3$ (7:3), 0 °C, 10 h, 81%, 95% ee, (**S**)-**5a**
with **44** (20 mol%): 50% KOH aq, PhMe-CHCl$_3$ (7:3), 0 °C, 24 h, 92%, 90% ee, (**S**)-**5a**

Poly(ethylene glycol)(PEG) supported cinchona PTCs were developed for the asymmetric alkylation by several groups (Scheme 6.13). The advantageous aspect of PEG includes that PEG is inexpensive, readily functionalized, and commercially available in different molecular weights. Moreover, it is readily soluble in many organic solvents and insoluble in a few other solvents, which enable to run a reaction under homogeneous catalytic conditions and to recover the catalyst as if it were bound to an insoluble matrix. The first report on the use of PEG-supported cinchona PTC in the alkylation was made by Benaglia and coworkers in 2003 [32]. The group prepared the PEG-supported derivatives of the Lygo–Corey PTCs by attaching the PEG moiety either on the oxygen at C9 or on the C6'–O of the quinoline ring and evaluated their catalytic efficiencies. The highest enantioselectivity (64% ee) was achieved with the C9–O supported PTC **45** at a low temperature of −78 °C. The Cahard group prepared the PEG-bound PTCs such as **46** in which PEG part is connected to the bridgehead nitrogen via benzoyl moiety [33]. From the systematic investigation of parameters affecting the reaction, cinchonidinium catalyst **46** was

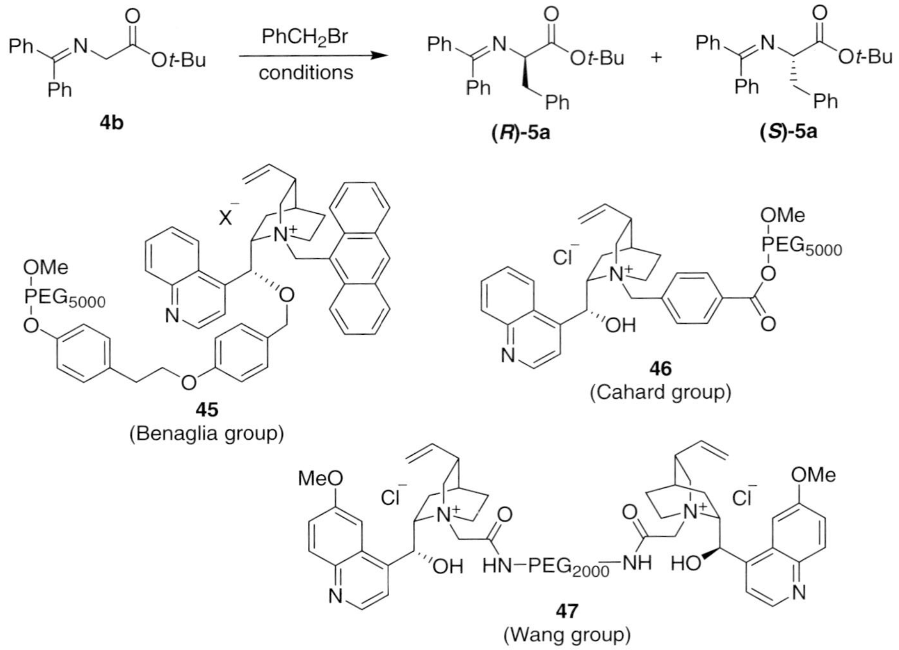

Scheme 6.13

with **45** (10 mol%): CsOH-H$_2$O, CH$_2$Cl$_2$, -78 °C, 60 h, 75%, 64% ee, **(S)-5a**
with **46** (10 mol%): 50% KOH aq, PhMe, 0 °C, 15 h, 84%, 81% ee, **(S)-5a**
with **47** (10 mol%): 1M KOH aq, r.t., 6 h, 98%, 83% ee, **(S)-5a**

found to be the best one for the asymmetric alkylation, and the nonpolar toluene solvent, aqueous potassium hydroxide base, and reaction temperature of 0 °C were determined for optimal reaction conditions. The water-soluble PEG-supported dimeric cinchona PTCs such as **47** were developed by Wang and coworkers [34]. The PTCs were prepared by reaction of diacetamido–PEG2000 chloride with natural cinchona alkaloid, and it was found that they showed better catalytic activity in the asymmetric alkylation of **4b** when the solvent is water rather than organic solvents. The interesting feature of these PTCs is that steady chemical yield along with enantioselectivity is guaranteed even after recovery and recycling several times. What makes this possible is that the acetamido-moiety-connected cinchona and PEG might be hardly disintegrated under the mild reaction conditions.

6.5.2
Alkylation of α-Monosubstituted α-Amino Acid Derivatives

Chiral α,α-dialkyl-α-amino acids (ααAAs), a class of nonproteinogenic amino acids, have been extensively studied because of their important role in the fields of synthetic, medicinal, and biological chemistry. Their quaternary chiral centers contribute not only to the molecular stability but also to the conformational preference, by inducing a preferable helical secondary structure of the peptide backbone, when incorporated into a peptide. Moreover, the biological activities of peptides containing ααAA can be maintained longer because of their resistance against enzymatic hydrolysis. Also, the ααAAs themselves are known to be powerful enzymatic inhibitors and useful synthetic building blocks via chemical transformations. Accordingly, the development of effective synthetic methods for chiral ααAAs is a very important and challenging subject in organic synthesis [35]. Since the O'Donnell group succeeded in the preparation of optically active α-alkylalanine derivatives using cinchona PTC in 1992, asymmetric phase-transfer catalysis has made distinctive contributions in this area (Scheme 6.14) [36]. The O'Donnell group obtained (R)-α-benzylated alanine derivative in enantiomeric excess by the asymmetric alkylation of 4-chlorobenzaldehyde imine of alanine *tert*-butyl ester (**48a**) with cinchoninium PTC **50**. From the examination of the different inorganic base system, the mixed solid base of KOH and K_2CO_3 was found to be the most effective. Lygo and coworkers accomplished the significantly improved enantioselectivity, compared with the O'Donnell's results, using their 9-anthracene-based PTC **51** [37]. The group emphasized that the mixed base employed in this reaction must be freshly prepared for reproducible results. Park, Jew, and coworkers performed systematic investigations to develop a higher enantioselectivity-producing reaction system using the PTC **29** [38]. The group found that 2-naphthyl moiety in the substrate **48b** provides a suitable steric bulkiness, and this naphthyl aldimine **48b** is superior to the other aryl aldimines including **48a** in terms of both chemical yield and enantioselectivity.

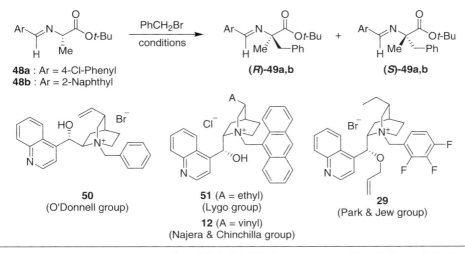

Scheme 6.14

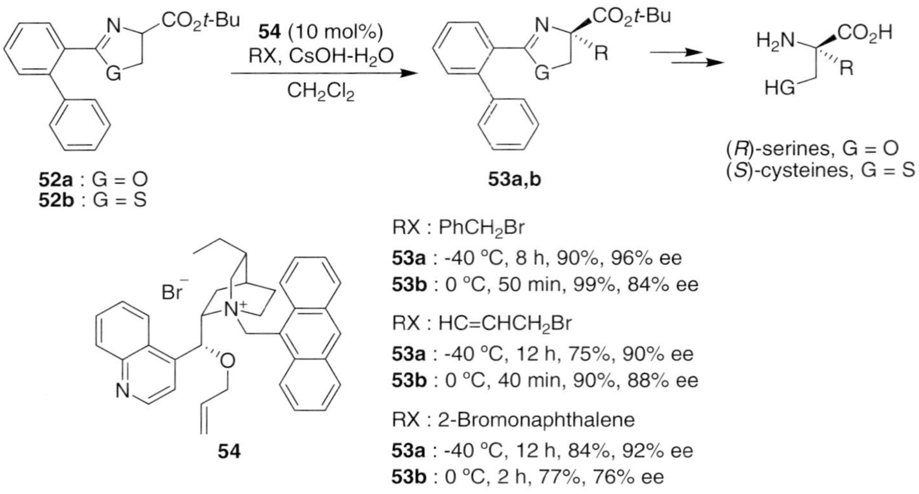

Scheme 6.15

52a: G = O
52b: G = S

RX : PhCH$_2$Br
53a : −40 °C, 8 h, 90%, 96% ee
53b : 0 °C, 50 min, 99%, 84% ee

RX : HC=CHCH$_2$Br
53a : −40 °C, 12 h, 75%, 90% ee
53b : 0 °C, 40 min, 90%, 88% ee

RX : 2-Bromonaphthalene
53a : −40 °C, 12 h, 84%, 92% ee
53b : 0 °C, 2 h, 77%, 76% ee

(R)-serines, G = O
(S)-cysteines, G = S

After inorganic base screening along with reaction temperature variation, solid rubidium hydroxide showed the best result at the reaction temperature of −35 °C. In 2007, additional fine-tuning of the reaction variables was performed by Nájera, Chinchilla, and coworker, from which the group was able to reduce the amount of the PTC **12** from 10 to 5 mol% without a significant loss of chemical and optical yields [39].

The group of Park and Jew developed the efficient reaction systems for the preparation of optically active α-alkylserine and α-alkylcysteine derivatives (Scheme 6.15) [40]. The group developed the oxazoline-based substrate for serines **52a** and the thiazoline-based substrate **52b** for cysteines, respectively. The oxazoline and thiazoline moieties fulfilled a dual function: the activation of the α-proton and the protection of both the amino and the side chain hydroxy groups. The *ortho*-biphenyl derivatives **52a** and **52b** were specifically designed for the cinchona PTC **54**, and solid cesium hydroxide monohydrate was chosen as an optimal base. The PTC **29** was also found to be as effective as the PTC **54** for the asymmetric alkylation of **52a**.

6.6
α-Substitution of Other Carbonyl Derivatives via PTC

6.6.1
α-Substitution of Monocarbonyl Compounds

Besides the asymmetric α-alkylation of α-amino acid derivatives described above, α-substitution of other types of carbonyl compounds using cinchona PTCs have been actively developed as well. In 1990, Vandewalle and coworker used the asymmetric alkylation of isotetralone derivative **55** with 1,5-dibromopentane using

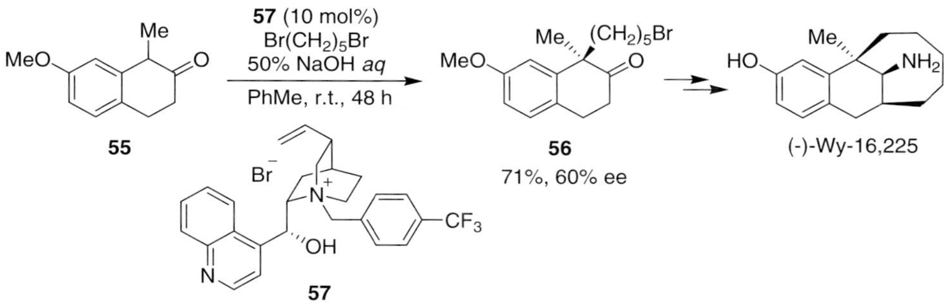

6.6 α-Substitution of Other Carbonyl Derivatives via PTC | 151

Scheme 6.16

cinchonidinium salt **57** under oxygen-free phase-transfer conditions in the course of the synthesis of analgesic agent (−)-Wy-16 225 (Scheme 6.16) [41].

In 1998, Corey and coworkers reported the alkylation of enolizable β,γ-unsaturated ester **58** in the presence of the catalyst **13** (Scheme 6.17) [42]. The α-alkylated products **59** were obtained in high optical purities with a broad variety of alkyl halides. Optically active tetrahydropyran derivatives **60** were prepared from the product **59b**.

In 1999, Arai, Shioiri, and coworkers reported the use of α-fluorotetralone **61** as the efficient substrate for the preparation of optically active fluoro compounds by asymmetric phase-transfer alkylation (Scheme 6.18) [43]. From the screening of the reaction variables, the group found that enantioselectivity was influenced by the aryl part of the catalyst, and the salt **63** possessing pentamethylphenyl moiety was the optimal PTC for this alkylation.

In 2004 and 2005, Andrus and coworkers reported a new approach to the synthesis of optically active α-alkyl-α-hydroxycarbonyl compounds by asymmetric catalytic

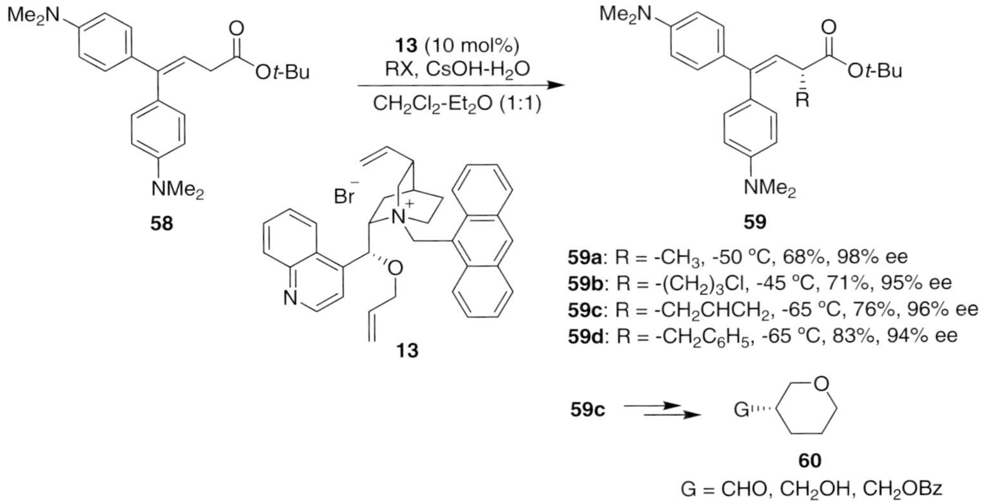

59a: R = -CH$_3$, -50 °C, 68%, 98% ee
59b: R = -(CH$_2$)$_3$Cl, -45 °C, 71%, 95% ee
59c: R = -CH$_2$CHCH$_2$, -65 °C, 76%, 96% ee
59d: R = -CH$_2$C$_6$H$_5$, -65 °C, 83%, 94% ee

60
G = CHO, CH$_2$OH, CH$_2$OBz

Scheme 6.17

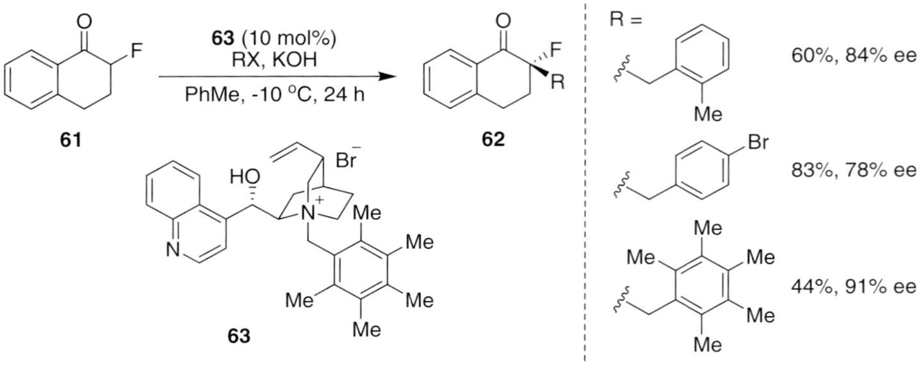

6 Cinchona-Catalyzed Nucleophilic α-Substitution of Carbonyl Derivatives

Scheme 6.18

phase-transfer glycolate alkylation (Scheme 6.19) [44]. The group performed the systematic investigation to search a suitable α-hydroxy carbonyl substrate as a glycolate surrogate with various modified acetophenones. Diphenylmethyloxy-2,5-dimethoxyacetophenone **64** was selected as the most suitable substrate. Highly optically enriched (S)-alkylated products **65** were obtained in the presence of the Park–Jew PTC **29** and solid CsOH·H$_2$O base in a mixed solvent of CH$_2$Cl$_2$ and n-hexanes (1 : 1) at −35 °C. Alkylated product **65** was elaborated to versatile glycolate intermediates **66** by the sequential bis-TMS peroxide mediated Baeyer–Villiger oxidation and selective transesterification reactions without loss of enantioselectivity. The ester product was purified to give a single enantiomer (>99% ee) by single recrystallization process.

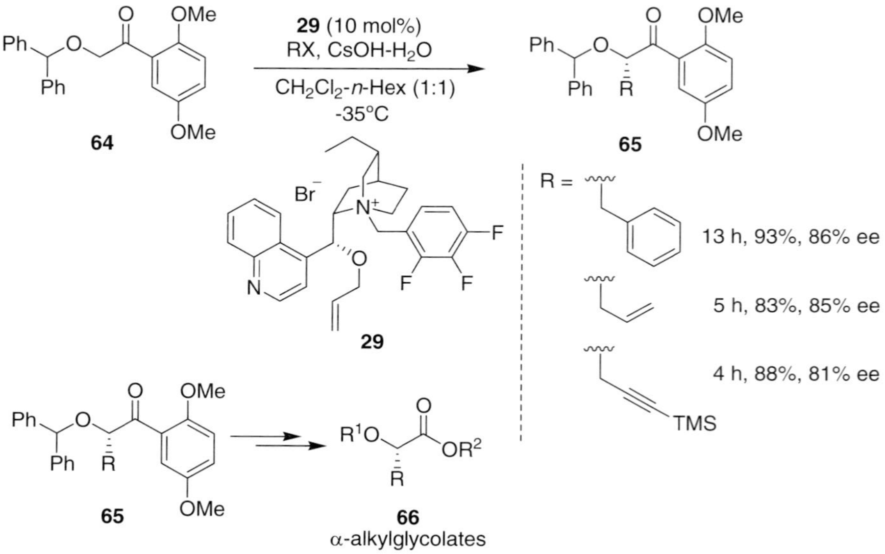

Scheme 6.19

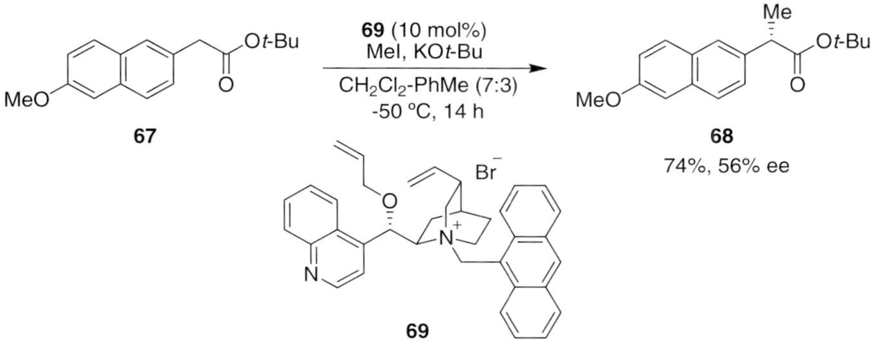

Scheme 6.20

In 2005, Ramachandran and coworker reported the asymmetric methylation of tert-butyl 2-(6-methoxynaphthalen-2-yl)acetate (**67**) using cinchoninium PTC **69** and strong base potassium tert-butoxide (Scheme 6.20) [45]. The optically active methylated product **68** can be converted to naproxen, a nonsteroidal anti-inflammatory drug, by hydrolysis of ester.

6.6.2
α-Substitution of β-Keto Carbonyl Compounds

Asymmetric alkylation of β-keto carbonyl derivatives under phase-transfer conditions is a very convenient and useful way to construct a chiral quaternary carbon center. In 2002, Dehmlow and coworkers reported the asymmetric benzylation of the cyclic β-ketoester, 2-(tert-butoxycarbonyl)cyclopentanone (**70**) in the presence of the cinchoninium PTC **72** in excellent chemical yield with 46% ee (Scheme 6.21) [46].

In 2004, Kim and coworker developed their own cinchona PTCs containing a specific bulky aryl moiety, 3,5-di-tert-butyl-4-methoxybenzyl group, such as **75**, and applied them to the asymmetric α-alkylation of the cyclic β-ketoesters, for example, 2-(ethoxycarbonyl)-1-indanone **73** ($n=0$) or 2-(ethoxycarbonyl)-1-tetralone **73** ($n=1$) as shown in Scheme 6.22 [47]. The enantioselectivities showed a considerable variation (44–99% ee) depending on the nature of electrophile.

In 2008, Dixon and coworkers reported catalytic enantioselective α-alkylations of β-keto carbonyl derivatives with aziridines by using a cinchona-derived catalyst **79** (Scheme 6.23) [48]. β-Keto carbonyl substrates **76** were reacted with N-protected

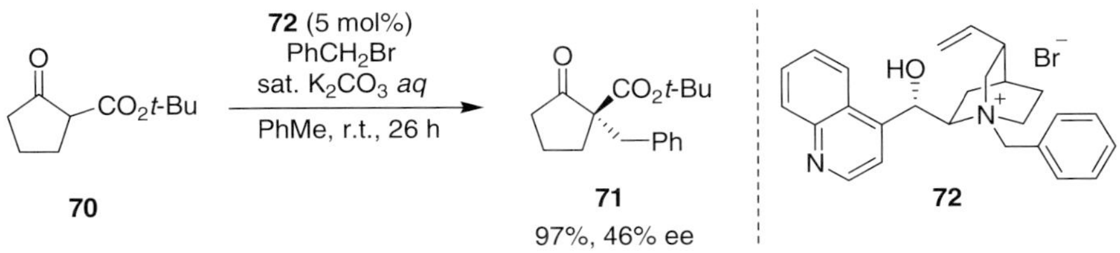

Scheme 6.21

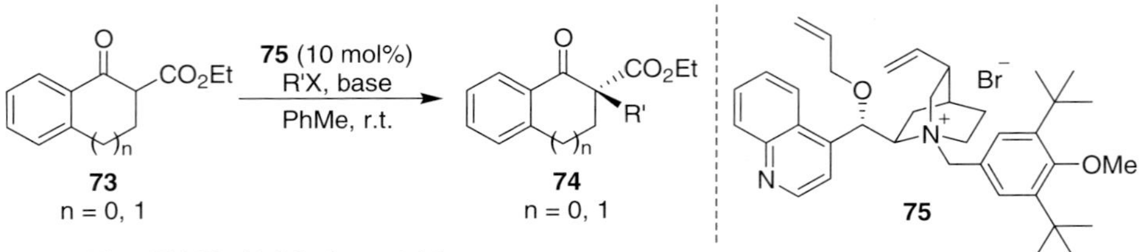

6 Cinchona-Catalyzed Nucleophilic α-Substitution of Carbonyl Derivatives

Scheme 6.22

n = 0, R' = -CH$_2$Ph, K$_2$CO$_3$ base, 72 h : 71%, 63% ee
n = 0, R' = -CH$_2$CHCH$_2$, K$_2$CO$_3$ base, 45 h : 70%, 57% ee
n = 1, R' = -CH$_2$Ph, KOH base, 6 h : 92%, 84% ee
n = 1, R' = -CH$_2$CH$_3$, KOH base, 29 h : 50%, 46% ee
n = 1, R' = -CH$_2$C$_6$H$_4$NO$_2$-p, KOH base, 13 h : 67%, 96% ee

aziridine **77** under phase-transfer catalytic conditions to afford the corresponding α-alkylated products **78** with high chemical and optical yields.

In 2005 and 2006, Jørgensen and coworkers reported the development of the first catalytic asymmetric nucleophilic aromatic substitution reaction of 2-(carboethoxy)cyclopentanone (**80**) with highly activated aromatic electrophiles

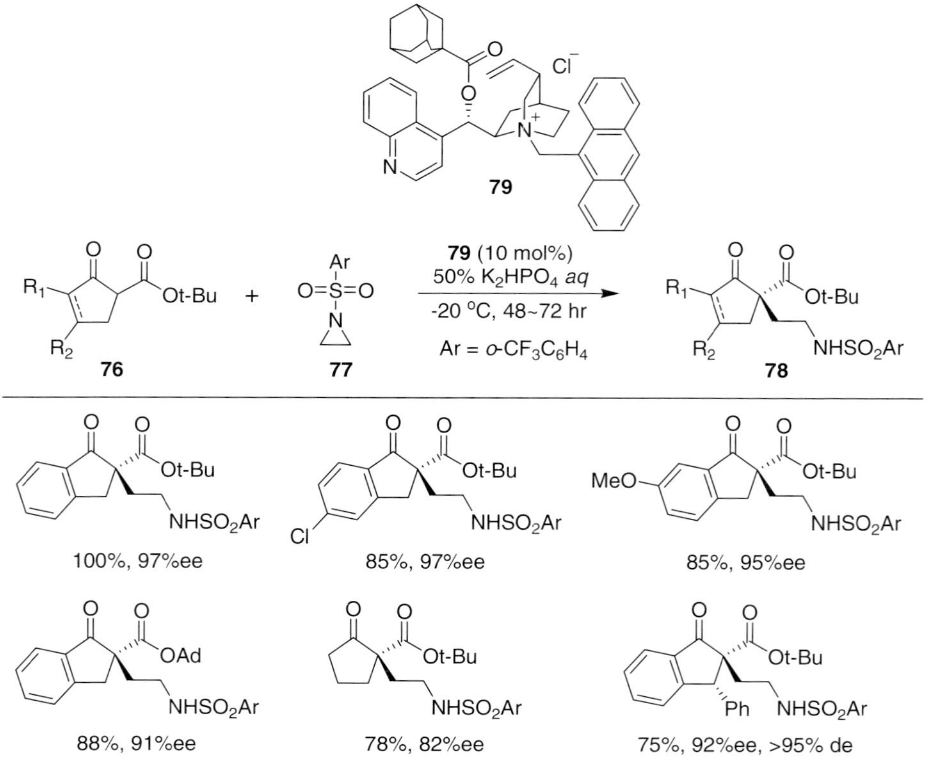

100%, 97%ee   85%, 97%ee   85%, 95%ee

88%, 91%ee   78%, 82%ee   75%, 92%ee, >95% de

Scheme 6.23

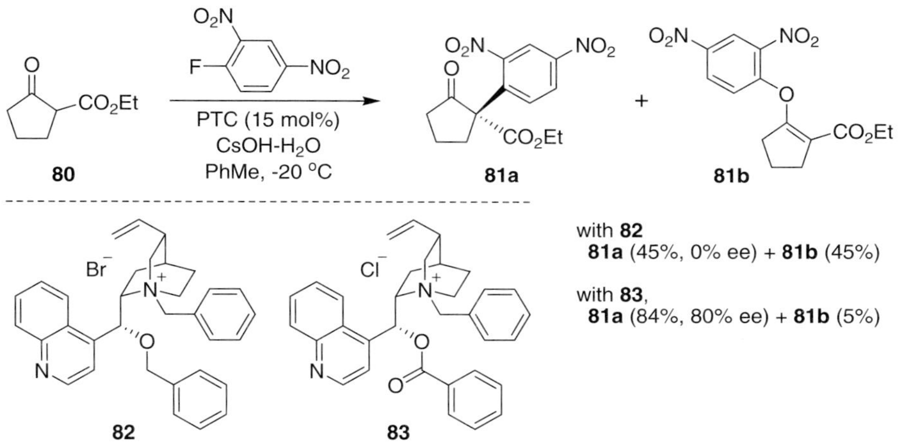

6.6 α-Substitution of Other Carbonyl Derivatives via PTC | 155

Scheme 6.24

such as 2,4-dinitrofluorobenzene under phase-transfer conditions (Scheme 6.24) [49]. A noteworthy difference in both regioselectivity (C-arylation versus O-arylation) and enantioselectivity was observed depending on the C9-O-substituent in the cinchona PTCs. The use of the O-benzoyl derivative **83** was found to be crucial to obtain C-substituted compound **81a** as the major product with high enantioselectivity. The authors proposed that the oxygen atom of the benzoyl group in **83** might stabilize an ion pair by hydrogen bonding, which significantly increased the C-arylation selectivity.

Jørgensen and coworkers also presented the first example of a catalytic enantioselective vinylic substitution in 2006 (Scheme 6.25) [50]. The hydrocinchoninium catalyst **79** incorporating a highly bulky substituent, 1-adamantoyl moiety, on the C9−O effectively catalyzed the introduction of α,β-unsaturated carbonyl groups to the α-position of the cyclic β-ketoester **70** via the addition–elimination-type substitution reaction, affording the vinylated product **85**. Generally high enantioselectivities (91–96% ee) were obtained with various β-chloropropenones **84**.

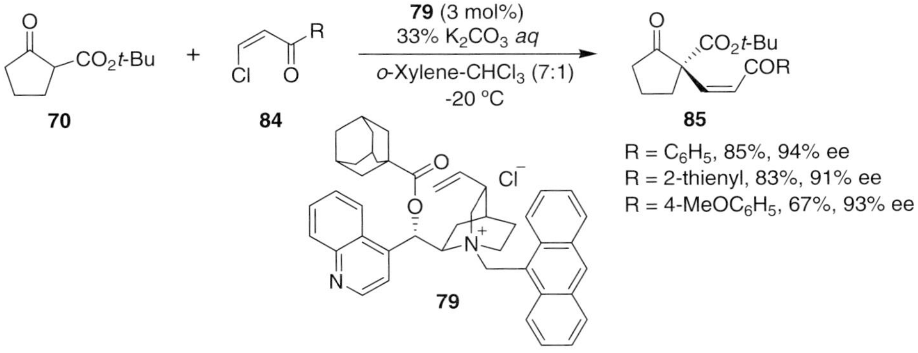

Scheme 6.25

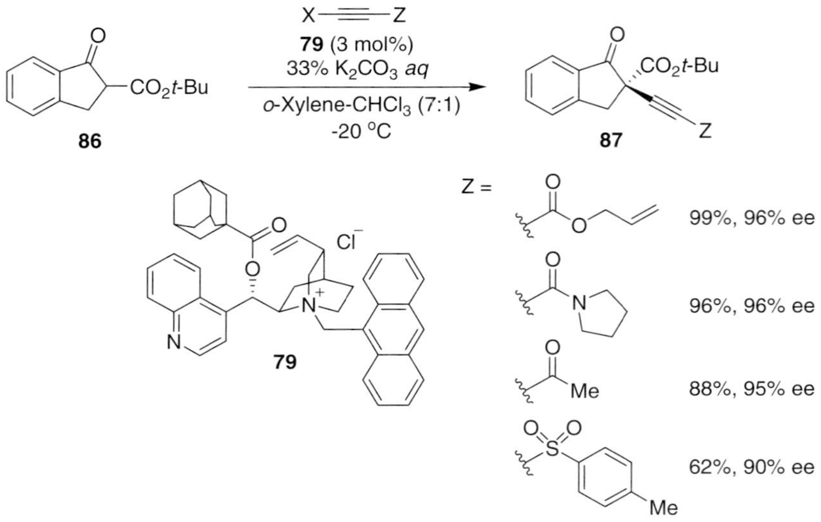

Scheme 6.26

In 2007, the same group also explored the direct α-alkynylation of the cyclic β-ketoester, 2-(*tert*-butoxycarbonyl)indanone (**86**), in the presence of the PTC **79** (Scheme 6.26) [51]. The alkynylated products **87** were obtained with excellent optical purities, and this transformation was extended to a wide range of substrates, including five-, six-, and seven-membered and aromatic conjugated β-ketoesters.

6.7
α-Heteroatom Substitution via PTC

The catalytic construction of heteroatom-substituted stereocenters has been an important and challenging task in asymmetric synthesis. Various synthetic methods for this issue have been developed. Among them, α-hydroxylation and α-halogenation via asymmetric phase-transfer catalytic reactions by cinchona quaternary ammonium salts are described here.

6.7.1
α-Hydroxylation of Carbonyl Derivatives

α-Hydroxylation of carbonyl derivatives was shown earlier than other heteroatom electrophiles in literature. In 1988, Shioiri and coworkers reported an approach for the synthesis of optically active useful α-hydroxy ketones including α-hydroxytetralone and α-hydroxyindanone from the corresponding aromatic cyclic ketones **88** using molecular oxygen as an electrophile along with triethylphosphite under phase-transfer conditions in the presence of **1** as catalyst (Scheme 6.27) [52].

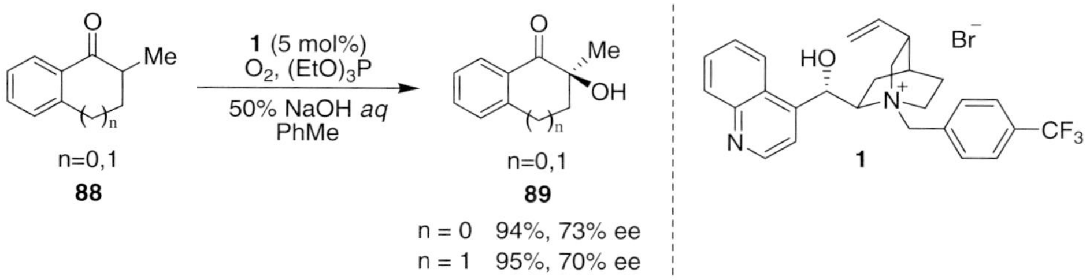

Scheme 6.27

A similar α-hydroxylation of 2-ethyltetralone was also performed by Dehmlow and coworkers [53].

6.7.2
α-Fluorination of Carbonyl Derivatives

Chiral organofluorine compounds are of importance because of their use in analytical, biological, medicinal, and polymer chemistry. Also, chiral organofluorine compounds containing a fluorine atom bonded directly to a stereogenic center have been applied to studies of enzyme mechanisms. In particular, fluorinated amino acids are of special interest for the design of new fluorine-containing peptides with unusual folding patterns, towing to H⋯F bonding or enzyme degradation resistance, which shows interesting biological properties. So, the development of effective asymmetric synthetic methodologies for α-fluorination has been popularly studied. In 2000, Shibata and coworkers reported direct fluorination of α-cyano ester **90** and trimethylsilylenol ethers **92** using stoichiometrically *in situ* generated N-fluoro-O(9)-acyl-hydroquinidinium salt **96** and N-fluoro-O(9)-acyl hydroquininium salt **97** to afford the corresponding α-fluorinated compounds **91**, **93**, respectively (Scheme 6.28) [54, 55]. The optimized N-fluoro-cinchona reagents **96**, **97** were adapted to α-fluorination of α-amino acid derivatives by Cahard and coworkers (Scheme 6.28) [56]. It was the first enantioselective electrophilic α-fluorination of phthaloylphenylglycinonitrile **94** for the synthesis of chiral α-fluorophenylglycine derivatives (up to 94% ee). Kim and coworkers reported the catalytic version of α-fluorination under phase-transfer conditions (Scheme 6.29) [57]. Treatment of β-keto esters **98** with N-fluorobenzenesulfonimide as the fluorine source in the presence of **100** under mild phase-transfer conditions afforded the corresponding α-fluoro-β-keto esters **99** in excellent yields with moderate enantiomeric excesses (up to 69% ee).

6.8
Nucleophilic α-Substitution of Carbonyl Derivatives via Non-PTC

Together with cinchona-PTC-mediated α-alkylations, the asymmetric nucleophilic α-substitution of carbonyl derivatives by using cinchona alkaloids as organocatalysts in nonbiphasic homogeneous conditions also have been extensively studied (e.g., arylation, hydroxylation, amination, hydroxyamination, and sulfenylation).

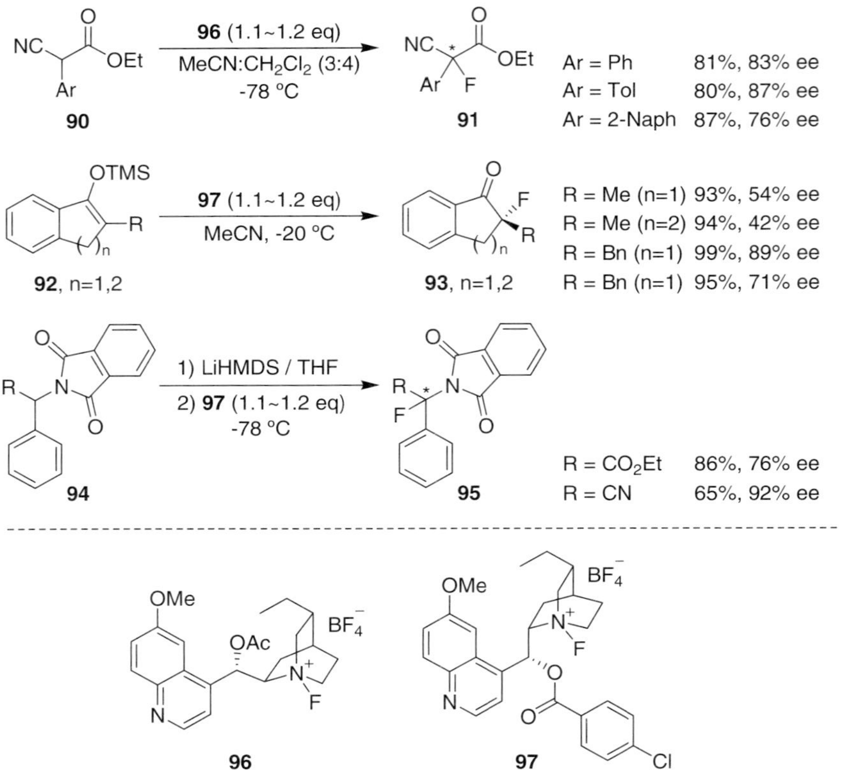

Scheme 6.28

6.8.1
α-Arylation of Carbonyl Derivatives

In 2007, Jørgensen and coworkers reported an asymmetric α-arylation of β-ketoester **101** under homogeneous condition using (−)-quinine as organocatalyst (Scheme 6.30) [58]. They demonstrated that the reaction of β-ketoesters **101** with 1,4-quinone **102** could be a good strategy of enantioselective α-arylation for aromatic

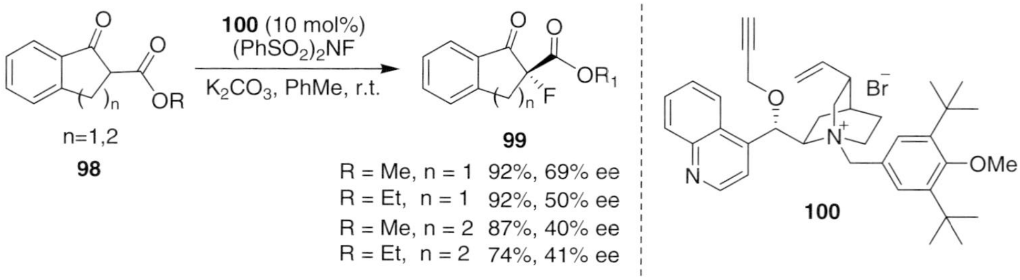

Scheme 6.29

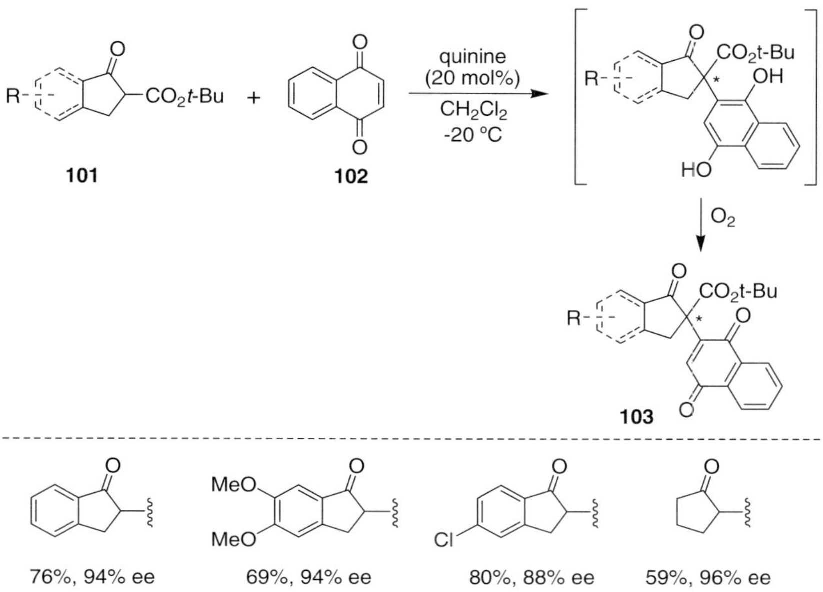
quinine
(20 mol%)
CH_2Cl_2
-20 °C
O_2
101
102
103
76%, 94% ee
69%, 94% ee
80%, 88% ee
59%, 96% ee

Scheme 6.30

compounds that contain electron-donating groups. This reaction can be very efficient one-pot synthetic method for complicated polycyclic and spiro chiral compounds (**103**) derived from various β-ketoesters and quinones.

6.8.2
α-Hydroxylation of Carbonyl Derivatives

In 2004, Jørgensen and coworkers disclosed α-hydroxylation of 1,3-dicarbonyl systems **104** (Scheme 6.31) [59]. They used hydroquinine (**106**) as the organocatalyst and cumyl hydroperoxide as the oxidant for the enantioselective α-hydroxylation of various β-keto esters **104** with enantioselectivities up to 80% ee. Optimization studies revealed that the *O*(9)-OH group of **106** is critical for good stereoselectivity, and the best enantioselectivity was obtained under CH_2Br_2 solvent at room temperature.

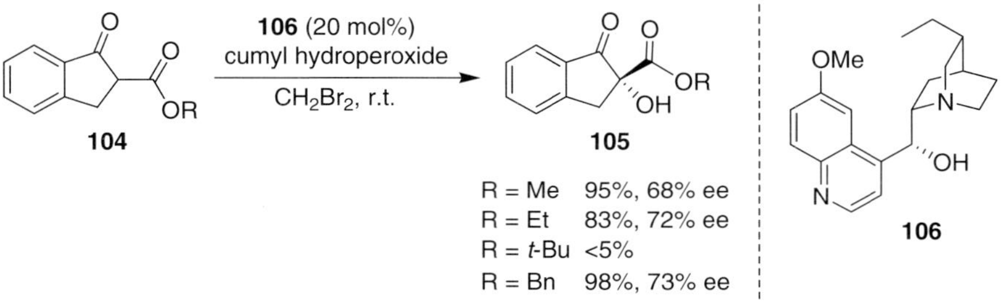
106 (20 mol%)
cumyl hydroperoxide
CH_2Br_2, r.t.
104
105
106
R = Me 95%, 68% ee
R = Et 83%, 72% ee
R = t-Bu <5%
R = Bn 98%, 73% ee

Scheme 6.31

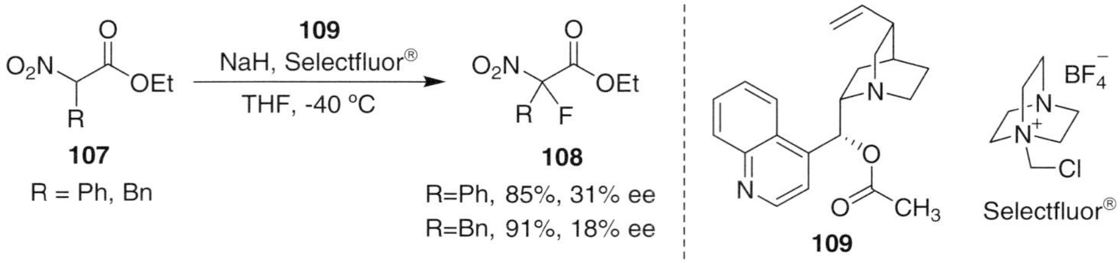

160 | *6 Cinchona-Catalyzed Nucleophilic α-Substitution of Carbonyl Derivatives*

Scheme 6.32

Furthermore, they demonstrated that the optically active α-hydroxy-β-keto esters **105** obtained undergo diastereoselective reduction giving anti-α,β-diols.

6.8.3
α-Halogenation of Carbonyl Derivatives

Stereoselective α-fluorination of α-nitro esters **107** was performed using Selectfluor® as a fluorinating agent and cinchona alkaloid catalyst **109** by Togni and coworkers (Scheme 6.32) [60]. Under the basic condition (NaH in THF at −40 °C) were obtained the α-fluorinated products **99** in high yield (up to 91%) with relatively low enantioselectivities (up to 31%).

Shibata and coworkers extended their enantioselective α-fluorination works [53, 54] to catalytic version using cinchona alkaloids–Selectfluor combination (Scheme 6.33) [61]. Acyl enol esters **110** were employed as substrates in the presence of **112** with Selectfluor and sodium acetate in CH_2Cl_2 to afford α-fluoroketones **111** (up to 53% ee).

This group continuously enlarged the scope of substrate to allylsilane, silyl enol ether **113** and oxindoles **115** for enantioselective catalytic α-fluorination (Scheme 6.34) [62]. They employed N-fluorobenzenesulfonimide (NFSI) as a fluorinating reagent with bis-cinchona alkaloid catalysts and excess base to provide the corresponding fluorinated compounds **114, 115** in excellent enantioselectivities up to 95% ee.

Cahard and coworkers prepared polystyrene-bound cinchona alkaloids (PS-CA, **119**), and successfully applied to enantioselective α-fluorination of silyl enol ether **117**. Soluble-phase PS-CA along with Selectfluor gave both good chemical and optical yields. Facile recovery of the PS-CA by solid/liquid separation allowed an

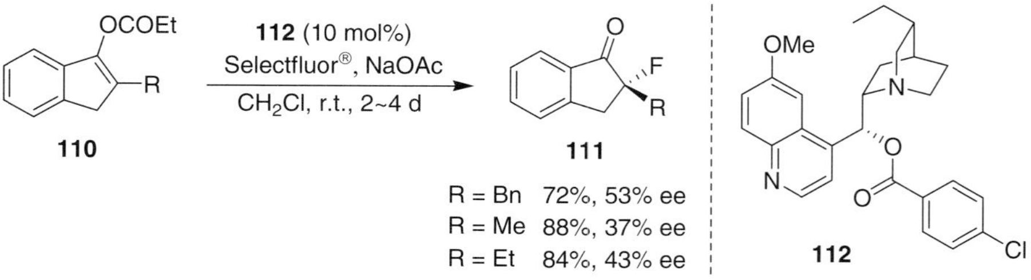

Scheme 6.33

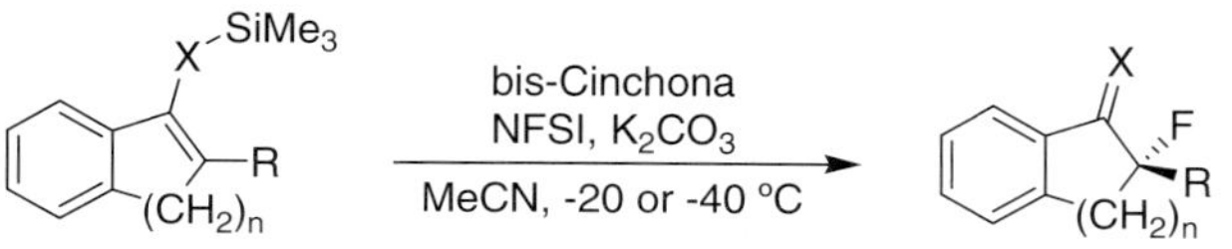

6.8 Nucleophilic α-Substitution of Carbonyl Derivatives via Non-PTC

113 n=1 or 2 → **114**

bis-Cinchona
NFSI, K$_2$CO$_3$
MeCN, -20 or -40 °C

X=CH$_2$, R=CH$_2$C$_6$H$_5$, (n=1) (DHQ)$_2$PYR, 75%, 94% ee
R=CH$_2$C$_6$H$_5$-p-Me, (n=1) (DHQ)$_2$PYR, 75%, 95% ee
R=CH$_2$C$_6$H$_5$-p-Cl, (n=1) (DHQ)$_2$PYR, 81%, 94% ee

X=O, R=CH$_2$C$_6$H$_5$, (n=2) (DHQ)$_2$PHAL, 82%, 82% ee
R=CH$_2$C$_6$H$_5$-p-Me, (n=2) (DHQ)$_2$PHAL, 79%, 86% ee
R=CH$_2$C$_6$H$_5$-p-Cl, (n=2) (DHQ)$_2$PHAL, 74%, 86% ee

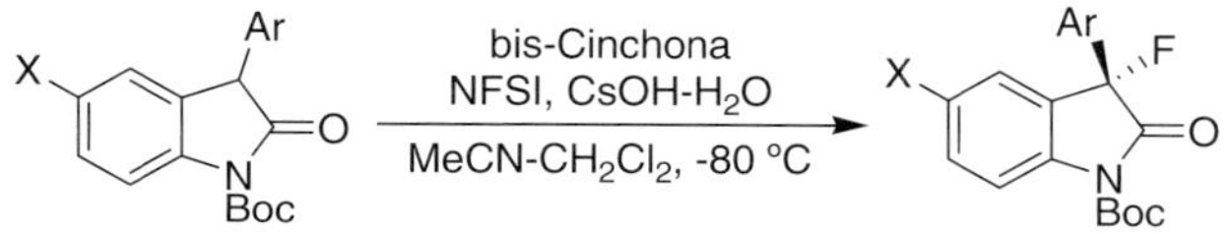

115 → **116**

bis-Cinchona
NFSI, CsOH-H$_2$O
MeCN-CH$_2$Cl$_2$, -80 °C

X=H,   Ar=Ph   (DHQ)$_2$AQN, 99%, 85% ee
X=H,   Ar=p-Tol (DHQ)$_2$AQN, 94%, 86% ee
X=Me,   Ar=p-Tol (DHQ)$_2$AQN, 86%, 84% ee
X=OMe, Ar=p-Tol (DHQ)$_2$AQN, 99%, 85% ee

Scheme 6.34

efficient recycling without any loss of enantioselectivity even in fourth uses (Scheme 6.35) [63].

In contrast to α-fluorination, direct α-chlorination and α-bromination were developed on the basis of ketene intermediate mechanism in the presence of O-benzoylquinine (**122**) by Lectka and coworkers in 2001 (Scheme 6.36) [64, 65]. Ketenes derived from acid chlorides **120** with either BEMP resin or proton sponge are added to **122** to form zwitterionic enolates, which are α-halogenated by perhaloquinones **123**, **124**. Finally, O-benzoylquinine moiety is substituted by haloaromatic phenolate anions, generated from **123** or **124**, to afford chiral α-halocarbonyl products **121** up to 99% ee.

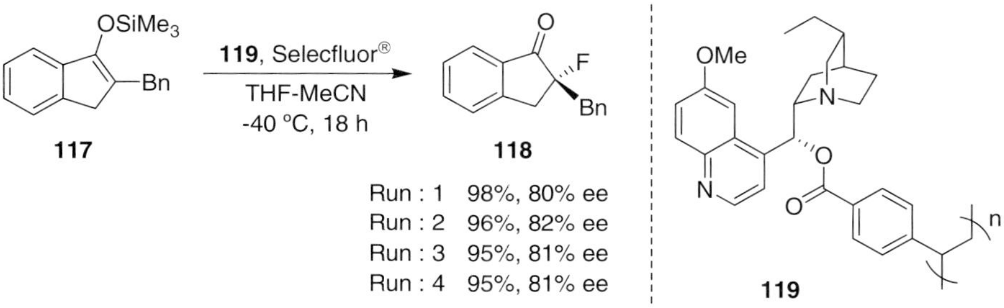

117 → **118**

119, Selecfluor®
THF-MeCN
-40 °C, 18 h

Run : 1 98%, 80% ee
Run : 2 96%, 82% ee
Run : 3 95%, 81% ee
Run : 4 95%, 81% ee

119

Scheme 6.35

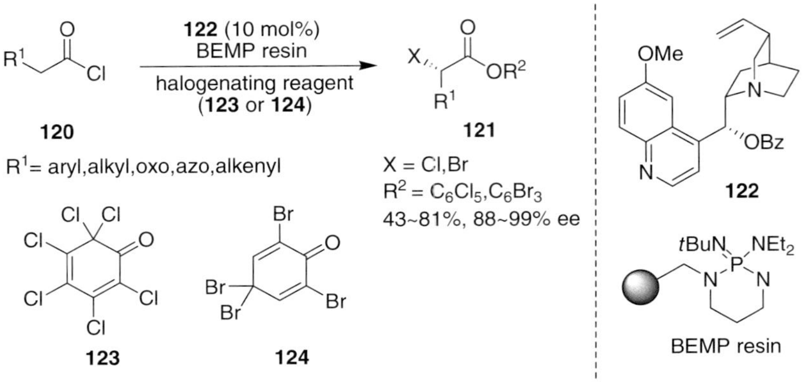

Scheme 6.36

In 2005, Lectka and coworkers also reported α-chlorination of acid halide by using polymer-supported cinchona catalyst via a column-based "flush and flow" system (Scheme 6.37) [66]. To a column of quinine-bound Wang resin **126** were added **120** and **123**, then the eluent (THF) was flowed by flushing to afford the corresponding α-chloroesters **125** up to 94% ee.

In 2006, Lectka and coworkers upgraded the enantioselective α-bromination with (R)-proline incorporated cinchona catalyst **129** and new bromination agent **127** (Scheme 6.38) [67]. It was notable that (R)-proline incorporated cinchona catalyst gave quite higher enantioselectivity (95% ee) than that of (S)-proline incorporated cinchona catalyst (88% ee) and their previous catalyst 122 in α-bromination of phenylacetyl chloride.

6.8.4
α-Amination of Carbonyl Derivatives

Catalytic asymmetric construction of nitrogen-substituted quaternary stereocenter has been another important and challenging task in asymmetric synthesis. The first α-amination using cinchona alkaloids was shown in 1,3-dicarbonyl substrate systems. In 2004, Pihko and coworkers showed enantioselective α-amination of cyclic β-ketoesters **130**, with dibenzyl azodicarboxylate (**131**) in the presence of

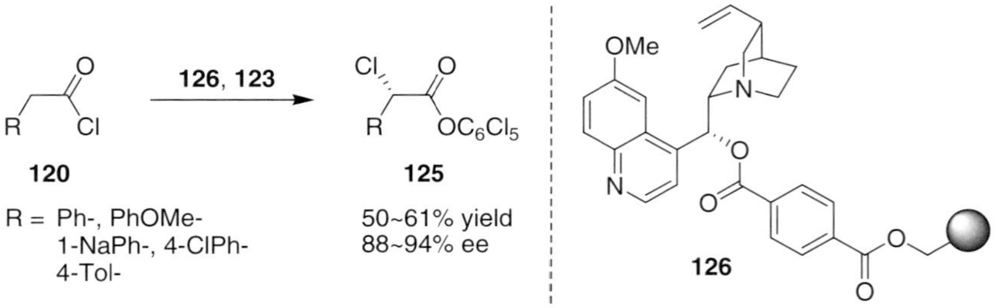

Scheme 6.37

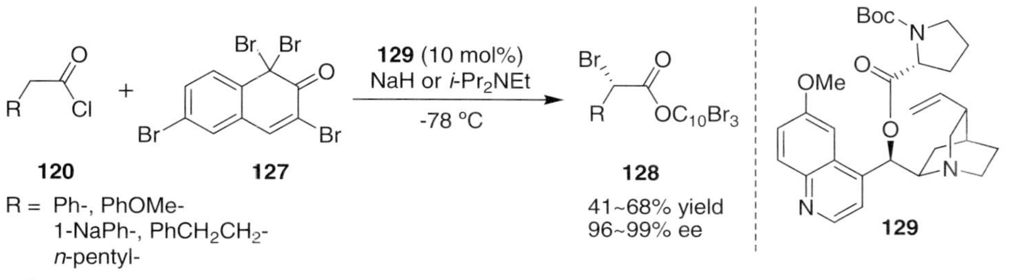

Scheme 6.38

cinchonidine or cinchonine as catalysts (Scheme 6.39) [68]. Highly enantioselective organocatalytic α-amination of α-aryl-α-cyanoacetates **133** was also reported by Jørgensen and coworkers. They accomplished highly enantioselective direct organocatalytic α-amination of **133** using di-*tert*-butyl azodicarboxylate (**134**) and catalytic amounts of α-isocupreidine (β-ICD, **136**) with enantioselectivities up to 98% ee (Scheme 6.40) [69]. Deng and coworkers extended the α-amination using their 6′-OH-modified bifunctional cinchona alkaloid **139** prepared from quinidine for the development of a highly enantioselective α-amination of α-aryl-α-cyanoacetates **137** (Scheme 6.41) [70].

Although there are several electrophilic α-amination methods using azodicarboxylates, Chen and coworkers reported the first α-amination of aromatic ketones (**140**) in 2007 (Scheme 6.42) [71]. They demonstrated that 9-amino-9-deoxyepicinchona alkaloid **143** was excellent organocatalyst for the direct enantioselective α-amination of aryl ketones **140** via enamine intermediate. *p*-Toluenesulfonic acid was used as

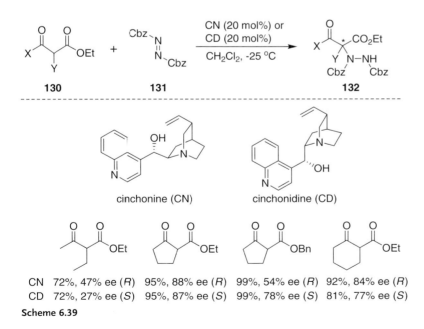

CN 72%, 47% ee (R)   95%, 88% ee (R)   99%, 54% ee (R)   92%, 84% ee (R)
CD 72%, 27% ee (S)   95%, 87% ee (S)   99%, 78% ee (S)   81%, 77% ee (S)

Scheme 6.39

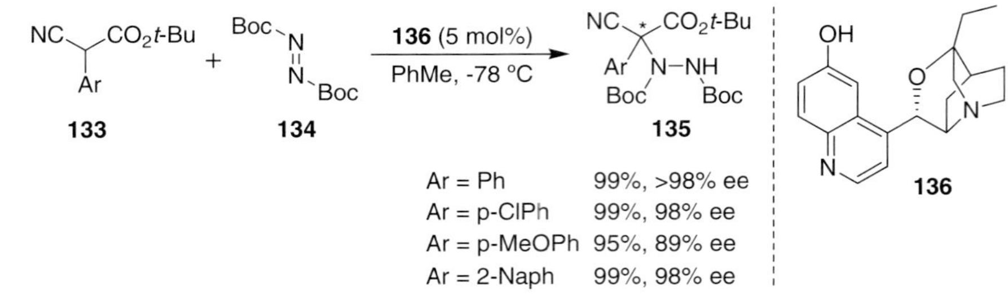
133
134
136 (5 mol%)
PhMe, -78 °C
135
136
Ar = Ph 99%, >98% ee
Ar = p-ClPh 99%, 98% ee
Ar = p-MeOPh 95%, 89% ee
Ar = 2-Naph 99%, 98% ee

Scheme 6.40

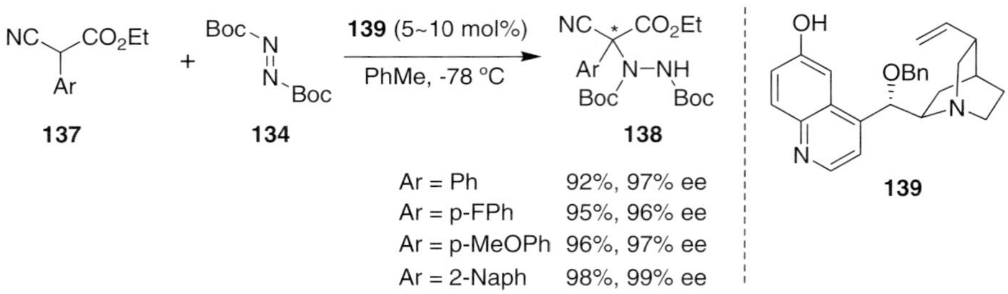
137
134
139 (5~10 mol%)
PhMe, -78 °C
138
139
Ar = Ph 92%, 97% ee
Ar = p-FPh 95%, 96% ee
Ar = p-MeOPh 96%, 97% ee
Ar = 2-Naph 98%, 99% ee

Scheme 6.41

acidic additive, and 4 Å molecular sieve was used for the removal of water during enamine formation of **140** and **141**. Both factors were very important to improve both chemical yield and enantioselectivity.

All the examples reported so far for the enantioselective organocatalytic aminoxylation (nucleophile attacks to the oxygen atom of the nitroso derivative) or hydroxyamination (nucleophile attacks to the nitrogen atom of the nitroso derivative) have employed an enamine intermediate and nitrosobenzene, but *O*-selectivity is obtained in majority. In 2007, a selective α-hydroxyamination was disclosed with cinchona alkaloids by Jørgensen and coworkers (Scheme 6.43) [72]. They developed the organocatalytic asymmetric addition of α-aryl-α-cyanoacetates **133** to nitrosobenzene (**144**) catalyzed by (−)-quinine in high yields and moderate enantioselectivities up to 59% ee.

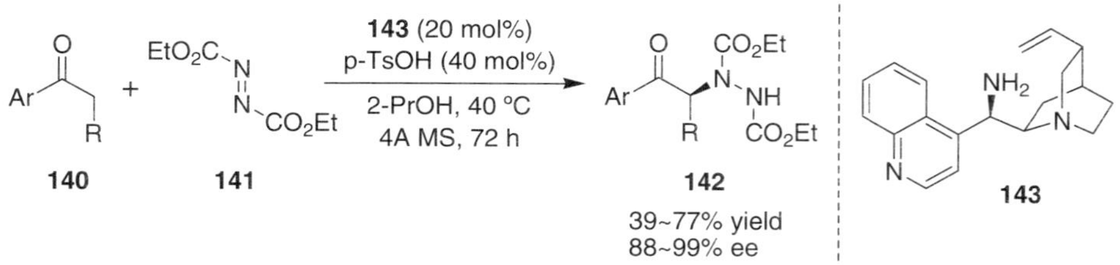
140
141
143 (20 mol%)
p-TsOH (40 mol%)
2-PrOH, 40 °C
4A MS, 72 h
142
39~77% yield
88~99% ee
143

Scheme 6.42

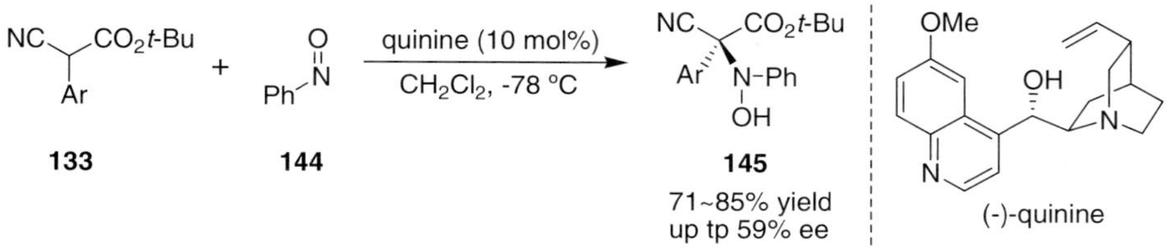

Scheme 6.43

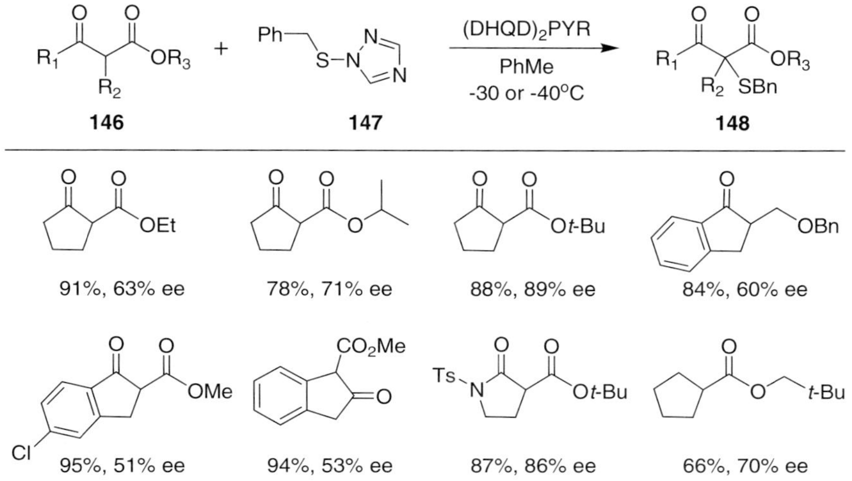

Scheme 6.44

6.8.5
α-Sulfenylation of Carbonyl Derivatives

In 2005, the application of cinchona alkaloid derivatives as catalysts for enantioselective α-sulfenylation of activated C−H bonds in lactones, lactams, and β-dicarbonyl compounds **146** by electrophilic sulfur reagent **147** was reported by Jørgensen and coworkers (Scheme 6.44) [73]. Optically active α-sulfenylated products **148** were obtained in good to excellent yields and up to 86% ee in the presence of 10 mol% $(DHQD)_2PYR$ in toluene solvent at −30 or −40 °C. Furthermore, the diastereoselective reduction of α-sulfenylated β-keto esters to give optically active α-sulfenylated β-hydroxy esters was also demonstrated.

6.9
Conclusions

The asymmetric α-substitution of carbonyl derivatives including imines of α-amino esters, achiral enolizable carbonyls, and cyclic β-ketoesters in the presence of chiral organocatalysts, especially chiral phase-transfer catalysts, is one of the powerful and efficient methods for the preparation of a variety of synthetically and/or biologically interesting organic compounds in optically active form. Cinchona alkaloids have played a very important role in the history of the development of catalytic asymmetric organic transformations as either various tertiary amine forms or their quaternary ammonium salt forms. As reviewed in this chapter, diverse types of cinchona alkaloid and cinchona alkaloidal ammonium salts have been developed as the highly efficient chiral catalysts for the asymmetric α-substitution of carbonyl derivatives. The unique structural feature and the ease of chemical modification as well as the availability of different forms of two pseudoenantiomers make the cinchona alkaloids be the first choice of resource to design new organic catalysts and reactions. A future challenge will be not only to explore the undiscovered talent of cinchona alkaloid as an organocatalyst but also to create a new efficient and practical organic reaction system.

References

1 For recent general reviews on asymmetric organocatalysis, see (a) Berkessel, A. and Gröger, H. (2005) *Asymmetric Organocatalysis – From Biomimetic Concepts to Applications in Asymmetric Synthesis*, Wiley-VCH Verlag GmbH, Weinheim; (b) B. List.(ed.) (2007) *Chem. Rev.*, **107** (12); (c) K.N. Houk and B. List (eds) (2004) *Acc. Chem. Res*, **37**, (8); (d) Ikunaka, M. (2007) *Org. Process Res. Dev.*, **11**, 495; (e) You, S.-L. (2007) *Chem. Asian J.*, **2**, 820; (f) Marion, N., Díez-González, S. and Nolan, S.P. (2007) *Angew. Chem. Int. Ed.*, **46**, 2; (g) Connon, S.J. (2006) *Chem. Eur. J.*, **12**, 5418; (h) List, B. (2006) *Chem. Commun.*, 819; (i) Seayad, J. and List, B. (2005) *Org. Biomol. Chem.*, **3**, 719; (j) Dalko, P.I. and Moisan, L. *Angew. Chem. Int. Ed*, (2004,) **43**, 5138.

2 For reviews on enamine catalysis, see (a) Mukherjee, S., Yang, J.W., Hoffmann, S., and List, B. (2007) *Chem. Rev.*, **107**, 5471; (b) List, B. (2004) *Acc. Chem. Res.*, **37**, 548; (c) List, B. (2002) *Tetrahedron*, **58**, 5572.

3 For reviews on organocatalysis via hydrogen bonding, see (a) Doyle, A.G., and Jacobsen, E.N. (2007) *Chem. Rev.*, **107**, 5713; (b) Akiyama, T., Itoh, J., and Fuchibe, K. (2006) *Adv. Syn. Catal.*, **348**, 999; (c) Taylor, M.S. and Jacobsen, E.N. (2006) *Angew. Chem. Int. Ed.*, **45**, 1520.

4 For reviews on organocatalysis via ion-pair, see (a) Dehmlow, E.V. and Dehmlow, S.S., (1993) *Phase Transfer Catalysis*, 3rd edn, Wiley-VCH Verlag GmbH, Weinheim; (b) Starks, C.M., Liotta, C.L., and Halpern, M.E. (1994) *Phase-Transfer Catalysis*, Chapman & Hall, New York; (c) Sasson, Y. and Neumann, R. (eds) (1997) *Handbook of Phase-Transfer Catalysis*, Blackie Academic & Professional, London; (d) Halpern, M.E.(ed.) (1997) *Phase-Transfer Catalysis*, (ACS Symposium Series 659), American Chemical Society, Washington, DC.

5 (a) Shioiri, T. (1997) in *Handbook of Phase-Transfer Catalysis* (eds Y. Sasson and R. Neumann), Blackie Academic & Professional, London, Chapter 14; (b) O'Donnell, M.J. (1998) *Phases – The Sachem Phase Transfer Catalysis Review*, (4), 5; (c) O'Donnell, M.J. (1999) *Phases – The Sachem Phase Transfer Catalysis Review*, (5), 5; (d) Nelson, A. (1999) *Angew. Chem. Int. Ed.*, **38**, 1583; (e) Shioiri, T. and Arai, S. (2000) in *Stimulating Concepts in Chemistry* (eds F. Vogtle, J.F. Stoddart, and M. Shibasaki), Wiley-VCH Verlag GmbH, Weinheim, p. 123; (f) O'Donnell, M.J. (2000) in *Catalytic Asymmetric Syntheses*,

2nd edn (ed. I. Ojima), Wiley-VCH Verlag GmbH, New York, Chapter 10; (g) O'Donnell, M.J. (2001) *Aldrichim. Acta*, **34**, 3; (h) Maruoka, K. and Ooi, T. (2003) *Chem. Rev.*, **103**, 3013; (i) O'Donnell, M.J. (2004) *Acc. Chem. Res.*, **37**, 506; (j) Lygo, B. and Andrews, B.I. (2004) *Acc. Chem. Res.*, **37**, 518.

6 For recent reviews on asymmetric phase-transfer catalysis, see (a) Maruoka, K.(ed.) (2008) *Asymmetric Phase Transfer Catalysis*, Wiley-VCH Verlag GmbH, Weinheim; (b) Maruoka, K. (2008) *Org. Process Res. Dev.*, **12**, 679; (c) Hashimoto, T. and Maruoka, K. (2007) *Chem. Rev.*, **107**, 5656; (d) Ooi, T. and Maruoka, K. (2007) *Aldrichim. Acta*, **40**, 77; (e) Ooi, T. and Maruoka, K. (2007) *Angew. Chem. Int. Ed.*, **46**, 4222; (f) Vachon, J. and Lacour, J. (2006) *Chimia*, **60**, 266; (g) Albanese, D. (2006) *Mini-Rev. Org. Chem.*, **3**, 195.

7 (a) O'Donnell, M.J., Esikova, I.A., Mi, A., Shullenberger, D.F., and Wu, S. (1997) in *Phase-Transfer Catalysis: Mechanisms and Synthesis* (ed. M.E. Halpern), (ACS Symposium Series 659), American Chemical Society, Washington,, DC; (b) Rabinovitz, M., Cohen, Y., and Halpern, M. (1986) *Angew. Chem. Int. Ed.*, **25**, 960.

8 Dolling, U.-H., Davis, P., and Grabowski, E.J.J. (1984) *J. Am. Chem. Soc.*, **106**, 446.

9 Hughes, D.L., Dolling, U.-H., Ryan, K.M., Schoenewaldt, E.F., and Grabowski, E.J.J. (1987) *J. Org. Chem.*, **52**, 4745.

10 O'Donnell, M.J. and Eckrich, T.M. (1978) *Tetrahedron Lett.*, **19**, 4625.

11 O'Donnell, M.J., Bennett, W.D., Bruder, W.A., Jacobsen, W.N., Knuth, K., LeClef, B., Polt, R.L., Boldwell, F.G., Mrozack, S.R., and Cripe, T.A. (1988) *J. Am. Chem. Soc.*, **110**, 8520.

12 (a) O'Donnell, M.J., Bennett, W.D. and Wu, S. (1989) *J. Am. Chem. Soc.*, **111**, 2353; (b) Lipkowitz, K.B., Cavanaugh, M.W., Baker, B., and O'Donnell, M.J. (1991) *J. Org. Chem.*, **56**, 5181.

13 O'Donnell, M.J., Wu, S., and Huffman, J.C. (1994) *Tetrahedron*, **50**, 4507.

14 (a) Lygo, B. and Wainwright, P.G. (1997) *Tetrahedron Lett.*, **38**, 8595; (b) Lygo, B., Crosby, J., Lowdon, T.R., and Wainwright, P.G. (2001) *Tetrahedron*, **57**, 2391; (c) Lygo, B., Crosby, J., Lowdon, T.R., Peterson, J.A., and Wainwright, P.G. (2001) *Tetrahedron*, **57**, 2403.

15 Corey, E.J., Xu, F., and Noe, M.C. (1997) *J. Am. Chem. Soc.*, **119**, 12414.

16 Lee, J.-H., Yoo, M.-S., Jung, J.-H., Jew, S.-s., Park, H.-g., and Jeong, B.-S. (2007) *Tetrahedron*, **63**, 7906.

17 (a) Jew, S.-s., Jeong, B.-S., Yoo, M.-S., Huh, H., and Park, H.-g. (2001) *Chem. Commun.*, 1244; (b) Park, H.-g., Jeong, B.-S., Yoo, M.-S., Park, M.-k., Huh, H., and Jew, S.-s. (2001) *Tetrahedron Lett.*, **42**, 4645; (c) Park, H.-g., Jeong, B.-S., Yoo, M.-S., Lee, J.-H., Park, M.-k., Lee, Y.-J., Kim, M.-J., and Jew, S.-s. (2002) *Angew. Chem. Int. Ed.*, **41**, 3036; (d) Park, H.-g., Jeong, B.-S., Yoo, M.-S., Lee, J.-H., Park, B.-s., Kim, M.G., and Jew, S.-s. (2003) *Tetrahedron Lett.*, **44**, 3497.

18 (a) Chinchilla, R., Mazón, P., and Nájera, C. (2002) *Tetrahedron: Asymmetry*, **13**, 927; (b) Chinchilla, R., Mazón, P., Nájera, C., and Ortega, F.J. (2004) *Tetrahedron: Asymmetry*, **15**, 2603.

19 (a) Siva, A. and Murugan, E. (2005) *J. Mol. Catal. A: Chem.*, **241**, 111; (b) Siva, A. and Murugan, E. (2005) *Synthesis*, 2927; (c) Siva, A. and Murugan, E. (2006) *J. Mol. Catal. A: Chem.*, **248**, 1.

20 Jew, S.-s., Yoo, M.-S., Jeong, B.-S., Park, I.Y., and Park, H.-g. (2002) *Org. Lett.*, **4**, 4245.

21 Yoo, M.-S., Jeong, B.-S., Lee, J.H., Park, H.-g., and Jew, S.-s. (2005) *Org. Lett.*, **7**, 1129.

22 Mazón, P., Chinchilla, R., Nájera, C., Guillena, G., Kreiter, R., Klein Gebbink, R.J.M., and van Koten, G. (2002) *Tetrahedron: Asymmetry*, **13**, 2181.

23 Elango, S., Venugopal, M., Suresh, P.S. and Elango, E. (2005) *Tetrahedron*, **61**, 1443.

24 Kumar, S. and Ramachandran, U. (2005) *Tetrahedron*, **61**, 7022.

25 Andrus, M.B., Ye, Z., and Zhang, J. (2005) *Tetrahedron Lett.*, **46**, 3839.

26 Wang, X., Lv, J., Wang, Y., and Wu, Y. (2007) *J. Mol. Catal. A: Chem.*, **276**, 102.

27 Li, Q., Li, L., Pei, W., Wang, S., and Zhang, Z. (2008) *Synth. Commun.*, **38**, 1470.

28 (a) Chinchilla, R., Mazón, P., and Nájera, C. (2000) *Tetrahedron: Asymmetry*, **11**, 3277; (b) Chinchilla, R., Mazón, P., and Nájera, C. (2004) *Adv. Synth. Catal.*, **346**, 1186.

29 Thierry, B., Plaquevent, J.-C., and Cahard, D. (2001) *Tetrahedron: Asymmetry*, **12**, 983.

30 Thierry, B., Perrard, T., Audouard, C., Plaquevent, J.-C., and Cahard, D. (2001) *Synthesis*, 1742.

31 (a) Shi, Q., Lee, Y.-J., Kim, M.-J., Park, M.-K., Lee, K., Song, H., Cheng, M., Jeong, B.-S., Park, H.-g., and Jew, S.-s. (2008) *Tetrahedron Lett.*, **49**, 1380; (b) Shi, Q., Lee, Y.-J., Song, H., Cheng, M., Jew, S.-s., Park, H.-g., and Jeong, B.-S. (2008) *Chem. Lett.*, **37**, 436.

32 Danelli, T., Annunziata, R., Benaglia, M., Cinquini, M., Cozzi, F., and Tocco, G. (2003) *Tetrahedron: Asymmetry*, **14**, 461.

33 Thierry, B., Plaquevent, J.-C., and Cahard, D. (2003) *Tetrahedron: Asymmetry*, **14**, 1671.

34 Wang, X., Yin, L., Yang, T., and Wang, Y. (2007) *Tetrahedron: Asymmetry*, **18**, 108.

35 For an excellent recent review on asymmetric synthesis of α,α-disubstituted α-amino acids, see Vogt, H. and Bräse, S., (2007) *Org. Biomol. Chem.*, **5**, 406.

36 O'Donnell, M.J. and Wu, S. (1992) *Tetrahedron: Asymmetry*, **3**, 591.

37 Lygo, B., Crosby, J., and Peterson, J.A. (1999) *Tetrahedron Lett.*, **40**, 8671.

38 Jew, S.-s., Jeong, B.-S., Lee, J.-H., Yoo, M.-S., Lee, Y.-J., Park, B.-s., Kim, M.G., and Park, H.-g. (2003) *J. Org. Chem.*, **68**, 4514.

39 Chinchilla, R., Nájera, C., and Ortega, F.J. (2007) *Eur. J. Org. Chem.*, 6034.

40 (a) Lee, Y.-J., Lee, J., Kim, M.-J., Kim, T.-S., Park, H.-g., and Jew, S.-s. (2005) *Org. Lett.*, **7**, 1557; (b) Kim, T.-S., Lee, Y.-J., Jeong, B.-S., Park, H.-g., and Jew, S.-s. (2006) *J. Org. Chem.*, **71**, 8276.

41 Nerinckx, W. and Vandewalle, M. (1990) *Tetrahedron: Asymmetry*, **1**, 265.

42 Corey, E.J., Bo, Y. and Busch-Petersen, J. (1998) *J. Am. Chem. Soc.*, **120**, 13000.

43 Arai, S., Oku, M., Ishida, T., and Shioiri, T. (1999) *Tetrahedron Lett.*, **40**, 6785.

44 (a) Andrus, M.B., Hicken, E.J., and Stephens, J.S. (2004) *Org. Lett.*, **6**, 2289; (b) Andrus, M.B., Hicken, E.J., Stephens, J.S., and Bedke, D.K. (2005) *J. Org. Chem.*, **70**, 3670.

45 Kumar, S. and Ramachandran, U. (2005) *Tetrahedron: Asymmetry*, **16**, 647.

46 Dehmlow, E.V., Düttmann, S., Neumann, B., and Stammler, H.-G. (2002) *Eur. J. Org. Chem.*, 2087.

47 Park, E.J., Kim, M.H., and Kim, D.Y. (2004) *J. Org. Chem.*, **69**, 6897.

48 Moss, T.A., Fenwick, D.R., and Dixon, D.J. (2008) *J. Am. Chem. Soc.*, **130**, 10076.

49 (a) Bella, M., Kobbelgaard, S., and Jørgensen, K.A. (2005) *J. Am. Chem. Soc.*, **127**, 3670; (b) Kobbelgaard, S., Bella, M., and Jørgensen, K.A. (2006) *J. Org. Chem.*, **71**, 4980.

50 Poulsen, T.B., Bernardi, L., Bell, M., and Jørgensen, K.A. (2006) *Angew. Chem. Int. Ed.*, **45**, 6551.

51 Poulsen, T.B., Bernardi, L., Alemán, J., Overgaard, J., and Jørgensen, K.A. (2007) *J. Am. Chem. Soc.*, **129**, 441.

52 Masui, M., Ando, A., and Shioiri, T. (1988) *Tetrahedron Lett.*, **29**, 2835.

53 Dehmlow, E.V., Düttmann, S., Neumann, B., and Stammler, H.-G. (2002) *Eur. J. Org. Chem.*, **24**, 2087.

54 Shibata, N., Suzuki, E., and Takeuchi, Y. (2000) *J. Am. Chem. Soc.*, **122**, 10728.

55 Shibata, N., Suzuki, E., Asahi, T., and Shiro, M. (2001) *J. Am. Chem. Soc.*, **123** (29), 7001.

56 Mohar, B., Baudoux, J., Plaquevent, J.-C., and Cahard, D. (2001) *Angew. Chem. Int. Ed.*, **40**, 4214.

57 Kim, D.Y. and Park, E.J. (2002) *Org. Lett.*, **4**, 545.

58 Alemán, J., Richter, B., and Jørgensen, K.A. (2007) *Angew. Chem. Int. Ed.*, **46**, 5515.

59 Acocella, M.R., Mancheno, O.G., Bella, M., and Jørgensen, K.A. (2004) *J. Org. Chem.*, **69**, 8165.

60 Ramírez, J., Huber, D.P., and Togni, A. (2007) *Synlett*, 1143.

61 Fukuzumi, T., Shibata, N., Sugiura, M., Nakamura, S., and Toru, T. (2006) *J. Fluorine Chem.*, **126**, 548.

62 Ishimaru, T., Shibata, N., Horikawa, T., Yasuda, N., Nakamura, S., Toru, T., and Shiro, M. (2008) *Angew. Chem. Int. Ed.*, **47**, 4157.

63 Thierry, B., Audouard, C., Plaquevent, J.-C., and Cahard, D. (2004) *Synlett*, 856.

64 Wack, H., Taggi, A.E., Hafez, A.M., Drury, W.J., III, and Lectka, T. (2001) *J. Am. Chem. Soc.*, **123**, 1531.

65 France, S., Wack, H., Taggi, A.E., Hafez, A.M., Wagerle, T.R., Shah, M.H., Dusich, C.L., and Lectka, T. (2004) *J. Am. Chem. Soc.*, **126**, 4245.

66 Bernstein, D., France, S., Wolfer, J., and Lectka, T. (2005) *Tetrahedron: Asymmetry*, **16**, 3481.

67 Dogo-Isonagie, C., Bekele, T., France, S., Wolfer, J., Weatherwax, A., Taggi, A.E., and Lectka, T. (2006) *J. Org. Chem.*, 71, 8946.

68 Pihko, P.M. and Pohjakallio, A. (2004) *Synlett*, 2115.

69 Saaby, S., Bella, M. and Jørgensen, K.A. (2004) *J. Am. Chem. Soc.*, **126**, 8120.

70 Liu, X., Li, H. and Deng, L. (2005) *Org. Lett.*, **7**, 167.

71 Liu, T.-Y., Cui, H.-L., Zhang, Y., Jiang, K., Du, W., He, Z.-Q., and Chen, Y.-C. (2007) *Org. Lett.*, **9**, 3671.

72 López-Cantarero, J., Cid, M.B., Poulsen, T.B., Bella, M., Ruano, J.L.G., and Jørgensen, K.A. (2007) *J. Org. Chem.*, **72**, 7062.

73 Sobhani, S., Fielenbach, D., Marigo, M., Wabnitz, T.C., and Jørgensen, K.A. (2005) *Chem. Eur. J.*, **11**, 5689.

7
Cinchona-Mediated Enantioselective Protonations
Jacques Rouden

7.1
Introduction

Conceptually, the enantioselective protonation is one of the oldest approaches to control the configuration of an asymmetric carbon. In most examples, it refers to the kinetically controlled C-protonation of a prostereogenic center such as an enolate or an enol and a tautomerization in the later case. This old concept, as a direct route for the preparation of optically active α-substituted carbonyl compounds (or electron-withdrawing groups), is still underdeveloped despite its synthetic potential and its apparent simplicity. Among the possible reasons, we can mention the following: the difficulty of controlling the approach of a proton toward a prostereogenic nucleophilic carbon due to its fast diffusion and small size, the problem of the O-protonation leading reversibly to the enol that can be tautomerized to the ketone without facial discrimination, the presence of two possible diastereoisomers Z and E of the enolate that can exhibit different enantiofacial selectivities, and the complexity of phenomena linked to metal aggregates, solvation, and complexation. All these mechanistic aspects have been carefully discussed in recent surveys [1].

In this chapter, we review the enantioselective protonation of enols/enolates where the asymmetry is brought by cinchona alkaloids, either the natural products or some analogues. The cinchona alkaloids may act as a "direct" protonating agent of enolates or as an acid–base bifunctional catalyst by first deprotonating the substrate to generate the enolate and then, as an acid, by reprotonating the carbanion.

Several methodologies have been developed to generate the prostereogenic intermediate necessary to achieve enantioselective protonation but all have in common a stable or transient species, enol or enolate, which is being protonated by a chiral proton source. In specific cases, it is difficult to determine the real structure of the intermediate obtained, enolate or enol or both, because of the lack of its characterization and precise mechanistic investigations.

Enantioselective protonations can be categorized into five groups depending on the type of reaction and substrate used to generate the key prochiral

Cinchona Alkaloids in Synthesis and Catalysis, Ligands, Immobilization and Organocatalysis
Edited by Choong Eui Song
Copyright © 2009 WILEY-VCH Verlag GmbH & Co. KGaA, Weinheim
ISBN: 978-3-527-32416-3

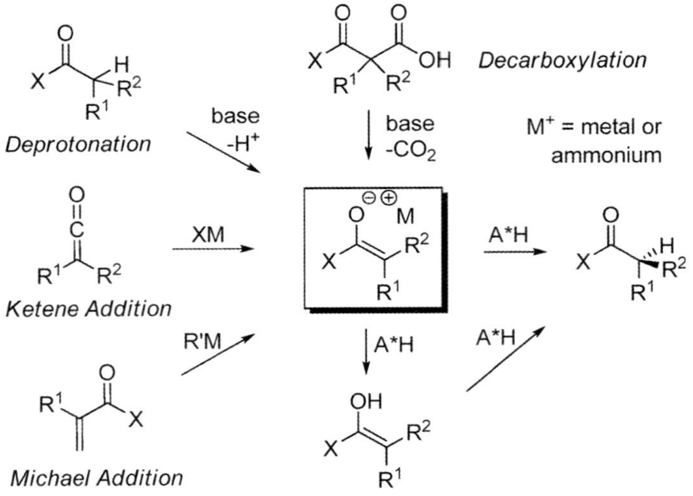

Scheme 7.1 Reactions for enolate generation.

intermediate (Scheme 7.1). The first group is devoted to the protonation of "preformed" enolates. In this group, the protonation of metal enolates generated by deprotonation of racemic substrates using a strong base will be distinguished from the desilylative protonation of silyl enol ethers and from the protonation of enamines. The second group deals with the nucleophilic additions on ketene derivatives catalyzed by organic bases such as tertiary amines. The third group describes nucleophilic additions on α,β-unsaturated carbonyl compounds. The fourth group reports on the decarboxylative protonation of malonyl or β-ketoacetic substrates. Finally, the last part gathers miscellaneous reactions involving asymmetric proton migrations mediated by cinchona alkaloids (not described in Scheme 7.1).

In this chapter, all research works directly in relation with "enantioselective protonations mediated by cinchona alkaloids" are exemplified by schemes. Several publications related to this type of chemistry, but not using cinchona alkaloids, are mentioned for comparison purposes but are not illustrated. Therefore, the reader can easily differentiate the studies involving cinchona alkaloids from other publications of interest for a better overall view of the research field.

7.2
Preformed Enolates and Equivalents

Among the methodologies listed in the introduction to generate the key enol/enolate intermediate, the enantioselective protonations of metal enolates, the so-called preformed enolates, or of their substitutes such as enamines or enol ethers have known, by far, the most intensive research development (Scheme 7.2).

On the asymmetric point of view, it can be summarized that the level of enantioselectivity obtained is highly dependent on the nature of the substrate,

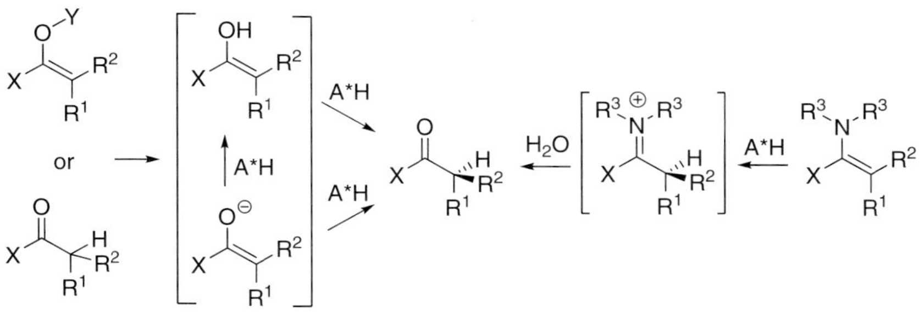

Scheme 7.2 Enols, enolates, and enamines for enantioselective protonations.

on the chiral source of proton, and on the experimental conditions (concentration, temperature, solvent, additives, and hydrolysis conditions), and no general trend has emerged. A survey of the literature data reveals that cinchona alkaloids have been little used although they display some of the structural requirements of the usual chiral protonating agents involved in these methodologies, in particular a proton linked to an oxygen atom such as an alcohol group and a tertiary amine. As a matter of fact, ephedrine and other β-amino alcohols have been frequently employed [2].

Since the seminal work of Lucette Duhamel [3] in 1976 describing what is the first "direct" asymmetric protonation of an enolate (in fact its enamine analogue), it is only in 1992 that Takeuchi et al. successfully used a cinchona alkaloid for the enantioselective protonation of a particular samarium enediolate under mild conditions [4]. Samarium diodide reduced benzil **1** into the corresponding enediolate **2**, which was then enantioselectively protonated by quinidine **3** at room temperature, affording (R)-benzoin **4** in 91% ee (Scheme 7.3). The presence of molecular oxygen was necessary to obtain high selectivities. However, the procedure was not catalytic as 3 equiv of quinidine **3** were needed. Moreover, only one substrate was described showing the limits of this procedure.

Recently, Levacher and coworkers developed the first organocatalytic enantioselective protonation of silyl enol ethers **5** using readily available cinchona alkaloids [5].

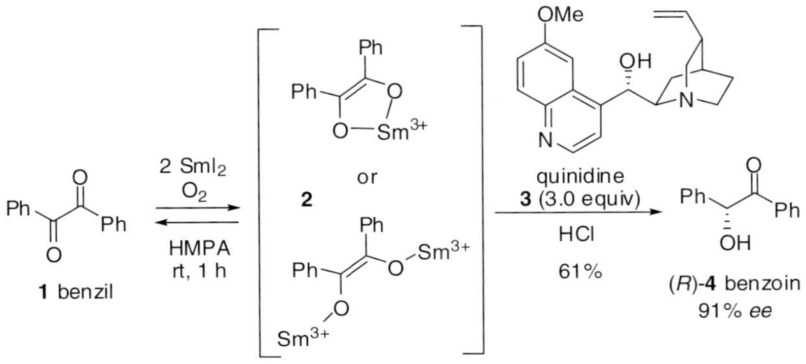

Scheme 7.3 Takeuchi's benzil-to-benzoin reductive asymmetric protonation [4].

Table 7.1 Levacher's organocatalytic enantioselective protonation of silyl enol ethers [5]

Entry	Silyl enol ethers	Yield (%)	ee (%)	Configuration[a]
1	**5a** $n=2$, $R^1=$ Me, $R^2=$ H	88	81	S
2	**5b** $n=2$, $R^1=$ Bn, $R^2=$ H	84	85	R
3	**5c** $n=2$, $R^1=$ Me, $R^2=$ OMe	98	81	nd
4	**5d** $n=2$, $R^1=$ Bn, $R^2=$ OMe	98	92	nd
5	**5e** $n=1$, $R^1=$ Me, $R^2=$ H	78	71	S
6	**5f** $n=1$, $R^1=$ Et, $R^2=$ H	76	74	S
7	**5g** $n=1$, $R^1=$ Bn, $R^2=$ H	78	64	nd

nd: not determined.
[a]Assigned by comparison with literature data.

Among the several screened cinchona derivatives, (DHQ)$_2$AQN **6** was selected for a full optimization of the conditions (solvent and temperature). The reaction was carried out in DMF with a combination of benzoyl fluoride and ethanol as a latent source of HF. Enantioselectivities up to 92% ee were obtained with tetralones trimethylsilyl enol ethers **5a–d** using 10 mol% of organocatalyst **6** (Table 7.1, entries 1–4). Indanone silyl enol ethers **5e–g** afforded the corresponding ketones **7e–g** with lower selectivities (64–74% ee, Table 7.1, entries 5–7) while the 2,2,6-trimethyl-cyclohexanone was obtained with only 58% ee, the reaction being conducted at −10 °C (data not shown in Table 7.1).

Shortly after, the same group published a study where readily available carboxylic acids, diacids, and N-protected amino acids were screened as proton sources [6]. The same substrates were used in the presence of citric acid instead of HF. This catalytic system displayed somewhat lower selectivity. For example, by using similar experimental conditions in the presence of citric acid at −10 °C, the enantioselective protonation of silyl enol ether **5c** afforded the corresponding ketone **7c** in excellent yield but lower enantioselectivity (up to 75% ee, Scheme 7.4, to be compared with entry 3, Table 7.1). However, upon further optimization, this process seems appealing in terms of simplicity, practicability, environmental concerns, and cost; therefore, adjustable for industrial use.

Since the first report by Duhamel, most of the enantioselective protonations were involving metal enolates and, therefore, the use of stoichiometric amounts of the chiral proton sources although catalytic versions are now emerging [7]. This subsection has summarized the three noticeable examples describing the

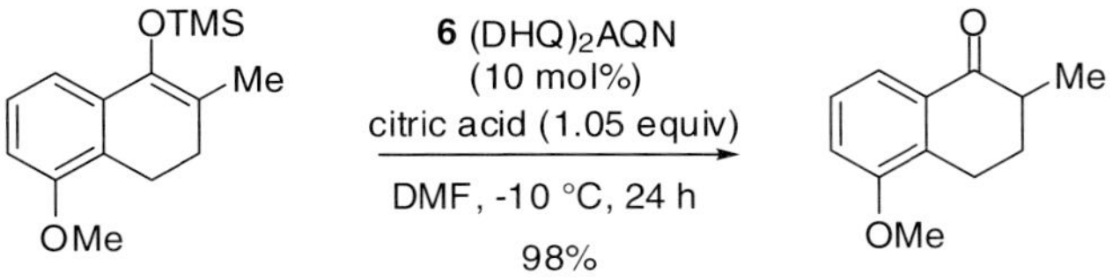

Scheme 7.4 Simplified organocatalytic protonation of a silyl enol ether [6].

enantioselective protonation of enolates or their equivalents using cinchona alkaloids. Today, with the advent of organocatalysis, numerous new analogues of these natural products have been synthesized, and some published works dealing with enantioselective protonations of enolates or their equivalents could be advantageously re-evaluated using these new structures as a readily available source of chirality.

7.3
Nucleophilic Addition on Ketenes

The nucleophilic addition on substituted ketenes is a well-known method to generate a prochiral enolate that can be further protonated by a chiral source of proton. Metallic nucleophiles are used under anhydrous conditions; therefore, the optically pure source of proton must be added then (often in a stoichiometric amount) to control the protonation. In the case of a protic nucleophile, an alcohol, a thiol, or an amine, the chiral inductor is usually present at the beginning of the reaction since it also catalyzes the addition of the heteroatomic nucleophile before mediating the enantioselective protonation (Scheme 7.5). The use of a chiral tertiary amine as catalyst generates a zwitterionic intermediate **B** by nucleophilic addition on ketene **A**, followed by a rapid diastereoselective protonation of the enolate to acylammonium **C**, and then the release of the catalyst via its substitution by the nucleophile ends this reaction sequence.

The pioneering studies in this field are done by Pracejus and coworkers in a series of papers published in the 1960s in which they studied the asymmetric addition/protonation of alcohols to ketenes [8]. They carried out a number of experiments to demonstrate the catalytic role of a tertiary amine on the nucleophilic

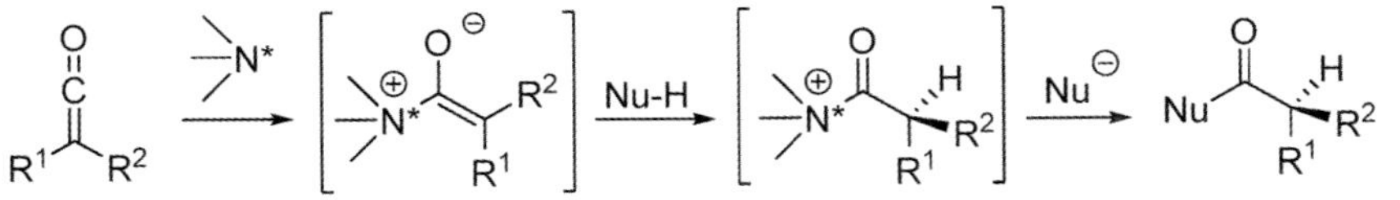

Nu: Heteroatomic Nucleophile

Scheme 7.5 Chiral tertiary amine catalyzed addition of heteroatomic nucleophiles on ketenes.

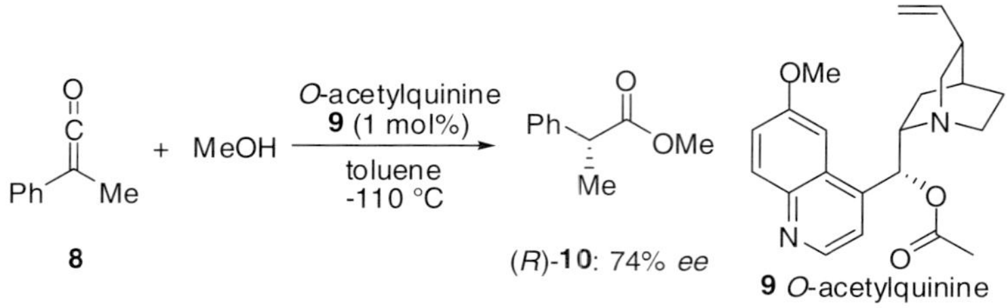

Scheme 7.6 Tandem methanol addition/asymmetric protonation on phenylmethylketene [8a].

addition as the rate determining step and the importance of the temperature on the selectivity-dependence protonating step. As a result of this important work, the best asymmetric protonation was observed in the addition of methanol on phenylmethylketene **8** catalyzed by the 9-O-acetylquinine **9** at −110 °C (Scheme 7.6). (R)-Methyl 2-phenylpropanoate **10** was isolated in 74% ee. Above this temperature, the selectivity dropped dramatically (<5% ee at −70 °C) [8a].

The same reaction (with identical substrates) was studied later by Nakamura and coworkers using polymer-bound cinchona alkaloids [9]. A poly-(9-O-acryloylcinchonine) **11** used as catalyst at −78 °C afforded the highest optical yield (Scheme 7.7, 35% ee). By comparison and under the same experimental conditions, the use of the homogeneous catalyst 9-O-propionylcinchonine afforded the product with only 4% ee, similar to that obtained by Pracejus at the same temperature. Comparative experiments carried out with similar catalysts derived from (S)-2-quinuclidinylmethanol showed the influence of the C8 configuration of the alkaloid in determining the configuration of the products.

The Simpkins's group developed a similar approach using prochiral silylketenes **12**, the silicium atom improving the stability of the starting material [10]. The nucleophilic addition of thiophenol **13** on trialkylsilylketenes **12** mediated by 10 mol% of 9-O-benzoylquinine **14** proceeded at low temperature to give α-silylthioesters **15** with enantiomeric excesses ranging from 79 to 93% ee (Scheme 7.8). 9-O-Benzoylquinidine **14** catalyzed the formation of the opposite optical isomer with the same level of enantioselectivity. Although the absolute configuration was not proven for all substrates, the presence of a phenyldimethylsilyl group instead of a trimethylsilyl reversed the selectivity induced by the catalyst, the (+)-enantiomers were produced in place of the (−)-products (comparison of (+)-**15d** with (−)-**15a–c**, Scheme 7.8). The synthetic potential of these chiral silylthioesters **15** was demonstrated by their straightforward transformations.

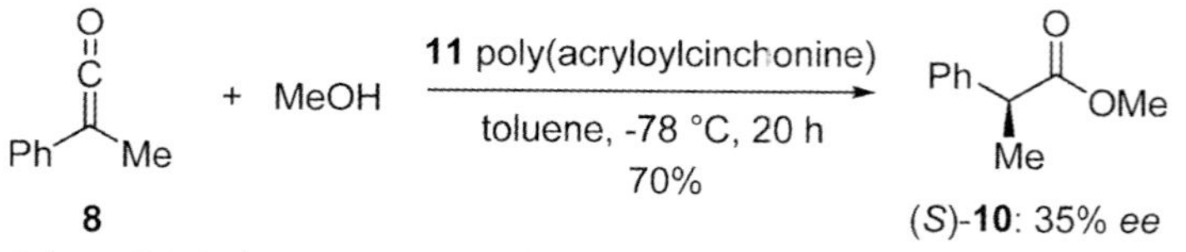

Scheme 7.7 Nakamura's supported catalysis of methanol addition on phenylmethylketene [9].

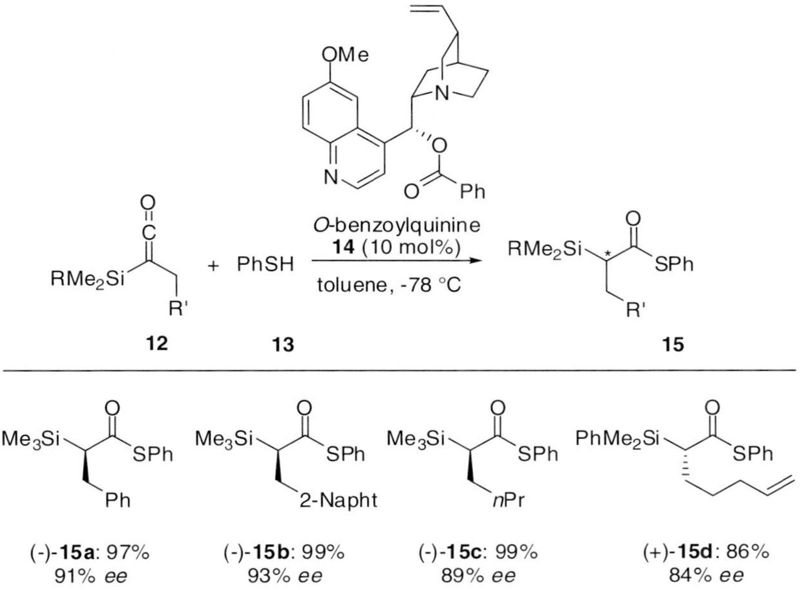

Scheme 7.8 Simpkins's thiophenol addition on silylalkylketenes [10].

The thioester group was readily converted into a ketone, an aldehyde, or a primary alcohol and the silyl moiety into a protected alcohol, all reactions preserving the stereochemistry of the products.

Pracejus also studied the tandem nucleophilic addition/diastereoselective protonation of optically pure (S)-phenylethylamine on phenylmethylketene [11]. With the aim of synthesizing amino acids and their derivatives, Calmes and coworkers reinvestigated the reaction of prochiral ketenes (generated *in situ* from acid chorides in the presence of a tertiary amine) with (R)-pantolactone, an α-hydroxylactone [12]. The authors have shown that the diastereoselectivity is dependent on the base used. Particularly, triethylamine and quinuclidine afforded complementary results and the diasteromeric ratios observed with quinuclidine suggest that the high stereoselections could be observed in nucleophilic additions to prochiral ketenes in the presence of cinchona alkaloids.

Despite the obvious potential of cinchona alkaloids as bifunctional chiral catalysts of the nucleophilic addition/enantioselective protonation on prochiral ketenes, no further contribution has appeared to date and only a few papers described this asymmetric reaction with other catalysts [13]. When the reaction is carried out with soft nucleophiles, the catalyst, often a chiral tertiary amine, adding first on ketene, is covalently linked to the enolate during the protonation. Thus, we can expect an optimal control of the stereochemical outcome of the protonation. This seems perfectly well suited for cinchona analogues and we can therefore anticipate successful applications of these compounds for this reaction in the near future.

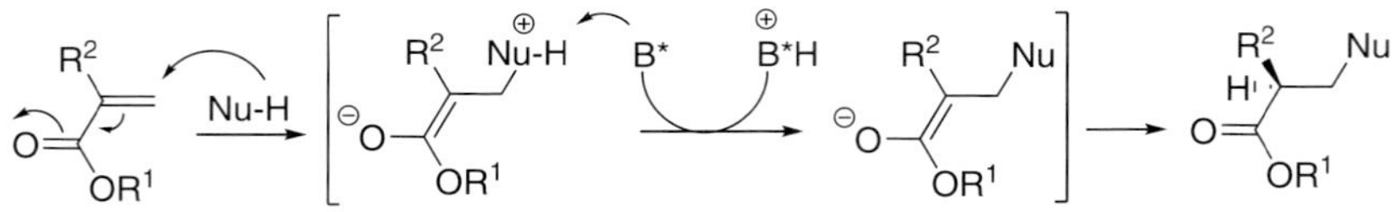

Scheme 7.9 Enantioselective protonation induced by Michael additions.

7.4
Michael Additions

The Michael addition of nucleophiles on α,β-unsaturated electron withdrawing groups, often carbonyl-containing functional groups, is a widely used reaction for the formation of C—C or C—heteroatom bonds. When the Michael acceptor bears a substituent on the α-position to the carbonyl, then an asymmetric carbon is created upon protonation of the transient enolate generated by the nucleophilic addition (Scheme 7.9).

Despite the importance of the Michael addition in organic synthesis, the tandem conjugate addition/enantioselective protonation has been little explored [14] and only a few publications have involved cinchona alkaloids as bifunctional catalysts **B*** for controlling the configuration of the chiral carbon created during protonation (Scheme 7.9).

Pracejus and coworkers reported the first Michael addition/enantioselective protonation mediated by cinchona alkaloids [15]. The authors put a special emphasis on the requirement of using chiral β-N,N-dialkylamino alcohol to achieve significant inductions. The addition of benzyl thiol **16** on 2-phthalimidoacrylate **17** catalyzed by 5 mol% of quinidine **3** gave the best selectivity (Scheme 7.10).

Kumar and coworkers described the Michael addition of thiophenol **13** to 2-phenylacrylates **19** using a catalytic amount of cinchona alkaloids [16]. Among the four natural alkaloids, quinine **20** and quinidine **3** afforded the best results with opposite enantioselectivity (Scheme 7.11). Methyl or isopropyl ester substrates **19a** and **19b** gave comparable selectivities whereas more sterically demanding esters (e.g., tBu or CH(iPr)$_2$) gave lower optical inductions. Based on the previous considerations and a computational analysis, the authors suggested a transition state

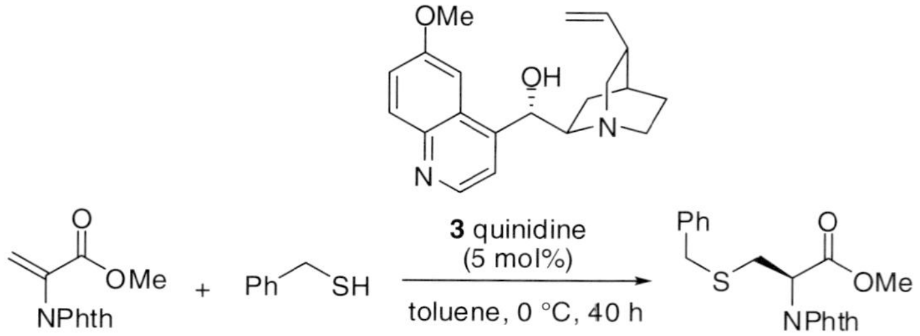

Scheme 7.10 Pracejus's benzyl thiol addition on α-substituted Michael acceptor [15].

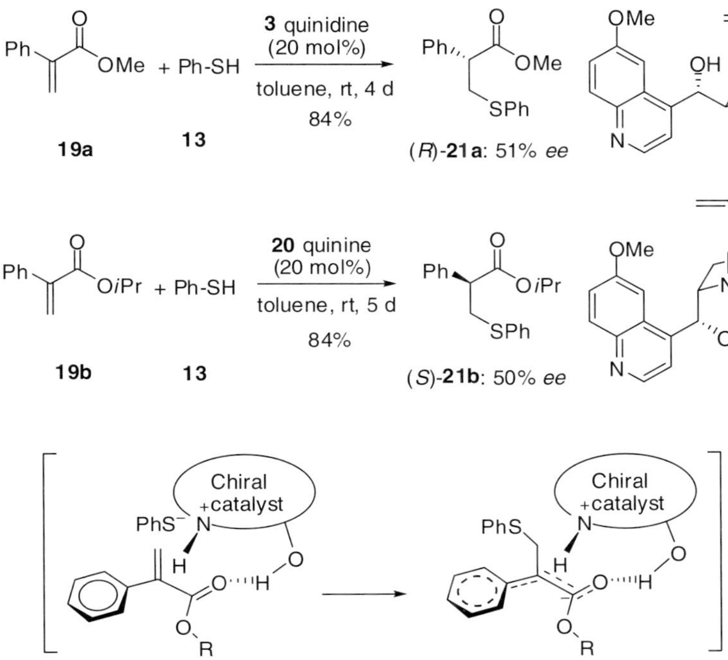

Scheme 7.11 Tandem thiophenol addition/enantioselective protonation on α-phenylacrylates [16].

featuring hydrogen bonding between the hydroxyl group of the catalyst and the carbonyl group of the ester, and an asymmetric transfer of a proton by the catalyst from the thiol group to the prochiral enolate (Scheme 7.11).

The synthetic utility of this reaction was further demonstrated by a three-step asymmetric synthesis of (S)-naproxen **22** an anti-inflammatory agent [16]. The thiol addition/asymmetric protonation on acrylate **19c** resulted in the Michael adduct (S)-**21c** in 46% ee followed by a Raney nickel mediated desulfurization and acid hydrolysis. (S)-Naproxen **22** was isolated in 85% ee after a single crystallization (Scheme 7.12).

Duhamel et al. have prepared a unique metastable enol **25** by Michael addition of thiobenzoic acid **23** on 2-benzylacrolein **24** at low temperature [17]. The enantioselective tautomerization of this configurationally pure enol **25** ($Z:E$ ratio >95:5) was performed at −70 °C in the presence of a stoichiometric amount of cinchonidine **26** to afford the corresponding enantioenriched aldehyde **27** in 71% ee (Scheme 7.13). N-Methyl ephedrine, having the same configuration as cinchonidine **26** for the amino alcohol moiety, produced the same enantiomer although with a lower selectivity (58% ee). No yield was given for both reactions.

Plaquevent and coworkers synthesized methyl dihydrojasmonate **28** using this methodology by performing the asymmetric Michael addition of dimethyl malonate **29** on 2-pentyl-2-cyclopentenone **30** [18]. The mechanism involved the tandem deprotonation of the malonate **29** using solid–liquid phase-transfer catalysis

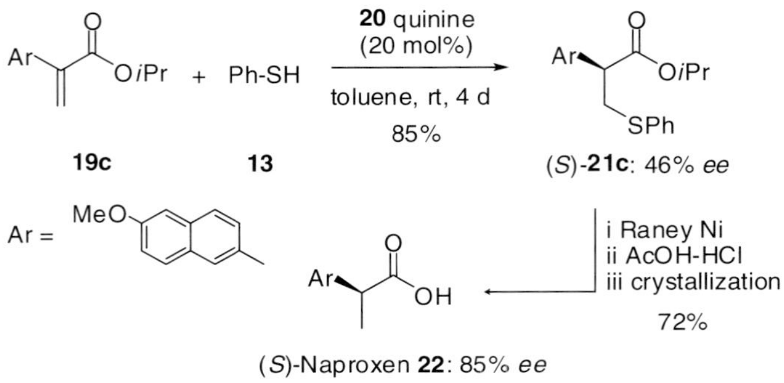

Scheme 7.12 Kumar's asymmetric synthesis of (S)-naproxen [16].

followed by enantioselective Michael addition on cyclopentenone **30** and ended by diastereoselective protonation. The reaction carried out under solvent-free conditions employed a quinine–ammonium derivative **31** as catalyst (Scheme 7.14). Due to a pseudoenantiomeric effect, the opposite enantiomer of the jasmonate precursor **32** was synthesized using the quinidine-based catalyst. Then, subsequent decarboxylation of the malonyl moiety using the Krapcho procedure afforded the expected compound **28** in high enantiomeric excess. No racemization was observed. The authors did not evaluate the direct involvement of the catalyst during the protonating step. However, since only the thermodynamic *trans*-diastereoisomer was obtained, we suspect a diastereoselective control of the protonation by the first asymmetric center created by the Michael addition and not by the catalyst. Then, owing to the basic experimental conditions, a thermodynamic equilibration of the product occurs.

More recently, Deng and coworkers developed an unprecedented transformation where two nonadjacent stereocenters were created with a high stereocontrol during the nucleophilic addition of a β-keto ester or a β-keto nitrile **33** onto 2-chloroacrylonitrile **34** (Scheme 7.15) [19]. The overall asymmetry of this tandem process was controlled by a 6′-OH cinchona alkaloids such as **35** that acted as a bifunctional

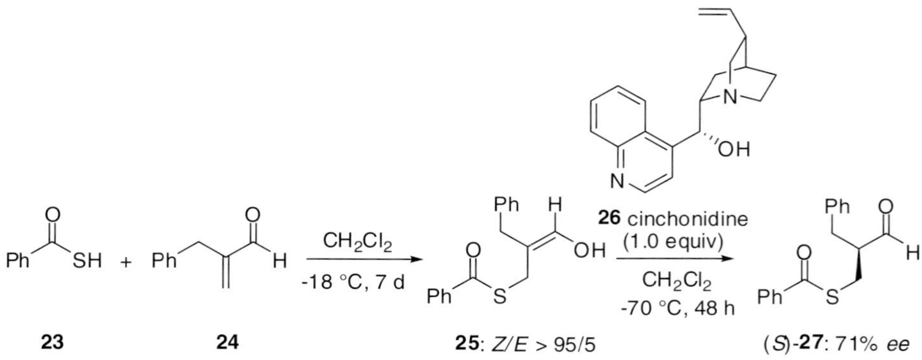

Scheme 7.13 Duhamel's unique enantioselective tautomerization of enol [17].

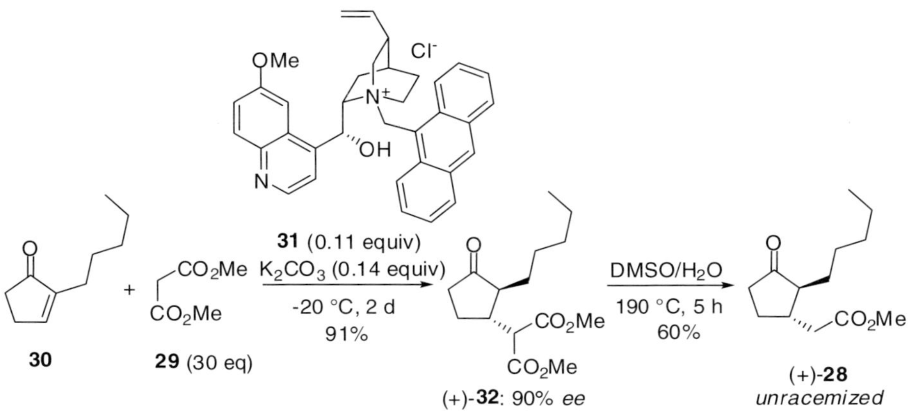

Scheme 7.14 Plaquevent's asymmetric synthesis of methyl dihydrojasmonate [18].

catalyst. The stereochemical outcome of this asymmetric tandem reaction resulted from a network of hydrogen-bonding interactions between the cinchona alkaloid, the reacting Michael donor and Michael acceptor in the nucleophilic step, and subsequently with the enol intermediate in the protonation step. As a base,

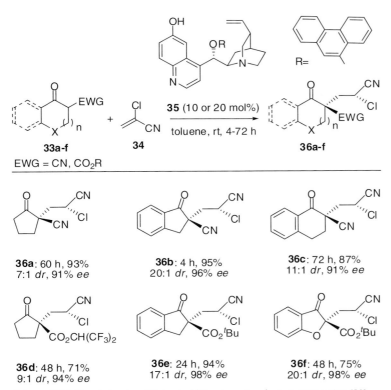

Scheme 7.15 Deng's tandem Michael addition/enantioselective protonation [19].

the catalyst deprotonated the β-keto ester or the β-keto nitrile thus controlling the chirality of the quaternary center created during the Michael addition. Then, as chiral acid (indeed the conjugated acid of the quinuclidine amine moiety), it carried out the enantioselective protonation of the prochiral enolate generated in the first step. Comparative experiments with DABCO excluded the possibility of a diastereocontrol of the protonation by the first stereocenter created during the conjugate addition.

During the course of this study, the model reaction of α-chloroacrylonitrile 34 with α-cyanocyclopentanone 33a revealed that the C9 substituent of the catalyst had a significant impact on both the enantioselectivity and diastereoselectivity of the reaction. Catalyst 35, which was the best catalyst with cyclic β-keto nitriles and β-keto esters 33, was ineffective for acyclic substituted α-cyanoacetates 37. Thus, the modification of the C9 substituent of the catalyst 38 (9-O-Ac instead of 9-O-phenanthrenyl) afforded significantly improved results (ee from 79 to 88% for R = Ph X = OEt), the highest dr and ee being obtained with the thiolesters 39b–c (dr up to 10 : 1 and ee up to 93%). This methodology was successfully applied to the synthesis of a precursor of enantioenriched (−)-manzacidin A 40, a biologically interesting natural product (Scheme 7.16).

In a second report describing similar chemistry and substrates, Deng and coworkers improved the selectivities at both stereocenters by using thiourea derivatives of quinine 41 and quinidine [20]. The reaction was carried under mild conditions at lower catalyst loading. Moreover, the results described in this paper (Scheme 7.17) were complementary of those obtained previously. Indeed, all four diastereoisomers of a given product can be synthesized by using quinine or quinidine derivatives. The model proposed by the authors involves the same type of dual activation as previously with a slight modification. In that case, the thiourea group

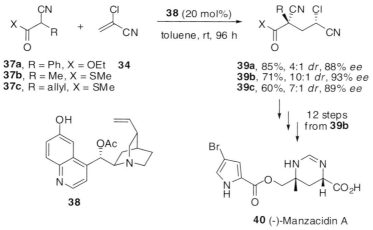

Scheme 7.16 Deng's Michael addition/enantioselective protonation with acyclic cyanoacetates [19].

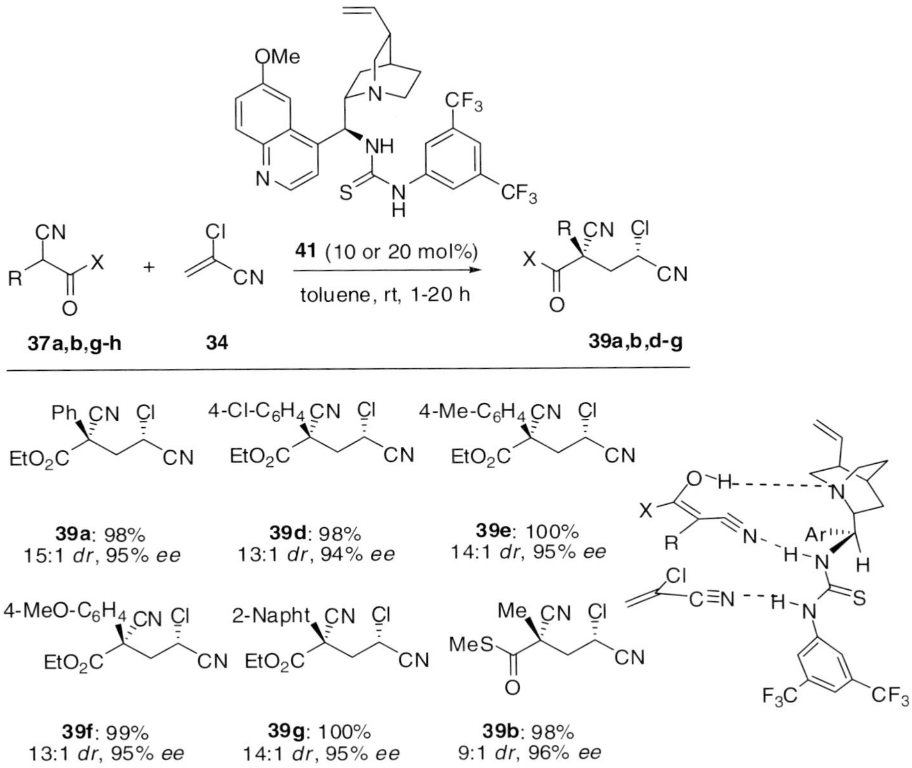

Scheme 7.17 Deng's optimization of Michael addition/enantioselective protonation with acyclic cyanoacetates [20].

makes simultaneously two hydrogen bonds, one with the Michael acceptor and the other with the nucleophile providing a higher rigidity of the supramolecular assembly of the transition state.

As mentioned at the beginning of this subsection, this methodology is an important tool for synthesis of carbonyl compounds with a stereogenic carbon center in the α-position. Only recently, this type of enantioselective protonation has been carried out using catalytic amounts of the chiral inductor. From this survey and the few references cited in the introduction not using cinchona alkaloids [14], the past decade has witnessed great progresses in the tandem Michael addition/enantioselective protonations. Despite a few synthetic applications that underline the scientific maturity of this area, many challenges still remain, in particular the Michael additions of nitrogen nucleophiles (primary and secondary alkylamines) or of "hard" enolates combined with efficient enantioselective protonations. In this relatively young area, there is no doubt that cinchona alkaloids and their analogues can play a decisive role as powerful catalysts.

7.5
Enantioselective Decarboxylative Protonation

Surprisingly the enantioselective decarboxylative protonation (EDP) of a malonate is an old reaction, which has been rarely recognized as an enantioselective protonation. This closely related reaction to the malonic acid synthesis has been recently reviewed [21]. The first example of this reaction in its enantioselective version, reported by Marckwald in 1904, is also probably the first example of an asymmetric reaction in the chemical history [22]. The chiral inductor used in this work was brucine, a readily available optically pure compound. Then, for more than 70 years, this reaction has received little attention. According to the generally accepted mechanism, an enol intermediate is generated by the unimolecular decomposition of the diacid or the ketoacid upon loss of CO_2 that proceeds through a cyclic proton transfer. The base-catalyzed mechanism follows probably a slightly different path involving an enolate intermediate generated first by a deprotonation of the carboxylic acid followed by a rapid loss of CO_2 (Scheme 7.18). Starting from a racemic or prochiral compound, the chirality is determined at the last stage of the reaction and is induced by an optically active source not covalently linked to the substrate, but intimately involved in the enantioselective formation of the C−H bond.

7.5.1
Copper-Catalyzed EDP

In the late 1980s, the group of Maumy described the first mild method for the EDP [23]. Prochiral phenylalkylmalonic acids **42** were decarboxylated at 60 °C in acetonitrile in the presence of a combination of cinchonidine **26** and copper (I) chloride. Low enantiomeric excesses of carboxylic acids **44** were obtained using more than 1 equiv of cinchonidine **26** (due to the acidic character of the product) with a substoichiometric amount of the copper salt (Scheme 7.19). EDP, performed with racemic hemimalonates **45** using catalytic amounts of the copper–cinchonidine system (20 and 40 mol%, respectively), afforded comparative levels of enantioselectivity (Scheme 7.20).

Brunner and coworkers showed that the amount of copper (I) chloride could be reduced to 3 mol% for the EDP of substituted malonic acid **42a**, using experimental

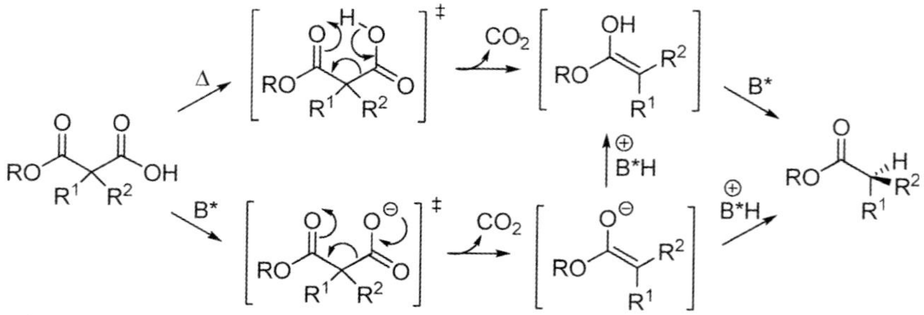

Scheme 7.18 Proposed mechanisms of enantioselective decarboxylative protonation.

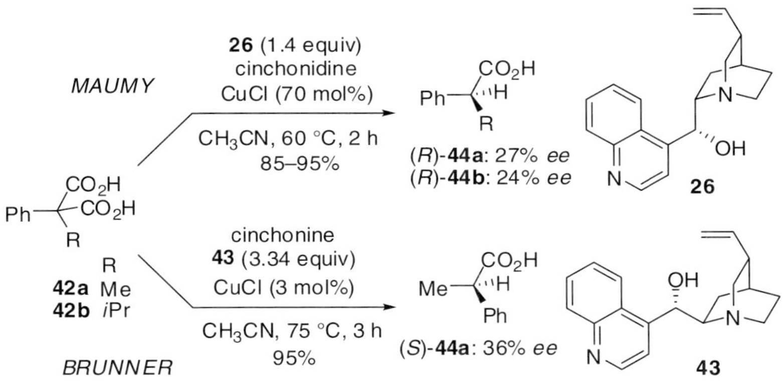

Scheme 7.19 Maumy's [23] and Brunner's [24] copper-catalyzed EDP with prochiral malonic acids.

conditions similar to those described by Maumy but a higher loading of cinchonine **43** [24]. The enantioselectivity reached 36% ee (Scheme 7.19). In the case of hemimalonate **45a**, Brunner obtained, with the combination copper-cinchonine **43** (3 and 8 mol%, respectively), the same level of enantioselectivity than Maumy using 40 and 20 mol% of copper chloride and cinchonidine **26** (Scheme 7.20). However, as described in Section 7.5.3, reinvestigation of this reaction showed that the role of copper is negligible.

7.5.2
Palladium-Catalyzed EDP

The research group of Muzart and Hénin studied extensively the palladium-catalyzed EDP of allyl- or benzyl-carboxylated compounds. Mainly two types of substrates, prochiral enol carbonates **A** and racemic β-keto esters **B**, were used to afford enols **C** as transient species [25]. In the presence of a chiral proton source, asymmetric protonation/tautomerization of enols led to enantioenriched ketones **D**

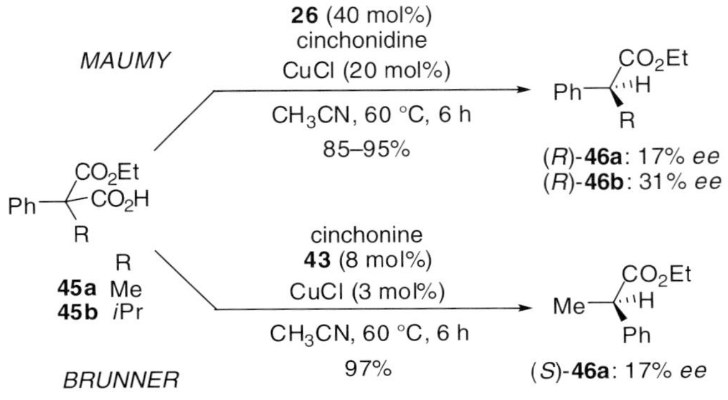

Scheme 7.20 Maumy's [23] and Brunner's [24] copper-catalyzed EDP with racemic hemimalonates.

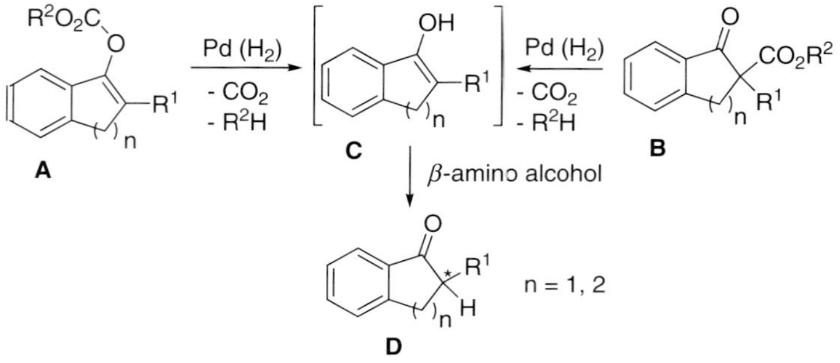

Scheme 7.21 Palladium-catalyzed tandem debenzylation/EDP.

(Scheme 7.21). A survey of different chiral proton sources showed that it was necessary to use a β-amino alcohol rather than an alcohol, an amine, or an acid as chiral inductor.

Initial studies [26] carried out with enol carbonates **A** or β-keto esters **B** afforded ketones **D** in high yields and selectivities up to 75% ee [27], using ephedrine or a β-amino alcohol derived from camphor. Surprisingly, cinchona alkaloids bearing the required β-amino alcohol moiety were not tested. It was necessary to wait till 2001 and later for publications describing the use cinchona alkaloids in these reactions, either the natural products or some analogues.

The hydrogenolysis/decarboxylation/asymmetric protonation reaction cascade of acyclic benzyl β-oxo-esters such as **47** catalyzed by Pd/C with H_2 in the presence of a catalytic amount of cinchonine **43** afforded the (S)-ketone **48** with enantioselectivities up to 75% ee, similar to previous results obtained with other β-amino alcohols. The reaction was carried out at room temperature in a short reaction time [28]. The best solvent for both yield and ee was ethyl acetate, compared with acetonitrile and THF. Comparative performances of cinchona alkaloids with other commonly used β-aminoalcohols are displayed on Scheme 7.22.

The palladium-catalyzed deprotection and EDP was then applied to the synthesis of 3-substituted 4-chromanones from cyclic β-keto esters **51** [29]. Among the amino alcohols tested, cinchonine **43** led to optically active chromanone **52** with a moderate 22% ee (Scheme 7.23), far from the enantioselectivity induced by aminoborneol (60% ee).

Access to optically active 2-fluoro-1-tetralone **53** was achieved using the same palladium-mediated cascade reaction [30]. The catalytic enantioselective decarboxylative protonation of 2-fluoro benzyl β-keto ester **54** in the presence of 30 mol% of quinine **20** afforded enantioenriched (S)-tetralone **53** in 65% ee (Scheme 7.24). The reaction was very sensitive to the nature of the palladium catalyst used. Furthermore, a minor amount of defluorinated product was observed. Several other cinchona derivatives were tested including analogues of cinchonine described by Brunner in organocatalytic EDP (see Section 7.5.3), but these chiral inductors afforded low selectivities (<30% ee).

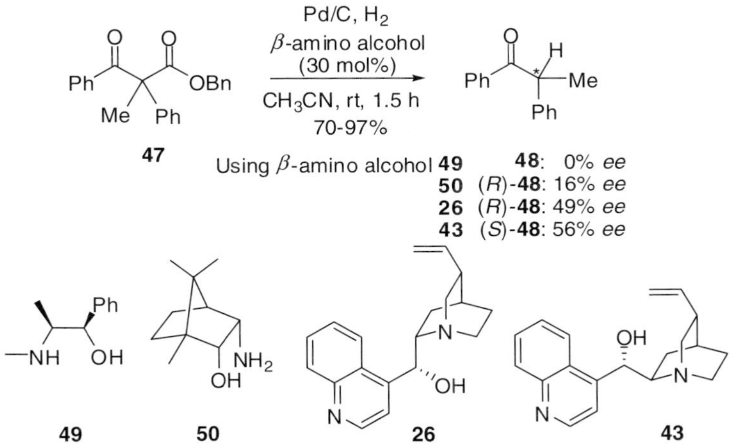

7.5 Enantioselective Decarboxylative Protonation

Scheme 7.22 Comparative performances of β-amino alcohols for the palladium-catalyzed tandem debenzylation/EDP of acyclic β-ketoesters [28].

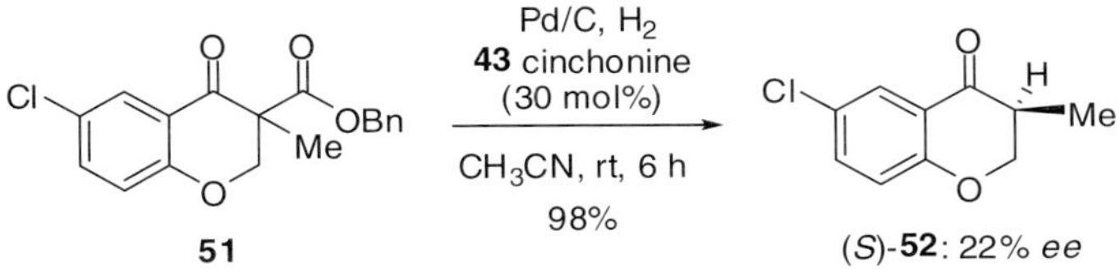

Reaction of **47** with Pd/C, H$_2$, β-amino alcohol (30 mol%), CH$_3$CN, rt, 1.5 h, 70–97%.

Using β-amino alcohol **49** — **48**: 0% ee
50 (R)-**48**: 16% ee
26 (R)-**48**: 49% ee
43 (S)-**48**: 56% ee

β-Amino alcohols: **49**, **50**, **26**, **43**.

Scheme 7.23 Muzart/Hénin's palladium-catalyzed tandem debenzylation/EDP of cyclic β-ketoesters [29].

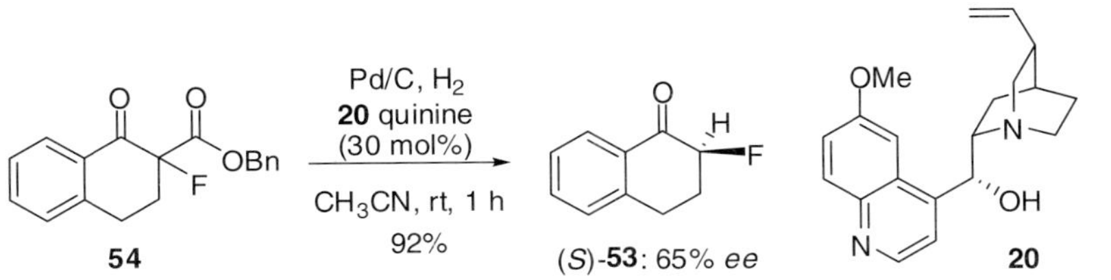

51 → with Pd/C, H$_2$, **43** cinchonine (30 mol%), CH$_3$CN, rt, 6 h, 98% → (S)-**52**: 22% ee

Scheme 7.24 Muzart/Hénin's asymmetric synthesis of fluorotetralone via debenzylative EDP [30].

54 → with Pd/C, H$_2$, **20** quinine (30 mol%), CH$_3$CN, rt, 1 h, 92% → (S)-**53**: 65% ee; **20**.

Next to Muzart's work, Baiker and coworkers reinvestigated the reaction parameters of the palladium-catalyzed EDP of cyclic β-keto esters in the presence of various chiral proton sources including cinchona alkaloids [31]. When working with benzyl ester **55a** as model compounds, they demonstrated the crucial effect of the solvent on the enantioselectivity of the reaction. In the palladium-catalyzed debenzylation of **55a** carried out at room temperature with hydrogen, the highest conversions but the lowest enantioselectivities were achieved in protic polar solvents

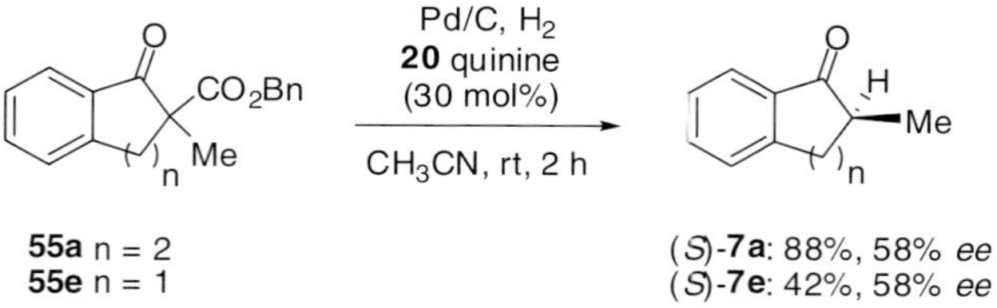

Scheme 7.25

55a n = 2
55e n = 1

(S)-**7a**: 88%, 58% ee
(S)-**7e**: 42%, 58% ee

Scheme 7.25 Baiker's reinvestigation of Muzart/Hénin's procedure [31, 32].

such as alcohols or acetic acid whereas polar aprotic solvents were more appropriate to observe good enantioselectivities with lower conversions. The best compromise was reached in acetonitrile using 30 mol% of quinine **20** (or quinidine to obtain the opposite enantiomer) affording (S)-tetralone **7a** in 58% ee and 88% yield after 2 h (Scheme 7.25). The same enantioselectivity was obtained with indanone **7e**. The selectivity was also shown to be dependent on the water content. Addition of one drop of water to the standard conditions decreased dramatically the ee of tetralone **7a**. The water content of the supported Pd could also contribute to the large differences observed among different palladium sources. Overall, the same level of selectivity was achieved compared to Muzart/Hénin's work. However, Baiker concluded from a thorough mechanistic study to an organocatalytic process by the cinchona alkaloid alone, the Pd being only necessary for the initial debenzylation of the ester [32].

7.5.3
Organocatalyzed EDP

Next to Brunner's work [24], Darensbourg et al. minimized the role of copper in the decarboxylation of malonic acids and esters [33]. The decarboxylation of phenylalkylmalonic acid derivatives occurred via a predissociation step involving metal–carboxylate bond cleavage, thus showing an inhibiting effect of copper on the carbon dioxide loss. Consistent with this mechanistic proposal, the rates of decarboxylation were greatly enhanced upon sequestering metal cation with chelating nitrogen bases.

Shortly later, Brunner demonstrated the uselessness of copper [34]. The reaction of hemimalonate **45a** carried out at room temperature in THF with 10 mol% of cinchonine **43** without copper afforded **46a** with 34% ee (Scheme 7.26), comparable to that obtained with copper but with a higher reaction rate. Beside its chiral inducting effect, the alkaloid also has a catalytic role in the decarboxylation process since the substrate is perfectly stable at room temperature in the absence of base.

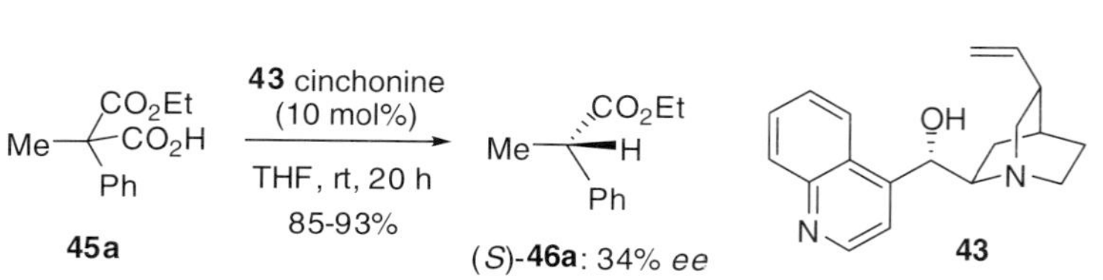

Scheme 7.26 Brunner's first organocatalyzed EDP [34].

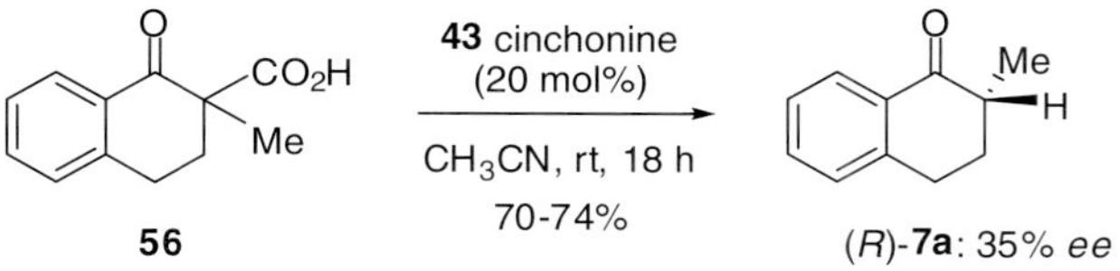

7.5 Enantioselective Decarboxylative Protonation

Scheme 7.27 Muzart/Hénin's organocatalyzed EDP [35].

Muzart and Hénin reached the same conclusions, that is, copper was useless and only the amine, as a base, catalyzed the decarboxylation [35]. Catalytic amounts of enantiopure amino alcohols triggered the EDP of β-keto acid **56** at room temperature in acetonitrile (Scheme 7.27). Among the screened amino alcohols, cinchonine **43** induced the best selectivity and 2-methyl-1-tetralone **7a** was obtained in 35% ee, equal to that obtained with ephedrine.

Enantioenriched β-hydroxyisobutyric acid **57** was synthesized using EDP catalyzed by cinchona alkaloids [36]. Under the best conditions, with a stoichiometric amount of cinchonidine **26**, β-hydroxyisobutyric ester **57** was obtained in low ee and moderate yield (Scheme 7.28). No enantioselectivity was observed when using 10 mol% of base.

Brunner and Schmidt developed the first efficient organocatalyzed EDP for the synthesis of a precursor **59** of (S)–naproxen [37]. A variety of bases were screened and cinchona alkaloids appeared the most efficient catalysts. About 30 analogues were synthesized and tested in the EDP of the aryl substituted 2-cyano propionic acid **60**. A selectivity up to 72% ee was reached by using 10 mol% of 9-*epi*-cinchonine benzamide derivative **61** in anhydrous THF just below room temperature (Scheme 7.29). The EDP of 2-cyano-2-phenylpropionic acid **62** carried out under the same conditions afforded 2-phenyl propionitrile **63** in slightly lower enantioselectivity (60% ee). A kinetic study showed that the enantioselectivity did not change with time and the plot conversion versus time was linear indicating zero-order behavior of acid. Therefore, no kinetic resolution was observed. Interestingly, subjecting both enantiomers of 2-cyanopropanoic derivative **62** or the racemic substrate to EDP, the same enantioselectivity was reached. Such results favor a two-step decarboxylation/protonation mechanism via planar intermediates and their stereoselective protonation (Scheme 7.18).

The EDP methodology was applied to the synthesis of enantioenriched proteinogenic amino acid precursors [38]. Starting from commercially available N-acetyl 2-aminomalonate, N-acetyl 2-alkyl-2-aminomalonic hemiesters **64** were easily

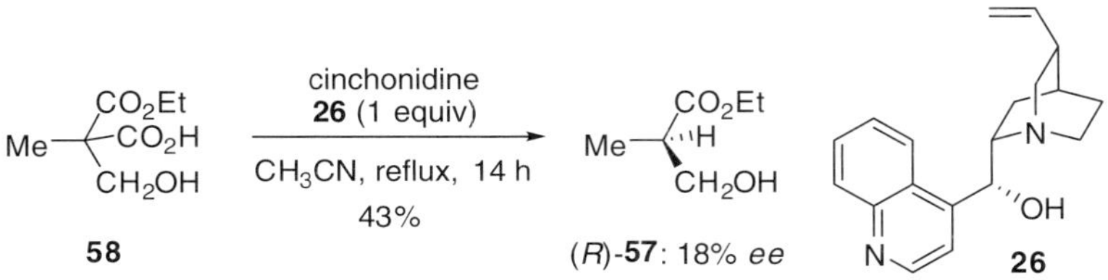

Scheme 7.28 β-Hydroxyisobutyric acid synthesis via EDP [36].

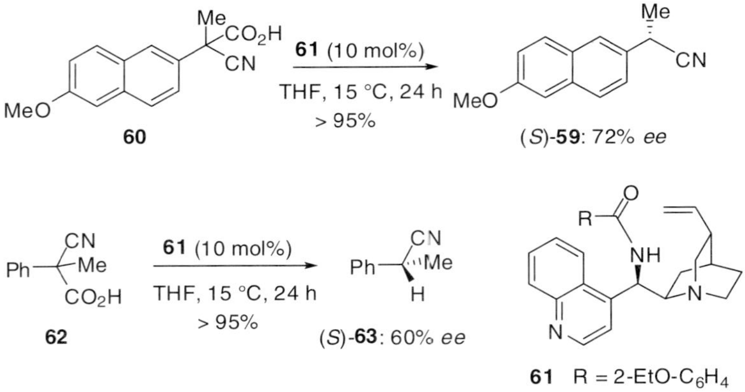

Scheme 7.29 Brunner's EDP of 2-cyanopropionic acid derivatives [37].

prepared and subjected to the EDP using 10 mol% of cinchona-based catalysts **65** in THF at 70 °C for 24 h (Scheme 7.30). As in the previous study, not more than 35 cinchona derivatives were tested and the highest selectivity (ee 71%) was achieved using 9-*epi*-cinchonine benzamide derivative **65b** with 2-benzyl-aminomalonate **64b** as substrate. Alanine acetamide **66a** was produced with slightly lower enantioselectivity using catalyst **65a**. Natural cinchona alkaloids afforded poor selectivities with such substrates (ee <10%, data not shown).

Synthesis of optically enriched N-protected ethyl pipecolate **67** was achieved using EDP catalyzed by cinchona alkaloids [39]. Numerous bases were screened and 10 mol% of 9-*epi*-cinchonine **68** afforded selectivity up to 71% ee with the N-benzoyl piperidino hemimalonate **69** on a multigram scale (Scheme 7.31), offering

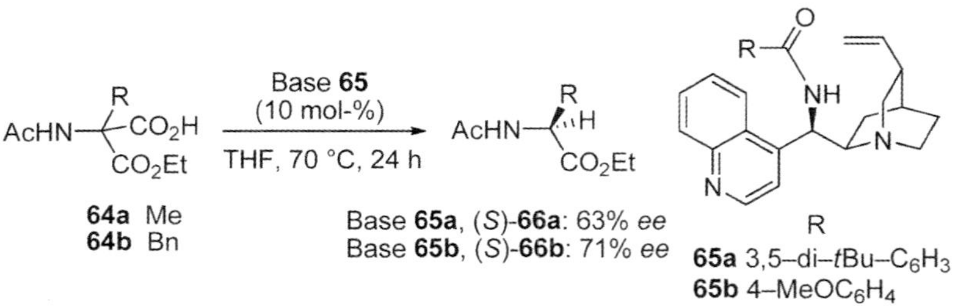

Scheme 7.30 Brunner's asymmetric synthesis of α-amino esters via EDP [38].

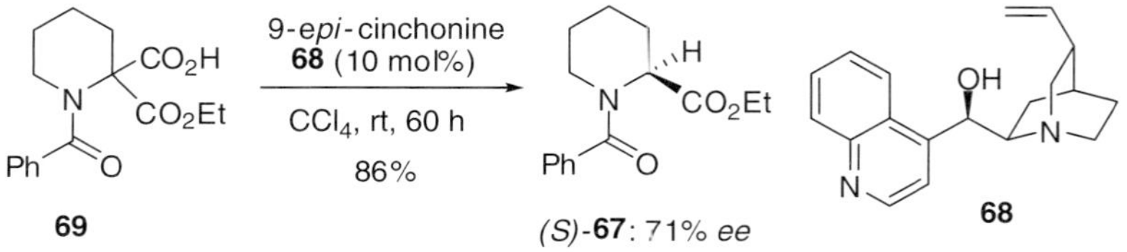

Scheme 7.31 Preparation of optically enriched pipecolate via EDP [39].

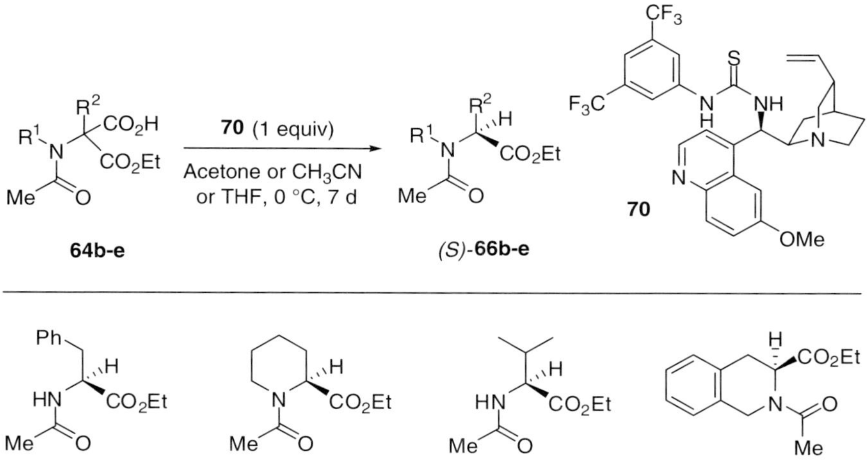

64b-e (S)-66b-e 70

(S)-66b: 86% 82% ee; (S)-66c: 90% 93% ee; (S)-66: 85% 89% ee; (S)-66e: 90% 86% ee

Scheme 7.32 Asymmetric synthesis of α-amino esters mediated by cinchona thiourea [41].

thus a practical alternative to the enantioselective protonation of the lithium enolate of an analogous substrate [40]. A careful study of the reaction parameters revealed that the selectivity was highly dependent on both the polarity of the solvent (toluene and carbon tetrachloride being the best ones) and the nitrogen substituent of the substrate. The temperature did not influence the enantioselectivity, only the rate of the reaction.

Optimization of the reaction conditions revealed that the highest selectivities were obtained in polar solvents at 0 °C using thiourea derivatives of cinchona alkaloids [41]. The scope of the reaction was studied and optically enriched cyclic and acyclic α-amino acids with 82–93% ee were prepared using 9-*epi*-quinidine thiourea derivative **70** (Scheme 7.32). As shown in the previous study and pointed out elsewhere by others [42], the 9-*epi* configuration of the catalysts induced the best enantioselectivities. The main drawbacks of the procedure are the long reaction time and the need of a stoichiometric amount of the chiral base **70** to maintain such a high level of selectivity (indeed the selectivity dropped to 70% ee when using 20 mol% of the catalyst). The (*R*)-enantiomers of the amino esters could be synthesized with the same level of selectivity using the quinine analogue thiourea.

The enantioselective decarboxylative protonation, which is directly related to the venerable malonic acid synthesis, has not shown the major advances in terms of selectivity that we could have expected for the oldest methodology used to carry out enantioselective protonations (Marckwald, 1904). The use of this reaction and, in particular, of its organocatalyzed version for the synthesis of compounds of biological significance is now emerging. This highlights the synthetic potential of this low-cost and operationally simple reaction, notably in the context of sustainable chemistry. However, this asymmetric reaction is not yet mature. Further investigations are required to fully understand and improve the selectivity to make this reaction general, practical, and useful for the synthesis of complex structures. In this context, cinchona alkaloids, which are already some of the most frequently used organocatalysts, will continue to demonstrate their usefulness for EDP.

7.6
Proton Migration

The last part of the chapter summarizes two specific reactions that can be viewed as a proton migration from an achiral sp^3 carbon to a prochiral sp^2 carbon, both reactions being catalyzed by cinchona derivatives.

Pète and coworkers reported on the enantioselective tautomerization of dienols produced by a Norrish type II photorearrangement of α,β-unsaturated carbonyl compounds [43]. Numerous substrates, catalysts and conditions were screened and the best enantioselectivities were obtained with chiral aminoalcohols derived from camphor. This enantioselective photodeconjugation carried out in dichloromethane at −40 °C with α,β-unsaturated ester **71** using 10 mol% of cinchonine **43** as chiral inductor afforded the deconjugated ester **72** in 41% ee and 40% isolated yield (with incomplete conversion, Scheme 7.33). The opposite enantiomer of ester **72** was produced under those conditions using cinchonidine **26** in 36% ee. The reaction proceeded through a dienol **73** generated by irradiation, which was then tautomerized by the dual action of both functional groups of the chiral β-amino alcohol. The transition state proposed by the authors involves the simultaneous deprotonation of the enol by the amine catalyst with the concomitant asymmetric protonation of the prochiral enolate carbon by the alcohol group of the catalyst.

A few years later, the same methodology was used with substituted arylketones [44]. Irradiation of α-disubstituted indanones, tetralones, and propiophenones bearing at least one hydrogen in the γ-position led to enol intermediates through a Norrish type II cleavage. In presence of catalytic amounts of optically active β-amino alcohols, enantioenriched arylketones were obtained with enantioselectivities up to 89% ee. Mechanistic and parameters investigations were carried out with indanone **74** and ephedrine as chiral inductor. Screening several β-amino alcohols including cinchona alkaloids, the photodeconjugation carried out at low temperature afforded moderate

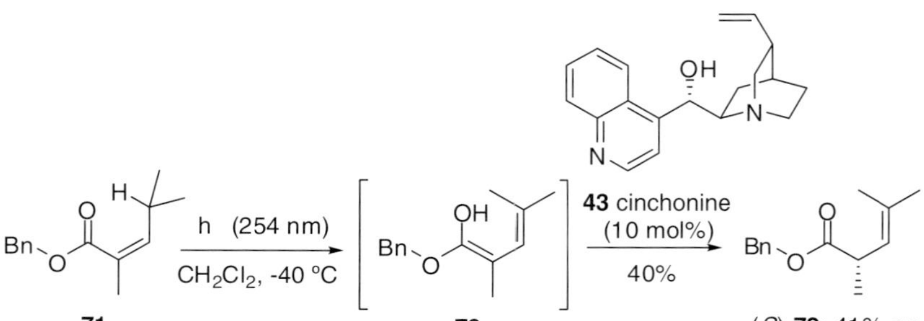

Scheme 7.33 Pète's photodeconjugation of α,β-unsaturated esters [43].

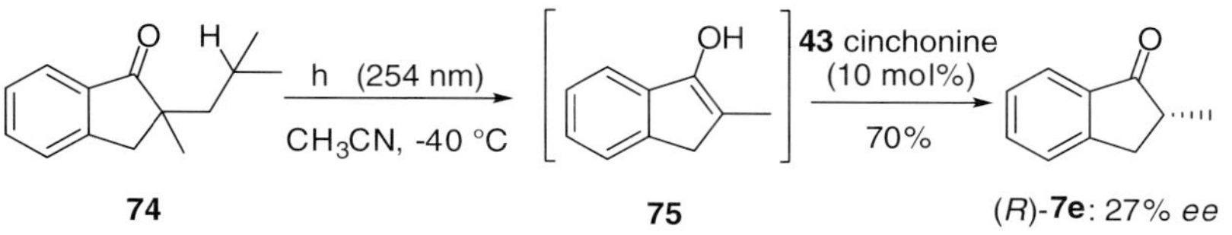

7.6 Proton Migration

Scheme 7.34 Photodeconjugation of α-disubstituted indanone [44].

ee for (R)-ketone **7e** via enol **75** in the presence of 10 mol% of cinchonine **43** (Scheme 7.34). Cinchonidine **26** gave the (S)-indanone **7e** with the same level of selectivity.

The Soloshonosk's group developed an original [1,3]-proton shift for the preparation of fluorinated amino acids [45]. This biomimetic synthesis of β-fluoroalkyl-β-amino acids from β-ketoesters **76** consisted in the thermodynamic equilibration of N-benzyl enamines **77** into N-benzylidene derivatives **78**, isomerized into **79** and then hydrolyzed to afford the free amino group. The proton shift was directed by the electron withdrawing character of fluoroalkyl groups and was catalyzed by an organic base, typically triethylamine or DBU. Initially, the reaction was intensively studied in its racemic version. Later, the enantioselective approach was investigated with a few fluorinated substrates and with only three organic chiral bases [46]. The reaction, carried out under solvent-free conditions at 100 °C using 10 mol% of cinchonidine **26**, afforded the optically enriched products **79** in selectivities up to 36% ee (Scheme 7.35).

Plaquevent and coworkers re-examined and sharply improved this method for a rapid access to enantioenriched β-trifluoromethyl-β-amino acid [47]. Nine cinchona-based catalysts were screened, and the best result was obtained using (DHQ)$_2$PHAL **80**. The reaction was performed starting with the p-nitrobenzyl enaminoester **81** at 80 °C and afforded the expected imine ester **82** in 90% isolated yield and 71% ee (Scheme 7.36). The authors put a special emphasis on the mechanistic aspect of the reaction using a deuterated substrate. According to the results, the deprotonation is both rate and asymmetric determining step.

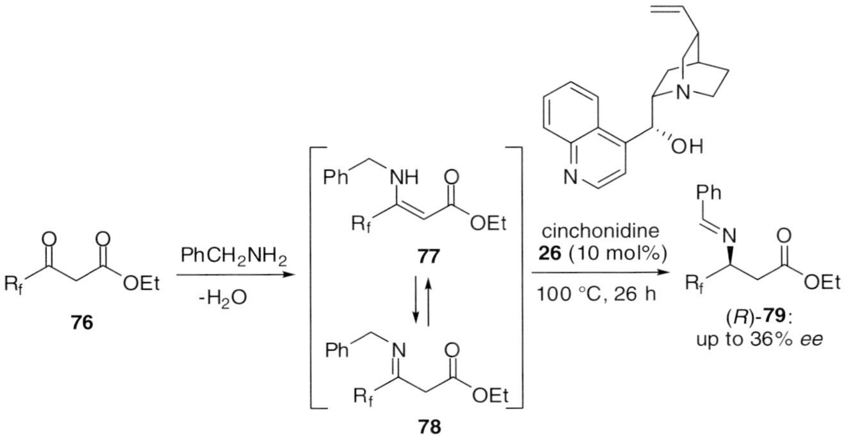

Scheme 7.35 Soloshonok's biomimetic asymmetric synthesis of β-amino esters [46].

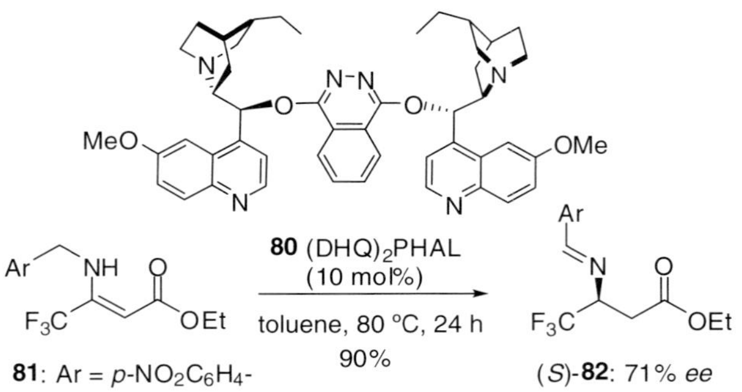

Scheme 7.36 Plaquevent's optimization of Soloshonosk's biomimetic procedure [47].

7.7
Summary of Cinchona-Mediated Enantioselective Protonations

Although in the past, the enantioselective protonation of enols/enolates has been regarded as a fundamental research subject, often employing stoichiometric amounts of the chiral inductor, the involvement of cinchona alkaloids has triggered the emergence of catalytic versions. Among the various methodologies listed in this chapter, the enantioselective decarboxylative protonation has seen the most intensive uses of these natural compounds or of their newly developed analogues. This is evident from the bifunctional nature of some of these compounds, the thiourea derivatives being the best ones although their catalytic performances need to be improved. Surprisingly, tandem Michael addition and nucleophilic addition on ketenes with enantioselective protonation, which seem best suited for the use of cinchona alkaloids, have not witnessed such development. With preformed metallic enolates, the anhydrous experimental conditions generally used to generate these species prevented the use of these natural products in catalytic amounts. Although very recent publications seem to prove the contrary, we can expect better advances in this area with neutral equivalents of enolates such as enol ethers or enamines.

Overall, cinchona alkaloids, which are already powerful (organo)catalysts in most major chemical reactions, will be expected to be major players in the enantioselective protonations of enols/enolates. This will persuade the synthetic chemist to incorporate these methodologies with confidence in total syntheses.

References

1 (a) Fehr, C. (1996) *Angew. Chem. Int. Ed. Engl.*, **35**, 2566–2587; (b) Eames, J. and Weerasooriya, N. (2001) *Tetrahedron: Asymmetry*, **12**, 1–24; (c) Duhamel, L., Duhamel, P., and Plaquevent, J.-C. (2004) *Tetrahedron: Asymmetry*, **15**, 3653–3691.

2 For a complete discussion on the chiral protonating agents, their structural and acidity requirements, see Duhamel, L., Duhamel, P., and Plaquevent, J.-C. (2004) *Tetrahedron: Asymmetry*, **15**, 3653–3691.

3 Duhamel, L. (1976) *C. R. Acad. Sci.*, **282**, 125–127.
4 Takeuchi, S., Miyoshi, N., Hirata, K., Hayashida, H., and Ohga, Y. (1992) *Bull. Chem. Soc. Jpn.*, **65**, 2001–2003.
5 Poisson, T., Dalla, V., Marsais, F., Dupas, G., Oudeyer, S., and Levacher, V. (2007) *Angew. Chem. Int. Ed.*, **46**, 7090–7093.
6 Poisson, T., Oudeyer, S., Dalla, V., Marsais, F., and Levacher, V. (2008) *Synlett*, 2447–2450.
7 (a) For the most recent examples of catalytic enantioselective protonations, see Cheon, C.H. and Yammamoto, H., (2008) *J. Am. Chem. Soc.*, **130**, 9246–9247; (b) Mitsuhashi, K., Ito, R., Arai, T., and Yanagisawa, A. (2006) *Org. Lett.*, **8**, 1721–1724; (c) Yanagisawa, A., Touge, T., and Arai, T. (2005) *Angew. Chem. Int. Ed.*, **44**, 1546–1548; (d) Reynolds, N.T. and Rovis, T. (2005) *J. Am. Chem. Soc.*, **127**, 16406–16407.
8 (a) Pracejus, H. (1960) *Liebigs Ann. Chem.*, **634**, 9–22; (b) Pracejus, H. and Tille, A. (1963) *Chem. Ber.*, **96**, 854–865; (c) Tille, A. and Pracejus, H. (1967) *Chem. Ber.*, **100**, 196–210; (d) Pracejus, H. and Kohl, G. (1969) *Liebigs Ann. Chem.*, **722**, 1–11.
9 (a) Yamashita, T., Yasueda, H., Miyauchi, Y., and Nakamura, N. (1977) *Bull. Chem. Soc. Jpn.*, **50**, 1532–1534; (b) Yamashita, T., Yasueda, H., Nakatani, N., and Nakamura, N. (1978) *Bull. Chem. Soc. Jpn.*, **51**, 1183–1185; (c) Yamashita, T., Yasueda, H., and Nakamura, N. (1979) *Bull. Chem. Soc. Jpn.*, **52**, 2165–2166.
10 Blake, A.J., Friend, C.L., Outram, R.J., Simpkins, N.S., and Whitehead, A.J. (2001) *Tetrahedron Lett.*, **42**, 2877–2881.
11 Pracejus, H. (1960) *Liebigs Ann. Chem.*, **634**, 23–29.
12 (a) Calmes, M., Escale, F., Glot, C., Rolland, M., and Martinez, J. (2000) *Eur. J. Org. Chem.*, 2459–2466; (b) Calmes, M., Escale, F., and Martinez, J. (2002) *Tetrahedron: Asymmetry*, **13**, 293–296.
13 For leading recent publications, see (a) Fehr, C., (2007) *Angew. Chem. Int. Ed.*, **46**, 7119–7121; (b) Schaefer, C. and Fu, G.C. (2005) *Angew. Chem. Int. Ed.*, **44**, 4606–4608; (c) Wiskur, S.L. and Fu, G.C. (2005) *J. Am. Chem. Soc.*, **127**, 6176–6177.
14 For recent examples, see (a) Hayashi, T., Senda, T., and Ogasawara, M., (2000) *J. Am. Chem. Soc.*, **122**, 10716–10717; (b) Nishimura, K., Ono, M., Nagaoka, Y., and Tomioka, K. (2001) *Angew. Chem. Int. Ed.*, **40**, 440–442; (c) Sibi, M.P., Petrovic, G., and Zimmerman, J. (2005) *J. Am. Chem. Soc.*, **127**, 2390–2391; (d) Li, B.-J., Jiang, L., Liu, M., Chen, Y.-C., Ding, L.-S., and Wu, Y. (2005) *Synlett*, 603–606; (e) Leow, D., Lin, S., Chittimalla, S.K., Fu, X., and Tan, C.-H. (2008) *Angew. Chem. Int. Ed.*, **47**, 5641–5645.
15 Pracejus, H., Wilcke, F.-W., and Hanemann, K. (1977) *J. Prakt. Chem.*, **319**, 219–229.
16 Kumar, A., Salunkhe, R.V., Rane, R.A., and Dike, S.Y. (1991) *J. Chem. Soc., Chem. Commun.*, 485–486.
17 Henze, R., Duhamel, L., and Lasne, M.-C. (1997) *Tetrahedron: Asymmetry*, **8**, 3363–3365.
18 Perrard, T., Plaquevent, J.-C., Desmurs, J.-R., and Hébrault, D. (2000) *Org. Lett.*, **2**, 2959–2962.
19 Wang, Y., Liu, X., and Deng, L. (2006) *J. Am. Chem. Soc.*, **128**, 3928–3930.
20 Wang, B., Wu, F., Wang, Y., Liu, X., and Deng, L. (2007) *J. Am. Chem. Soc.*, **129**, 768–769.
21 Blanchet, J., Baudoux, J., Amere, M., Lasne, M.-C., and Rouden, J. (2008) *Eur. J. Org. Chem.*, 5493–5506.
22 (a) Marckwald, W. (1904) *Ber. Dtsch. Chem. Ges.*, **37**, 349–354; (b) Marckwald, W. (1904) *Ber. Dtsch. Chem. Ges.*, **37**, 1368–1370.
23 Toussaint, O., Capdevielle, P., and Maumy, M. (1987) *Tetrahedron Lett.*, **28**, 539–542.
24 Brunner, H. and Kurzwart, M. (1992) *Monatsh. Chem.*, **123**, 121–128.
25 Detalle, J.-F., Riahi, A., Steinmetz, V., Hénin, F., and Muzart, J. (2004) *J. Org. Chem.*, **69**, 6528–6532.
26 (a) Hénin, F. and Muzart, J. (1992) *Tetrahedron: Asymmetry*, **3**, 1161–1164;

(b) Aboulhoda, S.J., Hénin, F., Muzart, J., Thorey, C., Behnen, W., Martens, J., and Mehler, T. (1994) *Tetrahedron: Asymmetry*, **5**, 1321–1326; (c) Aboulhoda, S.J., Létinois, S., Wilken, J., Reiners, I., Hénin, F., Martens, J., and Muzart, J. (1995) *Tetrahedron: Asymmetry*, **6**, 1865–1868; (d) Muzart, J., Hénin, F., and Aboulhoda, S.J. (1997) *Tetrahedron: Asymmetry*, **8**, 381–389; (e) Aboulhoda, S.J., Reiners, I., Wilken, J., Hénin, F., Martens, J., and Muzart, J. (1998) *Tetrahedron: Asymmetry*, **9**, 1847–1850.

27 Except in one particular example where 99% ee was obtained, see Muzart, J., Hénin, F., and Aboulhoda, S.J. (1997) *Tetrahedron: Asymmetry*, **8**, 381–389.

28 (a) Roy, O., Diekmann, M., Riahi, A., Hénin, F., and Muzart, J. (2001) *Chem. Commun.*, 533–534; (b) Roy, O., Riahi, A., Hénin, F., and Muzart, J. (2002) *Eur. J. Org. Chem.*, 3986–3994.

29 Roy, O., Loiseau, F., Riahi, A., Hénin, F., and Muzart, J. (2003) *Tetrahedron*, **59**, 9641–9648.

30 Baur, M.A., Riahi, A., Hénin, F., and Muzart, J. (2003) *Tetrahedron: Asymmetry*, **14**, 2755–2761.

31 Kukula, P., Matousek, V., Mallat, T., and Baiker, A. (2007) *Tetrahedron: Asymmetry*, **18**, 2859–2868.

32 Kukula, P., Matousek, V., Mallat, T., and Baiker, A. (2008) *Chem. Eur. J.*, **14**, 2699–2708.

33 (a) Darensbourg, D.J., Holtcamp, M.W., Khandelwal, B., and Reibenspies, J.H. (1994) *Inorg. Chem.*, **33**, 531–537; (b) Darensbourg, D.J., Holtcamp, M.W., Khandelwal, B., Klausmeyer, K.K., and Reibenspies, J.H. (1995) *Inorg. Chem.*, **34**, 2389–2398.

34 Brunner, H., Müler, J., and Spitzer, J. (1996) *Monatsh. Chem.*, **127**, 845–858.

35 Hénin, F., Muzart, J., Nedjima, M., and Rau, H. (1997) *Monatsh. Chem.*, **128**, 1181–1188.

36 Ryu, S.U. and Kim, S.Y.G. (1998) *J. Ind. Eng. Chem.*, **4**, 50–57.

37 Brunner, H. and Schmidt, P. (2000) *Eur. J. Org. Chem.*, 2119–2133.

38 Brunner, H. and Baur, M.A. (2003) *Eur. J. Org. Chem.*, 2854–2862.

39 (a) Rogers, L.M.-A., Rouden, J., Lecomte, L., and Lasne, M.-C. (2003) *Tetrahedron Lett.*, **44**, 3047–3050; (b) Seitz, T., Baudoux, J., Bekolo, H., Cahard, D., Plaquevent, J.-C., Lasne, M.-C., and Rouden, J. (2006) *Tetrahedron*, **62**, 6155–6165.

40 Martin, J., Lasne, M.-C., Plaquevent, J.-C., and Duhamel, L. (1997) *Tetrahedron Lett.*, **38**, 7181–7182.

41 Amere, M., Lasne, M.-C., and Rouden, J. (2007) *Org. Lett.*, **9**, 2621–2624.

42 McCooey, S.H. and Connon, S.J. (2005) *Angew. Chem. Int. Ed.*, **44**, 6367–6370.

43 (a) Piva, O., Mortezaei, R., Hénin, F., Muzart, J., and Pète, J.-P. (1990) *J. Am. Chem. Soc.*, **112**, 9263–9272; (b) Piva, O. and Pète, J.-P. (1990) *Tetrahedron Lett.*, **31**, 5157–5160.

44 Hénin, F., M'Boungou-M'Passi, A., Muzart, J., and Pète, J.-P. (1994) *Tetrahedron*, **50**, 2849–2864.

45 (a) Ono, T., Kuhkar, V.P. and Soloshonok, V.A. (1996) *J. Org. Chem.*, **61**, 6563–6569; (b) Soloshonok, V.A. and Berbasov, D.O. (2004) *J. Fluorine Chem.*, **125**, 1757–1763.

46 (a) Soloshonok, V.A. and Kuhkar, V.P. (1996) *Tetrahedron*, **52**, 6953–6964; (b) Soloshonok, V.A., Kirilenko, A.G., Galushko, S.V., and Kuhkar, V.P. (1994) *Tetrahedron Lett.*, **35**, 5063–5064.

47 Michaut, V., Metz, F., Paris, J.-M., and Plaquevent, J.-C. (2007) *J. Fluorine Chem.*, **128**, 500–506.

8
Cinchona-Catalyzed Nucleophilic 1,2-Addition to C=O and C=N Bonds
Hyeong Bin Jang, Ji Woong Lee, and Choong Eui Song

8.1
Introduction

The enantioselective nucleophilic addition of prochiral C=O and C=N moieties to the corresponding saturated chiral products is one of the most important stereoselective transformations on both the laboratory and the industrial scale. Although, over the past few decades, remarkable scientific achievements have been made in these research areas by using a variety of transitional metal-based catalysts, the sensitivity of the reaction to moisture and oxygen, as well as the toxic metal contamination of the product, usually restrict its practical application. Thus, currently, there is much interest in chiral organocatalysts, as they tend to be less toxic and more environmental friendly than traditional metal-based catalysts [1]. They are usually robust and thus tolerate moisture and oxygen, so that they usually do not demand any special reaction conditions.

The naturally occurring cinchona alkaloids (Figure 8.1), as described in other chapters of this book, have proven to be powerful organocatalysts in most major chemical reactions. They possess diverse chiral skeletons and are easily tunable for diverse catalytic reactions through different mechanisms, which make them privileged organocatalysts. The vast synthetic potential of cinchona alkaloids and their derivatives in the asymmetric nucleophilic addition of prochiral C=O and C=N bonds has also been well demonstrated over the last decade.

In this chapter, the current state of the art on the applications of cinchona alkaloids and their derivatives as chiral catalysts in the enantioselective nucleophilic addition of prochiral C=O and C=N bonds is discussed. The schemes exemplified in this chapter demonstrate the indispensable role of cinchona alkaloids as catalysts in these important research areas.

R = H
R' = CH=CH$_2$; Cinchonidine (**CD**)
R' = CH$_2$CH$_3$; Dihydrocinchonidine (**HCD**)

R = OMe
R' = CH=CH$_2$; Quinine (**QN**)
R' = CH$_2$CH$_3$; Dihydroquinine (**HQN**)

R = H
R' = CH=CH$_2$; Cinchonine (**CN**)
R' = CH$_2$CH$_3$; Dihydrocinchonine (**HCN**)

R = OMe
R' = CH=CH$_2$; Quinidine (**QD**)
R' = CH$_2$CH$_3$; Dihydroquinidine (**HQD**)

Figure 8.1 Structures of various naturally occurring cinchona alkaloids.

8.2
Aldol and Nitroaldol (Henry) Reactions

8.2.1
Aldol Reactions

The asymmetric aldol reaction represents the most versatile protocol for the preparation of optically enriched β-hydroxy ketones. During the last two decades, a number of observations have been made regarding asymmetric aldol and related reactions mediated by a cinchona-derived catalyst that affords high stereoselectivity.

8.2.1.1 Mukaiyama-Type Aldol Reactions

In 1993, the first cinchona-catalyzed enantioselective Mukaiyama-type aldol reaction of benzaldehyde with the silyl enol ether **2** of 2-methyl-1-tetralone derivatives was achieved by Shioiri and coworkers by using N-benzylcinchoninium fluoride (**1**, 12 mol%) [2]. However, the observed ee values and diastereoselectivities were low to moderate (66–72% for *erythro*-**3** and 13–30% ee for *threo*-**3**) (Scheme 8.1). The observed chiral induction can be explained by the dual activation mode of the catalyst, that is, the fluoride anion acts as a nucleophilic activator of the silyl enol ethers and the chiral ammonium cation activates the carbonyl group of benzaldehyde. Further investigations on the Mukaiyama-type aldol reaction with the same catalyst were tried later by the same [3] and another research group [4], but in all cases the enantioselectivities were too low for synthetic applications.

A marked improvement of the enantioselectivity was achieved by Corey and coworkers by using a bifluoride catalyst **4** [5]. Bifluoride derivatives such as **4** are usually less hygroscopic than monofluorides such as **1** and thus easier to handle. Using catalyst **4** (10 mol%), the silyl enol ethers **5** reacted with various aliphatic aldehydes smoothly at −70 °C via the aldol coupling reaction to afford the chiral β-hydroxy-α-amino esters **6** with moderate to good enantioselectivity for the *syn* adduct (72–89% ee) (Scheme 8.2). However, the diastereoselectiviy varied from 1 : 1 to 13 : 1 (*syn* : *anti*).

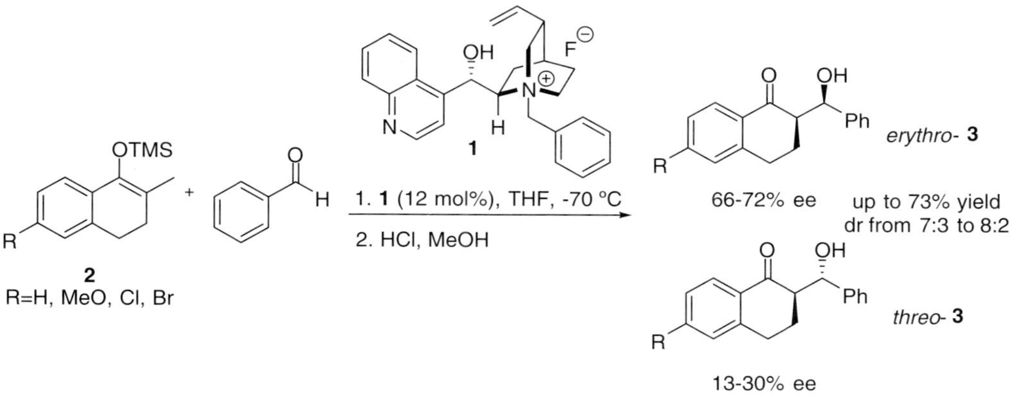

8.2 Aldol and Nitroaldol (Henry) Reactions | 199

Scheme 8.1

The structural variant **7** of Corey's bifluoride catalyst **4** was prepared later by Andrus and coworkers and applied as a catalyst (20 mol%) to the asymmetric Mukaiyama-type aldol reaction of aldehydes with the enol silylether **8** [6]. Excellent diastereoselectivity (up to >99/1) for the *syn*-aldol product **9** was achieved, especially with aromatic aldehydes. However, only moderate to good enantioselectivity (44–83% ee) was obtained (Scheme 8.3).

Recently, the Mukaiyama group demonstrated that the phenoxide anion could also be used as an activator of silyl enol ethers, instead of fluoride or bifluoride [7]. A similar approach to activate the Ruppert–Prakash reagent, TMSCF$_3$, was reported in the trifluoromethylation of carbonyls, which is discussed in Section 8.6 of this chapter [8]. Using **10** bearing the phenoxide anion as a catalyst (10 mol%), a range of aliphatic and aromatic aldehydes underwent smoothly the vinylogous Mukaiyama-type aldol

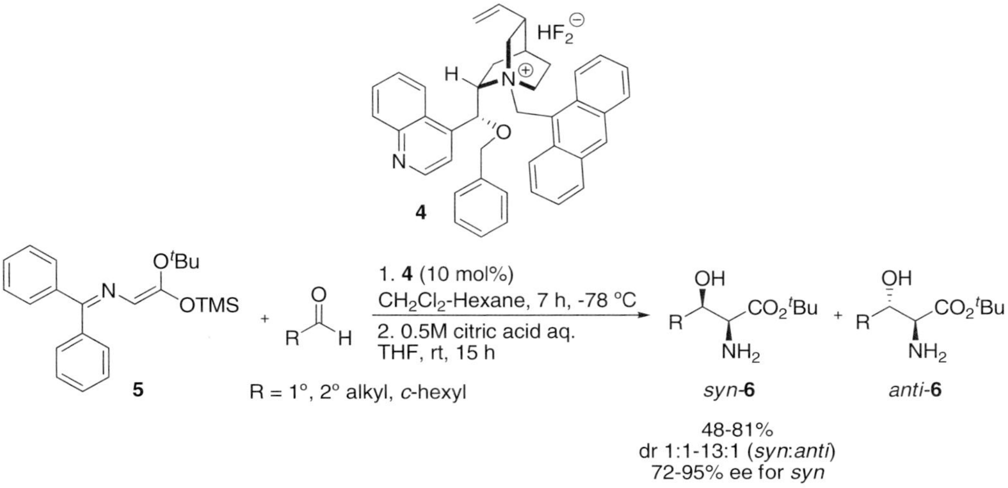

Scheme 8.2

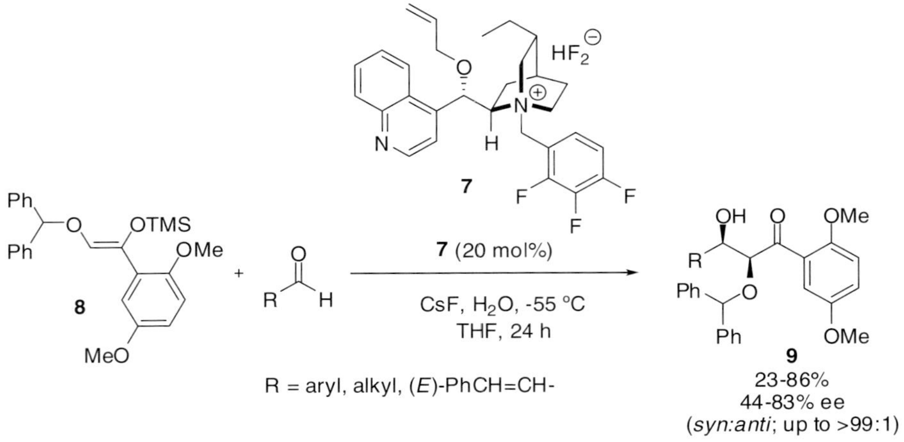

Scheme 8.3

reaction with the silyl enol ether **11** to give the *syn*-aldol product **12** as the major diastereomer with good to excellent ee values (80–97% ee) (Scheme 8.4).

8.2.1.2 Direct Aldol Reactions

As shown in Scheme 8.2, chiral β-hydroxy-α-amino acids can be obtained by the Mukaiyama-type aldol reaction of aldehydes with glycine-derived enol silyl ethers using cinchona-based quaternary ammonium salts. In 2004, Castle and coworkers [9] found that cinchona-based quaternary ammonium salts such as **13** are also able to catalyze the direct aldol reaction of aldehydes with the glycine donor **14** in the presence of a phosphazene base such as BTTP (*t*-butyliminotri(pyrrolidino)phos-

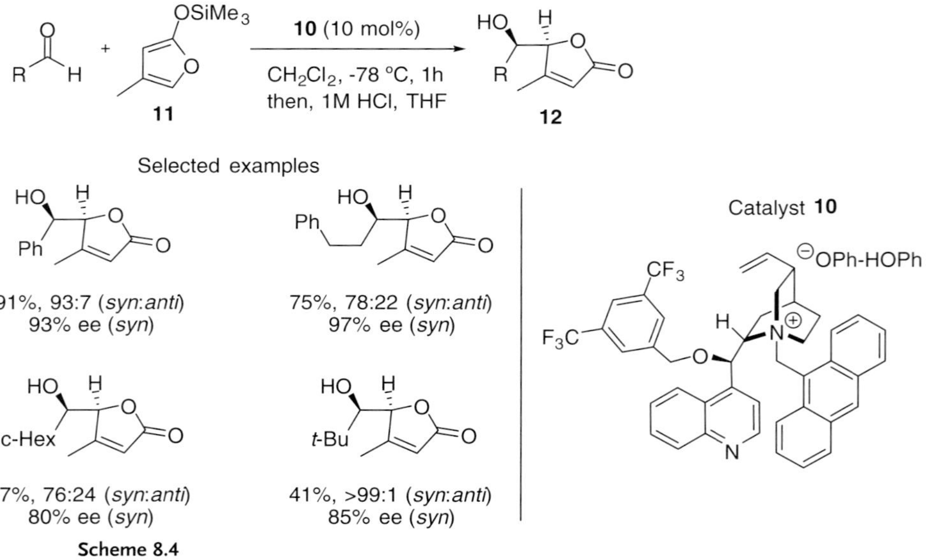

Scheme 8.4

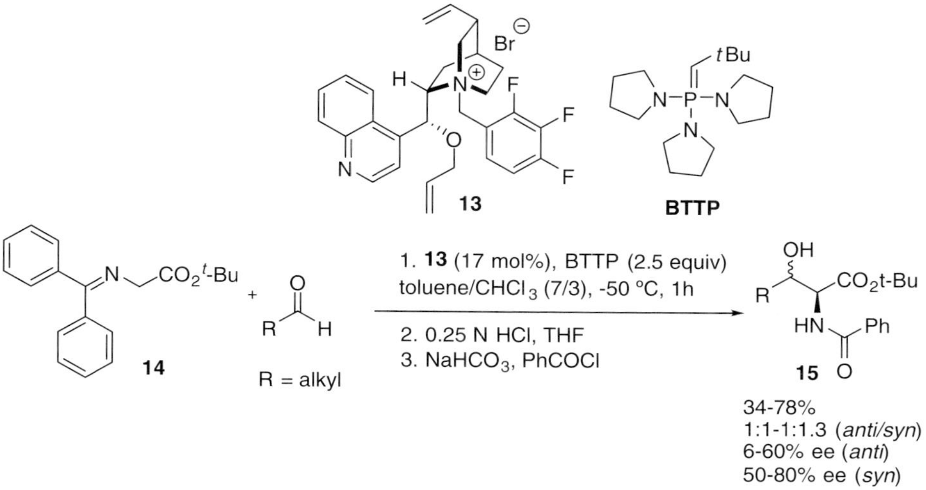

8.2 Aldol and Nitroaldol (Henry) Reactions

Scheme 8.5

phorane, 2.5 equiv), furnishing the β-hydroxy-α-amino esters **15**. However, the results obtained were not in the satisfactory range in terms of the chemical yield, diastereoselectiviy, and enantioselectivity (Scheme 8.5). Further studies [10] by the same group with different types of cinchona-based ammonium salts, however, did not improve the results to the level of preparative use.

In 2004, the group of Arai and Nishida found that cinchona-based ammonium salts such as **16** can catalyze the direct aldol reaction of diazoacetates **17** as an aldol donor with various aldehydes, affording the chiral α-diazo-β-hydroxyesters **18** which can be used as a precursor for the corresponding α-amino-β-hydroxy esters (Scheme 8.6) [11]. The obtained yields and ee values, however, were low to moderate. Especially, the electron-donating groups such as OMe in benzaldehyde had a strongly negative effect, resulting in a racemic product.

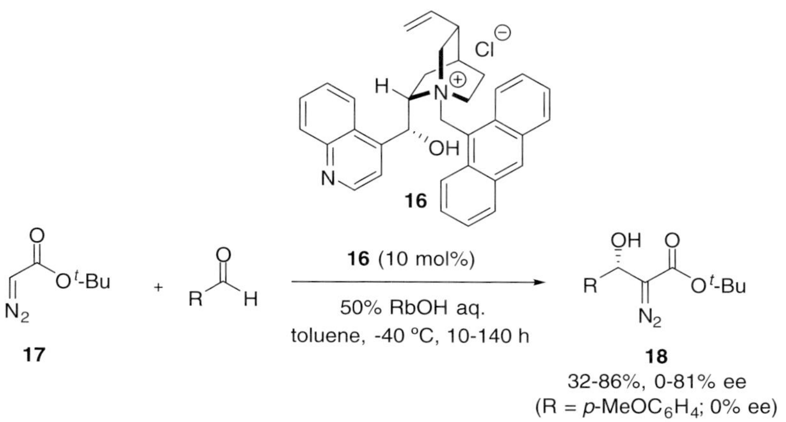

Scheme 8.6

As discussed above, all of the cinchona-based quaternary ammonium salts used as catalysts gave only poor to moderate diastereoselectivities and enantioselectivities for direct aldol reactions. Quite recently, a highly enantioselective, catalytic, direct aldol reaction was realized by adopting the enamine catalysis approach [12], in which 9-amino-*epi*-cinchona alkaloids are employed as aminocatalysts [13, 14].

In 2007, Liu and coworkers found that the direct aldol reaction of cyclic ketones with aromatic aldehydes could be catalyzed by the TFA salt of 9-amino-9-deoxy-*epi*-cinchonine (**19**) and its cinchonidine analogue (**20**) with high enantioselectivity [13]. In the presence of 10 mol% of catalyst **19** and 15 mol% of TfOH, a range of aromatic aldehydes underwent the aldol reaction with the cyclic ketone **21**, furnishing the aldol adducts **22** with good to excellent enantioselectivity (84–99% ee for *anti*). The opposite enantiomers of **22** could also be obtained with a similar level of ee using the cinchonidine-derived catalyst **20**. However, the diastereoselectivity varied from 1 : 1 to 9.2 : 1 (*syn*: *anti*) (Scheme 8.7). Electron-deficient aromatic aldehydes usually gave high chemical yields and ee values, whereas benzaldehyde, 1-naphthaldehyde, 4-MeO-benzaldehyde suffered from low chemical yields and somewhat lower ee values. Unfortunately, acyclic aldol donors such as acetone were shown to be unsuitable for this process. For example, the reaction of acetone with 4-nitrobenzaldehyde proceeded sluggishly to give a low chemical yield and ee (25%, 56% ee).

Later, the scope of this methodology was successfully extended to the intramolecular version by List and coworkers [14]. By employing 9-amino-9-deoxyepiquinine **24** as a catalyst (20 mol%) and an acid cocatalyst (AcOH, 60 mol%), 5-substituted-3-methyl-2-cyclohexene-1-ones (**26**) were obtained with high enantioselectivity (up to 94% ee) from the diketones **25** via the intramolecular aldol reaction (Scheme 8.8). The chiral enones **26** are valuable synthetic building blocks for the synthesis of many biologically important compounds (e.g., HIV-1 protease-inhibitive didemnaketals). The pseudoenantiomeric quinidine analogue **23** of **24** also provided the opposite

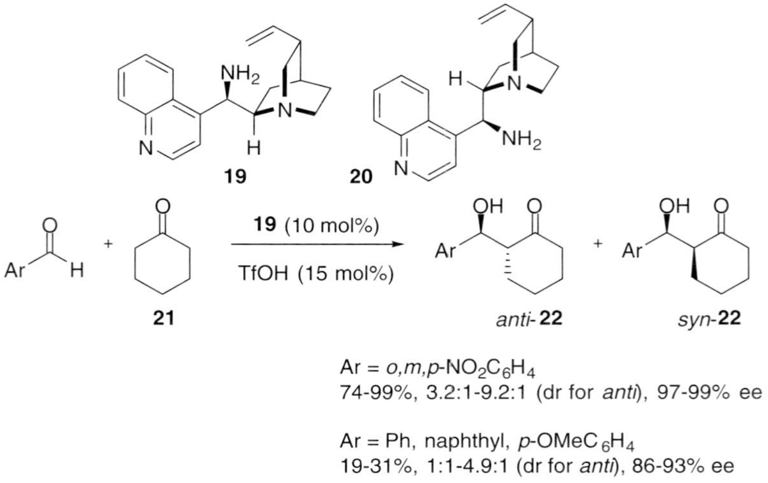
NH2
N
H
N
19
20
H
N
NH2
N
O
Ar
H
+
O
21
19 (10 mol%)
TfOH (15 mol%)
OH
O
Ar
anti-22
+
OH
O
Ar
syn-22
Ar = o,m,p-NO2C6H4
74-99%, 3.2:1-9.2:1 (dr for anti), 97-99% ee
Ar = Ph, naphthyl, p-OMeC6H4
19-31%, 1:1-4.9:1 (dr for anti), 86-93% ee

Scheme 8.7

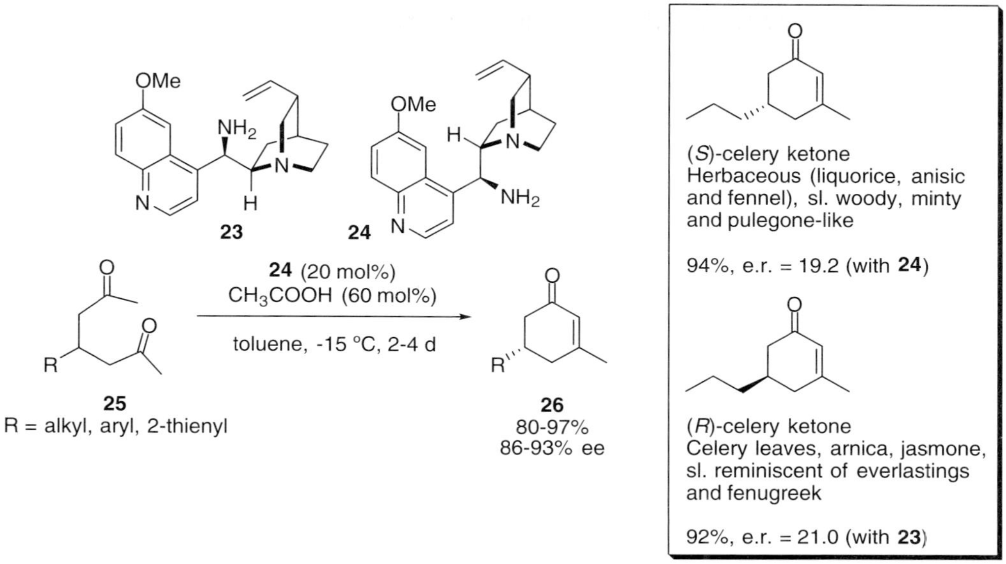

Scheme 8.8

enantiomers of **26** with the same level of enantiomeric excess. Of note, proline, the well-known amino catalyst, and the catalytic antibody 38C2 provided poor enantioselectivity for the same reaction (20 and 46% ee, respectively) [15]. It is also worth noting that the acidity of the acid cocatalyst was crucial to obtain good catalytic activity and enantioselectivity. The use of a stronger acid gave lower ee values. Acetic acid turned out to be the best cocatalyst in terms of the enantioselectivity (cf., R of **26** = n-pentyl; Cl_3CCO_2H (88% ee), CF_3CO_2H (81% ee), p-TsOH (66% ee), $PhCO_2H$ (89% ee), CH_3CO_2H (93% ee)). As has been proposed in the related aldol reactions catalyzed by a combination of diamine and protonic acid [16], the protonated quinuclidine moiety of catalysts **23** and **24** might act as a synergistic Brønsted acid for the direction and activation of the electrophilic carbonyl group by hydrogen bonding. This double activation model was supported by the fact that a significantly longer reaction time was required when only 1 equiv (i.e., 20 mol%) of acid to the catalyst was used.

The synthetic utility of this reaction was exemplified by the first enantioselective synthesis of both enantiomers of celery ketone **26** (R = nPr), a synthetic fragrance material with typical lovage and celery character (Scheme 8.8).

Very recently, Xiao and coworkers developed a new type of multifunctional cinchonidine-based catalyst, such as **27** having an additional proline moiety, for direct aldol reactions [17]. In the presence of an acid cocatalyst, the proline moiety of **27** can act as a Lewis base to activate the aldol donor via enamine formation. Moreover, both the NH proton of the amide linkage and the protonated tertiary amine at quinuclidine might act as a hydrogen-bonding donor to activate and direct the substrates. Using the cinchonidine-derived catalyst **27** (10 mol%) and acetic acid cocatalyst (20 mol%), the reaction of electron-deficient aromatic aldehydes with

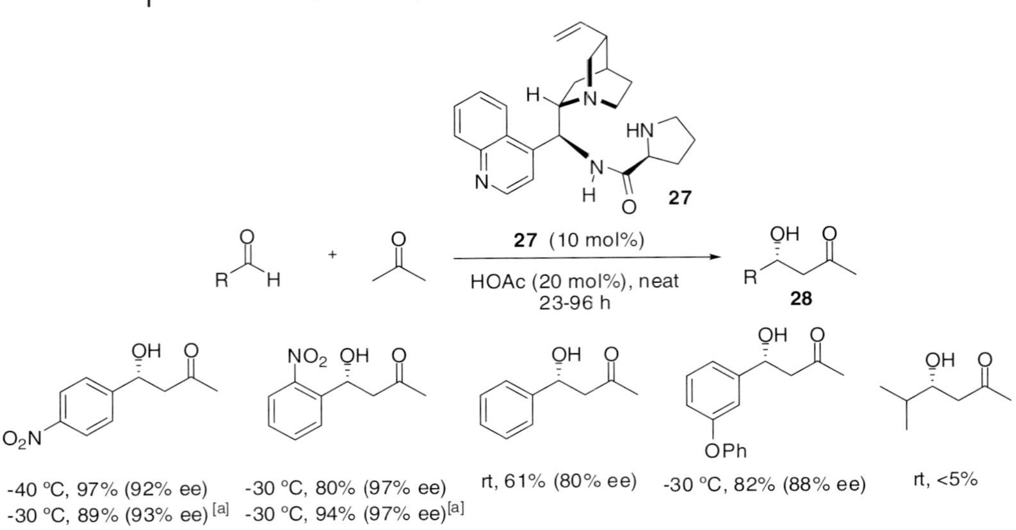

8 Cinchona-Catalyzed Nucleophilic 1,2-Addition to C=O and C=N Bonds

Scheme 8.9

Reaction: R-CHO + acetone → 28 (β-hydroxy ketone), with catalyst 27 (10 mol%), HOAc (20 mol%), neat, 23–96 h.

Products 28 with various R groups:

- R = 4-O$_2$N-C$_6$H$_4$: −40 °C, 97% (92% ee); −30 °C, 89% (93% ee) [a]
- R = 2-NO$_2$-C$_6$H$_4$: −30 °C, 80% (97% ee); −30 °C, 94% (97% ee) [a]
- R = C$_6$H$_5$: rt, 61% (80% ee)
- R = 3-PhO-C$_6$H$_4$: −30 °C, 82% (88% ee)
- R = iPr: rt, <5%

[a] Reactions were conducted with the cinchonine derived analogue of 27 to afford the opposite enantiomer.

acetone afforded the aldol product **28** with excellent ee values (up to 97% ee). However, benzaldehyde or electron-rich aromatic aldehydes gave lower enantioselectivities (80–88% ee). Moreover, aliphatic aldehydes (e.g., isobutyraldehyde) almost did not undergo the reaction (<5% conversion) (Scheme 8.9). The opposite enantiomers of **28** could also be obtained with the same degree of ee using the cinchonine analogue of **27** (Scheme 8.9).

2-Butanone **29** was also investigated as an aldol donor. The reactions of 2-butanone **29** with nitro-substituted benzaldehydes afforded the aldol product **30** with excellent stereoselectivity (up to 93:7 of dr and 98% ee). However, the reaction also occurred unselectively at the C1 position of the ketone, leading to the concomitant formation of **31** (Scheme 8.10).

Recently, Shibata, Toru and coworkers reported that modified bis-cinchona alkaloids such as (DHQD)$_2$PHAL (**32**) and (DHQ)$_2$PHAL (**33**), known as Sharpless

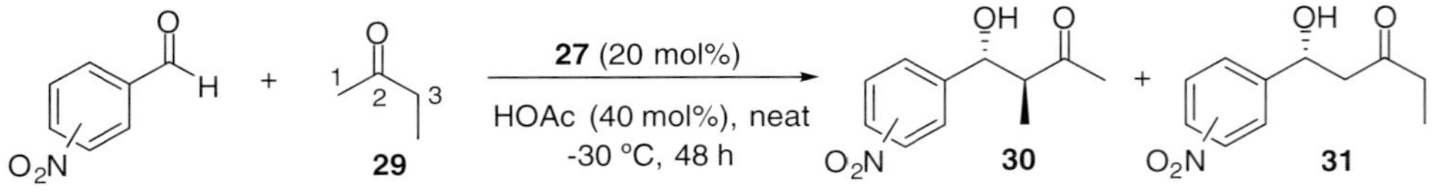

Scheme 8.10

Reaction: 2-NO$_2$/3-NO$_2$/4-NO$_2$-C$_6$H$_4$-CHO + 2-butanone **29** → **30** + **31**, with **27** (20 mol%), HOAc (40 mol%), neat, −30 °C, 48 h.

4-NO$_2$:
30 (53%, d.r. 93:7, 98% ee (anti)) + **31** (32%, 75% ee)
3-NO$_2$:
30 (71%, d.r. 93:7, 92% ee (anti)) + **31** (26%, 82% ee)
2-NO$_2$:
30 (39%, d.r. 96:4, 92% ee (anti)) + **31** (54%, 94% ee)

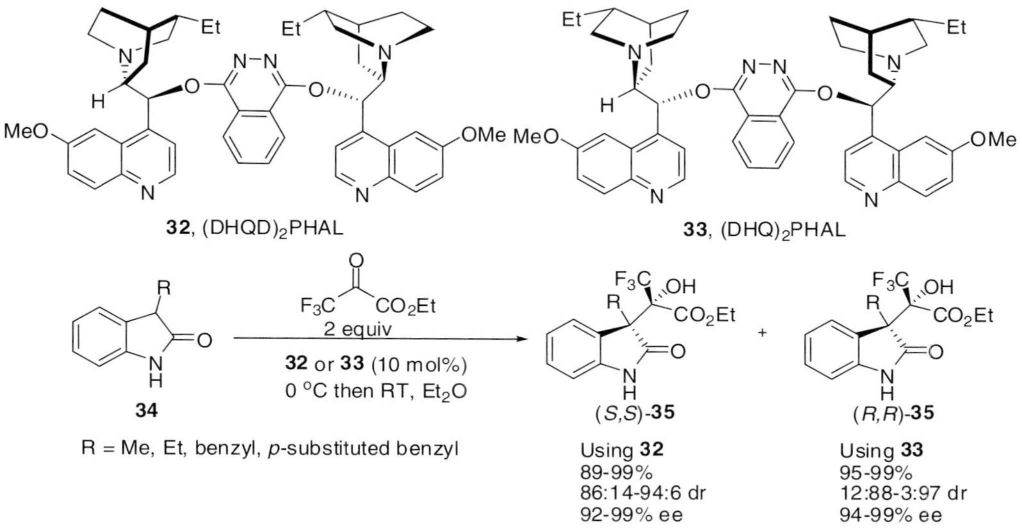

8.2 Aldol and Nitroaldol (Henry) Reactions | 205

Scheme 8.11

Reagents: **32** or **33** (10 mol%), F$_3$C-CO-CO$_2$Et (2 equiv), 0 °C then RT, Et$_2$O.

Substrate **34**: R = Me, Et, benzyl, p-substituted benzyl

Products (S,S)-**35** and (R,R)-**35**:

Using **32**: 89–99%, 86:14–94:6 dr, 92–99% ee
Using **33**: 95–99%, 12:88–3:97 dr, 94–99% ee

ligands, can serve as highly efficient catalysts for the aldol-type reaction of oxindoles with ethyl trifluoropyruvate [18]. This method provided the straightforward synthesis of both enantiomers of the trifluoromethylated oxindoles **35** having two stereogenic centers with high ee values (up to 99% ee) (Scheme 8.11). Various oxindoles with alkyl and aryl substituents were tolerated by this catalytic system. To rationalize the observed stereochemistry of the products, the authors proposed the dual function of quinuclidine (i) as a base for the deprotonation of the oxindole and (ii) as a Brønsted acid to activate the pyruvate via hydrogen bonding in the U-shaped reaction site supported by a π–π interaction with the quinoline ring to afford the (S,S) enantiomer with (DHQD)$_2$PHAL (**32**) catalyst (Figure 8.2).

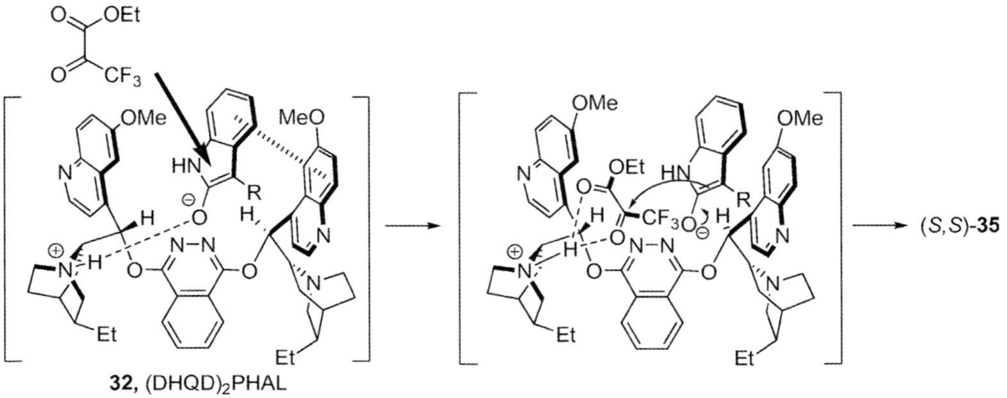

Figure 8.2 Proposed approximate structure of the substrate–catalyst (**32**) complex for the S,S-selective formation of adducts **35** in the aldol-type condensation of oxindoles with trifluoropyruvate.

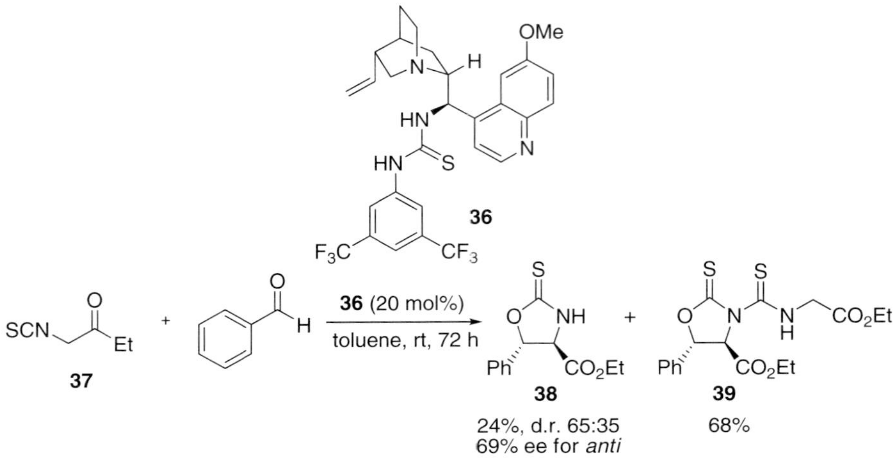

Scheme 8.12

Very recently, the quinidine-derived thiourea derivative **36** as a catalyst (20 mol%) was also applied to the enantioselective aldol addition of the α-isothiocyanato imide **37** to benzaldehyde affording the mixture of thiooxazolidinones **38** and **39** (Scheme 8.12) [19]. However, poor yields and only moderate enantiomeric excesses were observed. On the other hand, the same type of quinine-derived catalyst **40** (1 mol%) was proven to be highly reactive and enantioselective in the tandem thio-Michael-aldol reaction of 2-mercaptobenzaldehyde (**41**) with α,β-unsaturated oxazolidinone (**42**), furnishing benzothiopyranes (**43**) having three stereogenic centers in excellent stereoselectivity (>20:1 dr and 91–99% ee) (Scheme 8.13) [20]. The excellent stereoselectivity can be attributed to the multiple hydrogen bonding between the catalyst and the substrate.

8.2.2
Henry Reactions

The asymmetric nitroaldol (Henry) reaction can generate versatile chiral building blocks, for example, 1,2-amino alcohols, with readily available substances (i.e., aldehydes and ketones with nitroalkanes). The first asymmetric organocatalytic Henry reaction was achieved by Hiemstra and coworkers by using cinchona-derived catalysts such as **44** bearing a free hydroxyl group at the 6′-position. In the presence of 10 mol% of catalyst **44**, the reaction of various electron-deficient aromatic aldehydes with nitromethane at room temperature proceeded smoothly to afford the corresponding nitroaldol products **45** in high yield, but with only low ee values (up to 35% ee) (Scheme 8.14) [21].

Later, a dramatic jump in the ee values was achieved by the same research group by introducing the thiourea moiety at the 6′-position of **44** [22]. Using 10 mol% of the quinidine-derived catalyst **46**, the highly enantioselective Henry reaction between nitromethane and the aromatic aldehydes (electron-rich, electron-deficient, and

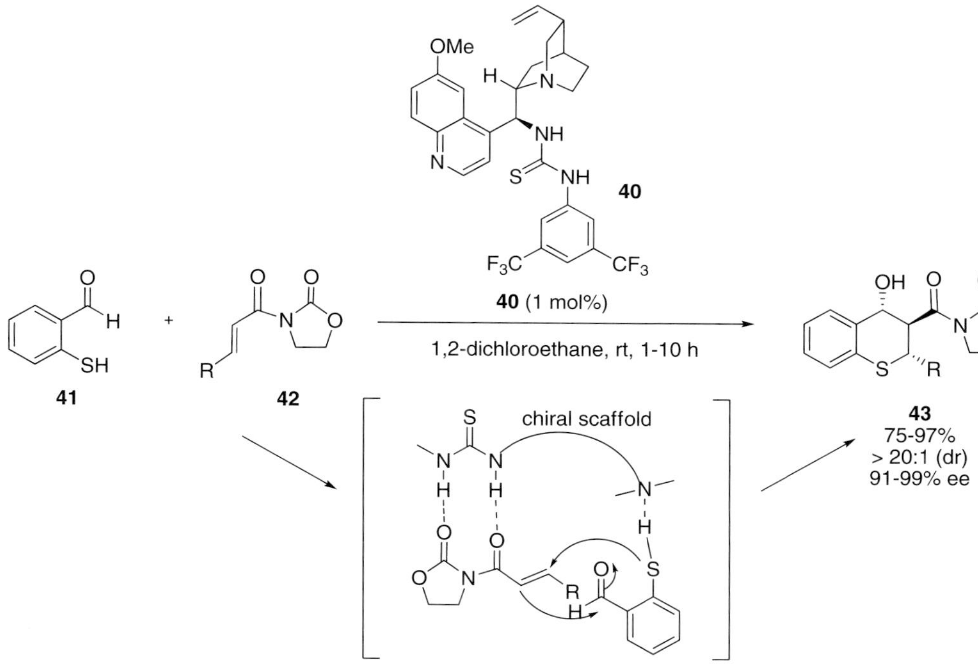

Scheme 8.13

hindered variants) was achieved in THF at −20 °C (up to 92% ee) (Scheme 8.15). The pseudoenantiomeric quinine-derived catalyst **47** provided the opposite enantiomers of **48** with similar ee values (up to 93%). However, disappointing results were obtained with aliphatic aldehyde substrates (up to 20% ee). The authors proposed a dual activation mode of the catalyst involving the activation of the aldehyde by the thiourea unit and deprotonation of the nitromethane acid by the quinuclidine base (Figure 8.3).

In 2006, Deng and coworkers reported that the quinidine and quinine derivatives, **49** and **50**, respectively, bearing a free OH group at the 6′-position catalyzed the enantioselective Henry reaction of α-ketoesters with nitromethane [23]. In the

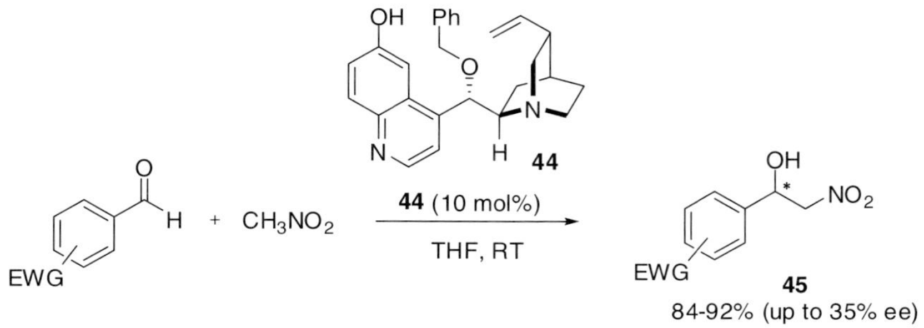

Scheme 8.14

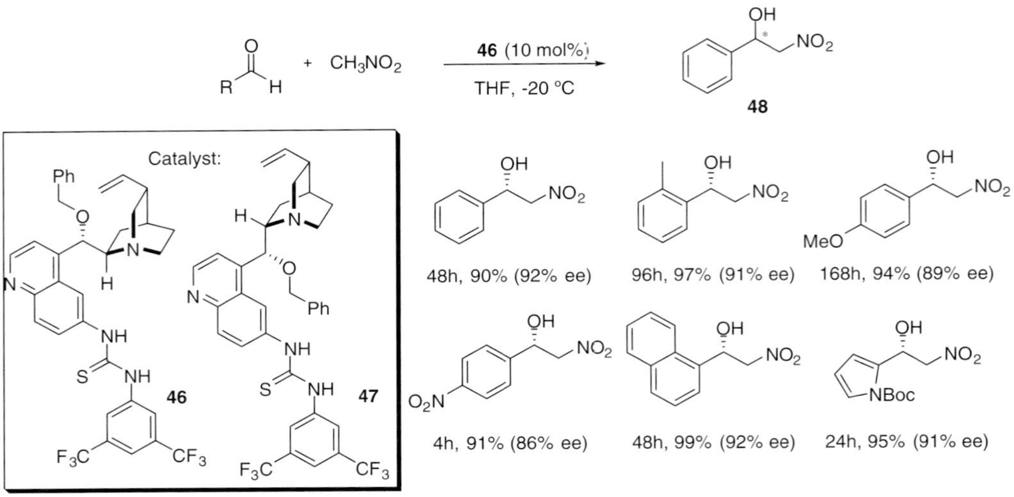

Scheme 8.15

presence of 5 mol% of the catalyst, the various α-ketoesters **51** underwent smoothly the nitroaldol reaction with nitromethane at −20 °C in CH_2Cl_2 to afford the chiral tertiary carbinols **52** in high yields (84–98%) and with good to excellent ee values (up to 90–97% ee) (Scheme 8.16). However, the same reaction with unactivated ketones such as acetophenone did not proceed at all.

Quite recently, Bandini, Umani-Ronchi and coworkers also reported the highly enantioselective Henry reaction of the various trifluoromethyl ketones **54** with nitromethane catalyzed by the C6′-hydroxy quinine derivatives **53** (5 mol%) [24]. Various aliphatic and aromatic ketones were smoothly converted to the desired tertiary carbinols **55** in high yields and ee values (up to 99%) without any significant electronic or steric demands (Scheme 8.17). The difluoroketones **56** proved just as useful as substrates (Scheme 8.18). Of note, the parent alkaloid, quinine, as a catalyst did not give rise to any asymmetric induction.

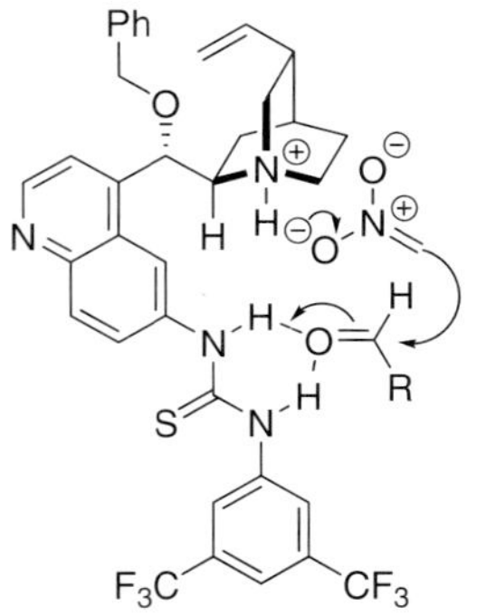

Figure 8.3 Proposed mode of reaction with catalyst **46**.

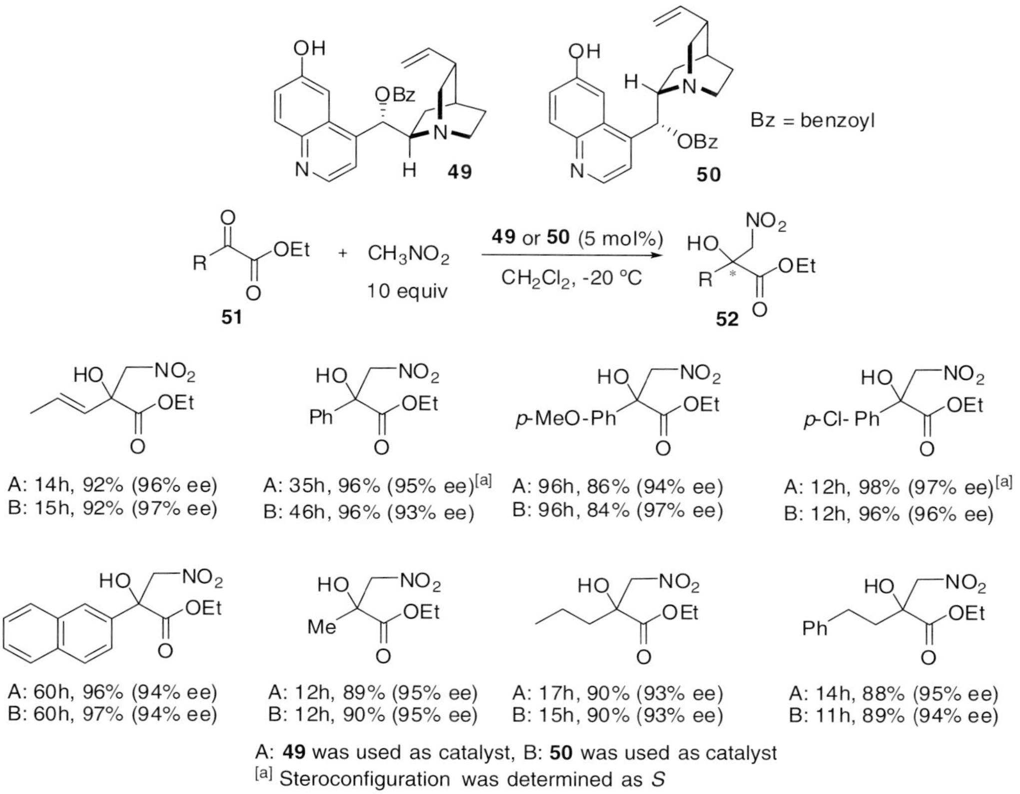

Scheme 8.16

Catalyst 49, 50 structures shown (Bz = benzoyl).

Reaction: 51 (R-CO-CO-OEt, α-ketoester) + CH₃NO₂ (10 equiv) → 52 (HO-C(NO₂CH₂)(R)-CO-OEt) with 49 or 50 (5 mol%), CH₂Cl₂, −20 °C.

Products (structure: HO-C(R)(CH₂NO₂)-C(O)OEt):

- R = propenyl: A: 14h, 92% (96% ee); B: 15h, 92% (97% ee)
- R = Ph: A: 35h, 96% (95% ee)[a]; B: 46h, 96% (93% ee)
- R = p-MeO-Ph: A: 96h, 86% (94% ee); B: 96h, 84% (97% ee)
- R = p-Cl-Ph: A: 12h, 98% (97% ee)[a]; B: 12h, 96% (96% ee)
- R = 2-naphthyl: A: 60h, 96% (94% ee); B: 60h, 97% (94% ee)
- R = Me: A: 12h, 89% (95% ee); B: 12h, 90% (95% ee)
- R = n-propyl: A: 17h, 90% (93% ee); B: 15h, 90% (93% ee)
- R = PhCH₂CH₂: A: 14h, 88% (95% ee); B: 11h, 89% (94% ee)

A: **49** was used as catalyst, B: **50** was used as catalyst
[a] Steroconfiguration was determined as *S*

8.3 Mannich and Nitro-Mannich Reactions

8.3.1 Mannich Reactions

Catalytic enantioselective Mannich reactions provide one of the most versatile approaches for the synthesis of optically active chiral amines. Recently, several organocatalytic protocols have been developed using the parent cinchona alkaloids or their derivatives.

In 2005, Schaus and coworkers found that the natural cinchona alkaloids such as cinchonine (**CN**) or cinchonidine (**CD**) themselves can serve as highly enantioselective catalysts (10 mol%) for the Mannich reaction of β-keto esters **57** with the various carbamate-protected aryl imines **58** [25]. Using either CN or CD, both enantiomers of the resulting secondary amine products **59** were obtained in excellent yields (up to 99%) and ee values (up to 96% ee) (Scheme 8.19). The extension of this protocol to encompass the use of the 2-substituted-1,3-dicarbonyl nucleophiles **60**

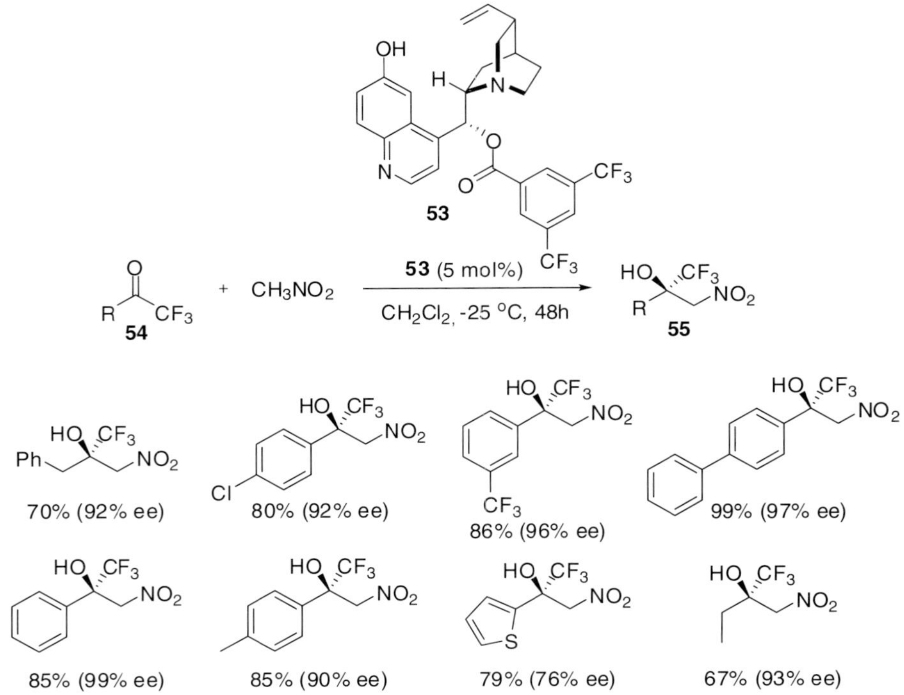

Scheme 8.17

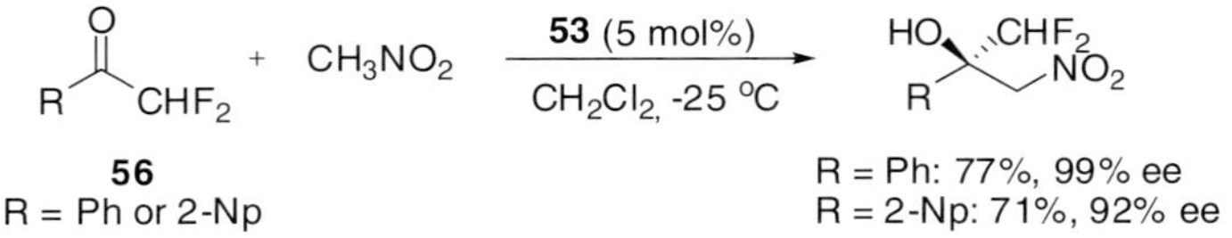

R-C(=O)-CF$_3$ (**54**) + CH$_3$NO$_2$ →[**53** (5 mol%), CH$_2$Cl$_2$, -25 °C, 48h] R-C(OH)(CF$_3$)-CH$_2$NO$_2$ (**55**)

Catalyst **53**: quinine derivative with 3,5-bis(trifluoromethyl)benzoate ester.

Selected products:
- Ph-C(OH)(CF$_3$)-CH$_2$NO$_2$: 70% (92% ee)
- 4-Cl-C$_6$H$_4$-C(OH)(CF$_3$)-CH$_2$NO$_2$: 80% (92% ee)
- 3-CF$_3$-C$_6$H$_4$-C(OH)(CF$_3$)-CH$_2$NO$_2$: 86% (96% ee)
- 4-biphenyl-C(OH)(CF$_3$)-CH$_2$NO$_2$: 99% (97% ee)
- benzyl-C(OH)(CF$_3$)-CH$_2$NO$_2$: 85% (99% ee)
- 4-Me-C$_6$H$_4$-CH$_2$-C(OH)(CF$_3$)-CH$_2$NO$_2$: 85% (90% ee)
- 2-thienyl-C(OH)(CF$_3$)-CH$_2$NO$_2$: 79% (76% ee)
- Et-C(OH)(CF$_3$)-CH$_2$NO$_2$: 67% (93% ee)

Scheme 8.18

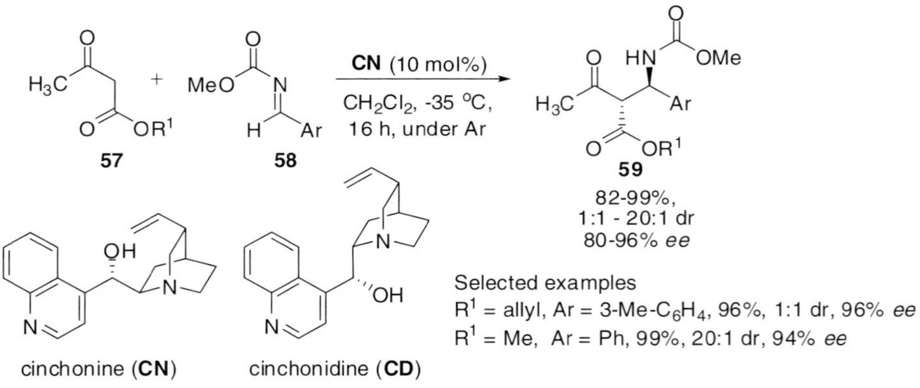

R-C(=O)-CHF$_2$ (**56**, R = Ph or 2-Np) + CH$_3$NO$_2$ →[**53** (5 mol%), CH$_2$Cl$_2$, -25 °C] R-C(OH)(CHF$_2$)-CH$_2$NO$_2$

R = Ph: 77%, 99% ee
R = 2-Np: 71%, 92% ee

Scheme 8.19

H$_3$C-C(=O)-CH$_2$-C(=O)-OR^1 (**57**) + MeO-C(=O)-N=CH-Ar (**58**) →[**CN** (10 mol%), CH$_2$Cl$_2$, -35 °C, 16 h, under Ar] H$_3$C-C(=O)-CH(C(=O)OR^1)-CH(Ar)-NH-C(=O)-OMe (**59**)

82-99%, 1:1 - 20:1 dr, 80-96% ee

cinchonine (**CN**)   cinchonidine (**CD**)

Selected examples
R^1 = allyl, Ar = 3-Me-C$_6$H$_4$, 96%, 1:1 dr, 96% ee
R^1 = Me, Ar = Ph, 99%, 20:1 dr, 94% ee

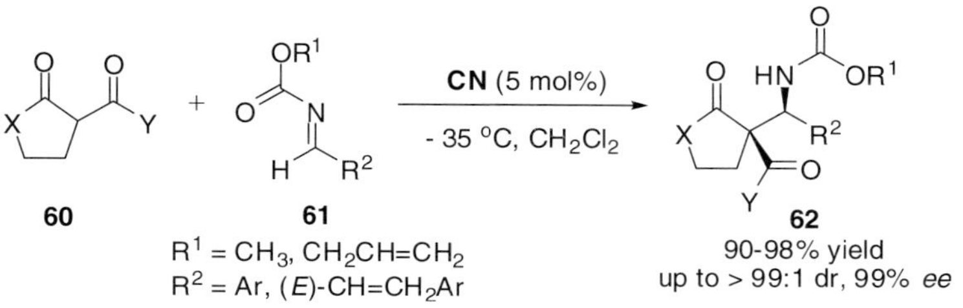

8.3 Mannich and Nitro-Mannich Reactions

Scheme 8.20

60 + 61 → 62

61: R¹ = CH₃, CH₂CH=CH₂; R² = Ar, (E)-CH=CH₂Ar

Conditions: CN (5 mol%), −35 °C, CH₂Cl₂

62: 90–98% yield, up to > 99:1 dr, 99% ee

allowed the formation of the α-quaternary carbon-bearing product **62** in yields of up to 98%, up to > 99 : 1 dr, and enantioselectivities of up to 99% ee (Scheme 8.20) [26]. The stereoselective control was attributed to the complexation of the chiral alkaloid with the nucleophile (Figure 8.4). In 2008, the same research group successfully applied this methodology to synthesize SNAP-7941, a potent melanin concentrating hormone receptor antagonist [27].

Bifunctional cinchona-based thioureas were also identified independently by the groups of Dixon [28], Deng [29], and Schaus [30] as highly efficient catalysts for the Mannich reactions of dicarbonyl pronucleophiles with carbamate-protected imines. In 2006, Dixon and coworkers [28] reported that in the presence of the cinchonine-derived thiourea **63**, the carbamate (Boc or Cbz)-protected aldimines **64** underwent the Mannich reaction with dialkylmalonates, affording the Mannich adducts **65** in high yields, with up to 97% ee (Scheme 8.21). This process was also applicable with similar efficiency to 1,3-diketones, β-ketoesters, and 2-substituted 1,3-dicarbonyl compounds. Shortly afterward, in a similar study, Deng and coworkers [29] and Schaus and coworkers [30] obtained comparable results using quinine or quinidine-derived catalyst analogues.

Molecular modeling studies performed by Schaus [30] to rationalize the observed sense of stereoselectivity revealed that the nucleophile (i.e., the conjugate base of the malonate) is hydrogen-bonded to both the thiourea moiety and the protonated quinuclidine catalyst and that one face of the nucleophile is blocked by the quinoline ring. Accordingly, the Re-face attack of the nucleophile on the Z-aldimine was proposed to produce the observed (S)-Mannich product (Figure 8.5).

Recently, Chen and coworkers demonstrated that the cinchonine- or cinchonidine-derived thioureas, **63** and **66**, respectively, were also able to catalyze the enantioselective

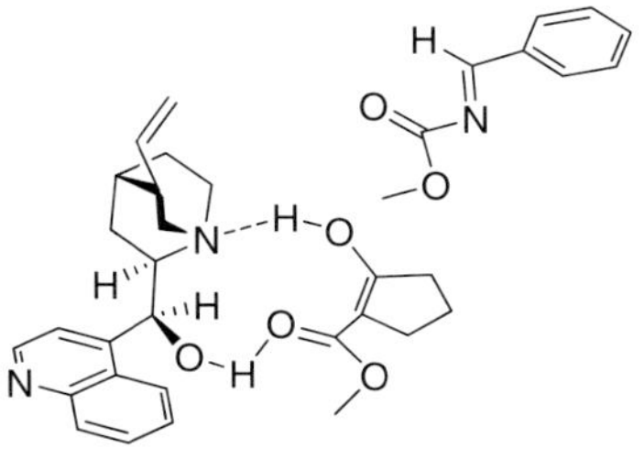

Figure 8.4 Proposed cinchonine/methyl 2- oxocyclopentane-carboxylate enol tautomer complex approaching the Re-face of methyl benzylidenecarbamate.

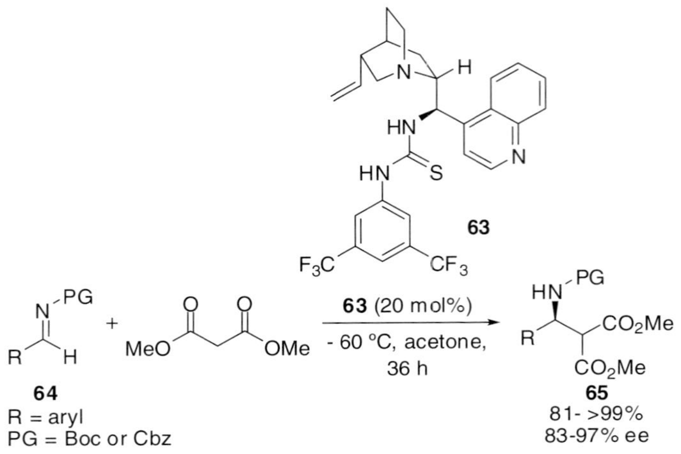

Scheme 8.21

64 + 63 (20 mol%), −60 °C, acetone, 36 h → 65
R = aryl
PG = Boc or Cbz
81->99%
83-97% ee

direct vinylogous Mannich reaction of the α,α-dicyanoolefin **67** as a nucleophilic partner and N-Boc benzaldimine [31]. This process exhibited exclusive γ-selectivity and antidiastereoselectivity to produce **68**. However, only moderate enantioselectivities (up to 61% ee) were achieved (Scheme 8.22). On the other hand, a modified analogue of Takemoto's catalyst was shown to be superior in this reaction.

Ricci, Pettersen, and coworkers reported that β-isocupreidine (β-ICD) derived from quinidine (**69**, 20 mol%) promotes the enantioselective decarboxylative addition of the malonic half thioesters **70** to the imines **71** to afford the protected β-amino thioesters **72**, which act as precursors for the preparation of β-amino acids [32] (Scheme 8.23). However, only low to moderate ee values (4–79% ee) were obtained.

As described above, N-protected aromatic imines can be successfully employed as substrates for Mannich reactions. However, although N-protected alkyl imines constitute another important class of imine substrates, their instability hampers their use. N-protected alkyl imines are spontaneously tautomerized into the corresponding enamines even at low temperature. In aza-Henry (nitro-Mannich) reactions, this problem was overcome independently by Palomo [33a] and the group of Herrera, Bernardi, and Ricci [33b] by using the stable α-amidosulfones **73** from

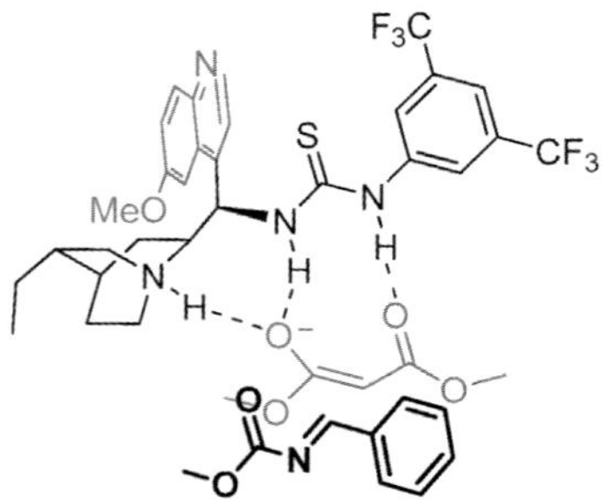

Figure 8.5 Proposed transition state model.

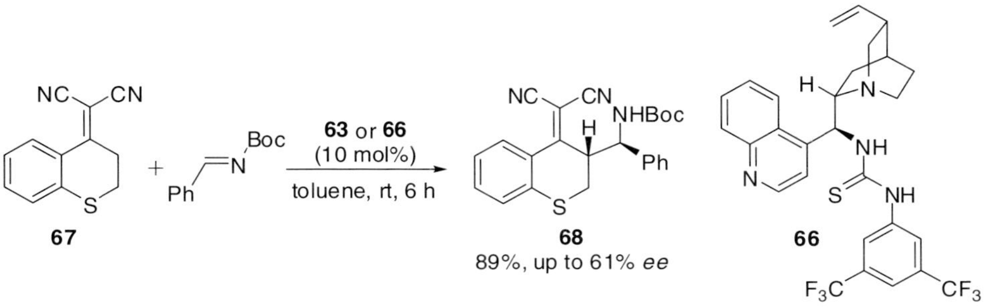

Scheme 8.22

which carbamate-protected imines can be generated *in situ* (see Section 8.3.2). In 2007, a highly enantioselective Mannich reaction with the *in situ* generation of carbamate-protected aliphatic and aromatic imines from the α-amidosulfones **73** catalyzed by the quinidine-based thiourea **36** or quinine-derived analogue **40** was reported by Deng and coworkers [34]. Both aliphatic and aromatic α-amido *p*-tolylsulfones **73** underwent the reaction with dibenzyl malonate to produce the products **74** in good to high yields and with up to 96% ee (Scheme 8.24).

Soon after, the groups of Ricci [35] and Schaus [36] also employed α-amidosulfones as stable imine precursors in cinchona-catalyzed Mannich reactions. Ricci and coworkers reported [35] that, under PTC conditions (toluene/aqueous K_2CO_3) using **75** as a catalyst (1 mol%), both the aliphatic and aromatic α-amido *p*-tolylsulfones **76** reacted with the malonates to afford the Mannich adducts **77** with high levels of enantioselectivity (85–99% ee) (Scheme 8.25). The subsequent decarboxylation/transesterification of **77** gave the corresponding β-amino acid derivatives without any alternation of the optical purities. The chiral dihydropyrimidones **80** were also successfully synthesized by Schaus and coworkers via the cinchonine catalyzed

Selected examples

R^1	R^2	yield (%)	% ee
p-$OMeC_6H_4$	p-$OMeC_6H_4$	52	68
p-$OMeC_6H_4$	$C_6H_5CH_2CH_2$	42	79
p-Cl-C_6H_4	p-$OMeC_6H_4$	55	65
p-Cl-C_6H_4	$C_6H_5CH_2CH_2$	47	73

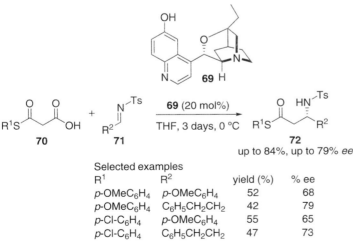

Scheme 8.23

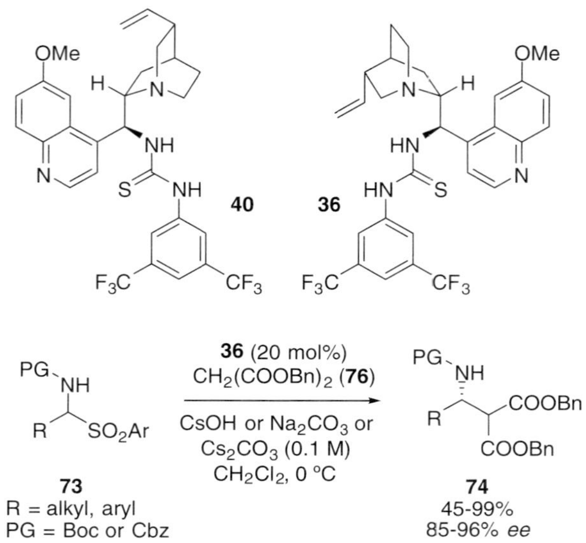

Scheme 8.24

Mannich reaction of the various diketones **78** with acyl amines generated *in situ* from the α-amidosulfones **79** (Scheme 8.26) [36].

Shibata, Toru, and coworkers developed a protocol for the highly enantioselective fluorobis(phenylsulfonyl)methylation of imines generated *in situ* from the α-amido sulfones **82** [37]. In the presence of the cinchona-derived PTC **81** (5 mol%), the asymmetric Mannich-type reaction of fluorobis(phenylsulfonyl)methane **83** with imines generated *in situ* from the α-amido sulfones **82** afforded the N-Boc α-fluorobisphenylsulfonyl amines **84** in high yields and with excellent enantioselectivities. The scope of the reaction was not limited to aromatic or heteroaromatic imines. Alkyl α-amidosulfones that have enolizable protons also gave good reactivities and

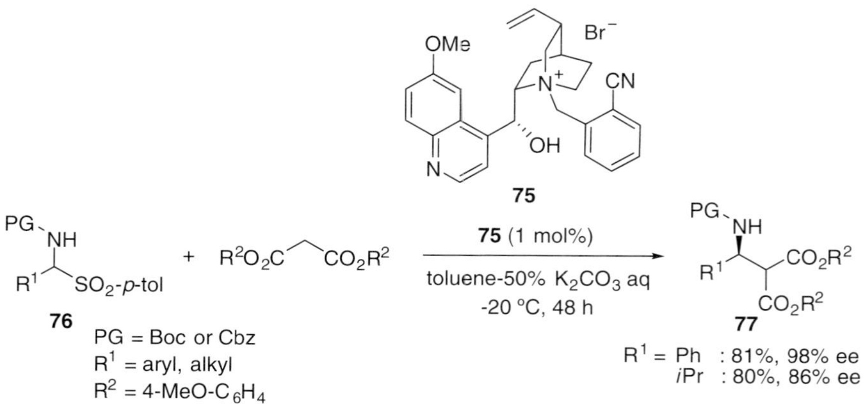

Scheme 8.25

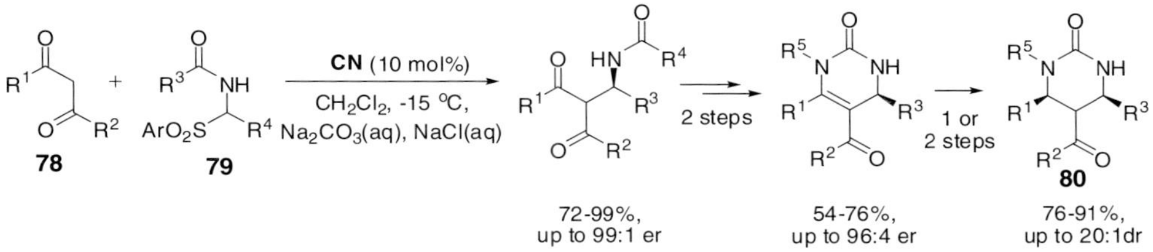

8.3 Mannich and Nitro-Mannich Reactions

Scheme 8.26

enantioselectivities (90–99% ee) (Scheme 8.27). The products **84** were successfully converted to the α-monofluoromethyl amines **85** by reductive desulfonylation.

8.3.2
Nitro-Mannich (Aza-Henry) Reactions

The catalytic asymmetric variants of the nitro-Mannich (aza-Henry) reaction have attracted a great deal of attention in recent years, partly because the resulting β-nitro amines can be readily derivatized into highly valuable chiral building blocks such as vicinal diamines. In 2005, the successful asymmetric nucleophilic addition of nitroalkanes to α-amido sulfones as imine surrogates was achieved independently by two research groups under phase-transfer conditions, using cinchona alkaloid derived quaternary ammonium salts as catalysts. As described previously, the α-amido sulfones are transformed into imines under the PTC conditions, that is, when using a phase-transfer catalyst in combination with an inorganic base.

Herrera, Bernardi, and coworkers demonstrated that the addition of nitromethane to a wide range of aromatic and aliphatic N-Boc imines generated *in situ* from the

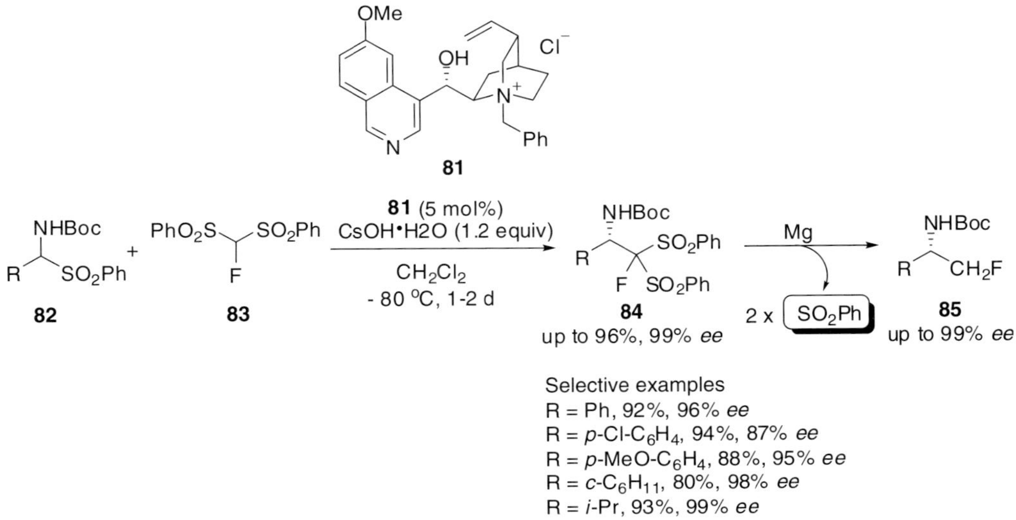

Scheme 8.27

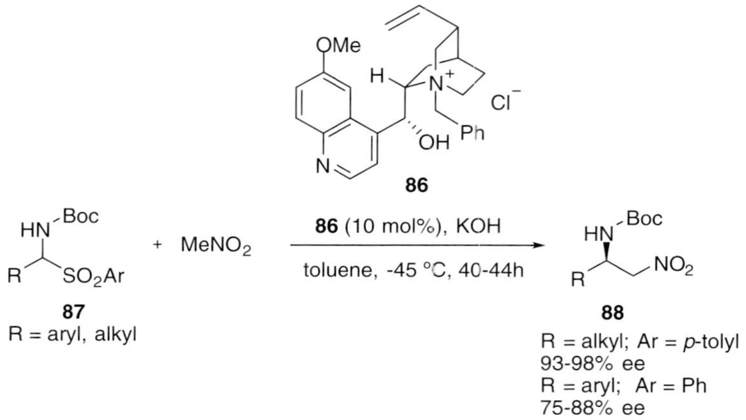

8 Cinchona-Catalyzed Nucleophilic 1,2-Addition to C=O and C=N Bonds

Scheme 8.28

α-amido sulfones **87** proceeded with high enantioselectivity under solid–liquid phase-transfer conditions by using the quinine-derived PTC **86** (10 mol%) as a catalyst and powdered KOH (5 equiv) as a base (Scheme 8.28) [38]. Remarkably, the aza-Henry reaction with enolizable aldehyde-derived azomethines gave an enantiomeric excess of more than 93% (93–98% ee).

In a similar study, Palomo and coworkers utilized CsOH·H$_2$O (1.2 equiv) as a solid base in combination with 12 mol% of quinine- (**86**) or cinchonidine- (**89**) derived quaternary ammonium chloride as a catalyst (12 mol%) for the reaction of nitromethane with a wide range of aromatic and aliphatic N-Boc imines generated *in situ* from the α-amido sulfones **90**, providing the corresponding aza-Henry adducts **91** in good yields and very high selectivities (Scheme 8.29) [39]. Moreover, they extended

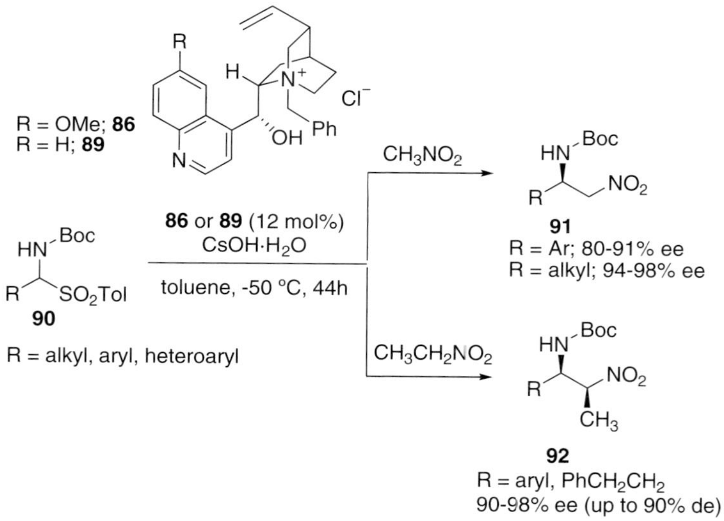

Scheme 8.29

the scope of this reaction to include nitroethane. The *syn*-adduct **92** was obtained predominantly with a high degree of enantioselectivity (Scheme 8.29). It is believed that in the above-discussed examples, the chiral PTC acts in multiple ways, first promoting the formation of the imine under mild conditions and then activating the nucleophile for asymmetric addition. Moreover, the free hydroxyl group of the catalysts also plays a key role in the activation of the substrate. The corresponding catalysts, whose alcohol group is in the form of a benzyl ether, displayed significantly lower efficiency (typically <10% conversion). This indicates that a hydrogen-bonding interaction between the free hydroxyl group and the nitro group's oxygen, facilitating nitronate formation, and/or between the hydroxyl and the azomethine's nitrogen, activating the electrophile and rigidifying the transition structure, plays a key role in this highly stereoselective transformation.

In addition to chiral PTCs, cinchona-based thioureas have also been proved to serve as catalysts for nitro-Mannich reactions. In 2006, Ricci and coworkers first reported that the quinine-based thiourea **40** (20 mol%) can catalyze the aza-Henry reaction between nitromethane and the *N*-protected imines **93** derived from aromatic aldehydes [40]. *N*-Boc-, *N*-Cbz-, and *N*-Fmoc protected imines gave the best results in terms of the chemical yields and enantioselectivities (up to 94% ee at −40 °C) (Scheme 8.30).

In a similar study, Schaus extended the scope of this reaction to include nitroethane [30]. The hydroquinine-derived thiourea catalyst **95** efficiently promotes the aza-Henry reaction of nitroethane with the acyl imines **96**, affording the *syn*-β-nitroamines **97** in good yields with enantioselectivities of 90–98% ee and *syn*-diastereoselectivities of up to 97% (Scheme 8.31).

Molecular modeling studies designed to rationalize the observed sense of stereoselectivity revealed that the nucleophile (i.e., the conjugate base of the nitroalkane) is linked by hydrogen-bond to both the thiourea moiety and the protonated quinuclidine catalyst and that one face of the nucleophile is blocked by the quinoline ring. According to the calculation, the *Re*-face attack of the nucleophile on the imine to

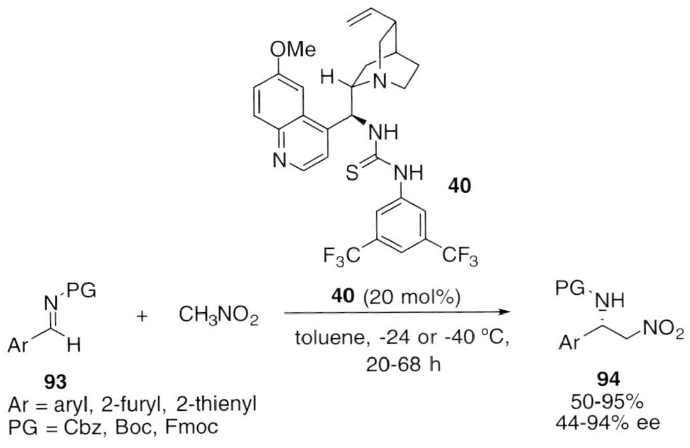

Scheme 8.30

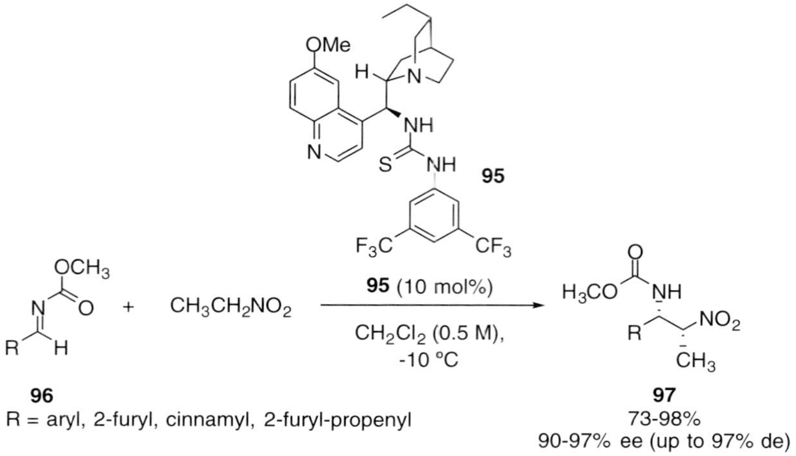
OMe
H
N
NH
N
S
NH
95
F3C
CF3
OCH3
N
O
R
H
+
CH3CH2NO2
95 (10 mol%)
CH2Cl2 (0.5 M),
-10 °C
H3CO
NH
NO2
R
CH3
96
R = aryl, 2-furyl, cinnamyl, 2-furyl-propenyl
97
73-98%
90-97% ee (up to 97% de)

Scheme 8.31

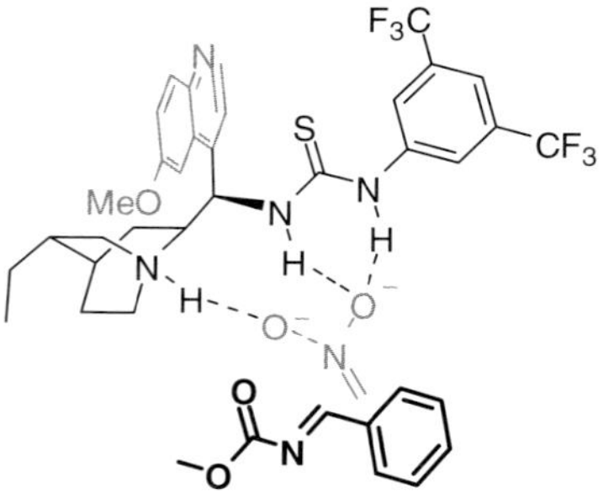
F3C
N
S
CF3
MeO
N
N
H
H
N
H
O
O
N
O
N
O

Figure 8.6 Proposed catalyst **95**/nitromethane complex approaching the *Re*-face of methyl benzylidenecarbamate.

produce the observed (*S*)-nitroamines is 1.6 kcal/mol more favorable than the *Si*-face attack (Figure 8.6).

8.4
Aldol- and Mannich-Related Reactions

8.4.1
Darzens Reactions

The Darzens reaction (tandem aldol-intramolecular cyclization sequence reaction) is a powerful complementary approach to epoxidation (see Chapter 5) that can be used for the synthesis of α,β-epoxy carbonyl and α,β-epoxysulfonyl compounds (Scheme 8.32). Currently, all catalytic asymmetric variants of the Darzens reactions are based on chiral phase-transfer catalysis using quaternary ammonium salts as catalysts.

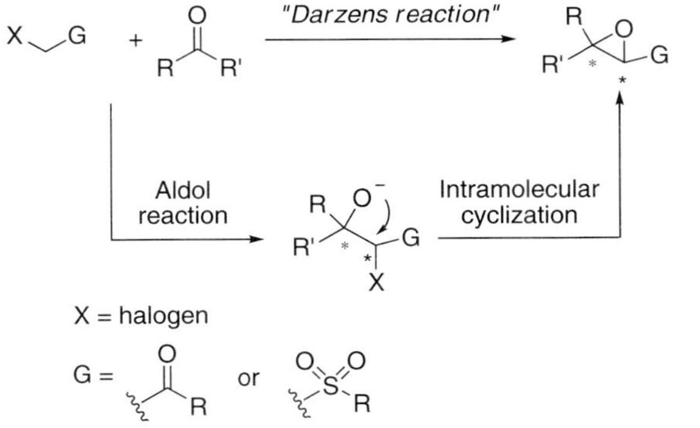

8.4 Aldol- and Mannich-Related Reactions

Scheme 8.32

The first catalytic asymmetric version of the Darzens reaction was achieved in 1978 by J. Hummelen and H. Wynberg [41]. The treatment of *p*-chlorobenzaldehyde and phenacylchloride with the strong base NaOH in the presence of the benzyl quininium chloride **86** as a chiral catalyst (6 mol%) afforded the *trans*-chalcone epoxide **98** in 68% yield. However, the optical yield achieved was only in the range of 7–9% ee (Scheme 8.33).

Two decades later, Arai and coworkers significantly improved the enantioselectivity by introducing a *para*-trifluoromethyl group onto the benzyl moiety of **86** [42]. Up to 86% ee was achieved in the reaction of the phenacyl chloride or racemic cyclic α-chloroketones **99** with several aliphatic and aromatic aldehydes in the presence of **100** (10 mol%) as a catalyst and LiOH as a base (Scheme 8.34). Interestingly, the *para*-methylbenzyl-incorporated analogue of **100**, which is sterically equivalent to the trifluoromethylated derivative **100**, afforded racemic *trans*-epoxides. This result strongly indicates that the electron density of the phenyl group plays an important role in the asymmetric induction. In addition, the use of the 9-OH protected PTC gave unsuccessful results, indicating that the hydrogen-bonding interaction between the OH-group of the catalyst and substrate is also crucial in the asymmetric induction step.

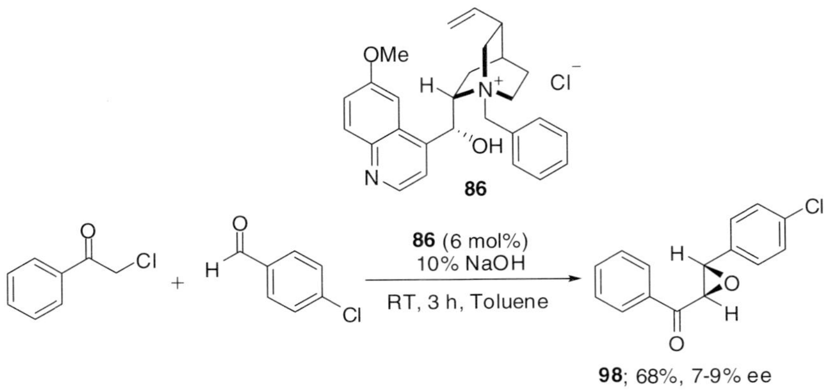

Scheme 8.33

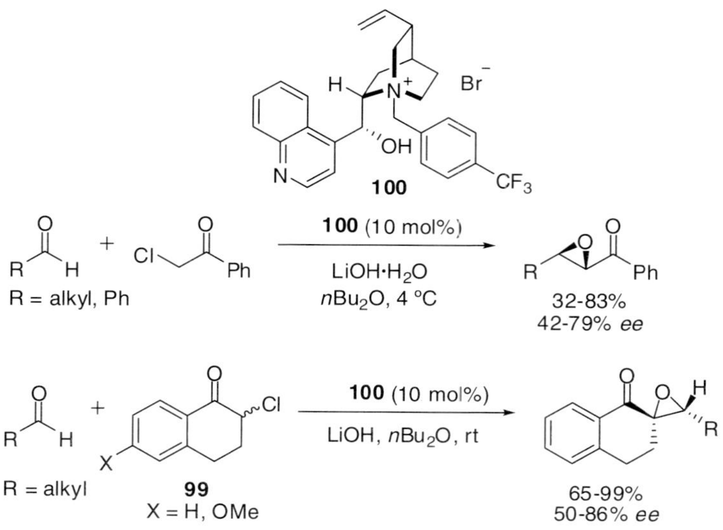

Scheme 8.34

[Scheme 8.34 shows catalyst 100 (a cinchona-derived quaternary ammonium bromide with OH and CF₃-benzyl groups) used in Darzens reactions:

Reaction 1: RCHO (R = alkyl, Ph) + Cl-CH₂-C(O)-Ph → epoxy ketone, using 100 (10 mol%), LiOH·H₂O, nBu₂O, 4 °C; 32–83% yield, 42–79% ee.

Reaction 2: RCHO (R = alkyl) + α-chlorotetralone derivative 99 (X = H, OMe) → spiro-epoxide, using 100 (10 mol%), LiOH, nBu₂O, rt; 65–99% yield, 50–86% ee.]

Later, Arai, and coworkers applied their PTC-mediated Darzens reaction conditions to the synthesis of optically active α,β-epoxysulfones [43]. The reaction of the chloromethyl phenyl sulfone 102 with aromatic aldehydes under PTC conditions (101 (10 mol%), KOH-toluene) at room temperature afforded the desired *trans*-epoxysulfones 103 in good yields and with moderate to good ee values (33–81%) (Scheme 8.35). On the other hand, aliphatic aldehydes gave the Darzens product as the racemates. The use of a Lewis acid additive (Sn(OTf)₂) improved the selectivity, but only slightly (up to 32% ee). Under PTC conditions, chloromethylsulfone is transformed into the corresponding carbanion, which then reacts with the

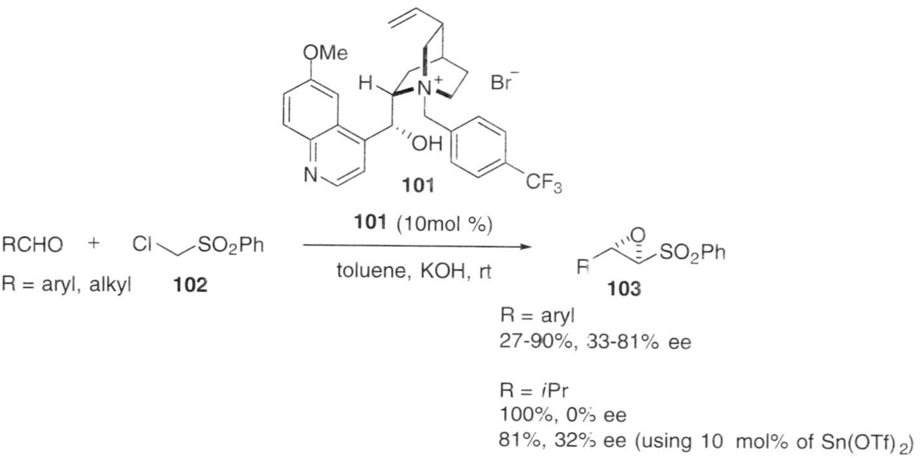

Scheme 8.35

RCHO + Cl-CH₂-SO₂Ph → epoxysulfone 103
R = aryl, alkyl   102

Conditions: 101 (10 mol %), toluene, KOH, rt

R = aryl: 27–90%, 33–81% ee
R = iPr: 100%, 0% ee
81%, 32% ee (using 10 mol% of Sn(OTf)₂)

Catalyst 101: cinchona-derived quaternary ammonium bromide bearing OMe, OH, and 4-CF₃-benzyl groups.

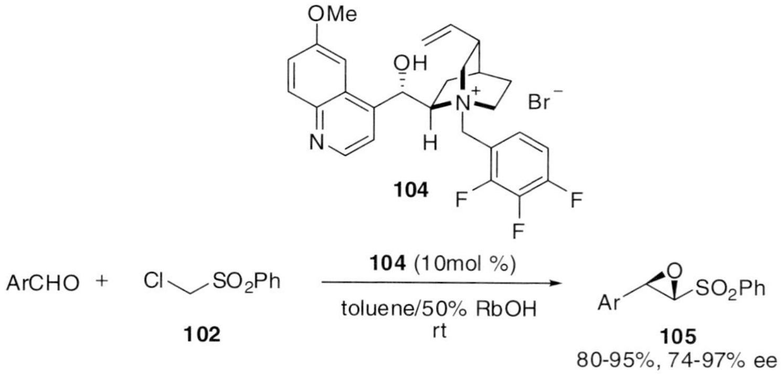

8.4 Aldol- and Mannich-Related Reactions

ArCHO + Cl–CH₂–SO₂Ph **102** →(104 (10 mol %), toluene/50% RbOH, rt)→ Ar–CH(–O–)CH(SO₂Ph) **105** 80-95%, 74-97% ee

Scheme 8.36

electrophilic carbonyl to give the Darzens products. Thus, the interaction between the reactive carbanion and the chiral quaternary ammonium cation may be responsible for the asymmetric induction.

Quite recently, Park and coworkers also examined the asymmetric Darzens reaction of chloromethyl phenyl sulfone **102** with various aromatic aldehydes using the 2,3,4-trifluorobenzyl-incorporated quinidinium bromide **104** as a chiral PTC and RbOH as a base [44]. The resulting epoxysulfones **105** were obtained in good chemical yields (81–95%) and with quite impressive enantioselectivity (up to 97% ee) (Scheme 8.36). The opposite enantiomer of **105** could also be obtained with a similar degree of ee using the quininium analogue of **104**. This significant improvement of the enantioselectivity compared with Wynberg's (**86**) or Arai's (**101**) catalyst was ascribed by the authors for the presence of the 2'-F moiety in the catalyst, which might help to form a more rigid conformation of the catalyst by coordinating with water via hydrogen bonding.

8.4.2
Morita–Baylis–Hillman Reactions and Aza-Morita–Baylis–Hillman Reactions

The Morita–Baylis–Hillman reaction and its aza-variant – the reaction of an electron-deficient alkene with an aldehyde (MBH) or an imine (aza-MBH) – provide a convenient route to highly functionalized allylic alcohols and amines. This reaction is catalyzed by simple amines or phosphines, which can react as a Michael donor with an electron-deficient alkene, generating an enolate intermediate. This intermediate in turn undergoes the aldol or Mannich reaction with electrophilic C=O or C=N bonds, respectively, to deliver allylic alcohols and amines.

8.4.2.1 Morita–Baylis–Hillman Reactions
One of the first practical catalytic asymmetric MBH reactions was reported by Hatakeyama and coworkers, who utilized β-isocupreidine (**69**, β-ICD), derived from quinidine, as a catalyst (10 mol%) [45a]. Enantioselectivities as high as 99% were achieved in the reaction of a variety of aliphatic and aromatic aldehydes with the very

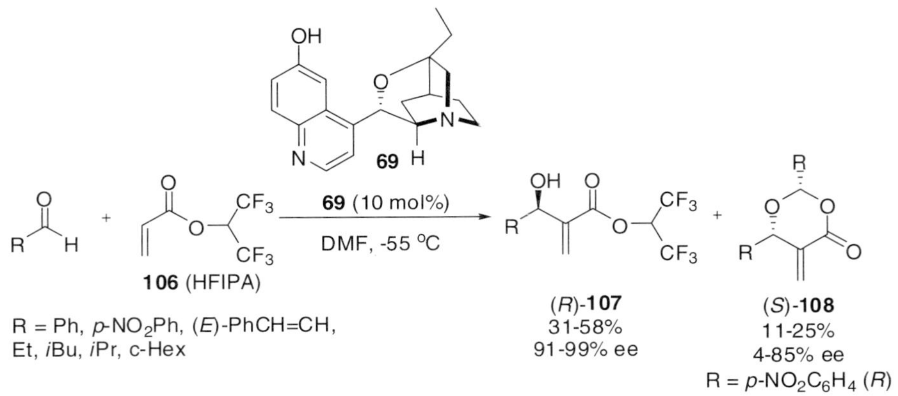
OH
O
N
69 H
R H
CF3
O
CF3
106 (HFIPA)
69 (10 mol%)
DMF, -55 °C
OH O
R
CF3
CF3
(R)-107
31-58%
91-99% ee
R
O O
R
O
(S)-108
11-25%
4-85% ee
R = p-NO2C6H4 (R)
R = Ph, p-NO2Ph, (E)-PhCH=CH,
Et, iBu, iPr, c-Hex

Scheme 8.37

electrophilic hexafluoroisopropyl acrylate (HFIPA, **106**) affording the (*R*)-enriched adducts **107** (Scheme 8.37) [45a]. The presence of a rigid quinuclidine ether moiety and a C6′-OH group was proven to be crucial for the activity and enantioselectivity, since the analogue of **69** having the 6′-OMe group gave almost no enantioselectivity (up to 10% ee). However, preparatively, the concomitant formation of the dioxanone derivatives **108** is somewhat disadvantageous, resulting in only moderate yields (40–58%) of **107**. Moreover, these side products are formed with the opposite sense of asymmetric induction and must be separated by tedious work-up procedures. However, a further careful study by the same authors revealed that the yields of the wanted product **107** can be to some extent improved by using the azeotropically dried β-ICD (**69**) [45b].

To rationalize the observed stereochemistry of the products and the key role of the phenolic OH group of β-ICD **69**, the authors proposed the plausible mechanism depicted in Scheme 8.38 [45a]. The Michael addition of **69** to **106** (HFIPA) forms the enolate **109**, which in turn undergoes the aldol reaction with the aldehyde to furnish an equilibrium mixture of several diastereomers. Among them are the two betaine intermediates, **110** and **111**, stabilized by intramolecular hydrogen bonding between the oxy anion and the phenolic OH, the conformations of which are nearly ideal for the subsequent elimination reactions (Scheme 8.38). The subsequent elimination process would produce the wanted (*R*)- or (*S*)-**107** with the regeneration of the catalyst. However, the antiperiplanar arrangement of the ammonium portion and α-hydrogen of the ester group in the subsequent E2 or E1cb reaction causes the intermediate **111** to suffer from severe steric interactions between the substituent R and the ester and quinuclidine moieties (see Newman projection **112**) and, thus, it reacts with a second aldehyde rather than undergoing elimination to form the (*S*)-dioxanone **108**. On the other hand, the intermediate **110** can undergo facile elimination to produce the (*R*)-ester **107** because of the lesser steric hindrance.

In work conducted by the same group, the utility of this methodology in synthetic activities was exemplified by the total synthesis of epopromycin B, a plant cell wall

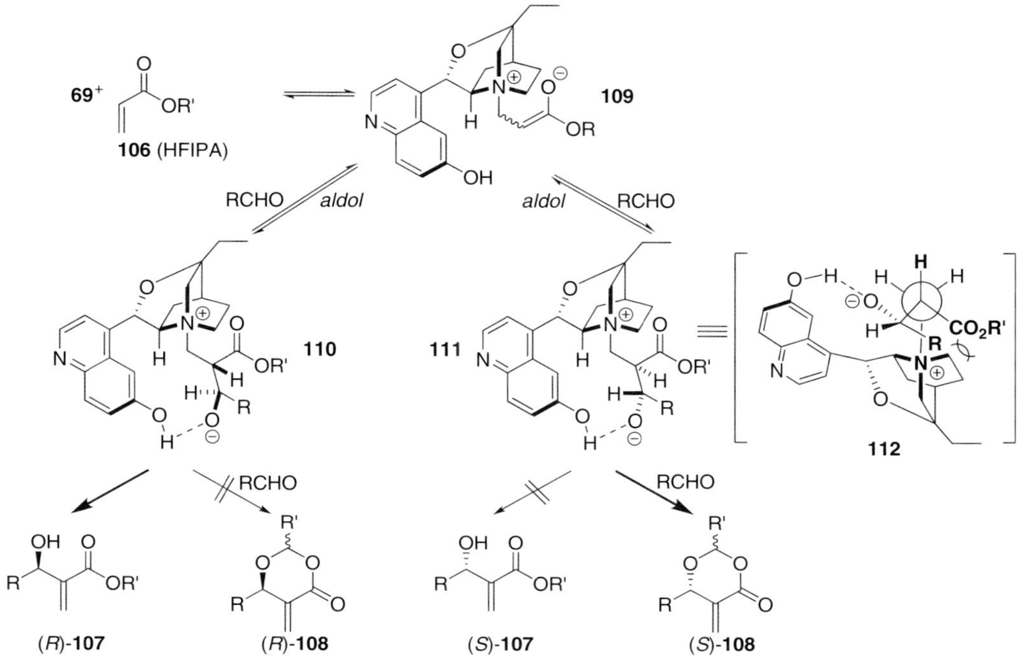

Scheme 8.38

synthesis inhibitor (Scheme 8.39) [46], and mycestericin E, a potent immunosuppressor [47] (Scheme 8.40) [47].

Later, Shi, and Jiang examined the performance of the catalyst **69** in the MBH reaction of aldehydes with methyl vinyl ketone (MVK) [48]. However, MVK was much

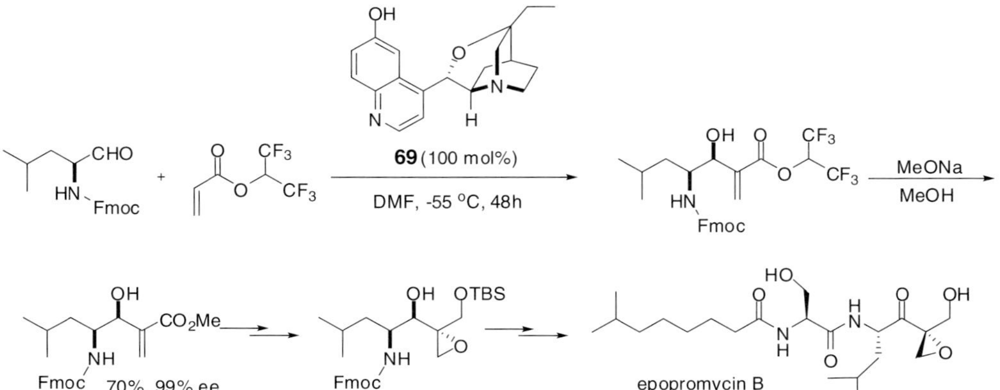

Scheme 8.39

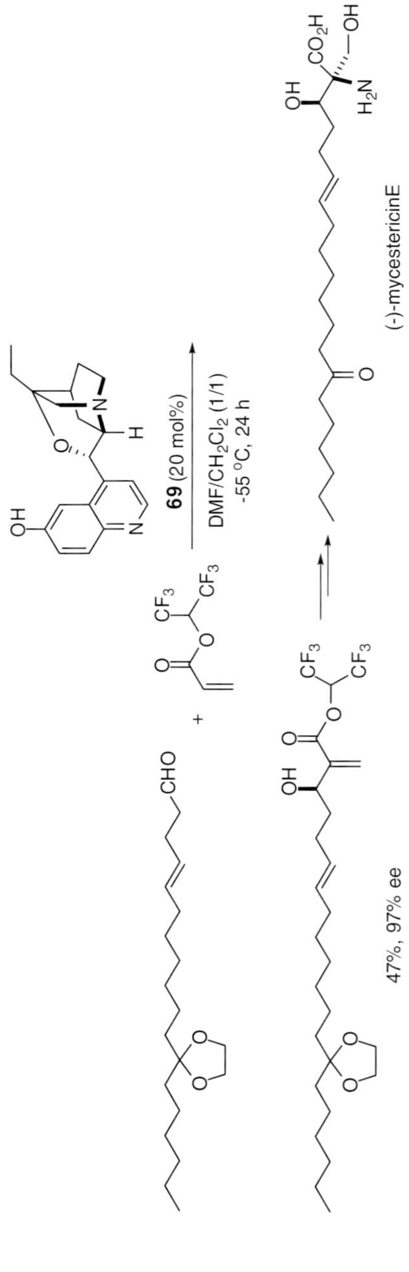

Scheme 8.40

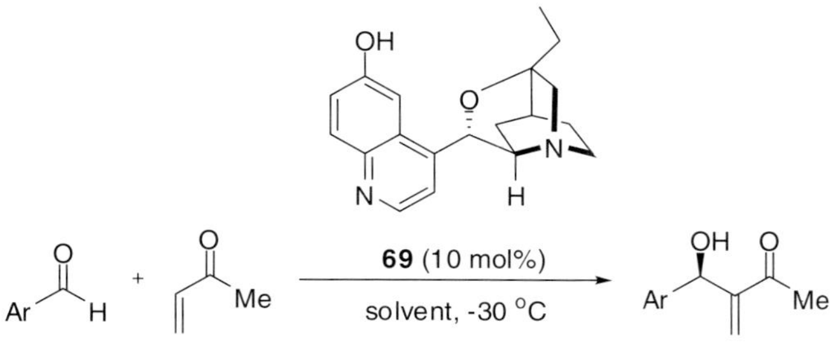

8.4 Aldol- and Mannich-Related Reactions

Scheme 8.41

Ar-CHO + CH$_2$=CH-C(O)-Me → (with 69 (10 mol%), solvent, -30 °C) → Ar-CH(OH)-C(=CH$_2$)-C(O)-Me

Ar = p-BrC$_6$H$_5$
78%, 25% ee
43%, 49% ee (using LiClO$_4$ additive)

less reactive than **106** (HFIPA) and, thus, the β-ICD-mediated addition of activated aromatic aldehydes to MVK required a higher temperature and longer reaction times, resulting in very low enantioselectivities (7–49% ee), although the use of Brønsted or Lewis acid additives to increase the reaction rate improved the yield and selectivity, but only slightly (Scheme 8.41).

The monocarboxylic acid salts of dimeric cinchona alkaloids such as (DHQD)$_2$PYR (**113**), (DHQD)$_2$PHAL (**32**), or (DHQD)$_2$AQN (**114**), known as Sharpless ligands, were also investigated for their catalytic efficiency in the asymmetric MBH reactions of methyl acrylate and electron-deficient aromatic aldehydes [49]. Conceptually, these compounds are bifunctional catalysts; in the presence of acid additives, one of the two amine functions of the dimers forms a salt and serves as an effective Brønsted acid, while the other tertiary amine of the catalyst acts as a nucleophile. However, these compounds provided only limited success. Salts derived from (DHQD)$_2$PYR (**113**) or (DHQD)$_2$PHAL (**32**) afforded trace amounts of the adducts **115**. Only salts of (DHQD)$_2$AQN (**113**) gave adducts such as **115** in ee values of up to 77%, albeit in very low yield (4–11%) (Scheme 8.42). Notably, without an acid, the reaction afforded the opposite enantiomer with slow conversion.

8.4.2.2 Aza-Morita–Baylis–Hillman Reactions

Asymmetric aza-MBH reactions were studied by the groups of Hatakeyama [50], Shi [51], and Adolfsson [52]. The results revealed that β-ICD (**71**) is also an efficient and general catalyst for the asymmetric aza-MBH reactions of activated aryl imines (e.g., N-diphenylphosphinoyl aryl imines and N-sulfonated aryl imines) with various activated olefins, such as HFIPA, MVK, ethyl vinyl ketone (EVK), acrolein, and acrylates. The diphenylphosphinoyl aryl imines **116** with HFIPA (**106**) in the presence of 10 mol% of β-ICD (**69**) can be smoothly transformed into the corresponding N-protected α-methylene-β-amino acid esters **117** in reasonable yields (42–97%) and moderate ee values (54–73%) (Scheme 8.43) [50]. Interestingly, the reaction proceeded with (S)-selectivity, while the analogous MBH reaction with aldehydes afforded the (R)-products (see Section 8.4.2.1, [45a]). Under the same reaction conditions, other N-sulfonated (N-Bz, N-Ts, and N-Ms) aryl imines also

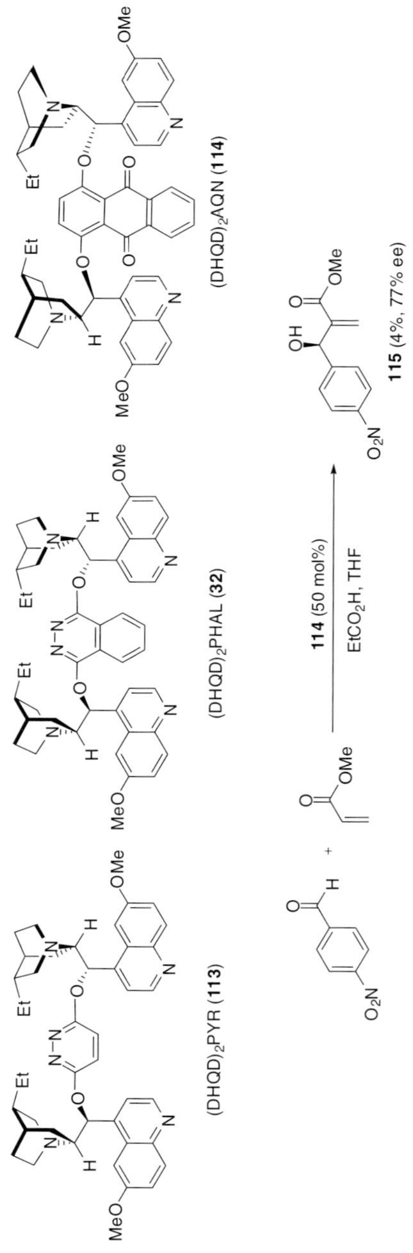

Scheme 8.42

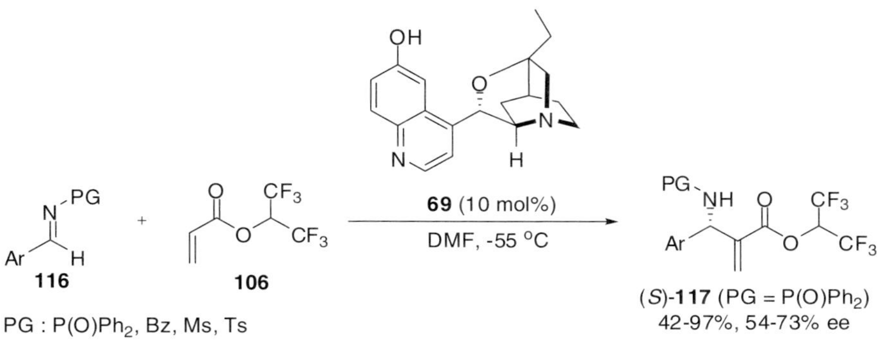

8.4 Aldol- and Mannich-Related Reactions

Scheme 8.43

PG : P(O)Ph$_2$, Bz, Ms, Ts

provided (S)-enriched products, albeit with lower enantioselectivity. The observed enantioselectivity was rationalized by a plausible mechanism governed by hydrogen bonding, similar to the mechanism depicted in Scheme 8.38.

Significantly higher ee values (up to 99%) for β-ICD (**69**)-catalyzed aza-MBH reactions were achieved by using MVK, EVK, acrolein, or methyl acrylate as an acceptor. All of these Michael acceptors afforded the corresponding adducts **118** in moderate to good yields and with good to excellent ee values (up to 99%) (Scheme 8.44) [51]. However, interestingly, the adducts **118** derived from MVK and EVK show the opposite configuration to those obtained with acrolein and methyl acrylate [51b]. The outcome of the absolute configuration of the adducts was rationalized on the basis of the mechanism depicted in Scheme 8.38, that is, different

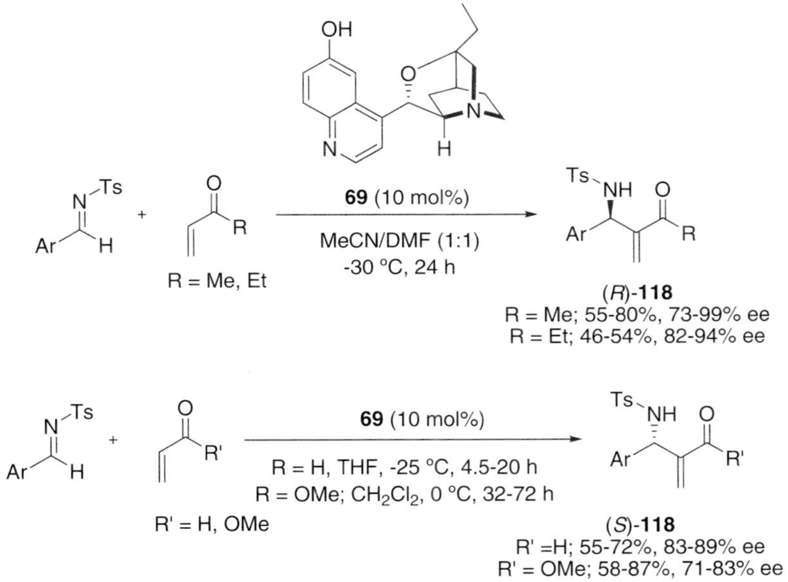

Scheme 8.44

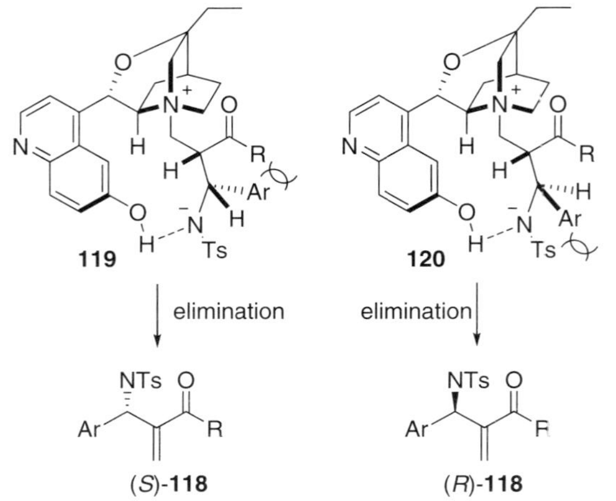

Scheme 8.45

rates in the elimination step of the two betaine intermediates **119** and **120**, stabilized by intramolecular hydrogen bonding between the amidate ion and the phenolic OH (Scheme 8.45).

8.4.3
Nucleophilic Addition of Ammonium Ketene Enolate to C=O or C=N Bonds

Cinchona alkaloids such as **121** possess a nucleophilic quinuclidine structure and can act as versatile Lewis bases to react with ketenes generated *in situ* from acyl halides in the presence of an acid scavenger. By acting as nucleophiles, the resulting ketene enolates can react intermolecularly [53] or intramolecularly [54] with electrophilic C=O or C=N bonds to deliver formal [2 + 2]-cycloadducts, such as chiral β-lactones or β-lactams, via aldol (or Mannich)-intramolecular cyclization sequence reactions (Scheme 8.46). The nucleophilic ammonium enolate can also react with energetic

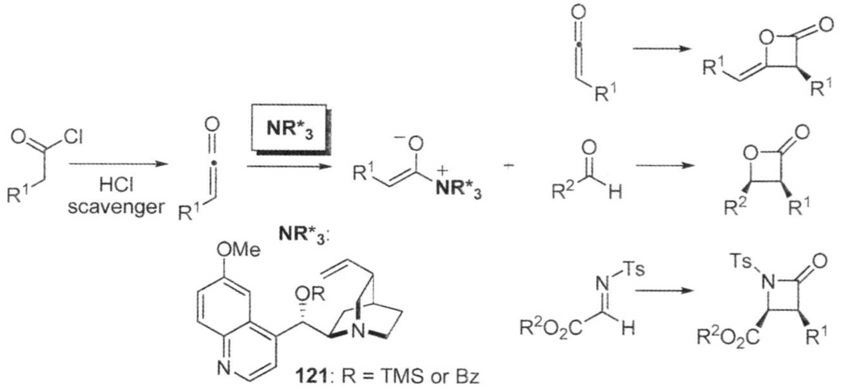

Scheme 8.46

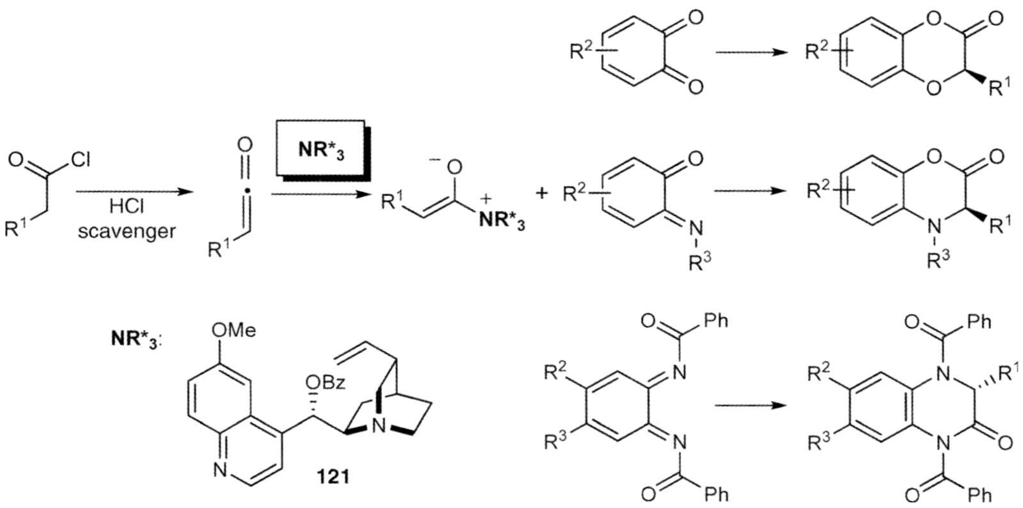

Scheme 8.47

heterodynes to produce [4 + 2] cycloadducts (Scheme 8.47). Further detailed information on these highly important reactions can be found in Sections 4.2.5 and 10.2.

8.5 Cyanation Reactions

8.5.1 Cyanohydrin Synthesis

Optically pure cyanohydrins serve as highly versatile synthetic building blocks. Much effort has, therefore, been devoted to the development of efficient catalytic systems for the enantioselective cyanation of aldehydes and ketones using HCN or trimethylsilyl cyanide (TMSCN) as a cyanide source [55]. More recently, cyanoformic esters (ROC(O)CN), acetyl cyanide ($CH_3C(O)CN$), and diethyl cyanophosphonate have also been successfully employed as cyanide sources to afford the corresponding functionalized cyanohydrins. In 1912, Bredig and Fiske discovered that the addition of HCN to benzaldehyde is accelerated by the alkaloids, quinine and quinidine, and that the resulting cyanohydrins are optically active and of opposite chirality [56]. However, the ee values obtained were very low (<10% ee). Furthermore, prolonged reaction times resulted in a loss of optical activity because of the racemization of the product. As mentioned in Chapters 1 and 4, this is, in fact, one of the earliest studied asymmetric organocatalysis reactions. Higher ee values of cyanohydrins in the same reaction were achieved under heterogeneous conditions using polymer-bound cinchona alkaloids [57]. Danda and coworkers reported that the polymers **123**, containing quinidine or quinine, catalyzed the asymmetric addition of HCN to 3-phenoxyben-

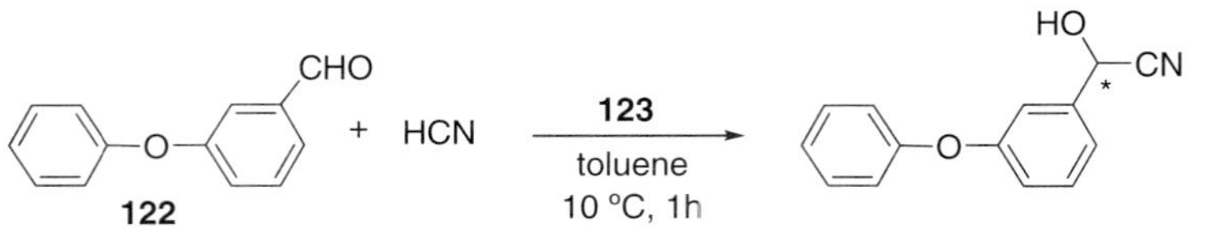

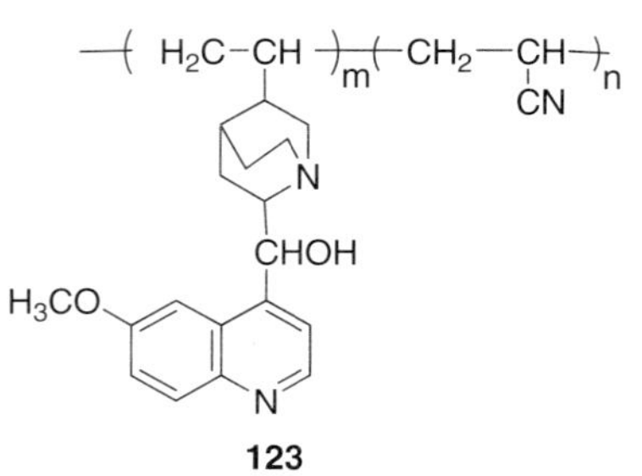

Scheme 8.48

catalyst	% yield	% ee
quinidine	97	22 (S)
quinine	93	5 (R)
123a	98	46 (S)
123b	97	20 (R)

123a; poly(quinidine-co-acrylonitrile)
123b; poly(quinine-co-acrylonitrile)

zaldehyde 122. Although the results achieved in these reactions were far from satisfactory for preparative purposes, interestingly, polymeric alkaloids exhibited higher enantioselectivities than those obtained with the corresponding monomeric alkaloids, quinidine and quinine (Scheme 8.48). For example, the polymer 123a containing quinidine gave the (S)-isomer of cyanohydrin in 98% yield and 46% ee, while the parent quinidine gave the (S)-isomer of cyanohydrin in 97% yield and 22% ee. The reaction mixture is gel-like in the case of polymeric alkaloids, whereas it is homogeneous in the case of the corresponding monomers. This gelation would be preferable both to increase the enantioselectivity and to decrease the racemization of the products.

The first preparatively useful cinchona-catalyzed cyanation methodology was developed by Deng and coworker in 2001. They demonstrated that the alkaloid derivatives, (DHQD)$_2$AQN (114) and (DHQD)PHN (124), and their pseudoenantiomeric analogues, (DHQ)$_2$AQN (125) and (DHQ)PHN (126), catalyzed the asymmetric addition of ethyl cyanoformate (EtOCOCN) to a variety of acyclic and cyclic dialkyl ketones. Both the α,α-dialkylated and α-acetal aliphatic ketones 127 were transformed to the tertiary cyanohydrin carbonates 128 in quite remarkable enantioselectivity and in reasonable yields (Scheme 8.49). Clearly, the pseudoenantiomeric catalyst pairs 114/125 and 124/126 afforded products of the opposite configuration. It is believed that the catalytic cycle is initiated by the quinuclidine base of the alkaloid reacting with ethyl cyanoformate to form the chiral cyanide/acylammonium ion pair 129, followed by the enantioselective addition of cyanide to the ketone and the subsequent alkoxycarbonylation of the resulting cyanoalkoxide 130 with the regeneration of the catalyst (Scheme 8.50) [58, 59].

Soon after, the same research group found that dimeric cinchona alkaloids such as (DHQ)$_2$AQN (125) and (DHQD)$_2$PHAL (33) can also be used as highly enantioselective organic Lewis base catalysts for the cyanosilylation of acetal ketones (131,

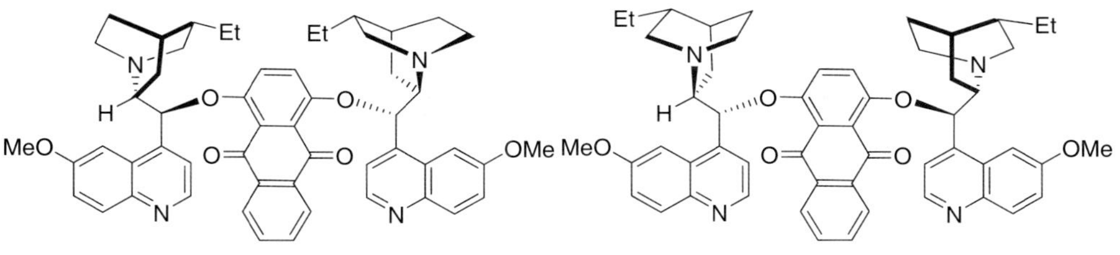

8.5 Cyanation Reactions | 231

(DHQD)₂AQN (**114**)

(DHQ)₂AQN (**125**)

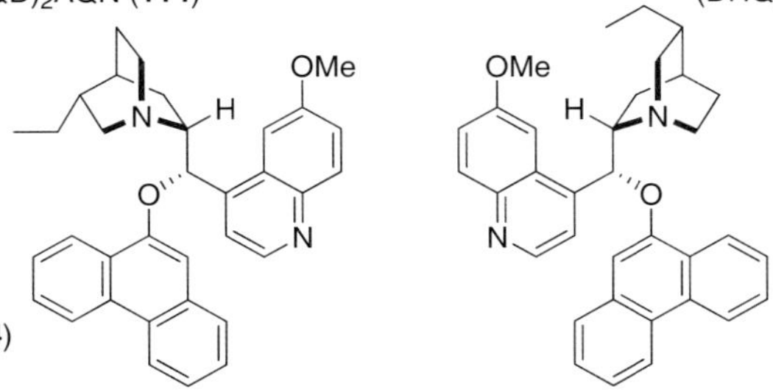

(DHQD)PHN (**124**)

(DHQ)PHN (**126**)

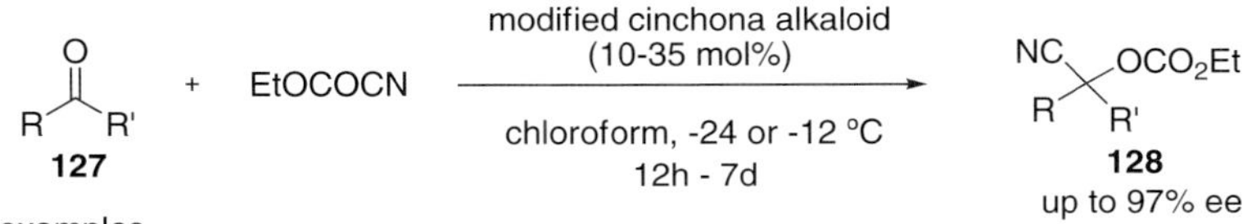

$$\underset{\textbf{127}}{\text{R}\overset{\text{O}}{\underset{}{\|}}\text{R'}} + \text{EtOCOCN} \xrightarrow[\text{chloroform, -24 or -12 °C}]{\text{modified cinchona alkaloid (10-35 mol%)}} \underset{\underset{\text{up to 97% ee}}{\textbf{128}}}{\text{NC}\underset{\text{R R'}}{\overset{}{\searrow}}\text{OCO}_2\text{Et}}$$

Selected examples

substrate	catalyst (mol%)	Time (d)	Conversion (%)	Yield (%)	ee (%)
Me,,, Me cyclopentanone	**114** (15)	2	68	66	97
	125 (15)	4	79	76	95
Me Me cyclohexanone	**114** (20)	4	65	62	91
	125 (30)	5	56	53	92
EtO,, EtO cyclopentanone	**124** (10)	7	quant.	99	94
	126 (30)	4	83	80	95
EtO EtO cyclohexanone	**124** (35)	5	82	78	96
n-PrO, Me, n-PrO	**114** (15)	2	68	66	97
	125 (15)	4	79	76	95
Me, Me, EtO OEt	**114** (20)	4	65	62	91
	125 (30)	5	56	53	92

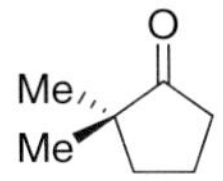

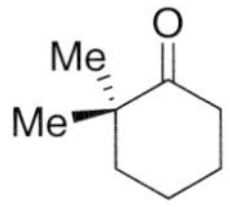

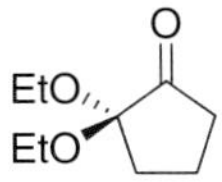

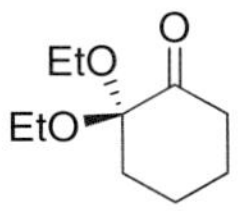

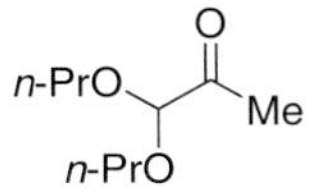

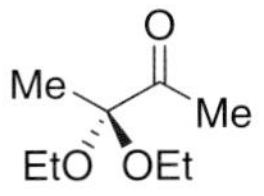

Scheme 8.49

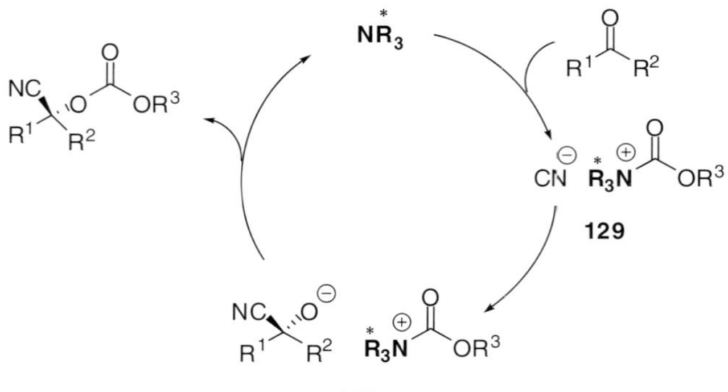

Scheme 8.50

α,α-dialkoxy ketones) [60]. Several α-acetal ketones **131** bearing a broad range of alkyl, aryl, alkenyl, and alkynyl substituents were converted to the corresponding cyanohydrin TMS ethers **132** with 90–98% ee at catalyst loadings of 2–20 mol% (Scheme 8.51). This methodology was used by the same group as the stereochemistry-defining step in the total synthesis of bisorbicillinolide, bisorbicillinol, and bisorbibutenolide [61].

More recently, Feng and coworkers [62] realized that the enantioselective cyanoformylation of aromatic aldehydes with ethyl cyanoformate, using a chiral quaternary ammonium salt and triethylamine. However, despite their optimization studies on the catalyst structure and reaction conditions, only moderate ee values (up to 72% ee) were obtained.

8.5.2
Strecker Synthesis

The catalytic asymmetric Strecker reaction represents one of the most direct and viable methods for the asymmetric synthesis of α-amino acids and their derivatives. Corey and coworkers reported the first successful use of a cinchona alkaloid derived chiral ammonium salt **133** as a catalyst in the asymmetric Strecker reactions of

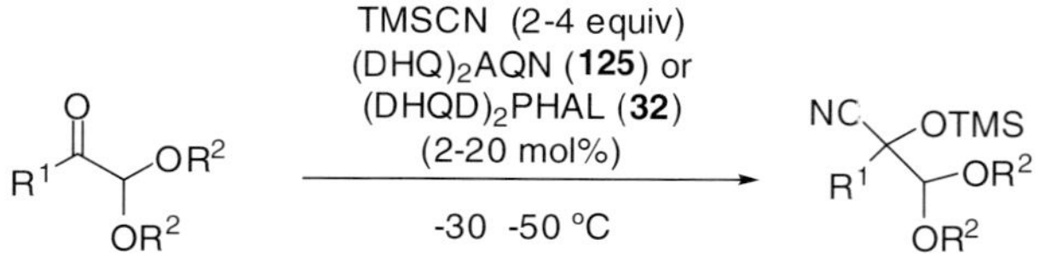

Scheme 8.51

131
R^1 = alkyl, aryl alkenyl, alkynyl
R^2 = Et, nPr

132
81-99%, 90-98% ee

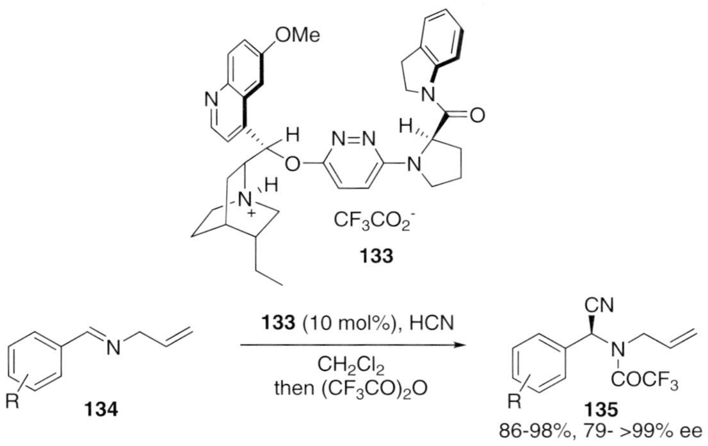

Scheme 8.52

the *N*-allyl aldimines **134** with HCN [63]. In the presence of 10 mol% of **133** and HCN, a variety of substituted aryl *N*-allyl aldimines **134** were transformed into the corresponding (*S*)-α-amino nitriles **135** in high yields and generally high ee values (86–98% yield, 79–99% ee) (Scheme 8.52). A computational study designed to rationalize the observed sense of enantioselectivity revealed that the cyanide ion should attack the *Re*-face of the imine hydrogen-bonded by the protonated quinuclidine moiety of **133** to produce the (*S*)-amino nitrile **135**, since the *Si*-face of the imine is blocked by the pyridazine linker (Figure 8.7).

Recently, the group of Herrera, Bernardi, and Ricci realized the enantioselective synthesis of protected α-amino nitriles from the corresponding α-amino sulfones **137**, which act as effective precursors for the *in situ* generation of imines, by the use of acetone cyanohydrin (**138**) as a cyanide source using quinine-derived PTC **136** [64]. The aminonitriles **139** were produced with broad generality in 50–88% ee (Scheme 8.53). However, a similar protocol using KCN and TMSCN resulted in a lower ee value.

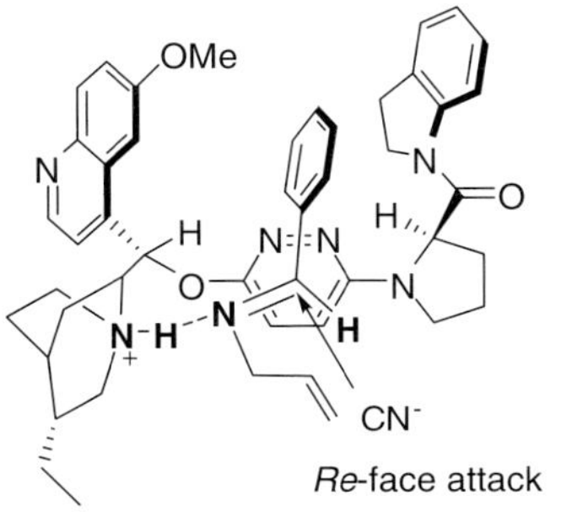

Figure 8.7 Proposed pre-transition-state assembly for the catalytic hydrocyanation of *N*-allylbenzaldimine.

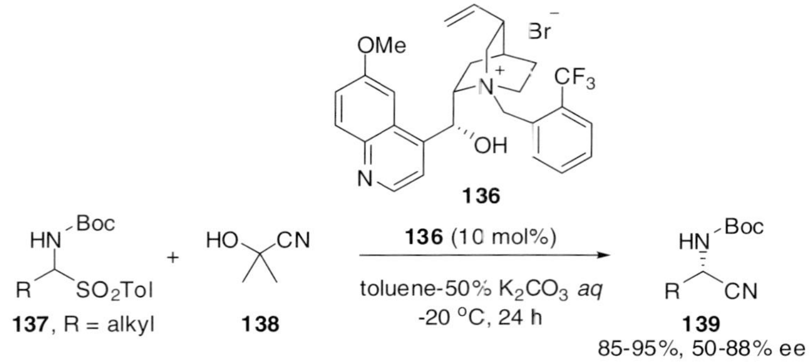

234 | 8 Cinchona-Catalyzed Nucleophilic 1,2-Addition to C=C and C=N Bonds

Scheme 8.53

137 (R = alkyl) + 138 → 139 (85–95%, 50–88% ee)

Conditions: 136 (10 mol%), toluene–50% K_2CO_3 aq, −20 °C, 24 h.

Catalyst 136: cinchona-derived ammonium salt with OMe quinoline, OH, CF_3-benzyl, Br^-.

Substrate 137: HN-Boc, R, SO_2Tol.
Substrate 138: HO-C(CN)-.
Product 139: HN-Boc, R, CN.

8.6
Trifluoromethylation

The introduction of a trifluoromethyl group with a strong electron-withdrawing ability can lead to significant changes in the physical, chemical, and biological properties of molecules because of the high liphophilic character of this group. Compounds with a trifluoromethyl group thus find wide application in the pharmaceutical field. Therefore, asymmetric approaches to the direct introduction of a trifluoromethyl group into prochiral aldehydes or ketones by a nucleophilic process represent an important synthetic strategy.

In 1994, the first enantioselective trifluoromethylation reaction was achieved with the Ruppert–Prakash reagent, $TMSCF_3$, in the presence of the cinchona-based quaternary ammonium fluoride 140 [65]. The chiral induction can arise from the dual activation mode of the catalyst, that is, the fluoride anion acts as the nucleophilic activator of $(TMS)CF_3$ and the chiral ammonium cation activates the carbonyl group of 141. However, the observed ee values of the obtained carbinols 142 do not exceed 51% and decrease considerably when nonaromatic carbonyl compounds (15% ee for $R^1 = n\text{-}C_7H_{15}$; $R^2 = H$) are used, which implies that π–π stacking interactions between the carbonyl compound and cinchoninium occur (Scheme 8.54).

In 2003, Caron and coworkers enhanced the stereoselectivity by introducing a bulky subunit at the quinuclidine nitrogen atom of cinchona alkaloids [66]. By using only 4 mol% of catalyst 143, the desired product 145 was obtained with up to 92% ee (Scheme 8.55). However, this catalyst proved not to be generally applicable to other carbonyl compounds.

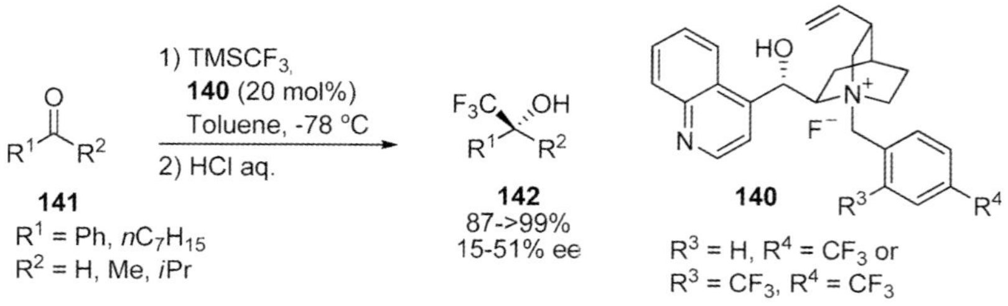

Scheme 8.54

141 ($R^1 = Ph, nC_7H_{15}$; $R^2 = H, Me, iPr$)

1) $TMSCF_3$, 140 (20 mol%), Toluene, −78 °C
2) HCl aq.

→ 142 (87–>99%, 15–51% ee), F_3C–C(OH)(R^1)(R^2)

Catalyst 140: cinchona-derived ammonium fluoride, $R^3 = H, R^4 = CF_3$ or $R^3 = CF_3, R^4 = CF_3$.

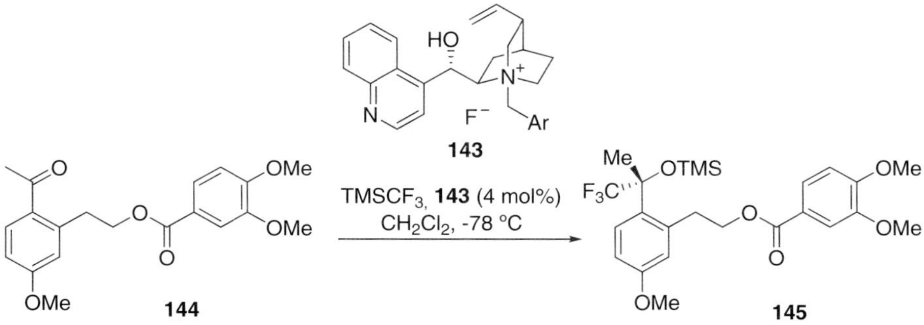

8.6 Trifluoromethylation

Scheme 8.55

Selected examples
Ar of **143**:
3,5-(MeO)$_2$C$_6$H$_3$; 98%, 83% ee
9-anthracenyl; 95%, 85% ee
1-naphthyl; 97%, 92% ee

Recently, Shibata, Toru, and coworkers demonstrated that a mixture of chiral ammonium bromides, **146** and **147**, and tetramethylammonium fluoride (TMAF) [67] or KF [68] can be directly used as a catalyst without the need for the prior isolation of the fluoride salts. By utilizing this catalyst system, the reactions of various aromatic ketones or aldehydes with (TMS)CF$_3$ were completed within a few hours to afford the corresponding products in moderate to high enantioselectivities (Schemes 8.56 and 8.57).

Recently, the groups of Mukaiyama [69] and Feng [70] independently demonstrated that aryl oxide anions such as phenoxide [69] or binaphthoxide [70] can be used instead of fluoride to activate (TMS)CF$_3$ as a Lewis base. Several types of aromatic ketones and aldehydes were smoothly trifluoromethylated within a few hours in the presence of cinchona-based quaternary ammonium salts bearing an aryl oxide anion **148** or **149**, affording the corresponding product in excellent yields and with moderate to high ee values (up to 87% ee) (Schemes 8.58 and 8.59).

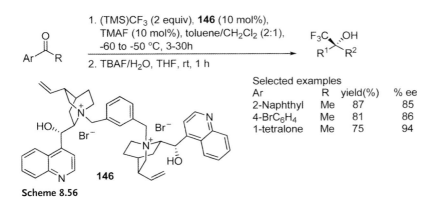

Selected examples

Ar	R	yield(%)	% ee
2-Naphthyl	Me	87	85
4-BrC$_6$H$_4$	Me	81	86
1-tetralone	Me	75	94

Scheme 8.56

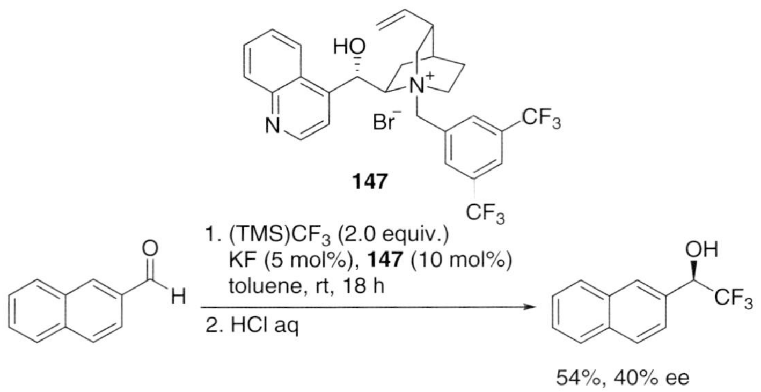
HO
N
N+
Br−
CF3
147
CF3
O
H
1. (TMS)CF3 (2.0 equiv.)
KF (5 mol%), 147 (10 mol%)
toluene, rt, 18 h
2. HCl aq
OH
CF3
54%, 40% ee

Scheme 8.57

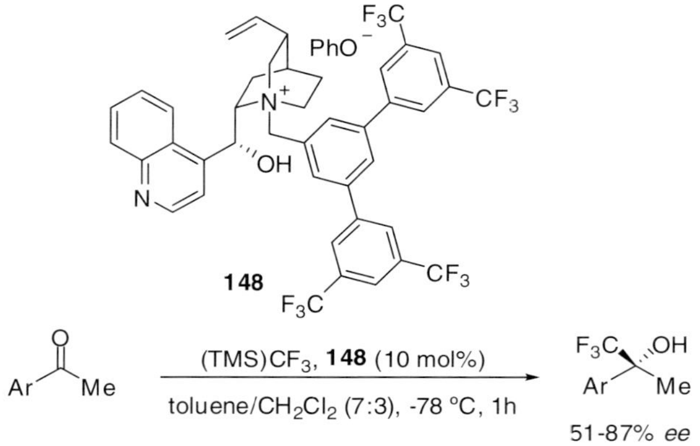
F3C
PhO−
N+
CF3
OH
N
148
CF3
F3C
O
Ar
Me
(TMS)CF3, 148 (10 mol%)
toluene/CH2Cl2 (7:3), -78 °C, 1h
F3C
OH
Ar
Me
51-87% ee

Scheme 8.58

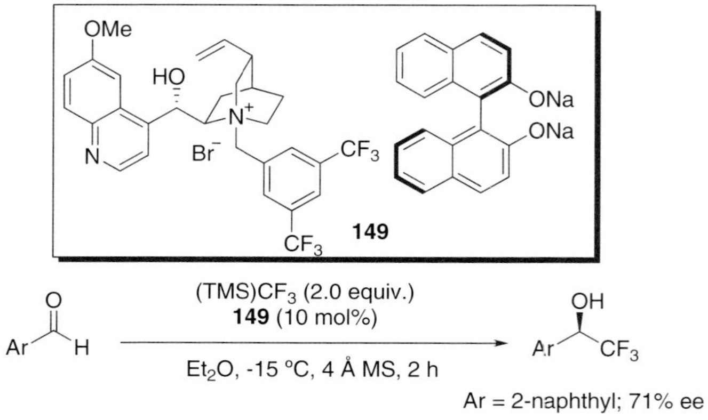
OMe
HO
N+
N
Br−
CF3
CF3
149
ONa
ONa
O
Ar
H
(TMS)CF3 (2.0 equiv.)
149 (10 mol%)
Et2O, -15 °C, 4 Å MS, 2 h
OH
Ar
CF3
Ar = 2-naphthyl; 71% ee

Scheme 8.59

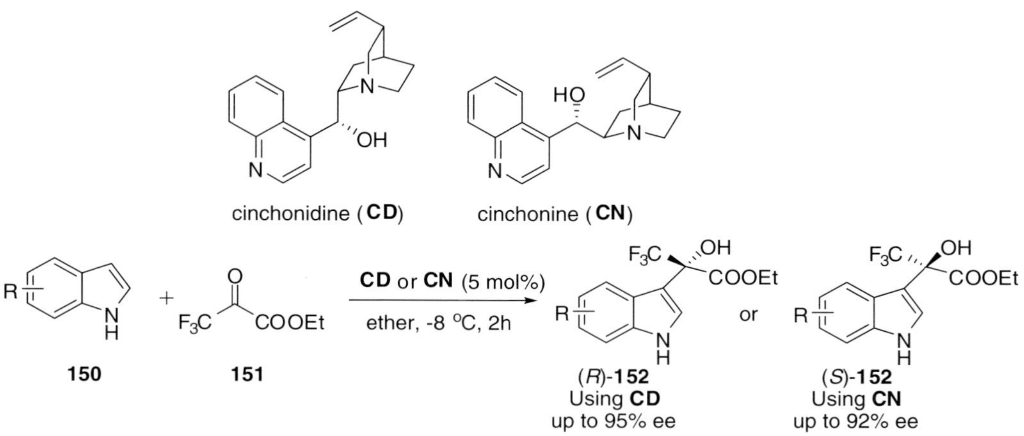

Scheme 8.60

8.7
Friedel–Crafts Type Alkylation

The electron-rich nature of indoles enables their Friedel–Crafts type 1,2-nucleophilic addition to carbonyls and imines, providing direct access to enantiomerically enriched indol derivatives. The first enantioselective cinchona-catalyzed 1,2-addition of indoles **150** to ethyl 3,3,3-trifluoropyruvate **151** was reported by Török and coworkers [71]. Natural cinchona alkaloids such as cinchonine and cinchonidine were shown to be highly effective in this reaction, providing excellent yields and enantioselectivities for both enantiomeric products **152** (up to 99% yield, 95% ee) (Scheme 8.60). However, the blocking of either the quinuclidine nitrogen or the 9-OH group led to dramatically reduced enantioselectivity (3–7% ee), strongly indicating the base–acid bifunctional activation mechanism of this reaction. Of note, N-methylindoles gave racemic products, also indicating the importance of the hydrogen-bonding interaction of the indole and catalyst for enantiodifferentiation.

As described above, although natural cinchona alkaloids provide high enantioselectivity, the substrate scope is narrow. These catalysts are only effective toward trifluoropyruvates. The substrate scope of carbonyls was dramatically broadened by using the bifunctional hydroquinine and hydroquinidine derived catalysts **153** and **154**, bearing the acidic C6′-OH group, which can promote the reaction through cooperative hydrogen-bonding catalysis [72]. In the presence of 10 mol% of **153** or **154**, a wide range of carbonyls such as alkynyl α-ketoesters, aryl α-ketoesters, and even aldehydes smoothly reacted with indoles, affording the corresponding products **155** in good to excellent yields and enantioselectivities (up to 99% ee) (Scheme 8.61). However, the parent cinchona alkaloids (quinidine and cinchonine) and cinchona-based thioureas such as **36** were much less effective in this reaction in terms of the reactivity and enantioselectivity.

Very recently, the enantioselective Friedel–Crafts alkylation of the simple phenols **157** with trifluoropyruvate was also accomplished by using the bifunctional quinidine

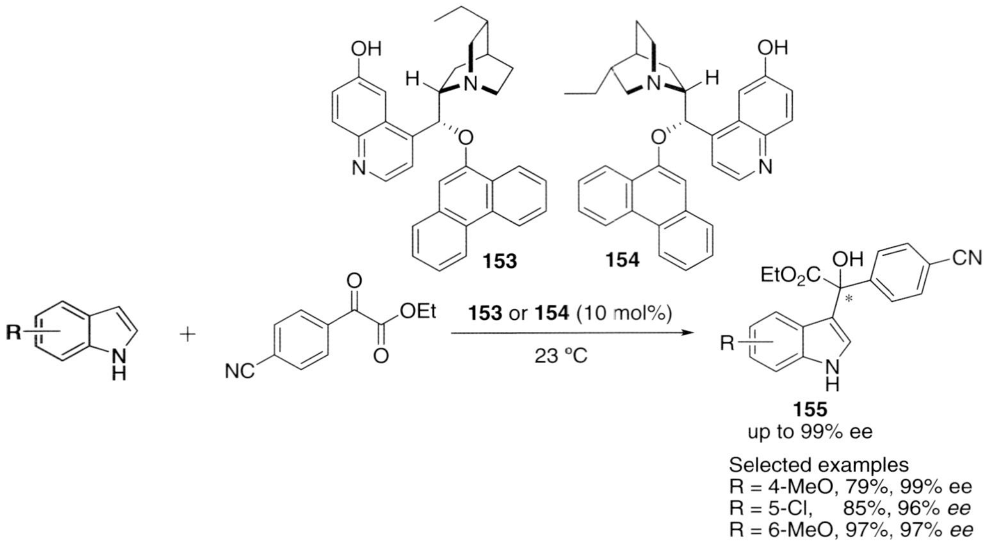

Scheme 8.61

derived catalyst **156** (10 mol%) [73]. Various phenols **157** were reacted para regioselectively with trifluoropyruvate in the presence of the catalyst **156**, affording the Friedel–Crafts alkylation product **158** in good to excellent yields (58–96%) and enantioselectivities (71–94% ee) (Scheme 8.62). The bifunctional thiourea derivative **40** was shown to be ineffective for this reaction (11% ee).

Deng and coworkers also found that the bifunctional quinidine- or quinine-based thioureas **36** and **40** effectively catalyze the asymmetric Friedel–Crafts reaction of indoles with a variety of aromatic and alkyl-substituted aldimine derivatives **159**. Both enantiomeric forms of the products **160** were obtained with uniformly high

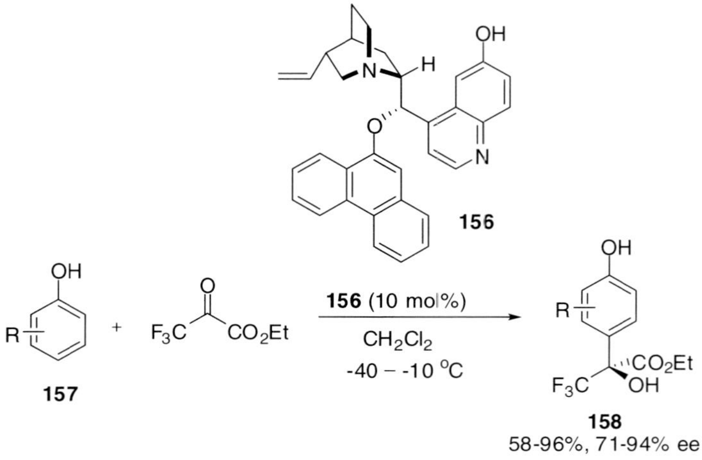

Scheme 8.62

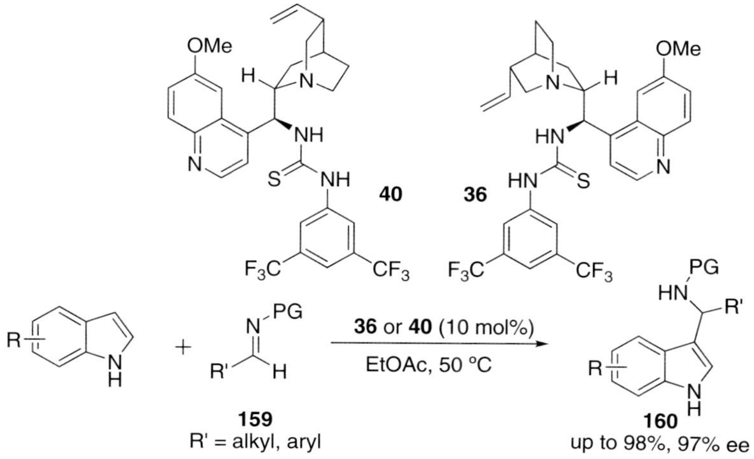

Scheme 8.63

levels of enantiomeric excess, regardless of the electronic characteristics of either reaction component (Scheme 8.63) [74]. Notably, N-Me indole was found to be inactive, indicating that 9-thiourea cinchona alkaloids activate the indole and the imine through hydrogen-bonding interactions.

The heterogeneous version of this reaction was quite recently reported by Guo and coworkers [75]. Highly interestingly, in the reaction of the indole with the N-Bs (Bs, benzene sulfonyl) phenyl imine **162**, enhanced enantioselectivity (99.2% ee at 40 °C) compared with that (92% ee at 50 °C [74], 93.2% ee at 40 °C [75]) obtained with the homogeneous analogue **40** was achieved using the mesoporous silica (SBA-15) supported *epi*-quinine thiourea catalyst **161**. The increased ee value can be ascribed to the confinement effect [76] of the support (Scheme 8.64).

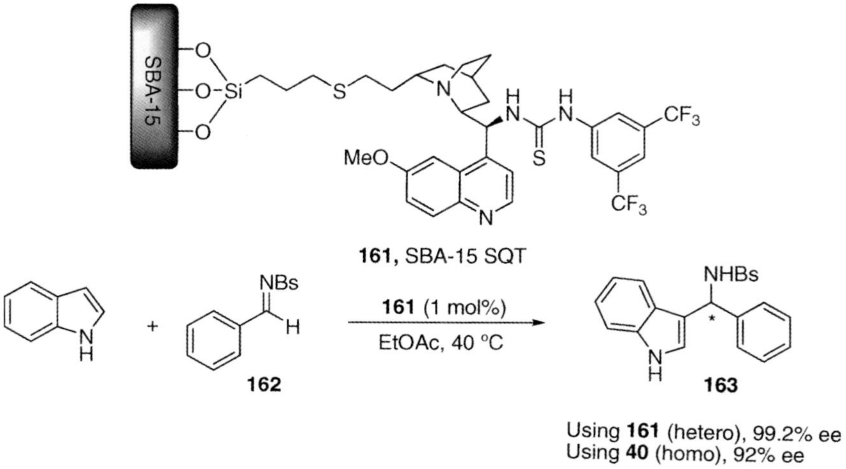

Scheme 8.64

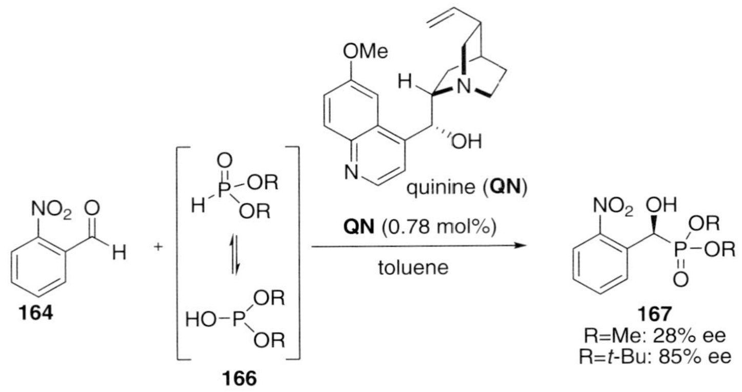

Scheme 8.65

Other types of cinchona-catalyzed enantioselective Friedel–Crafts type reactions such as the conjugate addition of naphthols to nitroalkenes [77] and Friedel–Crafts amination [78] are discussed in the respective sections of Chapters 6 and 9.

8.8
Hydrophosphonylation

Organophosphorous materials have attracted significant interest for their strong application potential toward bioactive compounds. By employing dialkyl phosphonates as nucleophiles, various aldehydes or imines could be converted to the desired α-hydroxy or α-amino phosphonates via 1,2-addition. It is known that the actual nucleophilic species is not the phosphonate form, but the phosphite, since the equilibrium between the phosphonate and phosphite is shifted toward the latter in the presence of a base [79]. In 1983, the first catalytic asymmetric hydrophosphonylation was achieved by using cinchona alkaloids as catalysts [80]. Wynberg and coworker reported that quinine (<1 mol%) catalyzes the reaction between an aldehyde 164 and a dialkyl phosphonate 165 to produce the optically active α-hydroxyphosphonate ester 167. When the aldehyde has an *ortho* substituent such as a nitro group, preferably one that aids in restricting the rotation of the aldehyde group, asymmetric induction takes place. The ee values increased from 28% with dimethyl phosphite to 80–85% with more hindered substrates such as di-*tert*-butyl phosphite (Scheme 8.65). The use of the pseudoenantiomeric quinidine furnished the opposite enantiomer of 167 with the same degrees of ee. Interestingly, however, O-acetyl quinine showed almost no reactivity or enantioselectivity, indicating that the free C9-hydroxy group might be critical for the catalysis and enantiodiscriminating step.

More than two decades after Wynberg's pioneering work, in 2006, Pettersen and Fini found that aromatic N-Boc-imines such as 168 also react with diethylphosphonate in the presence of catalytic quantities of quinine (10 mol%) to produce the α-aminophosphonates 169 in moderate to good yields and with up to

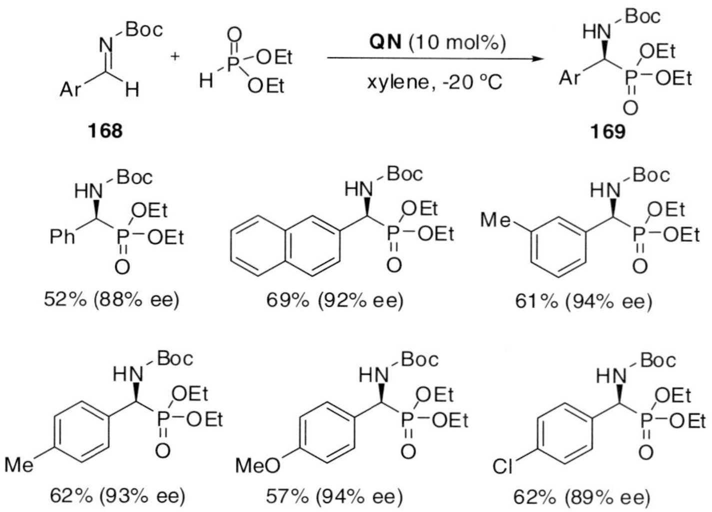

Scheme 8.66

Reaction: **168** (Ar-CH=N-Boc) + HP(O)(OEt)$_2$ → **169** (Ar-CH(NHBoc)-P(O)(OEt)$_2$), QN (10 mol%), xylene, −20 °C

Products:
- Ph: 52% (88% ee)
- 2-Naphthyl: 69% (92% ee)
- 3-Me-C$_6$H$_4$: 61% (94% ee)
- 4-Me-C$_6$H$_4$: 62% (93% ee)
- 4-MeO-C$_6$H$_4$: 57% (94% ee)
- 4-Cl-C$_6$H$_4$: 62% (89% ee)

94% ee (Scheme 8.66) [81]. However, quinidine (pseudoenantiomer of quinine) produced the enantiomer of **169** with significantly lower ee (48% ee). Moreover, similar to the observation made by Wynberg, C9-modified quinine (e.g., benzoyl) showed almost no reactivity for the hydrophosphonyl reaction, indicating the crucial role of the acidic-free C9-hydroxy group in the catalysis. Taking into account the above results, the authors proposed a plausible mechanism (Figure 8.9), in which imine is activated by hydrogen bonding with 9-OH group of catalyst and the phosphite is activated by the nitrogen of quinuclidine and guided to attack the imine [81]. It should be noted here that although this protocol afforded high enantioselectivity, a long reaction time was usually required (2–7 days). Under the same conditions, highly reactive N-tosylimine afforded products with low enantioselectivity.

To address the low enantioselectivity of reactive N-sulfonylimines, quite recently, Nakamura and Toru [82] employed the imines **170** protected with the 6-methyl-2-pyridylsulfonyl group, whose pyridine group might participate in the formation of a geometrically defined complex between the catalyst and imines, resulting in high chiral inducibility. A broad range of N-(6-methyl-2-pyridylsulfonyl)-imines **170** reacted

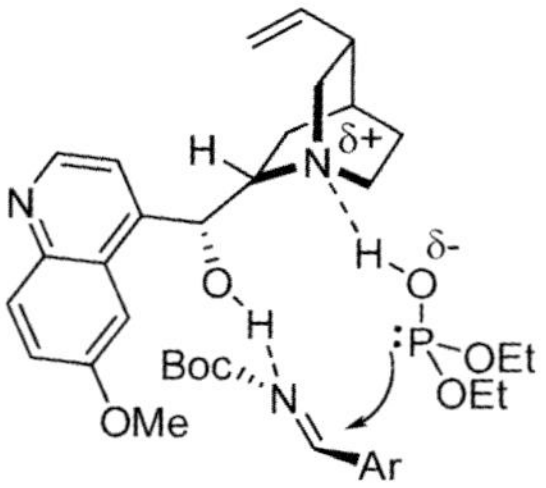

Figure 8.8 Proposed reaction mechanism.

very rapidly with diphenyl phosphite (**171**) in the presence of quinine, hydroquinine, or hydroquinidine at −40 or −78 °C. The desired chiral phosphonates **172** were obtained within a few hours (<1–6 h) in nearly quantitative yields (up to 99%) and with good to excellent enantioselectivity (82–98% ee) (Scheme 8.67). Furthermore, most of the products were crystalline and, thus, could be obtained in the enantiomerically pure form after a single crystallization. Here again, C9-protected quinine (e.g., acetyl) provided very low enantioselectivity (36% ee), indicating the crucial role of hydrogen bonding between the free C9-hydroxy group of the cinchona alkaloids and imines in the asymmetric induction. As expected by the authors, it was also observed that the presence of the pyridine ring at the sulfonyl group is critical for high enantioselectivity. Imines that do not possess the pyridine ring at the sulfonyl group were found to be

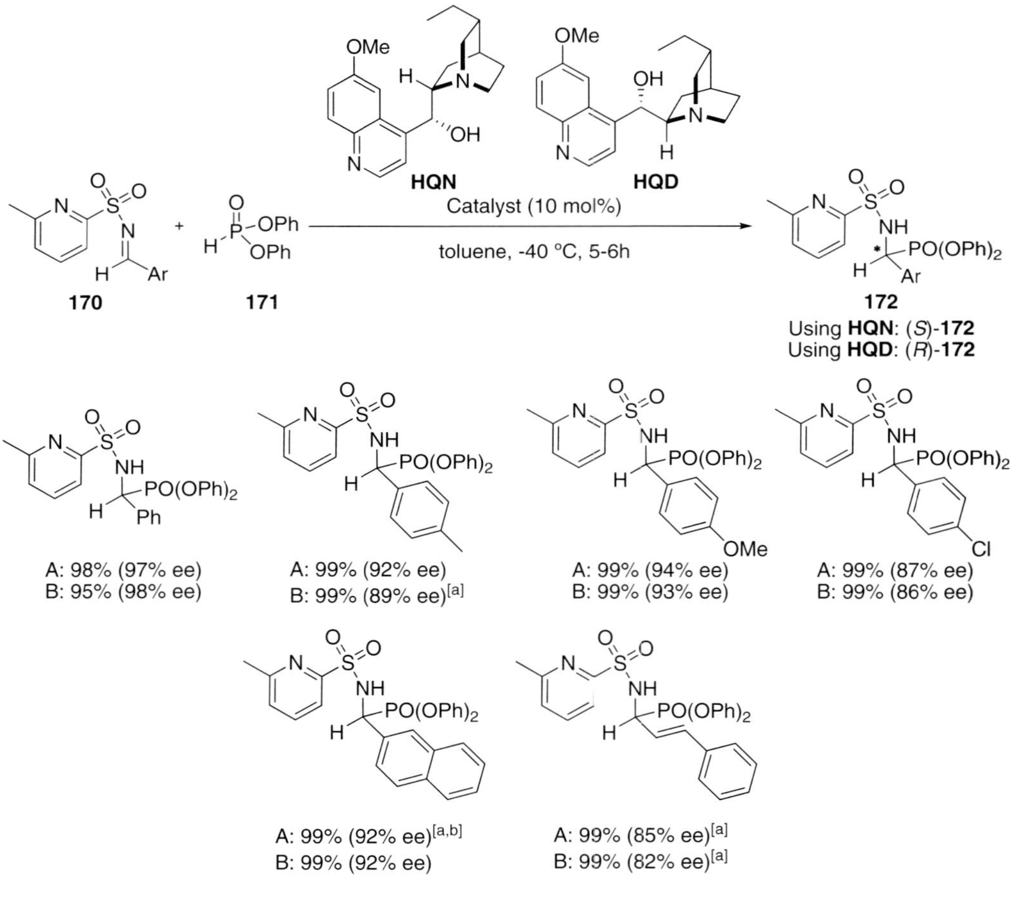

A: Using **HQN**; B: Using **HQD**
[a]Reaction time: less than 1h, [b]Quinine was used as catalyst instead of **HQN**
Scheme 8.67

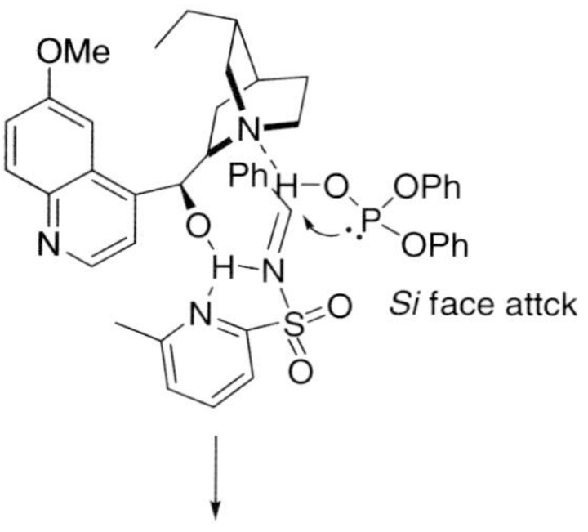

8.8 Hydrophosphonylation

Figure 8.9 Proposed reaction mechanism and model of transition state.

much less effective than **170** in terms of the reactivity and enantioselectivity. From the above consideration, the authors proposed a transition state model (Figure 8.9), in which the sulfonyl imine was directed via two hydrogen-bonding interactions and the phosphite is activated by the nitrogen of quinuclidine and guided to attack the *Si*-face of the imine.

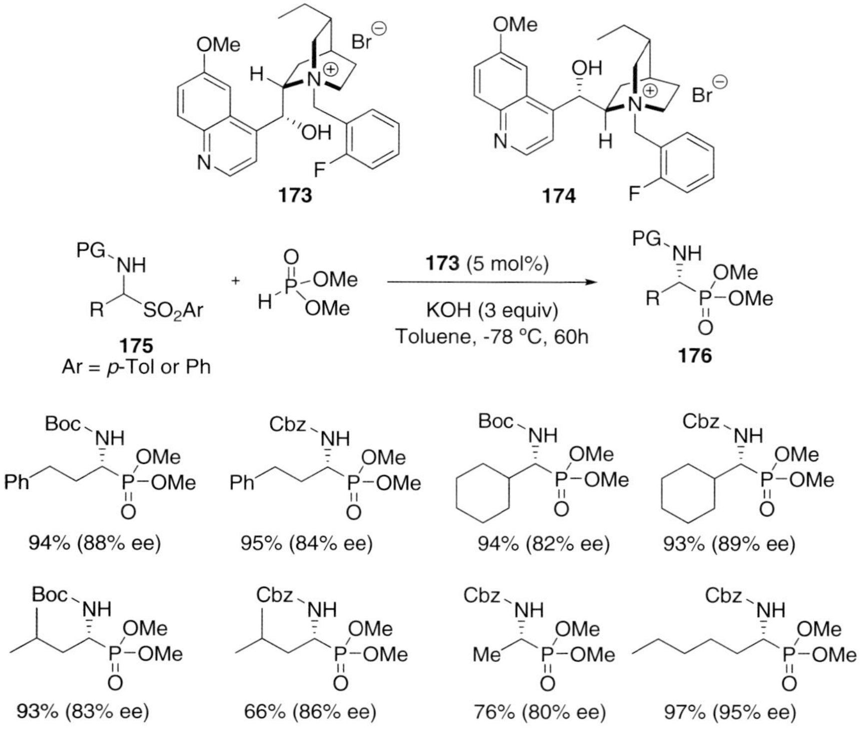

Scheme 8.68

Although the strategies for hydrophosphonylation discussed above gave satisfactory results in terms of the enantioselectivity, they do not appear to be suitable for linear unbranched imines, because of their instability, resulting from their imine–enamine tautomerism. Quite recently, Bernardi and Ricci reported that the use of phase-transfer catalysis in asymmetric hydrophosphonylation reactions with dimethyl phosphonate can allow a range of optically active α-amino phosphates to be obtained directly from α-amido sulfones derived from α- and β-branched, as well as linear unbranched aldehydes [83]. The use of α-amido sulfones **175** under PTC conditions, generating very reactive imines *in situ*, prevented the imine–enamine tautomerism. Thus, in the presence of 5 mol% of the hydroquinine-derived PTC **173** and KOH (3 equiv), the reaction of several amido sulfones **175** derived even from linear unbranched aldehydes with dimethyl phosphite at −78 °C and furnished the corresponding α-amino phosphates **176** in good to excellent yields and ee values (up to 99% yield and 95% ee) (Scheme 8.68). The opposite enantiomer of **176** was obtained with a similar level of ee using the hydroquinidine-derived PTC **174**.

8.9
Conclusions

This chapter presented the current state of the art on the applications of cinchona alkaloids and their derivatives as chiral catalysts in the asymmetric nucleophilic addition of prochiral C=O and C=N bonds. As shown in many of the examples discussed above, the vast synthetic potential of cinchona alkaloids and their derivatives in these reactions has been well demonstrated over the past few years. Cinchona-based organocatalysts possess diverse chiral skeletons and are easily tunable for diverse catalytic reactions through different mechanisms. Therefore, there is no doubt that the further development of cinchona-based organocatalysts for this major reaction type will continue to provide exciting results in the near future.

Acknowledgments

This work was supported by grants KRF-2008-005-J00701 (MEST) and R11-2005-008-00000-0 (SRC program of MEST/KOSEF).

References

1 (a) Berkessel, A. and Gröger, H. (2005) *Asymmetric Organocatalysis*, Wiley-VCH Verlag GmbH, Weinheim; (b) Dalko, P.I. (ed.) (2005) *Enantioselective Organocatalysis*, Wiley-VCH Verlag GmbH, Weinheim; (c) List, B. (ed.) (2007) *Chem. Rev*, **107**, 5413–5883 (special issue on organocatalysis); (d) Houk, K. N., and List, B.(eds) (2004) *Acc. Chem. Res.*, **37**, 487–631 (special issue on asymmetric organocatalysis).

2 Ando, A., Miura, T., Tatematsu, T., and Shioiri, T. (1993) *Tetrahedron Lett.*, **34**, 1507–1510.
3 Shiori, T., Bohsako, A., and Ando, A. (1996) *Heterocycles*, **42**, 93–97.
4 Bluet, G. and Campagne, J.-M. (2001) *J. Org. Chem.*, **66**, 4293–4298.
5 Horikawa, M., Busch-Petersen, J., and Corey, E.J. (1999) *Tetrahedron Lett.*, **40**, 3843–3846.
6 Andrus, M.B., Liu, J., Ye, Z., and Cannon, J.F. (2005) *Org. Lett.*, **7**, 3861–3864.
7 Nagao, H., Yamane, Y., and Mukaiyama, T. (2007) *Chem. Lett.*, **36**, 8–9.
8 (a) Nagao, H., Yamane, Y., and Mukaiyama, T. (2007) *Chem. Lett.*, **36**, 666–667; (b) Nagao, H., Kawano, Y., and Mukaiyama, T. (2007) *Bull. Chem. Soc. Jpn.*, **80**, 2406–2412; (c) Zhao, H., Qin, B., Liu, X., and Feng, X. (2007) *Tetrahedron*, **63**, 6822–6826.
9 Mettath, S., Srikanth, G.S.C., Dangerfield, B.S., and Castle, S.L. (2004) *J. Org. Chem.*, **69**, 6489–6492.
10 Ma, B., Parkinson, J.L., and Castle, S.L. (2007) *Tetrahedron Lett.*, **48**, 2083–2086.
11 (a) Arai, S., Hasegawa, K., and Nishida, A. (2004) *Tetrahedron Lett.*, **45**, 1023–1026; (b) Arai, S., Hasegawa, K., and Nishida, A. (2005) *Tetrahedron Lett.*, **46**, 6171; (c) Hasegawa, K., Arai, S., and Nishida, A. (2006) *Tetrahedron*, **62**, 1390–1401.
12 For a review, see Mukherjee, S., Yang, J.W., Hoffmann, S., and List, B. (2007) *Chem. Rev*, **107**, 5471–5569.
13 Zheng, B.-L., Liu, Q.-Z., Guo, C.-S., Wang, X.-L., and He, L. (2007) *Org. Biomol. Chem.*, **5**, 2913–2915.
14 Zhou, J., Wakchaure, V., Kraft, P., and List, B. (2008) *Angew. Chem. Int. Ed.*, **47**, 7656–7658.
15 (a) Agami, C. and Sevestre, H. (1984) *J. Chem. Soc., Chem. Commun.*, 1385–1386; (b) Agami, C., Platzer, N., and Sevestre, H. (1987) *Bull. Soc. Chim. Fr.*, **2**, 358–360; (c) List, B., Lerner, R.A., and Barbas, III, C.F. (1999) *Org. Lett.*, **1**, 59–61.
16 For a review, see Saito, S. and Yamamoto, H., (2004) *Acc. Chem. Res.*, **37**, 570–579.
17 Chen, J.-R., An, X.-L., Zhu, X.-Y., Wang, X.-F., and Xiao, W.-J. (2008) *J. Org. Chem.*, **73**, 6006–6009.
18 Ogawa, S., Shibata, N., Inagaki, J., Nakamura, S., Toru, T., and Shiro, M. (2007) *Angew. Chem. Int. Ed.*, **46**, 8666–8669.
19 Li, L., Klauber, E.G., and Seidel, D. (2008) *J. Am. Chem. Soc.*, **130**, 12248–12249.
20 Zu, L., Wang, J., Li, H., Xie, H., Jiang, W., and Wang, W. (2007) *J. Am. Chem. Soc.*, **129**, 1036–1037.
21 Marcelli, T., van der Hass, R.N.S., van Maarseveen, J.H., and Hiemstra, H. (2005) *Synlett*, 2817–2819.
22 (a) Marcelli, T., van der Hass, R.N.S., van Maarseveen, J.H., and Hiemstra, H. (2006) *Angew. Chem. Int. Ed.*, **45**, 929–931; (b) Hammar, P., Marcelli, T., Hiemstra, H., and Himo, F. (2007) *Adv. Synth. Catal.*, **349**, 2537–2548.
23 Li, H., Wang, B., and Deng, L. (2006) *J. Am. Chem. Soc.*, **128**, 732–733.
24 Bandini, M., Sinisi, R., and Umani-Ronchi, A. (2008) *Chem. Commun.*, 4360–4362.
25 Lou, S., Taoka, B.M., Ting, A., and Schaus, S.E. (2005) *J. Am. Chem. Soc.*, **127**, 11256–11257.
26 Ting, A., Lou, S., and Schaus, S.E. (2006) *Org. Lett.*, **8**, 2003–2006.
27 Goss, J.M. and Schaus, S.E. (2008) *J. Org. Chem.*, **73**, 7651–7656.
28 Tillman, A.L., Ye, J., and Dixon, D.J. (2006) *Chem. Commun.*, 1191–1193.
29 Song, J., Wang, Y., and Deng, L. (2006) *J. Am. Chem. Soc.*, **128**, 6048–6049.
30 Bode, C.M., Ting, A., and Schaus, S.E. (2006) *Tetrahedron*, **62**, 11499–11505.
31 Liu, T.-Y., Cui, H.-L., Long, J., Li, B.-J., Wu, Y., Ding, L.-S., and Chen, Y.-C. (2007) *J. Am. Chem. Soc.*, **129**, 1878–1879.
32 Ricci, A., Pettersen, D., Bernardi, L., Fini, F., Fochi, M., Herrera, R.P., and Sgarzani, V. (2007) *Adv. Synth. Catal.*, **349**, 1037–1040.
33 (a) Palomo, C., Oiarbide, M., Laso, A., and López, R. (2005) *J. Am. Chem. Soc.*, **127**, 17622–17623; (b) Fini, F., Sgarzani, V.,

Pettersen, D., Herrera, R.P., Bernardi, L., and Ricci, A. (2005) *Angew. Chem. Int. Ed.*, **44**, 7975–7978.

34 Song, J., Shih, H.-W., and Deng, L. (2007) *Org. Lett.*, **9**, 603–606.

35 Fini, F., Bernardi, L., Herrera, R.P., Pettersen, D., Ricci, A., and Sgarzani, V. (2006) *Adv. Synth. Catal.*, **348**, 2043–2046.

36 Lou, S., Dai, P., and Schaus, S.E. (2007) *J. Org. Chem.*, **72**, 9998–10008.

37 Mizuta, S., Shibata, N., Goto, Y., Furukawa, T., Nakamura, S., and Toru, T. (2007) *J. Am. Chem. Soc.*, **129**, 6394–6395.

38 Fini, F., Sgarzani, V., Pettersen, D., Herrera, R.P., Bernardi, L., and Ricci, A. (2005) *Angew. Chem. Int. Ed.*, **44**, 7975–7978.

39 Palomo, C., Oiarbide, M., Laso, A., and López, R. (2005) *J. Am. Chem. Soc.*, **127**, 17622–17623.

40 Bernardi, L., Fini, F., Herrera, R.P., Ricci, A., and Sgarzani, V. (2006) *Tetrahedron*, **62**, 375–380.

41 Hummelen, J. and Wynberg, H. (1978) *Tetrahedron Lett.*, **19**, 1089–1092.

42 (a) Arai, S. and Shioiri, T. (1998) *Tetrahedron Lett.*, **39**, 2145–2148; (b) Arai, S., Shirai, Y., Ishida, T., and Shioiri, T. (1999) *Chem. Commun.*, 49–50; (c) Arai, S., Shirai, Y., Ishida, T., and Shioiri, T. (1999) *Tetrahedron*, **55**, 6375–6386.

43 (a) Arai, S., Ishida, T., and Shioiri, T. (1998) *Tetrahedron Lett.*, **39**, 8299–8302; (b) Arai, S. and Shioiri, T. (2002) *Tetrahedron*, **58**, 1407–1413.

44 Ku, J.-M., Yoo, M.-S., Park, H.-g., Jew, S.-s., and Jeong, B.-S. (2007) *Tetrahedron*, **63**, 8099–8103.

45 (a) Iwabuchi, Y., Nakatani, M., Yokoyama, N., and Hatakeyama, S. (1999) *J. Am. Chem. Soc.*, **121**, 10219–10220; (b) Nakano, A., Kawahara, S., Akamatsu, S., Morokuma, K., Nakatami, M., Iwabuchi, Y., Takahashi, K., Ishihara, J., and Hatakeyama, S. (2006) *Tetrahedron*, **62**, 381–389.

46 Iwabuchi, Y., Sugihara, T., Esumi, T., and Hatakeyama, S. (2001) *Tetrahedron Lett.*, **42**, 7867–7871.

47 Iwabuchi, Y., Furukawa, M., Esumi, T., and Hatakeyama, S. (2001) *Chem. Commun.*, 2030–2031.

48 Shi, M. and Jiang, J.-K. (2002) *Tetrahedron: Asymmetry*, **13**, 1941–1947.

49 Mocquet, C.M. and Warriner, S.L. (2004) *Synlett*, 356–357.

50 Kawahara, S., Nakano, A., Esumi, T., Iwabuchi, Y., and Hatakeyama, S. (2003) *Org. Lett.*, **5**, 3103–3105.

51 (a) Shi, M. and Xu, Y.-M. (2002) *Angew. Chem. Int. Ed.*, **41**, 4507–4510; (b) Shi, M., Xu, Y.-M., and Shi, Y.-L. (2005) *Chem. Eur. J.*, **11**, 1794–1802.

52 Balan, D. and Adolfsson, H. (2003) *Tetrahedron Lett.*, **44**, 2521–2524.

53 (a) Wynberg, H. and Staring, E.G.J. (1982) *J. Am. Chem. Soc.*, **104**, 166–168; (b) Calter, M.A., Orr, R.K., and Song, W. (2003) *Org. Lett.*, **5**, 4745–4748; (c) Zhu, C., Shen, X., and Nelson, S.G. (2004) *J. Am. Chem. Soc.*, **126**, 5352–5353; (d) Taggi, A.E., Hafez, A.M., Wack, H., Yong, B., Drury, W.J., and Lectka, T. (2000) *J. Am. Chem. Soc.*, **122**, 7831–7832; (e) Taggi, A.E., Hafez, A.M., Wack, H., Yong, B., Ferraris, D., and Lectka, T. (2002) *J. Am. Chem. Soc.*, **124**, 6626–6635; (f) France, S., Shah, M.H., Weatherwax, A., Wack, H., Roth, J.P., and Lectka, T. (2005) *J. Am. Chem. Soc.*, **127**, 1206–1215; (g) Paull, D.H., Abraham, C.J., Scerba, M.T., Alden-Danforth, E., and Lectka, T. (2008) *Acc. Chem. Res.*, **41**, 655–663.

54 Cortez, G.S., Tennyson, R.L., and Romo, D. (2001) *J. Am. Chem. Soc.*, **123**, 7945–7946.

55 (a) North, M. (1993) *Synlett*, 807–820; (b) Effenberger, F. (1994) *Angew. Chem. Int. Ed.*, **33**, 1555–1564; (c) Gregory, R.J.H. (1999) *Chem. Rev.*, **99**, 3649–3682; (d) Mori, A. and Noue, I.S. (1999) in *Comprehensive Asymmetric Catalysis II* (eds E.N. Jacobsen, A. Pfaltz, and H. Yamamoto), Springer, Berlin, pp. 983–992.

56 Bredig, G. and Fiske, P.S. (1912) *Biochem. Z.*, **46**, 7.

57 Kobayashi, S., Tsuchiya, Y., and Mukaiyama, T. (1991) *Chem. Lett.*, **20**, 541–544.
58 Tian, S.-K. and Deng, L. (2001) *J. Am. Chem. Soc.*, **123**, 6195–6196.
59 Tian, S.-K. and Deng, L. (2006) *Tetrahedron*, **62**, 11320–11330.
60 Tian, S.-K., Hong, R., and Deng, L. (2003) *J. Am. Chem. Soc.*, **125**, 9900–9901.
61 Hong, R., Chen, Y., and Deng, L. (2005) *Angew. Chem. Int. Ed.*, **44**, 3478–3481.
62 Peng, D., Zhou, H., Liu, X., Wang, L., Chen, S., and Feng, X. (2007) *Synlett*, 2448–2450.
63 Huang, J. and Corey, E.J. (2004) *Org. Lett.*, **6**, 5027–5029.
64 Herrera, R.P., Sgarzani, V., Bernardi, L., Fini, F., Pettersen, D., and Ricci, A. (2006) *J. Org. Chem.*, **71**, 9869–9872.
65 Iseki, K., Nagai, T., and Kobayashi, Y. (1994) *Tetrahedron Lett.*, **35**, 3137–3138.
66 Caron, S., Do, N.M., Arpin, P., and Larivée, A. (2003) *Synthesis*, **11**, 1693–1698.
67 Mizuta, S., Shibata, N., Akiti, S., Fujimoto, H., Nakamura, S., and Toru, T. (2007) *Org. Lett.*, **9**, 3707–3710.
68 Mizuta, S., Shibata, N., Hibino, M., Nagano, S., Nakamura, S., and Toru, T. (2007) *Tetrahedron*, **63**, 8521–8528.
69 (a) Nagao, H., Yamane, Y., and Mukaiyama, T. (2007) *Chem. Lett.*, **36**, 666–667; (b) Nagao, H., Kawano, Y., and Mukaiyama, T. (2007) *Bull. Chem. Soc. Jpn.*, **80**, 2406–2412.
70 Zhao, H., Qin, B., Liu, X., and Feng, X. (2007) *Tetrahedron*, **63**, 6822–6826.
71 Török, B., Abid, M., London, G., Esquibel, J., Török, M., Mhadgut, S.C., Yan, P., and Prakash, G.K.S. (2005) *Angew. Chem. Int. Ed.*, **44**, 3086–3089.
72 Li, H., Wang, Y.-Q., and Deng, L. (2006) *Org. Lett.*, **8**, 4063–4065.
73 Zhao, J.-L., Liu, L., Gu, C.-L., Wang, D., and Chen, Y.-J. (2008) *Tetrahetron Lett.*, **49**, 1476–1479.
74 Wang, Y.-Q., Song, J., Hong, R., Li, H., and Deng, L. (2006) *J. Am. Chem. Soc.*, **128**, 8156–8157.
75 Yu, P., He, J., and Guo, C. (2008) *Chem. Commun.*, 2355–2357.
76 Thakur, S.S., Lee, J.E., Lee, S.H., Kim, J.M., and Song, C.E. (2008) in *Handbook of Asymmetric Heterogeneous Catalysis* (eds K. Ding and Y. Uozumi), Chapter 2, Wiley-VCH Verlag GmbH, Weinheim, pp. 25–72.
77 Liu, T.-Y., Cui, H.-L., Chai, Q., Long, J., Li, B.-J., Wu, Y., Ding, L.-S., and Chen, Y.-C. (2007) *Chem. Commun.*, 2228–2230.
78 Brandes, S., Bella, M., Kjærsgaard, A., and Jørgensen, K.A. (2006) *Angew. Chem. Int. Ed.*, **45**, 1147–1151.
79 Wiemer, D.F. (1997) *Tetrahedron*, **53**, 16609–16633.
80 (a) Wynberg, H. and Smaardijk, A.A. (1983) *Tetrahedron Lett.*, **24**, 5899–5900; (b) Smaardijk, A.A., Noorda, S., van Bolhuis, F., and Wynberg, H. (1985) *Tetrahedron Lett.*, **26**, 493–496.
81 Pettersen, D., Marcolini, M., Bernardi, L., Fini, F., Herrera, R.P., Sgarzani, V., and Ricci, A. (2006) *J. Org. Chem.*, **71**, 6269–6272.
82 Nakamura, S., Nakashima, H., Yamamura, A., Shibata, N., and Toru, T. (2008) *Adv. Synth. Catal.*, **350**, 1209–1212.
83 Fini, F., Micheletti, G., Bernardi, L., Pettersen, D., Fochi, M., and Ricci, A. (2008) *Chem. Commun.*, 4345–4347.

9
Cinchona-Catalyzed Nucleophilic Conjugate Addition to Electron-Deficient C=C Double Bonds

Ji Woong Lee, Hyeong Bin Jang, and Choong Eui Song

9.1
Introduction

Asymmetric conjugate addition reactions represent a fundamental approach to constructing densely functionalized products that are readily transformed into a variety of useful chiral intermediates. Since the initial studies [1] by Wynberg in the late 1970s and early 1980s on cinchona-catalyzed conjugate addition reactions, numerous examples of these reactions in which cinchona alkaloids (Figure 9.1) induce asymmetry have been reported. Today, cinchona alkaloids are classified as the most privileged chiral catalysts, promoting diverse types of conjugate addition reactions with excellent performances. In this chapter, the current state-of-the-art applications of cinchona alkaloids and their derivatives as chiral organocatalysts in enantioselective nucleophilic addition to electron-deficient C=C double bonds are discussed. The following section details developments in this area categorized according to the class of Michael acceptor substrate employed in the reaction, namely, α,β-unsaturated carbonyl substrates, nitroolefins, vinyl sulfones, and vinyl phosphates. The authors also focus on the mode of action of each catalytic system, with models being proposed.

9.2
Conjugate Reaction of α,β-Unsaturated Ketones, Amides, and Nitriles

9.2.1
Natural Cinchona Alkaloids as Catalysts

When the Michael donors have a sufficiently low pK_a, the Michael addition can be catalyzed by a base. The first catalytic asymmetric conjugate addition was achieved by Wynberg *et al.* in 1975 using cinchona bases [1a]. They performed the reaction of cyclic β-ketoesters such as **1** with methyl vinyl ketone in the presence of quinine and

9 Cinchona-Catalyzed Nucleophilic Conjugate Addition to Electron-Deficient C=C Double Bonds

R = H
R' = CH=CH$_2$; Cinchonidine (**CD**)
R' = CH$_2$CH$_3$; Dihydrocinchonidine (**HCD**)

R = OMe
R' = CH=CH$_2$; Quinine (**QN**)
R' = CH$_2$CH$_3$; Dihydroquinine (**HQN**)

R = H
R' = CH=CH$_2$; Cinchonine (**CN**)
R' = CH$_2$CH$_3$; Dihydrocinchonine (**HCN**)

R = OMe
R' = CH=CH$_2$; Quinidine (**QD**)
R' = CH$_2$CH$_3$; Dihydroquinidine (**HQD**)

Figure 9.1 Structures of various naturally occurring cinchona alkaloids.

obtained the enantiomerically enriched adducts 2 in nearly quantitative yields and moderate ee (up to 68% ee) [1a]. By using a cinchona alkaloid as a catalyst, a range of other donors such as other β-ketoesters, nitrosulfone, 1,3-diketones, α-cyanoesters, α-nitroesters [1], or aromatic thiols [1b, c] underwent the corresponding reactions with enones (Scheme 9.1). Mechanistic studies [1c] to establish the mode of action of the catalysts revealed that they operated bifunctionally; the C9-hydroxy group activates the electrophile through hydrogen bonding and a basic tertiary amine moiety activates the nucleophile via deprotonation (Figure 9.2). Reactions with catalysts lacking a free 9-OH group led to dramatically reduced reactivity and enantioselectivity, strongly supporting this dual activation mechanism.

Quite recently, the use of natural cinchona alkaloids as catalysts for the intramolecular oxo-Michael addition of o-tigloylphenol (**3**), furnishing chiral cis-2,3-dimethyl-4-chromanone **4**, which is a valuable intermediate for the synthesis of the anti-HIV-1 active coumarins, (+)-calanolide A (**5a**), and (+)-inophyllum B (**5b**), was reexamined by Ishikawa and coworkers (Scheme 9.2) [2]. The parent cinchona alkaloids,

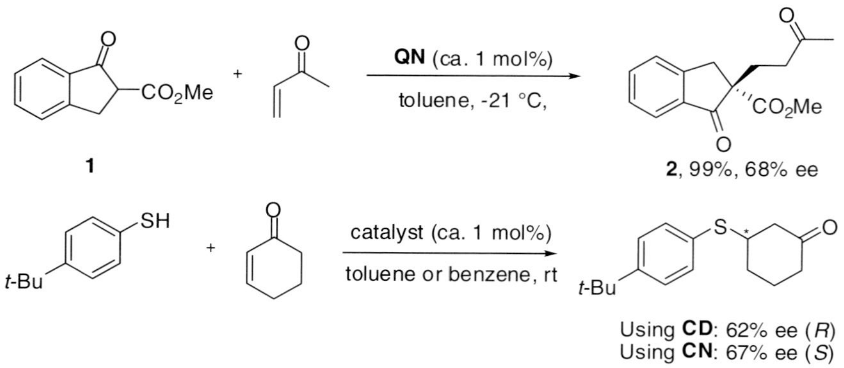

Scheme 9.1

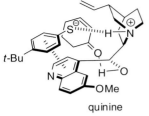

quinine

Figure 9.2 Transition-state model of quinine-catalyzed Michael addition of thiol to cyclohexenone.

quinine and dihydroquinine, showed remarkable enantioselectivity (up to 98% ee) in this reaction, albeit with low diastereoselectivity (about 40% ds).

For the asymmetric synthesis of the 2-substituted chromane **7** via the intramolecular Michael addition reaction of **6**, Merschaert et al. also employed natural cinchona alkaloids such as **HCD** as catalysts (Scheme 9.3) [3]. Here again, the 9-O functionalization and dehydroxylation of the natural alkaloid showed a large negative effect, indicating that the presence of the 9-OH group is needed to achieve both good kinetics and enantioselectivity. Moreover, C3 modifications of this parent alkaloid did not lead to any significant improvement in the results in terms of the enantioselectivity and catalytic activity.

The Jørgensen group also applied the parent cinchona alkaloids as catalysts to the aza-Michael addition of hydrazones **8** to cyclic enones **9** [4] and the asymmetric deconjugative Michael reaction of alkylidene cyanoacetates **10** with acrolein (**11**) [5]. However, only a moderate level of enantioselectivity was obtained in both reactions (Scheme 9.4). Of note, for the deconjugative Michael reaction, the delocalized allylic anion **12** could be generated via the deprotonation of **10** by the cinchona base and might attack the electrophilic enal at either the α- or the γ-position. However, in this study, only the α-adducts were produced.

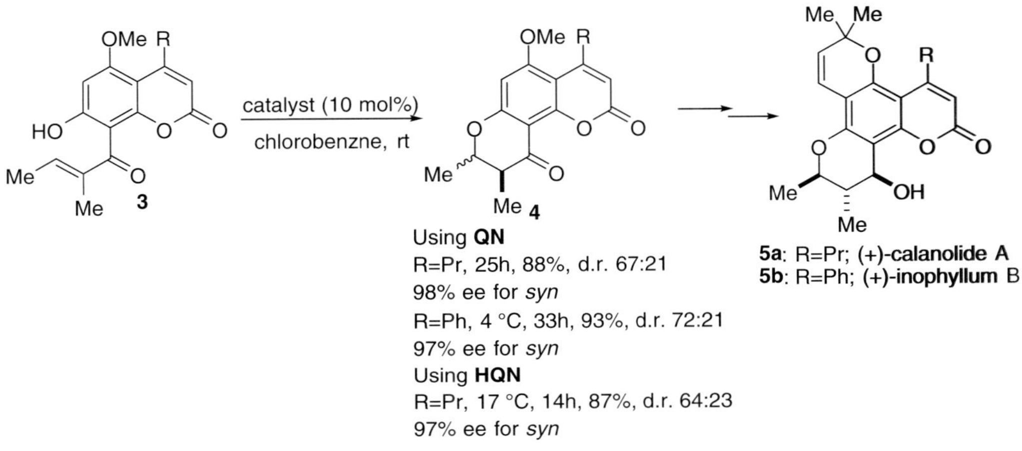

Scheme 9.2

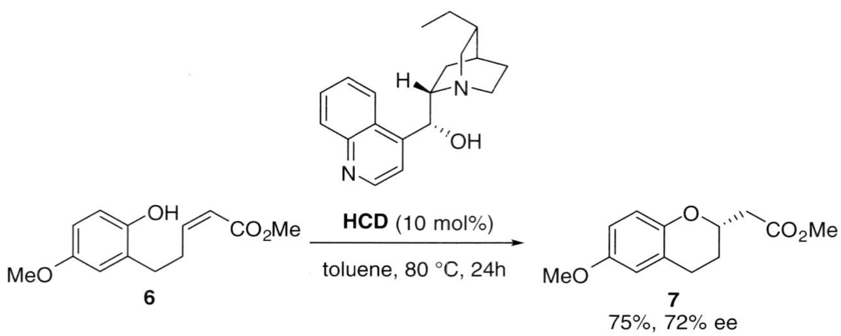

Scheme 9.3

9.2.2
PTC-Catalyzed Enantioselective Michael Addition Reactions

Asymmetric Michael additions can also be performed under phase-transfer conditions with an achiral base in the presence of a chiral quaternary ammonium salt as a phase-transfer agent. Conn and coworkers conducted the Michael addition of 2-propyl-1-indanone (13) to methyl vinyl ketone under biphasic conditions (aq 50% NaOH/toluene) using the cinchonine/cinchonidine-derived chiral phase-transfer catalysts (PTCs), 14a and 14b, as a catalyst (Scheme 9.5). However, only low to

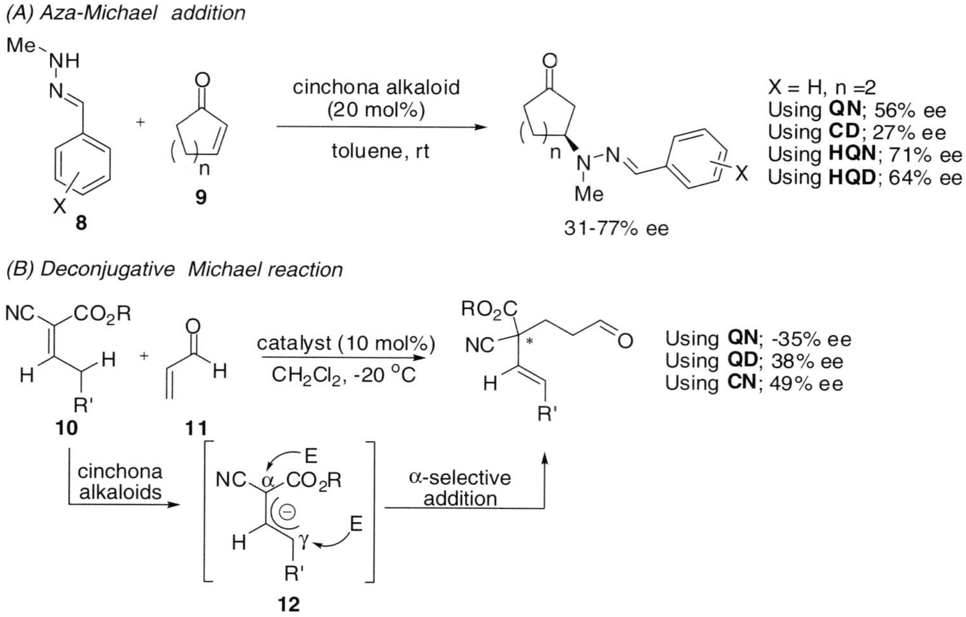

Scheme 9.4

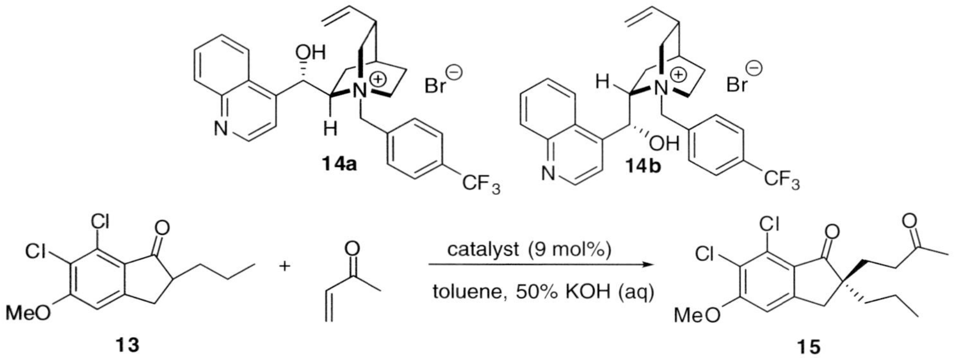

9.2 Conjugate Reaction of α,β-Unsaturated Ketones, Amides, and Nitriles

Scheme 9.5

Using **14a**; (S)-**15** (80% ee)
Using **14b**: (R)-**15** (40% ee)

moderate ee values were obtained (40–80% ee) [6]. Of note, the parent cinchona alkaloids are not basic enough to catalyze this reaction, since 2-alkylindanones contain a less acidic proton than the β-keto esters.

Corey's group achieved a dramatic improvement in the enantioselectivity in 1997 by using O(9)-allyl-N-(9-anthracenylmethyl) cinchonidinium bromide (**16**) as a phase-transfer catalyst. Under PTC conditions, the glycinate Schiff base of benzophenone **17** underwent addition reactions with various Michael acceptors such as α,β-unsaturated esters [7a], α,β-unsaturated ketones [7a], and acrylonitriles [7b], furnishing chiral α-functionalized glycinate Schiff bases in high yields and with excellent enantioselectivity (up to 99% ee) (Scheme 9.6). This protocol enabled the facile preparation of various functionalized chiral α-alkyl-amino acids. With methyl acrylate as an acceptor, the (S)-glutamic acid derivative **18** could be produced. The naturally occurring (S)-ornithine (**19**) could also be concisely synthesized as its dihydrochloride using acrylonitrile as an acceptor. By employing this protocol, all possible ^{13}C and ^{15}N isotopomers of L-lysine, L-ornithine, and L-proline were also prepared by Lugtenburg and coworkers [7c].

The activated allylic acetates **20** were also used by Ramachandran and coworkers as an acceptor in this reaction, which enabled the synthesis of various enantiomerically enriched glutamic acid derivatives **21** (Scheme 9.7) [8a]. O'Donnell and coworkers [8b] also conducted the Michael addition of the glycinate Schiff base of benzophenone **17** to a range of Michael acceptors **22** using **16** as a catalyst. However, they employed an organic-soluble base (BEMP) instead of an inorganic base. Generally, good to high ee values (76–89% ee) were obtained (Scheme 9.8).

Later on, the substrate scope of this methodology was extensively explored by Corey's group using the structural analogues **23** and **24** of the above-discussed catalyst **16**. A range of Michael donors such as nitromethane [9b] and silyl enolethers [9c] were successfully applied using the catalyst **23** or **24** (Scheme 9.9). In the case where the β-alkyl chalcones **25** with a proton at C(γ) were used as acceptors, self-dimerization occurred under PTC conditions to produce the chiral 1,5-dicarbonyl

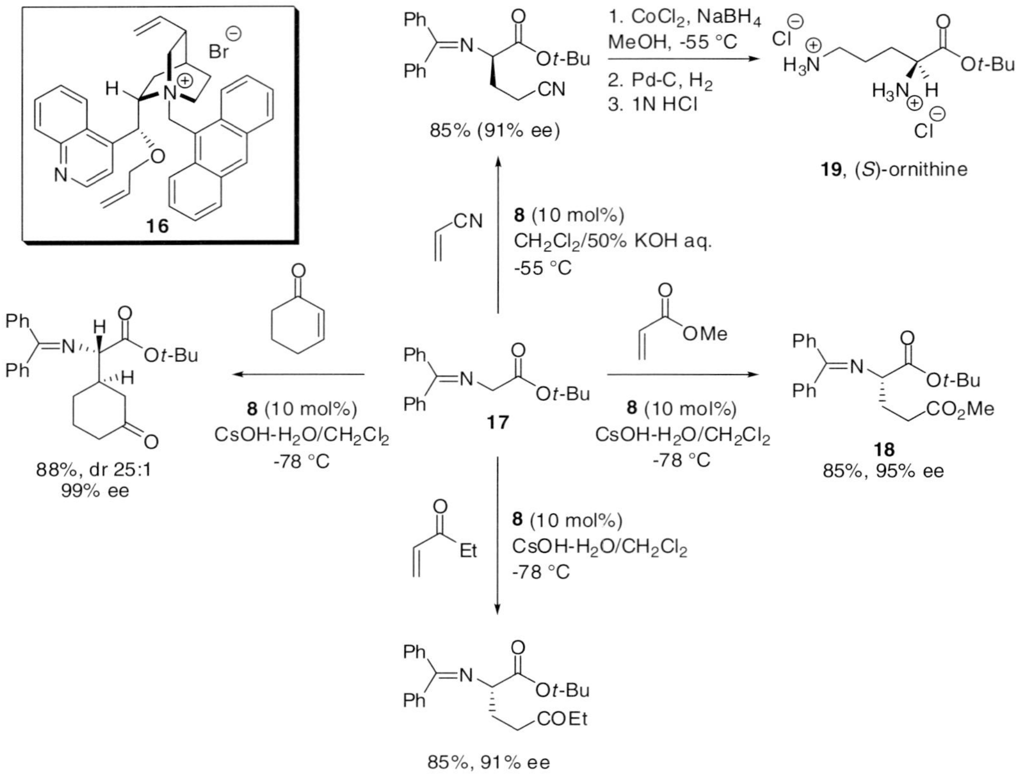

Scheme 9.6

compounds **26** with good to excellent enantioselectivity [9a]. This dimerization occurs through a Michael addition–double bond transposition sequence (Scheme 9.10), triggered by the abstraction of a γ-proton. This protocol provides simple access to the chiral α-alkylated γ-ketoacids **27**. Recently, the self-dimerization of cyclic enones under PTC conditions was also reported by Bella and coworkers [10].

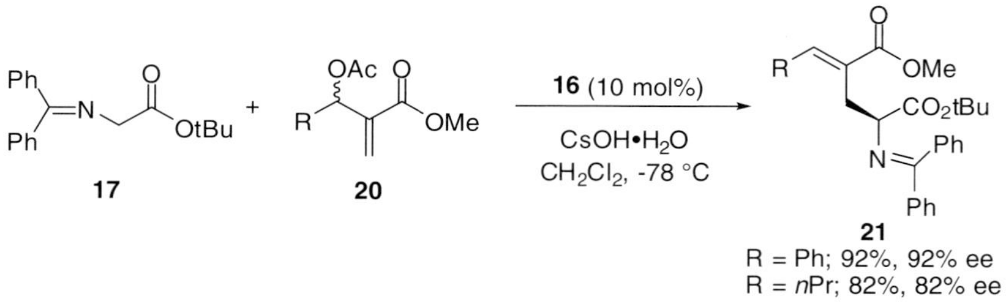

Scheme 9.7

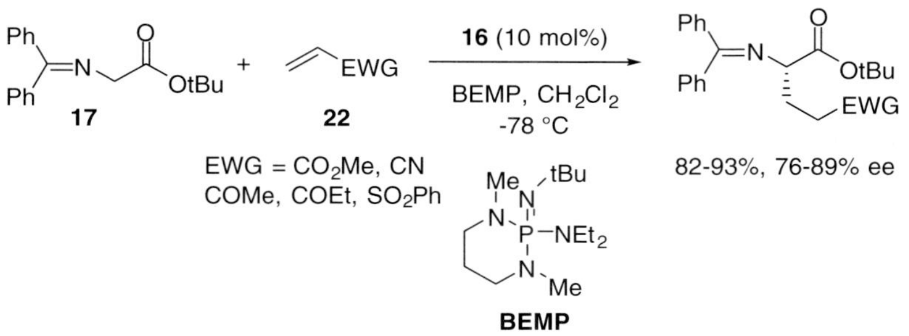

9.2 Conjugate Reaction of α,β-Unsaturated Ketones, Amides, and Nitriles

Scheme 9.8

Corey's methodology was also successfully applied by Plaquevent and coworkers to the synthesis of chiral methyl dihydrojasmonate (**31**), which has a jasmine-like odor. The reaction of α-pentyl cyclopentanone (**29**) and dimethyl malonate under PTC conditions (11 mol% of **28**, K_2CO_3 (0.16 equiv.) at −20 °C) gave the key intermediate **30** for the synthesis of methyl dihydrojasmonate **31** in 91% yield and with 90% ee (Scheme 9.11) [11].

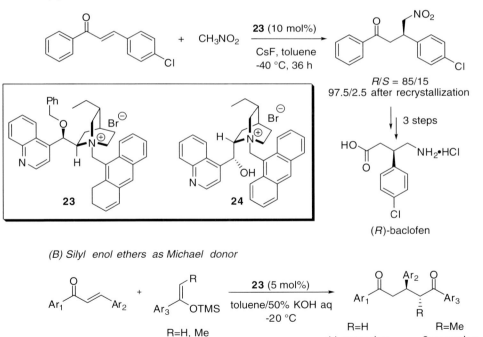

Scheme 9.9

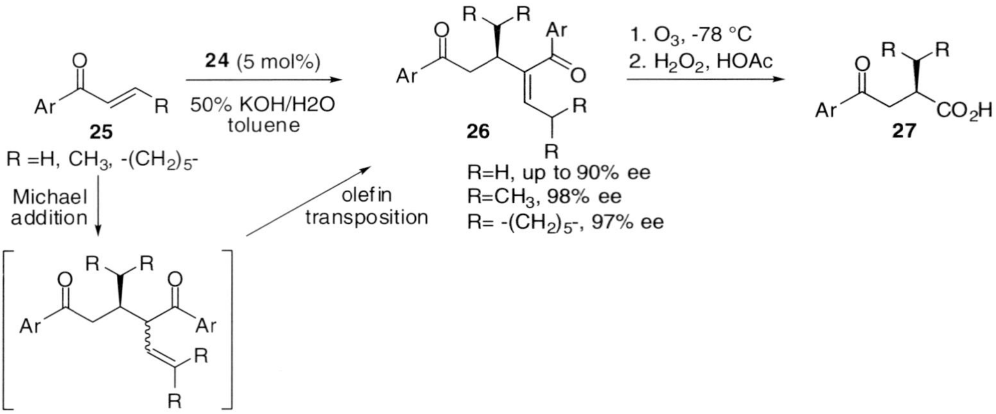

Scheme 9.10

Silyl enolates are also useful Michael donors in the asymmetric tandem Michael lactonization reaction. In 2006, the Mukaiyama group demonstrated that the phenoxide anion could be used instead of fluoride as a nucleophilic activator of silyl enolates [12]. Using the cinchonidine-derived PTC **32** bearing the phenoxide anion (5 mol%), a range of silyl enolates **33–35** smoothly underwent the reaction with the chalcone **36** to give the corresponding 3,4-dihydropyran-2-ones **37–39**, respectively, in up to 98% yield, 96% ee, and dr >99 : 1 (Scheme 9.12). Interestingly, the cinchonidine-derived PTC **40** bearing a free C9−OH group and highly bulky substituent to quinuclidine's nitrogen atom produced the opposite enantiomers to those that were obtained with catalyst **32** [12d] (Scheme 9.13).

In 2007, Jørgensen and coworkers reported the first example of the asymmetric 1,6-addition of β-ketoesters to electron-poor δ-unsubstituted dienes using the cinchona-derived PTCs, **42a,b** (3 mol%) (Scheme 9.14) [13]. Various electron-poor δ-unsubstituted dienes **44** having ketones, esters, and sulfones as substituents

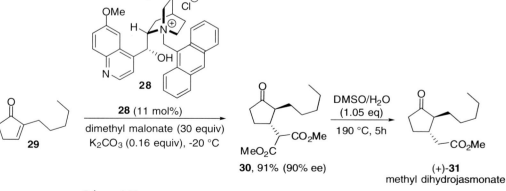

Scheme 9.11

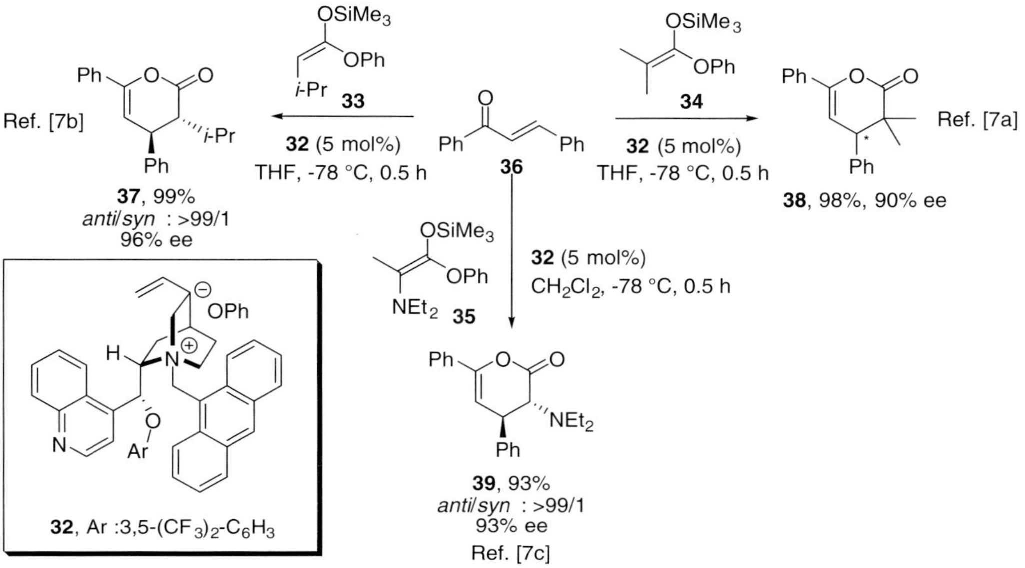

9.2 Conjugate Reaction of α,β-Unsaturated Ketones, Amides, and Nitriles

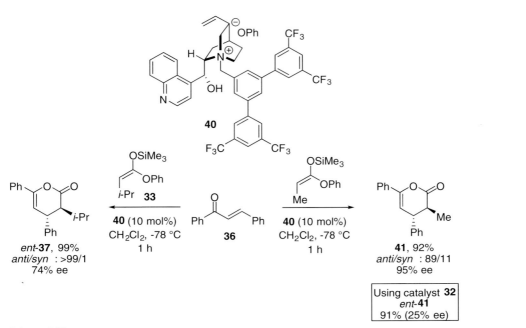

Scheme 9.12

Scheme 9.13

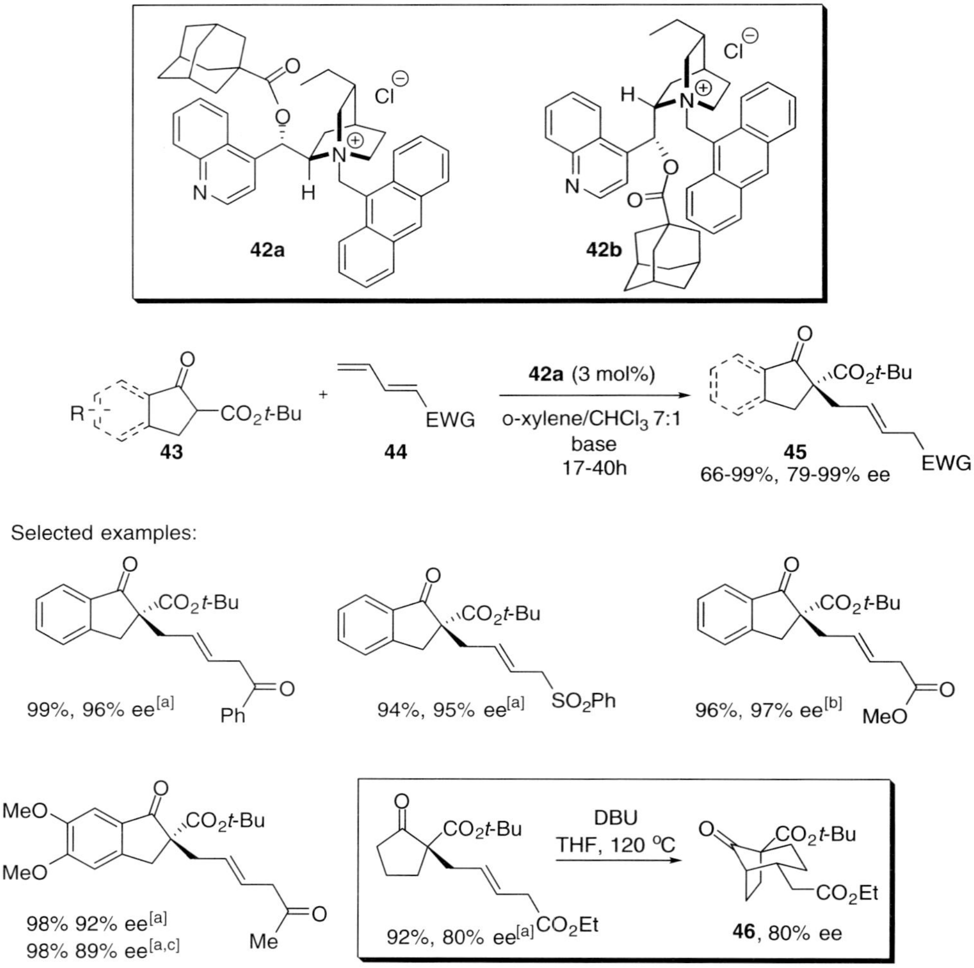

258 | *9 Cinchona-Catalyzed Nucleophilic Conjugate Addition to Electron-Deficient C=C Double Bonds*

Scheme 9.14

[a] 50% K_2HPO_4, -20 °C, [b] 40% Cs_2CO_3, 4 °C, [c] The opposite enantiomer obtained using **42b**

underwent conjugate addition reactions with the various β-keto esters **43** affording the 1,6-adducts **45** in high chemical yield and excellent enantioselectivity (up to 99% yield and 97% ee). Bicyclo[3.2.1]octan-8-one **46** could be obtained as a single diasteromer via the DBU-promoted isomerization and intramolecular Michael addition of the 1,6-adduct (see the box in Scheme 9.14).

The glycinate Schiff base of benzophenone **17** was also shown to be a suitable Michael donor for the asymmetric 1,6-addition to the activated dienes **44** having ketones, esters, and sulfones as substituents. Using Corey's phase-transfer catalyst, **16**, the corresponding allylated products **47** were obtained as a single *E*-isomer with high enantioselectivity (from 92 to 98% ee). The synthetic utility of this reaction

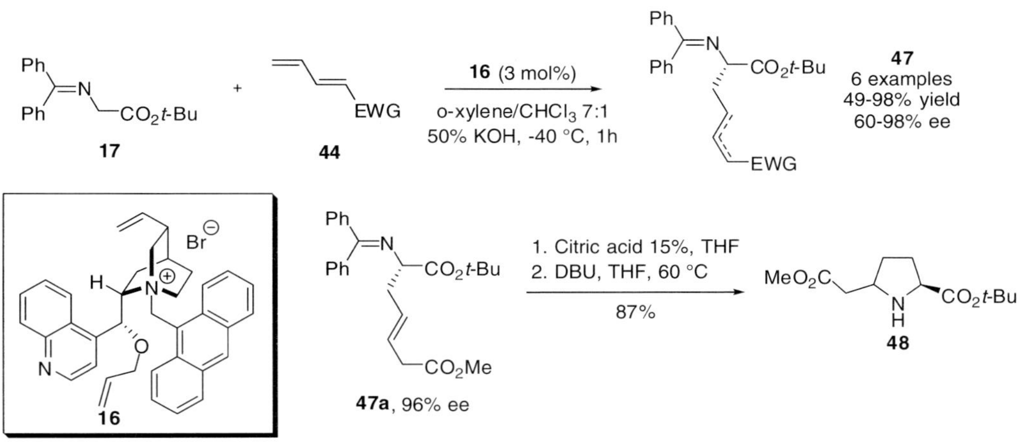

9.2 Conjugate Reaction of α,β-Unsaturated Ketones, Amides, and Nitriles | 259

Scheme 9.15

was demonstrated by the two-step transformation of the allylated adduct **47a** into the 2,5-*cis*-pyrrolidine (**48**) with the diastereomerically pure form, via the hydrolysis of the imine followed by DBU-promoted double-bond isomerization and cyclization (Scheme 9.15).

Very recently, the use of the electron-deficient allenic esters and ketones **50** as acceptors was also reported by the same group [14]. Under PTC conditions (**42a** (3 mol%), o-xylene/CHCl$_3$, K$_2$CO$_3$ aq), the cyclic β-ketoester **49** underwent an addition to the electron-deficient allenic esters and ketones **50**, giving the corresponding β,γ-unsaturated (isolated alkene) carbonyl compounds **51** with excellent enantioselectivity (up to 96% ee, Scheme 9.16). The Michael adduct **51** could be transformed into the optically active hexahydrobenzopyranone **51a** and **51b** with a 2 : 1 (**51a** : **51b**) diastereomeric ratio via a simple one-step procedure (Scheme 9.17).

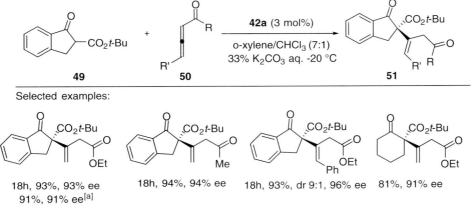

[a] The opposite enantiomer was obtained by using **42b** as catalyst

Scheme 9.16

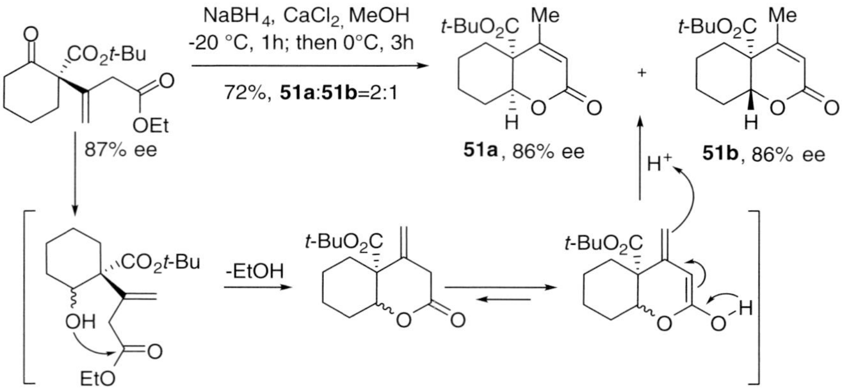

Scheme 9.17

However, the use of the glycinate Schiff base imine (**17**) as a Michael donor provided low to moderate enantioselectivity (60–88% ee).

In 2007, Jørgensen and coworkers also reported the "*anti*-Michael" reaction of the cyclic β-ketoesters **53** with the sulfone group-substituted acrylonitrile **54** under PTC conditions (**52** (6 mol%), CHCl$_3$, aq Cs$_2$CO$_3$ or K$_3$PO$_4$) [15]. As depicted in Scheme 9.18, the sulfone group of the acceptor directed the nucleophile and then is removed to afford the *anti*-Michael (α-addition Morita–Baylis–Hillman-like) adducts **55** in variable yields (42–90%) and ee values (60–94% ee) (Scheme 9.18).

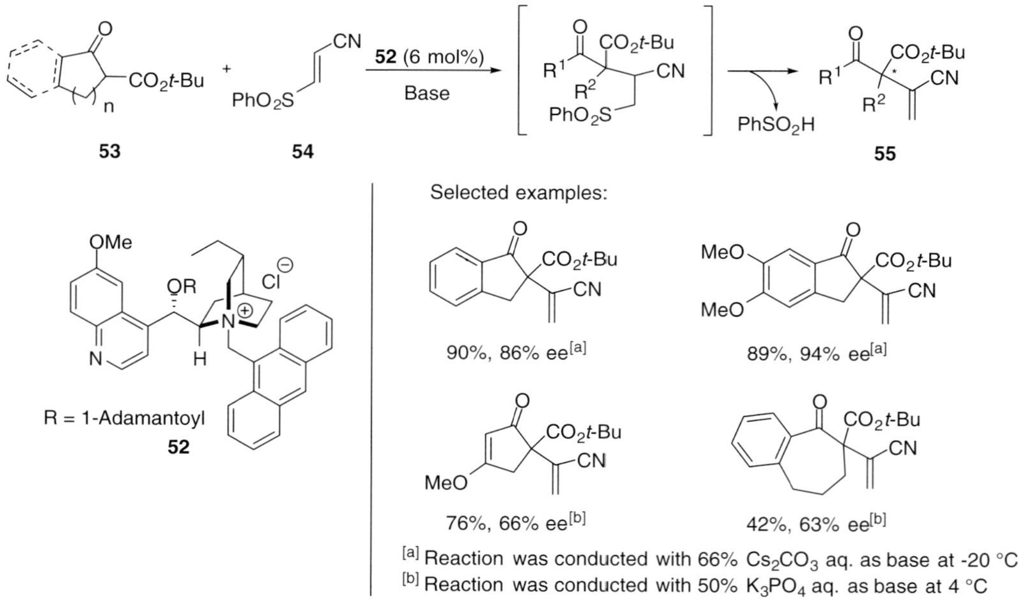

[a] Reaction was conducted with 66% Cs$_2$CO$_3$ aq. as base at -20 °C
[b] Reaction was conducted with 50% K$_3$PO$_4$ aq. as base at 4 °C

Scheme 9.18

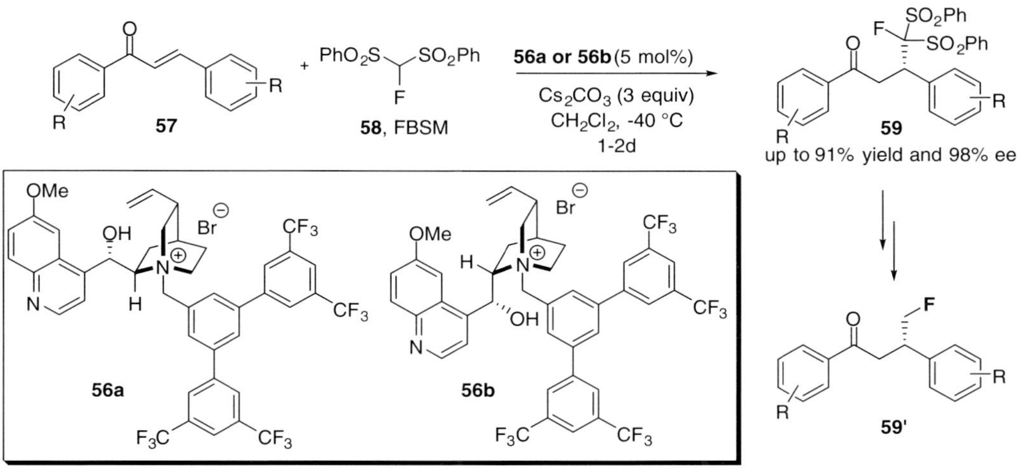

9.2 Conjugate Reaction of α,β-Unsaturated Ketones, Amides, and Nitriles

Scheme 9.19

In 2008, fluorobis(phenylsulfonyl)methane (FBSM, **58**), a synthetic equivalent of the monofluoromethide species, was successfully employed as a Michael donor by the Shibata group [16]. Under phase-transfer conditions using the quinidine-derived PTC **56a** with a sterically demanding substituent at the quinuclidine nitrogen, FSBM (**58**) was added to the *trans*-chalcones **57**, affording the corresponding adducts **59** with high to excellent ee values (up to 98% ee, Scheme 9.19). The use of the quinine-derived analogue **56b** furnished the opposite enantiomer of **59** (*ent*-**59**) with the same degree of ee. The conjugate addition adducts could be readily converted into the corresponding monofluoromethylated derivatives **59′** via reductive desulfonylation as a key step without racemization.

Computational studies designed to rationalize the observed sense of stereoselectivity revealed that the chalcone is linked to the 9-OH group of the catalyst by hydrogen bonding. The aromatic π–π interactions between the quinoline ring of catalyst **56a** and chalcone further stabilize the transition-state structure and direct the FBSM to approach from the *Re*-face of the chalcone, affording the *R*-isomer of **59**. The *Si*-face is blocked by the bulky parts of the benzyl substituent of the catalyst (Figure 9.3).

9.2.3
Non-PTC-Catalyzed Enantioselective Michael Addition Reactions

In 2004, Jørgensen and coworkers demonstrated that the Sharpless bis-cinchona alkaloid (DHQ)$_2$PHAL (**60**) could be a promising catalyst for the enantioselective conjugate addition of 1,3-diketones **62** to alkynones **61** [17]. For both aromatic and aliphatic alkynones **61**, the addition of the β-diketones **62** proceeded in high yields and good to excellent enantioselectivity (up to 95% ee), giving the addition products **63** as a mixture of *E/Z*-isomers that can then be selectively isomerized to the more stable *E*-isomer (>50 : 1) by treating it with a catalytic amount of Bu$_3$P (or I$_2$) without affecting the ee values (Scheme 9.20).

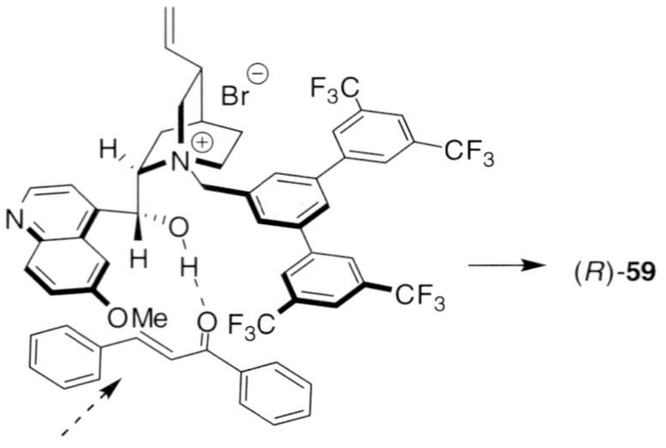

FBSM (58)

Figure 9.3 A proposed transition-state assembly for the enantioselective Michael addition of FBSM (**58**) to **57** catalyzed by **56a**.

In 2006, Deng and coworkers reported that quinine/quinidine-derived catalysts (**64a,b**) bearing a free OH group at the C6′-position and bulky phenanthryl moiety at the 9(O)-position quite efficiently promoted the Michael addition of the α-substituted β-ketoesters **65** to the α,β-unsaturated ketones **66** (Scheme 9.21) [18]. The reaction with as little as 1.0 mol% of catalyst **64** afforded excellent stereoselectivity and chemical yields (up to 98% ee with quantitative yield) for a wide range of both donors and acceptors.

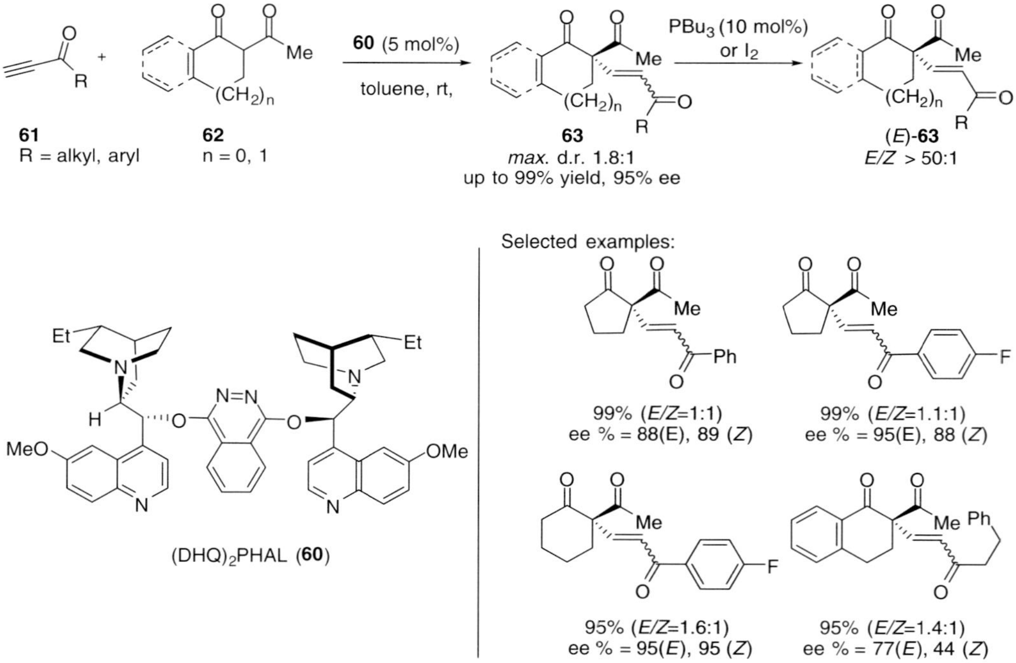

Scheme 9.20

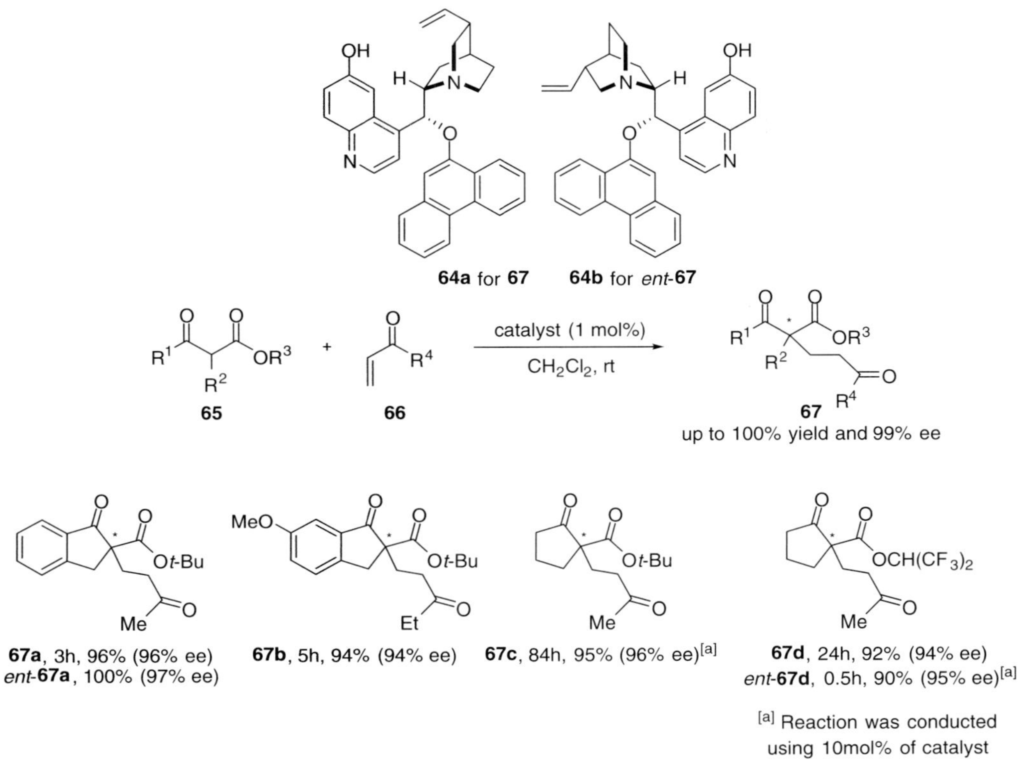

9.2 Conjugate Reaction of α,β-Unsaturated Ketones, Amides, and Nitriles | 263

64a for **67**   **64b** for ent-**67**

65 + **66** → (catalyst (1 mol%), CH$_2$Cl$_2$, rt) → **67**
up to 100% yield and 99% ee

67a, 3h, 96% (96% ee)
ent-**67a**, 100% (97% ee)

67b, 5h, 94% (94% ee)

67c, 84h, 95% (96% ee)[a]

67d, 24h, 92% (94% ee)
ent-**67d**, 0.5h, 90% (95% ee)[a]

[a] Reaction was conducted using 10 mol% of catalyst

Scheme 9.21

In a similar study by Dixon [19], acrylic esters, thioesters, and N-acryloyl pyrrole were also identified as effective acceptors in the Michael addition with β-ketoesters catalyzed by **64b**. Enantiomeric excesses of up to 98% and yields of up to 96% were reported (Scheme 9.22).

Deng also extended the substrate scope of the C6′-hydroxycinchona alkaloid catalysts **64a** and **64b** to the enals **69** in which the latter act as acceptors [20]. It is known that enals rapidly undergo decomposition to form insoluble oligomers or polymers in the presence of nucleophilic catalysts such as DABCO, quinuclidine, and β-isocupreidine (β-ICD), and so on. In contrast to these nucleophilic catalysts, the C6′-hydroxycinchona alkaloids **64** did not promote such polymerization, and, thus, in the presence of **64** (0.1–10 mol%), the wanted Michael addition of the β-keto esters **68** to the enals **69** rapidly went to completion to produce the 1,4-adducts **70** in excellent yields (up to 99%) and stereoselectivity (up to 98% ee and >25 : 1 ds) (Scheme 9.23). Moreover, the reactivity of catalysts **64** was remarkable for this reaction and, thus, the catalytic amount could be reduced to 0.1 mol% without any significant loss of enantioselectivity. All of these results indicate that the alkaloid catalysts **64** act not as nucleophilic catalysts but as general base catalysts.

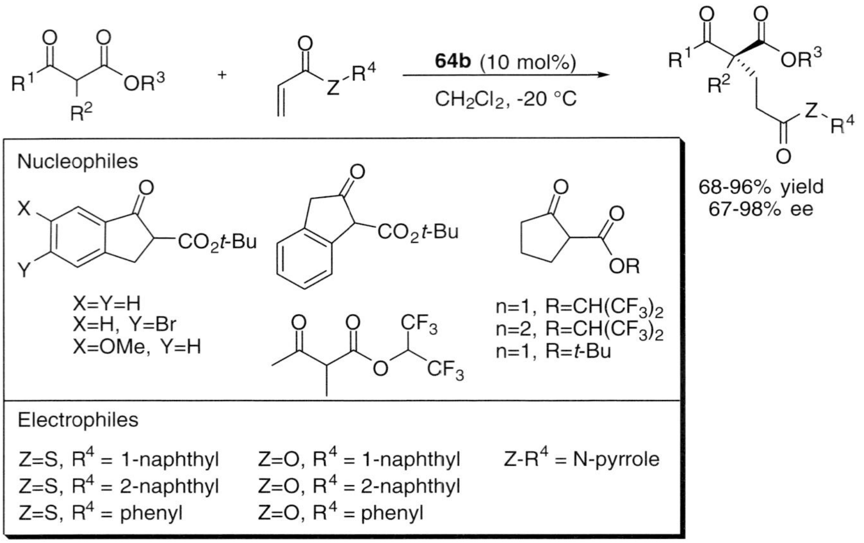

Scheme 9.22

However, the catalysts **64** were found to be ineffective for the α-substituted-α-cyanoacetate donors **72** (up to 48% ee). The same authors overcame this problem by introducing the new 6′-OH cinchona alkaloid catalyst **71** that afforded excellent yields (90–100%) and a synthetically useful level of enantioselectivities (80–95% ee) for conjugate addition to acrolein with the α-cyanoacetates **72** bearing a range of α-aryl and α-heteroaryl groups (Scheme 9.24).

Deng also successfully applied the C6′–OH catalysts **64a,b** to the asymmetric tandem conjugate addition–protonation reactions of the α-cyanoketones or

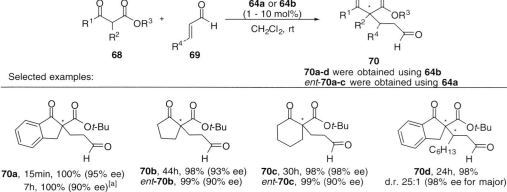

Scheme 9.23

[a] Reaction was conducted using 0.1 mol% of catalyst

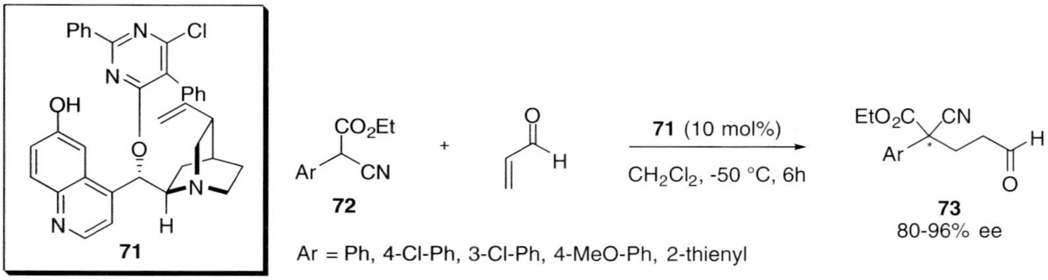

9.2 Conjugate Reaction of α,β-Unsaturated Ketones, Amides, and Nitriles

Scheme 9.24

β-ketoesters **74** to the α-chloro acrylonitrile **75**, creating 1,3-tertiary-quaternary stereocenters [21]. The reactions of various cyclic α-cyanoketones and β-ketoesters **74** with **75** proceeded in 71–95% yield to afford the corresponding adducts **76** in 7–35 : 1 dr, and the major diastereomers were produced in 91–99% ee (Scheme 9.25). This methodology was used by the same group as the stereochemistry-defining step in the total synthesis of manzacidin A (**77**) (Scheme 9.26).

Mechanistically, this catalytic reaction proceeds via enantioselective Michael addition and the subsequent protonation of the transient enol intermediate in a stereoselective manner (Scheme 9.27). Thus, the authors proposed that the catalysts serve as a dual-function catalyst for this tandem reaction; namely, the stereochemical outcome of this tandem reaction resulted from a network of hydrogen-bonding interactions between the catalyst with the reacting donor and acceptor in the addition step and, subsequently, with the putative enol intermediate (**78**) in the protonation step (Scheme 9.28).

In 2005, Chen and coworkers found that the *epi*-cinchonidine/cinchonine-derived thiourea catalysts, **79a,b**, can serve as highly active promoters of the Michael addition of thiophenol to the α,β-unsaturated imide **80**; however, the reaction proceeded with low enantioselectivity (up to 17% ee) (Scheme 9.28) [22].

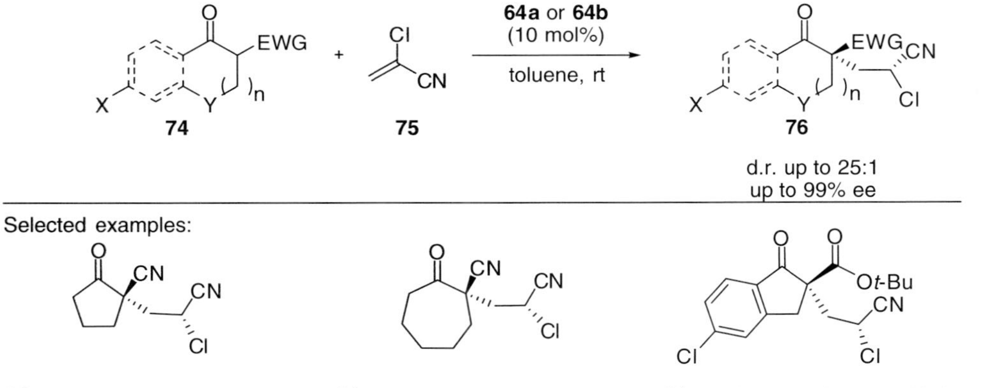

64a: 96h, 89%, d.r. 5:1, 85% ee   **64a**: 72h, 92%, d.r. 25:1, 94% ee   **64a**: 24h, 82%, d.r. 13:1, 98% ee
64b: 60h, 93%, d.r. 7:1, 91% ee   **64b**: 24h, 88%, d.r. 25:1, 95% ee   **64b**: 24h, 81%, d.r. 20:1, 99% ee

Scheme 9.25

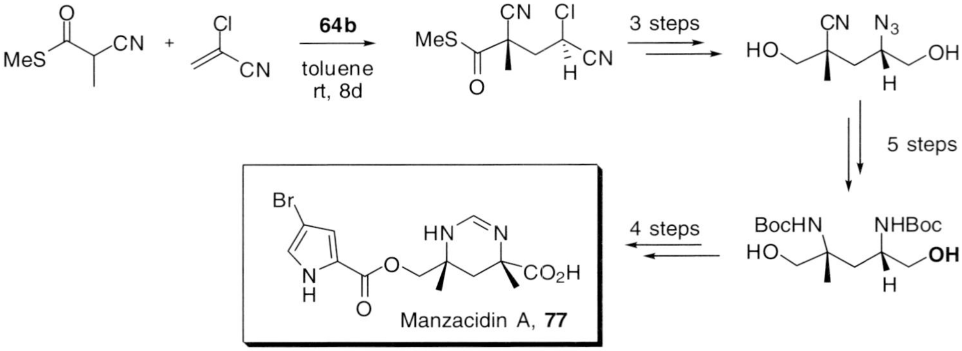

Scheme 9.26

A short time later, Soós and coworkers reported that the *epi*-quinine/quinidine-derived thiourea derivatives **81** and **82** efficiently catalyzed the enantioselective addition of nitromethane to chalcones [23]. Using 10 mol% of the *epi*-dihydroquinine thiourea **81a**, the asymmetric Michael reaction of nitromethane with chalcones produced the Michael adduct **84** in excellent yields and ee values (93%, 96% ee, Scheme 9.29). Moreover, when the reaction temperature was increased to 50–75 °C, the catalyst loading could be reduced to 0.5 mol% without any significant loss of yield or enantiocontrol. The key to this success is their ability to simultaneously activate both the nucleophile and the electrophile and to control their encounter in a well-defined chiral environment. Of note, the stereochemistry of the C9-*epi* catalyst is critical for promoting catalysis. The analogue **83** having the natural cinchona alkaloid stereochemistry exhibited no catalytic activity (quinine itself was also inactive). By employing this methodology, enantioenriched (*R*)-rolipram (**86**) was concisely prepared from the reaction of nitromethane and α,β-unsaturated *N*-acylpyrrole **85** (Scheme 9.30) [23b].

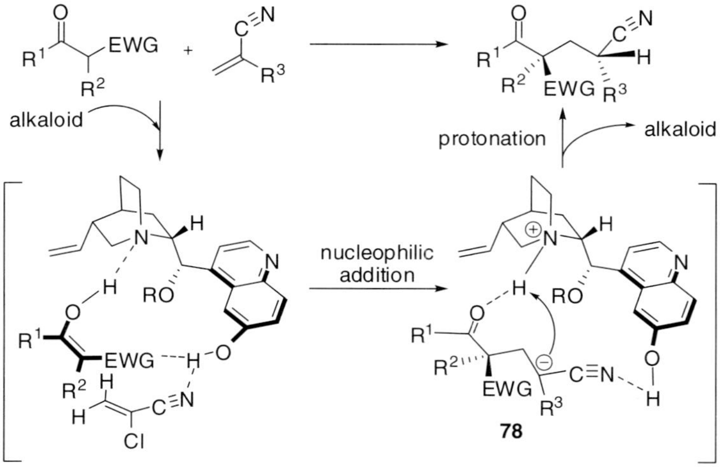

Scheme 9.27

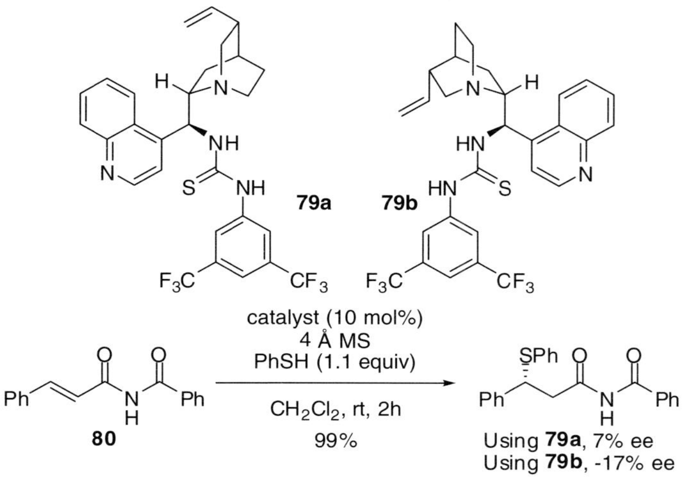

9.2 Conjugate Reaction of α,β-Unsaturated Ketones, Amides, and Nitriles | 267

Scheme 9.28

Reaction of **80** with PhSH (1.1 equiv), catalyst (10 mol%), 4 Å MS, CH$_2$Cl$_2$, rt, 2h, 99%.
Using **79a**, 7% ee
Using **79b**, −17% ee

Soon afterward, a broad spectrum of Michael donors and Michael acceptors were successfully employed in the thiourea-catalyzed Michael addition [24–27]. Selected examples are outlined in Schemes 9.31–9.33.

In 2006, Wang and coworkers reported the asymmetric Michael addition of a broad spectrum of nucleophiles to chalcones (**25**) using the thiourea catalyst **81a** [24].

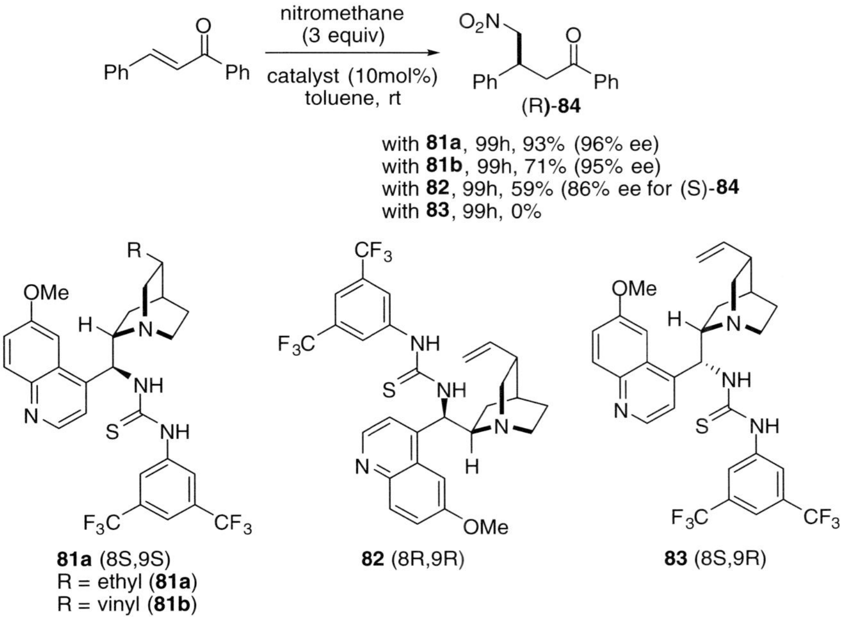

Reaction of chalcone (PhCH=CHC(O)Ph) with nitromethane (3 equiv), catalyst (10 mol%), toluene, rt, giving (R)-**84**.

with **81a**, 99h, 93% (96% ee)
with **81b**, 99h, 71% (95% ee)
with **82**, 99h, 59% (86% ee for (S)-**84**
with **83**, 99h, 0%

81a (8S,9S)
R = ethyl (**81a**)
R = vinyl (**81b**)

82 (8R,9R)

83 (8S,9R)

Scheme 9.29

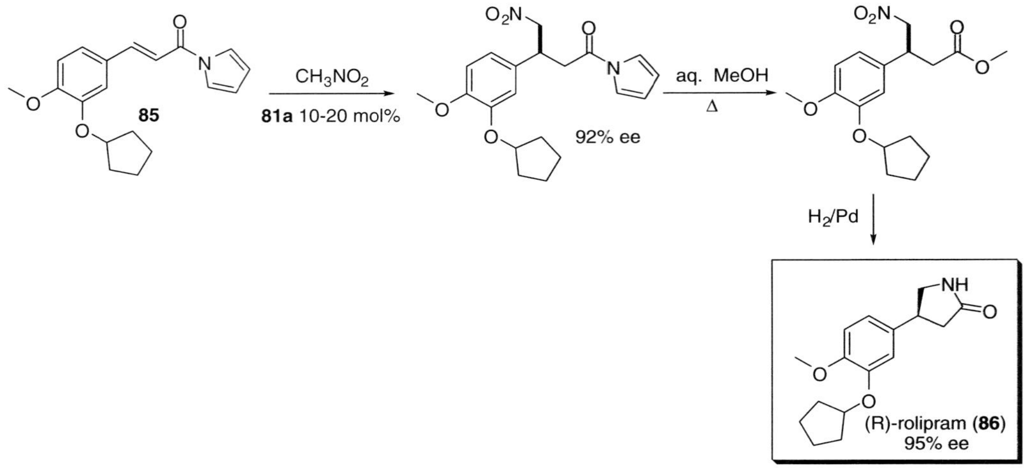

9 Cinchona-Catalyzed Nucleophilic Conjugate Addition to Electron-Deficient C=C Double Bonds

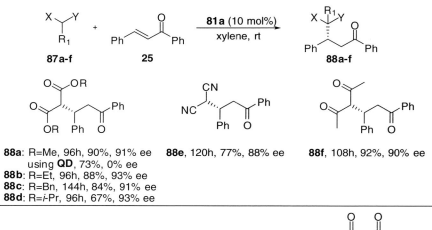

Scheme 9.30

The reactions afforded excellent enantioselectivity (up to 98% ee) and high yields for diverse donors, including malonate esters, ketoesters, 1,3-diketones, nitroesters, and 1,3-dinitriles (**87a–f**). In contrast to the results obtained with thiourea catalysts, the natural quinidine catalyst afforded only the racemic product (Scheme 9.31). The

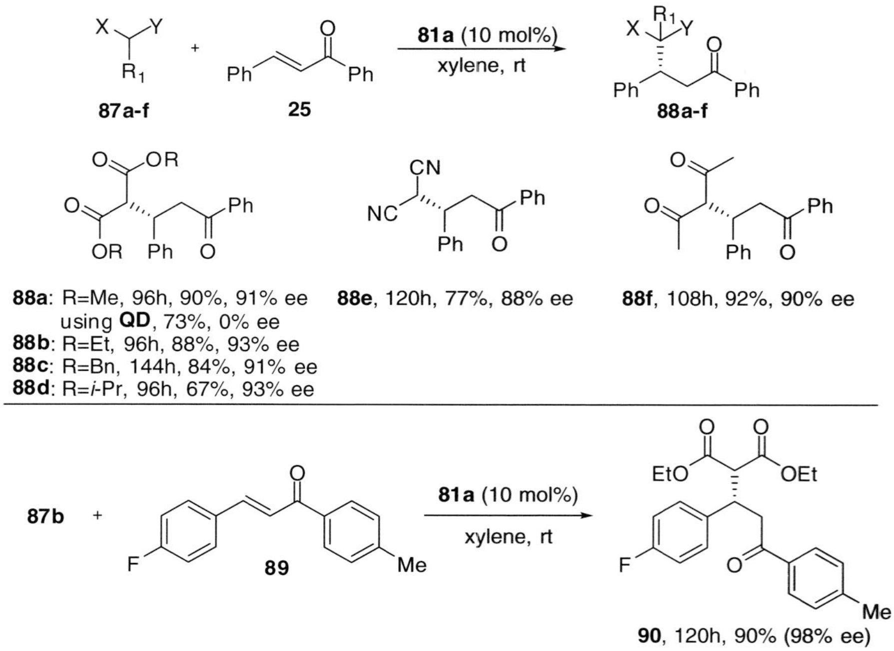

88a: R=Me, 96h, 90%, 91% ee using **QD**, 73%, 0% ee
88b: R=Et, 96h, 88%, 93% ee
88c: R=Bn, 144h, 84%, 91% ee
88d: R=i-Pr, 96h, 67%, 93% ee

88e, 120h, 77%, 88% ee

88f, 108h, 92%, 90% ee

90, 120h, 90% (98% ee)

Scheme 9.31

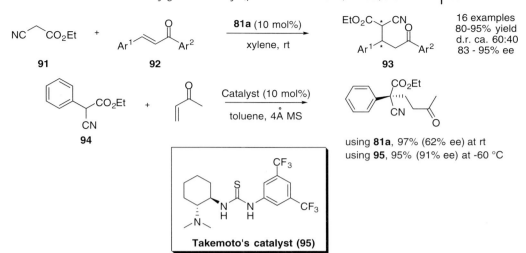

Scheme 9.32

cyano acetates **91** were also examined as nucleophilic partners by Chen and coworkers [25, 26]. In the presence of the cinchona-based thiourea catalyst **81a** and **81b** (10 mol%), ethyl cyanoacetate **91** reacted with various chalcones **92** to afford the Michael adducts **93** in high yields (80–95%) and with high to excellent

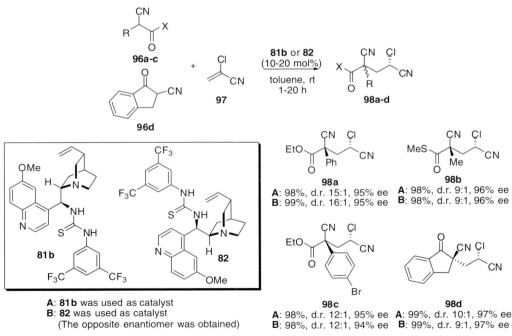

Scheme 9.33

enantioselectivities for both syn/anti-diastereomers (83–95% ee) (Scheme 9.32) [26], while the reaction of the α-substituted cyanoacetate (**94**) with methyl vinyl ketone (MVK) gave only modest enantioselectivity (up to 62% ee) [25]. On the other hand, Takemoto's catalyst **95** was shown to be superior in the reaction of α-substituted cyanoacetate with MVK (91% ee). Deng also investigated the catalytic efficiency of quinine/quinidine-derived thiourea catalysts (**81b,82**) for the asymmetric tandem conjugate addition–protonation reactions of α-cyanoesters (**96a–c**) or α-cyanoketones **96d** with electron-deficient acrylonitrile (**97**), creating 1,3-tertiary-quaternary stereocenters [27]. The reactions with the catalyst (10–20 mol%) afforded high stereoselectivity (9–25 : 1 dr, 94–99% ee) and excellent product yields (94–100%) for a wide range of cyclic α-substituted cyanoketones and acyclic α-substituted cyanoesters (Scheme 9.33).

The thiourea catalyst (**99**, 10 mol%) was also identified by Scheidt and coworkers as a highly efficient catalyst for the enantio- and diastereoselective intramolecular oxo-Michael addition of *ortho*-acyl phenol **100** to give the 3-carboxyflavanones **101** with excellent levels (up to 94% ee) of stereocontrol (Scheme 9.34). The subsequent acid-catalyzed *tert*-butyl ester deprotection/decarboxylation sequence could also be carried out in one pot to give the flavanone derivatives **102** [28]. It was suggested by the authors that the quinuclidine group of the catalyst was able to function as a general base catalyst to deprotonate phenol, while the thiourea group was able to simultaneously activate the β-ketoester moiety by hydrogen bonding. This intramolecular bifunctional mechanism was supported by control experiments; the monofunctional catalyst N,N-bis-trifluoromethylphenyl thiourea failed to promote the reaction. Moreover, the use of either quinine or a combination of quinine and thiourea afforded very low enantioselectivity (below 23% ee).

Wang and coworkers found that the quinine-derived thiourea catalyst **81b** (1 mol%) was also highly reactive and enantioselective for the tandem thio-Michael-aldol reaction of various 2-mercaptobenzaldehydes (**103**) with α,β-unsaturated oxazolidinones (**104**), furnishing benzothiopyranes (**105**) with three stereogenic centers in

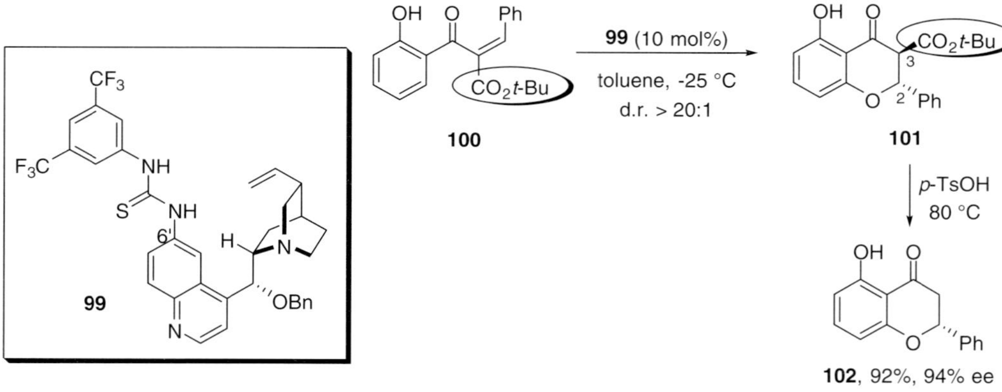

Scheme 9.34

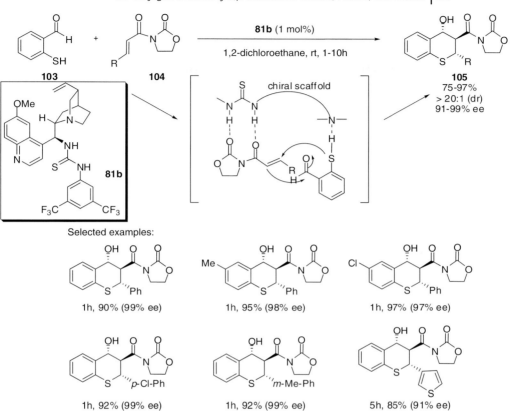

Scheme 9.35

excellent stereoselectivity (>20:1 dr and 91–99% ee) (Scheme 9.35) [29a]. The excellent stereoselectivity can be attributed to multiple hydrogen bonding between the catalyst and the substrate.

Later on, the same group developed a highly stereoselective Michael–Michael cascade process catalyzed by the quinine-based thiourea **81b** [29b]. In the presence of a low catalyst loading (2 mol%), the reaction of *trans*-3-(2-mercaptophenyl)-2-propenoic acid ethyl ester (**106**) with a wide range of nitroalkenes (**107**) afforded thiochromanes (**108**) having three new stereocenters with high stereoselectivity (dr >30:1, up to 99% ee, Scheme 9.36).

It was suggested that this reaction proceeds via the DKR-mediated Michael-retro-Michael–Michael–Michael cascade reaction pathway; the initially formed Michael adduct undergoes a DKR process in the presence of catalyst **81b**, where the deprotonation of the highly acidic proton of **109** by the quinuclidine base of the catalyst leads to a reversible and stereoselective retro-Michael–Michael–Michael process. This proposal was supported by the observation that the reaction of racemic **109** with nitrostyrene under identical conditions depicted in Scheme 9.36

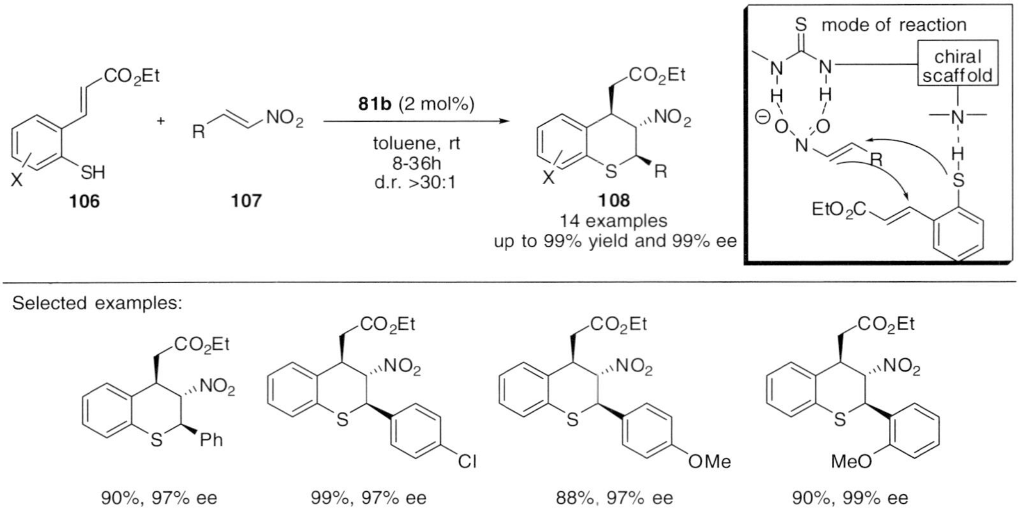

272 | *9 Cinchona-Catalyzed Nucleophilic Conjugate Addition to Electron-Deficient C=C Double Bonds*

Scheme 9.36

Selected examples:
90%, 97% ee; 99%, 97% ee; 88%, 97% ee; 90%, 99% ee

afforded the adduct **110** in 94% yield and excellent stereoselectivity (95% ee, dr>30:1), and with the same configuration as that observed from the direct reaction of **106** with nitrostyrene (Scheme 9.37).

In 2008, Deng developed an efficient asymmetric Michael addition of the aza-Michael donor **112** to α,β-unsaturated ketones by adopting the iminium catalysis approach [30]. In the presence of C9-*epi*-amino cinchona alkaloids (**111**, **112a,b**) and acid cocatalyst (20–40 mol%), the addition of **114** to a variety of alkyl- and aryl-substituted α,β-unsaturated ketones **113** proceeded smoothly via the iminium intermediate and afforded the corresponding adducts in good to outstanding enantioselectivity (up to 99% ee) (Scheme 9.38). However, excessively long reaction times were usually required at ambient temperature. Heating the reaction to 40 °C resulted in more acceptable rates without any major erosion of the yields or enantioselectivity.

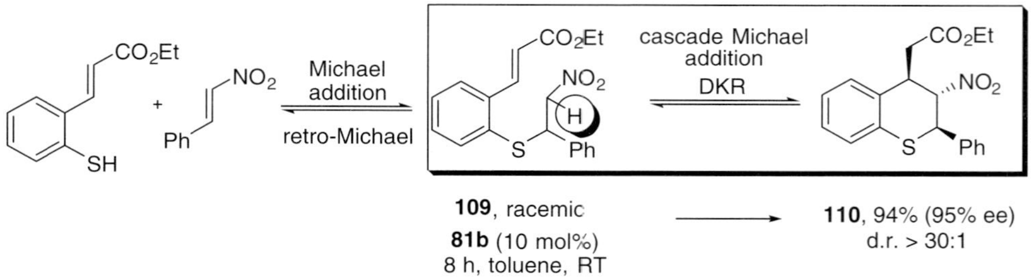

Scheme 9.37

9.2 Conjugate Reaction of α,β-Unsaturated Ketones, Amides, and Nitriles

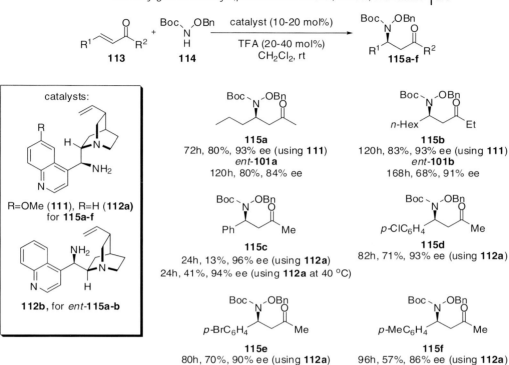

Scheme 9.38

Chen and coworkers also successfully applied the TFA salt of the 9-*epi*-amino-9-deoxycinchona alkaloids **111** as a chiral organocatalyst to domino Michael–Michael reactions [31]. A variety of α,β-dicyanoalkenes acting as Michael donors underwent the reaction with the α,β-unsaturated ketones **117** (see also Section 10.4). For example, the highly enantioselective desymmetrization of prochiral α,α-dicyanoalkenes (**116**) from 4-substituted cyclohexanones was achieved via tandem Michael–Michael addition reactions with α,β-unsaturated ketones (**117**), giving the bicyclic products **118** with two new C–C bonds and four stereogenic centers including one quaternary carbon center in a single operation with a high level of stereoselectivity (dr >99 : 1, up to >99% ee, Scheme 9.39).

In 2008, Ye and coworkers also developed a new type of multifunctional cinchonidine-based catalyst, such as **119** having an additional primary amine moiety, for the Michael addition of nitroalkane to cyclic enones [32]. In the presence of an acid cocatalyst, the primary amine moiety of **119** can act as a Lewis base to activate the Michael acceptor via iminium formation. The catalysts **119a** and **119b** (5 mol%) provided quite excellent enantioselectivity (up to 98% ee) for the Michael addition of nitroalkanes to cyclohexenone (Scheme 9.40). The observed retardation of the reaction rate and the opposite sense of enantioselectivity obtained with the catalyst **119b** indicated the importance of the configuration of the cyclohexane

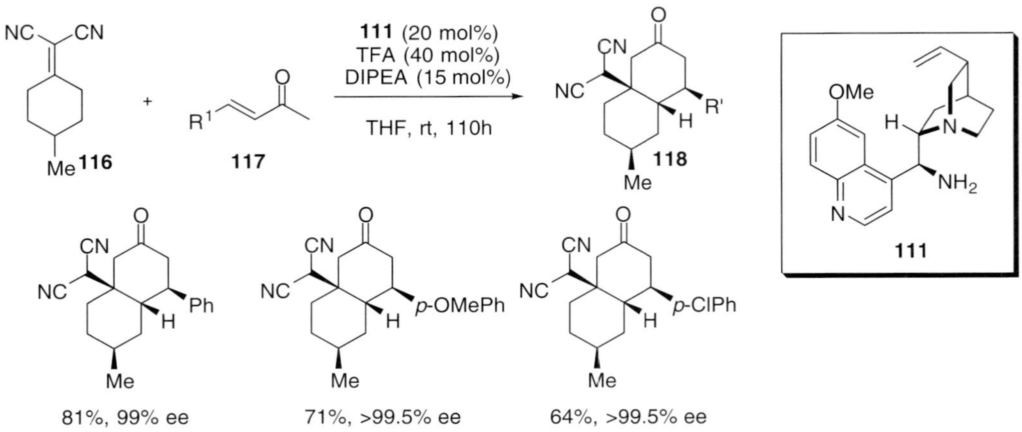

Scheme 9.39

diamine for the asymmetric catalysis in this case. The reaction with a nitroethane also proceeded in high enantioselectivity, albeit with a low diastereoselectivity (**121**).

9.3
Conjugate Addition of Nitroalkenes

The nitro group is a powerful electron-withdrawing substituent, and this property dominates the chemistry of all molecules containing this functional group. Especially, nitroalkenes, being markedly electron deficient, readily undergo addition reactions with many different nucleophiles. Recently, a number of observations have been made of asymmetric conjugate additions mediated by a cinchona alkaloid-derived catalyst that afford high enantioselectivity and diastereoselectivity, in which nitroalkenes are involved as a Michael acceptor.

In 2004, Deng and coworkers reported the first preparatively useful results for the catalytic asymmetric Michael additions of malonates to nitroolefins. They found that

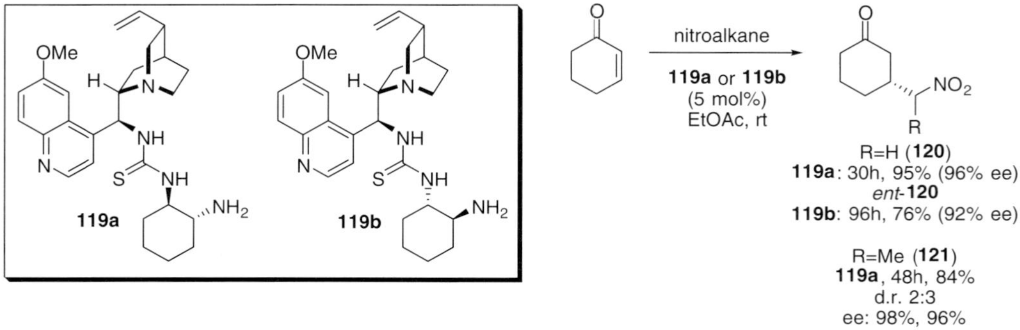

Scheme 9.40

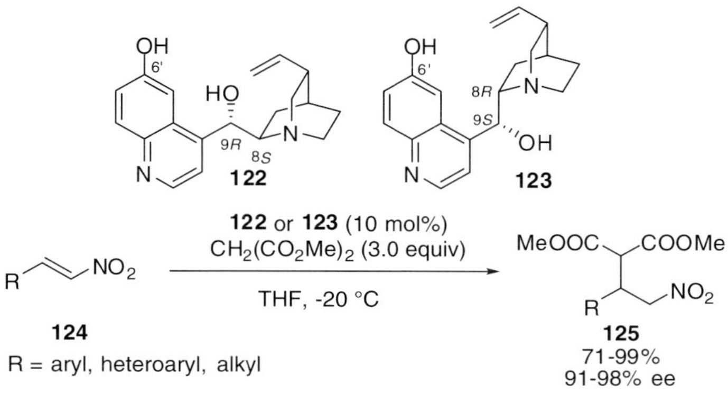

9.3 Conjugate Addition of Nitroalkenes

Scheme 9.41

readily accessible 6′-demethylated quinine/quinidine, **122** and **123**, respectively, could serve as highly effective promoters of asymmetric Michael addition processes [33]. In the presence of either **122** or **123** (10 mol%), a range of nitroalkenes **124** bearing aryl, heteroaryl, and alkyl groups with varying electronic and steric properties underwent the reaction with dimethyl malonate, affording the conjugate adducts **125** in good to excellent yields (71–99%) and ee values (91–98%) (Scheme 9.41). However, the parent alkaloids, quinine, and quinidine, provided much lower reaction rates and enantioselectivities (e.g., 16% ee from the reaction of dimethyl malonate with the *trans*-phenyl nitroalkene in the presence of quinidine). This result indicates that, together with the quinuclidine nitrogen, the phenolic OH group plays an important role in the stabilization and organization of the transition-state assembly of this reaction.

Soon afterward, the same group demonstrated that the quinidine- and quinine-derived catalysts, **122** and **123**, respectively, also accept an exceptionally wide range of trisubstituted carbon Michael donors **126** [34]. Outstanding diastereoselectivity (>98 : 2) and quantitative enantioselectivity (exceeding 99% ee) were obtained with most of the trisubstituted carbon donors that were examined **126** (e.g., cyclic **126a** and acyclic β-ketoesters **126b,c**, 2-substituted 1,3-diketones **126d**, α-nitroesters **126e**, and α-cyanoesters **126f** (Scheme 9.42). Of note, β-isocupreidine (β-ICPD) **128**, a conformationally rigid analogue of **122**, was also identified as an efficient catalyst for the same reactions and, interestingly, exhibited remarkably similar dr and ee profiles to those obtained using **122**. In the light of these observations, the authors suggested that the catalyst **122** or **123** might exist as a gauche-open active conformer in the transition state (Figure 9.4).

A range of other Michael donors such as nitroalkanes [35], anthrone [36], and *N*-heterocycles [37] were also shown to be applicable to the above-discussed catalyst system, that is, catalysts bearing a 6′-OH group such as **122** and **123**. The enantioselective conjugate addition reaction of a range of cyclic and acyclic nitroalkanes **129** with aromatic and heteroaromatic nitroolefins **130** under neat conditions afforded the synthetically useful 1,3-dinitro compounds **131** in good yields (70–82%) and with good degrees of enantioselectivity (67–88% ee) using **122** (10 mol%)

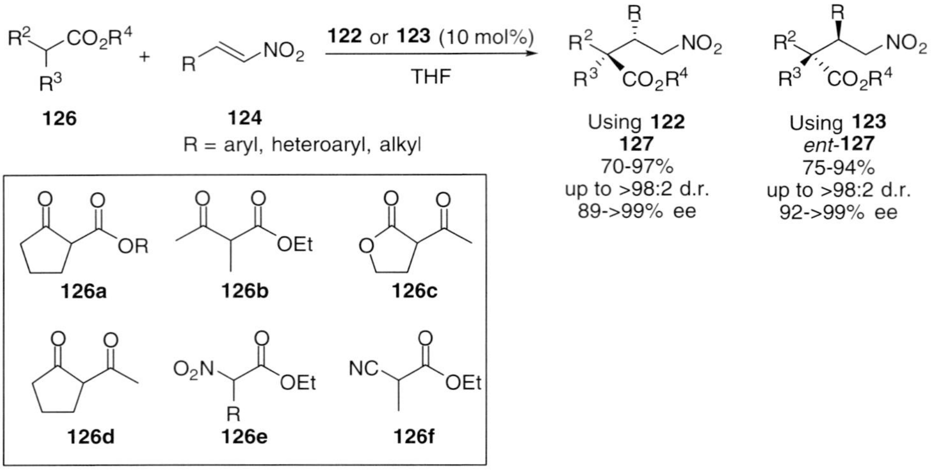

Scheme 9.42

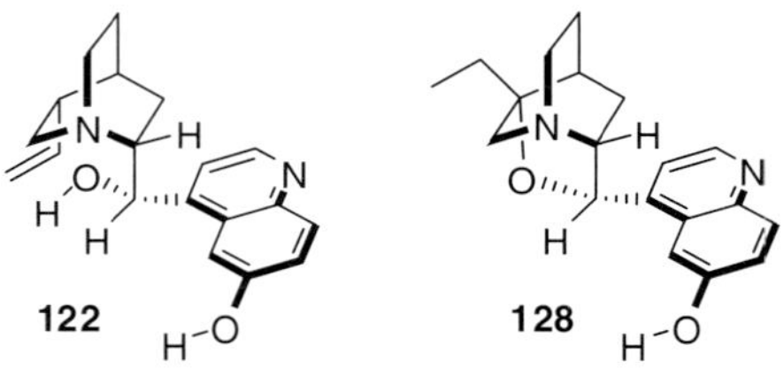

[Scheme showing reaction of 126 + 124 with 122 or 123 (10 mol%) in THF giving 127 (using 122: 70-97%, up to >98:2 d.r., 89->99% ee) and ent-127 (using 123: 75-94%, up to >98:2 d.r., 92->99% ee). R = aryl, heteroaryl, alkyl. Substrates shown: 126a, 126b, 126c, 126d, 126e, 126f.]

Figure 9.4 Gauche-open conformer of **122** and the structure of **128**.

(Scheme 9.43) [35]. However, no reaction occurred in the case of less reactive aliphatic nitroolefins. The use of anthrone **133** as the nucleophilic component in the presence of the O-benzoylated derivative **132** (5 mol%) of **122** also provided excellent yields (91–99%) and enantiomeric excesses (up to 99%) (Scheme 9.44) [36]. The Michael additions of an N-heterocycle as a nitrogen nucleophile to an electron-deficient nitroalkene also took place using catalyst **122** (10 mol%). A range of N-heterocycles (triazole and tetrazole) and nitroolefins could be employed in this process, producing the adducts with good to high enantiomeric excesses (up to 94% ee) (Scheme 9.45) [37].

In 2005, the groups of Connon [38] and Dixon [39] independently reported that *epi*-cinchona-based (thio)urea derivatives can serve as excellent bifunctional organocatalysts

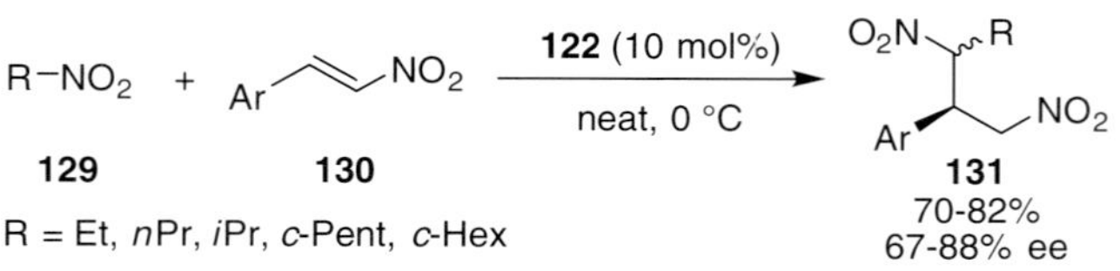

Scheme 9.43

R–NO$_2$ (**129**) + Ar–NO$_2$ (**130**) → (with **122** (10 mol%), neat, 0 °C) → **131** (70-82%, 67-88% ee)

R = Et, nPr, iPr, c-Pent, c-Hex

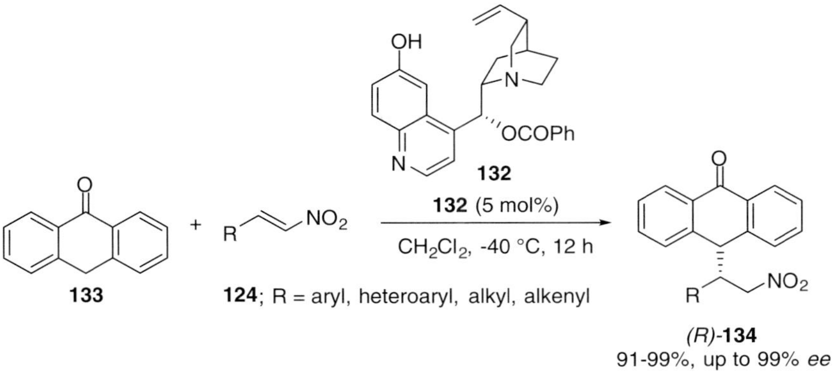

Scheme 9.44

for the asymmetric addition of malonates to a variety of nitroalkenes. Using 2–5 mol% of the *epi*-dihydroquinine- and *epi*-dihydroquinidine-derived thiourea derivatives, **81a** and **81a′**, respectively, dimethyl malonate allowed the conversion of both electronically activated and deactivated β-nitrostyrenes **130** to give the enantioselective 1,4-adducts **137** in high yield (91–95%) and excellent enantioselectivity (87–99% ee) (Scheme 9.46). The high activity of the thiourea catalyst was highlighted by the synthesis of the thienyl nitroalkane **137** (Ar = 2-thienyl) in 92% yield and 94% ee using as little as 0.5 mol% of the catalyst under mild reaction conditions (0 °C, 46 h). However, aliphatic nitroolefins afforded slightly lower enantioselectivities (75–86% ee). Of note, the 9-*epi* catalyst stereochemistry is critical for obtaining high levels of enantioselectivity. Analogues having the natural cinchona alkaloid stereochemistry gave much lower ee values. The synergistic cocatalytic effect of the thiourea moiety and the quinuclidine nitrogen of **81a** and **81a′** was proposed.

Soon afterward, various types of carbon [40–44], oxygen [45], and phosphorous [46] Michael donors were successfully employed in the thiourea-catalyzed addition to nitroalkenes. In the presence of the bifunctional *epi*-9-amino-9-deoxy cinchonine-based thiourea catalyst **79a**, the 5-aryl-1,3-dioxolan-4-ones **138** bearing an acidic α-proton derived from mandelic acid derivatives and hexafluoroacetone were identified by Dixon and coworkers as effective pronucleophiles in diastereo- and enantioselective Michael addition reactions to nitrostyrenes **124** [40]. While the diastereoselectivity obtained exceeded 98%, the enantiomeric excess recorded

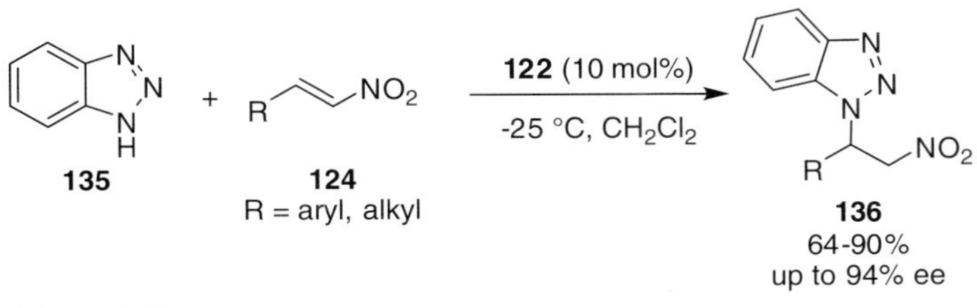

Scheme 9.45

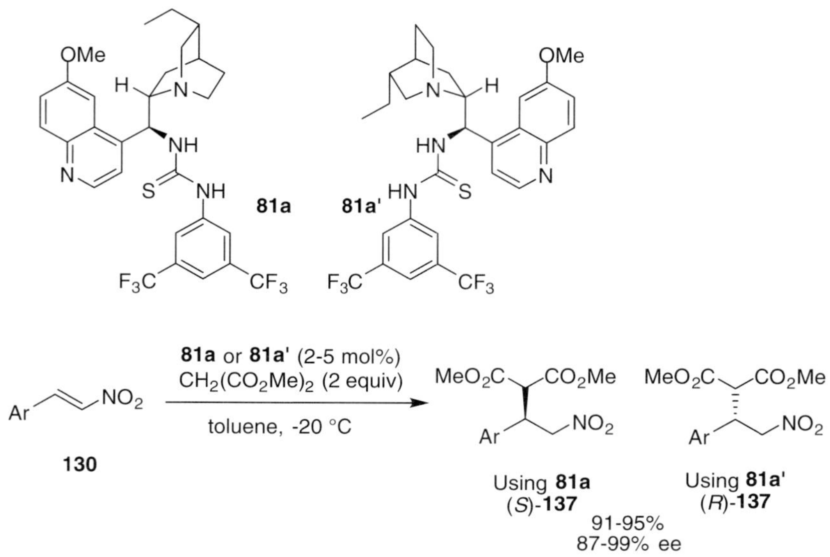

Scheme 9.46

was in the range of 60–89% (Scheme 9.47). The obtained adducts **139** can be easily converted to the chiral α-hydroxy acid derivatives.

In 2008, Jørgensen and coworkers reported that oxazolones **140** could also smoothly undergo addition to nitroalkenes **124** in the presence of bifunctional cinchona-based thiourea organocatalysts such as **81a** and **81a'**. The reaction with 4-phenyloxazolones (**140**, R^1 = Ph) took place at C4 affording the adducts **141** with moderate to good ee values (up to 83% ee). On the other hand, 4-alkyloxazolones (R^1 = Me, i-Bu) were added to the C2-position of **140**, furnishing **142** with a moderate to high enantioselectivity (up to 92% ee) (Scheme 9.48). The adducts **141** can be used

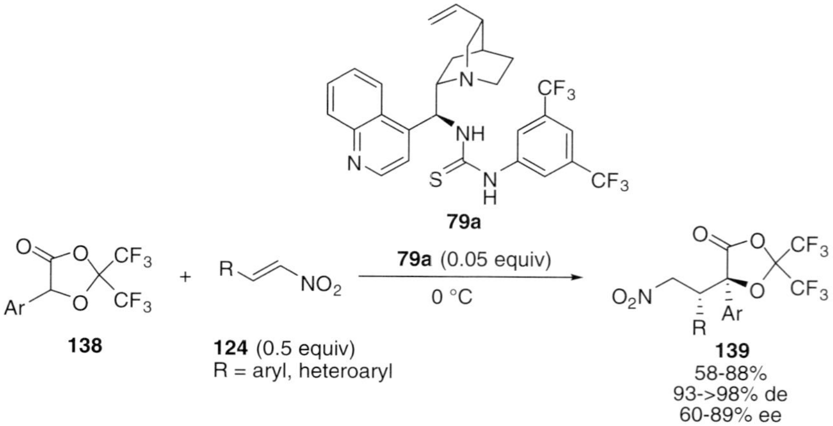

Scheme 9.47

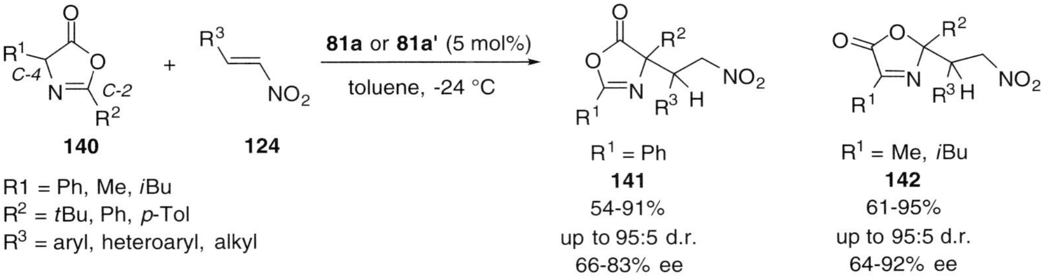

9.3 Conjugate Addition of Nitroalkenes | 279

Scheme 9.48

140: R^1 = Ph, Me, iBu; R^2 = tBu, Ph, p-Tol; R^3 = aryl, heteroaryl, alkyl

124 + 140 → 81a or 81a' (5 mol%), toluene, −24 °C

141: R^1 = Ph, 54–91%, up to 95:5 d.r., 66–83% ee

142: R^1 = Me, iBu, 61–95%, up to 95:5 d.r., 64–92% ee

as precursors for the synthesis of optically active α,α-disubstituted α-amino acids [41].

The *epi*-quinine urea **81b** was also found by Wennemers to promote an asymmetric decarboxylation/Michael addition between thioester **143** and **124** to afford the product **144** in good yield and high enantioselectivity (up to 90% ee) (Scheme 9.49). Here, malonic acid half-thioesters serve as a thioester enolate (i.e., enolate Michael donors). This reaction mimics the polyketide synthase-catalyzed decarboxylative acylation reactions of CoA-bound malonic acid half-thiesters in the biosynthesis of fatty acids and polyketides. The authors suggested, analogously with the enzyme system, that the urea moiety is responsible for activating the deprotonated malonic acid half-thioesters that, upon decarboxylation, react with the nitroolefin electrophile simultaneously activated by the protonated quinuclidine moiety (Figure 9.5) [42].

The use of naphthols **145** as the carbon nucleophilic reaction component in Friedel–Crafts type Michael addition reactions was also reported in 2007 by Chen and coworkers [43]. In this system, the pronucleophile is activated by the quinuclidine unit of bifunctional cinchona-based thiourea catalysts such as **81a'**. A range of aryl- and alkyl-substituted nitroalkene derivatives **124** were applicable to this system. The corresponding adducts **146** were obtained with 85–95% ee at low temperature

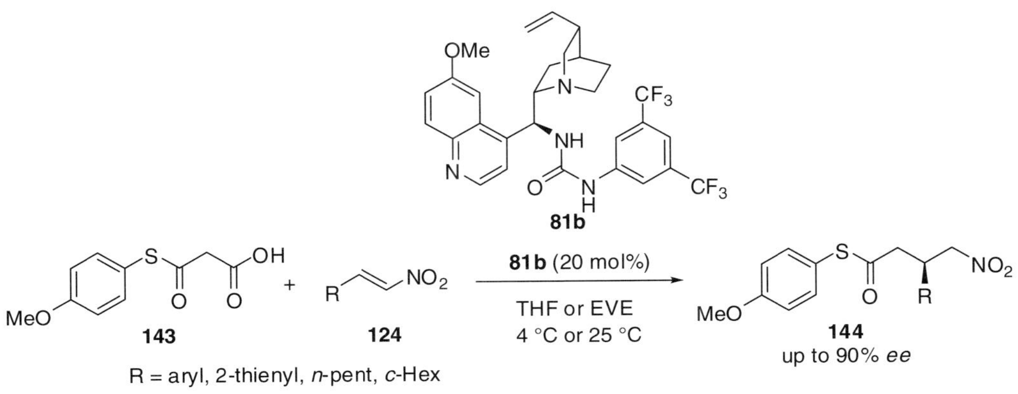

143 + 124 → 81b (20 mol%), THF or EVE, 4 °C or 25 °C → 144, up to 90% ee

R = aryl, 2-thienyl, n-pent, c-Hex

Scheme 9.49

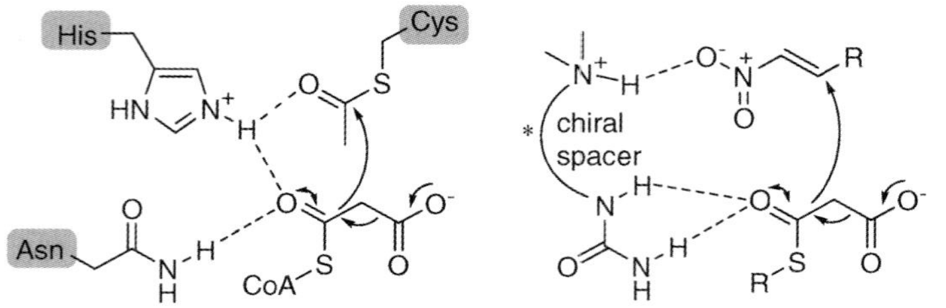

Figure 9.5 Left: activation of MAHT in the active site of polyketide synthases; right: schematic representation of the catalyst design.

($-50\,°C$) along with the dimerization products **147**, which become the major product when the reaction time is extended (Scheme 9.50).

In 2006, Scheidt and coworkers [44] reported the first enantioselective direct nucleophilic addition of the silylated thiazolium salt **148**, a precursor of the equivalent acyl anion, to nitroalkene **149** in the presence of tetramethylammonium fluoride (TMAF) and stoichiometric amounts of quinine-based thiourea **81b**, producing the chiral β-nitroketone **150** in 67% yield and with 74% ee (Scheme 9.51). The acyl anion equivalent **152** can be generated by the desilylation of **148** with TMAF, followed by the 1,2-H shift of the resulting alkoxide **151**. The observed asymmetric induction indicates that there is a strong interaction between the thiourea and the nitroalkene during the carbonyl anion addition step.

In 2007, Jørgensen and coworkers demonstrated that the bifunctional thiourea-cinchona alkaloid catalysts **81b** also promoted the enantioselective addition of oximes **153** as oxygen nucleophiles to nitroolefins **124** giving the adduct **154** in good yield with a high level of enantioselectivity (Scheme 9.52) [45]. The obtained adduct **154** can be converted to the optically active aliphatic nitro- or aminoalcohols. It is believed that the

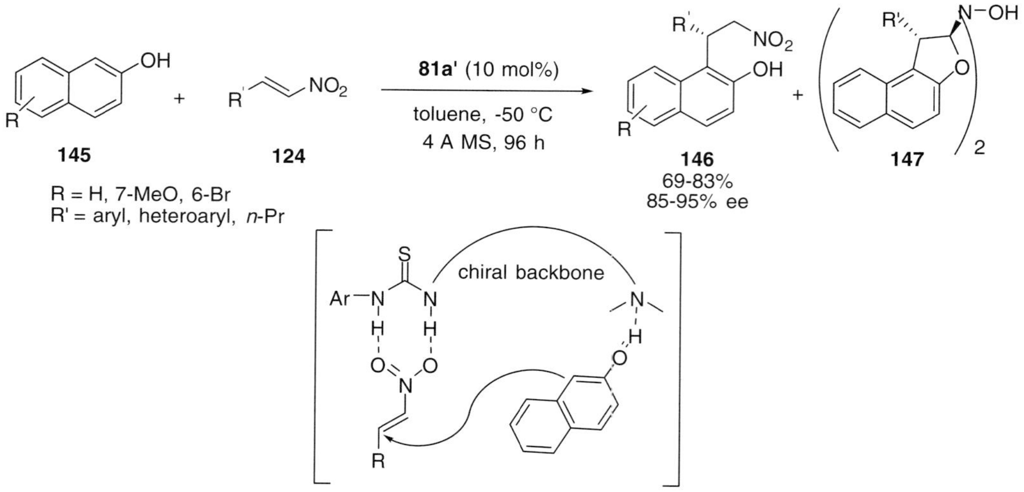

Scheme 9.50

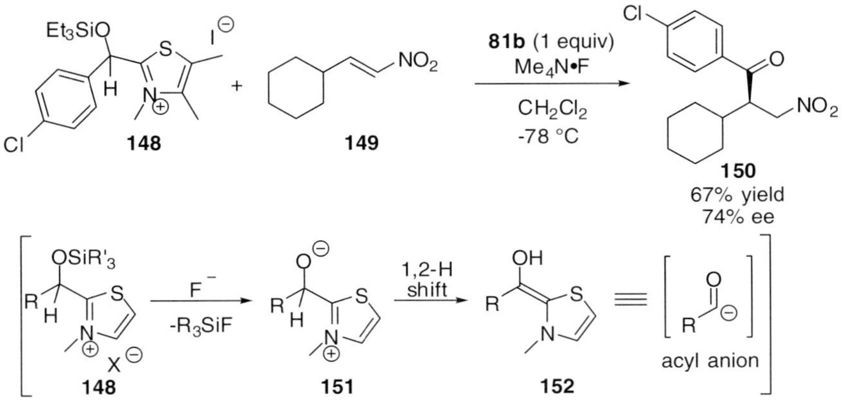

Scheme 9.51

thiourea-cinchona alkaloids bring about the simultaneous activation of both the oxime and the nitroalkene, through hydrogen-bonding interactions. However, the scope of this method is limited to nitroolefin substrates with aliphatic substitutents. Nitrostyrenes are not applicable due to the existence of a competitive retro-Michael process.

The hydroquinine thiourea catalyst 81a was also identified by Melchiorre and coworkers as a promising catalyst for the asymmetric addition of secondary phosphines 155 to aromatic nitroalkenes 130 furnishing 156 (obtained after protection with *in situ* formed borane) [46]. However, the obtained ee values were only moderate (up to 66% ee) (Scheme 9.53).

As described above, cinchona-based (thio)ureas have proven to be highly efficient H-bond donor catalysts. In 2008, Rawal and coworkers developed a highly promising new family of cinchona-based H-bond donor catalysts such as 157 by replacing the thiourea moiety of cinchona-based thiourea catalysts with the squaramide unit [47]. The squaramide moiety of 157 is able to form two H-bonds to a reactant due to the more accessible reaction site and fixed *syn*-orientation of the NH-protons. Using only 0.5 mol% of the cinchonine-derived squaramide catalyst 157, various Michael donors 158 and nitroalkenes 130 were smoothly converted to the desired adducts 159 in excellent yield and ee values (up to 99% ee) (Scheme 9.54).

In 2007, Connon and McCooey developed highly efficient, asymmetric *syn*-selective addition reactions of enolizable carbonyl compounds to nitroolefins by adopting the enamine catalysis approach [48]. The 9-*epi*-amino cinchona alkaloid derivative (160, 9-*epi*-DHQDA) as an aminocatalyst promoted the addition of a variety

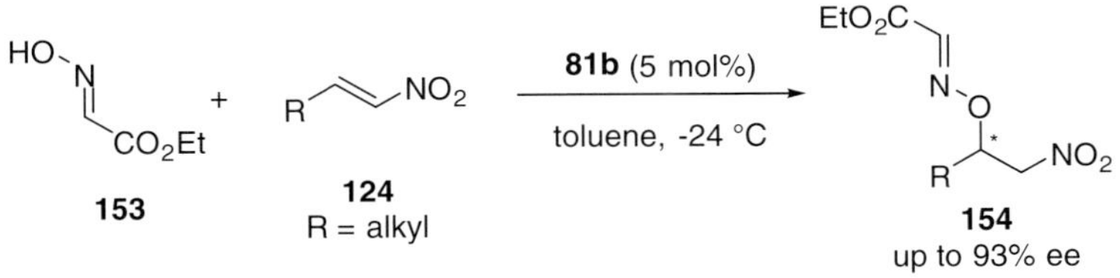

Scheme 9.52

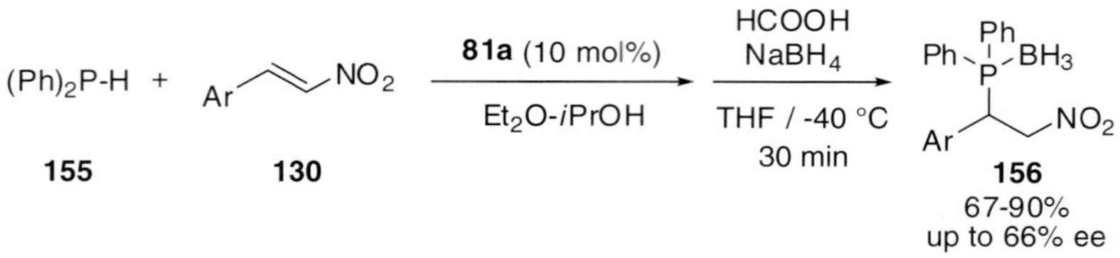

Scheme 9.53

(Ph)₂P-H + Ar–CH=CH–NO₂ → [81a (10 mol%), Et₂O-iPrOH; HCOOH, NaBH₄, THF / -40 °C, 30 min] → Ph–P(Ph)(BH₃)–CH(Ar)–CH₂–NO₂

155 **130** **156**
67-90%
up to 66% ee

of ketones **161** (acyclic/cyclic) to **124** to give the *syn* adducts **162** in good to outstanding enantio- and diastereoselectivity (Scheme 9.55). The reaction of acyclic ketones with **124** gave relatively higher stereoselectivities than cyclic ketones.

By employing the primary amine catalyst **160**, Zhong and coworkers developed the tandem Michael–Henry reaction of ketones with nitroalkenes to provide highly functionalized chiral hexanes and pentanes with high diastereo- and enantioselectivity [49]. The selected examples depicted in Scheme 9.56 show that, in the presence of **160** (10–15 mol%), various Michael donors and nitroalkenes smoothly underwent the tandem reaction with almost quantitative yield and extremely high enantioselectivity with the complete diastereoselectivity of the products. Further details of this reaction can be seen in Section 10.4.

Sharpless bis-cinchona alkaloids such as [DHQD]₂PYR (**163a**) have proved to serve as highly efficient catalysts for the asymmetric vinylogous Michael addition of the electron-deficient vinyl malonitriles **164** as the nucleophilic species to nitroolefins **124** [50]. This process exhibited exclusive γ-regioselectivity and high diastereo- and enantioselectivity. Only the *anti*-products **165** were observed in all reactions (Scheme 9.57). Of note, 1-tetralone did not react with nitroolefins under these

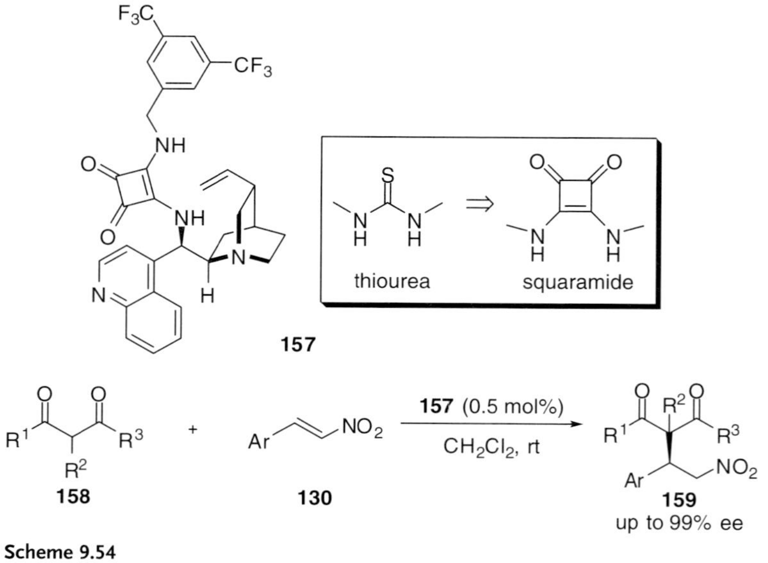

157

R¹–C(O)–CR²(R³)–C(O)– **158** + Ar–CH=CH–NO₂ **130** →[157 (0.5 mol%), CH₂Cl₂, rt]→ R¹–C(O)–C(R²)(R³)–C(O)–CH(Ar)–CH₂–NO₂ **159**
up to 99% ee

Scheme 9.54

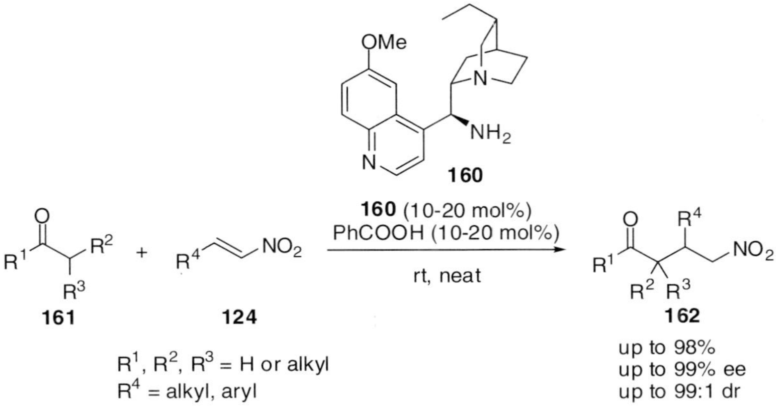

9.3 Conjugate Addition of Nitroalkenes | 283

Scheme 9.55

R¹, R², R³ = H or alkyl
R⁴ = alkyl, aryl

conditions, which implies that malonitriles can serve as a masking group and as an activating group.

The bis-cinchona alkaloid, [DHQD]$_2$PYR (**163a**, 20 mol%), was also employed by Jørgensen and coworkers for the asymmetric conjugate addition of the azide **166** to the α,β-unsaturated nitro compounds **124** to obtain the 1,2-diamino scaffold from the 2-azido-1-nitro compounds **167** that were formed [51]. (DHQD)$_2$PYR (**163a**) provided a good yield, but only a moderate level of enantioselectivity (up to 62% ee) (Scheme 9.58). It was also found that the proton donor that was added had a major influence on the reaction course, not only on the enantiomeric excess but also on the face-selectivity.

Natural cinchona alkaloids were recently examined as catalysts for Michael addition reactions. In the presence of a natural cinchona alkaloid such as cinchonine (**CN**), 2-hydroxy-1,4-naphthoquinone **168** as the nucleophilic component was added

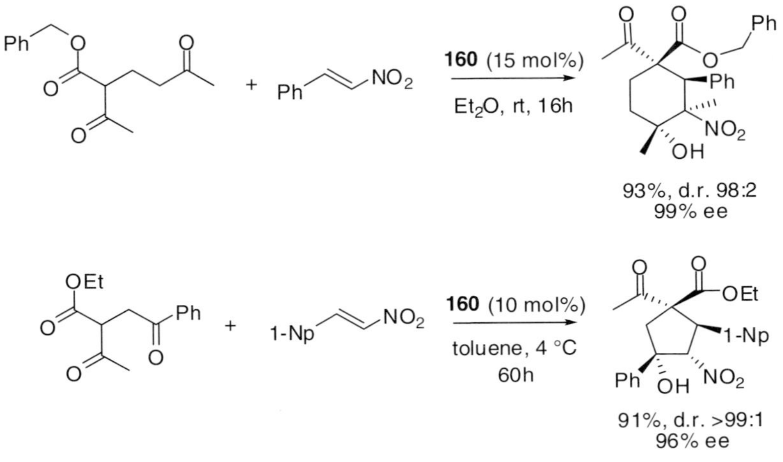

Scheme 9.56

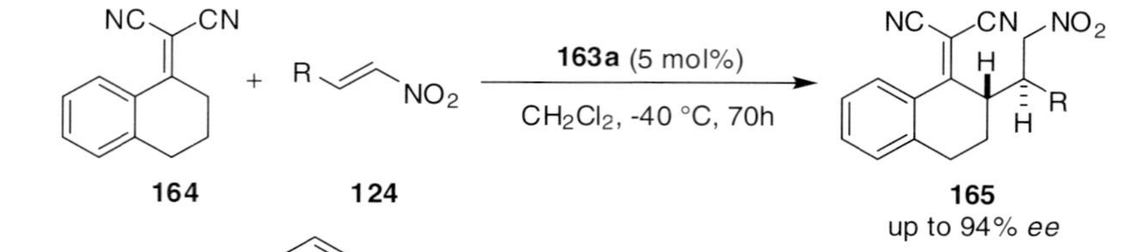

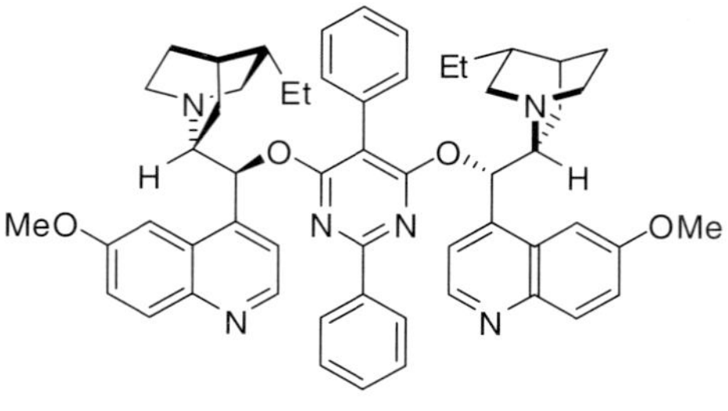

Scheme 9.57

R	Yield(%)	ee (%)
4-BrC$_6$H$_4$	90	91
2-thiophenyl	93	94
4-MeOC$_6$H$_4$	93	94

Selected example

[DHQD]$_2$PYR (163a)

to nitrostyrene **169** affording the adduct **170** with modest enantioselectivities (up to 64% ee) (Scheme 9.59) [52]. Quinine (**QN**) catalyzed the enantioselective Michael addition of diphenyl phosphate (**171**) to aromatic nitrostyrenes **124** in xylene affording the adducts **172** in good yields and with good to high enantioselectivities (up to 88% ee) [53]. The obtained adducts **172** can be chiral precursors for the synthesis of α-substituted β-aminophosphonates. However, relatively low ee values (45–63% ee) were also observed for aliphatic nitroalkenes (Scheme 9.60).

9.4
Conjugate Addition of Vinyl Sulfones and Vinyl Phosphates

The Michael addition of carbon nucleophiles to vinyl sulfones is a useful reaction for synthetic and medicinal chemistry. In 2005, Deng and coworkers reported the first catalytic, asymmetric conjugate additions of α-substituted α-cyanoacetates to vinyl sulfones using the cinchona alkaloid-derived catalysts **64a** and **64b** [54]. In the presence of 20 mol% of **64a** or **64b**, the α-cyanoacetates **173** with a range of aryl and heteroaryl groups of varying electronic and steric properties were smoothly added to phenylvinyl sulfone **174** (Ar = Ph), providing the conjugate adducts **175** in excellent enantioselectivity (up to 97% ee) and good to excellent yields (Scheme 9.61).

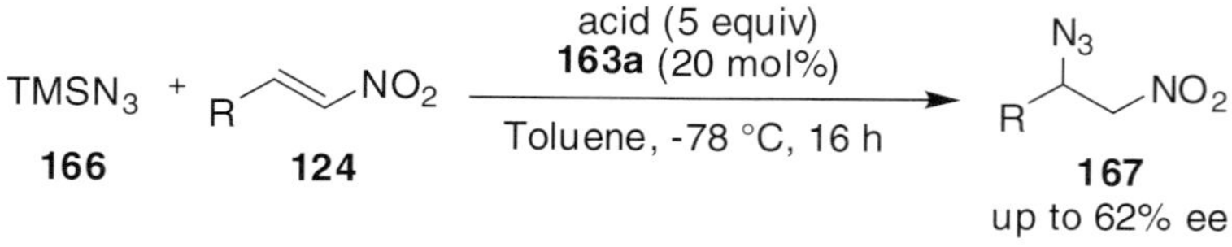

acid: AcOH or 2,4,6-trimethoxybenzoic acid

Scheme 9.58

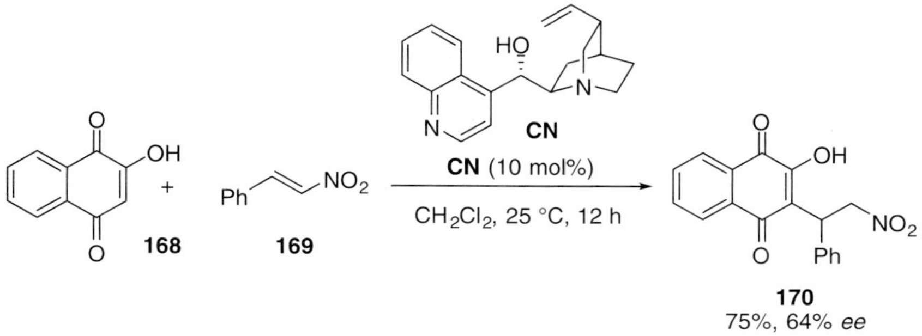

9.4 Conjugate Addition of Vinyl Sulfones and Vinyl Phosphates | 285

Scheme 9.59

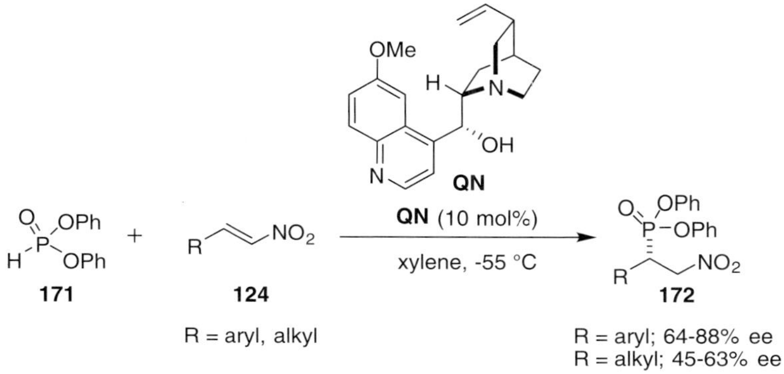

Scheme 9.60

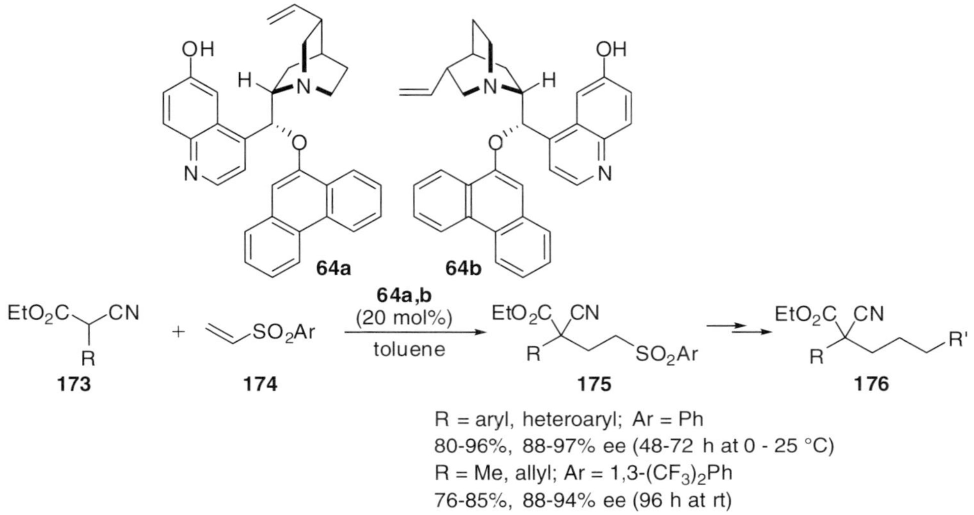

R = aryl, alkyl

R = aryl; 64–88% ee
R = alkyl; 45–63% ee

Scheme 9.61

R = aryl, heteroaryl; Ar = Ph
80–96%, 88–97% ee (48–72 h at 0 – 25 °C)
R = Me, allyl; Ar = 1,3-(CF$_3$)$_2$Ph
76–85%, 88–94% ee (96 h at rt)

However, compared to their aryl congeners, α-alkyl α-cyanoacetates were significantly less active as Michael donors, resulting in very poor yields. This problem was overcome by introducing an electron-withdrawing substituent on the aromatic ring of **174** to enhance the electrophilicity of the vinyl sulfones. With 3,5-bis-(trifluoromethyl)phenyl vinyl sulfone **174** (Ar = 1,3-$(CF_3)_2$Ph), 100% conversion after 96 h at room temperature was achieved with good to excellent enantioselectivity (up to 94% ee). The 1,4-adduct was applied to the asymmetric synthesis of the biologically significant α,α-disubstituted amino acids **176**.

In 2006, Chen and coworkers reported that cinchona-based thioureas (**79a** or **81b**) serve as catalysts for the Michael addition of α-phenyl cyanoacetate (**94**) to phenyl vinyl sulfone (**177**) at room temperature, affording the addition product **178**. Nearly quantitative yields were obtained. However, the obtained ee values were only moderate (43–54% ee) (Scheme 9.62) [55].

Very recently, Lu and coworkers successfully applied aminocatalysis via the enamine intermediate to the Michael addition of cyclic ketones to vinyl sulfones **181** [56]. In the presence of the cinchonidine-derived primary amine salt **179**, the Michael reactions between vinyl sulfones **181** and cyclic ketones **180** proceeded smoothly, affording the desired adduct **182** in very high yield and with excellent enantioselectivity (up to 97% ee) (Scheme 9.63). They also successfully applied this methodology to the synthesis of sodium cyclamate. However, this protocol gave poor yields and ee values for acyclic ketones.

9.4.1
Vinyl Phosphate

In 2008, Jørgensen and coworkers reported the first example of the asymmetric conjugate addition of β-ketoesters **184** to the vinyl bisphosphonate esters **183** catalyzed by the parent cinchona alkaloids (20 mol%) such as dihydroquinine and dihydroquinidine [57]. High yields and enantioselectivities (up to 99% ee) were achieved for a wide range of indanone-based β-ketoesters as well as various 5-tert-butyloxycarbonyl

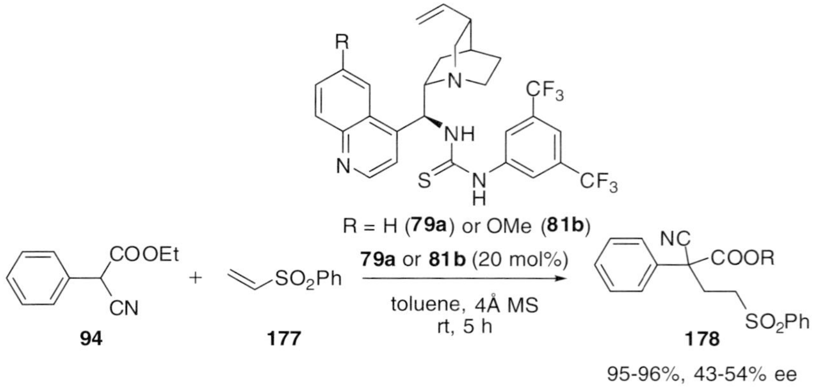

Scheme 9.62

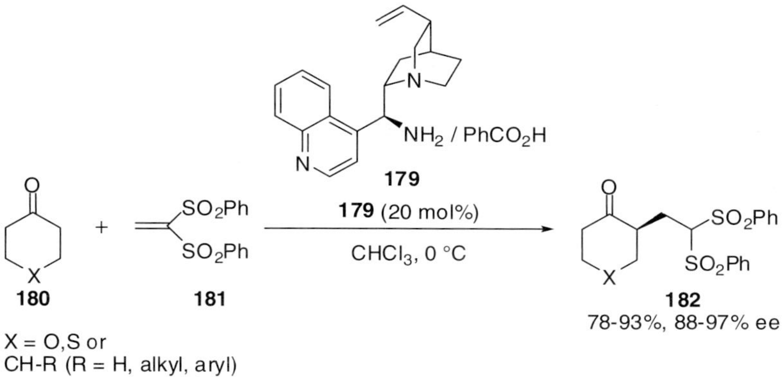
NH2 / PhCO2H
179
SO2Ph
SO2Ph
179 (20 mol%)
CHCl3, 0 °C
SO2Ph
SO2Ph
180
181
182
78-93%, 88-97% ee
X = O,S or
CH-R (R = H, alkyl, aryl)

Scheme 9.63

cyclopentenones (Scheme 9.64). The 1,4-addition products **185** could be easily converted to synthetically valuable vinyl phosphonate derivatives by Horner–Wadsworth–Emmons olefination. To explain the observed sense of enantioselectivity, the plausible transition-state model depicted in Figure 9.6 was proposed. In the less-favored intermediate (Figure 9.6), steric repulsion between the quinoline part of the catalyst and the enolate of the β-ketoester is proposed to disfavor this reaction path.

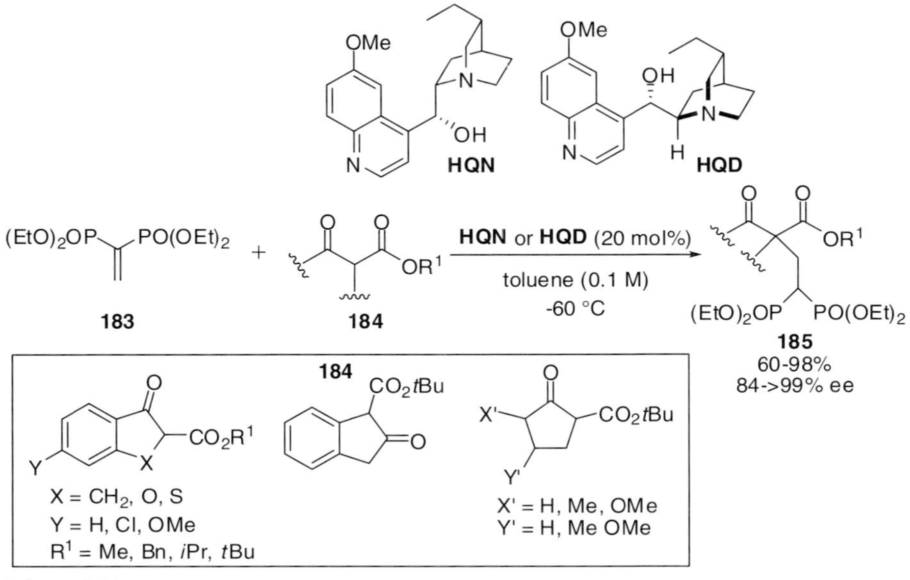
OMe
OH
HQN
OMe
OH
HQD
(EtO)2OP
PO(OEt)2
OR1
HQN or HQD (20 mol%)
toluene (0.1 M)
-60 °C
OR1
(EtO)2OP
PO(OEt)2
183
184
185
60-98%
84->99% ee
184
CO2R1
X = CH2, O, S
Y = H, Cl, OMe
R1 = Me, Bn, iPr, tBu
CO2tBu
CO2tBu
X' = H, Me, OMe
Y' = H, Me OMe

Scheme 9.64

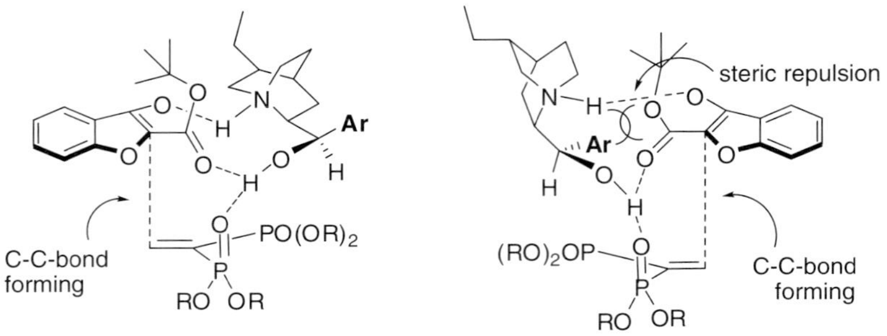

Figure 9.6 Proposed approach of the intermediates to the bisphosphonate.

9.5
Cyclopropanation and Other Related Reactions

9.5.1
Cyclopropanation

Cyclopropane units are valuable synthetic building blocks for the synthesis of biologically active molecules. For this reason, there is an abundance of literature on the synthesis of this unique group of compounds [58]. The first organocatalytic, asymmetric cyclopropanation reaction was achieved in 1999 by Arai and coworker by using chiral PTCs derived from cinchona alkaloids [59]. The results obtained, however, were far from satisfactory for synthetic use in terms of yields and enantiomeric excesses (up to 83% ee). The first preparatively useful enantioselective cyclopropanation process was developed in 2003 by Gaunt and coworkers by using cinchona alkaloid derivatives as chiral base catalysts [60]. Using quinine (**163b** or **186a**) or quinidine (**163a** or **186b**) derivatives as organocatalysts (10–20 mol%), both enantiomers of a range of chiral-functionalized cyclopropanes **187** could be obtained with excellent diastereo- and enantioselectivity (up to 97% ee) (Scheme 9.65). Mechanistically, this catalytic process proceeds via the ammonium ylide intermediate **189**, generated from the reaction of the α-halocarbonyl compounds **188** with the amine catalyst **186** (or **163**). The Michael addition of the ammonium ylide **189** to the α,β-unsaturated carbonyls **190** and the subsequent cyclization affords the cyclopropane product, while regenerating the catalyst **186** (or **163**) (Scheme 9.66) [60].

Later on, the scope of this method was successfully extended to the intramolecular version by the same research group. Using the quinine- and quinidine-derived catalysts, **186a,b** and **191a,b**, respectively, the chiral [4,1,0]bicycloheptanes **192** were obtained as a single diastereomer with excellent ee values (usually over 95% ee). Although only moderate yields were obtained with the catalysts **186a** or **186b**, the catalysts **191** bearing the methyl group at the C2′-position of the quinoline ring provided much higher yields (Scheme 9.67) [61]. The increased yields obtained with **191** were attributed to the inhibition of alkylation occurring at the quinoline N atom with halocarbonyl substrates due to the steric hindrance (Figure 9.7).

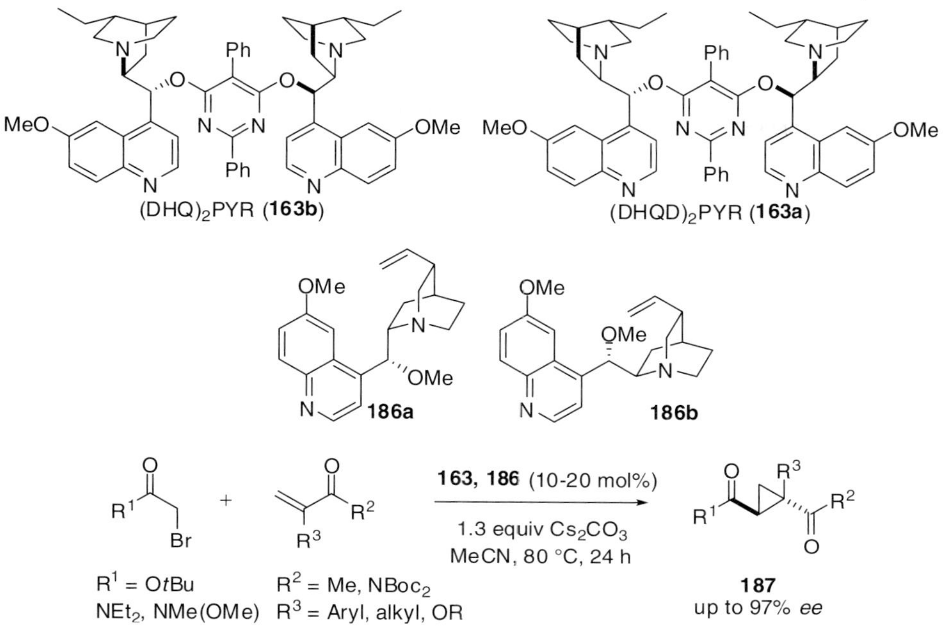

9.5 Cyclopropanation and Other Related Reactions

Scheme 9.65

In 2006, Ohkata and coworkers found that the natural cinchonidine (**CD**) functions as a Brønsted base catalyst (1 mol%) in the reaction between chloromethyl ketones **193** and β-substituted methylidenemalononitriles **194** to furnish the corresponding tetrasubstituted *trans*-cyclopropanes **195** with enantioselectivities of up to 82% ee (Scheme 9.68) [62]. The 9-OH-protected derivatives provided almost no enantioselectivity, indicating that the hydrogen bonding is crucial for stereoinduction.

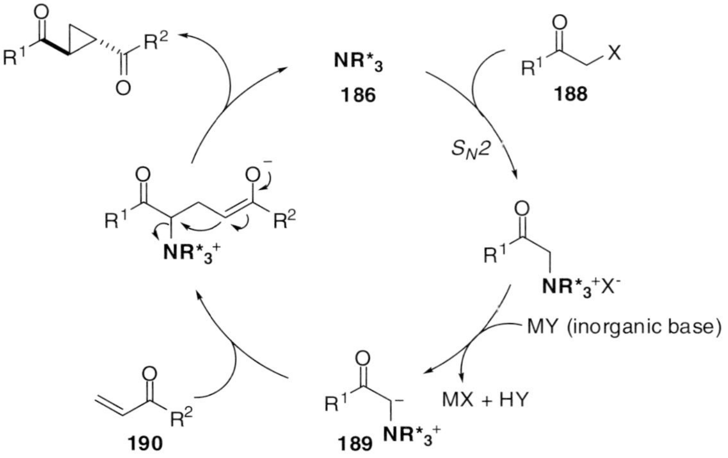

Scheme 9.66

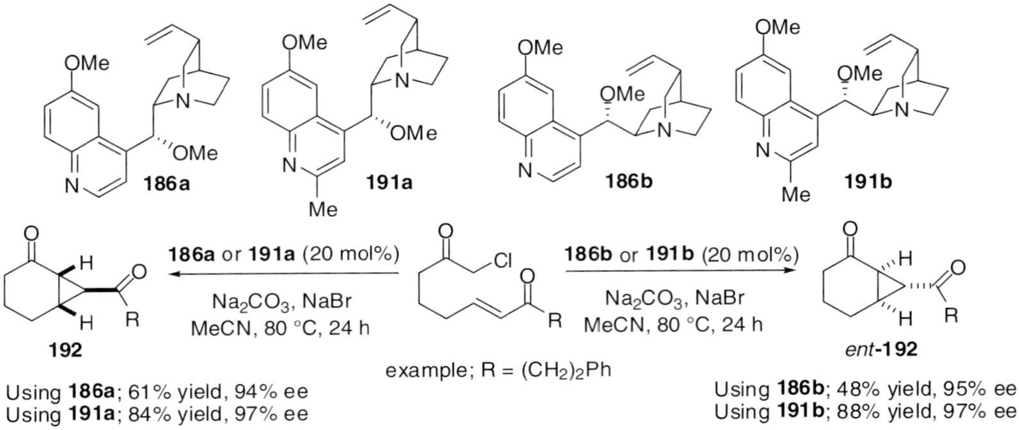

9 Cinchona-Catalyzed Nucleophilic Conjugate Addition to Electron-Deficient C=C Double Bonds

186a or 191a (20 mol%)
Na$_2$CO$_3$, NaBr
MeCN, 80 °C, 24 h

186b or 191b (20 mol%)
Na$_2$CO$_3$, NaBr
MeCN, 80 °C, 24 h

example; R = (CH$_2$)$_2$Ph

192
Using **186a**; 61% yield, 94% ee
Using **191a**; 84% yield, 97% ee

ent-192
Using **186b**; 48% yield, 95% ee
Using **191b**; 88% yield, 97% ee

Scheme 9.67

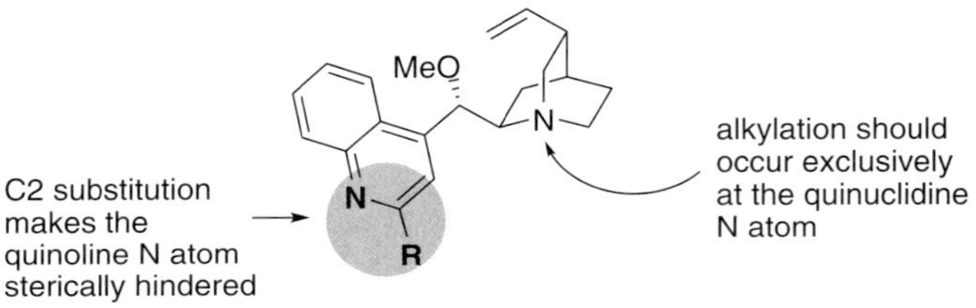

C2 substitution makes the quinoline N atom sterically hindered

alkylation should occur exclusively at the quinuclidine N atom

Figure 9.7 Modified catalysts **191** with substitution at the C2' position on the quinoline ring.

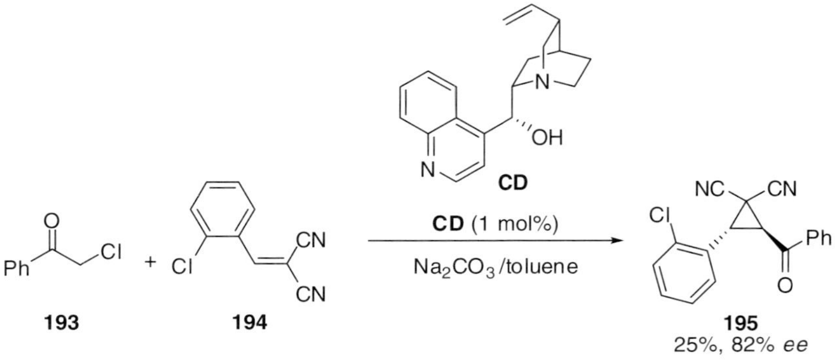

193 + **194** → **CD** (1 mol%), Na$_2$CO$_3$/toluene → **195**
25%, 82% ee

Scheme 9.68

In the same year, Connon and coworkers [63] reported that the chiral bifunctional cinchona alkaloid-based thiouea **81a** is also able to catalyze the addition of dimethyl chloromalonate **196** to nitroolefins **124**, leading to the Michael adduct that cyclizes to form the cyclopropane **197** in the presence of DBU. Almost single diastereomeric nitrocyclopropanes (>98% de) were obtained in good yields. However, the enantioselectivity obtained with this type of catalyst was poor (≤47% ee) (Scheme 9.69).

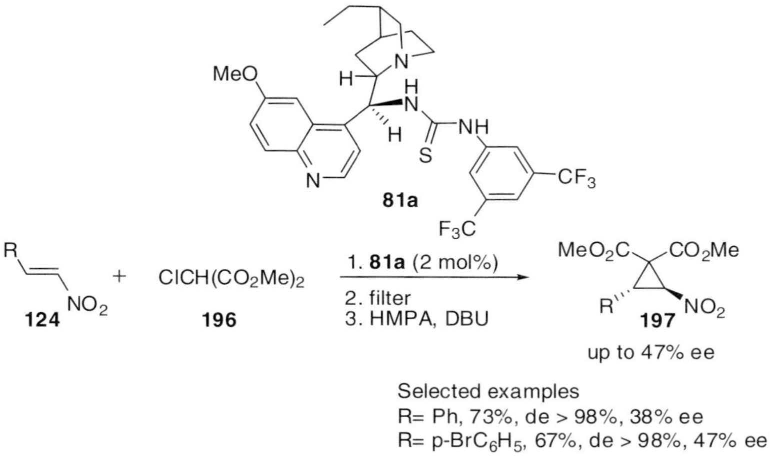

9.5 Cyclopropanation and Other Related Reactions

Scheme 9.69

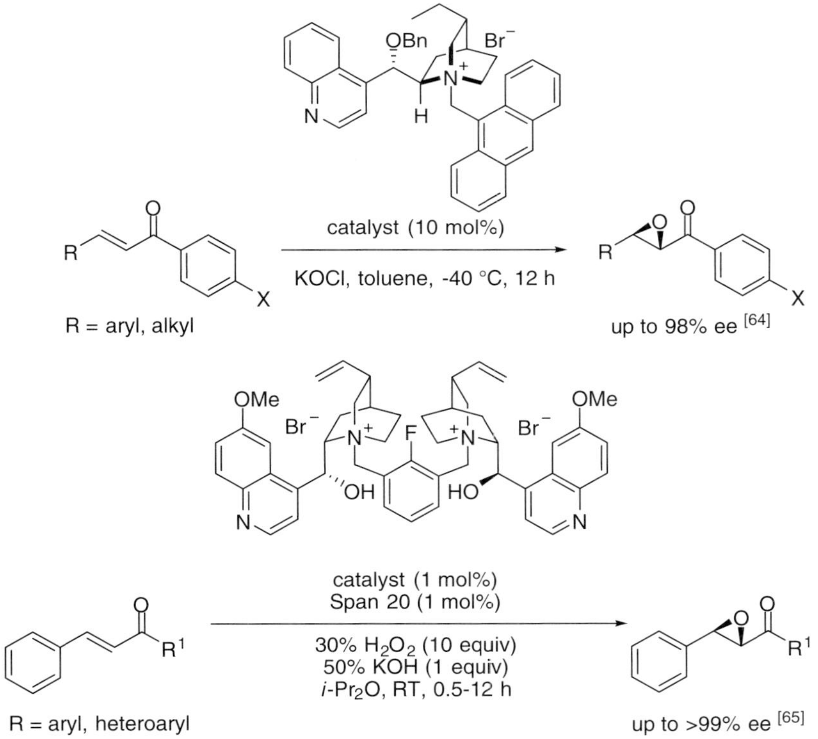

Selected examples
R= Ph, 73%, de > 98%, 38% ee
R= p-BrC$_6$H$_5$, 67%, de > 98%, 47% ee

up to 47% ee

up to 98% ee [64]

up to >99% ee [65]

R = aryl, heteroaryl

Scheme 9.70 [64, 65].

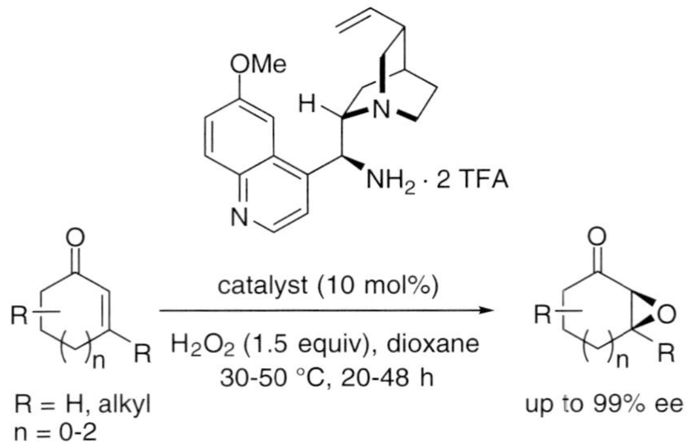

Scheme 9.71 [66].

9.5.2
Epoxidation and Aziridination

In the presence of cinchona derivatives as catalysts, peroxides or hypochlorites as Michael donors react with electron-deficient olefins to give epoxides via conjugate addition–intramolecular cyclization sequence reactions. Two complementary methodologies have been developed for the asymmetric epoxidation of electron-poor olefins, in which either cinchona-based phase-transfer catalysts or 9-amino-9(deoxy)-*epi*-cinchona alkaloids are used as organocatalysts. Mechanistically, in these two

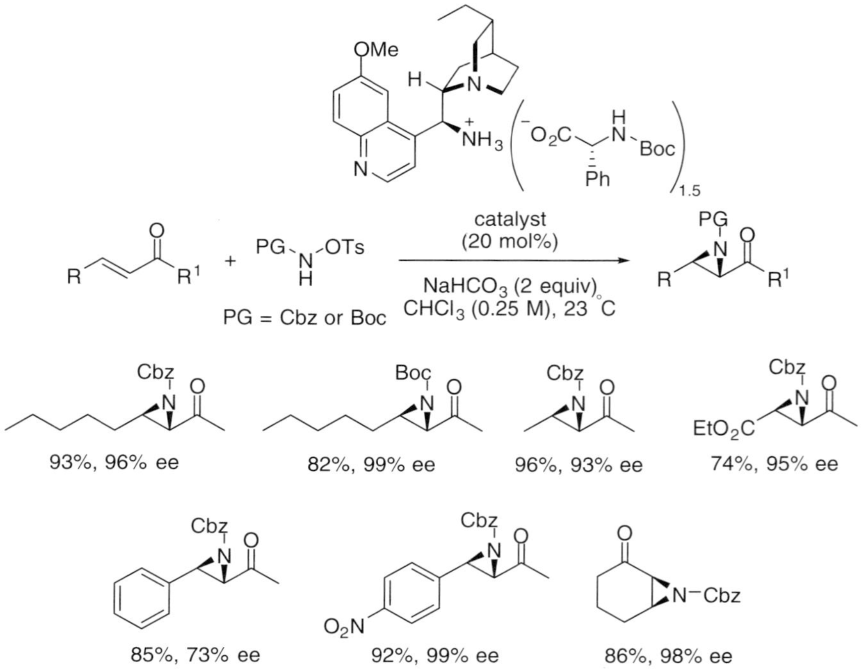

Scheme 9.72 [67].

methodologies, the reaction proceeds via a chiral ion pairing mechanism and iminium catalysis, respectively. Cinchona-based PTCs have been shown to be highly efficient for the enantioselective epoxidation of acyclic enones (Scheme 9.70). On the other hand, iminium catalysis using cinchona-based bifunctional primary amine catalysts has been proved to be suitable for the epoxidation of cyclic enone systems (Scheme 9.71).

Similarly, N-hydroxamates act as nitrogen-transfer reagents to give N-arylaziridines via conjugate addition. However, there are few examples where good enantioselectivity has been achieved using chiral PTCs. Here too, a simple and highly versatile method for asymmetric aziridination has been achieved by adopting iminium catalysis (Scheme 9.72). More detailed information on these highly important reactions can be found in Section 5.2.

9.6
Conclusions

This chapter presented the current state-of-the-art applications of cinchona alkaloids and their derivatives as chiral inducers in asymmetric conjugate addition reactions. As shown in a number of the examples discussed, cinchona alkaloids are classified as the most privileged chiral catalysts, promoting diverse types of conjugate addition reactions with excellent performances. Nevertheless, these reactions are still attractive targets for catalyst development, since numerous types of Michael donors and acceptors participate in them. Therefore, there is no doubt that the further development of cinchona-based organocatalysts for this major reaction category will continue to provide more exciting results in the near future.

Acknowledgments

This work was supported by grants KRF-2008-005-J00701(MOEHRD), R11-2005-008-00000-0 (SRC program of MEST/KOSEF) and R31-2008-000-10029-0 (WCU program).

References

1 (a) Wynberg, H. and Helder, R. (1975) *Tetrahedron Lett.*, **16**, 4057–4060; (b) Hermann, K. and Wynberg, H. (1979) *J. Org. Chem.*, **44**, 2238–2244; (c) Hiemstra, H. and Wynberg, H. (1981) *J. Am. Chem. Soc.*, **103**, 417–430.

2 (a) Tanaka, T., Kumamoto, T. and Ishikawa, T. (2000) *Tetrahedron: Asymmetry*, **11**, 4633–4637; (b) Sekino, E., Kumamoto, T., Tanaka, T., Ikeda, T., and Ishikawa, T. (2004) *J. Org. Chem.*, **69**, 2760–2767.

3 Merschaert, A., Delbeke, P., Daloze, D., and Dive, G. (2004) *Tetrahedron Lett.*, **45**, 4697–4701.

4 Perdicchia, D. and Jørgensen, K.A. (2007) *J. Org. Chem.*, **72**, 3565–3568.

5 Bell, M., Frisch, K., and Jørgensen, K.A. (2006) *J. Org. Chem.*, **71**, 5407–5410.

6 Conn, R.S.E., Lovell, A.V., Karady, S., and Weinstock, L.M. (1986) *J. Org. Chem.*, **51**, 4710–4711.

7 (a) Corey, E.J., Noe, M.C., and Xu, F. (1998) *Tetrahedron Lett.*, **39**, 5347–5350; (b) Zhang, F.-Y. and Corey, E.J. (2000) *Org. Lett.*, **2**, 1097–1100; (c) Siebum, A.H.G., Tsang, R.K.F., van der Steen, R., Raap, J., and Lugtenburg, J. (2004) *Eur. J. Org. Chem.*, 4391–4396.

8 (a) Ramachandran, P.V., Madhi, S., Bland-Berry, L., Reddy, M.V.R., and O'Donnell, M.J. (2005) *J. Am. Chem. Soc.*, **127**, 13450–13451; (b) O'Donnell, M.J., Delgado, F., Domínguez, E., de Blas, J., and Scott, W.L. (2001) *Tetrahedron: Asymmetry*, **12**, 821–828.

9 (a) Zhang, F.-Y. and Corey, E.J. (2004) *Org. Lett.*, **6**, 3397–3399; (b) Corey, E.J. and Zhang, F.-Y. (2000) *Org. Lett.*, **2**, 4257–4259; (c) Zhang, F.-Y. and Corey, E.J. (2001) *Org. Lett.*, **3**, 639–641.

10 Ceccarelli, R., Insogna, S., and Bella, M. (2006) *Org. Biomol. Chem.*, **4**, 4281–4284.

11 Perrard, T., Plaquevent, J.-C., Desmurs, J.-R., and Hébrault, D. (2000) *Org. Lett.*, **2**, 2959–2962.

12 (a) Tozawa, T., Yamane, Y., and Mukaiyama, T. (2006) *Chem. Lett.*, **35**, 56–57; (b) Tozawa, T., Yamane, Y., and Mukaiyama, T. (2006) *Chem. Lett.*, **35**, 360–361; (c) Nagao, H., Yamane, Y., and Mukaiyama, T. (2006) *Chem. Lett.*, **35**, 1398–1399; (d) Mukaiyama, T., Nagao, H., and Yamane, Y. (2006) *Chem. Lett.*, **35**, 916–917.

13 Bernardi, L., López-Cantarero, J., Niess, B., and Jørgensen, K.A. (2007) *J. Am. Chem. Soc.*, **129**, 5772–5778.

14 Elsner, P., Bernardi, L., Salla, G.D., Overggard, J., and Jørgensen, K.A. (2008) *J. Am. Chem. Soc.*, **130**, 4897–4905.

15 Alemán, J., Reyes, E., Richter, B., Overgaard, J., and Jørgensen, K.A. (2007) *Chem. Commun.*, 3921–3923.

16 Furukawa, T., Shibata, N., Mizuta, S., Nakamura, S., Toru, T., and Shiro, M. (2008) *Angew. Chem. Int. Ed.*, **47**, 8051–8054.

17 Bella, M. and Jørgensen, K.A. (2004) *J. Am. Chem. Soc.*, **126**, 5672–5673.

18 Wu, F., Li, H., Hong, R., and Deng, L. (2006) *Angew. Chem. Int. Ed.*, **45**, 947–950.

19 Ligby, C.L. and Dixon, D.J. (2008) *Chem. Commun.*, 3798–3800.

20 Wu, F., Hong, R., Khan, J., Liu, X., and Deng, L. (2006) *Angew. Chem. Int. Ed.*, **45**, 4301–4305.

21 Wang, Y., Liu, X., and Deng, L. (2006) *J. Am. Chem. Soc.*, **128**, 3928–3930.

22 Li, B.-J., Jiang, L., Liu, M., Chen, Y.-C., Ding, L.-S., and Wu, Y. (2005) *Synlett*, 603–606.

23 (a) Vakulya, B., Varga, S., Csámpai, A., and Soós, T. (2005) *Org. Lett.*, **7**, 1967–1969; (b) Vakulya, B., Varga, S., and Soós, T. (2008) *J. Org. Chem.*, **73**, 3475–3480.

24 Wang, J., Li, H., Zu, L., Jiang, W., Xie, H., Duan, W., and Wang, W. (2006) *J. Am. Chem. Soc.*, **128**, 12652–12653.

25 Liu, T.-Y., Li, R., Chai, Q., Long, J., Li, B.-J., Wu, Y., Ding, L.-S., and Chen, Y.-C. (2007) *Chem. Eur. J.*, **13**, 319–327.

26 Gu, C.-l., Liu, L., Sui, Y., Zhao, J.-L., Wang, D., and Chen, Y.-J. (2007) *Tetrahedron: Asymmetry*, **18**, 455–463.

27 Wang, B., Wu, F., Wang, Y., Liu, X., and Deng, L. (2007) *J. Am. Chem. Soc.*, **129**, 768–769.

28 Biddle, M.M., Lin, M., and Scheidt, K.A. (2007) *J. Am. Chem. Soc.*, **129**, 3830–3831.

29 (a) Zu, L., Wang, J., Li, H., Xie, H., Jiang, W., and Wang, W. (2007) *J. Am. Chem. Soc.*, **129**, 1036–1037; (b) Wang, J., Xie, H., Li, H., Zu, L., and Wang, W. (2008) *Angew. Chem. Int. Ed.*, **47**, 4177–4179.

30 Lu, X. and Deng, L. (2008) *Angew. Chem. Int. Ed.*, **47**, 7710–7713.

31 Kang, T.-R., Xie, J.-W., Du, W., Feng, X., and Chen, Y.-C. (2008) *Org. Biomol. Chem.*, **6**, 2673–2675.

32 Li, P., Wang, Y., Liang, X., and Ye, J. (2008) *Chem. Commun.*, 3302–3304.

33 Li, H., Wang, Y., Tang, L., and Deng, L. (2004) *J. Am. Chem. Soc.*, **126**, 9906–9907.

34 Li, H., Wang, Y., Tang, L., Wu, F., Liu, X., Guo, C., Foxman, B.M., and Deng, L. (2005) *Angew. Chem. Int. Ed.*, **44**, 105–108.

35 Wang, J., Li, H., Zu, L., Jiang, W., and Wang, W. (2006) *Adv. Synth. Catal.*, **348**, 2047–2050.
36 Shi, M., Lei, Z.-Y., Zhao, M.-X., and Shi, J.-W. (2007) *Tetrahedron Lett.*, **48**, 5743–5746.
37 Wang, J., Li, H., Zu, L., and Wang, W. (2006) *Org. Lett.*, **8**, 1391–1394.
38 McCooey, S.H. and Connon, S.J. (2005) *Angew. Chem. Int. Ed.*, **44**, 6367–6370.
39 Ye, J., Dixon, D.J., and Hynes, P.S. (2005) *Chem. Commun.*, 4481–4483.
40 Hynes, P.S., Stranges, D., Stupple, P.A., Guarna, A., and Dixon, D.J. (2007) *Org. Lett.*, **9**, 2107–2110.
41 Alemán, J., Milelli, A., Cabrera, S., Reyes, E., and Jørgensen, K.A. (2008) *Chem. Eur. J.*, **14**, 10958–10966.
42 Lubkoll, J. and Wennemers, H. (2007) *Angew. Chem. Int. Ed.*, **46**, 6841–6844.
43 Liu, T.-Y., Cui, H.-L., Chai, Q., Long, J., Li, B.-J., Wu, Y., Ding, L.-S., and Chen, Y.-C. (2007) *Chem. Commun.*, 2228–2230.
44 Mattson, A.E., Zuhl, A.M., Reynolds, T.E., and Scheidt, K.A. (2006) *J. Am. Chem. Soc.*, **128**, 4932–4933.
45 Dinér, P., Nielsen, M., Bertelsen, S., Niess, B., and Jørgensen, K.A. (2007) *Chem. Commun.*, 3646–3648.
46 Bartoli, G., Bosco, M., Carlone, A., Locatelli, M., Mazzanti, A., Sambri, L., and Melchiorre, P. (2007) *Chem. Commun.*, 722–724.
47 Malerich, J.P., Hagihara, K., and Rawal, V.H. (2008) *J. Am. Chem. Soc.*, **130**, 14416–14417.
48 McCooey, S.H. and Connon, S.J. (2007) *Org. Lett.*, **9**, 599–602.
49 (a) Tan, B., Chua, P.J., Li, Y., and Zhong, G. (2008) *Org. Lett.*, **10**, 2437–2440; (b) Tan, B., Chua, P.J., Zeng, X., Lu, M., and Zhong, G. (2008) *Org. Lett.*, **10**, 3489–3492.
50 Xue, D., Chen, Y.-C., Wang, Q.-W., Cun, L.-F., Zhu, J., and Deng, J.-G. (2005) *Org. Lett.*, **7**, 5293–5296.
51 Nielsen, M., Zhuang, W., and Jørgensen, K.A. (2007) *Tetrahedron*, **63**, 5849–5854.
52 Zhou, W.-M., Liu, H., and Du, D.-M. (2008) *Org. Lett.*, **10**, 2817–2820.
53 Wang, J., Heikkinen, L.D., Li, H., Zu, L., Jiang, W., Xie, H., and Wang, W. (2007) *Adv. Synth. Catal.*, **349**, 1052–1056.
54 Li, H., Song, J., Liu, X., and Deng, L. (2005) *J. Am. Chem. Soc.*, **127**, 8948–8949.
55 Liu, T.-Y., Long, J., Li, B.-J., Jiang, L., Li, R., Wu, Y., Ding, L.-S., and Chen, Y.-C. (2006) *Org. Biomol. Chem.*, **4**, 2097–2099.
56 Zhu, Q., Cheng, L., and Lu, Y. (2008) *Chem. Commun.*, 6315–6317.
57 Capuzzi, M., Perdicchia, D., and Jørgensen, K.A. (2008) *Chem. Eur. J.*, **14**, 128–135.
58 Pellissier, H. (2008) *Tetrahedron*, **64**, 7041–7095.
59 Arai, S., Nakayama, K., Ishida, T., and Shioiri, T. (1999) *Tetrahedron Lett.*, **40**, 4215–4218.
60 (a) Papageorgiou, C.D., Ley, S.V., and Gaunt, M.J. (2003) *Angew. Chem. Int. Ed.*, **42**, 828–831; (b) Papageorgiou, C.D., Cubillo de Dios, M.A., Ley, S.V., and Gaunt, M.J. (2004) *Angew. Chem. Int. Ed.*, **43**, 4641–4644.
61 (a) Bremeyer, N., Smith, S.C., Ley, S.V., and Gaunt, M.J. (2004) *Angew. Chem. Int. Ed.*, **43**, 2681–2684; (b) Johansson, C.C.C., Bremeyer, N., Ley, S.V., Owen, D.R., Smith, S.C., and Gaunt, M.J. (2006) *Angew. Chem. Int. Ed.*, **45**, 6024–6028.
62 Kojima, S., Suzuki, M., Watanabe, A., and Ohkata, K. (2006) *Tetrahedron Lett.*, **47**, 9061–9065.
63 McCooey, S.H., McCabe, T., and Connon, S.J. (2006) *J. Org. Chem.*, **71**, 7494–7497.
64 Corey, E.J. and Zhang, F.-Y. (1999) *Org. Lett.*, **1**, 1287–1290.
65 Jew, S.-s., Lee, J.-H., Jeong, B.-S., Yoo, M.-S., Kim, M.-J., Lee, Y.-J., Lee, J., Choi, S.-h., Lee, K., Lah, M.S., and Park, H.-g. (2005) *Angew. Chem. Int. Ed.*, **44**, 1383–1385.
66 Wang, X., Reisinger, C.M., and List, B. (2008) *J. Am. Chem. Soc.*, **130**, 6070–6071.
67 Pesciaioli, F., De Vincentiis, F., Galzerano, P., Bencivenni, G., Bartoli, G., Mazzanti, A., and Melchiorre, P. (2008) *Angew. Chem. Int. Ed.*, **47**, 8703–8706.

10
Cinchona-Catalyzed Cycloaddition Reactions
Yan-Kai Liu and Ying-Chun Chen

10.1
Introduction

Asymmetric pericyclic reaction represents one of the most straightforward protocols to access enantioenriched cyclic compounds and triggers continuing interest in organic synthesis. Fruitful results have been achieved by the catalysis of chiral metal complexes over the past decades [1]. On the other hand, recently small organic molecules have also contributed a lot to this area owing to the rapid development of asymmetric organocatalysis [2].

Cinchona alkaloids are readily available natural chiral compounds and have a long history to be utilized as organocatalysts in asymmetric catalysis [3, 4]. They are multifunctional, tunable, and more importantly, they could promote a diversity of reactions through different catalytic mechanisms, which make them privileged catalysts in organocatalysis. In this chapter, the applications of cinchona alkaloids and their derivatives for asymmetric cycloaddition reactions after 2000, especially for the construction of a variety of five- and six-membered cyclic compounds, are discussed.

10.2
Asymmetric Cycloadditions Catalyzed by Quinuclidine Tertiary Amine

Cinchona alkaloids possess a nucleophilic quinuclidine structure and can perform as versatile Lewis bases to react with ketenes generated *in situ* from acyl halides in the presence of an acid scavenger. The resulting ketene enolates can react with electrophilic C=O or C=N bonds to deliver chiral β-lactones [5] or β-lactams [6], respectively, in a [2 + 2] cycloaddition manner, which is discussed in Chapter 5 in detail. Gaunt et al. also developed practical "one pot" cyclopropanation processes mediated by the modified cinchona alkaloids via ammonium ylide intermediates [7]. Although the catalytic strategy has been well established, the utilization of ammonium enolate based [4 + 2] cycloaddition is rare probably because of the relative unreactivity of the

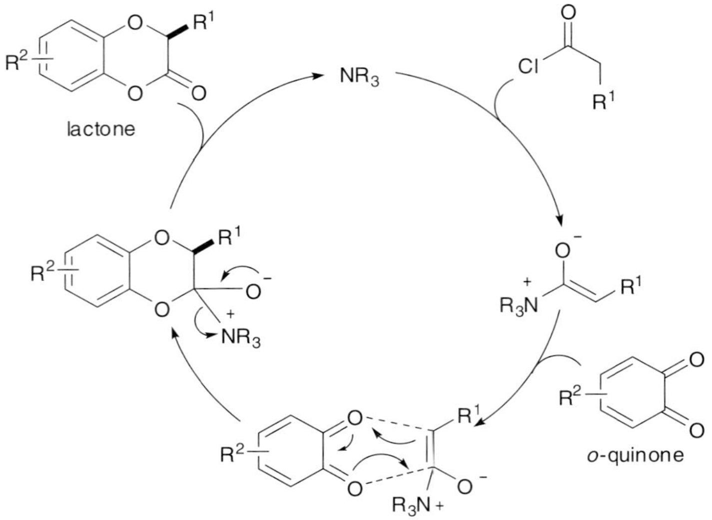

Scheme 10.1 Catalytic cycle for the [4 + 2] cycloaddition of ketene enolate and o-quinone.

enolates themselves toward various heterodienes. Recently, Lectka et al. have developed some elegant inverse electron demand [4 + 2] cycloadditions of ketenes with more energetic heterodienes by the catalysis of modified cinchona alkaloids.

As illustrated in Scheme 10.1, a tertiary amine base would react with the *in situ* generated ketenes to afford zwitterionic enolates. Then these intermediates would react with electrophilic o-quinones to deliver lactone adducts after the release of the tertiary amine catalyst [8].

By using 10 mol% of benzoylquinidine (BQD, **1a**), a variety of acid chlorides reacted efficiently with o-chloranil at −78 °C in the presence of nonnucleophilic Hünig base (*i*Pr$_2$EtN, DIPEA) (Scheme 10.2). The desired lactones **2** were produced in up to 91% yield with excellent enantiopurity. On the other hand, o-bromanil and 9,10-phenanthrene quinone were also successfully applied in this cycloaddition reaction.

The o-chloranil-derived lactones can be readily converted to chiral α-oxygenated carboxylic acid derivatives through methanolysis followed by CAN oxidation. In each case, the optical activity could be fully preserved (Scheme 10.3).

Following the same strategy, further improvements in similar catalytic asymmetric [4 + 2] cycloaddition reaction have been made by Lectka group. The cyclic 1,4-benzoxazinones **3** (Scheme 10.4) that rely on the highly enantioselective [4 + 2] cycloaddition of o-benzoquinone imides with chiral ketene enolates were efficiently constructed, which can be derivatized *in situ* to provide α-amino acid derivatives in good to excellent yields and with virtual enantiopurity [9].

Substrates for this type of cycloadditions are not limited to o-quinones and o-benzoquinone imides but can be extended to o-benzoquinone diimides also. In a preliminary study, it was found that some unidentified byproducts as well as the desired quinoxalinone were formed, which made the reaction sluggish and resulted

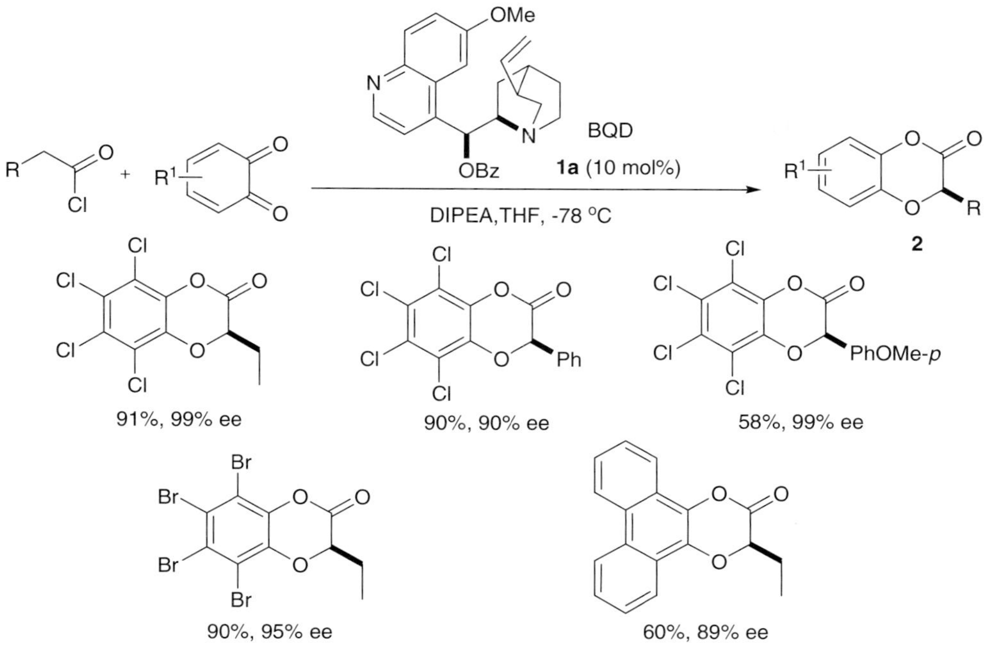

10.2 Asymmetric Cycloadditions Catalyzed by Quinuclidine Tertiary Amine

Scheme 10.2 [4 + 2] cycloaddition of ketene enolates and o-quinones.

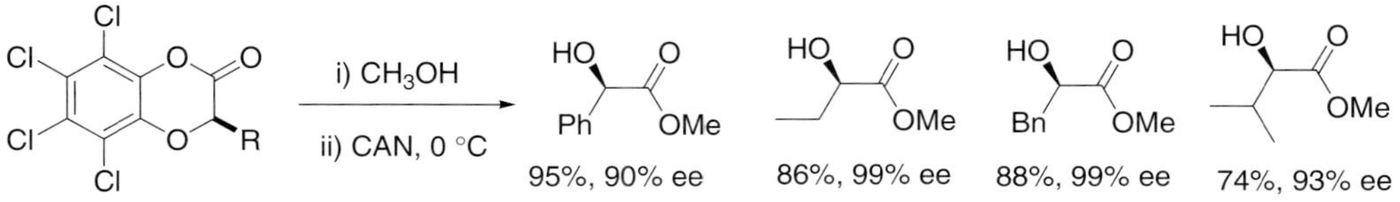

Scheme 10.3 Synthesis of chiral α-hydroxyl carboxylic acids.

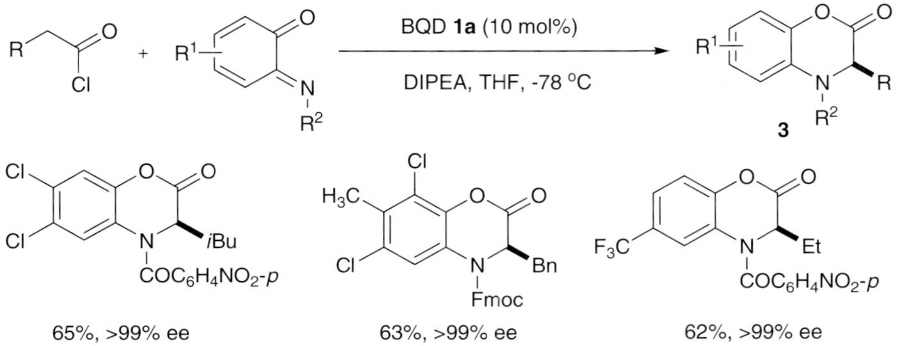

Scheme 10.4 [4 + 2] cycloaddition of o-benzoquinone imides with ketene enolates.

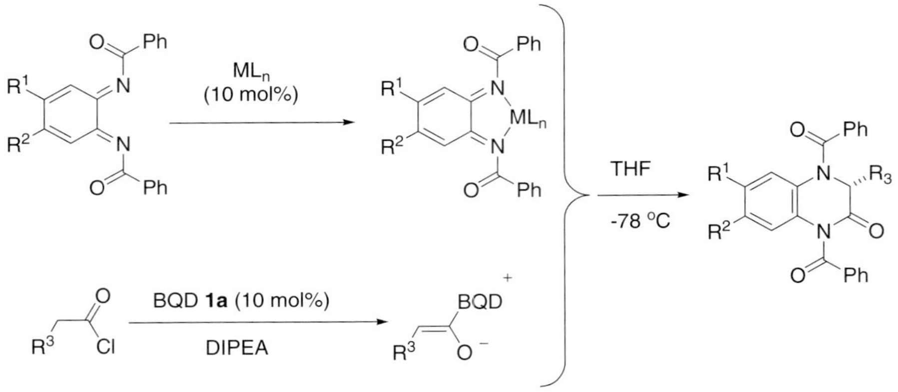

Scheme 10.5 Bifunctional [4 + 2] cycloaddition of o-benzoquinone diimides and ketene enolates.

in a lower yield. Breakthrough resulting in high yields was achieved when a Lewis acid was applied as cocatalyst, which could increase the electrophilicity of the diimide without interfering with the nucleophilic enolate (Scheme 10.5) [10].

With $Zn(OTf)_2$ (10 mol%) as the cocatalyst, the 3,4-dihydroquinoxalin-2-one products **4** were obtained in remarkably increased yields and with excellent enantioselectivities (>99% ee) in each case (Table 10.1, entries 1–12). It is worth to note that the opposite enantiomer was obtained with similarly high enantioselectivity and yield for the cycloaddition reaction when benzoylquinine (BQN, **1b**) was used instead of BQD **1a** (entry 13).

Later, Lectka et al. reported a detailed synthetic and mechanistic study of unusual [4 + 2] cycloaddition of ketene enolates and o-quinones by the bifunctional catalysis of cinchona alkaloids BQD **1a** (or BQN **1b**) and Lewis acids. The undertaken investigations based on the integration of experimental and calculated data itself demonstrated a surprising cooperative LA/LB interaction on a ketene enolate. It showed that the reaction of o-quinone undergoes a mechanistic "switch" in which the mode of activation changes from Lewis acid (LA) complexation of the quinone to metal complexation of the chiral ketene enolate.

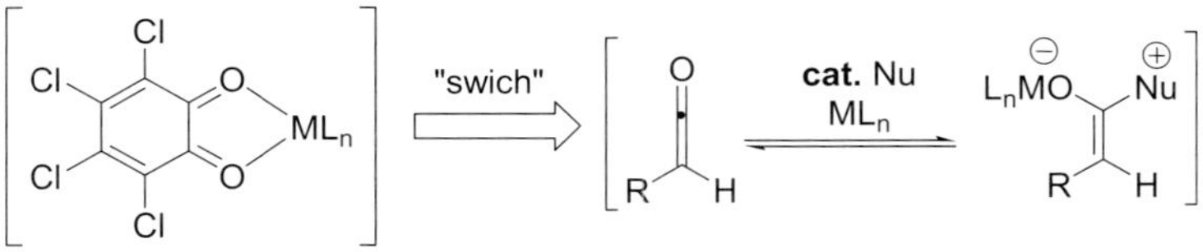

A large number of Lewis acids had been screened, and trans-$(Ph_3P)_2PdCl_2$ had been found to be the best Lewis acid to give the highest yield. As outlined in Scheme 10.6, two main possibilities might be broached: (i) formation of a metal enolate lowers the activation barrier for carbon–oxygen bond formation and (ii) the metal cocatalyst stabilizes the enolate, thus increasing its concentration and perhaps enhancing its chemoselectivity.

Table 10.1 Synthesis of 3,4-dihydroquinoxalin-2-one products.

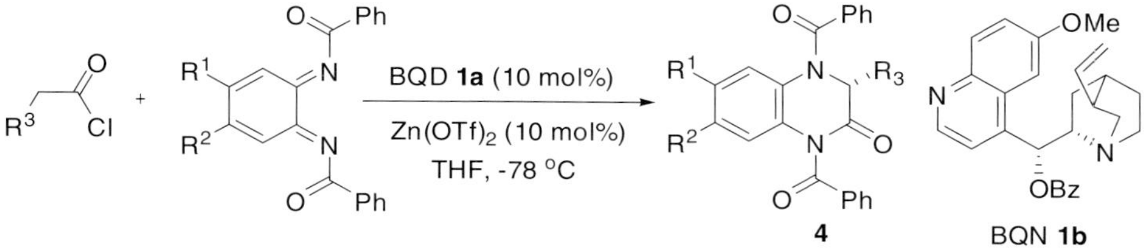

Entry	R¹	R²	R³	Yield 4 (%)	ee (%)
1	Cl	Cl	Et	82	>99
2	Cl	Cl	CH_2SMe	84	>99
3	Cl	Cl	Bn	85	>99
4	Cl	Cl	CH_2phthalimide	93	>99
5	H	H	iPr	71	>99
6	H	H	i-Bu	73	>99
7	CF_3	H	Et	81	>99
8	CF_3	H	Bn	77	>99
9	CF_3	H	p-Br-Bn	78	>99
10	COPh	H	Et	84	>99
11	COPh	H	i-Bu	84	>99
12	COPh	H	$(CH_2)_3Cl$	87	>99
13[a]	COPh	H	p-Br-Bn	69	>99

BQN **1b** was used.

The catalytic system has been efficiently applied to the sequential [4 + 2] cycloaddition/ring opening reaction to afford highly enantioenriched α-hydroxylated carbonyl derivatives in excellent yields (Scheme 10.7). A variety of important classes of compounds, such as α-hydroxy-γ-lactone, α-hydroxy-γ-lactam, factor Xa inhibitor,

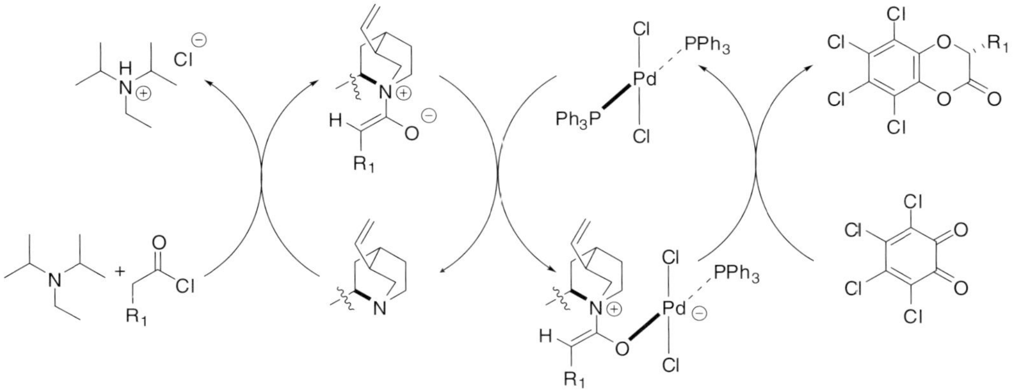

Scheme 10.6 Proposed bifunctional mechanism for the Pd(II) cocatalyzed cycloaddition.

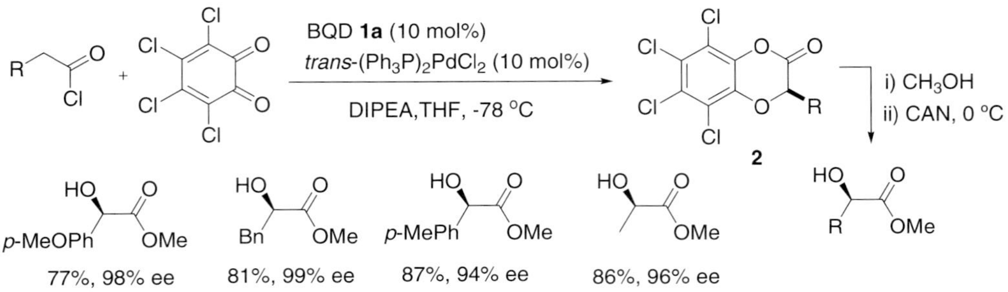

Scheme 10.7 Synthesis of α-hydroxy esters.

Reagents/conditions: BQD **1a** (10 mol%), trans-(Ph$_3$P)$_2$PdCl$_2$ (10 mol%), DIPEA, THF, −78 °C; then i) CH$_3$OH, ii) CAN, 0 °C.

Products:
- p-MeOPh, OMe: 77%, 98% ee
- Bn, OMe: 81%, 99% ee
- p-MePh, OMe: 87%, 94% ee
- (H), OMe: 86%, 96% ee
- R, OMe

and α-hydroxy DAPT, could be synthesized in high yield and enantioselectivity by simply modifying the acid chlorides [11].

Another catalytic application of chiral ketene enolates to [4 + 2]-type cyclizations was the discovery of their use in the diastereoselective and enantioselective syntheses of disubstituted thiazinone. Nelson and coworkers described the cyclocondensations of acid chlorides and α-amido sulfones as effective surrogates for asymmetric Mannich addition reactions in the presence of catalytic system composed of O-TMS quinine **1c** or O-TMS quinidine **1d** (20 mol%), LiClO$_4$, and DIPEA. These reactions provided chiral Mannich adducts masked as cis-4,5-disubstituted thiazinone heterocycles **5**. It was noteworthy that the *in situ* formation of enolizable N-thioacyl imine electrophiles, which could be trapped by the nucleophilic ketene enolates, was crucial to the success of this reaction. As summarized in Table 10.2, the cinchona-catalyzed ketene-N-thioacyl-imine cycloadditions were generally effective for a variety of alkyl-substituted ketenes and aliphatic imine electrophiles (≥95% ee, ≥95% cis : trans) [12].

However, cycloaddition involving aryl α-amido sulfone afforded modestly attenuated yields due to competing formation of the β-lactam adducts, possessing the unanticipated *trans* diastereoselection that was not observed for alkyl imine electrophiles (Eq. (10.1)).

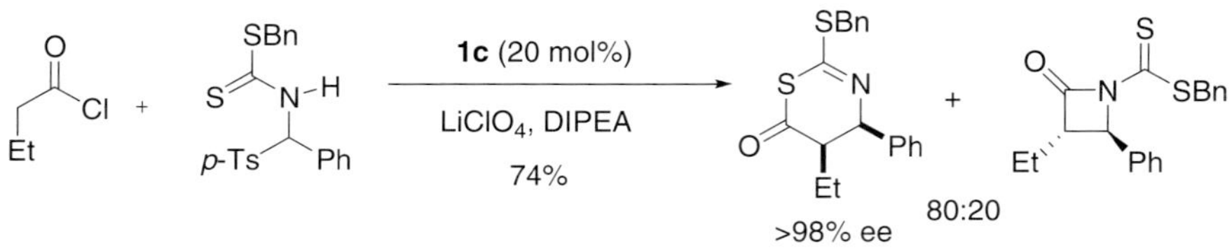

Et–C(O)Cl + p-Ts–CH(Ph)–N(H)–SBn → [**1c** (20 mol%), LiClO$_4$, DIPEA] → thiazinone (S,S-N,Ph,Et; 74%, >98% ee) + β-lactam (80:20)

(10.1)

The utility of the chiral cycloadducts was revealed by the ring opening processes available to the thiazinone heterocycles. Dipeptide derivative was directly prepared through amine-mediated ring opening reaction, suggestive of the function of thiazinone as an activated ester surrogate. In addition, hydride-mediated thiazinone reduction led to the enantioenriched β-amino aldehyde derivative (Scheme 10.8).

The applications of cinchona-catalyzed ketene enolates can be extended to α,β-unsaturated aliphatic acyl halides. Peters et al. presented a new concept for the synthesis of α,β-unsaturated δ-lactones, which are subunits of a number of natural and unnatural products that display a wide range of biological activity. They proposed

10.2 Asymmetric Cycloadditions Catalyzed by Quinuclidine Tertiary Amine

Table 10.2 Cinchona alkaloid catalyzed ketene-N-thioacyl imine [4 + 2] cycloadditions.[a]

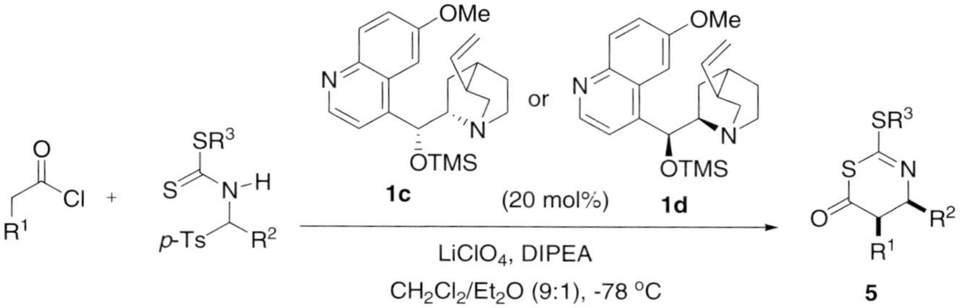

Entry	R^1	R^2	R^3	Yield 5 (%)	ee (%) (cis : trans)
1	Me	CH$_2$CH$_2$Ph	Et	76	>98 (95 : 5)
2	Me	C$_6$H$_{11}$	Et	75	>98 (95 : 5)
3	Me	CH$_2$OBn	Et	51	>98 (>97 : 3)
4	CH$_2$Ph	CH$_2$CH$_2$Ph	Et	65	>98 (>97 : 3)
5	Me	CH$_2$CH$_2$CH$_3$	Bn	67	98 (>97 : 3)
6	Et	CH$_2$CH(CH$_3$)$_2$	Bn	72	>98 (>97 : 3)
7	iPr	CH$_2$CH$_2$Ph	Bn	63	>98 (>97 : 3)
8	Me	CH$_2$CH(CH$_3$)$_2$	Bn	74	>95 (>97 : 3)

[a]Entries 1–7, catalyst **1c**; entry 8, catalyst **1d**.

that substituted vinylketenes should be formed *in situ* by dehydrohalogenation of α,β-unsaturated acid chlorides, which could be trapped by a cinchona base, quinidine O-TMS ether **1d**, to generate the zwitterionic dienolates. Subsequent [4 + 2] cycloaddition with aldehydes, by either a stepwise or a concerted mechanism, would afford the α,β-unsaturated lactones [13].

The expected cycloaddition of α,β-unsaturated acid chloride and chloral indeed occurred in the presence of tertiary amine. The yields could be significantly improved by adding a Lewis acid cocatalyst, Sn(OTf)$_2$, which would facilitate the deprotonation of acid chloride and activate the aldehyde substrate. More satisfactory results could be obtained when acid chloride was added slowly by syringe pump to avoid massive

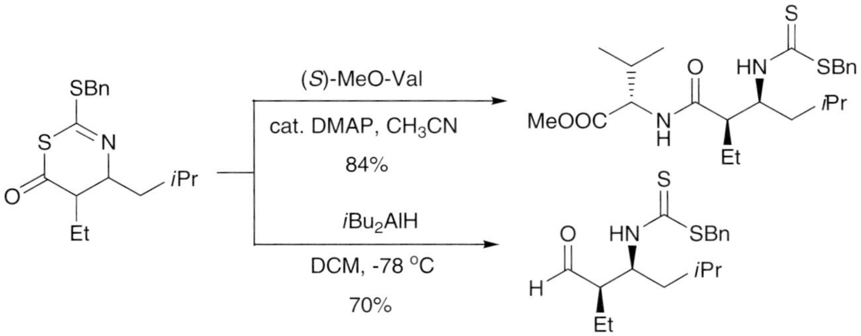

Scheme 10.8 Transformations of thiazinone.

Table 10.3 Preparation of α,β-unsaturated δ-lactones.

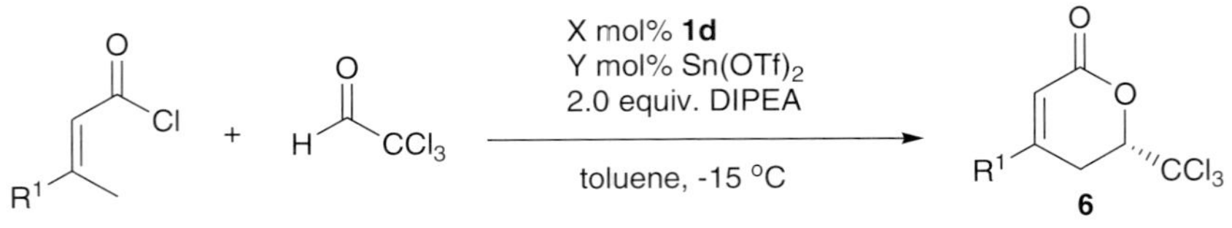

Entry	R¹	X	Y	Yield 6 (%)	ee (%)
1	iPr	20	10	78	82
2	Et	20	10	60	54
3	cHex	20	10	75	83
4	t-Bu	40	20	80	95
5	Et₃Si	40	20	43	96
6	n-Bu₃Si	100	30	61	97

polymerization. As summarized in Table 10.3, an array of δ-lactones **6** was generally attained in good yields, with up to 97% ee. The enantioselectivity depended primarily upon the steric bulk of R¹ of acid chlorides, while with the unbranched Et substituent the ee was only moderate (Table 10.3, entry 2).

The trichloromethyl moiety in lactones is a synthetically versatile functional group that can be converted into several valuable functionalities, which are outlined in Scheme 10.9.

Except for the classical [4 + 2] reaction of the ketene enolates catalyzed by the quinuclidine tertiary amine group, a new type of enantioselective asymmetric [4 + 2] annulation between electron-deficient heterodienes and acetylene dicarboxylates had been reported by Waldmann and Kumar group (Scheme 10.10).

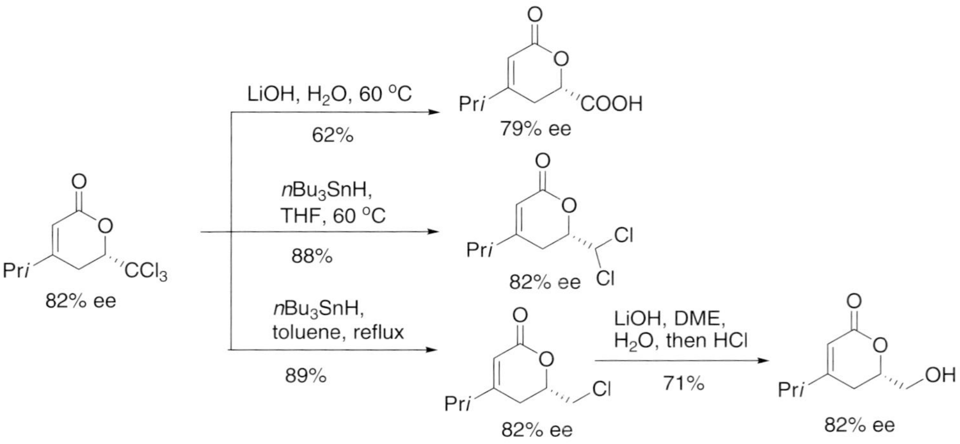

Scheme 10.9 Synthetic modifications of the trichloromethyl group.

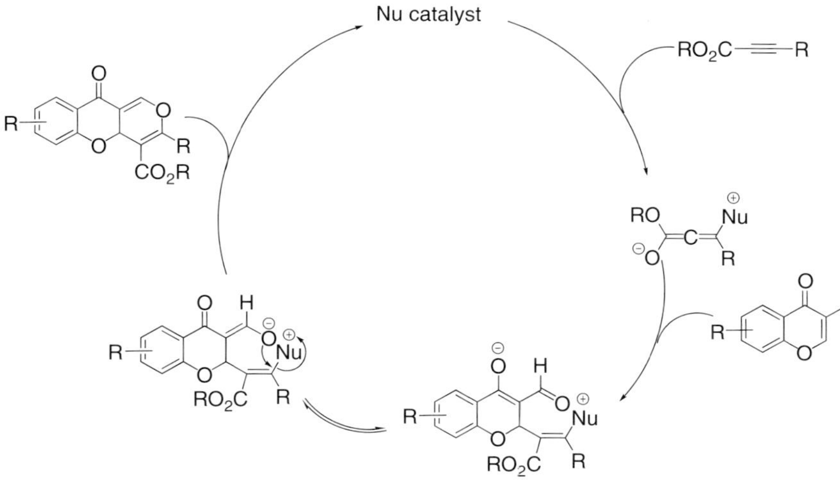

10.2 Asymmetric Cycloadditions Catalyzed by Quinuclidine Tertiary Amine

Scheme 10.10 Postulated mechanism for nucleophilic catalysis.

In the presence of the newly developed chiral organocatalyst **1e**, the ring annulation reaction of acetylene dicarboxylates and 3-formylchromones successfully gave rise to tricyclic benzopyrones **7** in good yield (46–91%) and with enantioselectivity of 81–87% ee (Table 10.4) [14].

Chen and coworkers published a formal [3 + 3]-type reaction to give highly substituted cyclohexenes **8**. This domino process consists of an allylic–allylic alkylation of an α,α-dicyanoalkene derived from 1-indanone and Morita–Baylis–Hillman carbonates, following an intramolecular Michael addition, by employing dual organocatalysis of commercially available modified cinchona alkaloid (DHQD)$_2$AQN **1f** (hydroquinidine (anthraquinone-1,4-diyl) diether) and (S)-BINOL. The cyclic adducts

Table 10.4 Asymmetric [4 + 2] annulation reaction.

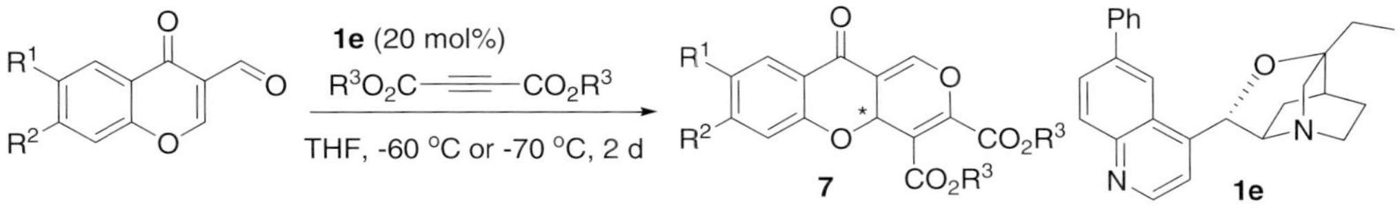

Entry	R^1	R^2	R^3	ee (%)	Yield (%)
1	H	H	Me	83	91
2	Cl	H	Me	82	78
3	iPr	H	Me	81	52
4	H	H	Et	87	52
5	Cl	Me	Me	81	55
6	Br	H	Et	85	46

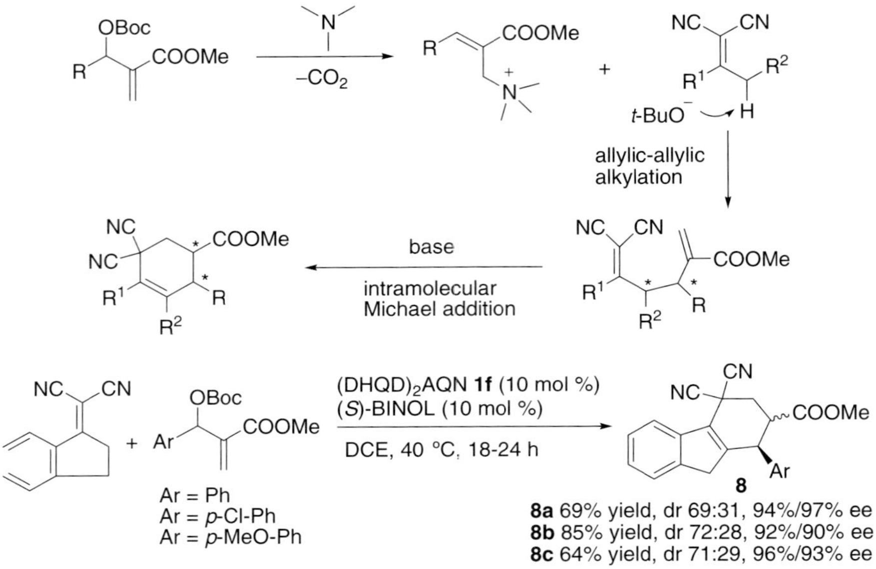

Scheme 10.11 Formal [3 + 3]-type reaction to access cyclohexenes.

were obtained with good to high yield (64–85%), excellent enantioselectivities (92–96% ee), and modest dr values (Scheme 10.11).

It should be noted that the intramolecular Michael addition did not occur under the above catalytic conditions when other α,α-dicyanoalkenes were applied. However, the cyclic product **8d** could be attained with the catalysis of a stronger base DBU (Eq. (10.2)) [15].

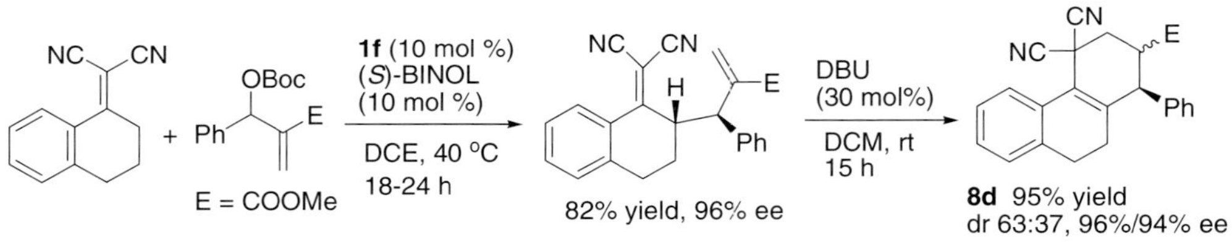

(10.2)

Apart from acting as effective Lewis base catalysts, the quinuclidine structure of cinchona alkaloids can also participate in the other cycloaddition reaction by a different catalytic mechanism. Calter *et al.* described an interesting asymmetric "interrupted" Feist–Bénary reaction between ethyl bromopyruvates and cyclohexadione. They proposed that the protonated cinchona alkaloid would perform as a Brønsted acid to form hydrogen-bonding interaction with α-ketoester moiety, rendering it more electrophilic toward attack by either the enol or enolate of cyclohexandione. Then intramolecular alkylation would afford the formal [3 + 2] cycloadduct (Scheme 10.12) [16].

As outlined in Table 10.5, excellent yield and enantioselectivity were obtained in the reaction of simple ethyl bromopyruvate and cyclohexadione catalyzed by a quinidine

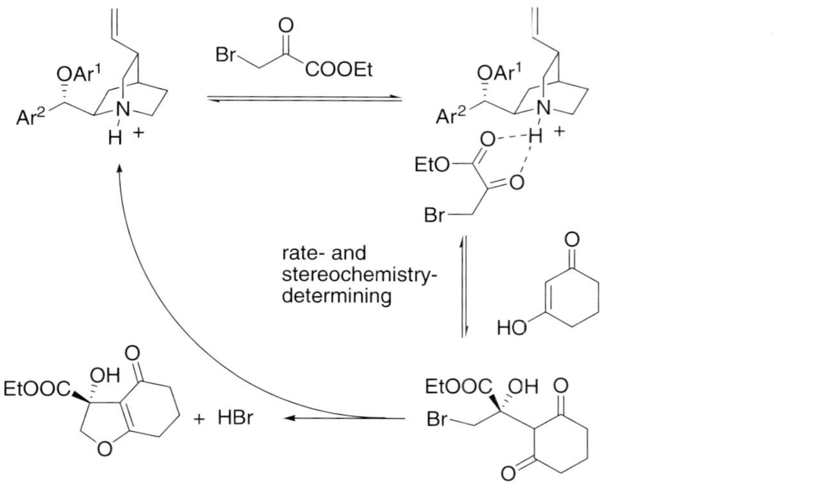

Scheme 10.12 Catalytic mechanism for the asymmetric interrupted Feist–Bénary reaction.

derivative **1g** (10 mol%) in addition to a slight excess of proton sponge (PS) (Table 10.5, entry 1). On the other hand, further modified conditions should be applied in the reactions of racemic, secondary β-bromo-α-ketoesters. As these enantiomers do not interconvert under the reaction conditions, a necessary condition for dynamic kinetic resolution, they found that the interconversion of the enantio-

Table 10.5 Asymmetric "interrupted" Feist–Bénary reaction.

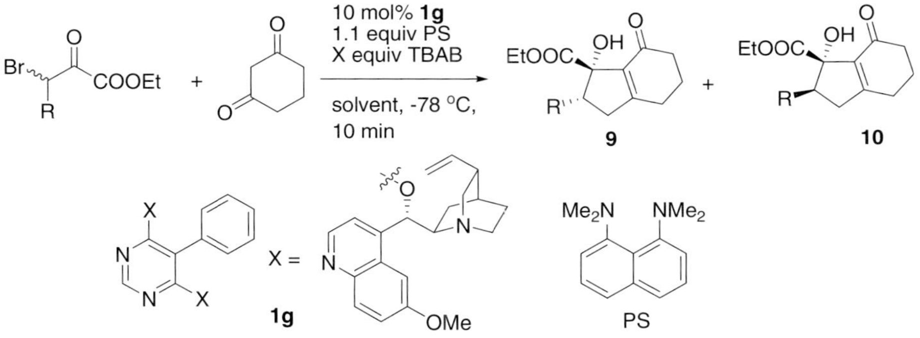

Entry	R	Solvent	X	9:10	Yield 9 (%)	ee (%)
1	H	CH_2Cl_2	0	—	98	91
2	Me	THF	0.5	98:2	95	94
3	nPent	THF	0.5	96:4	96	93
4	I-Bu	THF	0.5	97:3	94	96
5	Ph	THF	0.5	96:4	94	93

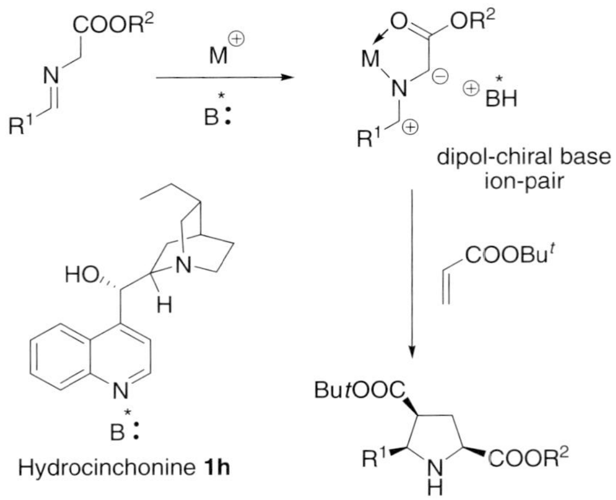

Scheme 10.13 1,3-Dipolar cycloaddition of azomethine ylides activated by a metal salt and a chiral base.

mers could be efficiently accelerated by adding 0.5 equiv of tetrabutylammonium bromide (TBAB) by a S_N2 mechanism in a polar, aprotic solvent THF. Some bicyclic adducts **9** were produced in high yields and almost complete diastereoselectivities and enantioselectivities (entries 2–5).

Azomethine ylides are very important 1,3-dipoles, and they are usually used to react with alkenes leading to the formation of the highly substituted pyrrolidine derivatives [17]. A novel and practical process for the 1,3-dipolar cycloaddition of azomethine ylides with alkenes had been reported by Jørgensen and coworkers [18]. They proposed that a dipol-chiral base ion pair would be generated between α-imino ester–metal complex and a cinchona alkaloid, and subsequent cycloaddition with dipolarophile would take place in a stereoselective manner (Scheme 10.13).

Studies of different cinchona alkaloids as the chiral Brønsted bases and a metal salt showed that hydrocinchonine **1h** (Scheme 10.13) and AgF were the most effective combination. The scope of the [3 + 2] cycloaddition of azomethine ylides and alkenes was investigated. Selected examples are shown in Scheme 10.14. High yields and moderate enantioselectivities were obtained from a variety of α-imino esters. It was worth mentioning that most of the pyrrolidine derivatives **11** obtained from *tert*-butyl acrylate were solids that could be enantiomerically enriched by crystallization.

10.3
Asymmetric Cycloadditions Catalyzed by Bifunctional Cinchona Alkaloids

Natural cinchona alkaloids are bifunctional aminoalcohol compounds. They can catalyze some asymmetric reactions through concerted activation modes, which could generally provide better stereocontrol. In 1996, Nakatani and coworkers

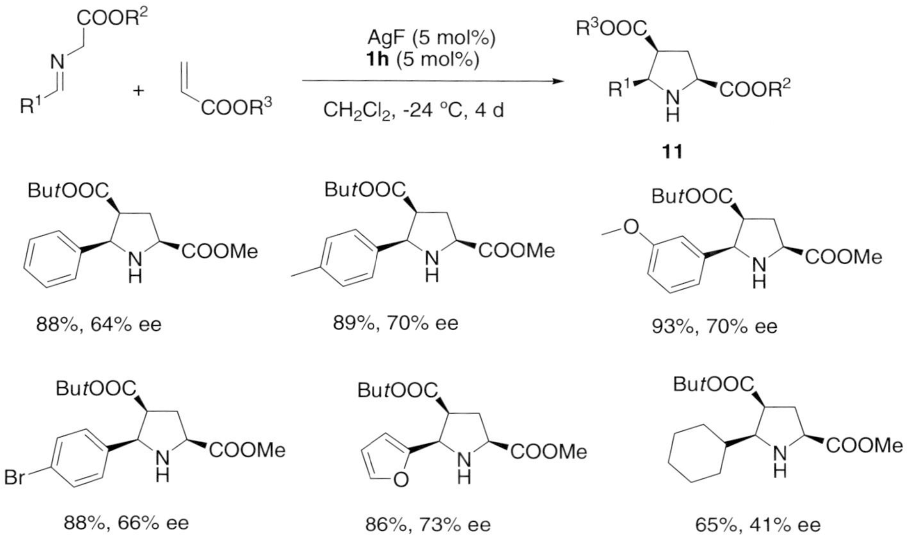

Scheme 10.14 1,3-Dipolar cycloaddition of azomethine ylides and *tert*-butyl acrylate.

explored the early work in Diels–Alder reaction of 3-hydroxy-2-pyrone with N-methylmaleimide catalyzed by natural cinchona alkaloids, while modest enantioselectivity was afforded in the presence of stoichiometric amounts of cinchonidine **1i**. The formation of rigid complex between 3-hydroxy-2-pyrone and cinchonidine was important to achieve better enantiocontrol (Scheme 10.15) [19].

Recently, more efforts have been devoted to the design of novel bifunctional organocatalysts derived from natural cinchona alkaloids, which have been successfully applied in a number of asymmetric reactions. Deng *et al.* promoted the extensive

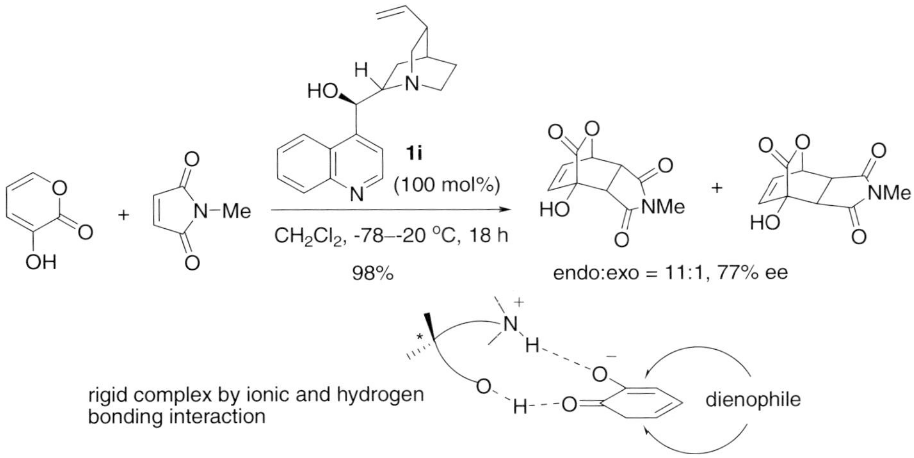

Scheme 10.15 Diels–Alder reaction of 3-hydroxy-2-pyrone with N-methylmaleimide.

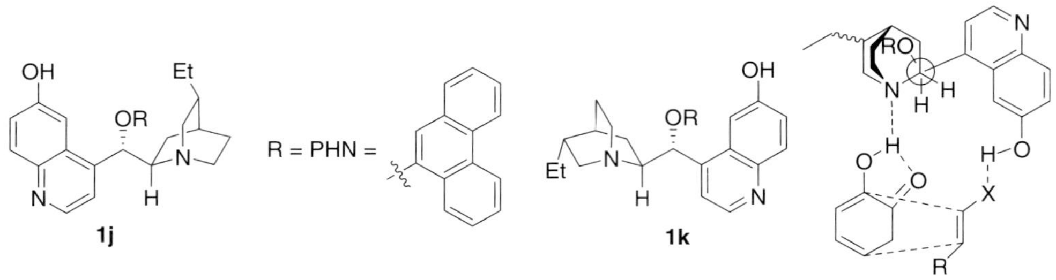

Scheme 10.16 Bifunctional catalysis for Diels–Alder reaction of 2-pyrone.

work on the Diels–Alder reaction of 3-hydroxy-2-pyrones with dienophiles [20]. The 6′-OH derivatives **1j** (from quinidine) or **1k** (from quinine) could simultaneously raise the energy of the HOMO of 3-hydroxy-2-pyrone and lower the energy of the LUMO of dienophiles, and it afforded significantly better catalytic efficiency than that by unmodified cinchona alkaloid (Scheme 10.16).

The bifunctional catalysts **1j** and **1k** were found to tolerate a broad variety of pyrones and dienophiles. As shown in Table 10.6, the reactions proceeded well from 76 : 24 to 93 : 7 dr, and the major diastereoisomers **12** were obtained mostly in greater than 90% ee. Even a relatively unreactive dienophile 3-methylbut-3-en-2-one could be employed, thereby generating optically active chiral building blocks containing two adjacent quaternary stereocenters (Table 10.6, entry 4).

However, when the dienophile was replaced with unsaturated nitrile, **1j** (or **1k**) was proved to be ineffective. Other bifunctional catalysts, 9-thiourea cinchona alkaloids **1l** and **1m** [21], which were prepared from 9-amino-9-deoxyepiquinidine and 9-amino-

Table 10.6 Diels–Alder reaction with **1j** and **1k** (in parentheses).

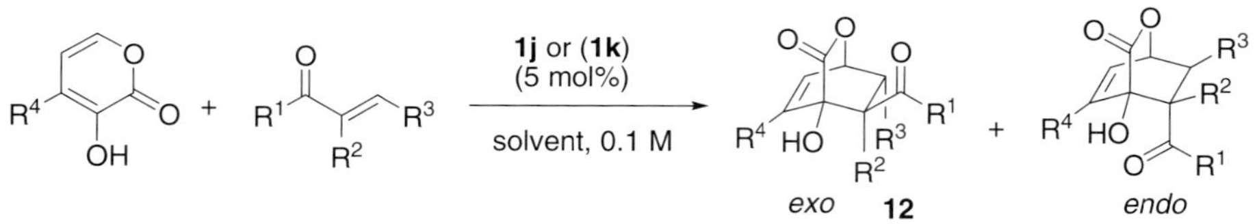

Entry	R^4	R^1, R^2, R^3	Solvent	T (°C)	exo : endo	Yield[a] (%)	ee (%)
1	H	Ph, H, COOEt	Et$_2$O	rt	93 : 7 (94 : 6)	87 (90)	94 (87)
2	H	p-BrPh, H, COOEt	Et$_2$O	rt	91 : 9 (94 : 6)	91 (93)	91 (83)
3	H	Ph, H, PhCO	Et$_2$O	rt	93 : 7	100[b]	90
4[c]	H	Me, Me, H	Et$_2$O	rt	24 : 76 (26 : 74)	65 (63)	91 (90)
5	Ph	Ph, H, COOEt	Et$_2$O	0	95 : 5	84	85
6	Me	Ph, H, COOEt	Et$_2$O	0	88 : 12	87	82
7	Cl	Ph, H, COOEt	EtOAc	rt	86 : 14	77	84

[a]For major diastereomers.
[b]Yield of *exo* product.
[c]10 mol% catalyst was used.

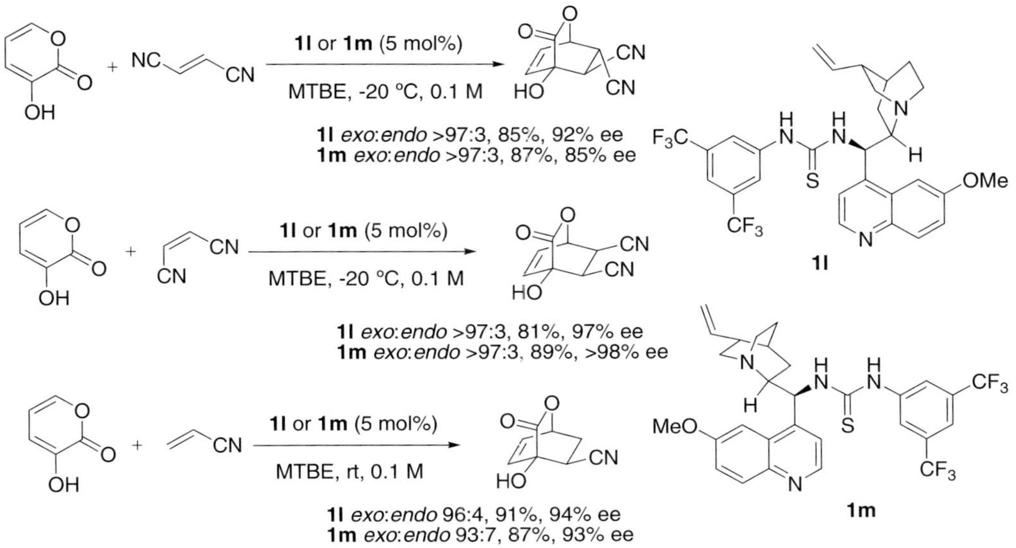

10.3 Asymmetric Cycloadditions Catalyzed by Bifunctional Cinchona Alkaloids | 311

Scheme 10.17 Diels–Alder reaction catalyzed by bifunctional thiourea catalysts.

9-deoxyepiquinine, respectively, had to be used to get high enantioselectivity and diastereoselectivity [20] (Scheme 10.17).

Zhang et al. investigated the asymmetric 1,3-dipolar cycloaddition of *tert*-butyl 2-(diphenylmethyleneamino)acetate and nitroalkenes promoted by bifunctional thiourea compounds derived from cinchona alkaloids, affording chiral pyrrolidine derivatives **13** with multisubstitutions. Catalyst **1m** delivered the best results in terms of catalytic activity, diastereoselectivity and enantioselectivity. Nevertheless, only moderate ee values could be obtained while the diastereoselectivities were generally good (Scheme 10.18) [22].

Optically pure 2,3-dihydropyrroles are important unsaturated heterocyclic compounds because of their application as chiral building blocks in the organic synthesis and the total synthesis of natural products. However, the asymmetric organocatalytic synthesis of chiral 2,3-dihydropyrroles is scarce. Highly diastereo- and enantioselective syntheses of 2,3-dihydropyrroles **14** by the base catalyzed asymmetric

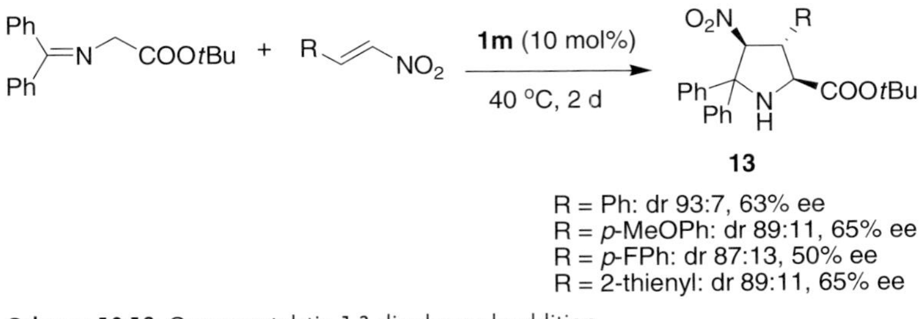

R = Ph: dr 93:7, 63% ee
R = *p*-MeOPh: dr 89:11, 65% ee
R = *p*-FPh: dr 87:13, 50% ee
R = 2-thienyl: dr 89:11, 65% ee

Scheme 10.18 Organocatalytic 1,3-dipolar cycloaddition.

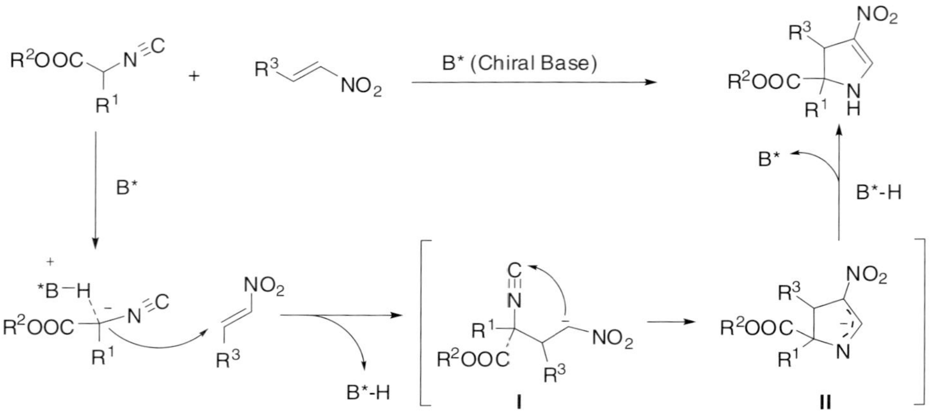

Scheme 10.19 Formal [3 + 2] cycloaddition of isocyanoesters to nitroalkenes.

formal [3 + 2] reaction of isocyanoesters to nitroolefins had been designed by Gong and coworkers [23]. They proposed that chiral base could promote an asymmetric Michael addition of isocyanoesters to nitroalkenes by activating the acidic α-carbon atom of isocyanoesters to generate intermediates **I**. Subsequent intramolecular cyclization reactions of intermediates **I** afforded precursor 1,2-dihydropyrroles **II**, which may be converted into dihydropyrrole after protonation (Scheme 10.19).

Various cinchona alkaloids and derivatives have been investigated in the asymmetric cycloaddition of isocyanoesters and nitroolefins. With regard to the type of organocatalysts investigated, the catalyst **1n**, as a bifunctional organocatalyst bearing tertiary amine and 6′-OH structures, had served as an efficient and highly stereoselective organocatalyst. 6′-Hydroxy group could introduce H-bond into the catalytic process, which can significantly promote the catalytic efficiency. Dichloromethane was found to be very useful as solvent, while chloroform or toluene was used as low yield (9–31%) was obtained. Given attention to the reaction rate and stereoselectivity, the reaction temperature had been locked at 35 °C. A broad range of nitroolefins and isocyanoesters were proceeding in the presence of a catalytic amount (20 mol%) of catalyst **1n**. Good to high yields and excellent enantioselectivity in the range 90–>99% ee were usually obtained. Some selected examples are shown in Scheme 10.20.

10.4
Asymmetric Cycloaddition Reactions Catalyzed by Cinchona-Based Primary Amines

In recent years, the *in situ* generation of chiral iminium ions from a chiral secondary amine and the Michael acceptor, mostly α,β-unsaturated aldehydes and ketones, is used as a powerful strategy for a range of asymmetric cycloaddition reactions [24]. However, the chiral α,β-unsaturated iminium ions, provided by reversible condensation from α,β-unsaturated ketones with secondary amine, could not always work well in this asymmetric approach, probably because of poor generation of the corresponding iminium cations [25].

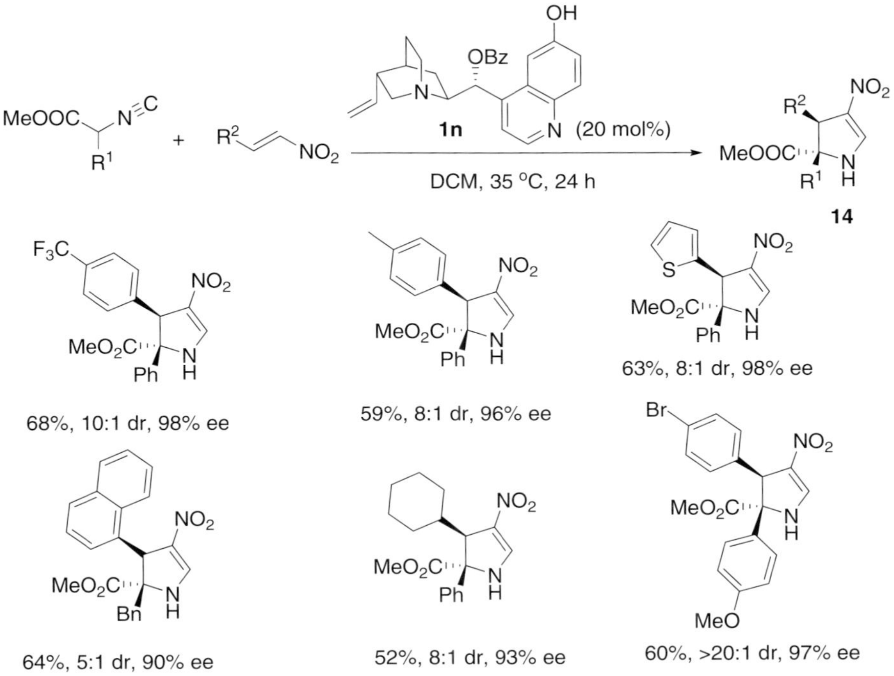

10.4 Asymmetric Cycloaddition Reactions Catalyzed by Cinchona-Based Primary Amines | 313

Scheme 10.20 Synthesis of chiral 2,3-dihydropyrroles.

Chen and coworkers have reported a new domino Michael–Michael addition reaction between α,α-dicyanoalkene [26] derived from cyclohexanone and benzylideneacetone, resulting in a stepwise [4 + 2]-type cycloaddition to afford almost enantiopure bicyclic adduct **15**. In contrast to the completely inert function of secondary ammonium salt, a primary amine, 9-amino-9-deoxyepiquinine **1o** [27], in combination with trifluoroacetic acid, was found to be highly efficient in the activation of the α,β-unsaturated ketone by tandem iminium–enamine catalysis (Scheme 10.21) [28].

A variety of α,α-dicyanoalkenes derived from aryl ketones have also been extensively explored under the same catalytic conditions, and in general the vinylogous Michael adducts were obtained due to the steric hindrance in the following enamine catalysis by primary amine **1o** [28]. Nevertheless, an interesting domino Michael–Michael–*retro*-Michael reaction was observed for α,α-dicyanoalkenes derived from acetophenone and propiophenone, giving a facile process to chiral 2-cyclohexen-1-one derivatives. It was noteworthy that a kinetic resolution was observed in the intramolecular Michael addition step (Scheme 10.22).

Chen *et al.* have further applied the TFA salt of 9-amino-9-deoxyepiquinine **1o** as chiral organocatalyst to the desymmetrization of prochiral α,α-dicyanoalkenes from 4-substituted cyclohexanones via domino Michael–Michael addition reactions with α,β-unsaturated ketones. These reactions exhibited high synthetic efficacy and bicyclic products **16** with two new C−C bonds, and four stereogenic centers

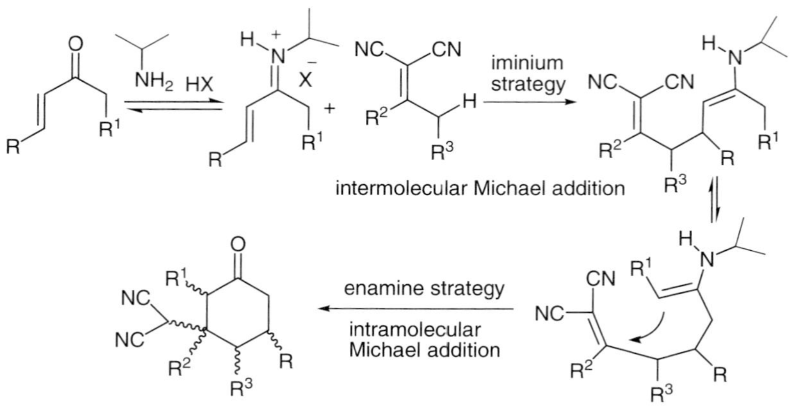

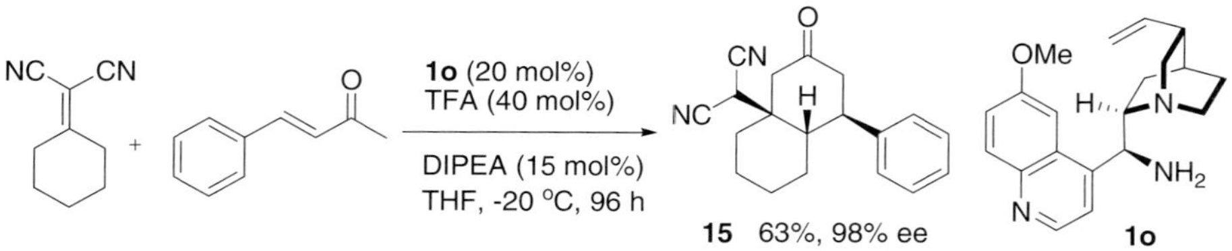

Scheme 10.21 Domino Michael–Michael addition by iminium–enamine catalysis.

including one quaternary carbon center were assembled in a single operation with high level of stereoselectivity. Selected examples were outlined in Table 10.7 (dr >99:1, up to >99% ee) [29].

Based on the experiences on the Diels–Alder reaction of 2-pyrones [20], Deng et al. have investigated the stereoselective Diels–Alder reaction of 2-pyrones and α,β-unsaturated ketones. They also found that primary amines **1o** and **1p** derived from quinine and quinidine, respectively, were the optimal iminium catalysts (5 mol%). The acid additive has a crucial effect on the efficiency of the reaction. TFA (20 mol%) gave the best results, and the Michael reaction, as a side reaction, could also be prohibited.

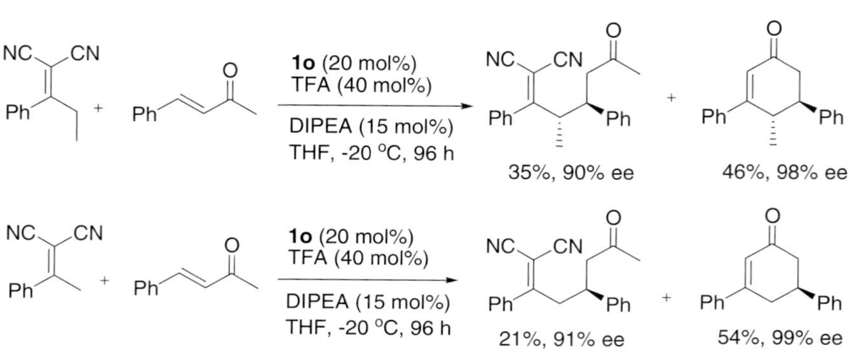

Scheme 10.22 Vinylogous Michael addition and domino Michael–*retro*-Michael reaction.

10.4 Asymmetric Cycloaddition Reactions Catalyzed by Cinchona-Based Primary Amines

Table 10.7 Desymmetrization of prochiral α,α-dicyanoalkenes via domino Michael–Michael addition.

Entry	R	R^1	Yield 16 (%)	ee (%)
1	Me	Ph	81	99
2	Me	p-CH$_3$OPh	71	>99
3	Me	p-ClPh	64	>99
4	Me	m-ClPh	72	95
5	Me	1-Np	45	98
6	Me	2-Thienyl	60	97
7	Ph	Ph	61	95
8	t-Bu	Ph	71	>99
9	BzO	Ph	60	98

On the other hand, lower reaction temperature (0–30 °C) was indispensable to decrease the background reaction. Under these conditions, a wide range of other α,β-unsaturated ketones and substituted 2-pyrones had been converted into bicyclic chiral compounds **17** in high yield, diastereomeric ratio, and enantiomeric excess (Table 10.8). Interestingly, the authors noted that, in contrast to 2-pyrone, electron-rich dienes bearing neither a hydrogen-bond acceptor nor donor such as cyclopentadiene and cyclohexadiene were inactive for the Diels–Alder reaction with benzylideneacetone catalyzed by **1p** and TFA. They propose that the activation of 2-pyrone by the multifunctional amine **1p** is also required for the D–A reaction to occur [30].

Both Michael reaction and the Henry reaction represent powerful C–C bond forming processes, which could provide important tools for the construction of highly functionalized carbon skeletons. Recently, cinchona alkaloids and their derivatives have been identified as efficient bifunctional organocatalysts in asymmetric Michael reaction and Henry reaction. On the basis of the above concept, an asymmetric tandem Michael–Henry reaction was successfully achieved by Zhong and coworkers [31]. In pioneering work, several cinchona alkaloid and derivative catalysts had been screened in this Michael–Henry reaction, involving a nitroolefin and carbon nucleophiles. However, the thiourea bifunctional catalysts such as **11** and **1m** [21], which are effective in asymmetric Michael reactions, did not give the satisfiying results. In contrast, the primary amine, 9-amino-9-deoxyepiquinine **1o**, provided the best results. Zhong *et al.* have proposed an interesting catalytic mode for their reaction: the nitroolefins are assumed to interact with the primary amine moiety of **1o** via double hydrogen bonding interaction, thus enhancing the electrophilic character of the reacting carbon center. The carbanion (adjacent to the nitro group)

316 | *10 Cinchona-Catalyzed Cycloaddition Reactions*

Table 10.8 Asymmetric Diels–Alder reaction of 2-pyrones and α,β-unsaturated ketones.

Entry	R	R^1; R^2	T (°C)	dr (exo:endo)	Yield (%) exo + endo/exo	ee (%)
1[a]	H	Ph Me	0	80:20	87/69	98
2[a]	H	p-ClC$_6$H$_4$ Me	0	82:18 79:21	91/74 89/69	96 98
3	H	p-MeOC$_6$H$_4$ Me	0	81:19 84:16	95/74 68/57	96 97
4	H	2-Thienyl Me	0	80:20	85/67	99
5	H	2-Furyl Me	0	80:20	71/56	99
6	H	Me Me	−30	84:16	92/77	99
7	H	(CH$_2$)$_3$Cl Me	−20	76:24	83/63	96
8	H	H Et	−20	97:3	99/96	99
9	H	H nPen	−20	97:3	98/91	96
10	Ph	Ph Me	0	83:17	83/63	96
11	Me	Ph Me	0	80:20	60/48	96
12	Cl	p-BrC$_6$H$_4$ Me	0	67:33	63/42	90

[a]Data in parentheses were obtained with **1o**.

generated from the Michael addition then attacks the carbonyl group to afford Henry products.

As outlined in Scheme 10.23, this catalytic strategy could be applied for the synthesis of multifunctionalized cyclohexanes **18** and cyclopentanes **19** with two quaternary stereocenters in excellent enantioselectivities (up to >99% ee) and high diastereoselectivities (93 : 7–99 : 1).

Zhong *et al.* also developed a domino double Michael reaction to access multi-substituted cyclopetanes **20** with excellent enantioselectivities (90–99% ee) and high diastereoselectivities (95 : 5–>99 : 1) catalyzed by 9-amino-9-deoxyepiquinine **1o** (Scheme 10.24). A similar concerted activation mode was proposed [32].

In 2000, MacMillan *et al.* developed the first amine-catalyzed 1,3-dipolar cycloaddition of α,β-unsaturated aldehydes and nitrones by LUMO-activation [33]. However,

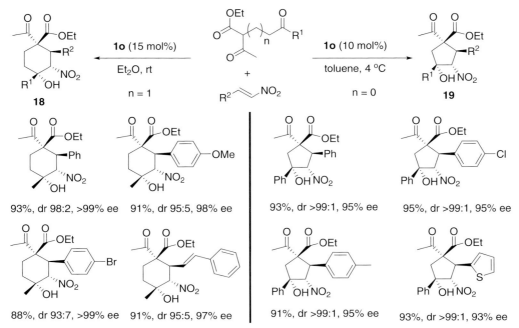

Scheme 10.23 Construction of multifunctionalized cyclohexanes and cyclopentanes.

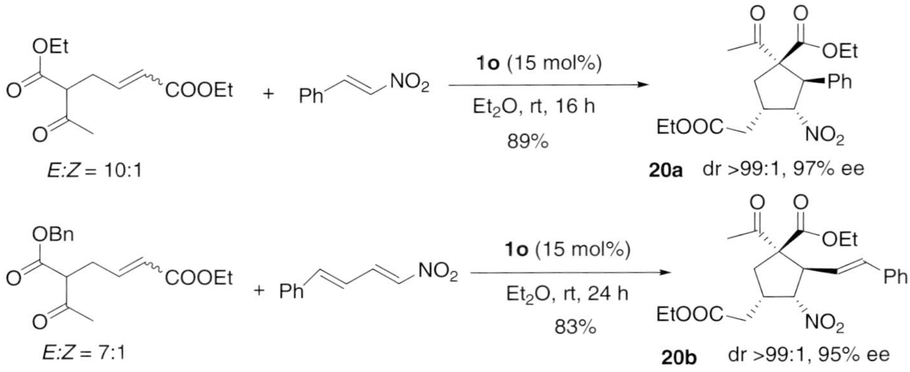

Scheme 10.24 Domino double Michael reaction.

the organocatalytic 1,3-dipolar cycloaddition of enones is rare, probably because of the lack of suitable amine catalysts. Chen et al. found that 9-amino-9-deoxyepiquinine **1o** (10 mol%) in combination with acid (20 mol%) smoothly catalyzed the 1,3-dipolar cycloaddition of azomethine imine and 2-cyclohexen-1-one to give the tricyclic product **21a** with excellent diastereoselectivity (dr >99 : 1) at 20 °C. Unfortunately, only moderate ee values were obtained. Later, it was established that 6′-hydroxy-9-amino-9-deoxyepiquinine **1q**, a newly designed multifunctional primary amine, could serve as a more prominent iminium catalyst in the reaction of 2-cyclohexen-1-one and azomethine imine (Scheme 10.25) [34].

It is clear, as the authors predicted, that the free hydroxy group is critical in the catalytic process. Both the reactivity and the enantioselectivity were dramatically improved when the reaction was catalyzed by **1q** salt. On the other hand, another appearance could also support that a hydrogen bond might be formed from the 6′-OH group with the carbonyl group of dipole. When 4 Å MS was added to the reaction solution to remove the trace amount of water that generated during the formation of an active iminium intermediate, so that the influencing factor of the expected hydrogen bonding might be excluded. As a result, the ee value was raised obviously. The transition state shown was proposed by the

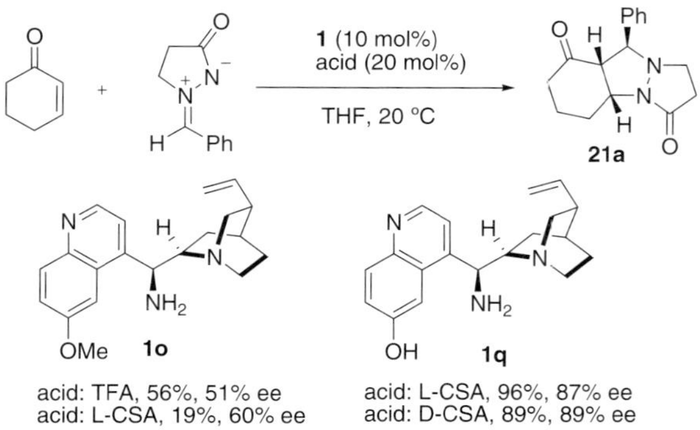

Scheme 10.25 Primary amine catalyzed 1,3-dipolar cycloaddition of cyclic enone.

10.4 Asymmetric Cycloaddition Reactions Catalyzed by Cinchona-Based Primary Amines | 319

Chen's group. Both functional groups of the 9-amino-9-deoxyepicinchona alkaloids catalyst play an important role in the asymmetric catalytic process by coordinating to 2-cyclohexene-1-one and azomethine imine.

re-face attack   endo-selectivity

The authors have further screened the catalytic conditions, and satisfactory results for both enantioselectivity and yield were obtained by using a bulky acid, triisopropylbenzenesulfonic acid (TIPBA), at higher temperature (40 °C). A variety of azomethine imines reacted directly with cyclic enones, forming the desired tricyclic products in good yield (72–99%) and ee (86–95%), while having excellent diastereoselectivity (all >99 : 1) (Table 10.9). Moreover, 6′-OH-9-amino-9-deoxyepi-quinidine **1r**

Table 10.9 Asymmetric 1,3-dipolar cycloaddition of cyclic enones and azomethine imines.

Entry	n	R	t (h)	Yield 21 (%)	ee (%)
1	1	Ph	36	89	90
2	1	p-ClC$_6$H$_4$	60	73	92
3	1	o-ClC$_6$H$_4$	18	80	94
4	1	p-MeOC$_6$H$_4$	36	99	92
5	1	2-Furyl	96	99	95
6	1	iPr	40	76	91
7	1	nPr	40	76	87
8	0	Ph	60	78	90
9	2	p-MeOC$_6$H$_4$	60	76	93
10[a]	1	m-ClC$_6$H$_4$	60	72	−90
11[a]	1	Cyclohexyl	40	83	−85
12[a]	0	p-MeOC$_6$H$_4$	40	75	−90

[a] **1r** was used instead of **1q**.

from quinidine was also prepared, and good results were gained while the products had the opposite configuration (entries 16–19). Hence, both enantiomers of the cycloaddition products could be readily attainable. However, low ee values (<30%) were obtained with acyclic enones.

10.5
Asymmetric Cycloaddition Catalyzed by Cinchona-Based Phase-Transfer Catalysts

Drawing from their success with catalytic [4 + 2] cycloaddition, Lectka group developed another highly enantioselective cycloaddition of o-quinone methide (o-QM) with silyl ketene acetals, using a chiral cinchona alkaloid derived ammonium, N-(3-nitrobenzyl)quinidinium fluoride 1s, as a precatalyst. The free hydroxyl group of the cinchona alkaloid moiety was crucial to high optical induction. A variety of silyl ketene acetals had been screened to afford the cycloadducts 22 with good ee (72–90%) and excellent yield (84–91%) (Scheme 10.26) [35].

γ-Lactam moieties, fused to carbocyclic rings, are common in bioactive natural products. The preparation of optically active γ-lactams by asymmetric synthesis is also resulted in a number of inspiring enantioselective chemical syntheses. Jørgensen and coworkers [36] had demonstrated a novel use of the organocatalytic enantioselective vinylic substitution reaction for the single-step construction of C5 quaternary 3-halo-3-pyrrolin-2-ones (Scheme 10.27).

A key to the reaction is the use of the stereochemically well-defined α,β-dihalogenated acrylate ester as the electrophile in the substitution process. Under the control of a chiral phase-transfer catalyst, enantioselective C−C bond formation by the stereospecific substitution of the chlorine atom, with retention of configuration,

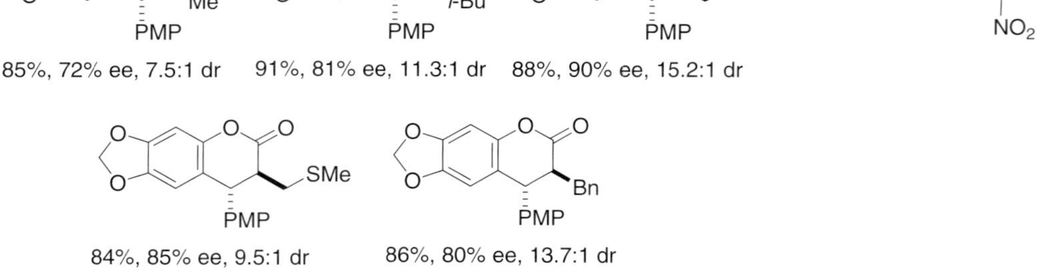

Scheme 10.26 Cycloaddition of o-quinone methide with silyl ketene acetal.

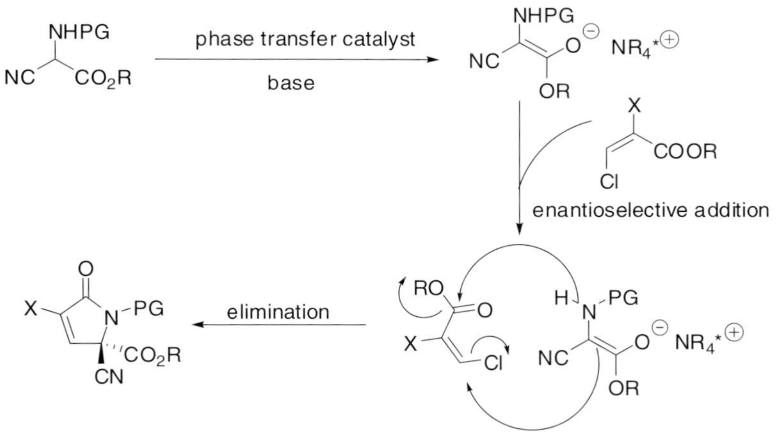

10.5 Asymmetric Cycloaddition Catalyzed by Cinchona-Based Phase-Transfer Catalysts

Scheme 10.27 Preparation of γ-lactams.

by a 1,2-dinucleophile followed by immediate ring closure results in the selective formation of 3-halo-3-pyrrolin-2-ones.

This protocol was subsequently optimized and developed for high enantioselective variants using the dihydrocinchonine-derived phase-transfer catalyst **1t** or **1u**. Enantiopure products **23** were easily obtained after recrystallization with high yields (Table 10.10).

The iodo-substituted products were obtained as single enantiomers in 63–73% yield by recrystallization of the crude reaction mixture. The chloro-substituted product was unfortunately not crystalline, but was isolated in 90% yield and with 85% ee after column chromatography. It is worth to note that the overall process is very practical, scalable, and chromatography-free.

The optically active γ-lactam can be easily modified. It is possible to introduce different substituents at C3 by standard palladium catalyzed Suzuki coupling, Sonogashira coupling, or Negisi coupling reactions (Scheme 10.28) [36].

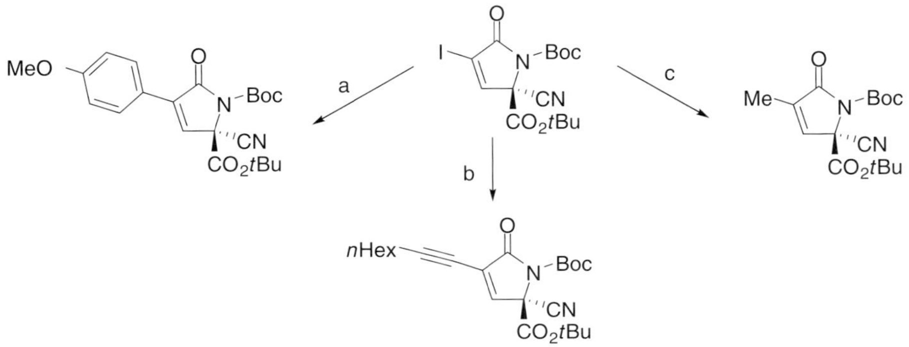

Scheme 10.28 Different coupling reactions with 3-iodo-3-pyrrolin-2-one.

Table 10.10 The synthetic scope for the preparation of 3-halo-3-pyrrolin-2-ones.

Entry	Catalyst	X	ee (%) (crude)	Yield 23 (%)	ee (%) (configuration)
1	1t	I	79	73	>99 (R)
2	1ru	I	78	68	>99 (S)
3	1t	I	89	71	>99 (R)
4	1u	I	73	62	>99 (S)
5	1t	I	91	63	>99 (R)
6	1t	I	80	64	>99 (R)
7	1t	Cl	85	90	85 (R)

10.6
Conclusion

As discussed above, the vast synthetic potential of cinchona alkaloids and their derivatives in the asymmetric cycloadditions has been well demonstrated over the past few years. Cinchona-based organocatalysts possess diverse chiral skeletons and are multifunctional. There is no doubt that the further development of cinchona-based organocatalysts and their catalyzed cycloaddition reactions will continue to provide exciting results in the near future.

References

1. Maruoka, K. (2000) in *Catalytic Asymmetric Synthesis* (ed. I. Ojima), Wiley-VCH Verlag GmbH, New York, Chapter 8, p. 467.

2. (a) Dalko, P.I. and Moisan, L. (2004) *Angew. Chem. Int. Ed.*, **43**, 5138; (b) Berkessel, A. and Gröger, H. (2005) *Asymmetric Organocatalysis*, Wiley-VCH Verlag GmbH, Weinheim, (c) Shen, J. and Tan, C.-H. (2007) *Org. Biomol. Chem.*, **6**, 3229.

3. (a) Hermann, K. and Wynberg, H. (1979) *J. Org. Chem.*, **44**, 2238; (b) Hiemstra, H. and Wynberg, H. (1981) *J. Am. Chem. Soc.*, **103**, 417.

4. For recent reviews of cinchona alkaloids based organocatalysis, see (a) Tian, S.-K., Chen, Y., Hang, J., Tang, L., McDaid, P., and Deng, L., (2004) *Acc. Chem. Res.*, **37**, 621; (b) Gaunt, M.J. and Johansson, C.C.C. (2007) *Chem. Rev.*, **107**, 5596; (c) Chen, Y.-C. (2008) *Synlett*, 1919; (d) Bartoli, G. and Melchiorre, P. (2008) *Synlett*, 1759.

5. (a) Wynberg, H. and Staring, E.G.J. (1982) *J. Am. Chem. Soc.*, **104**, 166; (b) Calter, M.A., Orr, R.K., and Song, W. (2003) *Org. Lett.*, **5**, 4745; (c) Zhu, C., Shen, X., and Nelson, S.G. (2004) *J. Am. Chem. Soc.*, **126**, 5352.

6. (a) Taggi, A.E., Hafez, A.M., Wack, H., Yong, B., Drury, W.J., and Lectka, T. (2000) *J. Am. Chem. Soc.*, **122**, 7831; (b) Taggi, A.E., Hafez, A.M., Wack, H., Yong, B., Ferraris, D., and Lectka, T. (2002) *J. Am. Chem. Soc.*, **124**, 6626; (c) France, S., Shah, M.H., Weatherwax, A., Wack, H., Roth, J.P., and Lectka, T. (2005) *J. Am. Chem. Soc.*, **127**, 1206.

7. (a) Papageorgiou, C., Ley, S.V., and Graunt, M.J. (2003) *Angew. Chem. Int. Ed.*, **42**, 828; (b) Bremeyer, N., Smith, S.C., Ley, S.V., and Granut, M.J. (2004) *Angew. Chem. Int. Ed.*, **43**, 2681; (c) Papageorgiou, C.D., Cubillo de Dios, M.A., Ley, S.V., and Graunt, M.J. (2004) *Angew. Chem. Int. Ed.*, **43**, 4641.

8. Bekele, T., Shah, M.H., Wolfer, J., Abraham, C.J., Weatherwax, A., and Lectka, T. (2006) *J. Am. Chem. Soc.*, **128**, 1810.

9. Wolfer, J., Bekele, T., Abraham, C.J., Dogo-Isonagie, C., and Lectka, T. (2006) *Angew. Chem., Int. Ed.*, **45**, 7398.

10. (a) Abraham, C.J., Paull, D.H., Scerba, M.T., Grabinski, J.W., and Lectka, T. (2006) *J. Am. Chem. Soc.*, **128**, 13370; (b) For an account, see Paull, D.H., Abraham, C.J., Scerba, M.T., Alden-Danforth, E., and Lectka, T., (2008) *Acc. Chem. Res.*, **41**, 655.

11. Abraham, C.J., Paull, D.H., Bekele, T., Scerba, M.T., Dudding, T., and Lectka, T. (2008) *J. Am. Chem. Soc.*, **130**, 17085.

12. Xu, X., Wang, K., and Nelson, S.G. (2007) *J. Am. Chem. Soc.*, **129**, 11690.

13. Tiseni, P.S. and Peters, R. (2007) *Angew. Chem., Int. Ed.*, **46**, 5325.

14 Waldmann, H., Khedkar, V., Dückert, H., Schürmann, M., Oppel, I.M., and Kumar, K. (2008) *Angew. Chem. Int. Ed.*, **47**, 6869.

15 Cui, H.-L., Peng, J., Feng, X., Du, W., Jiang, K., and Chen, Y.-C. (2009) *Chem. Eur. J.*, **15**, 1574.

16 Calter, M.A., Phillips, R.M., and Flaschenriem, C. (2005) *J. Am. Chem. Soc.*, **127**, 14566.

17 For reviews, see (a) Coldham, I. and Hufton, R., (2005) *Chem. Rev.*, **105**, 2765; (b) Nájera, C. and Sansano, J.M. (2005) *Angew. Chem. Int. Ed.*, **44**, 6272; (c) Stanley, L.M. and Sibi, M.P. (2008) *Chem. Rev.*, **108**, 2887.

18 Alemparte, C., Blay, G., and Jørgensen, K.A. (2005) *Org. Lett.*, **7**, 4569.

19 Okamura, H., Nakamura, Y., Iwagawa, T., and Nakatani, M. (1996) *Chem. Lett.*, 193.

20 Wang, Y., Li, H., Wang, Y.-Q., Liu, Y., Foxman, B.M., and Deng, L. (2007) *J. Am. Chem. Soc.*, **129**, 6364.

21 For a review on cinchona-derived thiourea catalysts, see Connon, S.J. (2008) *Chem. Commun.*, 2499.

22 Xue, M.-X., Zhang, X.-M., and Gong, L.-Z. (2008) *Synlett*, 691.

23 Guo, C., Xue, M.-X., Zhu, M.-K., and Gong, L.-Z. (2008) *Angew. Chem. Int. Ed.*, **47**, 3414.

24 For a review on iminium catalysis, see Erkkila, A., Majander, I., and Pihko, P.M., (2007) *Chem. Rev.*, **107**, 5416.

25 Ishihara, K. and Nakano, K. (2005) *J. Am. Chem. Soc.*, **127**, 10504.

26 For the applications of α,α-dicyanoalkenes as versatile vinylogous synthons, see (a) Poulsen, T.B., Alemparte, C., and Jørgensen, K.A., (2005) *J. Am. Chem. Soc.*, **127**, 11614; (b) Xue, D., Chen, Y.-C., Cun, L.-F., Wang, Q.-W., Zhu, J., and Deng, J.-G. (2005) *Org. Lett.*, **7**, 5293; (c) Liu, T.-Y., Cui, H.-L., Long, J., Li, B.-J., Wu, Y., Ding, L.-S., and Chen, Y.-C. (2007) *J. Am. Chem. Soc.*, **129**, 1878.

27 Brunner, H., Bügler, J., and Nuber, B. (1995) *Tetrahedron: Asymmetry*, **6**, 1699.

28 Xie, J.-W., Chen, W., Li, R., Zeng, M., Du, W., Yue, L., Chen, Y.-C., Wu, Y., Zhu, J., and Deng, J.-G. (2007) *Angew. Chem. Int. Ed.*, **46**, 389.

29 Kang, T.-R., Xie, J.-W., Du, W., Feng, X., and Chen, Y.-C. (2008) *Org. Biomol. Chem.*, **6**, 2673.

30 Sing, R.P., Bartelson, K., Wang, Y., Su, H., Lu, X., and Deng, L. (2008) *J. Am. Chem. Soc.*, **130**, 2422.

31 (a) Tan, B., Chua, P.J., Li, Y., and Zhong, G. (2008) *Org. Lett.*, **10**, 2437; (b) Tan, B., Chua, P.J., Zeng, X., Lu, M., and Zhong, G. (2008) *Org. Lett.*, **10**, 3489.

32 Tan, B., Shi, Z., Chua, P.J., and Zhong, G. (2008) *Org. Lett.*, **10**, 3425.

33 Jen, W.S., Wiener, J.J.M., and MacMillan, D.W.C. (2000) *J. Am. Chem. Soc.*, **122**, 9874.

34 Chen, W., Du, W., Duan, Y.-Z., Wu, Y., Yang, S.-Y., and Chen, Y.-C. (2007) *Angew. Chem. Int. Ed.*, **46**, 7667.

35 Alden-Danforth, E., Scerbe, M.T., and Lectka, T. (2008) *Org. Lett.*, **10**, 4951.

36 Poulsen, T.B., Dickmeiss, G., Overgaard, J., and Jørgensen, K.A. (2008) *Angew. Chem. Int. Ed.*, **47**, 4687.

11
Cinchona-Based Organocatalysts for Desymmetrization of *meso*-Compounds and (Dynamic) Kinetic Resolution of Racemic Compounds

Ji Woong Lee, Hyeong Bin Jang, Je Eun Lee, and Choong Eui Song

11.1
Introduction

The desymmetrization of *meso*-compounds is a very useful method of preparing enantioenriched molecules from achiral compounds in a single step. In dynamic kinetic resolution (DKR), it is also possible to convert the racemic reactant with 100% completion because both (reactant) enantiomers engage in chemical equilibrium and exchange. In this way, the faster reacting enantiomer is replenished in the course of the reaction at the expense of the slower reacting one. Over the last few decades, remarkable scientific achievements have been made in these research areas by using a variety of enzymatic and nonenzymatic approaches. Among them, in many cases, the naturally occurring cinchona alkaloids and their derivatives have been shown to be superior in terms of both their catalytic activity and their selectivity. Some reactions (e.g., the alcoholytic desymmetrization of cyclic anhydrides) have already reached the level of large-scale synthetic practicability as regards the catalyst activity and enantioselectivity. The key structural feature responsible for their successful utility is the presence of the Lewis basic quinuclidine nitrogen and Lewis acidic substituents at the C9-position, such as hydroxyl, thiourea, and sulfonamide groups.

In this chapter, we attempt to review the current state of the art in the applications of cinchona alkaloids and their derivatives as chiral organocatalysts in these research fields. In the first section, the results obtained using the cinchona-catalyzed desymmetrization of different types of *meso*-compounds, such as *meso*-cyclic anhydrides, *meso*-diols, *meso*-endoperoxides, *meso*-phospholene derivatives, and prochiral ketones, as depicted in Scheme 11.1, are reviewed. Then, the cinchona-catalyzed (dynamic) kinetic resolution of racemic anhydrides, azlactones and sulfinyl chlorides affording enantioenriched α-hydroxy esters, and N-protected α-amino esters and sulfinates, respectively, is discussed (Schemes 11.2 and 11.3).

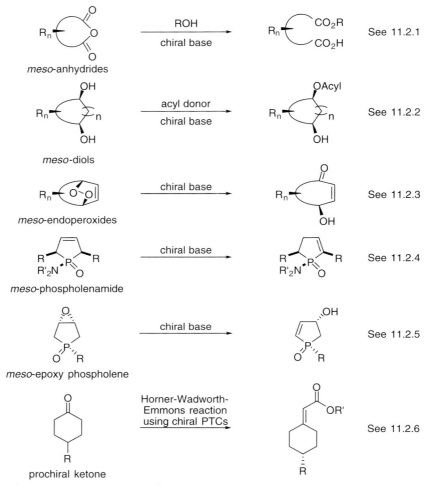

Scheme 11.1 Desymmetrization of *meso*-compounds.

11.2
Desymmetrization of *meso*-Compounds

11.2.1
Desymmetrization of *meso*-Cyclic Anhydrides

There has been considerable interest in the stereoselective ring opening of *meso*-cyclic anhydrides. The stereoselective alcoholysis of these anhydrides is particularly attractive as the resulting hemiesters are used as versatile intermediates in the construction of many bioactive compounds [1]. Much effort has, therefore, been devoted to the development of efficient enzymatic and nonenzymatic catalytic systems for this reaction [2]. Among the stereoselective catalysts developed to date,

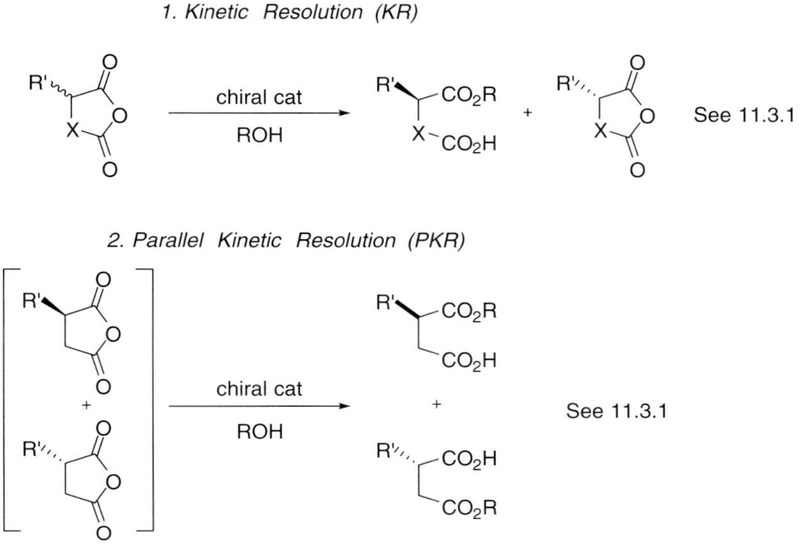

11.2 Desymmetrization of meso-Compounds

1. Kinetic Resolution (KR)

2. Parallel Kinetic Resolution (PKR)

Scheme 11.2 Kinetic resolution of racemic compounds.

cinchona-based catalysts have been regarded as one of the most powerful chirality inducers.

In 1985, the first nonenzymatic catalytic process was reported by Oda and coworkers who found that the natural cinchona alkaloids 1–4 catalyzed the methanolytic

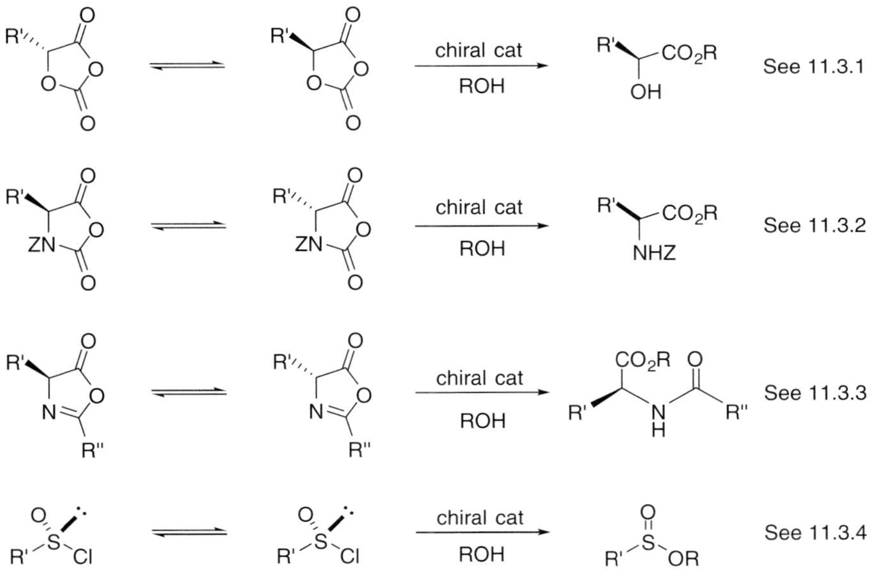

Scheme 11.3 Dynamic kinetic resolution of racemic compounds.

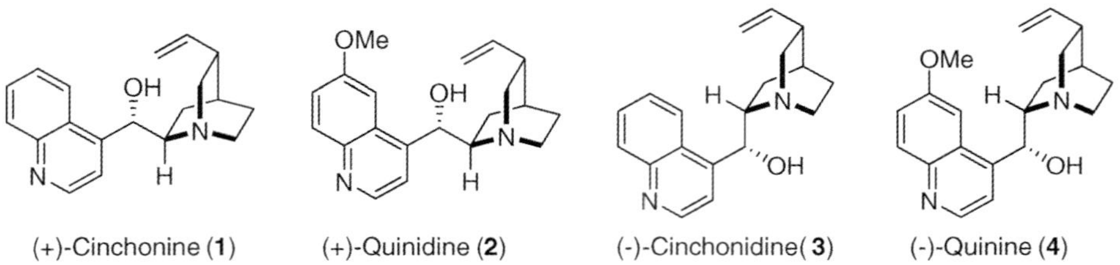

(+)-Cinchonine (1)   (+)-Quinidine (2)   (−)-Cinchonidine (3)   (−)-Quinine (4)

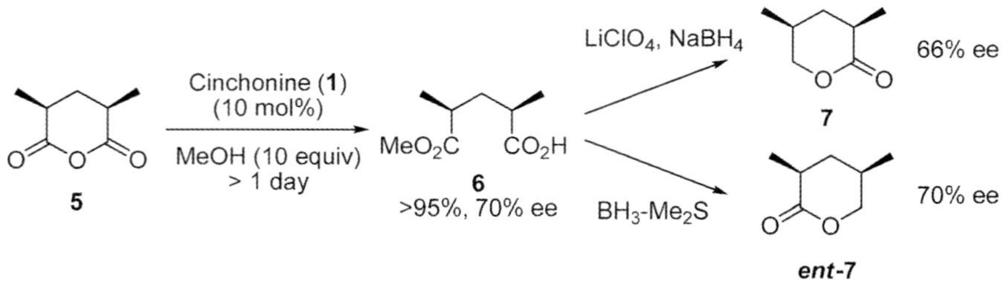

Scheme 11.4

desymmetrization of cis-2,4-dimethyl glutaric anhydride (5) [3a]. The use of cinchonine (1) (10 mol%) as a catalyst provided the hemiester (6) in >95% yield and 70% ee. Its pseudoenantiomer, cinchonidine (3), provided the antipodal hemiester in 64% ee. The obtained hemiester 6 could be easily converted into either the lactone 7 or its enantiomer *ent*-7 by the selective reduction of the carbonyl groups (Scheme 11.4) [3b].

On the basis of the significant kinetic isotope effect ($k_{MeOH}/k_{MeOD} = 2.3$) in the cinchonine (1, 10 mol%)-catalyzed ring opening of cis-2,4-dimethylglutaric anhydride by methanol (20 equiv) in toluene, Oda proposed the general base catalysis mechanism depicted in Scheme 11.5 [3a, b]. Another comparative study on the reaction rates (k_{obs}) obtained with different bases (cinchonidine (2.26×10^{-3}), quinuclidine (2.26×10^{-3}), and quinoline (4.34×10^{-5})) and with different

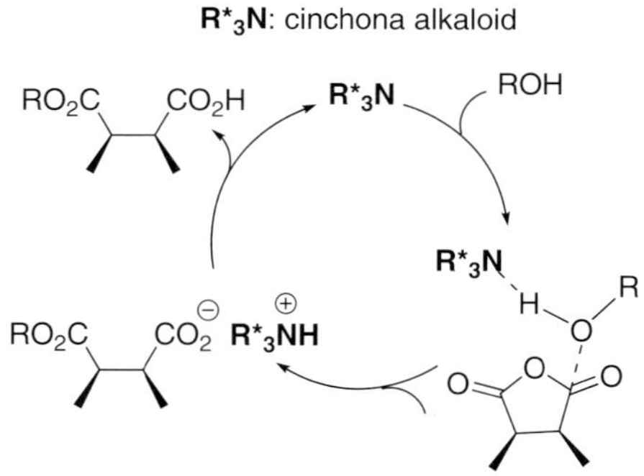

Scheme 11.5 General base catalysis mechanism proposed by Oda.

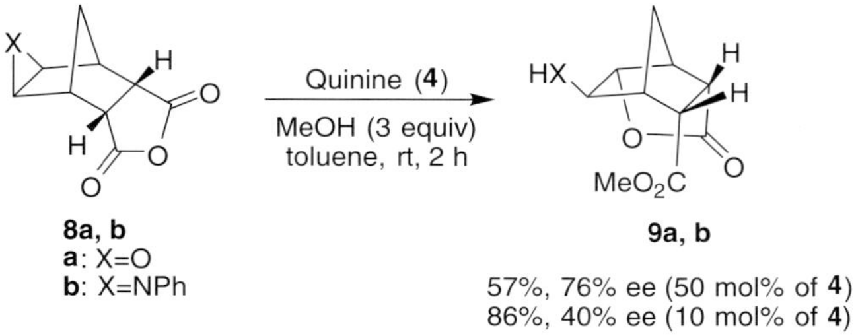

Scheme 11.6

8a, b
a: X=O
b: X=NPh

9a, b
57%, 76% ee (50 mol% of **4**)
86%, 40% ee (10 mol% of **4**)

nucleophiles (MeOH (36 h), EtOH (10 d) and i-PrOH (no reaction)) supported this mechanism [3b].

Shortly thereafter, Aiken and coworkers also reported that quinine (**4**) could be used as a catalyst (50 mol%) to promote the methanolytic desymmetrization of the *meso*-epoxy anhydride **8a** to give the lactone **9a** in 57% yield and 76% ee (Scheme 11.6) [4]. Lowering the reaction temperature to 0 or −30 °C did not result in any increase in selectivity. *meso*-Aziridine anhydride **8b** was also tested under similar reaction conditions, but a lower enantioselectivity (40% ee) was obtained (Scheme 11.6).

A more efficient protocol for the enantioselective alcoholysis of *meso*-cyclic anhydrides mediated by cinchona alkaloids was reported by Bolm and coworkers in 1999 and subsequent publications. Their optimization study on the reaction conditions (reaction temperature and solvent system) revealed that, in the presence of 110 mol% of quinidine (**2**), a range of different *meso*-cyclic anhydrides underwent alcoholysis to give methyl hemiesters in 61–99% yield and 85–99% ee with 3 equiv of MeOH in toluene/CCl$_4$, at −55 °C [5]. Under identical conditions, quinine (**4**) provided the opposite enantiomers with comparable enantioselectivities (75–99% ee). However, reducing the amount of quinidine to 10 mol% resulted in a significantly lower yield and dramatic loss of enantioselectivity. However, by using base additives to reactivate the protonated cinchona alkaloids, the catalytic protocol was accomplished [6]. After intensively screening various tertiary amines to evaluate their efficacy as a reactivator, pempidine (**10**) was found to be the optimal one. Thus, in the presence of quinidine (10 mol%) and pempidine (1 equiv), the methanolytic desymmetrization of a variety of *meso*-cyclic anhydrides at −55 °C afforded the corresponding hemiesters in excellent yields and ee values (Scheme 11.7). A further systematic optimization study revealed that benzyl alcohol and toluene were the optimum nucleophile and solvent, respectively. Under these conditions, the hemibenzyl esters could be obtained with high enantioselectivity (up to 99%) [7]. In addition, the crystallinity of many of the benzyl hemiesters was advantageous in terms of the product purification. However, in Bolm's protocol, a low temperature is crucial for achieving an optimum enantioselectivity, consequently resulting in too long a reaction time (6 days).

In 2000, Deng and coworkers [8] found that, in the absence of a stoichiometric achiral base such as **10**, commercially available bis-cinchona alkaloids such as the

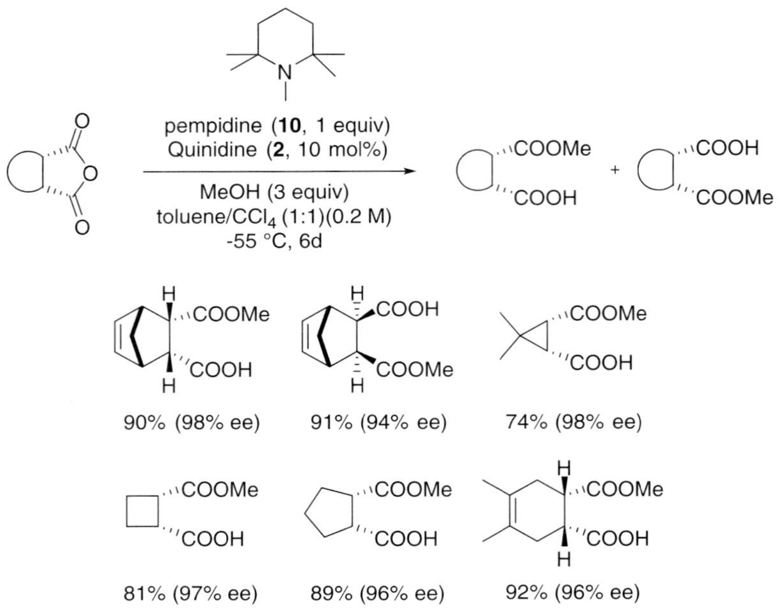

Scheme 11.7

anthraquinone-bridged dimers of dihydroquinidine (DHQD)$_2$AQN (**11**) and dihydroquinine (DHQ)$_2$AQN (**12**) – known as Sharpless ligands – can effectively catalyze the alcoholytic desymmetrization reactions of *meso*-cyclic anhydrides. At a catalyst loading of 5–30 mol%, a broad range of cyclic anhydrides were converted to the corresponding hemimethyl esters in excellent yields and enantiomeric excesses (up to 98% ee) (Scheme 11.8). However, the dihydroquinine-derived (DHQ)$_2$AQN (**12**) always gave slightly lower enantioselectivity than the quinidine analogue **11** [8]. Although this approach represents a significant scientific achievement, as in the case of Bolm's protocol, a low reaction temperature (−20 to −40 °C) proved crucial to achieving an optimal enantioselectivity, consequently resulting in quite a long reaction time (43–120 h) that may restrict its use in large-scale applications. Moreover, it was reported later by the research scientists of the Daiso Co. Ltd., Japan that the dimeric cinchona alkaloid catalysts **11** and **12** are not readily available on an industrial scale, because of the difficulty in synthesizing them [9].

To overcome some of the drawbacks of the previously reported protocols, in 2008, the Song and Connon groups independently investigated the potential of bifunctional cinchona-based (thio)ureas such as **13** [10] as catalysts in the methanolysis of *meso*-anhydrides [11]. It was assumed by the authors that the quinuclidine group of the catalyst would be able to function as a general base catalyst and activate the nucleophile (alcohol), while the thiourea group would be able to simultaneously activate the electrophile (anhydride) by double hydrogen bonding. Consistent with this assumption, in the presence of the thiourea catalyst **13** (1–10 mol%), a range of substrates, including mono-, bi-, and tricyclic anhydrides,

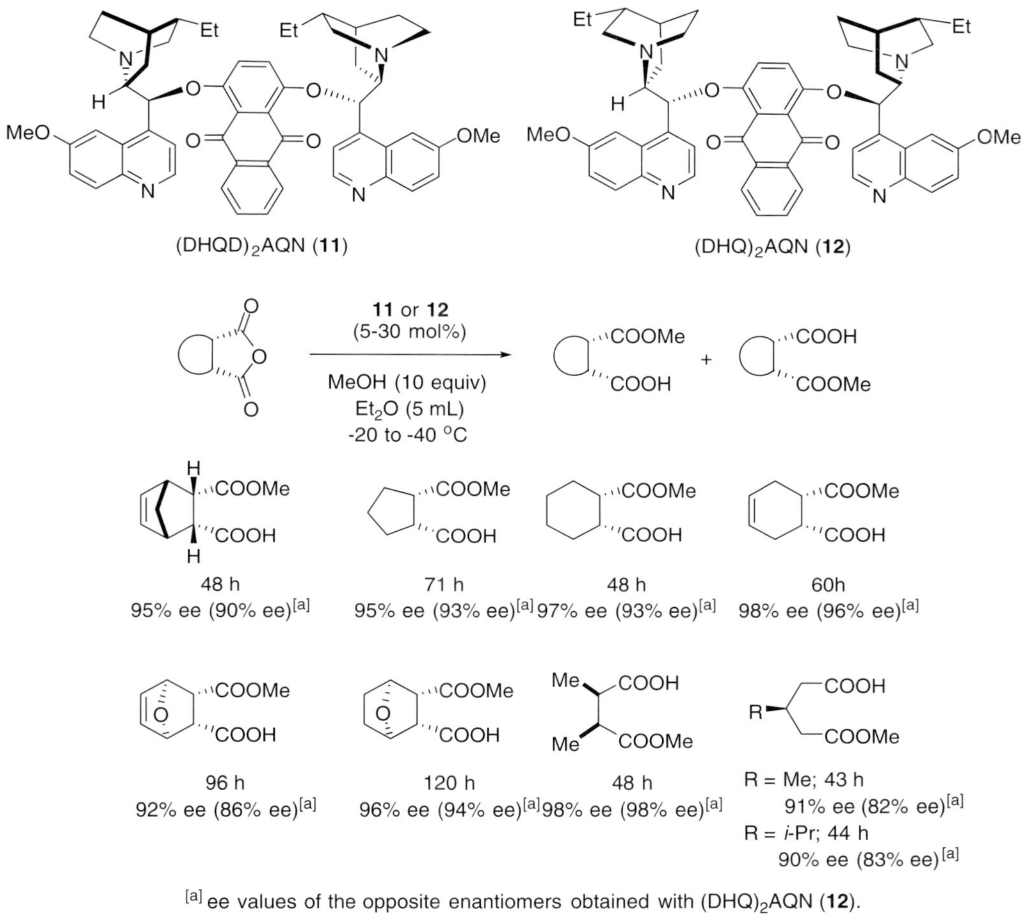

11.2 Desymmetrization of meso-Compounds

Scheme 11.8

[a] ee values of the opposite enantiomers obtained with (DHQ)$_2$AQN (**12**).

were found to be very smoothly converted at room temperature to the corresponding hemiesters in excellent yields and ee values (up to 97% ee) (Scheme 11.9). More recently, Connon and coworkers also presented the thiolysis of 3-methylglutaric anhydride with the same catalyst to afford 98% yield and 92% ee values [12]. To explain the observed sense of stereoselectivity and increased reaction rate, Song proposed the plausible transition-state structure **14** (Figure 11.1) in which the quinuclidine nitrogen and thiourea moiety work cooperatively to stabilize the transition state.

However, highly interestingly, unusual concentration and temperature effects on the enantioselectivity were observed by Song and coworkers [11a]. As shown in Figure 11.2, the enantioselectivity in the methanolytic desymmetrization reaction of the *meso*-cyclic anhydride **15** increases with increasing dilution of the reaction mixture and on raising the reaction temperature from −20 to 20 °C.

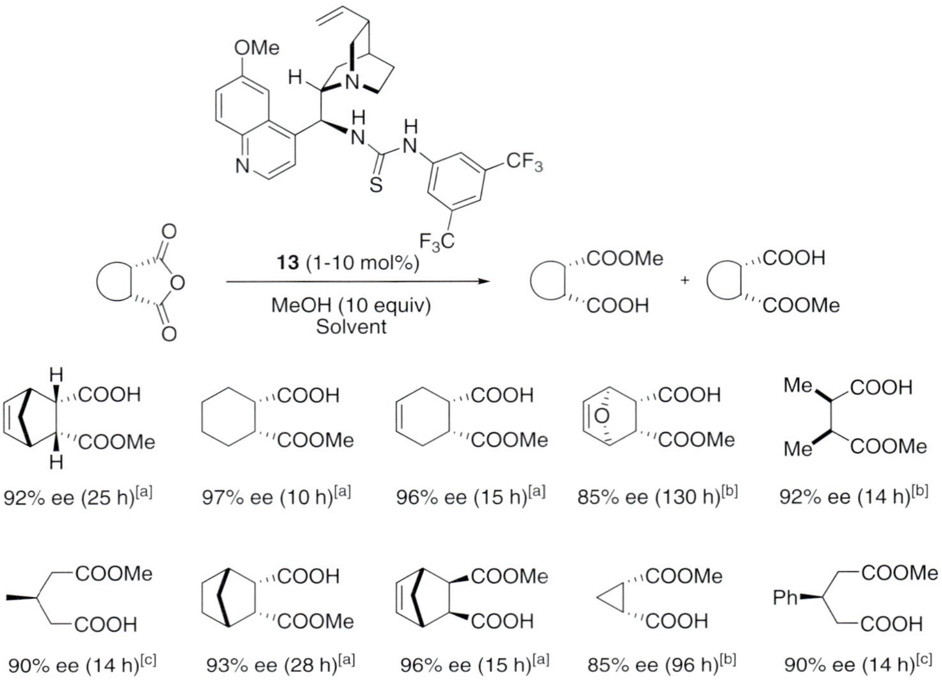

Scheme 11.9

Catalyst **13** (cinchona-based thiourea with 3,5-bis(trifluoromethyl)phenyl group)

Reaction: cyclic anhydride + MeOH (10 equiv), **13** (1–10 mol%), Solvent → monomethyl ester (COOMe/COOH) + regioisomer (COOH/COOMe)

Substrate scope (ee and reaction time):

- norbornene diacid monoester: 92% ee (25 h)[a]
- cyclohexane: 97% ee (10 h)[a]
- cyclohexene: 96% ee (15 h)[a]
- epoxide: 85% ee (130 h)[b]
- Me,Me dimethyl: 92% ee (14 h)[b]
- acyclic methyl: 90% ee (14 h)[c]
- norbornane: 93% ee (28 h)[a]
- bicyclic: 96% ee (15 h)[a]
- cyclopropane: 85% ee (96 h)[b]
- Ph-substituted: 90% ee (14 h)[c]

[a] From ref [11(a)], In 1,4-dioxane (12.5 mM), **13** (10 mol%), rt.
[b] From ref [11(b)], In MTBE (15 mM), **13** (1 mol%), rt.
[c] From ref [11(b)], In MTBE (7.5 mM), **13** (1 mol%), 0 °C.

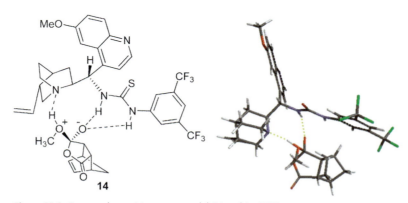

Figure 11.1 Proposed transition-state model **14** and its DFT-computed structure (B3LYP at 6-31G* level).

11.2 Desymmetrization of meso-Compounds

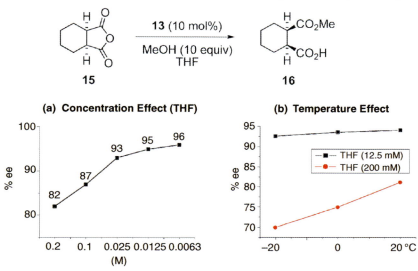

Figure 11.2 Effect of (a) concentration and (b) temperature on the enantioselectivity in the methanolysis of **15** catalyzed by **13**.

In the light of these experimental results, Song and coworkers [11a] suggested that the cinchona-based thiourea catalyst may exist mainly in the dimeric (or higher order) form by self-association at high concentrations and at low temperature. On the other hand, under dilute conditions and at ambient temperature, the catalyst could exist mainly in the monomeric form that is responsible for the high enantioselectivity. It is well known that, according to their X-ray crystal structures, monofunctional and bifunctional (thio)ureas form aggregates in the solid state through H-bonding between the (thio)urea N–H groups and the (thio)urea sulfur or oxygen atom in an intermolecular fashion [13]. Moreover, a recent NMR spectroscopic study also showed that the thiourea **13** exists as a dimer, even in solution [14].

As described above, although cinchona-based thioureas such as **13** are excellent catalysts for alcoholysis, they suffer from self-aggregation, resulting in lower reactivity and strong dependency of the enantioselectivity on the concentration and temperature. To address the self-association issue of bifunctional thioureas, Song and coworkers quite recently developed the new bifunctional cinchona-based sulfonamide catalyst **17** that showed no appreciable effect of the concentration or temperature on the reactivity and enantioselectivity in the methanolysis of *meso*-anhydrides, as shown in Figure 11.3 [15]. Thus, even in relatively highly concentrated conditions (0.1 M), the methanolytic desymmetrization of a variety of mono-, bi-, and tri-cyclic anhydrides gave excellent yields and ee values (up to 98%) with catalyst **17** (5–10 mol%) within a significantly shorter reaction time than that observed for the other reported catalytic systems (Scheme 11.10). For example, the desymmetrization of **15** with 10 mol% of **17** proceeded unprecedentedly fast; the reaction was completed

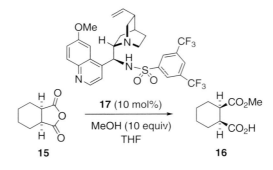

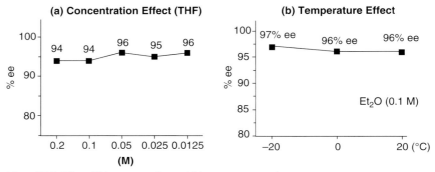

Figure 11.3 Effect of (a) concentration and (b) temperature on the enantioselectivity in the methanolysis of **15** catalyzed by **17**.

within 1 h, affording the chiral hemiester **16** with excellent ee (96% ee). Reducing the amount of catalyst to 0.5 mol% still resulted in excellent catalytic activity and enantioselectivity (e.g., 93% ee) (Scheme 11.10).

The comparative data in Scheme 11.11 highlights the superior catalytic efficiency of the sulfonamide **17** over the thiourea catalyst **13**. Under the same reaction conditions (0.5 mmol of **15**, 5 mmol of MeOH, 5 ml of Et$_2$O, and 1 mol% of catalyst), the sulfonamide catalyst **17** gave excellent enantioselectivity (95% ee) while the thiourea catalyst **13** gave only moderate ee (62%). The stereoselectivity of **13** can be increased to 88% ee only under highly dilute conditions, but at the expense of the reactivity [15].

Very recently, Song and coworkers also developed a promising polymer-supported analogue **18** of the cinchona-based sulfonamide organocatalyst **17** [16]. The styrene-divinylbenzene copolymeric catalyst **18** showed excellent activity and enantioselectivity (up to 97% ee) in the heterogeneous methanolytic desymmetrization of a range of *meso*-cyclic anhydrides. All reactions with **18** (10 mol%) were completed within a few hours to give the corresponding hemiesters in quantitative yields and excellent ee values (up to 97% ee) (Scheme 11.12). All of the products were obtained in the pure form by the simple filtration of the resin **18** followed by the evaporation of the volatiles. Moreover, the indefinite stability of the polymeric sulfonamide **18** under

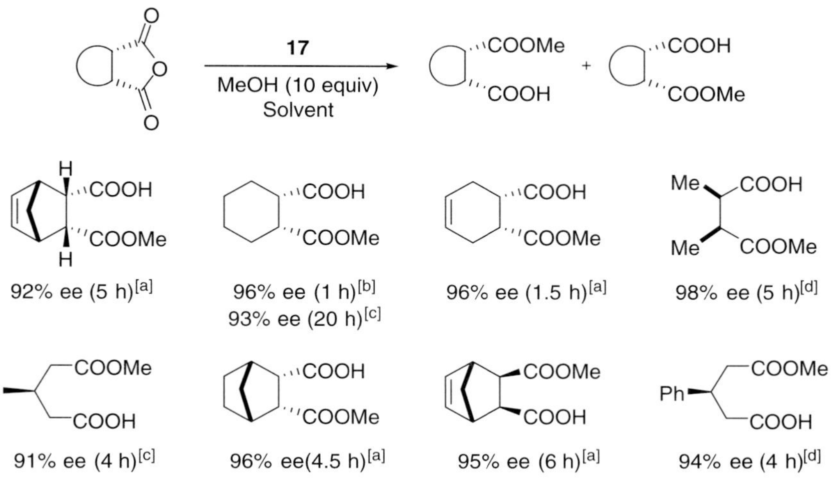

11.2 Desymmetrization of meso-Compounds

[meso-anhydride] →(17, MeOH (10 equiv), Solvent)→ [cyclohexene-COOMe/COOH] + [cyclohexene-COOH/COOMe]

Product	ee / time
bicyclic COOH/COOMe (norbornene)	92% ee (5 h)[a]
cyclohexane COOH/COOMe	96% ee (1 h)[b]; 93% ee (20 h)[c]
cyclohexene COOH/COOMe	96% ee (1.5 h)[a]
Me,Me-COOH/COOMe	98% ee (5 h)[d]
iPr-COOMe/COOH	91% ee (4 h)[c]
norbornane COOH/COOMe	96% ee (4.5 h)[a]
norbornene COOMe/COOH	95% ee (6 h)[a]
Ph-COOMe/COOH	94% ee (4 h)[d]

[a] In Et$_2$O (0.1 M), 5 mol% of catalyst was used at RT.
[b] In Et$_2$O (0.1 M), 10 mol% of catalyst was used at RT.
[c] In Et$_2$O (0.1 M), 0.5 mol% of catalyst was used.
[d] In MTBE, 10 mol% of catalyst was used at -20 °C.

Scheme 11.10

catalytic conditions allowed for its repeated recycling without any loss of turnover time or enantioselectivity even after the tenth run. Therefore, this process seems appealing in terms of its simplicity, practicability, environmental concerns, and cost and is therefore suitable for industrial use.

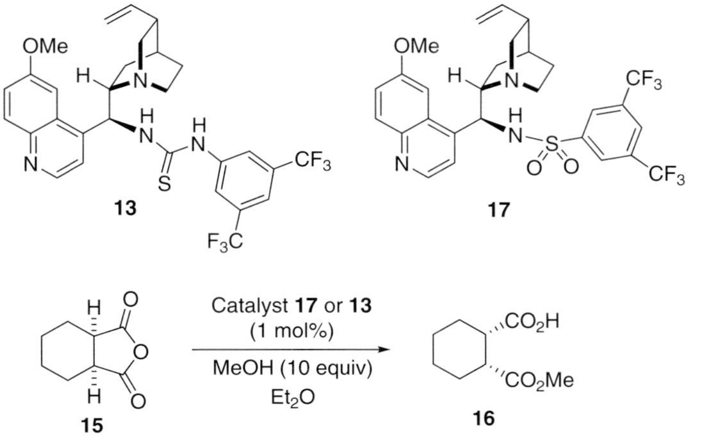

[Structures of catalysts 13 and 17 — cinchona-alkaloid-derived thiourea (13) and sulfonamide (17) bearing 3,5-bis(CF$_3$)phenyl groups]

[15 (cis-cyclohexane-1,2-dicarboxylic anhydride)] →(Catalyst 17 or 13 (1 mol%), MeOH (10 equiv), Et$_2$O)→ [16: CO$_2$H/CO$_2$Me]

Using 17; 95% ee (6 h in Et$_2$O (0.1 M))
Using 13; 62% ee (6 h in Et$_2$O (0.1 M))
       88% ee (20 h in 0.0125 M Et$_2$O)

Scheme 11.11 Methanolytic desymmetrization of **15** with catalysts **17** and **13**.

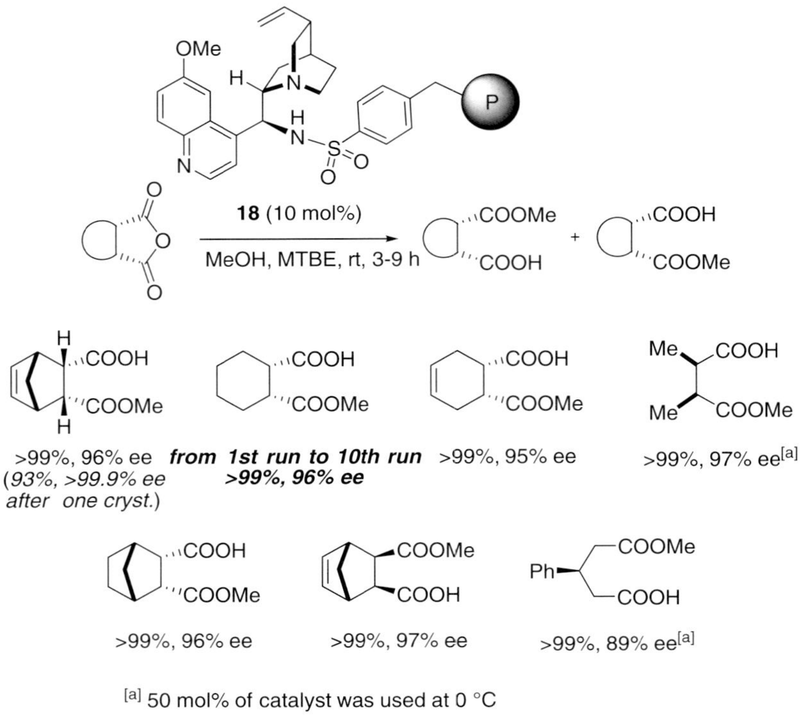
OMe
H
N
H
N
S
O
O
P
18 (10 mol%)
MeOH, MTBE, rt, 3-9 h
COOMe
COOH
COOH
COOMe
>99%, 96% ee
(93%, >99.9% ee after one cryst.)
from 1st run to 10th run
>99%, 96% ee
>99%, 95% ee
>99%, 97% ee[a]
>99%, 96% ee
>99%, 97% ee
>99%, 89% ee[a]
[a] 50 mol% of catalyst was used at 0 °C

Scheme 11.12

11.2.1.1 Applications

The cinchona-catalyzed alcoholysis of *meso*-anhydrides has been successfully applied to the synthesis of key intermediates for a variety of industrially interesting biologically active compounds. Some selected examples are summarized in Scheme 11.13. More detailed information on the synthetic application of this reaction is available in the recent comprehensive review of this topic by Bolm *et al.* [1a].

11.2.2 Desymmetrization of *meso*-Diols

The enantioselective discrimination of one of the hydroxyl groups of *meso*-diols can give chiral monoprotected diols, which serve as versatile intermediates for asymmetric organic synthesis. In addition to the enzymatic methods, a number of chemical approaches have been reported using chiral 1,2-diamine catalysts, chiral phospholane-based catalysts, planar chiral DMAP derivatives, and oligopeptide-based catalysts [2, 28]. Surprisingly, however, relatively a few publications are devoted to this reaction with cinchona-based organocatalysts.

The first cinchona-mediated desymmetrization of *meso*-diols was reported by Duhamel and coworker. Using an excess amount of *O*-benzoylquinidine (**19**, 2 equiv)

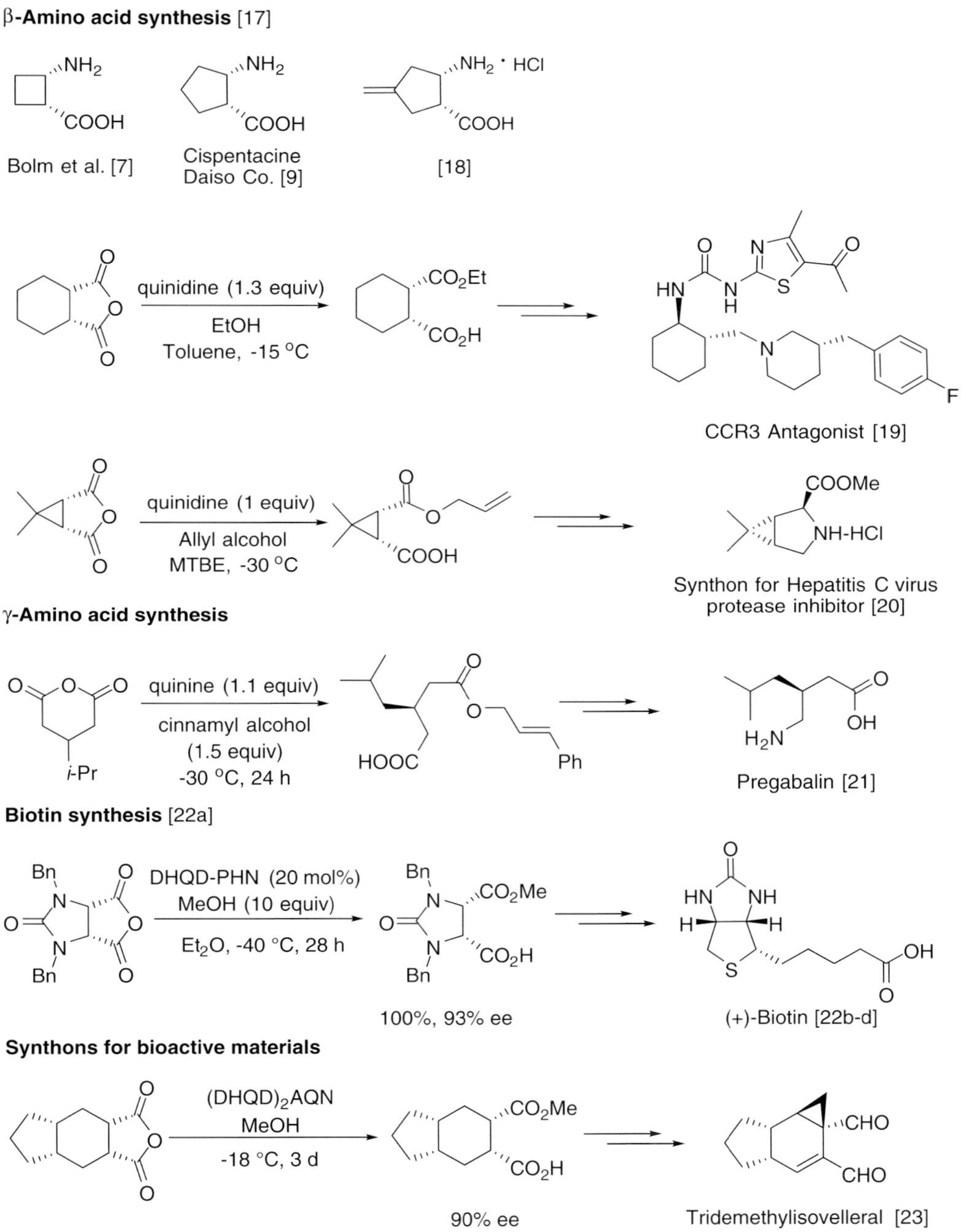
β-Amino acid synthesis [17]
NH2
COOH
NH2
COOH
NH2 · HCl
COOH
Bolm et al. [7]
Cispentacine
Daiso Co. [9]
[18]
quinidine (1.3 equiv)
EtOH
Toluene, -15 °C
CO2Et
CO2H
CCR3 Antagonist [19]
quinidine (1 equiv)
Allyl alcohol
MTBE, -30 °C
COOH
COOMe
NH-HCl
Synthon for Hepatitis C virus
protease inhibitor [20]
γ-Amino acid synthesis
quinine (1.1 equiv)
cinnamyl alcohol
(1.5 equiv)
-30 °C, 24 h
i-Pr
HOOC
Ph
H2N
OH
Pregabalin [21]
Biotin synthesis [22a]
Bn
DHQD-PHN (20 mol%)
MeOH (10 equiv)
Et2O, -40 °C, 28 h
CO2Me
CO2H
100%, 93% ee
HN
NH
OH
(+)-Biotin [22b-d]
Synthons for bioactive materials
(DHQD)2AQN
MeOH
-18 °C, 3 d
CO2Me
CO2H
90% ee
CHO
CHO
Tridemethylisovelleral [23]

Scheme 11.13 Applications of alcoholytic desymmetrization of *meso*-anhydrides.

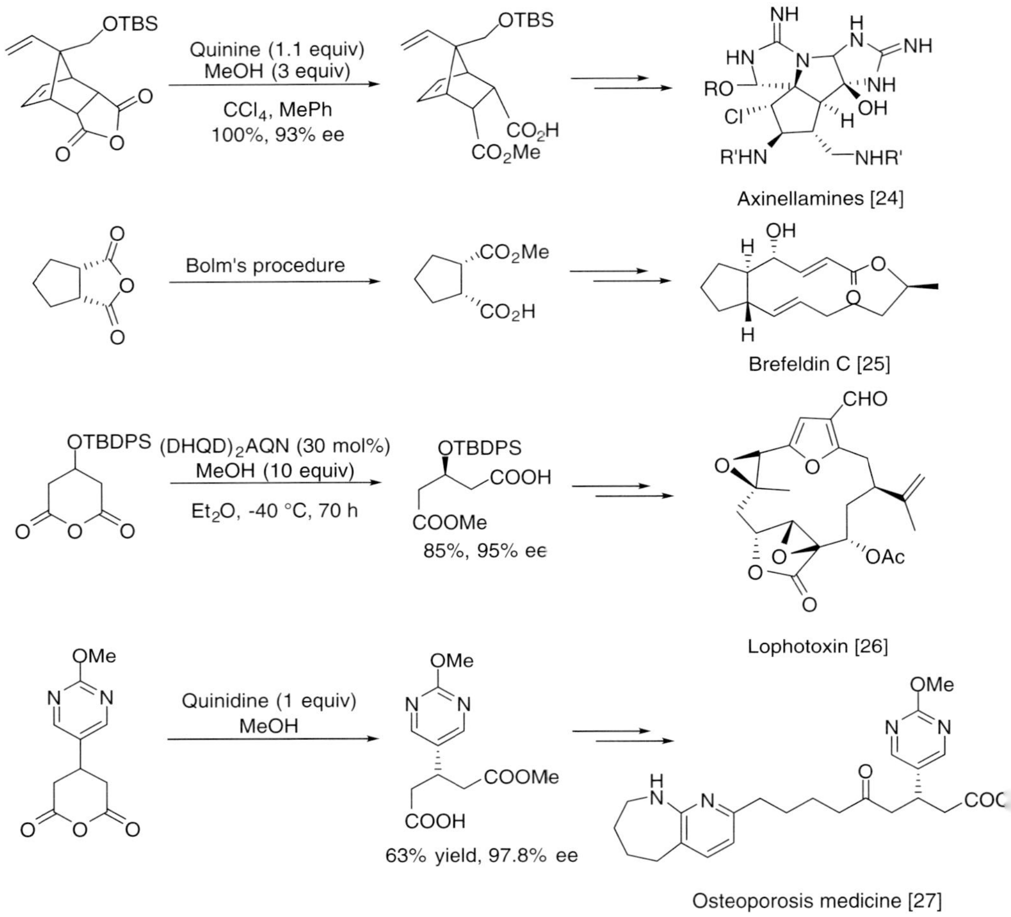

Scheme 11.13 (Continued).

[29] and benzoyl chloride as an acyl donor, cis-2-cyclopentene-1,4-diol (**20**) was converted to the monobenzoylated product **21** with up to 47% ee. However, the isolated yield (8%) was disappointingly low (Scheme 11.14).

Significantly improved results were obtained in 2003 by employing the phosphinite derivative **22** of cinchonine as a catalyst, which can be obtained as a mixture with the undesired corresponding phosphinate from the reaction of cinchonine (**1**) and chlorodiphenylphosphane [30a]. The desymmetrization reaction of a range of 1,2-meso-diols with benzoyl chloride in the presence of 30 mol% of the phosphinite **22** as a mixture containing about 15% phosphinate afforded the corresponding monobenzoylated diol in excellent yields and ee values (up to 94% ee) (Scheme 11.15). It was postulated by the authors that the reaction is initiated by the activation of the

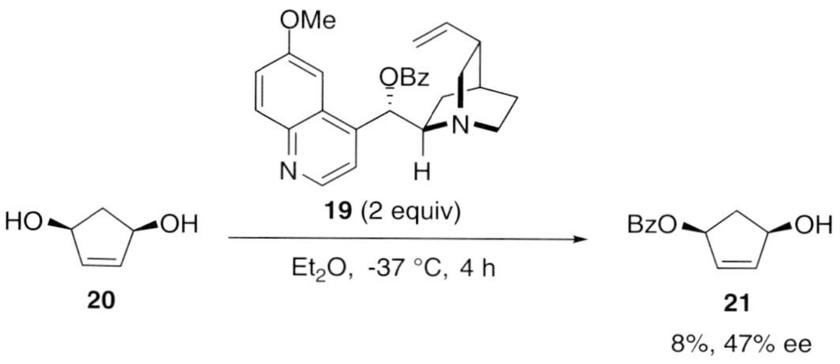

11.2 Desymmetrization of meso-Compounds

Scheme 11.14

HO—[cyclopentene]—OH → BzO—[cyclopentene]—OH
20   19 (2 equiv), Et₂O, −37 °C, 4 h   **21**
8%, 47% ee

acylating reagent (benzoyl chloride) by the Lewis basic trivalent phosphinite group. A subsequent report by the same group demonstrated that the quinidine-derived phosphinite **23** (30 mol%) was also highly effective as a catalyst for the desymmetrization of meso-1,3- and 1,4-diols, affording the monoacylated products with up to 99% ee (Scheme 11.16) [30b].

The catalytic efficiency of the diamines, **24** and **25**, derived from the "truncated cinchona alkaloids," quincorine and quincoridine, respectively, for the desymmetrization of meso-1,4-diols was also investigated by Kündig and coworkers [31a, b]. Both pseudoenantiomers **24** and **25** efficiently catalyzed the desymmetrization of the meso-complex **26** with benzoyl chloride, giving the enantiopure monobenzoylated Cr(CO)₃ complexes, **27** and **ent-27**, respectively, with up to 99% ee (Scheme 11.17). This process will provide easy access to new planar chiral complexes.

Later, Kündig and coworkers extended the substrate scope of catalysts **24** and **25** from meso-1,4-diol complexes such as **26** to simple meso-1,2-diols **28** [31c]. In the presence of 2 mol% of the catalyst, all of the tested cyclic and acyclic 1,2-meso-diols, except for substrates incorporating phenyl groups, were efficiently desymmetrized to

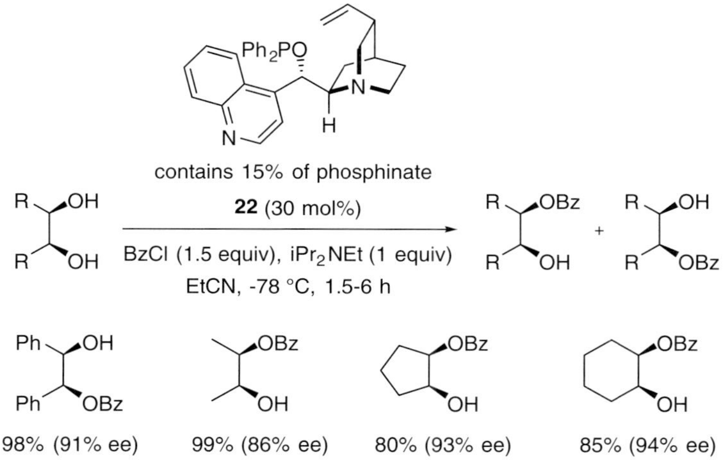

R(OH)—R(OH) → R(OBz)—R(OH) + R(OH)—R(OBz)

contains 15% of phosphinate
22 (30 mol%)
BzCl (1.5 equiv), iPr₂NEt (1 equiv)
EtCN, −78 °C, 1.5–6 h

Ph(OH)—Ph(OBz)   (OBz)(OH)   cyclopentane-OBz/OH   cyclohexane-OBz/OH
98% (91% ee)   99% (86% ee) 80% (93% ee)   85% (94% ee)

Scheme 11.15

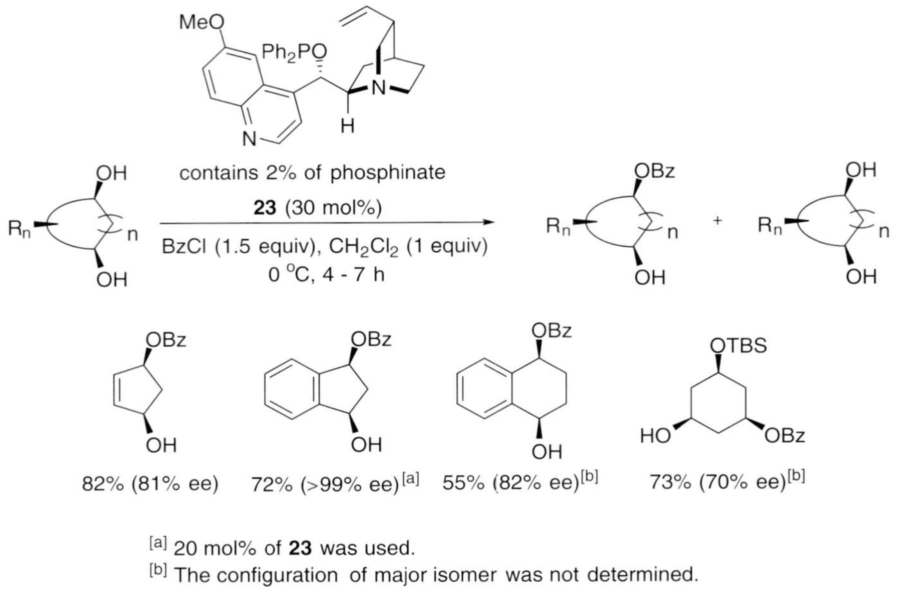
MeO
Ph2PO
N
H
contains 2% of phosphinate
23 (30 mol%)
BzCl (1.5 equiv), CH2Cl2 (1 equiv)
0 °C, 4 - 7 h
OH
Rn
n
OBz
OTBS
HO
82% (81% ee)
72% (>99% ee)[a]
55% (82% ee)[b]
73% (70% ee)[b]
[a] 20 mol% of 23 was used.
[b] The configuration of major isomer was not determined.

Scheme 11.16

give the monobenzoylated product with good to excellent enantioselectivities (up to 97% ee) (Scheme 11.18). However, substrates incorporating phenyl groups afforded products with much lower selectivities (13–83% ee), which was attributed by the authors to their low solubility under the reaction conditions employed (see examples in the box in Scheme 11.18). The authors proposed that the observed high ee values can be attributed partly to the kinetic resolution of the monobenzoylated product **29** due to the faster second benzoylation of the minor (−)-monobenzoylated product to

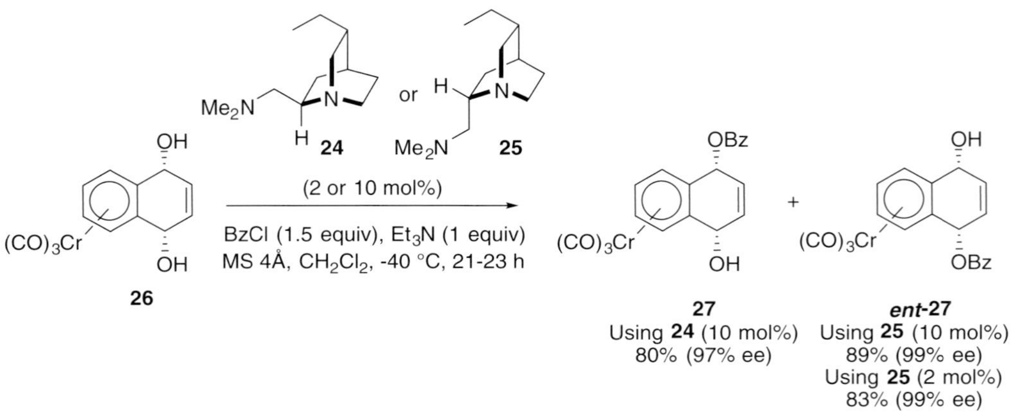
Me2N
N
H
24
or
25
(2 or 10 mol%)
BzCl (1.5 equiv), Et3N (1 equiv)
MS 4Å, CH2Cl2, -40 °C, 21-23 h
OH
OBz
(CO)3Cr
26
27
Using 24 (10 mol%)
80% (97% ee)
ent-27
Using 25 (10 mol%)
89% (99% ee)
Using 25 (2 mol%)
83% (99% ee)

Scheme 11.17

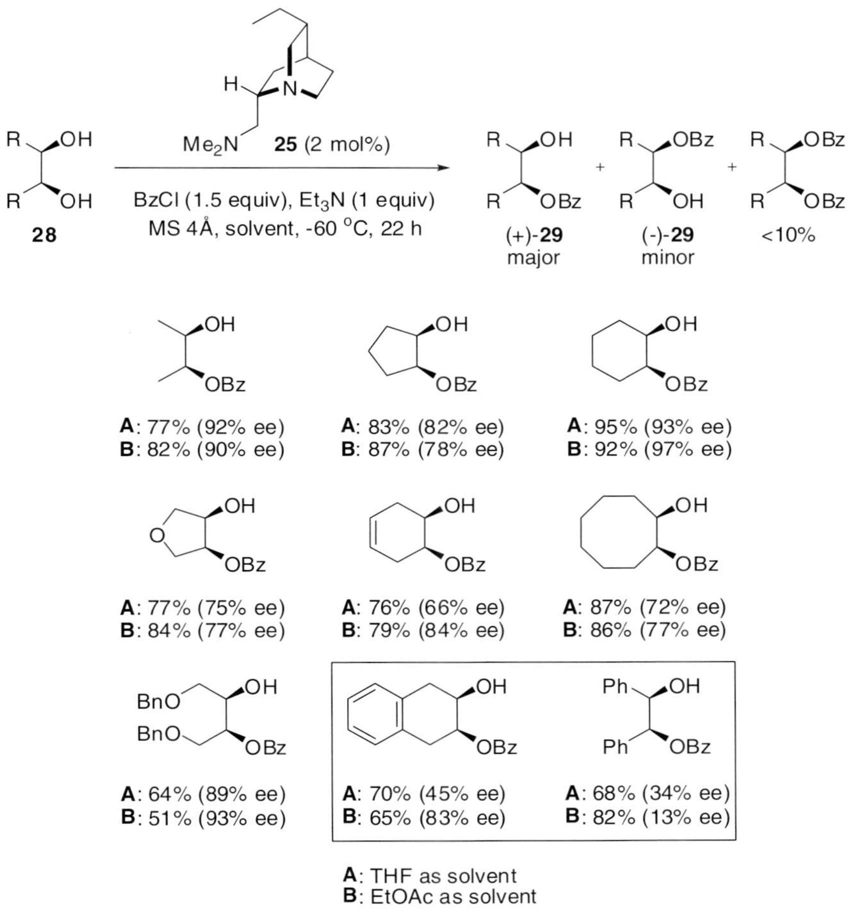

11.2 Desymmetrization of meso-Compounds

Scheme 11.18

A: THF as solvent
B: EtOAc as solvent

the dibenzoylated compounds, consequently increasing the ee value of the major (+)-monobenzoylated product (Scheme 11.19).

11.2.3
Desymmetrization of *meso*-Endoperoxides

The enantiomerically enriched 4-hydroxyenones **32** are an important class of chiral building blocks. This class of molecules has generally been obtained by the enzymatic or nonenzymatic catalytic desymmetrization of *meso*-cyclic diols **31** (or their derivatives) that can be prepared from the *meso*-endoperoxides **30** [32]. It is also well known that achiral bases such as NEt$_3$ can promote the Kornblum–DeLaMare rearrangement of the *meso*-endoperoxides **30** [33]. This process is believed to proceed via the E2

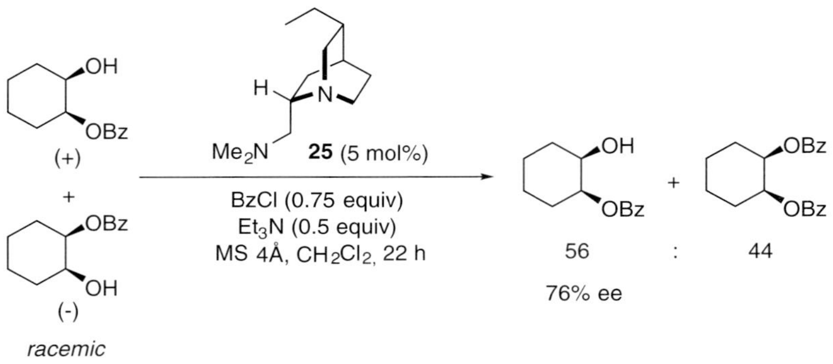

11 Cinchona-Based Organocatalysts for Desymmetrization

Scheme 11.19

mechanism that is initiated by the abstraction of the acidic α-proton of the peroxides **30** by a base to give the 4-hydroxyenones **32** (Scheme 11.20)[34].

In 2006, Toste and coworkers reported the first enantioselective Kornblum–DeLaMare rearrangement based on the desymmetrization of the *meso*-endoperoxides **30** catalyzed by cinchona alkaloids, which provides a more direct route to the γ-hydroxyenones **32** than any of the other previously developed methods [35]. Initial attempts to achieve the desymmetrization of the endoperoxide **33** affording **34** with monomeric and dimeric cinchona alkaloid catalysts gave disappointing results (e.g., quinidine (51%, 10% ee), O-acetyl quinidine (50%, 14% ee), and (DHQD)$_2$PHAL (38%, 32% ee)). A dramatic increase in both the yields (99%) and the ee values (99% ee) was obtained by using the quinidine-derived catalyst **35** or **36** bearing a free C6′–OH and an acetyl (**35**) or phenanthryl (**36**) group at the C9–O atom (Scheme 11.21). The pseudoenantiomeric 9-O-acetyl-6′OH-quinine analogue also provided the corresponding enantiomer with similar ee (97% ee). A variety of other substrates could also be desymmetrized to give the corresponding γ-hydroxyenones with moderate to excellent ee values (up to 99% ee) (Scheme 11.21). To explain the increased rate and enantioselectivity observed with the catalyst having a free C6′–OH, the authors proposed the transition-state model depicted in Figure 11.4,

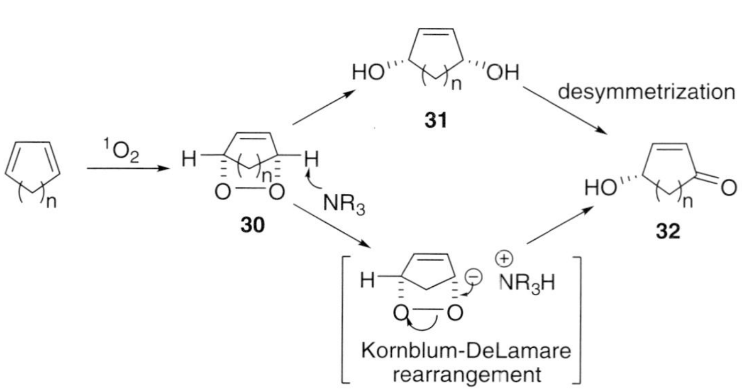

Scheme 11.20

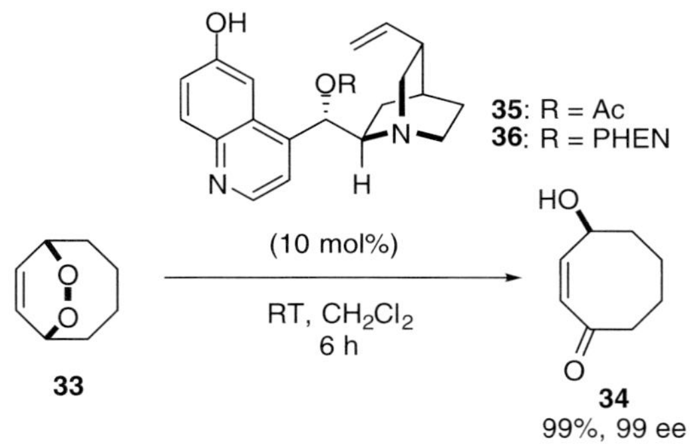

11.2 Desymmetrization of meso-Compounds

Scheme 11.21

Another set of selected examples (using 5 or 10 mol% of **35**)

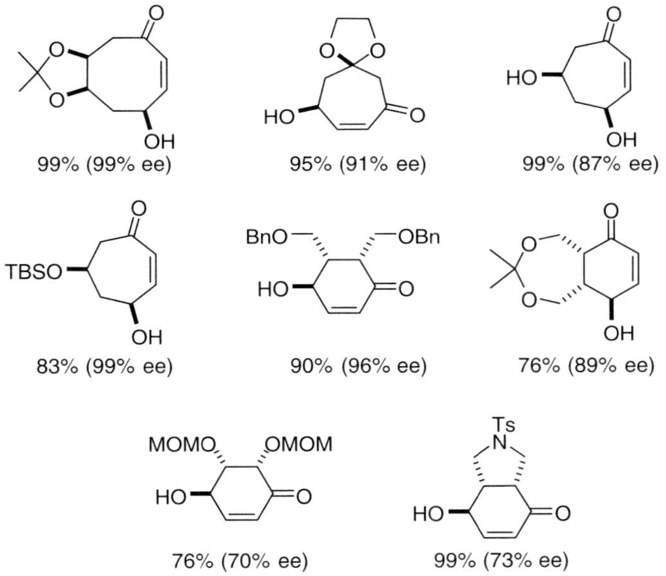

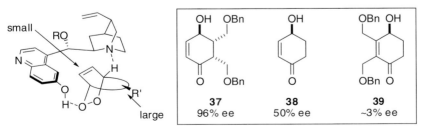

Figure 11.4 Proposed model for enantiodiscrimination step.

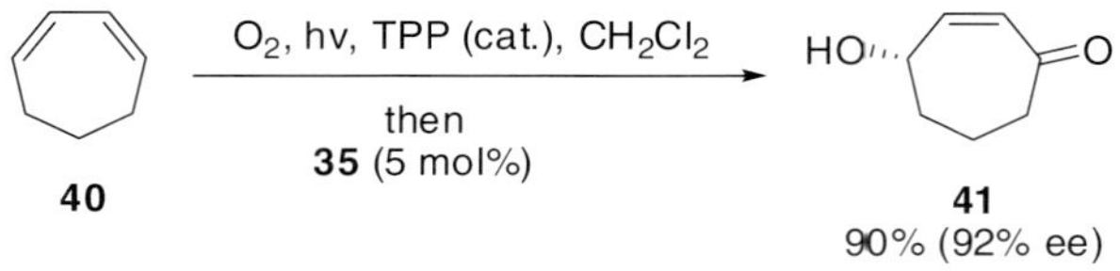

Scheme 11.22

in which the quinuclidine nitrogen and hydroxyl group simultaneously interact with the endoperoxide such that the olefin moiety is positioned in the sterically demanding pocket encompassed by the quinoline ring. This model affords a reasonable explanation for the dramatically different ee values of the three cyclohexenones **37–39** (see, examples in the box in Figure 11.4). Moreover, the large primary kinetic isotope effect ($k_D/k_H \sim 5$) of this reaction is consistent with the quinuclidine base-initiated E2 elimination mechanism.

The preparative usefulness of this protocol was highlighted by the one-pot synthesis of the γ-hydroxyenone **41** by the 1,4-dioxygenation of 1,3-cycloheptadiene (**40**) with singlet oxygen, followed by the **35**-catalyzed desymmetrization of the resulting *meso*-endoperoxides. The desired γ-hydroxyenone was obtained in 90% yield with 92% ee (Scheme 11.22).

11.2.4
Desymmetrization of *meso*-Phospholenes via Alkene Isomerization

Optically active phospholanes are considered as chiral building blocks for the synthesis of 2,5-dialkylphospholane ligands such as DuPHOS for transition metal catalysis. Quite recently, the catalytic asymmetric synthesis of phospholenes was realized for the first time via the cinchona-catalyzed double-bond isomerization of the *meso*-phospholene amide **42** [36]. In the presence of 10 mol% of cinchonine (**1**), the 3-phospholene amide **42** underwent an irreversible double-bond shift to the conjugated chiral 2-phospholene amide **43** with up to 83% ee (95% ee after recrystallization) (Scheme 11.23). The pseudoenantiomeric cinchonidine (**3**) gave the corresponding opposite enantiomer with a slightly lower ee value (72% ee). The obtained 2-phospholene product **43** can be used as a synthetic precursor of the phospholane ligand after hydrolysis and diastereoselective hydrogenation.

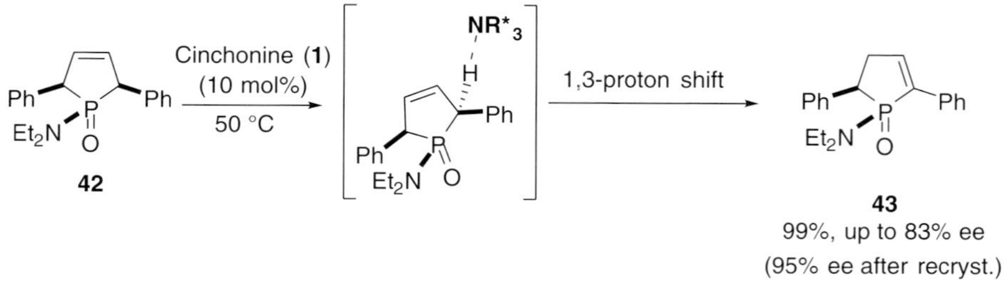

Scheme 11.23

11.2.5
Desymmetrization of *meso*-Epoxy Phospholenes to Allyl Alcohols via Rearrangement

The desymmetrization of *meso*-epoxides to enantiomerically enriched allylic alcohols via asymmetric rearrangement has been recognized as a highly useful strategy in the field of asymmetric synthesis [37]. The chiral lithium base-promoted rearrangement reaction is well established [38]. Recently, Pietrusiewicz and coworkers reported that cinchona alkaloids can also serve as effective chiral bases in the enantioselective rearrangement of the 3-phospholene epoxide **44** to the allylic alcohol **45** [39]. Among the parent cinchona alkaloids examined in this study, quinidine (**2**) (50 mol%) showed the best enantioselectivity (52% ee) (Scheme 11.24). However, the reactivity was too low for preparative use. After 90 days, a yield of 41% of the rearranged product was obtained. Notably, the free OH at the C9-position is crucial for asymmetric induction. O-protected quinidine was completely ineffective. The model proposed by the authors (Figure 11.5) can explain the crucial role of the free OH group at the C9-position in determining both the reactivity and the observed stereochemical outcome. Due to the hydrogen-bond interaction between the alkaloid hydroxyl group and epoxide phosphoryl moiety, the quinuclidine base abstracts the more easily accessible anti-β-proton, unlike in the case of lithium amide bases. Of note, it is accepted that the lithium amide base-catalyzed rearrangement proceeds via initial lithium coordination to the epoxy oxygen followed by a *syn*-β-proton.

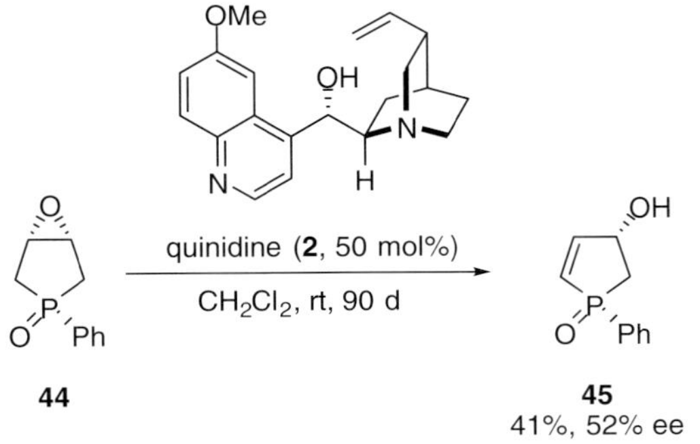

Scheme 11.24

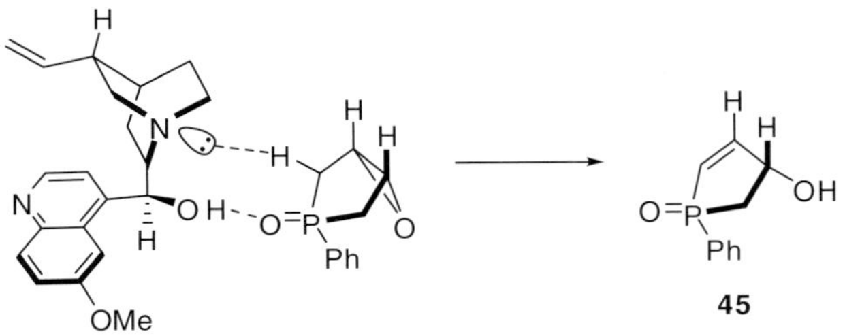

Figure 11.5 Preferred quinidine approach to epoxide.

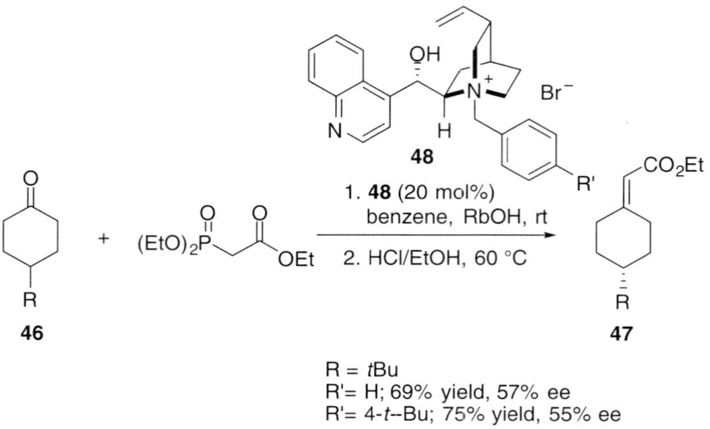

Scheme 11.25

R = tBu
R'= H; 69% yield, 57% ee
R'= 4-t-Bu; 75% yield, 55% ee

11.2.6
Desymmetrization of Prochiral Ketones by Means of Horner–Wadsworth–Emmons Reaction

The enantioselective desymmetrization of prochiral ketones of the type **46** by means of the Horner–Wadsworth–Emmons reaction is an elegant approach to the synthesis of the enantioenriched enone products **47**. Enones such as **47** bearing a "remote" stereogenic center are difficult to prepare by other methods. The first catalytic asymmetric Horner–Wadsworth–Emmons reaction was realized by using cinchona-based PTCs as organocatalysts [40]. Up to 55–57% ee values and satisfactory yields were obtained using catalytic amounts (20 mol%) of the N-benzylcinchoninium salts **48** and RbOH as a base (Scheme 11.25).

11.3
(Dynamic) Kinetic Resolution of Racemic Compounds

11.3.1
(Dynamic) Kinetic Resolution of Racemic Cyclic Anhydrides

In 2001, Deng and coworkers found that nucleophilic catalysts such as $(DHQD)_2AQN$ (**11**) can also affect the parallel kinetic resolution of racemic anhydrides by alcoholysis, that is, the two substrate enantiomers are converted into regioisomeric esters. A variety of monosubstituted racemic succinic anhydrides were converted in the presence of $(DHQD)_2AQN$ (**11**, 15–20 mol%) and allyl alcohol as a nucleophile to both regioisomeric hemiesters **49** and **50** with synthetically reliable ee values (up to 98% ee) and yields (Scheme 11.26) [41]. The obtained regioisomeric 2- or 3- aryl succinates could be converted to the β- and α-aryl-γ-butylolactones, which constitute valuable synthons for various pharmaceuticals.

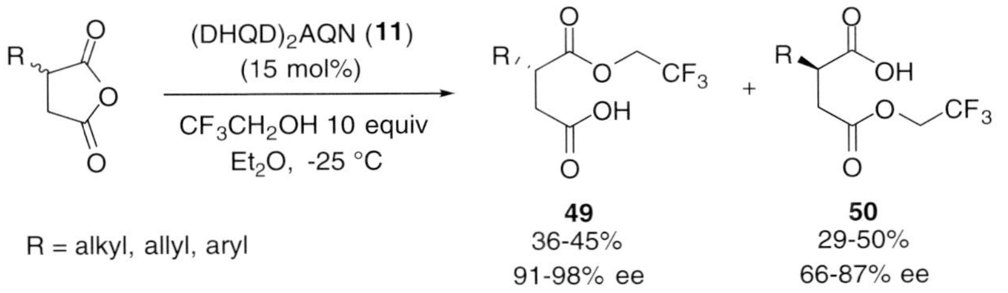

Scheme 11.26

In the same period, Deng and coworkers also reported that the racemic 5-alkyl 1,3-dioxolane-2,4-diones **51** undergo kinetic resolution in the presence of alcohols (ethanol or allyl alcohol) and dimeric cinchona alkaloids such as (DHQD)$_2$AQN (**11**, 10 mol%) [42]. A range of 5-alkyl dioxolanes **51** were highly enantioselectively converted to the (R)-esters (R)-**52** (selectivity factors up to 133) (Scheme 11.27). The hydrolysis of the reaction mixture converts the remaining (S)-**53** to the (S)-acid (s)-**53** that can be easily separated from the (R)-ester **52** by extraction (Scheme 11.27).

When the 5-aryl 1,3-dioxolanes **54** were employed instead of the 5-alkyl substrates **51**, the additional racemization of the dioxolanes occurred, due to the higher acidity of the α-CH, thereby enabling dynamic kinetic resolution. As shown in Scheme 11.28, the highly enantioenriched (up to 96% ee) α-hydroxy acid esters **55** were obtained in reasonable yields (up to 86%). However, unsatisfactory results (60–62% ee) were obtained with substrates bearing an *ortho*-substituted benzene ring at the 5-position (see examples in the box in Scheme 11.28). By following the course of the reaction, the authors observed that the high ee values (85–90% ee) at the initial stage of this reaction significantly dropped to 60–62% ee values after the completion of the reactions. This observation indicates that the alcoholysis actually occurred highly enantioselectively. However, the rate of the racemization reaction was not sufficient for an efficient DKR process due to the steric hindrance of the *ortho*-substituents.

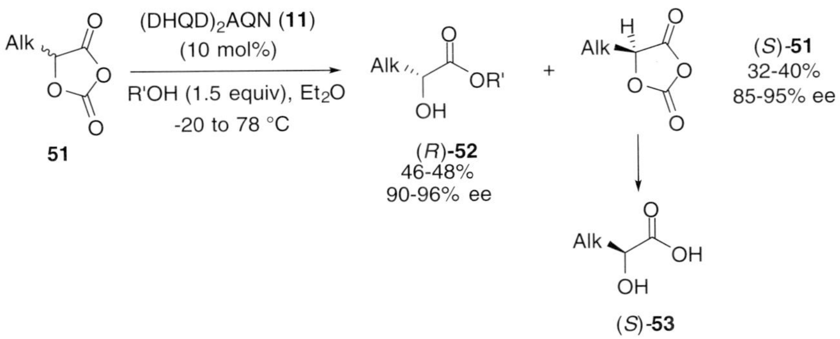

Scheme 11.27

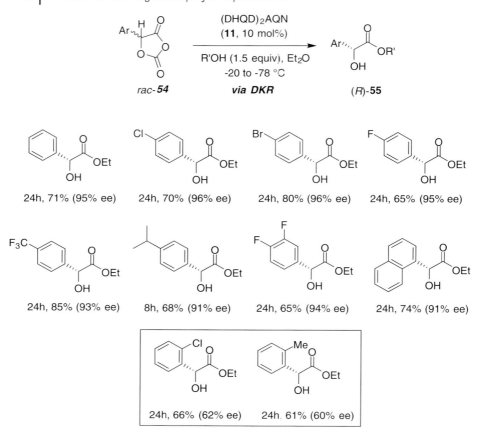

Scheme 11.28

11.3.2
(Dynamic) Kinetic Resolution of Racemic N-Cyclic Anhydrides

The kinetic resolution of the N-urethane-protected N-carboxyanhydrides (UNCAs) 56 by methanolysis in the presence of (DHQD)$_2$AQN (11) was also reported by Deng and coworkers for the synthesis of enantiopure α-amino acids [43a]. At low temperatures (from −40 to −78 °C), a range of UNCAs 56 were successfully resolved to give the (R)-N-protected α-amino esters 57 (up to 97% ee) and the remaining (S)-56 (up to 98% ee) (Scheme 11.29).

Their subsequent investigation revealed that, at increased temperature, the racemization of the 5-aryl N-carboxy anhydrides 58 proceeds sufficiently fast to enable their dynamic kinetic resolution. Using 20 mol% of (DHQD)$_2$AQN (11) and allyl alcohol as a nucleophile, enantiomeric excesses of >90% and almost quantitative yields were achieved for several aryl N-carboxy anhydrides (Scheme 11.30) [43b]. The obtained enantioenriched amino esters 59 could be simply transformed to the corresponding amino acids with an unchanged ee value by hydrogenolysis of the allyl group.

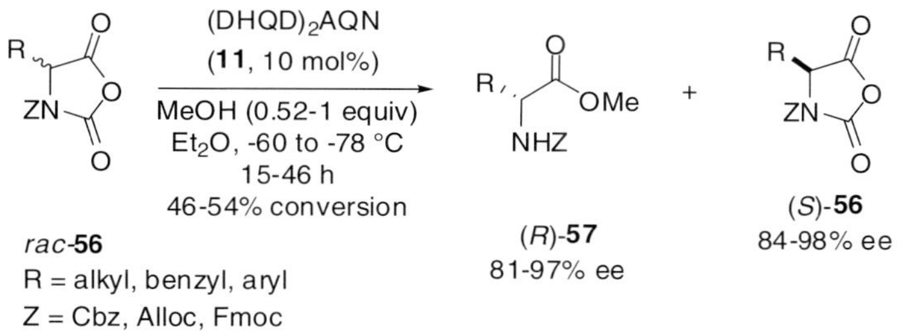

11.3 (Dynamic) Kinetic Resolution of Racemic Compounds

Scheme 11.29

rac-56, R = alkyl, benzyl, aryl; Z = Cbz, Alloc, Fmoc

Conditions: (DHQD)$_2$AQN (11, 10 mol%), MeOH (0.52–1 equiv), Et$_2$O, −60 to −78 °C, 15–46 h, 46–54% conversion

Products: (R)-57, 81–97% ee; (S)-56, 84–98% ee

As depicted in Scheme 11.31, the authors proposed a general base catalysis mechanism in which dimeric cinchona alkaloids such as (DHQD)$_2$AQN (11) act as a dual activator (i.e., (i) the racemization of UNCA by α-deprotonation and (ii) the activation of the nucleophile (R′OH)).

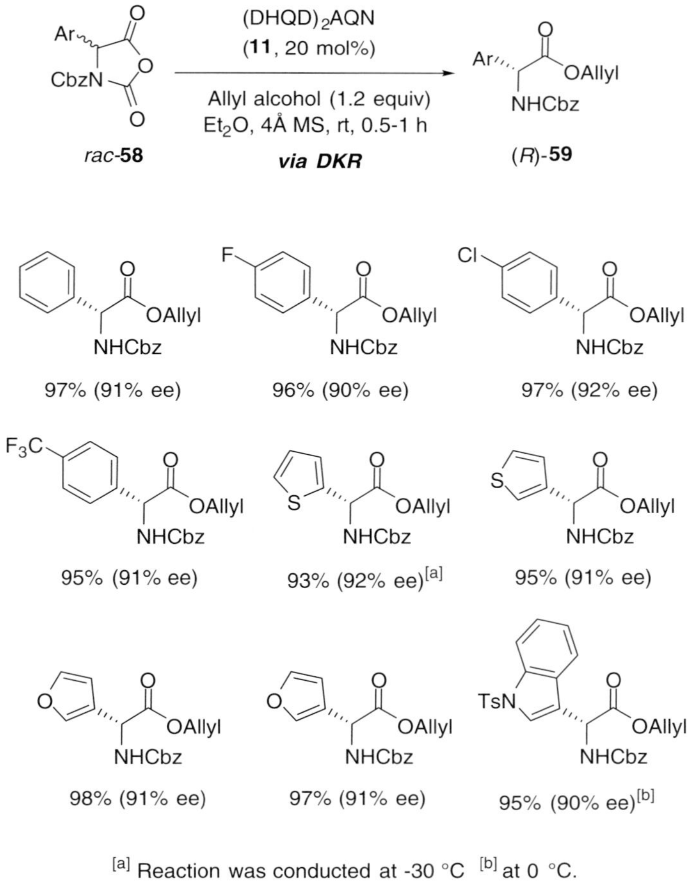

rac-58 → (R)-59 via DKR

Conditions: (DHQD)$_2$AQN (11, 20 mol%), Allyl alcohol (1.2 equiv), Et$_2$O, 4Å MS, rt, 0.5–1 h

Products (Ar-CH(NHCbz)-C(O)OAllyl):

- Ph: 97% (91% ee)
- 4-F-C$_6$H$_4$: 96% (90% ee)
- 4-Cl-C$_6$H$_4$: 97% (92% ee)
- 4-F$_3$C-C$_6$H$_4$: 95% (91% ee)
- 2-thienyl: 93% (92% ee) [a]
- 3-thienyl: 95% (91% ee)
- 2-furyl: 98% (91% ee)
- 3-furyl: 97% (91% ee)
- N-Ts-indolyl: 95% (90% ee) [b]

[a] Reaction was conducted at −30 °C [b] at 0 °C

Scheme 11.30

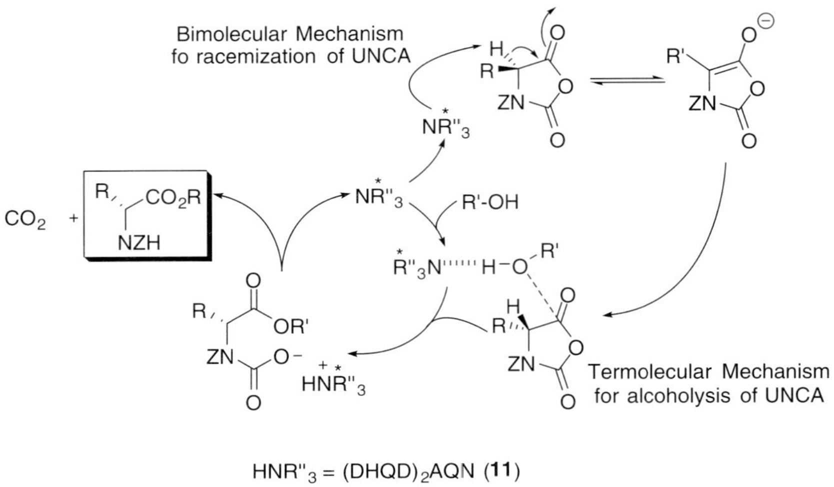

Scheme 11.31 Dual catalytic motion of modified bis-cinchona alkaloid.

11.3.3
Dynamic Kinetic Resolution of Racemic Azlactones

The catalytic dynamic kinetic resolution of the α-substituted azlactones **60** by alcoholysis is also an elegant shortcut to various amino acids. Several strategies involving enzymes, Ti-based complexes, cyclic dipeptides, and chiral DMAP nucleophilic catalysts have been employed to bring about the selective DKR of azlactones [44]. However, most of these methods suffer from either limited substrate scope and/or long reaction times. In 2004, Berkessel and coworkers obtained somewhat promising results for the DKR of the azlactones **60** using bifunctional Takemoto-type catalysts bearing both a Lewis acidic (thio)urea moiety and a Brønsted basic tertiary amine group [44]. In most cases, the N-benzoyl amino acid esters were obtained with moderate to high ee values. It was proposed that the catalysts operate via the simultaneous activation of both the electrophile and the nucleophile by hydrogen bonding and general base catalysis [45]. Quite recently, Connon and coworkers also demonstrated that bifunctional cinchona-based ureas as catalysts can efficiently promote the DKR reactions of the azlactones **60** using an alcohol nucleophile [12]. Screening the catalytic efficiency of various cinchona-based (thio)urea catalysts revealed that hydroquinine urea **61** (10 mol%) was the optimum catalyst. This catalyst is relatively insensitive to the steric bulk of the substrate, allowing a range of both unhindered and hindered azlactones **60** to undergo DKR in CH_2Cl_2 at $-20\,°C$, affording the N-benzoylated (S)-amino esters **62** with a high level of enantioselectivity (78–88% ee) (Scheme 11.32). The robust nature of catalyst **61** was demonstrated by the one-pot DKR from N-benzoyl racemic alanine **63** that gave the corresponding enantioenriched allyl ester **64** in 94% yield and with 88% ee (Scheme 11.33).

11.3.4
Catalytic Sulfinyl Transfer Reaction via Dynamic Kinetic Resolution of Sulfinyl Chlorides

Asymmetric sulfinyl transfer reactions are of significant interest, since they can be applied to the synthesis of chiral sulfur-containing compounds, which are valuable synthons for pharmaceuticals and natural products [46]. Moreover, chiral sulfoxides are versatile chiral auxiliaries for asymmetric organic synthesis. A number of studies

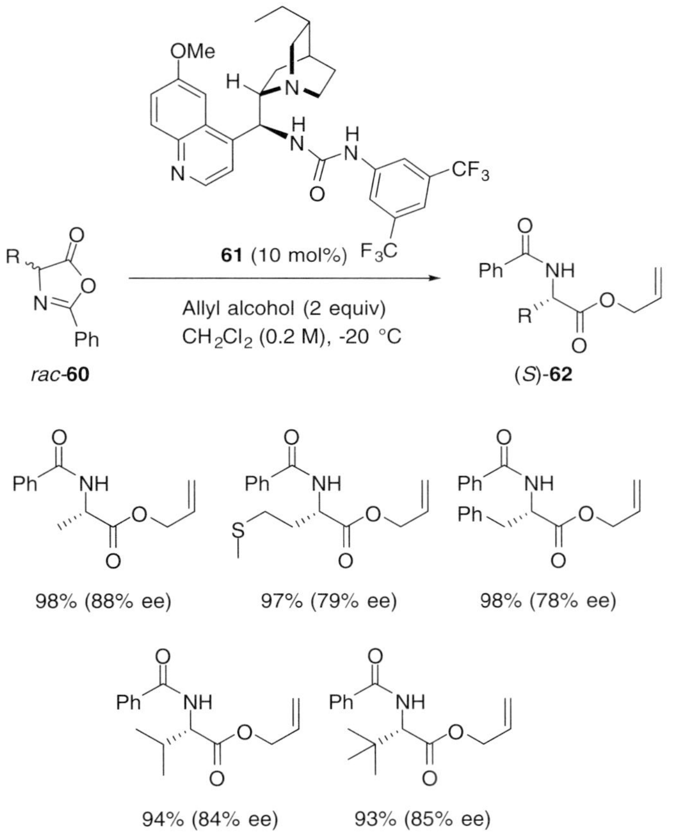

Scheme 11.32

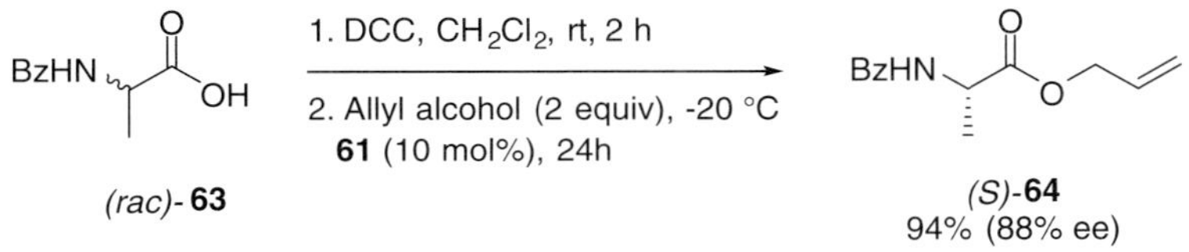

Scheme 11.33

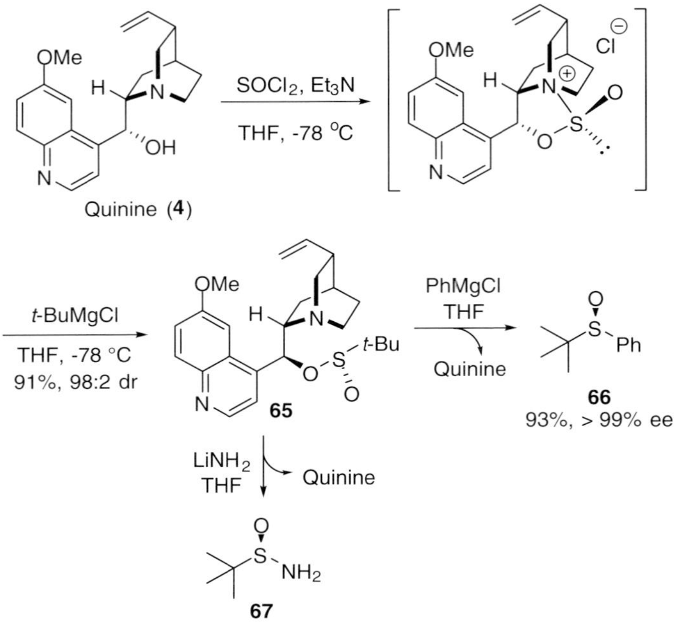

Scheme 11.34

were devoted to this field, but practical and general catalytic sulfinyl transfer reactions using organocatalysts did not appear until the beginning of the 2000s [47]. In 2004, Ellman et al. reported the first catalytic, dynamic kinetic resolution of t-butyl sulfinyl chloride with arylmethyl alcohols using an oligopeptide-based catalyst to afford the sulfinate esters with up to 81% ee [48]. Subsequently in 2005, Senanayake et al. prepared chiral sulfinyl transfer agents such as **65** by treating quinine (**4**) or quinidine (**2**) with thionyl chloride and tert-BuMgCl (Scheme 11.34) [49]. The obtained tert-butanesulfinate **65** could be successfully used as a chiral auxiliary by reacting it with Grignard reagents or lithium amide to prepare the chiral sulfoxides **66** or sulfonamide **67** (up to 99% ee) (Scheme 11.34).

Similarly, in 2005, Shibata and coworkers also reported that N-sulfinylammonium salts such as **72**, generated *in situ* via a cinchona alkaloid **68** or **69**/racemic sulfinyl chloride (**70**) combination, serve as highly enantioselective sulfinylating agents for a range of achiral alcohols to give synthetically valuable chiral sulfinates **71** [50]. Both enanatiomers of the sulfinates **71** could be accessed in excellent enantiopurity by using either of the pseudoenantiomeric quinine or quinidine derivatives (e.g., quinidine acetate (**68**) or hydroquinine acetate (**69**)) (Scheme 11.35). Moreover, this reaction proceeds via the dynamic kinetic resolution of the racemic sulfinyl chlorides, resulting in a maximum 100% theoretical yield, since the sulfinyl stereocenter of the N-sulfinylammonium salts **72** can be easily epimerized (Scheme 11.35). Thus, the enantioselectivity depends on the relative reaction rate of the alcohol with the diastereomeric sulfinylammonium salts **72**.

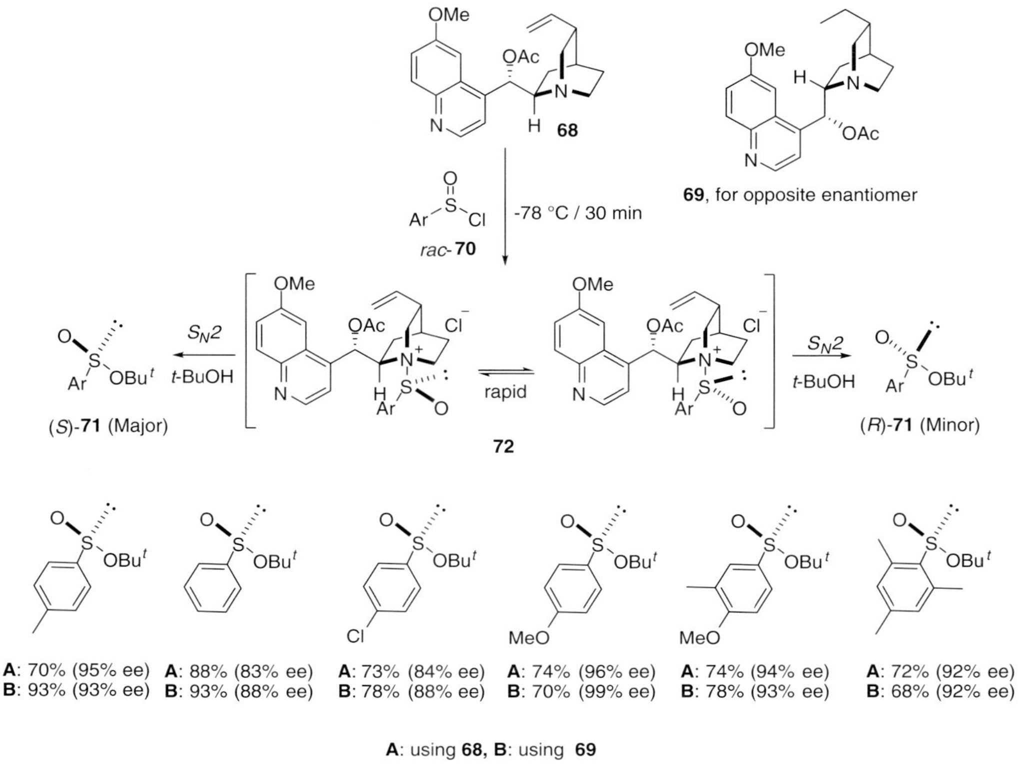

11.3 (Dynamic) Kinetic Resolution of Racemic Compounds

Scheme 11.35

Soon after, a catalytic version of the above stoichiometric asymmetric sulfinyl transfer reaction based on a cinchona alkaloid/sulfinyl chloride was developed by the same group. The additional use of an inorganic (Na_2CO_3, K_2CO_3) or organic base (proton sponge, i-Pr_2EtN) effectively promoted the catalytic turnover of the cinchona alkaloid by capturing the liberated HCl and, thus, reactivating the quinuclidine nitrogen of the alkaloid [51]. In the presence of a catalytic amount (10 mol%) of quinidine 1-naphthoate (**73**) or quinine 1-naphthoate and an additional base (2.5 equiv), 9-fluorenol was effectively sulfinylated at −40 °C to furnish the corresponding sulfinates **76** with up to 95% ee. Reducing the catalyst loading from 10 to 1 mol% did not lead to a decrease in the ee value although the reaction time was prolonged (Scheme 11.36).

At around the same time, Ellman and coworkers [52] also reported similar results for cinchona-catalyzed sulfinyl transfer reactions to those obtained by the Shibata group [51]. They found that the parent cinchona alkaloids effectively catalyzed the sulfinyl transfer reaction of *tert*-butanesulfinyl chloride (*rac*-**74**) and a variety of benzyl alcohols in the presence of a proton sponge (2.5 equiv). For example, in the presence of 10 mol% of quinidine (**2**) and 2.5 equivalents of a proton sponge, sulfinyl transfer with 2,4,6-trichlorobenzyl alcohol (**77**) provided the sulfinate ester **78** in 92% yield and with 90% ee (Scheme 11.37).

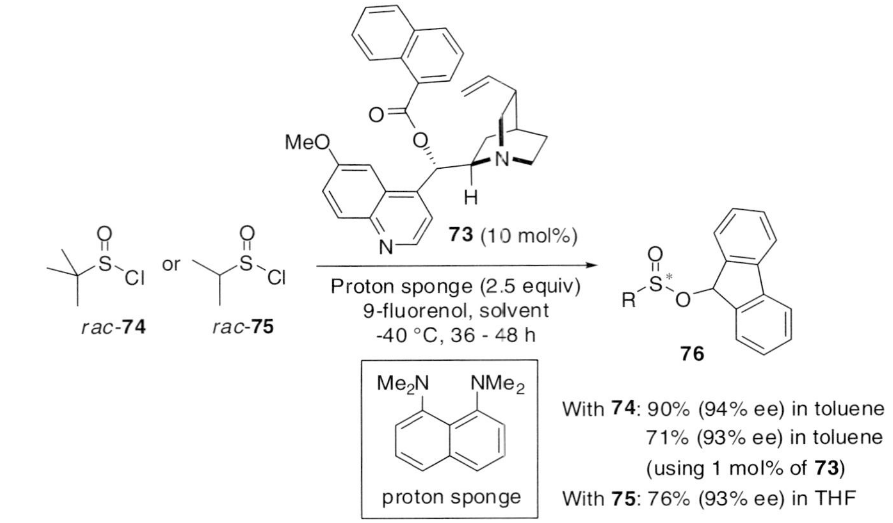

Scheme 11.36

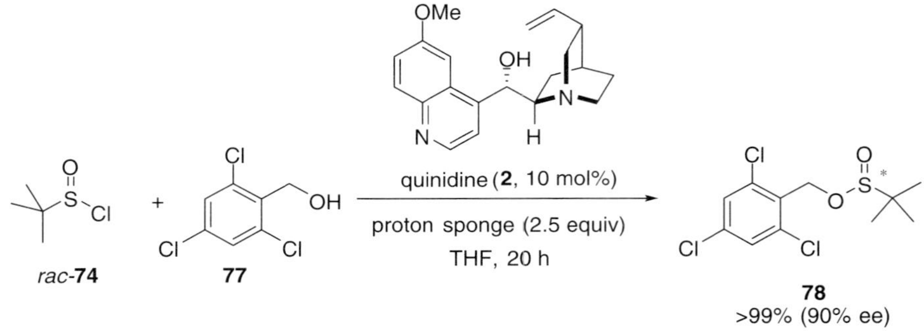

Scheme 11.37

11.4
Conclusions

This chapter presented the current stage of development in the desymmetrization of *meso*-compounds and (dynamic) kinetic resolution of racemic compounds in which cinchona alkaloids or their derivatives are used as organocatalysts. As shown in many of the examples discussed above, cinchona alkaloids and their derivatives effectively promote these reactions by either a monofunctional base (or nucleophile) catalysis or a bifunctional activation mechanism. Especially, the cinchona-catalyzed alcoholytic desymmetrization of cyclic anhydrides has already reached the level of large-scale synthetic practicability and, thus, has already been successfully applied to the synthesis of key intermediates for a variety of industrially interesting biologically active compounds. However, for other reactions, there is still room for improvement in terms of the general substrate scope, chemical yield, and product optical purity. More systematic studies designed to understand the details of the asymmetric induction step should be performed that will lead to more efficient and practical catalyst systems. There is no doubt that such studies will be conducted and thus significant further progress can be expected in the not too distant future.

Acknowledgments

This work was supported by grants KRF-2008-005-J00701 (MOEHRD), R11-2005-008-00000-0 (SRC program of MEST/KOSEF) and R31-2008-000-10029-0 (WCU program).

References

1 (a) Atodiresei, J., Schiffers, I., and Bolm, C. (2007) *Chem. Rev.*, **107**, 5683–5712; (b) Chen, Y., McDaid, P., and Deng, L. (2003) *Chem. Rev.*, **103**, 2965–2984; (c) Tian, S.-K., Chen, Y., Hang, J., Tang, L., McDais, P., and Deng, L. (2004) *Acc. Chem. Res.*, **37**, 621–631; (d) Berkessel, A. and Gröger, H. (2005) *Asymmetric Organocatalysis*, Wiley-VCH Verlag GmbH, Weinheim, Chapter 13; (e) Dalko, P I. (2005) *Enantioselective Organocatalysis* (ed. P.I. Dalko), Wiley-VCH Verlag GmbH, Weinheim, pp. 312–321; (f) Spivey, A.C. and Andrews, B.I. (2001) *Angew. Chem. Int. Ed.*, **40**, 3131–3134.

2 For a recent review on enzymatic desymmetrization reactions, see García-Urdiales, E., Alfonso, I., and Gotor, V., (2005) *Chem. Rev.*, **105**, 313–354; for a review on nonenzymatic desymmetrization reaction, see Willis, M.C., (1999) *J. Chem. Soc., Perkin Trans. I*, 1765–1784.

3 (a) Hiratake, J., Yamamoto, Y., and Oda, J. (1985) *J. Chem. Soc., Chem. Commun.*, 1717–1719; (b) Hiratake, J., Inagaki, M., Yamamoto, Y., and Oda, J. (1987) *J. Chem. Soc., Perkin Trans. I*, 1053–1058.

4 (a) Aitken, R.A., Gopal, J., and Hirst, J.A. (1988) *J. Chem. Soc., Chem. Commun.*, 632–634; (b) Aitken, R.A. and Gopal, J. (1990) *Tetrahedron: Asymmetry*, **1**, 517–520.

5 Bolm, C., Gerlach, A., and Dinter, C.L. (1999) *Synlett*, 195–196.

6 Bolm, C., Schiffers, I., Dinter, C.L., and Gerlach, A. (2000) *J. Org. Chem.*, **65**, 6984–6991.

7 (a) Bolm, C., Schiffers, I., Atodiresei, I., and Hackenberger, C.P.R. (2003) *Tetrahedron: Asymmetry*, **14**, 3455–3467; (b) Bolm, C., Schiffers, I., Dinter, C.L., Difrére, L., Gerlach, A., and Raabe, G. (2001) *Synthesis*, 1719–1730; (c) Bolm, C., Atodiresei, I., and Schiffers, I. (2005) *Org. Synth.*, **82**, 120–124.

8 Chen, Y.G., Tian, S.-K., and Deng, L. (2000) *J. Am. Chem. Soc.*, **122**, 9542–9543.

9 Ishii, Y., Fujimoto, R., Mikami, M., Murakami, S., Miki, Y., and Furukawa, Y. (2007) *Org. Process Res. Dev.*, **11**, 609–615.

10 For a recent review on cinchona-based (thio)ureas, see Connon, S.J. (2008) *Chem. Commun*, 2499–2510.

11 (a) Rho, H.S., Oh, S.H., Lee, J.W., Lee, J.Y., Chin, J., and Song, C.E. (2008) *Chem. Commun.*, 1208–1210; (b) Peschiulli, A., Gun'ko, Y., and Connon, S.J. (2008) *J. Org. Chem.*, **73**, 2454–2457.

12 Peschiulli, A., Quigley, C., Tallon, S., Gun'ko, Y.K., and Connon, S.J. (2008) *J. Org. Chem.*, **73**, 6409–6412.

13 Etter, M.C., Urbañczyk-Lipkowska, Z., Zia-Ebrahimi, M., and Panunto, T.W. (1990) *J. Am. Chem. Soc.*, **112**, 8415–8426.

14 TárKányi, G., Király, P., Varga, S., Vakulya, B., and Soós, T. (2008) *Chem. Eur. J.*, **14**, 6078–6086.
15 Oh, S.H., Rho, H.S., Lee, J.W., Lee, J.E., Youk, S.H., Chin, J., and Song, C.E. (2008) *Angew. Chem. Int. Ed.*, **47**, 7872–7875.
16 Youk, S.H., Oh, S.H., Rho, H.S., Lee, J.E., Lee, J.W., and Song, C.E. (2008) *Chem. Commun.*, 2220–2222.
17 (a) For cyclo β-amino acids, see Fülöp, F. (2001) *Chem. Rev.*, **101**, 2181–2204, (b) for β-peptides, see Cheng, R.P., Gellman, S.H., and DeGrado, W.F., (2001) *Chem. Rev.*, **101**, 3219–3232.
18 Mittendorf, J., Arold, H., Fey, P., Matzke, M., Militzer, H.-C., and Mohrs, K.-H. (1995) Ger. Offen. DE 44 00 749 A1.
19 Yue, T.-Y., McLeod, D.D., Albertson, K.B., Beck, S.R., Deerberg, J., Fortunak, J.M., Nugent, W.A., Radesca, L.A., Tang, L., and Xiang, C.D. (2006) *Org. Process Res. Dev.*, **10**, 262–271.
20 Park, J., Sudhakar, A., Wong, G.S., Chen, M., Weber, J., Yang, X., Kwok, D.-I., Jeon, I., Raghavan, R.R., Tamarez, M., Tong, W., and Vater, E.J. (2004) PCT Int. Appl. WO 2004113295 A1.
21 Hameršak, Z., Stipetić, I., and Avdagić, A. (2007) *Tetrahedron: Asymmetry*, **18**, 1481–1485.
22 (a) De Clercq, P.J. (1997) *Chem. Rev.*, **97**, 1755–1792; (b) Choi, C., Tian, S.-K., and Deng, L. (2001) *Synthesis*, 1737–1741; (c) Dai, H.-F., Chen, W.-X., Zhao, L., Xiong, F., Sheng, H., and Chen, F.-E. (2008) *Adv. Synth. Catal.*, **350**, 1635–1641; (d) Huang, J., Xiong, F., and Chen, F.-E. (2008) *Tetrahedron: Asymmetry*, **19**, 1436–1443.
23 Röme, D., Johansson, M., and Sterner, O. (2007) *Tetrahedron Lett.*, **48**, 635–638.
24 Starr, J.T., Koch, G., and Carreira, E.M. (2000) *J. Am. Chem. Soc.*, **122**, 8793–8794.
25 Archambaud, S., Aphecetche-Julienne, K., and Guingant, A. (2005) *Synlett*, 139–143.
26 (a) Wipf, P. and Grenon, M. (2006) *Can. J. Chem.*, **84**, 1226–1241; (b) Wipf, P. and Soth, M.J. (2002) *Org. Lett.*, **4**, 1787–1790.
27 Keen, S.P., Cowden, C.J., Bishop, B.C., Brands, K.M.J., Davies, A.J., Dolling, U.H.,
Lieberman, D.R., and Stewart, G.W. (2005) *J. Org. Chem*, **70**, 1771–1779.
28 (a) Rendler, S. and Oestreich, M. (2008) *Angew. Chem. Int. Ed.*, **47**, 248–250; (b) Berkessel, A. and Gröger, H. (2005) *Asymmetric Organocatalysis*, Wiley-VCH Verlag GmbH, Weinheim, Chapter 13. 3.
29 Duhamel, L. and Herman, T. (1985) *Tetrahedron Lett.*, **26**, 3099–3102.
30 (a) Mizuta, S., Sadamori, M., Fujimoto, T., and Yamamoto, I. (2003) *Angew. Chem. Int. Ed.*, **42**, 3383–3385; (b) Mizuta, S., Tsuzuki, T., Fujimoto, T., and Yamamoto, I. (2005) *Org. Lett.*, **7**, 3633–3635.
31 (a) Kündig, E.P., Lomberget, T., Bragg, R., Poulard, C., and Bernardinelli, G. (2004) *Chem. Commun.*, 1548–1549; (b) Kündig, E.P., García, A.E., Lomberget, T., and Bernardinelli, G. (2006) *Angew. Chem. Int. Ed.*, **45**, 98–101; (c) Kündig, E.P., Garcia, A.E., Lomberget, T., Garcia, P.P., and Romanens, P. (2008) *Chem. Commun.*, 3519–3521.
32 Zhao, Y., Rodrigo, J., Hoveyda, A.H., and Snapper, M.L. (2006) *Nature*, **443**, 67–70, and references cited therein.
33 (a) Kornblum, N. and DeLaMare, H.E. (1951) *J. Am. Chem. Soc.*, **73**, 880–881; (b) Hagenbuch, J.-P. and Vogel, P. *J. Chem. Soc., Chem. Commun.*, (1980) 1062–1063.
34 (a) Kelly, D.R., Bansal, H., and Morgan, J.J.G. (2002) *Tetrahedron Lett.*, **43**, 9331–9333; (b) Mete, E., Altundas, R., Secen, H., and Balci, M. (2003) *Turk. J. Chem.*, **27**, 145–153.
35 Staben, S.T., Linghu, X., and Toste, F.D. (2006) *J. Am. Chem. Soc.*, **128**, 12658–12659.
36 Hintermann, L. and Schmitz, M. (2008) *Adv. Synth. Catal.*, **350**, 1469–1473.
37 Berkessel, A. and Gröger, H. (2005) *Asymmetric Organocatalysis*, Wiley-VCH Verlag GmbH, Weinheim, Chapter 13. 4.
38 For a review, see (a) Magnus, A.S., Bertilsson, S.K., and Andersson, P.G., (2002) *Chem. Soc. Rev.*, **31**, 223–229; (b) Eames, J. (2002) *Eur. J. Org. Chem.*, 393–401.
39 (a) Pietrusiewicz, K.M., Koprowski, M., and Pakulski, Z. (2002) *Tetrahedron:*

Asymmetry, **13**, 1017–1019; (b) Pakulski, Z., Koprowski, M., and Pietrusiewicz, K.M. (2003) *Tetrahedron*, **59**, 8219–8226.
40 Arai, S., Hamaguchi, S., and Shioiri, T. (1998) *Tetrahedron Lett.*, **39**, 2997–3000.
41 Chen, Y. and Deng, L. (2001) *J. Am. Chem. Soc.*, **123**, 11302–11303.
42 Tang, L. and Deng, L. (2002) *J. Am. Chem. Soc.*, **124**, 2870–2871.
43 (a) Hang, J., Tian, S.-K., Tang, L., and Deng, L. (2001) *J. Am. Chem. Soc.*, **123**, 12696–12697; (b) Hang, J., Li, H., and Deng, L. (2002) *Org. Lett.*, **4**, 3321–3324.
44 Berkessel, A., Cleemann, F., Mukherjee, S., Müller, T.N., and Lex, J. (2005) *Angew. Chem. Int. Ed.*, **44**, 807–811, and references cited therein.
45 Berkessel, A., Mukherjee, S., Müller, T.N., Cleemann, F., Roland, K., Brandenburg, M., Neudörfl, J.-M., and Lex, J. (2006) *Org. Biomol. Chem.*, **4**, 4319–4330.
46 Carreño, M.C. (1995) *Chem. Rev.*, **95**, 1717–1760.
47 For leading reference on the preparation of chiral sulfoxides, see Fernández, I. and Khiar, N., (2003) *Chem. Rev.*, **103**, 3651–3705.
48 Evans, J.W., Fierman, M.B., Miller, S.J., and Ellman, J.A. (2004) *J. Am. Chem. Soc.*, **126**, 8134–8135.
49 Lu, B.Z., Jin, F., Zhang, Y., Wu, X., Wald, S.A., and Senanayake, C.H. (2005) *Org. Lett.*, **7**, 1465–1468.
50 Shibata, N., Matsunaga, M., Nakagawa, M., Fukuzumi, T., Nakamura, S., and Toru, T. (2005) *J. Am. Chem. Soc.*, **127**, 1374–1375.
51 Shibata, N., Matsunaga, M., Fukuzumi, T., Nakamura, S., and Toru, T. (2005) *Synlett*, **11**, 1699–1702.
52 Peltier, H.M., Evans, J.W., and Ellman, J.A. (2005) *Org. Lett.*, **7**, 1733–1736.

Part Three
Organic Chemistry of Cinchona Alkaloids

12
Organic Chemistry of *Cinchona* Alkaloids
Hans Martin Rudolf Hoffmann and Jens Frackenpohl

12.1
Introduction

During the last two decades, cinchona alkaloids have emerged as powerful chiral auxiliaries leading to well-known landmark developments in asymmetric synthesis already described in preceding chapters; but more recently, these alkaloids themselves have been shown to undergo some remarkable transformations and skeletal shifts that are rapidly widening the outlook in the chemistry of cinchona bases.

By way of introduction, let us look at one of the oldest reactions of cinchona chemistry, namely, the Pasteur cleavage to quinotoxine [1]. On treatment with acid, both the quinine **1** and quinidine **2** pair as well as the cinchonidine and cinchonine pair are cleaved to give quinicine (quinotoxine) **3** and cinchonicine **4** (cinchotoxine), respectively (Scheme 12.1). Three stereocenters are lost and 3,4-difunctionalized piperidines are formed.

Consider a possible mechanism. At first sight, the reaction looks innocuous and almost trivial – yet one has to ask why the alkaloid 1,2-amino alcohol is broken up at all. A priori, one can envisage several possibilities (Scheme 12.2):

1. **A 1,2-hydride shift**: As formulated (Scheme 12.2), the reaction could be considered an aza-analogous pinacol–pinacolone rearrangement. However, it seems doubtful that the $C=N^+<$ heterolysis on its own is efficient enough to pull off this self-oxidation/reduction.

2. **Conventional β-elimination with deprotonation at C9**: Since a hydroxyl proton such as C9–OH is generally more mobile and acidic than the C9–H proton, it is not immediately obvious why the C–H proton should come off at all.

3. **Tautomerization with concomitant protonation of the bridgehead Nsp^3 and the less basic Nsp_2 nitrogen; generation of an enol i, which is also an enamine, and $C8-N^+R_3$ heterolysis**: The pyridine-type heterocycle is an electron-attracting moiety as exemplified by the spontaneous low-temperature enolization of 1-azabicyclic [3.2.2]ketones **115** and **116** described below (Scheme 12.43).

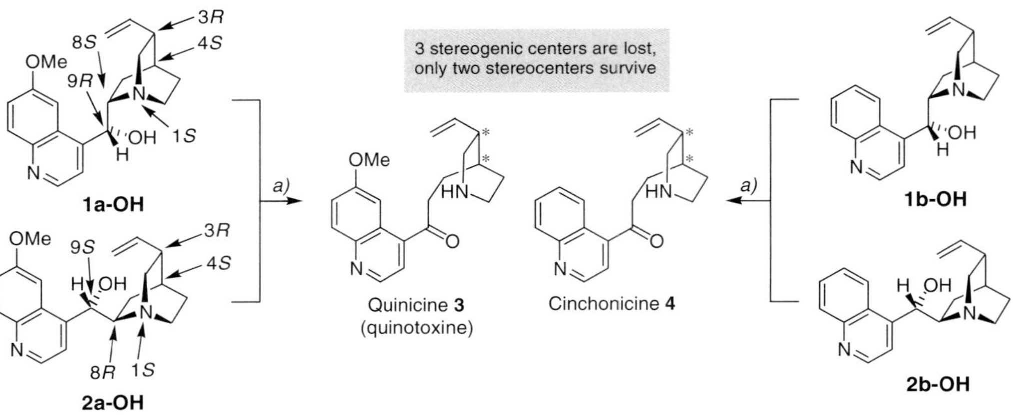

Scheme 12.1 The quinotoxine cleavage (L. Pasteur, 1853).
Reagents and conditions: (a) water–acetic acid (13 : 1), 100 °C, 35 h.

Deprotonation at C9 with extended conjugation seems feasible because the resulting enol *i* is sterically stabilized and of a type known to exist transiently [2]. Tautomer *i* is an activated oxyallyl system [3], which is built into a 1,5-*N*,*N*-acetal.

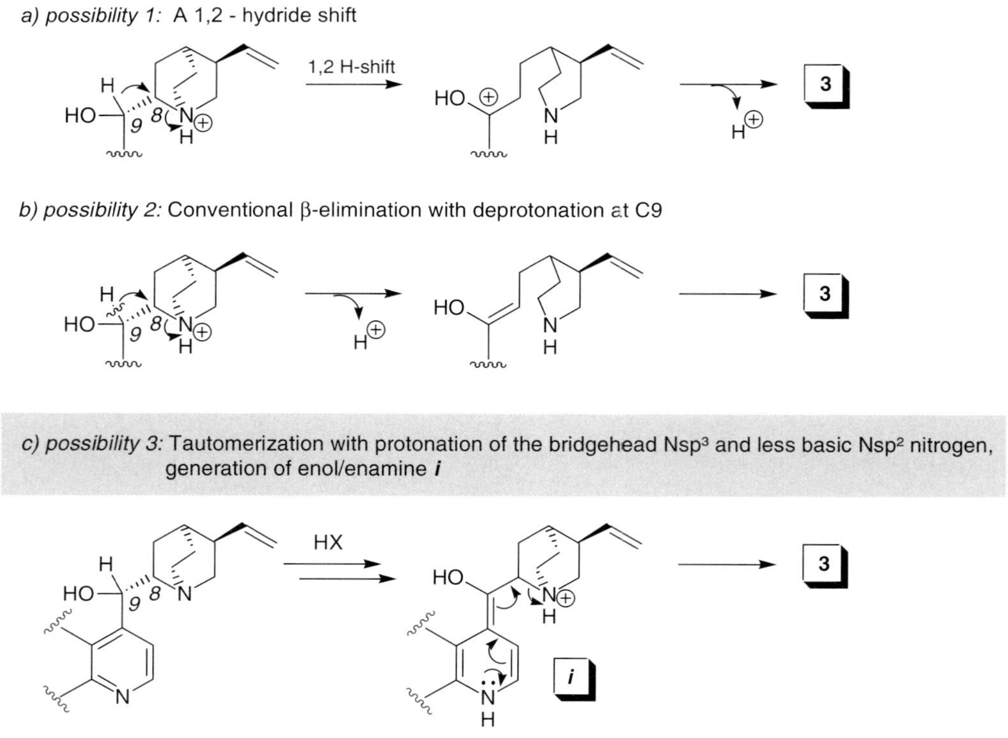

a) possibility 1: A 1,2 - hydride shift

b) possibility 2: Conventional β-elimination with deprotonation at C9

c) possibility 3: Tautomerization with protonation of the bridgehead Nsp³ and less basic Nsp² nitrogen, generation of enol/enamine *i*

Scheme 12.2 From quinine to quinotoxine – possible routes.

Once species *i* is generated, the remote enamine nitrogen exerts high nucleophilic pressure at C8 and the observed C−N$^+$R$_3$ fission becomes plausible. It is of interest that a recent preparation of quinotoxine from quinine uses the 1912 Biddle [1h] procedure with water and little acetic acid (13 : 1) as a medium. Apparently, it does not pay to go over the top with respect to acidity. Acetate as a nucleophile and base will assist the required tautomeric change and proton jumps. Overall, the quinoline nitrogen serves as an Achilles' heel as it does in the QCI and QCD formation (Scheme 12.5). The role of the proton for the generation of quinotoxine can probably be taken on by an organometallic Lewis acid also (cf. the following discussion).

Thanks to quinicine **3** and cinchonicine **4** Pasteur achieved the first separation of racemic tartaric acid. This resolution is considered a milestone in organic chemistry. Conversely, the transformation of quinotoxine **3** into quinine in three steps is part of the (formal) Rabe–Kindler/Woodward–Doering synthesis of quinine as has recently been reaffirmed by R.M. Williams and his group (cf. Section 11.5.3). Meroquinene ester **7** is a 3,4-disubstituted piperidine formed from quininone **5** (cinchoninone) and base in the presence of ^{3}O$_2$. In this reaction, an activated bridgehead lactam is a key intermediate that is opened by KOBu^t. Three stereocenters are lost and only two stereocenters survive in the course of this transformation (Scheme 12.3) [4].

It has been suggested that cinchona bases, although most valuable chiral auxiliaries (e.g., the AD reaction [5] and asymmetric phase-transfer reactions [6]), "are very unlikely to find application as (chiral) building blocks" [7]. The recent developments outlined herein will change this view.

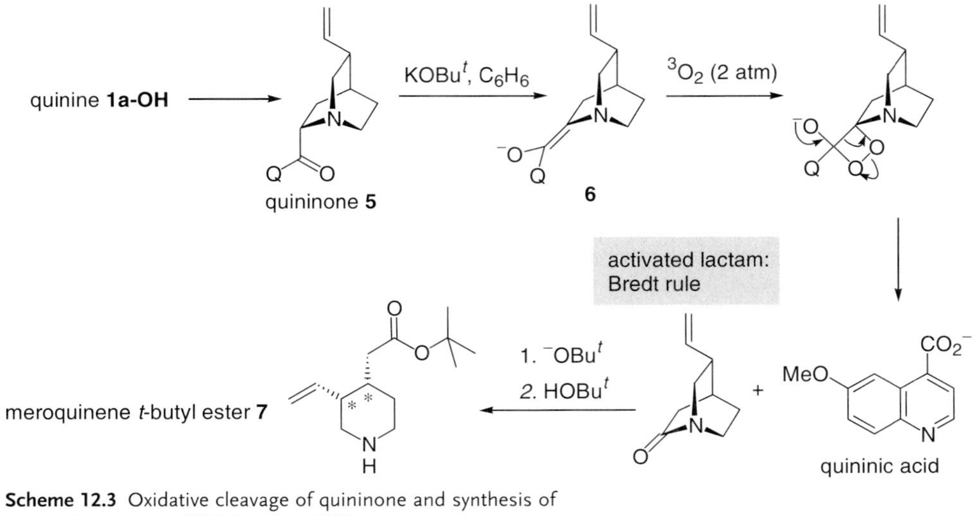

Scheme 12.3 Oxidative cleavage of quininone and synthesis of meroquinene ester (W.E. Doering, 1946).

12.2
Preparation of Quincorine and Quincoridine: Discovery of a Novel Cleavage Reaction of Cinchona Alkaloids

Quinine and quinidine have been transformed into quincorine (QCI) and quincoridine (QCD), respectively, via cleavage of the C4'–C9 bond of the natural products. These two bicyclic 1,2-amino alcohols contain four intact chiral centers, each including the 1S-configured bridgehead nitrogen. Like quinine and quinidine, they are available in enantiomerically pure form, and like their parent precursors, they are "pseudoenantiomeric": a hypothetical replacement of the vinyl group by hydrogen in QCI and QCD leaves carbon C2 as a single stereocenter. Our route to QCI and QCD is outlined in Scheme 12.4 [8]. 6'-Methoxyquinoline is formed from both quinine and quinidine.

Experiments with LiAlD$_4$ and other evidence suggest a unique sequence of elementary steps and reactivity embracing a variety of transient intermediates from nucleophilic, radical to electrophilic (Scheme 12.5).

Thus, while nucleophiles are believed to generally attack C2 of pyridine, LiAlD$_4$ and added secondary alcohol are visualized to furnish an alkoxy aluminum deuteride, which docks onto the secondary hydroxy group of the alkaloid (**2a-OH** → **i**). Hydrogen evolved is swept out by nitrogen. Thanks to the lone-pair back-donation from the alkoxy oxygens to aluminum (Scheme 12.5) that the intramolecular attack by

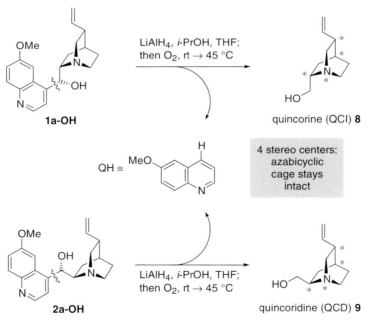

Scheme 12.4 Invention and innovation in natural product chemistry: quincorine and quincoridine from cinchona alkaloids (H.M.R. Hoffmann, 1996).

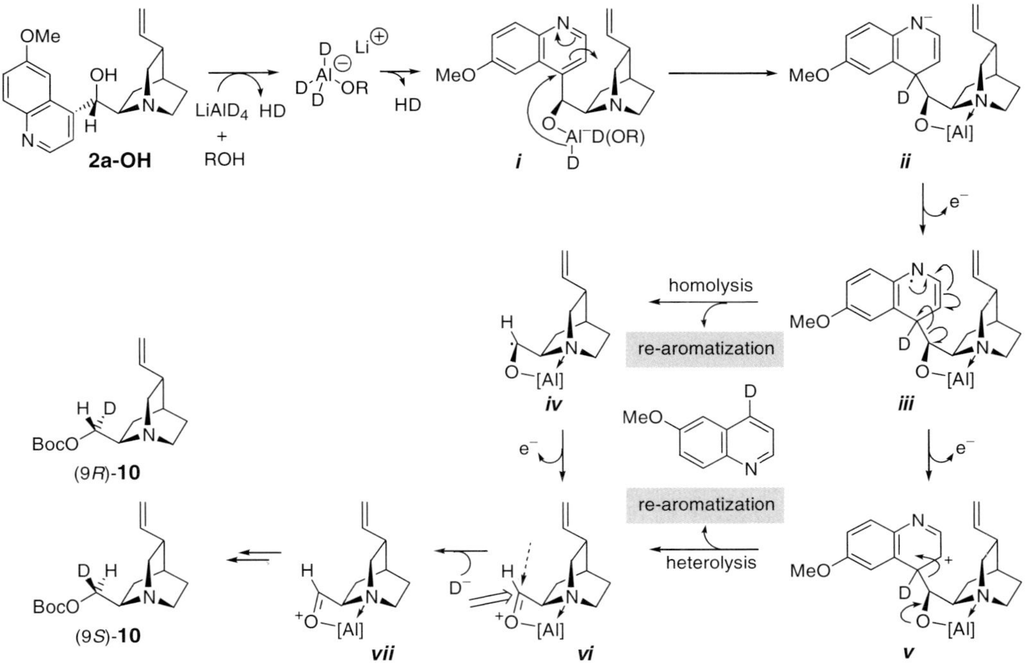

12.2 Preparation of Quincorine and Quincoridine: Discovery of a Novel Cleavage Reaction

Scheme 12.5 *En route* to quincoridine. Proposed mechanism of QCD formation from quinidine.

deuteride ion at carbon C4' is facilitated. An azaallyl anion (*i* → *ii*) is generated that undergoes one-electron oxidation to form an intermediate azaallylic radical (*ii* → *iii*). Homolysis of this radical affords neutral, deuterated 6-methoxyquinoline and aluminum-complexed ketyl radical *iv*. A second, one-electron oxidation is assumed to generate the electrophilic aldehyde (*iv* → *vi*). Alternatively, but considered less likely, azaallylic radical *iii* undergoes one-electron oxidation to azaallylic cation *v*, which would have to undergo a heterolytic fragmentation to aldehyde *vi* and 6'-methoxyquinoline. Aldehyde *vi* undergoes diastereoselective deuteration via an ionic reaction (*vi* → *vii*). Intramolecular complexation of aluminum by the bridgehead nitrogen is thought to restrict rotation of the formyl group. After Boc protection, deuterated diastereomer (9*R*)-10 is isolated in excess, in accord with nucleophilic attack from the more accessible side.

Thus, in addition to the familiar and classical degradation of cinchona bases, there is scope for unusual reactivity and new selectivity, involving radical, radical ion-SET, and ionic chemistry with sequential oxidation and also hydrogenation steps, all in a single flask! The transformation takes place under mild conditions with high functional group selectivity. Protecting and activating groups are not necessary. The process has been carried out on a 100 kg scale in a 5000 l vessel, thus turning QCI and QCD into readily available chiral building blocks (Figure 12.1). A potentially dangerous reaction is started and sustained by controlled access of oxygen (Scheme 12.5). The reaction mixture is kept below 50 °C and does not ignite.

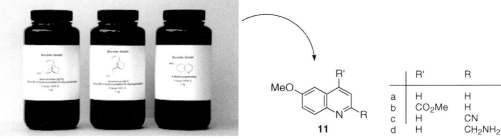

Figure 12.1 Commercially available OCI, QCD, and quinolines from Buchler Chininfabrik GmbH, Braunschweig, Germany.

The other product is 6′-methoxyquinoline **11a** that is obtained in high purity as almost colorless crystals, with only a slight yellow tinge. Previously, this heteroaromatic material was available commercially only as a brown mass. Derivatives with substituents at C2 and C4 have been prepared by nucleophilic addition to the pyridine nucleus (Figure 12.1).

12.3
Transformations of the Quinoline Moiety

Pyridine is an aromatic 6π electron heterocycle, which is isoelectronic with benzene, but electron deficient. Nucleophiles thus add almost invariably to carbon C2 of the imine-like C=N double bond. Perhaps the best known nucleophilic addition is the Chichibabin reaction with sodium amide in liquid ammonia, giving 2-aminopyridine. Reactions of the quinoline moiety of cinchona alkaloids can be more complex. Although expected 2′-addition can be achieved easily with organolithium reagents to yield **13** (Scheme 12.6) [9], LiAlH$_4$, for example, has been shown to attack C4′ en route to quincorine and quincoridine (Schemes 12.4 and 12.5). C4′ selectivity is due to chelation of aluminum by the C9−OH oxygen.

More recently, C4′−C bonds were formed upon reactions of cinchona alkaloids with an excess of Grignard reagents at elevated temperature [10]. The addition is terminated in unexpected fashion, namely, by the loss of heteroaromaticity and

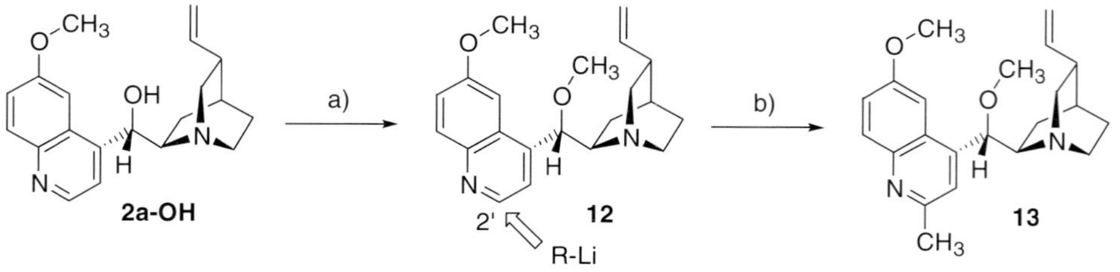

Scheme 12.6 Substitution at the C2′-position of the quinoline moiety with organolithium reagents. Reagents and conditions: (a) KH, MeI, THF; (b) MeLi, Et$_2$O, −78 °C−rt, then air or I$_2$, 72%.

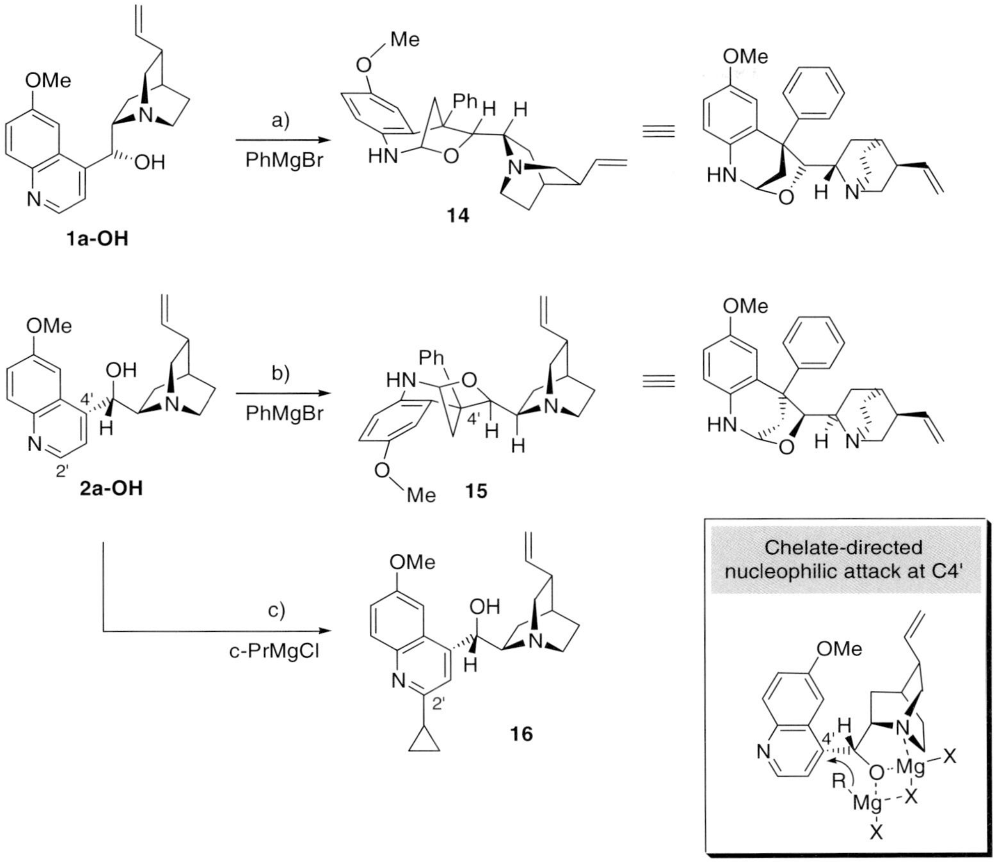

Scheme 12.7 Nucleophilic addition of Grignard reagents at C4′ – formation of [3.2.1]bicyclic N,O-acetals (L. Hintermann et al., 2007). Reagents and conditions: (a) PhMgBr, toluene, 70 °C, 3.5 h, 35%; (b) PhMgBr, toluene, 70 °C, 4 h, 53%; (c) c-PrMgCl, toluene, 65 °C, 20 h.

collapse to a [3.2.1]bicyclic N,O-acetal (Scheme 12.7). The outcome of this surprising reaction has been validated by X-ray crystallography. Aryl Grignard reagents with either electron-donating or electron-accepting groups were readily transformed into N,O-acetals **14** and **15**. Vinyl Grignard reagents also added to the quinoline nucleus, as did primary Grignard reagents and even isopropyl magnesium chloride. Highly sterically demanding nucleophiles yielded the "classical" 2′-alkylated alkaloid **16** as the major product instead.

Thus, the Grignard reagent can add to the less accessible 4′-position of the quinoline moiety rather than the 2′-position, and single diastereomers were isolated from all reactions (e.g., **14** and **15**). Both observations were explained by assuming a group transfer within a metal chelate (Scheme 12.7). The choice of toluene as solvent for not breaking up those metal clusters is a key factor for obtaining satisfactory results in the conjugate addition. The various adducts offer ample scope for further studies. It is worthy of note that adducts **14** and **15** are a "pseudoenantiomeric" pair, i.e. replacement of vinyl by hydrogen generates an enantiomeric pair. The two bridge head carbons of the [3.2.1] bicycle cannot be inverted independently.

Another variation of the quinoline moiety of the cinchona alkaloid scaffold involves a modification of the ether at C6′ [11]. To obtain novel phase-transfer catalysts, the methyl ether of quinidine **2a-OH** was cleaved to yield phenol **17-OH**. Subsequent Cs_2CO_3-mediated alkylation gave only quinidine analogues **18** that were quaternized to yield desired catalysts **19** (Scheme 12.8). Alkylation at C9 was not observed. Phenol **17-OH**, also named cupreidine, can be used as a precursor for the synthesis of various functionalized cinchona organocatalysts (Scheme 12.9). Thioureas **22** were synthesized by transforming the C6′−OH group into an amino functionality via Buchwald amination of the intermediate triflate **20** as key step [12]. Subsequent base-free coupling of **21** with isothiocyanates afforded the target catalysts **22** [13]. This four-step reaction sequence proceeded in high yield, also on a multigram scale. The phenolic C6′−OH also allows functionalization of the α-position, giving rise to C5′-substituted cinchona alkaloids. Quinidine-derived phenol **17** can be aminated in α-position by a simple addition of di-*tert*-butyl azodicarboxylate (D*t*-BuAD) in dichloromethane at ambient temperature [14]. Two diastereomeric hydrazides **23a** and **23b** could be isolated, and the principle of atropisomerism was successfully introduced into cinchona alkaloid chemistry. Interestingly, atropisomers **23a** and **23b** showed a pronounced difference in their polarity.

Functionalized oxazatwistanes with phenolic OH group and nonnatural configuration at C3 (see Schemes 12.19 and 12.21) have shown their utility as organocatalysts in a growing number of reactions [15], underscoring the importance of fundamental research in cinchona alkaloids and the careful investigation of substituent effects.

12.4
Basic Transformations of the Vinyl Side Chain

Conversion of the natural product quinine into its alkyne analogue **26** was reported at the beginning of the twentieth century [16], but to our surprise, no further work on this potentially interesting and useful class of compounds had come forward since then. Similarly, transformations such as oxidative cleavage or degradation to the ketone functionality at carbon C3 had remained unexplored until the 1990s.

12.4.1
Alkyne Cinchona Alkaloids, Their Derivatives, and Basic Transformations

We have recently converted the ethenyl group of the cinchona alkaloid into an ethynyl group by a bromination–double-dehydrobromination sequence [17, 18]. Bromination of the natural products **1**, **2** in CCl_4 provided the corresponding 10,11-dibromo-cinchona alkaloids in quantitative yield as a yellowish precipitate. The double dehydrobromination to afford alkynes **26** and **27** was studied under a variety of

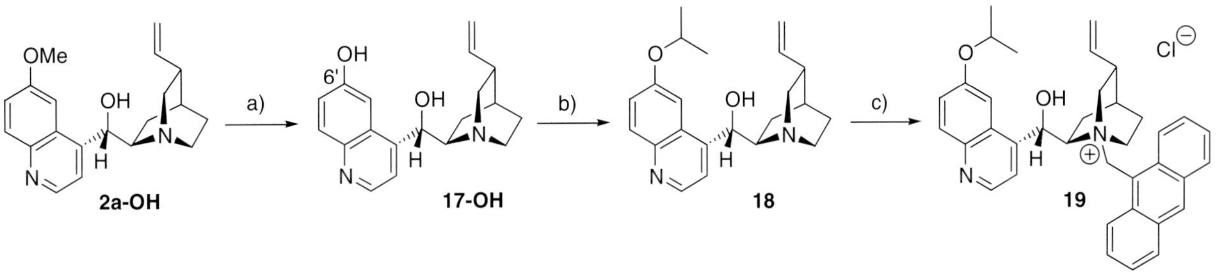

Scheme 12.8 Preparation of C6'-modified phase-transfer catalysts. Reagents and conditions: (a) NaH, EtSH, DMF, 110 °C, 16 h, 90%; (b) Cs$_2$CO$_3$, 2-bromopropane, DMF, 60 °C, 40 h, 87%; (c) 9-chloromethylanthracene, THF, reflux, 16 h, 67%.

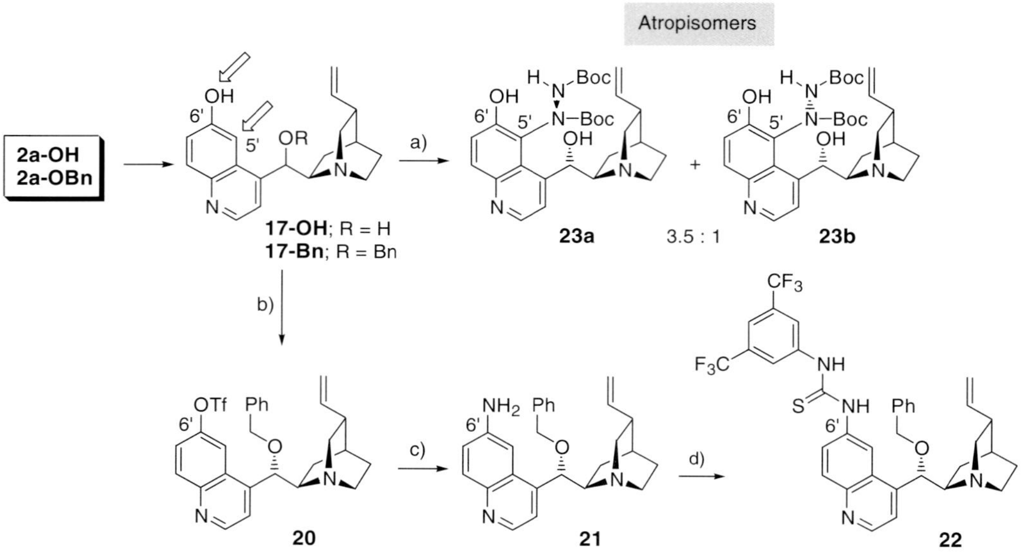

Scheme 12.9 Synthesis of novel cinchona alkaloid organocatalysts with modified quinoline moiety. Reagents and conditions: (a) Dt-BuAD, CH$_2$Cl$_2$, rt, 2 h, 93%; (b) PhNTf$_2$, DMAP, CH$_2$Cl$_2$; (c) 1. Pd(OAc)$_2$, BINAP, Cs$_2$CO$_3$, THF, Ph$_2$C=NH, 2. citric acid, THF, H$_2$O; (d) 3,5-(CF$_3$)$_2$PhNCS, THF.

conditions [19], but ordinarily, the yields did not exceed 50%. We therefore carried out the dehydrohalogenation in two steps. Addition of Et$_3$N provided vinylic bromides **24** and **25** in quantitative yield. Using the highly hygroscopic and lipophilic methyltrioctylammonium chloride (aliquat 336) [20] in catalytic amount together with powdered KOH simplified the second dehydrobromination, and alkyne formation was feasible under mild conditions (Scheme 12.10) [17]. All alkynes prepared were more polar and basic than parent natural cinchona alkaloids with vinyl side chain. Thus, it became clear early on that the remote ethynyl side chain would have a significant impact on chemical and physical properties in contrast to related dihydro analogues [21]. No inherent instability of these alkynes was observed.

Simple 10,11-didehydro QCD and 10,11-didehydro QCI are also accessible using this reliable two-step procedure. Thus, QCI, which solidifies at room temperature when pure, forms alkyne **28** with a change of chirality descriptor from R to S at carbon C5. QCI-derived alkyne **28** crystallizes readily and also sublimes on warming, forming beautiful colorless needles, reaching from wall to wall of the reaction vessel [17].

Thanks to the terminal alkyne function, a wide range of Pd-catalyzed sp^2–sp cross-coupling reactions are feasible for natural product derivatives **26** and **27** as well as for QCI and QCD alkynes involving vinylic, aromatic, and heteroaromatic centers [18]. Sonogashira coupling of the cinchona alkyne and coupling partners in the presence of (Ph$_3$P)$_2$PdCl$_2$ (0.05 equiv) and CuI (0.1 equiv) in a mixture (1 : 1) of Et$_3$N and THF proved to be suitable (e.g., **30a–b**, Scheme 12.11). QCI-alkynes in Scheme 12.11 are models for the related alkaloids [11, 22].

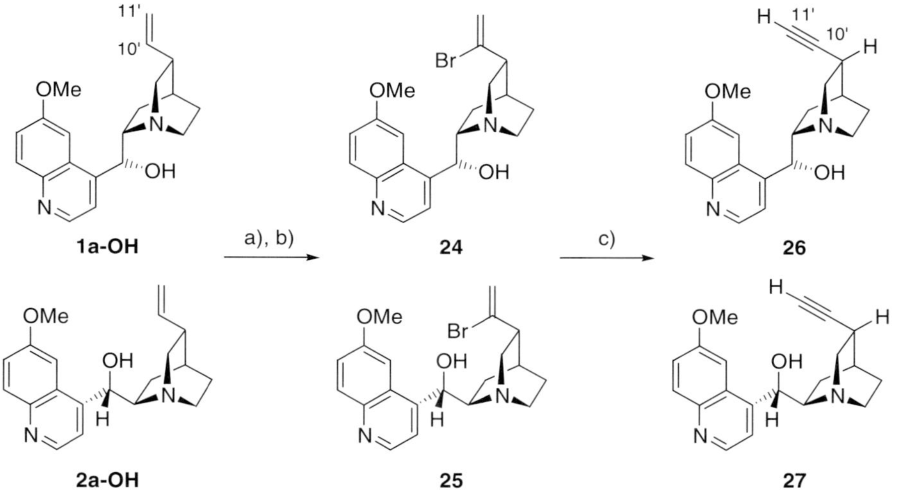

Scheme 12.10 Synthesis of 10,11-didehydro-quinine and -quinidine. Reagents and conditions: (a) Br$_2$, CCl$_4$, 0 °C, 2 h; (b) Et$_3$N, CHCl$_3$, rt, 14 h; (c) KOH, aliquat 336, THF, rt, 20 h.

Corresponding coupling with aniline derivatives, followed by cyclization, affords conjugate indole–cinchona alkaloids (**31**) [23]. Furthermore, cobalt-catalyzed cycloadditions such as the Pauson–Khand reaction proceed with high yields and moderate diastereoselectivities (e.g., **32**, Scheme 12.11). Terminal alkyne **28** is easily transformed into its iodinated analogue **33** [17], which can be hydrogenated to afford Z-vinyl iodide **34**, a further versatile building block for cross-coupling reactions. We have also used Sonogashira coupling to prepare enediynes, for example, enediyne **35**. Although enediyne **35** is not part of a strained ring, it undergoes a Bergman cycloaromatization at moderately elevated temperature (60–70 °C) in CHCl$_3$ to form **36** in 86% yield (Scheme 12.11) ([18]; I. Neda, personal communication).

12.4.1.1 The Ethynyl Group is Anything but a Spectator Substituent

All alkyne alkaloids prepared including 10,11-didehydro QCI and QCD were found to be crystalline and more polar on chromatography than the parent ethenyl (vinyl) alkaloids [17, 18, 24]. Increased polarity/basicity is consistent with the observation that Os-mediated AD reactions proceed significantly more slowly with alkyne ligands **37** (Figure 12.2) [25]. The efficiency of alkyne cinchona alkaloids in AD reactions was evaluated with both standard substrates and challenging targets, such as homoallylic alcohols and bryostatin C-ring lactol segment **38**. In the case of substituted C-ring lactols, a higher selectivity (although lower rates) was observed with novel alkyne AD-ligands, thus overcoming selectivity problems inherent to the use of commercial AD-mixes. Furthermore, novel quaternized cinchonidine-based phase-transfer catalysts **39** were synthesized upon replacement of the vinyl group by a set of alkyne moieties. As shown in the AD reaction, the asymmetric synthesis of β-hydroxy-α-amino acids via aldol reaction proceeds with lower rates but higher selectivity when using catalysts **39** [26].

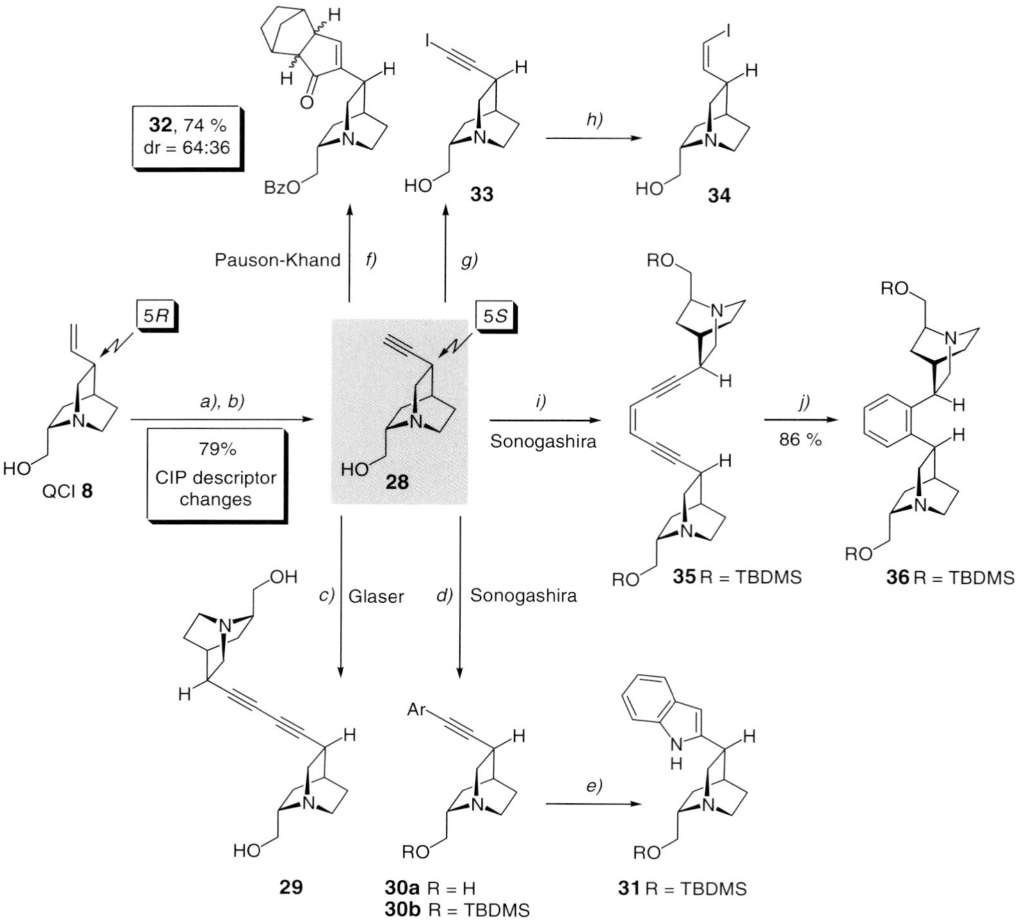

Scheme 12.11 Synthesis and transformations of QCl-alkyne **28**. Reagents and conditions: (a) 1. Br$_2$, CCl$_4$, 0 °C, 2 h; 2. Et$_3$N, CHCl$_3$, rt, 14 h; (b) KOH, aliquat 336, THF, reflux, 7 h; (c) (Ph$_3$P)$_2$PdCl$_2$, CuI, I$_2$, Et$_3$N, THF, 16 h; (d) (Ph$_3$P)$_2$PdCl$_2$, CuI, aryl or vinyl halide, Et$_3$N, THF, 16 h; (e) 1. TFAA (6 equiv), Et$_3$N (3 equiv), THF, −78 °C → 0 °C, 1 h; 2. imidazole (2.6 equiv), LiCl (1.2 equiv), Pd(OAc)$_2$ (5 mol%), DMF, 6 h, 35%; with 40 mol% of Pd(OAc)$_2$ the yield increases to 57%; (f) 1. Co$_2$(CO)$_8$, CH$_2$Cl$_2$, rt, 90 min; 2. NMO, alkene, CH$_2$Cl$_2$/THF (1 : 1), rt, 5 h; (g) I$_2$, morpholine, toluene, 70 °C, 5 h; (h) p-TsNHNH$_2$, NaOAc, THF, H$_2$O, 65 °C, 4 h; (h) (Ph$_3$P)$_2$PdCl$_2$, CuI, 1,2-dichloroethene (0.5 equiv), Et$_3$N/THF (1 : 1), 16 h; (i) CHCl$_3$, 60–70 °C, 4 h.

A priori, the 1-azabicyclo[2.2.2]octane helix may be twisted clockwise or counterclockwise and may also adopt a higher energy C$_{3v}$ conformation in which the three ethano bridges are fully eclipsed. Interaction of the nitrogen lone pair with the three antiperiplanar σ bonds is expected to be maximized in the C$_{3v}$ conformer (Scheme 12.19), which contains two fused boat (rather than twist) azacyclohexanes. The conformational energy (or A value) of the "slim" −C≡CH group is considerably

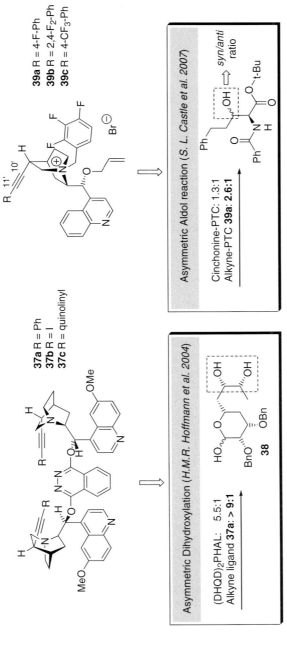

Figure 12.2 Impact of side chain modification on stereocontrol of cinchona alkaloid ligands and catalysts.

	Didehydroquinidine	Didehydro-QCI	Didehydro-QCD
N1-C2-C3-C4	11.4°	6.1°	18.3°
N1-C6-C5-C4	13.2°	11.0°	16.2°
N1-C7-C8-C4	12.3°	9.7°	15.3°
Σ	36.9°	26.8°	49.8°

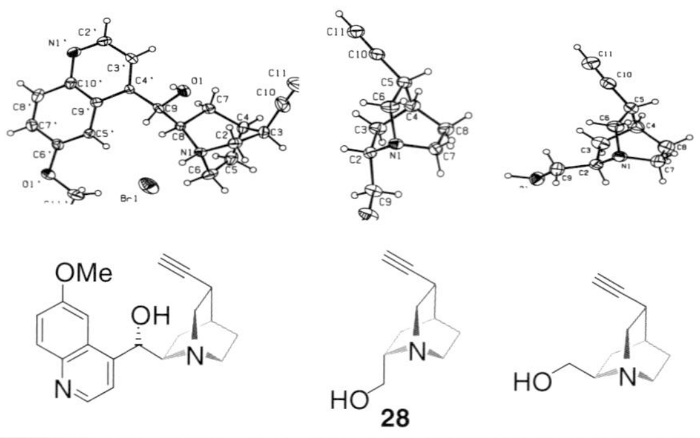

Figure 12.3 X-ray crystal structure of alkynes derived from cinchona alkaloids. Torsion angles defining twist sense, cf. Scheme 12.19.

less (1.71–2.18 kJ/mol) than that of the sterically more demanding $-CH=CH_2$ (6.23–7.00 kJ/mol) and $-CH_2-CH_3$ (7.49 kJ/mol) groups [27]. As a result, 1,3-*syn* repulsion at C3 and C5 will be less in the alkyne bicyclics than that in the ethyl and ethenyl analogues.

From a variety of X-ray crystal studies (e.g., didehydro-QCI/QCD and didehydroquinidine), we have established that alkyne quinuclidines are *twisted* consistently less than the vinyl and ethyl analogues. Overall torsion angles, that is, the sum of the three torsion angles, vary from 26 to 49° compared with up to 65° for ethyl and 60° for ethenyl derivatives, respectively (Figure 12.3). Fine-tuning of nitrogen polarity (and basicity), which may be important for catalysis and perhaps pharmacological activity, is thus possible ([28]; R. Wartchow, personal communication).

12.4.2
Fluorination of the Vinyl Side Chain

Introducing fluorine substituents into biologically active compounds is a strategy that is frequently used in medicinal chemistry or agrochemical research [29]. Fluorination of the C10-C11 side chain of quinine and quinidine can be achieved by the reaction of C9-O-acetylated **1a-OAc** and **2a-OAc** in the superacid system HF-SbF5 in the presence of HCl and CCl₄. With HCl as crucial additive for protonating both nitrogen atoms, geminal difluorination of quinine at C10 can be achieved, however, with epimerization at the adjacent bridging carbon C3. Difluorinated epimers **40a** and **40b** are formed as 1 : 1 mixtures (Scheme 12.12) [30]. Treatment of unprotected alkaloids with superacids affords tricyclic derivatives as only products (cf. sections below). Upon addition of an excess of H_2O_2 to $HF\text{-}SbF_5$ as a source of an OH^+ equivalent acetylated quinine, **1a-OAc** is transformed into corresponding C10-ketone **42** and fluorohydrin **41**, again both as epimeric mixtures [31]. As only one F-atom is introduced at C10, a new stereocenter is generated increasing the number of separable diastereomers. Remarkably, in the quinidine series of both transformations, rearranged [3.2.1]azabicyclic analogues **45** or **46** could be isolated in considerable yield alongside expected fluorinated epimers **43** and **44** [32].

Starting from acetylenic quinidine acetate **27-OAc**, the corresponding gem-difluoro derivative **43** was obtained as a single product. In this case, consecutive reactions such as epimerization and rearrangement to the thermodynamically more stable [3.2.1]azabicyclic analogue **46** were not observed [33].

12.4.3
Oxidation and Oxidative Cleavage of the Vinyl Group

Although early synthetic efforts in the late nineteenth century focused on the vinylic side chain of the cinchona alkaloids, functionalization remained a challenge as epimerization at carbon C3 (cinchona numbering) could occur (cf. Scheme 12.12) [34]. Dihydroxylation of protected quinidine under two-phase conditions furnishes an inseparable 1:1 mixture of epimeric diols **47a** and **47b**. Periodate cleavage under standard and other conditions invariably provides an epimeric mixture of C10-aldehydes **48a** and **49** (Scheme 12.13). For the preparation of epimerically pure acetylated aldehyde **48a**, it was crucial to first coat wet silica gel with $NaIO_4$ and then add the diol at room temperature [35]. In this fashion, the yield of aldehyde **48a** was also improved. Even when the modified aldehyde synthesis was applied to QCI **8** and QCD **9**, epimerization could not be inhibited entirely. In these cases, epimeric mixtures of 82:18 and 77:23 were obtained.

Aldehyde **48a** is a versatile intermediate for the preparation of various epimerically pure quinidine derivatives such as alcohol **50** via $NaBH_4$ reduction and amines **51** via reductive amination (Scheme 12.13). Deprotection of **48a** provided ε-hydroxyaldehyde **48b**. In the case of an acetyl-protecting group, deprotection was very easy due to intramolecular base catalysis [36]. Subsequent TIPS-triflate-promoted cyclization furnished a remarkably stable silylated hemiacetal **52** with defined configuration of carbon C10 [35]. Not only the epimerically pure C10-aldehydes **48a–b** but also the corresponding C10,C11-diol **47** provides access to a wide variety of diastereomerically pure quinidine derivatives [37]. Diastereomeric quinidine-based diols **53a** and **53b** were easily separated by column chromatography after selective O11-tosylation. Starting from monotosylated C10,C11-diol **53a** and **53b**, corresponding diastereomerically pure epoxides and 1,2-amino alcohols are accessible in satisfactory yield (82–85% overall, Scheme 12.13).

A second option for cleaving the vinylic side chain of cinchona alkaloids is the direct oxidation with $KMnO_4$ in HCl/MeOH affording corresponding C10 esters **54** and **55** in fair yield (Scheme 12.14). In an initial study, the ester functionality of **55** was converted into corresponding ketones **56** via a Weinreb amide intermediate.

12 Organic Chemistry of Cinchona Alkaloids

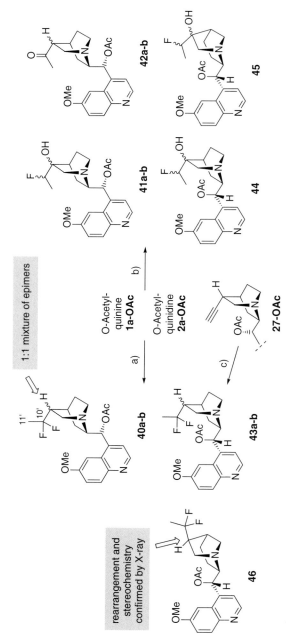

Scheme 12.12 Fluorination of the vinylic side chain: reactions of quinine and quinidine in superacids. Reagents and conditions: (a) 1. HF-SbF$_5$ (7:1), H$_2$O$_2$ (80%), −35 °C, 3 min; 2. ice water, Na$_2$CO$_3$; (b) 1. HF-SbF$_5$ (7:1), HCl, CCl$_4$, −30 °C, 10 min; 2. ice water, Na$_2$CO$_3$; (c) 1. HF-SbF$_5$ (7:1), −30 °C, 10 min; 2. ice water, Na$_2$CO$_3$.

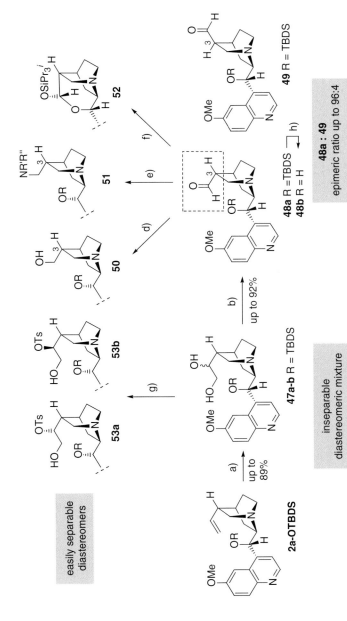

Scheme 12.13 Oxidative cleavage of the vinyl side chain *en route* to C10-aldehydes. Reagents and conditions: (a) OsO$_4$ (cat.), K$_2$CO$_3$, K$_3$[Fe(CN)$_6$], *t*-BuOH/H$_2$O (1 : 1), rt, 4 h; (b) NaIO$_4$, CH$_2$Cl$_2$, SiO$_2$, rt, 1 h; (c) TBAF, THF, 10 h, rt; (d) NaBH$_4$, MeOH, rt, 10 min; (e) ZnCl$_2$, NaBH$_3$CN, amine, MeOH, rt 24 h; (f) *i*-Pr$_3$SiOTf, 2,6-lutidine, CH$_2$Cl$_2$, 0–25 °C, 14 h; (g) *p*-TsCl, Et$_3$N, CH$_2$Cl$_2$, 24 h, rt.

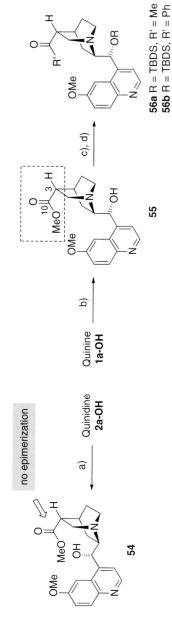

Scheme 12.14 Oxidative cleavage of the vinyl side chain via formation of C10-esters. Reagents and conditions: (a) 1. KMnO$_4$, H$_2$SO$_4$ (2N), 2 h, rt; 2. MeOH, conc. HCl (cat.), 4 d, rt; (b) 1. KMnO$_4$, H$_2$SO$_4$ (2N), 1 h, rt; 2. MeOH, conc. HCl (cat.), 1 d, rt; (c) 1. TBDMS-OTf, 2,6-lutidine, CH$_2$Cl$_2$, 0 °C – rt, 10 h; 2. Me$_2$AlCl, MeONHMe-HCl, CH$_2$Cl$_2$, 16 h, rt; (d) R-Li, THF, −78 °C – rt, 2 h.

In contrast to the oxidative cleavage of the C10-C11-moiety, hydroboration followed by oxidation furnishes C11-functionalized quinine and quinidine derivatives. As an illustration, only transformations in the quinidine series are shown (cf. Scheme 12.14). C11-aldehyde **58** was obtained either upon direct oxidation of the borane species with PCC/SiO$_2$ or via an improved stepwise procedure including (i) oxidation with Me$_3$NO-2 H$_2$O to yield the terminal alcohol **57** and (ii) subsequent Dess–Martin oxidation [35]. Oxidation and esterification of alcohol **57** using Jones' reagent and MeOH/HCl gave the C11-ester **59**, which is a suitable precursor for the synthesis of cinchona alkaloid macrocycles (Scheme 12.15) [38].

As is shown for quinidine C10-aldehyde **48**, its C11-homologue **58** forms an eight-membered acetal **61** upon treatment with TIPS-triflate [35]. Despite all the attention cinchona alkaloids have received in the field of asymmetric synthesis (cf. preceding chapters), there has only been limited use of these building blocks in supramolecular chemistry. A notable exception includes the use as chiral resolution reagent, where quinine forms an inclusion complex with binaphthols [39]. In recent years, Sanders et al. have intensively studied the macrolactonization of cinchona alkaloid esters. Cyclization of the quinine-based C11-ester, under thermodynamic control, gives mainly a cyclic tris-lactone [40], whereas quinidine-derived ester **59** is preorganized to

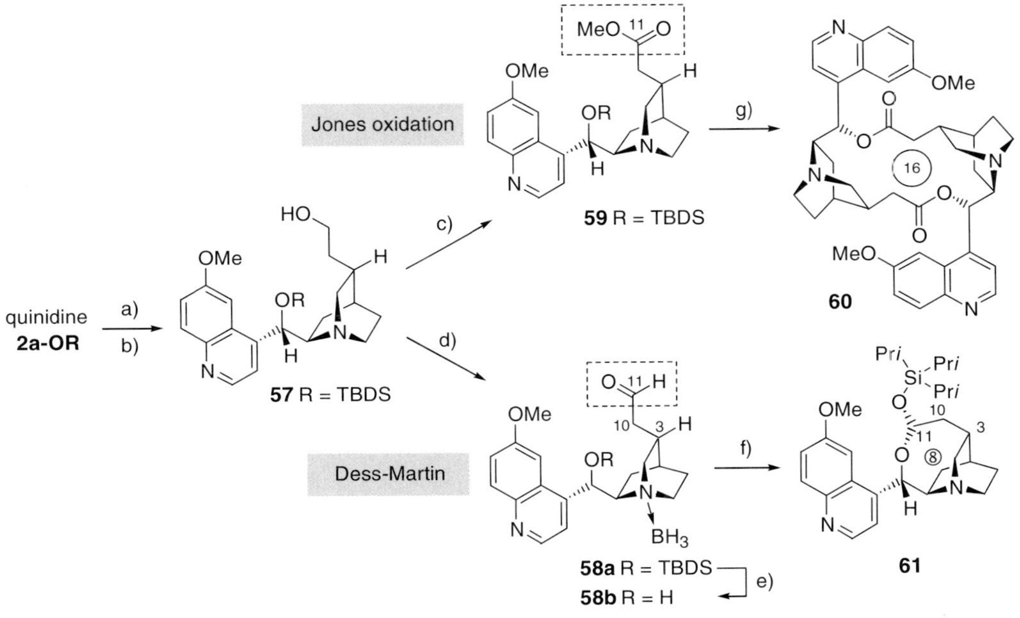

Scheme 12.15 Oxidation of the vinyl side chain at carbon C11. Reagents and conditions: (a) BH$_3$ (4 equiv), THF, 0 °C, 6 h, 97%; (b) Me$_3$NO-2H$_2$O, 100 °C; (c) 1. Jones reagent, acetone, rt; 2. HCl (conc.), MeOH, rt; (d) PCC/SiO$_2$ or Dess–Martin periodinane, CH$_2$Cl$_2$, 0 °C – rt, 5 h, 76%; (e) TBAF, THF, 10 h, rt; (f) i-Pr$_3$SiOTf (1.3 equiv), 2,6-lutidine (2 equiv), CH$_2$Cl$_2$, 0 °C – rt, 14 h; (g) 1. TBAF, THF, 10 h, rt; 2. KOMe, 18-crown-6, toluene, reflux.

form a medium-sized bis-lactone **60** (Scheme 12.15) [41]. At present, the various ring systems are accessible on a small scale only, high dilution being required for cyclization.

12.4.3.1 Oxidative Functionalization of Quincorine 8 and Quincoridine 9

Oxidation of 1,2-amino alcohols to α-amino acids is well known to be nontrivial and remains a challenging task especially if the basic amino nitrogen is not masked or protected [42]. Jones oxidation allows modification of the QCI and QCD side arm at carbon C9 affording C9-esters **62** and **63** selectively in acceptable yields (Scheme 12.16) [43]. Furthermore, the desired products can be isolated in epimerically pure form and also enantiopure. To our surprise, KMnO$_4$-mediated oxidation led to selective cleavage of the vinylic side chain of quincorine **8** and quincoridine **9**. The unprotected primary C9 alcohol function of **64** and **65** and the configuration at carbon C5 survived. 1,6-Dicarboxylic esters **66** and **67**, which are potential linkers, were formed only as side products in changing yield (10–20%).

As monosubstituted quinuclidines play an important role in medicinal chemistry, methods most frequently used for the preparation of 3-substituted quinuclidines

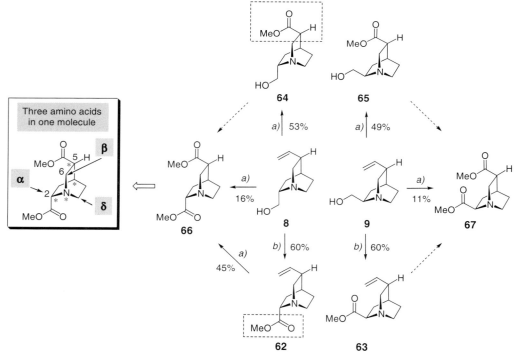

Scheme 12.16 Selective oxidation of the side arms of QCI and QCD. Reagents and conditions: (a) 1. KMnO$_4$ (2 equiv), H$_2$SO$_4$, H$_2$O, 0 °C → rt, 2. MeOH, cat. HCl, rt, 3 d; (b) 1. Jones reagent (3.6 equiv), acetone, 0 °C → reflux, 3 d, 2. MeOH, cat. HCl, rt, 3 d.

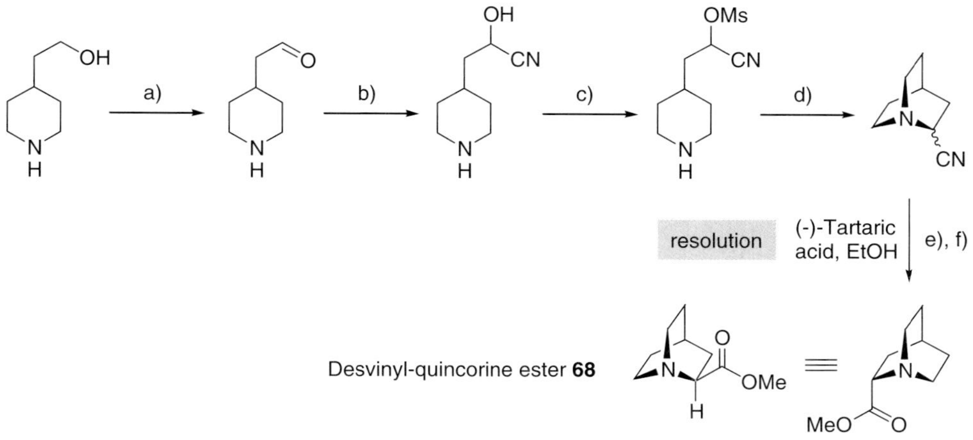

Scheme 12.17 Synthesis of enantiopure 5-quinuclidine-2-carboxylic acid ester. Reagents and conditions: (a) 1. Boc$_2$O, 2. DMSO, (COCl)$_2$, Et$_3$N, rt; (b) NaCN, HCl, H$_2$O; (c) MsCl, Et$_3$N, CH$_2$Cl$_2$; (d) 1. F$_3$CCO$_2$H, CH$_2$Cl$_2$, 2. Et$_3$N; (e) (+)-tartaric acid, EtOH, H$^+$; (f) 1. (COCl)$_2$, 2. MeOH, 3. Ba(OH)$_2$.

began with 3-quinuclidinyl methyl ester and 3-quinuclidinone. Likewise, quincorine- and quincoridine-based esters represent a novel class of seminatural, conformationally constrained amino acid esters. Due to their low molecular weight, their compact and rigid structure and their chemodifferentiated side arms, amino acid esters **62–65** are versatile chiral building blocks and novel acid soluble spacers.

To our knowledge, one alternative route to simple enantiopure quinuclidine-2-carboxylic acid has recently been described by Corey who assembled target molecule **68** whereas racemic **68** was first synthesized several decades ago by Prelog and [44]. Parent quinuclidine-2-carboxylic acid ester **68** that is structurally related to proline and pipecolinic acid was obtained from commercial 4-(2-hydroxyethyl)-piperidine in six chemical steps including one tartaric acid-mediated resolution (Scheme 12.17) [45]. A cyanoactivated intramolecular S$_N$2 reaction delivered the strained [2.2.2]bicyclic system. The cyano group serves as a handle of further functionality and elaboration.

12.4.4
Degradation of the Vinyl Side Chain: Synthesis of Cinchona Alkaloid Ketones

Metabolism of quinidine proceeds at several sites including oxygenation of the vinyl side chain [46]. With a view to preparing new metabolites with both natural and nonnatural, that is, *epi*-configuration at C3 as well as other enantiopure materials, von Riesen first developed a simple and reliable degradation procedure of the vinyl group to yield corresponding C3-ketones [47]. Hydrobromination of quinidine **2a-OH** with 48% HBr as performed in earlier work [48] is incomplete undoubtedly due to competing protonation of basic sites and lone pairs within the molecule. However, quinidine was found to dissolve in aqueous fuming 62% HBr (>6 equiv) completely to a clear, homogeneous solution. Apart from diastereomeric seven-ring ethers **79a** and **79b** (monohomotwistanes: cf. Section 12.5.1 and Scheme 12.20), the desired secondary bromide **70** was isolated as separable diastereomeric mixture [35, 49]. Base-mediated elimination of C10-bromide **70** with DBU furnished internal alkene **72**. The shift of olefinic double bond of parent cinchona alkaloids has recently attracted more interest. Isomerization via rhodium on alumina has been reported [50].

From a preparative standpoint, protection of the weak acidic OH group by an acyl or silyl protecting group not only facilitated separation and isolation but also improved the base-mediated elimination with DBU. The whole sequence was thus applied to O-protected quinine and related QCI and QCD (Scheme 12.18).

Subsequent dihydroxylation was carried out advantageously under two-phase conditions. No significant self-induction of diastereomer formation was observed by the AD-type ligand-cum-substrate (ratio of 1,2-diols: 1.2 : 1). Subsequent $NaIO_4$-mediated diol cleavage afforded the desired acetylated rubanone **74b** in 86% isolated yield on a 20 g scale. Given this reliable synthetic route, the door is open to further elaboration such as the Mannich α-aminomethylation of the carbonyl system. Thus, seminatural cinchona alkaloids have become accessible. Selected examples of possible transformations are discussed in the following section.

The conversion of QCI **8** and QCD **9** into their corresponding keto analogues **77** and **78** has been optimized following a pathway similar to the one described above and carried out on a multigram scale in 30–35% overall yield [51].

12.5
Selected Novel Transformations of the Quinuclidine Moiety of Cinchona Alkaloids

12.5.1
Cage Helicity and Further Consequences: Nucleophilic Attack on Quinuclidin-3-ones

1-Azabicyclo[2.2.2]octane represents a simple helix that, a priori, can be right-handed or left-handed (Scheme 12.19).

In contrast to cinchona alkaloids with ethyl or ethenyl side chain, where pseudoenantiomerism is often reflected by opposite twist sense, the corresponding alkyne derivatives are predominantly twisted clockwise. Similarly, both QCI-alkyne **28** and QCD-alkyne trace out a right-handed helix. However, strong counterclockwise twist is enforced in a suitably bridged oxazatwistane that can be seen to be a frozen conformation of quinidine (Schemes 12.19 and 12.46) and that has been used as asymmetric catalyst in Baylis–Hillman reactions (Scheme 12.20a). In fact, Hatakeyama has shown that semirigid chiral aminophenol **80-OH** as catalyst in the reaction of aldehydes with an activated acrylate afforded more than 90% ee that can be regarded as a major step forward in the study of the asymmetric Baylis–Hillman reaction [52]. A series of other hydroxylated amines derived from cinchona alkaloids has been screened, but only oxazatwistane **80-OH** proved to be a suitable ligand, in

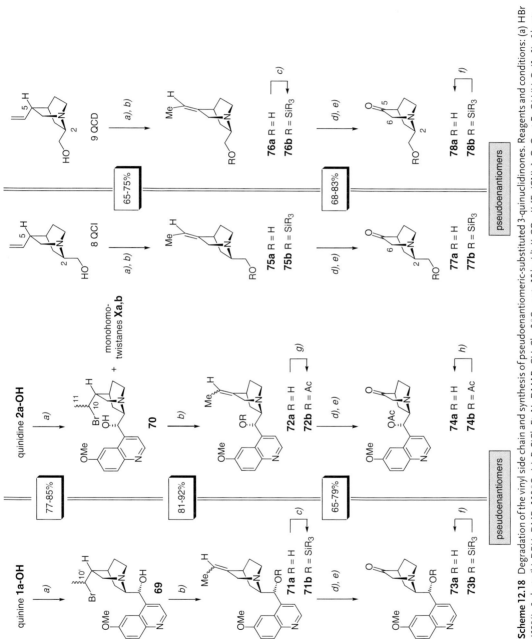

Scheme 12.18 Degradation of the vinyl side chain and synthesis of pseudoenantiomeric-substituted 3-quinuclidinones. Reagents and conditions: (a) HBr (62%), 3 d, rt; (b) DBU, DMF, 110 °C; (c) R₃SiCl, Et₃N, DMAP, CH₂Cl₂, 0 °C → rt, 10 h; (d) K₂CO₃, K₃[Fe(CN)₆], OsO₄, t-BuOH/H₂O (1 : 1), 6 h, rt; (e) NaIO₄, t-BuOH/H₂O (1 : 1), 6 h, rt; (f) TBAF, THF, 0 °C → rt, 12 h; (g) AcCl, Et₃N, CH₂Cl₂, rt, 16 h; (h) K₂CO₃, MeOH, rt, 1 h.

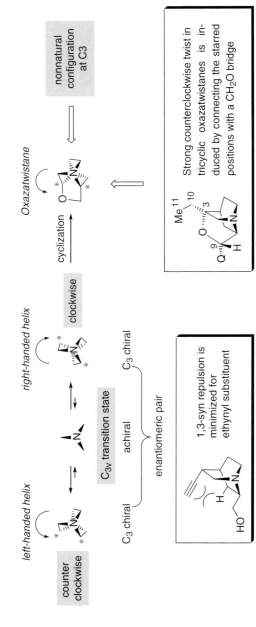

Scheme 12.19 Quinuclidines as a simple helix. Helicity of 1-azabicyclo[2.2.2]octane and twist sense of quinidine-derived oxazatwistanes.

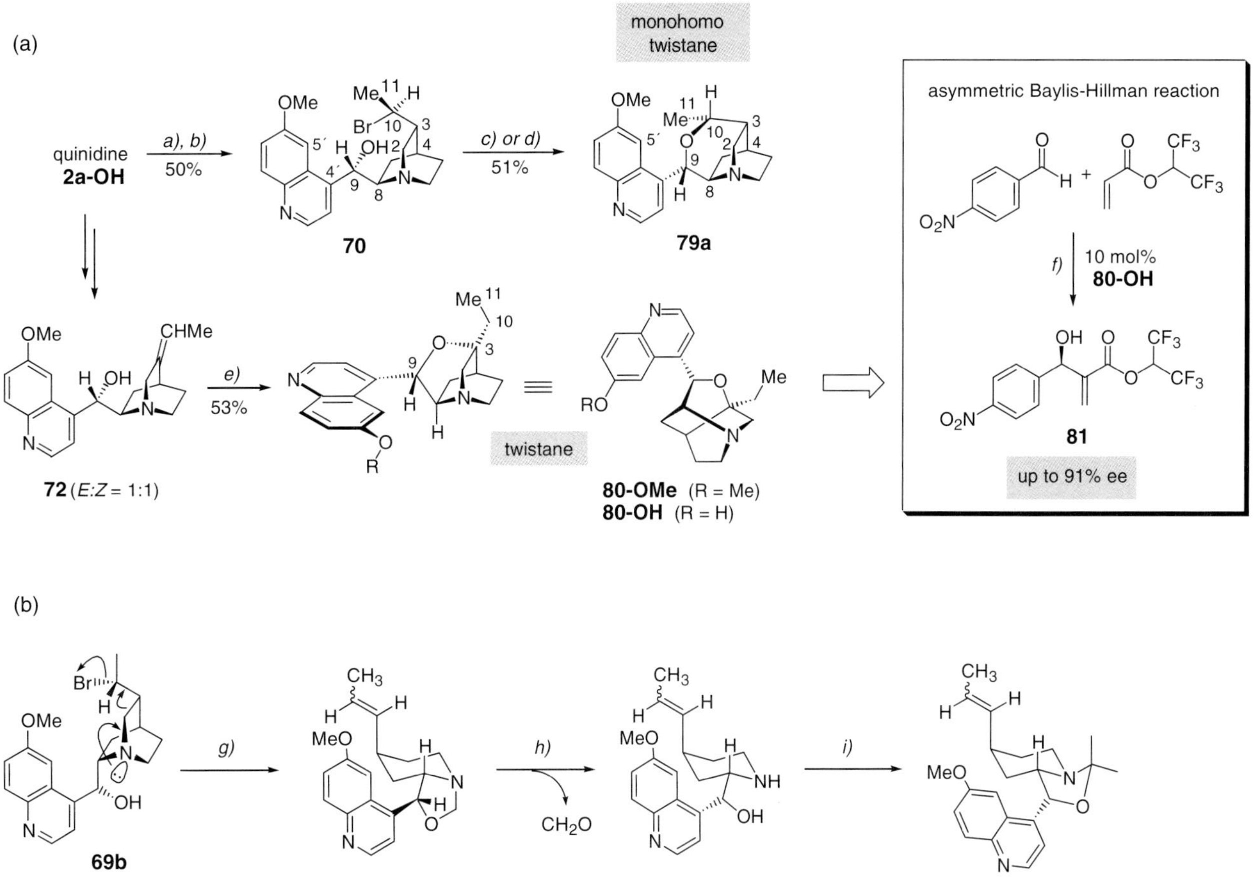

12.5 Selected Novel Transformations of the Quinuclidine Moiety of Cinchona Alkaloids | 385

Scheme 12.20 (a) Synthesis and application of quinidine-derived oxazatwistanes in the asymmetric Baylis–Hillman coupling; (b) Grob fragmentation of bromodihydroquinine **69**. Reagents and conditions: (a) HBr (62%), 3 d, rt; (b) KOH (25%), NaHCO$_3$; (c) 1. AgBF$_4$, THF, 0 °C → rt, 4 h; 2. Et$_3$N, KI, MeOH; (d) NaH, THF, 0 °C → rt, 2 d; (e) 1. KBr, H$_3$PO$_4$, 3 d, 100 °C; 2. KOH (25%), NaHCO$_3$; (f) DMF, −55 °C; (g) NaHCO$_3$, aq EtOH, reflux, 11 h; (h) (CO$_2$H)$_2$, aq EtOH, reflux, 80 min; (i) acetone, rt, 8 wk.

which the phenolic hydroxy group functions as a proton donor within the transition-state aggregate [53]. Models show – what may not immediately be obvious – that the heteroaromatic moiety can rotate forward and that the phenolic OH group together with the tertiary amine can enter into two-point contacts. The same twistane catalyst has been used recently for the catalytic asymmetric protonation of fluoro-enolic species derived from β-keto esters [54]. Similar to novel QCI- and QCD-based ligands, oxazatwistanes are easily prepared in few steps from readily available starting material [35]. Active catalyst **80-OH** is obtained from *iso*-quinidine **72** via acid-mediated cyclization in fair yield. Nonetheless, even seemingly simple transformations have to be monitored carefully: Homotwistane **79a** was initially obtained as a highly polar side product in the course of the hydrobromination of quinidine (Scheme 12.18) [46] and is also formed upon treatment of unprotected quinidine with the superacid system $HF\text{-}SbF_5$ [32].

Grob fragmentation of precursor **70** has been suppressed in favor of twistane and homotwistane formation by masking the nitrogen lone pair (with Ag^+ or H^+). Under S_N1-like conditions and at elevated temperature, cinchona-derived γ-amino bromides (e.g., **70** and its quinine analogue **69**) undergo Grob fragmentation with intramolecular capture of the intermediate iminium ion by the C9–OH group to give oxazabicyclo[4.3.0]-nonanes. Upon transacetalization with acetone, a cyclic N,O-acetal was isolated and characterized (Scheme 12.20b) [55]. Fragmentation is also a potential side reaction in the double dehydrobromination of 10,11-dibromo-1-azabicycles (cf. Schemes 12.10 and 12.11, **1a-OH** → **26**). The formation of desired alkyne **26** is facilitated by generating the intermediate vinylic bromide **24** under mild conditions and then using strong base to enforce the second, more difficult, dehydrobromination, again at as low a temperature as possible.

3-Quinuclidinone and derived 3-quinuclidinols represent fundamental pharmacophoric leads and have been of long-standing interest in medicinal chemistry in the continued quest for selective high-affinity ligands for neuronal receptors [56]. Substituted chiral quinuclidin-3-ones have been prepared by oxidative degradation of the vinyl side chain of QCI and QCD that was carried out by (i) double-bond shift and (ii) dihydroxylation and subsequent 1,2-diol cleavage (cf. Scheme 12.18). Because of their low molecular weight (mw = 155) and their compact bicyclic structure, both diastereomeric quinuclidin-5-ones **77** and **78** are attractive homochiral building blocks for asymmetric synthesis, pharmacology, and combinatorial chemistry. Substrate control of stereochemistry in reactions of ketones **77** and **78** with nucleophiles is a challenge greater than substrate control with sterically more encumbered parent cinchona alkaloids. We have established face selectivity and directed nucleophilic attack preferentially toward the supposedly more hindered *endo*-π-face of the carbonyl group, giving predominantly functionalized quinuclidinols ***anti*-82** and ***anti*-83** with natural configuration at carbon C5 and in diastereomeric excess of up to 97% (Scheme 12.21) [51].

π-Face selectivity, especially in the simple QCD series, is unprecedented and strongly depends on the size of the remote O-protecting group: it is highest with bulky O-protecting groups, involving a 1,7-stereoinduction (the bulk of the trityl group is evident considering that hexaphenylethane does not exist). The origin of

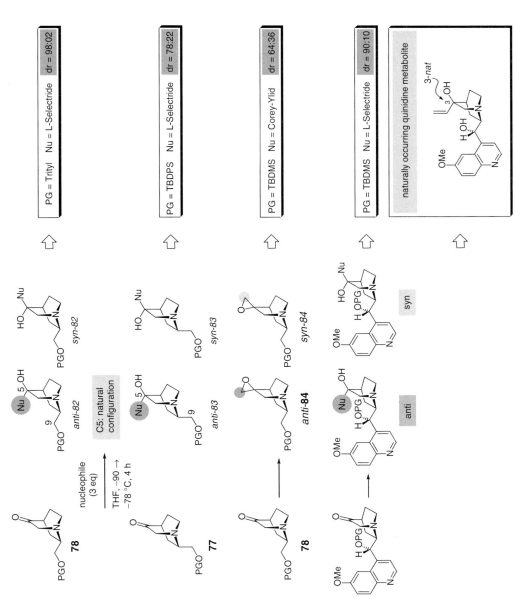

Scheme 12.21 Synthesis of substituted 2-hydroxymethyl quinuclidin-5-ols. Masochistic nucleophilic attack on quinuclidinone helix.

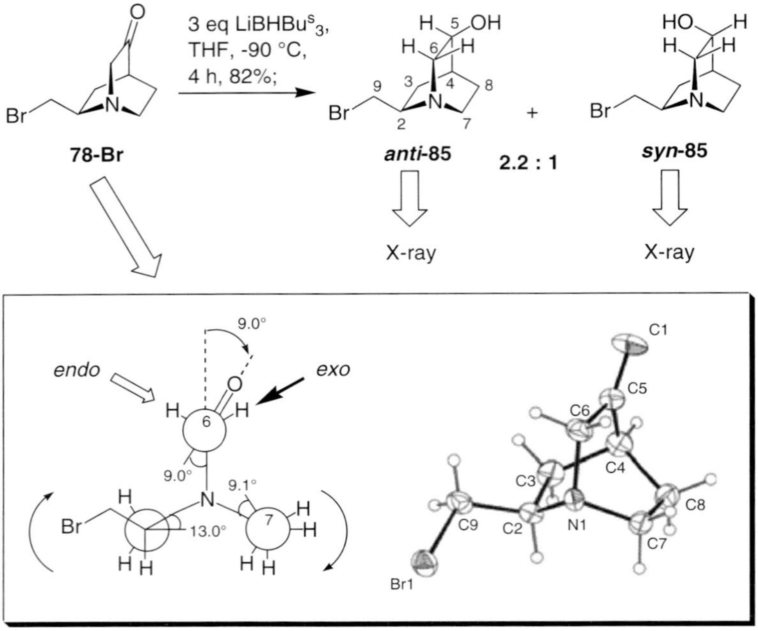

Scheme 12.22 Investigation of the origins of face selectivity: the impact of right-handed helicity.

π-facial selectivity is proposed to be due to helicity and torsional strain rather than electronic. The X-ray crystal structure of 2-bromomethylquinuclidin-5-one **78-Br** shows that its azabicyclic helix is twisted clockwise (Scheme 12.22) [57].

Thus, the "masochistic nucleophilic attack" from the supposedly more hindered face is favored affording quinuclidinols with natural configuration at C5. The right-handed helicity is conserved during preferred nucleophilic addition from the *endo* face: the fully eclipsed high-energy C_{3v}-like conformation is bypassed *en route* to the major products **anti-82**, **anti-82**, and **anti-85**. Stereochemical assignments at carbon C5 have been confirmed by NOE experiments and X-ray analyses of **anti-85** and **syn-85**.

Similarly, addition of vinyl magnesium bromide to the C3 carbonyl group of quinidine-based rubanone provides the known major quinidine metabolite directly (see also Scheme 12.21) [58]. Again, a surprising diastereoselectivity of up to 7:1 in favor of attack from the sterically more hindered π-face was observed (Scheme 12.21). As before, the quinidine-derived ketone traces out a right-handed helix. Chelation of magnesium by heteroatoms may help to increase diastereoselectivity but is now believed not to be decisive. π-Facial selectivity is observed upon reactions not only with hydride, Grignard, or organolithium reagents but also with Corey's sulfur ylide reagent affording epoxide **anti-84** as major product (Scheme 12.21).

12.5.2
Transformations at Carbon C6 and Formation of Bridgehead Bicyclic Lactams

A large number of quinuclidine-based lead structures have been synthesized in the 1990s [59]. Because of the high impact of stereochemistry on therapeutic and catalytic activity, we have investigated substrate-induced stereoselective functionalizations of cinchona alkaloid ketones not only at carbonyl C3 (cf. Section 11.5.1) but also at carbon C2 α to the nitrogen with the aim of extending the cinchona skeleton. Straightforward aldol reaction of 3-quinuclidinone with aromatic aldehydes have for some time been known to give single Z-isomers [60], whereas Z-selectivity is partially lost in corresponding reactions of cinchona alkaloids. Aldol reactions of silylated quinidine-based ketone **74c** furnished **86a–b** as E/Z mixtures, especially at elevated temperatures (Scheme 12.23). Furthermore, the corresponding reaction of acetylated rubanone **74a** proceeded Z-selectively but was accompanied by intramolecular cyclization: intramolecular acetyl shift and subsequent 6-exo-tet cyclization lead to novel tricyclic ketone **87** [61].

This unexpected cyclization prompted us to explore chemistry involving quinuclidine carbon C2 in detail. A careful evaluation of side products obtained en route to rubanone showed that tricyclic derivative **88** with an N,O-acetal function at carbon C2 was formed upon treatment of intermediate **72b** with chloramine T under O-deacetylating conditions (Scheme 12.24) [47].

As a starting point for functionalizing carbon C2 in a more targeted way, Langer first α-brominated acylated rubanones **74b–e** in satisfactory yield under optimized conditions and with complete diastereoselectivity (Scheme 12.25). The bromine entered the molecule away from the quinolyl residue, and the resulting α-halogenated ketones **89a–d** were converted in a single step into tricyclic N,O-acetals **90** containing a masked 1,2,3-tricarbonyl functionality [62]. This one-pot conversion

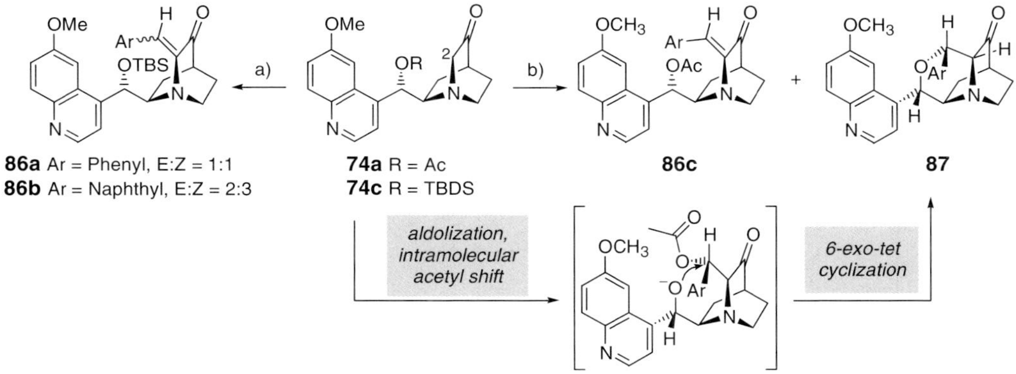

Scheme 12.23 Surprising transformations of rubanones: access to cinchona alkaloids functionalized at carbon C6. Reagents and conditions: (a) K_2CO_3, aryl aldehyde, MeCN; (b) Na_2CO_3, aryl aldehyde, THF.

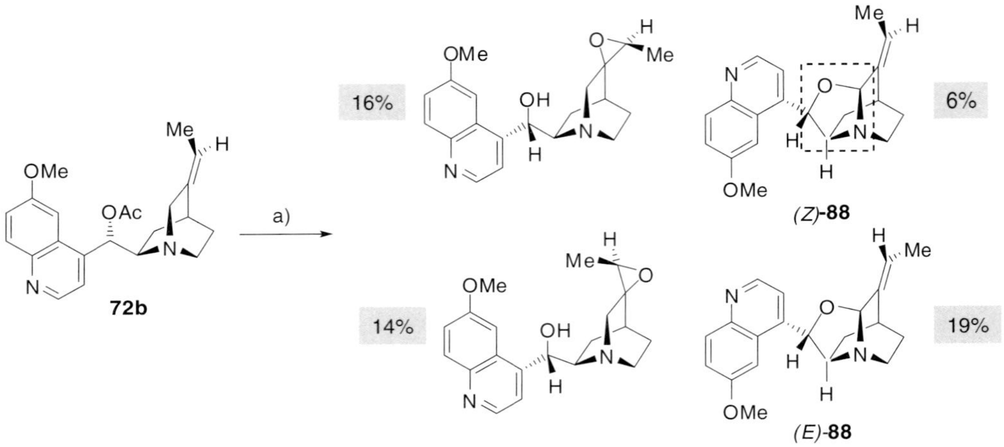

Scheme 12.24 Formation of tricyclic N,O-acetals as side products. Reagents and conditions: (a) 1. TsNCl⁻Na⁺, H₂O, H₂SO₄, acetone, 2. K₂CO₃, THF.

mediated by NaI and an excess of sodium azide includes an intramolecular acyl shift. The formation of an N,O-acetal at carbon C2 suggested that further 2-quinuclidinone analogues of cinchona alkaloids with a lactam functionality should be accessible.

5-Vinyl-2-quinuclidinone was originally postulated as an intermediate *en route* from quininone **5** to meroquinene ester **7** (Scheme 12.3). It is of general interest because amide resonance in the bicyclic lactam is impeded or blocked out altogether by the Bredt rule. However, early attempts to synthesize 5-vinyl-2-quinuclidinone were not successful. To introduce the desired lactam functionality in cinchona alkaloids, we developed a mild 1,2-carbonyl transposition. Treating α-halogenated ketones **89a** with various cyclic amines (e.g., pyrrolidine, piperidine, or morpholine) in solvent CH₂Cl₂ furnished a series of O-deprotected lactams **91** in fair yields (exemplified in Scheme 12.25 by morpholine) [63]. 2,3-Diamino oxirane **A** has been postulated as a reactive intermediate. Epoxide opening at room temperature is accompanied by a proton shift giving *endo*-configurated **endo-91a** under kinetic control. Extended reaction times at slightly elevated temperature or base treatment gave the *exo*-epimer, which was not stable but cyclized to 4-oxa-1-azatwistan-2-one **92**. The two-step conversion of protected rubanones **74** into corresponding 2-quinuclidinone derivatives represents a concise 1,2-carbonyl transposition and provides a short route to bridgehead lactam alkaloids. More recently, the sensitive parent unsubstituted 2-quinuclidinone **94** was generated by the Schmidt–Aubé reaction of azido ketone **93** under nonaqueous conditions and in the gas phase together with its corresponding [3.2.1] azabicyclic analogue **95** [64]. The twisted amide shows enhanced basicity comparable to that of secondary and tertiary amines, and has been captured and stabilized efficiently by N-protonation (Scheme 12.26).

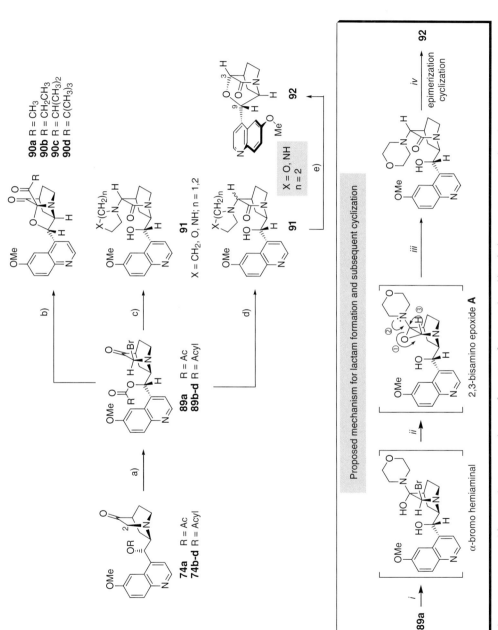

Scheme 12.25 Targeted functionalization at carbon C6: formation of N,O-acetals and bicyclic lactams. Reagents and conditions: (a) Br$_2$ (3.0 equiv), PBr$_3$ (1.2 equiv), HOAc, rt; (b) NaN$_3$ (excess), NaI (1 equiv), DMF, 115 °C; (c) sec. amine (10 equiv), CH$_2$Cl$_2$, rt, 10 h; (d) sec. amine (10 equiv), CH$_2$Cl$_2$, rt, 72 h; (e) KOH, H$_2$O, Bu$_4$NI, 40 °C.

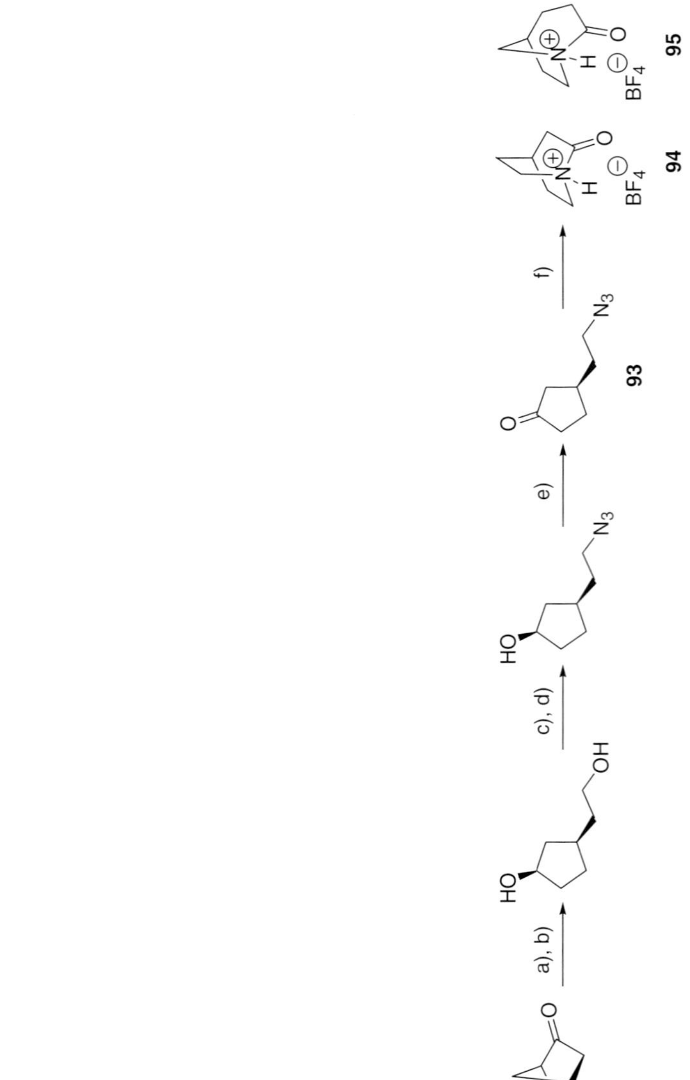

Scheme 12.26 Synthesis of parent 2-quinuclidinone. Reagents and conditions: (a) *m*-CPBA, NaHCO$_3$, CH$_2$Cl$_2$, rt; (b) LiAlH$_4$, Et$_2$O, rt; (c) TsCl, Et$_3$N, CH$_2$Cl$_2$, rt; (d) NaN$_3$, DMF, 70 °C; (e) Dess–Martin periodinane, CH$_2$Cl$_2$, rt; (f) HBF$_4$, Et$_2$O, rt.

12.5.3
Functionalization of Other Quinuclidine Carbons, for Example, C5 and C7

Synthetic access to quinidine positions other than C2 and C3 is more challenging. Treatment of quinine in superacid, for example, delivers a tricyclic derivative **96** via nucleophilic ring closure between the C9–OH oxygen and a carbocation at C5 (Scheme 12.27). However, cyclization occurs only subsequent to a rearrangement of the azabicyclo[2.2.2]octane unit into a thermodynamically more stable [3.2.1]-system [65].

7-Hydroxyquinine **97** was prepared recently in the course of a novel total synthesis approach toward cinchona alkaloids (Scheme 12.27). Interestingly, studies on stereocontrolled synthesis of quinine, apart from that of Uskokovic [66], were not published until 2001 and 2004, again underlining the fact that cinchona alkaloid

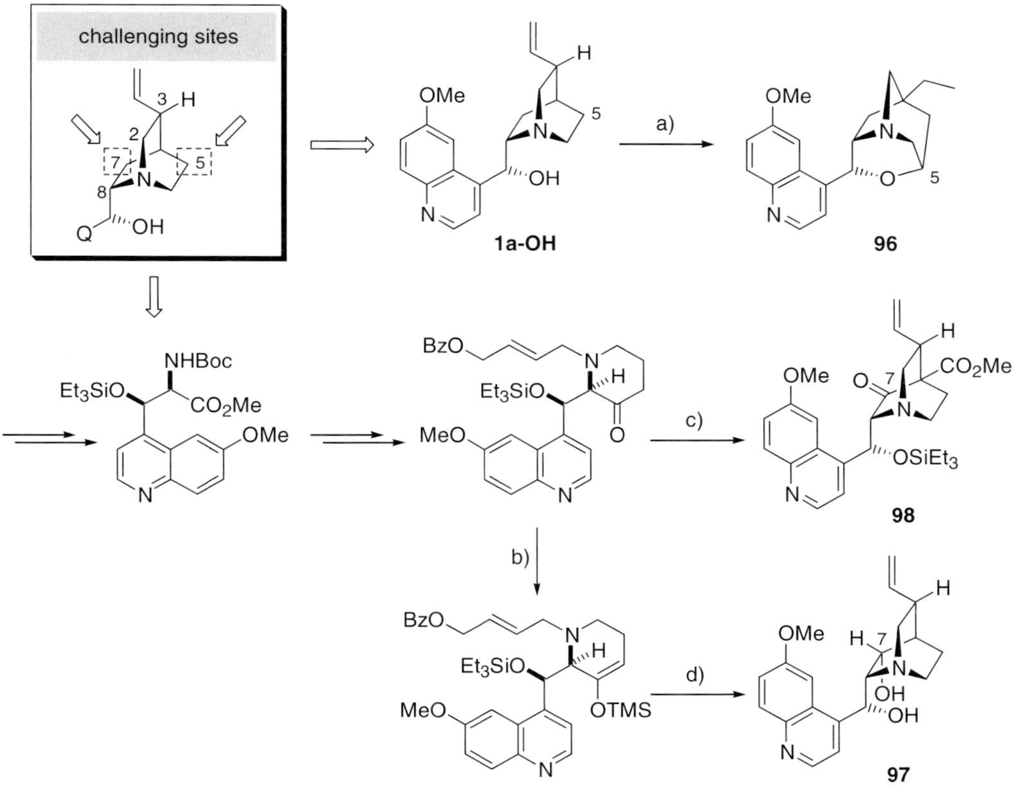

Scheme 12.27 Approaches to challenging positions C5 and C7 of the quinuclidine moiety. Reagents and conditions: (a) HF/SbF$_5$ (7:1), −30 °C; (b) TMSCl, LiHMDS, −78 °C; (c) 1. CNCO$_2$CH$_3$, NaHMDS, 2. Bu$_3$SnOMe, Pd$_2$(dba)$_3$, P(2-furyl)$_3$; (d) 1. Bu$_3$SnF, Pd$_2$(dba)$_3$, P(2-furyl)$_3$, 2. DIBAL-H.

chemistry remains a challenge [67]. The complete sequence toward **97** comprises more than 20 synthesis steps and proceeds via two key steps, (i) an asymmetric aldol reaction that establishes stereochemistry at carbon C8 and C9, followed by (ii) construction of the 3-vinyl-quinuclidine core by C3–C4 ring closure using a tailor-made intramolecular Pd-mediated allylic alkylation, with excellent regio- and diastereoselectivity [68]. Corresponding 7-quinuclidinone **98** could also be prepared. Interestingly, **97** was evaluated against two strains of plasmodium and found to be inactive.

Due to the high number of steps, challenging stereocontrol, and low overall yield, the known total synthesis of cinchona alkaloids cannot compete with the extraction of the natural products from readily available cinchona bark and any subsequent semisynthetic modifications.

12.6
Nucleophilic Substitution at Carbon C9

12.6.1
Unusual Steric Course of Solvolysis of C9-Activated Alkaloids: Access to C9-*epi*-Configured Stereoisomers

Solvolytic displacements at saturated carbon are among the best known and most investigated reactions of organic chemistry. It is generally accepted that the steric course of solvolysis ranges from complete inversion of configuration to substantial and even complete racemization, depending on the stability and lifetime of any ion pair and carbocation intermediates (S_N1 rule of Ingold and Hughes). In the past, hydrocarbons with tosyloxy leaving groups have usually been investigated in solvolyses. Substrates with a neighboring basic amino function as in cinchona alkaloids give rise to novel reactivity. The well-defined conformations of cinchona alkaloids and the presence or absence of the remote 6′-methoxy group also have to be considered (see Section 11.7.2). Methanolysis of O-tosylated quinine **1a-OTs** (which has to be prepared first of all) may appear to be hopelessly complex (Scheme 12.28), and even during the heydays of solvolytic work and controversy in the 1960s, C9-functionalized cinchona alkaloids were completely ignored.

We found that a simple change of leaving group (from **1a-OTs** to **1a-OMs**) is advantageous in many nucleophilic substitutions at C9. Mesylates could be handled more easily and dissolved readily in water in the presence of tartaric acid (1 equivalent), which serves as a proton donor and helps solubilize the alkaloid by multiple hydrogen bonding. Upon heating to 100 °C, completely inverted, nonnatural 9-*epi*-quinine ***epi*-1a-OH** was obtained as the only product in a *spot-to-spot* reaction and in good yield (Scheme 12.29a) [69].

Similarly, 9-*epi*-quinine mesylate ***epi*-1a-OMs** was submitted to aqueous hydrolysis in the presence of tartaric acid at reflux. In this case, 9-*epi*-quinine ***epi*-1a-OH** was again formed, however, with complete *retention* of configuration! In an analogous

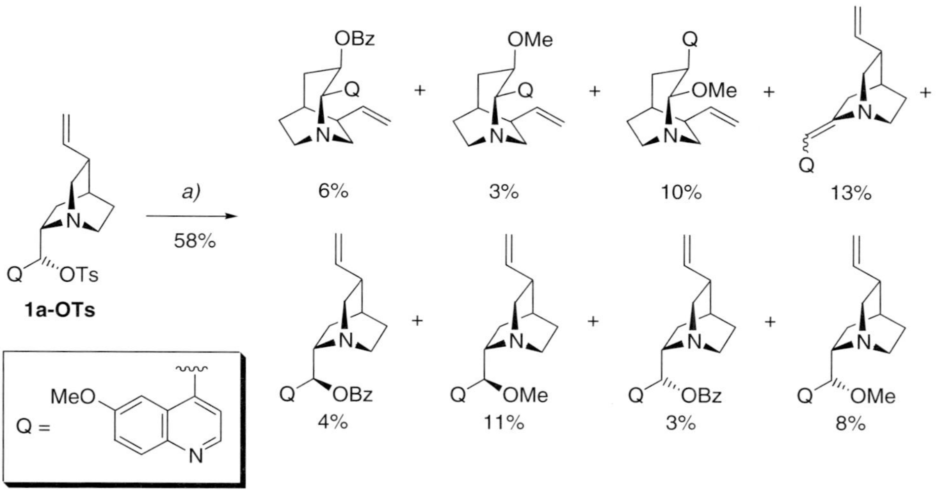

Scheme 12.28 Methanolysis of O-tosylated quinine **1a-OTs**.
Reagents and conditions: (a) C₆H₅CO₂Na (2 equiv), MeOH, reflux, 12 h.

sequence, pseudoenantiomeric quinidine **2a-OH** was converted into its mesylate and was hydrolyzed, furnishing nonnatural 9-*epi*-quinidine **epi-2a-OH** with complete *inversion* of configuration. However, hydrolysis of O-mesylated *epi*-quinidine **epi-2a-OMs** proceeded with complete retention of configuration. The results from the quinidine series (Scheme 12.29b) support the quinine work (Scheme 12.29a). Changing the leaving group made no difference. Both C9-*epi*-bromides and C9-*epi*-mesylates gave the alcohol with retention of configuration, whereas quinine and quinidine, the products of inversion, were not formed in this case. Put another way, 9-*epi*-quinine **epi-1a-OH** is formed from both 9-*nat* and 9-*epi* precursors **1a-OMs** and **epi-1a-OMs**. Similarly, 9-*epi*-quinidine **epi-2a-OH** arises from both 9-*nat* and 9-*epi*-precursors **2a-OMs** and **epi-2a-OMs**.

We have suggested that this surprising finding is due to conformational control and molecular recognition of solvent water via hydrogen bonding in the presence of tartaric acid. C9-*epi*-quinine is visualized to adopt a conformation containing a hydrophilic pocket with the leaving group and the (protonated) bridgehead nitrogen. The pocket is open to retentive solvolysis in an S$_N$i-type reaction involving a C$^+$X$^-$ ion pair (Scheme 12.30) [66]. The binding motif in the *epi* series is reminiscent of the postulated chelation of QCI-derived **8-OMs** by excess Li$^+$ (Scheme 12.34). For the C9-*nat* configured alkaloids, on the other hand, leaving group X and the protonated amino nitrogen are on opposite faces of the preferred ground-state conformation (Scheme 12.45), and inversion of configuration is strongly favored. We have used the inversion procedure to prepare all four nonnatural *epi*-configured cinchona alkaloids: *epi*-quinine (**epi-1a**), *epi*-quinidine (**epi-1b**), *epi*-cinchonidine (**epi-2a**), and *epi*-cinchonine (**epi-2b**).

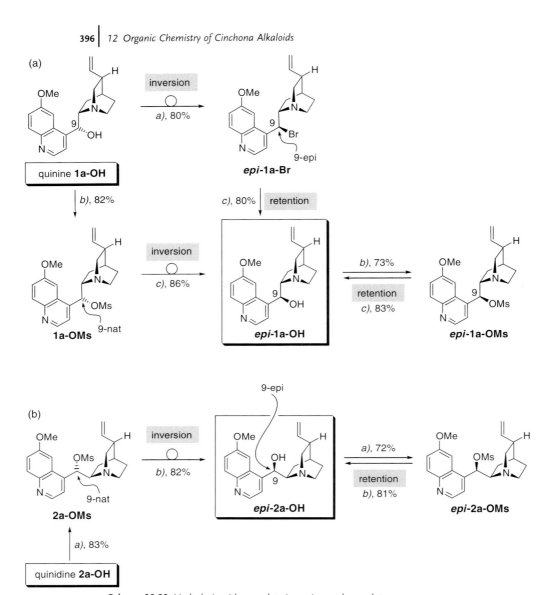

Scheme 12.29 Hydrolysis with complete inversion and complete retention of configuration (tartaric acid present). (a) Activated quinine; (b) activated quinidine. Reagents and conditions: (a) PPh$_3$, CBr$_4$, toluene, 20 °C, 12 h; (b) MsCl, Et$_3$N, THF, reflux, 4 h; (c) tartaric acid, H$_2$O, reflux, 0.5 h.

12.6.2
Replacing the C9-Hydroxy Group by Alternative Substituents

Mesylates of both native and *epi*-alkaloids are suitable precursors for further nucleophilic displacements at carbon C9. In contrast to solvolytic displacements

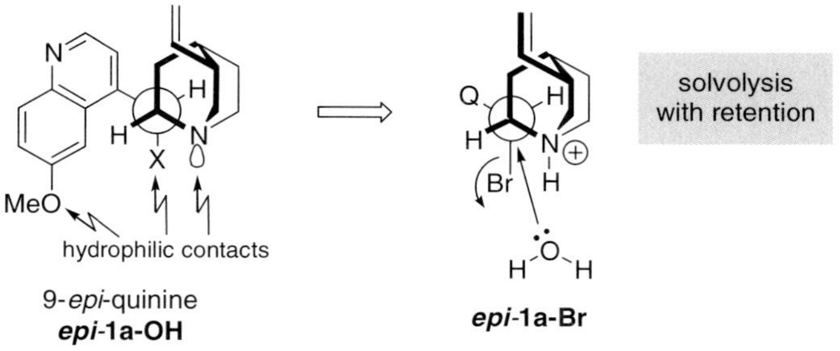

12.6 Nucleophilic Substitution at Carbon C9

Scheme 12.30 Rationalizing retention of configuration – in the rotamer shown, the bulky 1-azabicyclo[2.2.2]octane moiety faces the single H(C9) hydrogen.

outlined above (Scheme 12.29a,b), reactions of both **nat-2a-OMs** and **epi-2a-OMs** with strongly nucleophilic thiolates proceeded with complete inversion of configuration to yield thioethers **99** (Scheme 12.31). However, the steric environment at C9 remains precarious with respect to S_N2 reactions. For example, with more bulky 2-methyl-2-propanethiolate, nucleophilic displacement could not be carried out [70]. Quinidine derivatives **99** and corresponding selenides were tested as *N,S*-donating chiral ligands in the asymmetric allylic alkylation [71]. In accordance with other investigations described below, the 9-*epi*-configured alkaloid ligands were found to provide desired compounds with enhanced enantioselectivity.

9-Chloro-substituted quinine and quinidine derivatives were converted with suitable Grignard reagents in fair yield into their corresponding C9-vinyl and C9-aryl analogues. Again, both *nat-* and *epi-*configured precursors **nat-2a-Cl** and **epi-2a-Cl** yielded the same product **epi-100** with *epi-*configuration at C9 [72]. Thus, retention

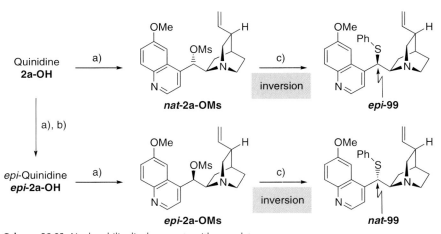

Scheme 12.31 Nucleophilic displacements with complete inversion of configuration. Reagents and conditions: (a) MsCl, Et$_3$N, THF, reflux, 4 h; (b) tartaric acid, H$_2$O, reflux, 0.5 h; (c) PhSNa, toluene, DMF, rt, 24 h.

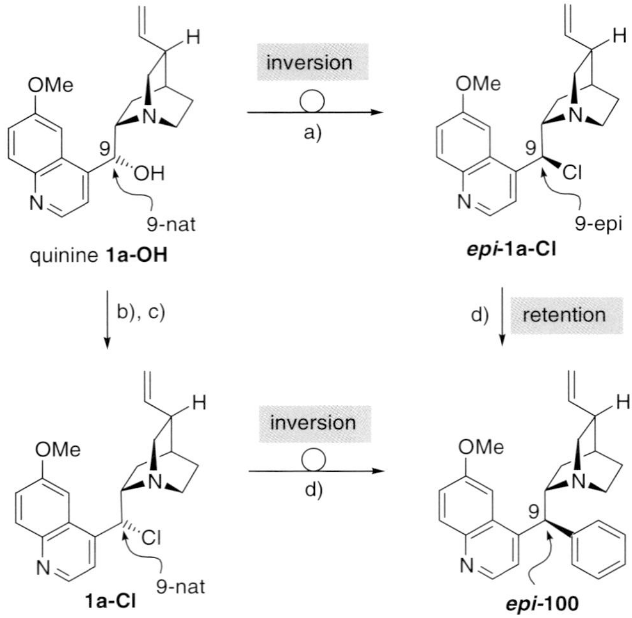

Scheme 12.32 Grignard reactions at carbon C9 with complete inversion and complete retention. Reagents and conditions: (a) SOCl$_2$; (b) MsCl, Et$_3$N, toluene, rt; (c) 1. tartaric acid, H$_2$O, reflux, 0.5 h; 2. SOCl$_2$; (d) PhMgBr (1.4 equiv), Et$_2$O, toluene, reflux, 4 h.

due to an S$_N$i-type mechanism is not limited to solvolytic displacements described above (cf. Scheme 12.29a and b) appears to be a more general feature of nonnatural 9-*epi*-configured cinchona alkaloids. In contrast, Grignard reaction of 9-*nat*-chlorides proceeded with complete inversion to afford ***epi*-100**. Stereoretention can be rationalized by coordination of the quinuclidine nitrogen to magnesium (instead of the proton as in Scheme 12.30) directing attack at stereocenter C9. Reactions with an excess of vinyl Grignard reagents analogous to (Scheme 12.32) also afforded C2'-vinylated byproducts.

Attention to C9-amino analogues of cinchona alkaloids has grown significantly in recent years as urea- and thiourea-substituted derivatives have emerged as readily accessible, robust, and tunable bifunctional organocatalysts for a wide range of transformations [73]. All published approaches toward organocatalyst systems **103** follow a synthetic route described by Brunner *et al.* in the mid-1990s [74]. The corresponding cinchona alkaloid (e.g., quinine) is allowed to react with hydrazoic acid in a Mitsunobu reaction to yield the corresponding azide **101** with clean inversion at carbon C9. Reduction of the azide either with triphenylphosphine or via hydrogenation furnishes 9-*epi*-amine **102**. Desired organocatalyst **103** is thereafter formed in good yield upon direct reaction with an isocyanate or a thioisocyanate (Scheme 12.33) [75]. In a similar straightforward

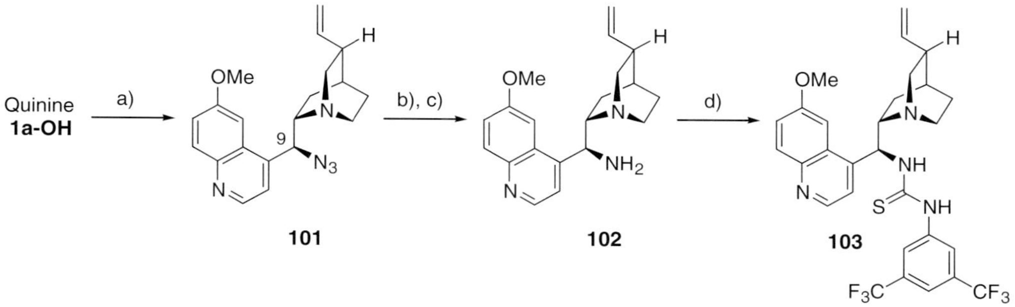

Scheme 12.33 9-*epi*-Aminoquinine as key intermediate for the synthesis of novel bifunctional organocatalysts. Reagents and conditions: (a) PPh$_3$, DIAD, HN$_3$, benzene, THF, rt; (b) 1. PPh$_3$, THF, 2. H$_2$O or (c) H$_2$, Pd/C, MeOH, rt, 2 h; (d) isothiocyanate, CH$_2$Cl$_2$, 0 °C – rt, 12 h.

manner, 9-*epi*-amine **102** can be alkylated via NaBH$_3$CN-mediated reductive amination.

Although the privileged cinchona alkaloid core structure with natural configuration at C9 has been the basis for ligand design in the past, C9-*epi*-configured organocatalysts can exhibit decidedly superior activity and enantioselectivity compared with their 9-*nat* analogues [76]. Fine-tuning of the catalyst by replacing the urea by a thiourea moiety has improved its efficacy further, especially in asymmetric Michael additions or Henry reactions. The relationship between catalyst activity and the relative orientation of the thiourea and quinuclidine moieties strongly indicates that a bifunctional mechanism is involved.

12.6.3
Transformations of QCI and QCD at Carbon C9: Access to a Novel Class of Small Molecule Ligands

Introduction of a nucleophile at carbon C9 of primary alcohols QCI **8** and QCD **9** may look simple, but it often requires optimization. S, P, and also Se have been introduced via their highly nucleophilic monoanions (Scheme 12.34) [77]. The N-C8-C9-O backbone conformation that allows free access to C9 from the rear is usually a minor rotamer only and S$_N$2 displacements may be sluggish. However, bidentate interaction with excess Li$^+$ in solvent dioxan (but not in THF) is suggested to change the rotameric equilibrium and to "open up" C9 to backside attack. The situation appears to parallel that in some carbohydrates where S$_N$2 displacements on 1,2-alkoxy halides can also be recalcitrant. Preparation of the iodomethyl derivative **104-I**, which enters into the "first cinchona rearrangement" (see the following discussion), thus proceeds via the derived mesylate with LiI (approximately 3 equiv) in refluxing dioxane in good yield (Scheme 12.34) [78].

Carbon–carbon bond formation at C9 with cyanide ion was feasible after careful optimization. Solvent toluene, CsF, and crown ether 18-C-6 were essential for

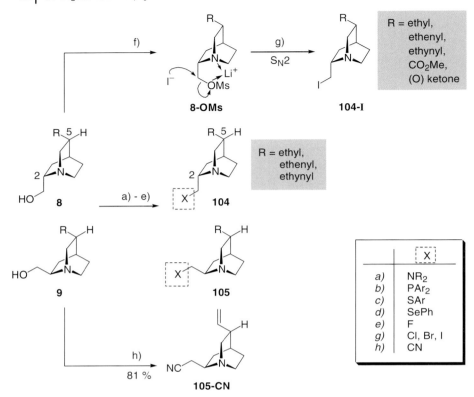

Scheme 12.34 S_N2 displacements in C9-activated QCI and QCD. Reagents and conditions: (a) R = H: PPh$_3$, DEAD, HN$_3$, THF, 0 °C → reflux, 2.5 h; R = Ar: 1. MsCl, Et$_3$N, DCM; 2. phosphide, THF, reflux; (b) 1. MsCl, Et$_3$N, DCM; 2. phosphide, THF, reflux; (c) 1. MsCl, Et$_3$N, DCM; 2. NaSAr, THF, reflux; (d) 1. MsCl, Et$_3$N, DCM; 2. KSePh, THF, reflux; (e) n-Bu$_4$NF (1 M soln. in THF); (f) MsCl, Et$_3$N, DCM; (g) LiX (X = Cl, Br, I), dioxane, reflux; (h) KCN, toluene, CsF, and crown ether 18-C-6.

displacements with KCN to afford, for example, **105-CN**. Omission of any of these three components (change of solvents, other cesium salts such as Cs$_2$CO$_3$, and the absence of crown ether) was not promising, however, *tetra-n* butylammonium can be used [74]. Related fluorides **104-F** and **105-F** and corresponding amines have also been prepared via standard methods [74, 79]. Interestingly, fluorine derivative **104-F** is highly volatile and has a characteristic odor. The Staudinger protocol with triphenylphosphine and trifluoroacetic acid provided 1,2-diamines directly from corresponding azides. A variant (PPh$_3$, 3-ClC$_6$H$_4$CO$_2$H) afforded a related *m*-chlorobenzamide, which is less water soluble and readily isolable [74]. Furthermore, QCI reacts spontaneously with dichloromethane, the nucleophilic bridgehead nitrogen being quaternized, apparently with some assistance from the hydroxymethyl group that enters hydrogen bonding to the leaving group (Scheme 12.35) [80].

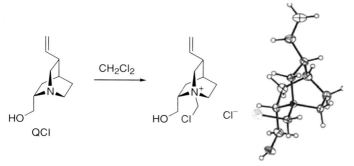

Scheme 12.35 Spontaneous quaternization of QCI in dichloromethane.

In contrast, diamine **104-NH$_2$** can be stored in CH$_2$Cl$_2$. Neda *et al.* have shown that upon contact with CO$_2$ (from the air) or with CO$_2$ in ether, **104-NH$_2$** reacts spontaneously to form the primary ammonium carbamate salt **106** (2 : 1 adduct) as a colorless crystalline solid. The neat salt is quite stable. Upon heating above 120 °C *in vacuo*, CO$_2$ is removed and pure 1,2-diamine **104-NH$_2$** can be recovered and used directly thereafter. In this simple fashion, even traces of QCI can be removed from **104-NH$_2$** without recourse to HPLC, distillation, or other traditional techniques of separation and purification (Scheme 12.36) [81].

Further applications of QCD- and QCI-derived bidentate ligands are beginning to emerge. Although hydrosilylations with QCI- and QCD-derived phosphine ligands afford only low ee values (up to 54%), the asymmetric Kumada–Corriu coupling with 2H-QCI *phos* (**104-PPh$_2$**) proceeds to afford up to 85% ee (Scheme 12.37) [82]. Recently, a collection of transition metals and the QCI- and QCD-diamine ligands has been evaluated in the catalytic asymmetric transfer hydrogenation of aromatic ketones, a benchmark reaction of asymmetric catalysis. In the course of a catalyst screening, it was found that complexes of QCI- and QCD-diamines **104-NH$_2$** and **105-NH$_2$** with oxophilic [IrCl(COD)$_2$] under phosphine-free conditions were the most active catalysts capable of reducing a broad range of aromatic ketones with excellent conversion and good enantioselectivities (up to 95% ee) (Scheme 12.37) [83]. Either pseudoenantiomeric diamine ligand was effective, providing ready access to enantiomeric secondary alcohol. These are the best selectivities for QCI- or

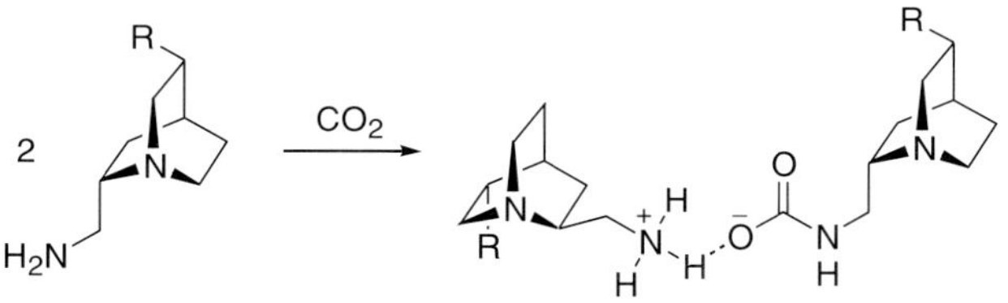

Scheme 12.36 Ammonium carbamate **106**. A stable crystalline carbamate salt from 1,2-diamine **104-NH$_2$**. (I. Neda, 2004).

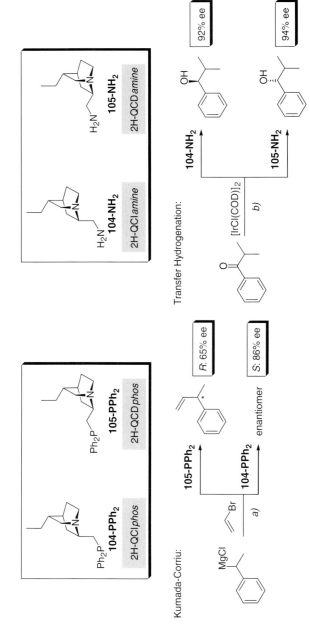

Scheme 12.37 Asymmetric catalysis with QCI- and QCD-derived phosphine and amino ligands (M. Lemaire, 2001; P. I. Arvidsson, 2003). Reagents and conditions: (a) NiCl$_2$, QCI/QCD-ligands 104-PPh$_2$ or 105-PPh$_2$, vinyl bromide, Et$_2$O, 12 h; (b) QCI/QCD-ligands 104-NH$_2$ or 105-NH$_2$, [IrCl(COD)]$_2$, i-PrOH, i-PrOK, 10 h.

QCD-derived ligands in an asymmetric transformation reported so far. Their low molecular weight (mw = 168) and their commercial availability, in both pseudoenantiomeric forms, suggest that these ligands with N-chiral bridgehead nitrogen will find further practical use in asymmetric synthesis. In fact, desymmetrization of a series of cyclic and acyclic *meso*-diols by the N-chiral TMEDA equivalents **104-NMe$_2$** and **105-NMe$_2$** has been reported very recently [84].

12.7
Novel Rearrangements of the Azabicyclic Moiety

12.7.1
First Cinchona Rearrangement

As yet, the 1-azabicyclo[3.2.2]nonane scaffold has been encountered in nature only rarely, for example, in tronoharine, vincathicine, and communesin B, in contrast to the 1-azabicyclo[2.2.2]octane moiety of cinchona alkaloids. A short and simple access to this particular bicyclic system is thus of high interest for exploring its biological and chemical properties.

The silver benzoate-mediated reaction of nonnatural 9-*epi*-quinine with bromide as leaving group (*epi*-**1a-Br**) in methanol furnishes a cage-expanded 1-azabicyclo [3.2.2]nonane **107-OMe** with a methoxy group α to the bridgehead nitrogen. This rearrangement is stereoelectronically favorable, involving a shift of the (*nonprotonated*) nitrogen lone pair, nucleophilic shift of carbon C7 to C9, and stereocontrolled external capture of a strained, nonplanar bridgehead iminium ion by the nucleophile. In the observed products, the quinolyl and methoxy group adopt a quasi-*trans*-2,3-diequatorial orientation. In the analogous sequence, quinidine with C9-*epi*-configured chlorine as leaving group provides the pseudoenantiomeric α-methoxy amine **108-OMe** (Scheme 12.38) [85]. All five chirality centers of the starting materials *epi*-**1a-Br** and *epi*-**2a-Cl** are neatly transposed into the five stereocenters of the cage-expanded product.

More recently, the "first cinchona cage expansion" has also been realized for quincorine and quincoridine (Scheme 12.39) [78]. There exists a substantial barrier to flipping of the tight three-carbon bridge and the limits of the Bredt rule are being tested. The rearrangement proceeds stereoselectively under mild S$_N$1-like conditions and tolerates additional functionality such as carbonyl and ester groups, as well as terminal alkynes. Evidence for the stereochemical outcome was provided by NOE experiments and X-ray analysis [86]. In the presence of nucleophiles, Lewis acid-mediated cleavage of α-amino ethers **109-OMe** and **110-OMe** affords a variety of enantiopure-substituted 1-azabicyclo[3.2.2]nonanes that are formed in S$_N$1-like reactions and with complete stereocontrol, that is, 100% retention. There is no leakage into pseudoenantiomeric **109-Nu** *en route* to product **110-Nu** or vice versa. The chiral carbocation is pyramidalized with imperfect orbital overlap, and model calculations suggest that the carbenium carbon is above the trigonal plane by about 36°. A variety of Lewis acids and nucleophiles were screened for preparing **109-Nu**

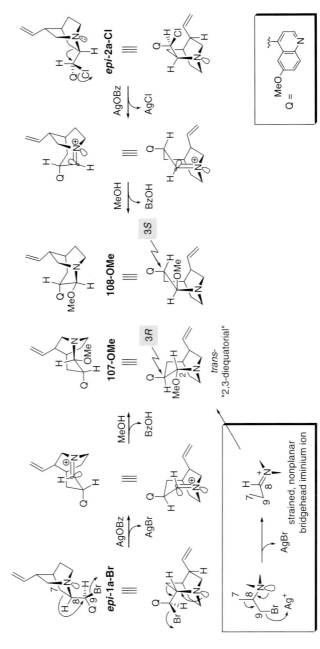

Scheme 12.38 "First cinchona cage expansion" of 9-*epi*-configured bromoquinine *epi*-**1a-Br** and *epi*-**2a-Cl** mediated by silver ion.

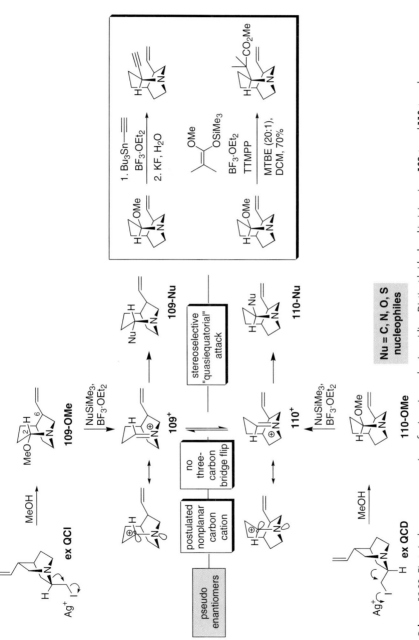

Scheme 12.39 First cinchona cage expansion of quincorine and quincoridine. Distinct bridgehead iminium ions **109+** and **110+** and their stereospecific capture by nucleophiles.

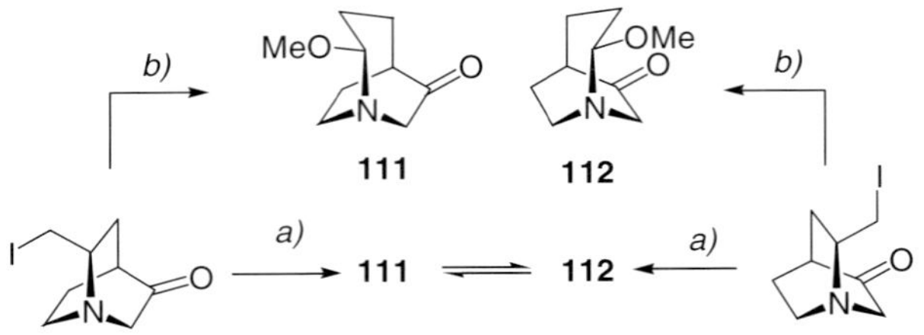

Scheme 12.40 Ring expansion of 2-iodomethylquinuclidin-5-ones, a further molecular diversity connector. Reagents and conditions: (a) AgOTf (1.1 equiv), MeOH, 50 °C; (b) AgOBz (1.1 equiv), MeOH, 50 °C.

and **110-Nu**. At first, simple alkylation, vinylation, and ethynylation proved to be difficult. Me$_3$SiCN provides new Strecker-type α-amino nitriles, whereas new β-amino acid esters can be prepared directly by two-carbon homologation using silylated ketene acetal (Nu=Me$_2$C=CO$_2$SiR$_3$) and BF$_3$–OEt$_2$ as Lewis acid with TTMPP [*tris*(trimethoxyphenyl) phosphine] as cocatalyst [78b]. Dead-end complexation of the bridgehead amino nitrogen by the Lewis acid and a narrow trajectory of nucleophilic attack of the iminium ion probably impede some of the desired displacements.

Even QCI- and QCD-derived 2-iodomethyl ketones afford ring-expanded 1-azabicyclo[3.2.2]nonan-6-ones **111** and **112**, respectively, although in a more sluggish reaction. Surprisingly, in view of the high stereoselectivity exhibited by the other silver triflate-mediated cage expansions, partial epimerization was observed upon treatment of iodo ketones with AgOTf (**111** → 80:20 and **112** → 68:32). Silver benzoate-promoted rearrangement of iodo ketones, however, gave epimerically pure ketones **111** and **112** (Scheme 12.40).

The first cinchona rearrangement is also feasible without Ag$^+$, provided stringent stereochemical and experimental conditions are fulfilled. In the preferred conformation of ***epi*-1a-OMs**, the C9-OMs leaving group and migrating C7-C8 σ-bond are antiperiplanar (Scheme 12.41). Under optimized conditions with NaOBz as a buffer (for the liberated methanesulfonic acid), α-amino ether **107-OMe** was formed in 81%

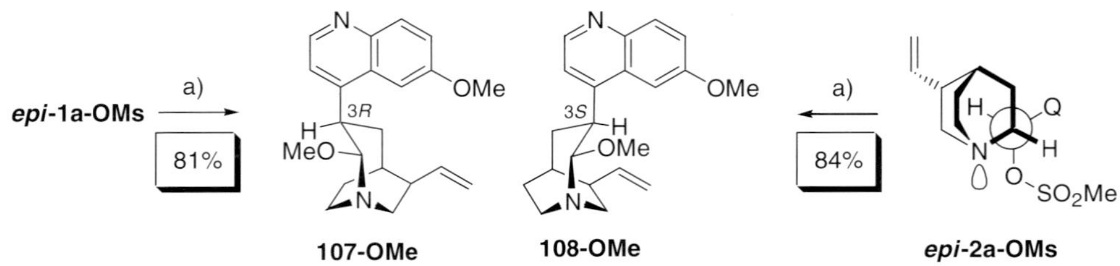

Scheme 12.41 First cinchona cage expansion of *epi*-configured O-mesylated quinine ***epi*-1a-OMs** and quinidine ***epi*-2a-OMs**. Preparation of α-amino ethers **107-OMe** and **108-OMe**. Reagents and conditions: (a) NaOBz (2 equiv), MeOH, rt, 2 h.

yield (Scheme 12.41), whereas pseudoenantiomeric α-amino ether **108-OMe** could be obtained in respectable 84% yield.

12.7.2
Second Cinchona Rearrangement

As shown in Scheme 12.28, methanolysis of O-tosylated quinine (**1a-OTs**) delivered a host of products that had to be chromatographed carefully and identified individually after separation. In fact, it would have been tempting to stop further investigation at this stage. However, the *cinch* bases (6'-R = H) cinchonidine **1b-OH** and cinchonine **2b-OH** with C9-*nat* configuration behave differently. Especially in solvent trifluoroethanol and also in water, the C9 mesylates of *cinch* bases undergo cage expansion to produce **113-OBz** and **114-OBz**, that is, β-functionalized amines that arise side-by-side with the products of stereoretentive, cage-conserving solvolysis (Scheme 12.42).

Alternatively, the cage-expanded alcohols **113-OH** ($3R$ configuration) and **114-OH** ($3S$ configuration) have been prepared in a simple complementary fashion from the C9-*nat* mesylates by heating at 100 °C in pure water. The hydrolysis of the mesylates can be considered *green chemistry*, since methanesulfonic acid is inexpensive, biodegradable, and toxicologically harmless. Oxidation provides the corresponding amino ketones **115** and **116**, which have a relatively acidic α-proton and are equilibrated spontaneously below room temperature: enolization is a rational pathway and also consistent with deuterium uptake at carbon C2 when the ketones are dissolved in solvent MeOD (Scheme 12.43) [87]. Thus, the pseudoenantiomeric barrier between cage-expanded 1,2-amino alcohol **113-OH** and **114-OH** can be broken easily after oxidation to ketone.

We have formulated the "second cage expansion" via nitrogen-bridged cation *viii* (Scheme 12.42) as an intermediate. Note that this is not a conventional tetrahedral ammonium ion. The 2-electron 3-center 9-cinchona cation has a nonclassical character because four ligands around the bridging nitrogen are constrained in one hemisphere. The nitrogen lone pair is in the second imaginary hemisphere and represents a fifth valence (the bulky Q' moiety points away from the azabicyclic core and toward the observer). The delocalized 9-cinchona cation *viii* (Scheme 12.42) can be compared with delocalized nonclassical 2-norbornyl cation *ix* [88] in which bridging of pentacoordinated carbon is constrained by the nortricyclic scaffold. Cation *viii* accounts for reversibility of the cage expansion and retention of configuration (= double inversion) upon nucleophilic attack of carbon C9 (Scheme 12.42, box).

In general, cinch bases (6'-R = H) are more prone to bridging (cf. *viii*) than quinine and quinidine (6'-R = OMe), where the methoxy donor facilitates electron delocalization and reduces electron demand at C9. Migrating aptitude of the bridgehead nitrogen also depends on conformation, solvent ionizing power, and pH (nonprotonated bridgehead nitrogen). The second cage expansion is open to many variations and has been carried out in the presence of added nucleophiles under S_N1-like conditions. With added *tetra-n*-butylammonium thiocyanate cage expansion is especially efficient (Scheme 12.44a; see Ref. [90]). With added azide ion (Scheme 12.44b), a cascade reaction occurs that is terminated by an intramolecular 1,3-dipolar Huisgen

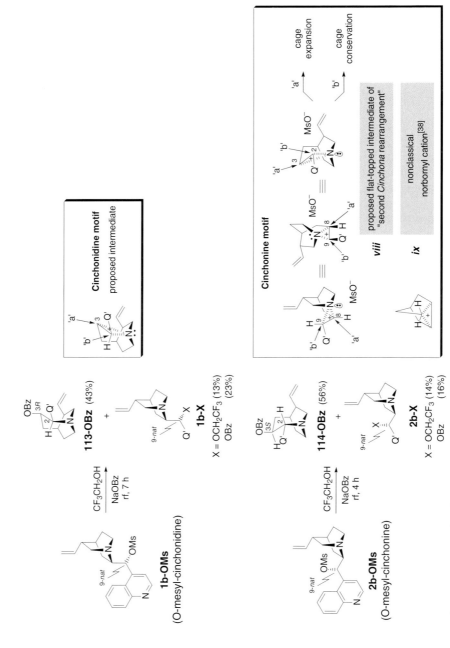

Scheme 12.42 "Second cinchona rearrangement." Effect of a change to O-mesylated *cinch* base (6'-R = H) and solvent trifluoroethanol.

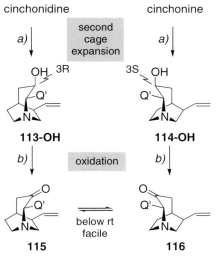

Scheme 12.43 Preparation of bicyclic amino ketones **115** and **116**. Oxidation breaks the pseudoenantiomeric barrier. Reagents and conditions: (a) 9-BBN, MeOH, H$_2$O, 16 h; (b) 1. (COCl)$_2$, DMSO, CH$_2$Cl$_2$, −78 °C, 2. Et$_3$N, rt, 1 h.

cycloaddition, giving triazoles (**117** and **118**), which are of general interest as antifungals. Remarkably, the second rearrangement is potentially reversible as has clearly been demonstrated in Scheme 12.44a and b [89]. For example, cage-expanded 1,5-diazatricyclo[4.4.1.0^{3,8}]undecane derivative **118** is formed from **27-OMs** under S$_N$1-like conditions. Equally, isomeric mesylate **114-OMs**, which contains a quasiequatorial and stereoelectronically favorable mesyloxy leaving group, forms tricyclic **118** with cage conservation and also homotwistane **117** with cage contraction. Thus, route A and route B are strikingly convergent with respect to the product spectrum. Skeleton and stereochemistry of both **27-OMs** and **114-OMs** are linked by the second cinchona rearrangement and an N-bridged cation such as *viii* (Scheme 12.42). Termination by *trans*-annular 1,3-dipolar cycloaddition is, of course, not feasible in the pseudoenantiomeric cinchonidine (quinine) series. In this case, the C9-azide moiety and the ethynyl (vinyl) acceptor are remote from each other.

A general cage expansion and solvolysis scheme together with a set of general guidelines for these baffling cinchona rearrangements has recently been put forward and is in excellent agreement with the experimental evidence accumulated by us so far (Scheme 12.45) [90].

12.8
General Experimental Hints and Obstacles

Although analytical chemistry has made giant strides forward in the last 100 years, separation by column chromatography can still be difficult. Certain cinchona

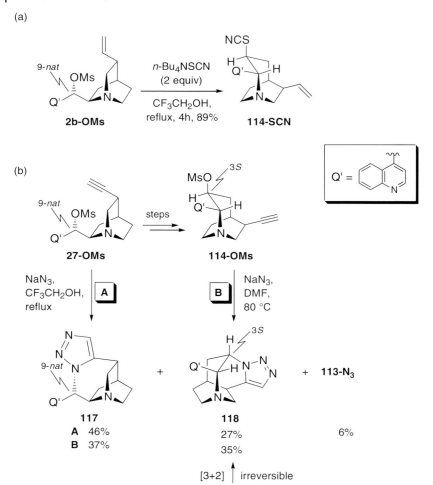

Scheme 12.44 Cascade reactions of **27-OMs** and **114-OMs** with azide ion under strongly ionizing conditions.

products stick so firmly to the column material, especially to silica gel, that they can hardly be eluted. Addition of triethyl amine or ammonia to the eluent may be necessary and helpful. Unless a precise mass balance is carried out, these products will be overlooked. The 9-*epi*-derivatives seem to be particularly prone to

12.8 General Experimental Hints and Obstacles

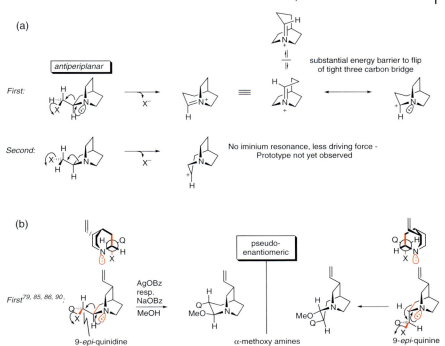

Migratory aptitude depends on configuration at C9, conformation, stereoelectronic factors, leaving group ability, solvation and acidity (protonation of bridgehead nitrogen).
In H_2O + *tartaric acid* the [2.2.2]azabicyclic cage is conserved (6'-R = OMe):

9-*epi* quinine-X ⇌ 9-*nat* quinine-X   (100% retention / 100% inversion)
9-*epi*-quinine: Pseudo-enantiomeric quinidines react analogously.

Extension to *cinch* bases feasible, although there is leakage into second rearrangement, depending on concentration of tartaric acid.[91] Stereoretentive hydrolysis with cage conservation ocurs with C9 *epi*-configured substrates: In the preferred conformation and in the presence of tartaric acid the (protonated) amino group and the hydrophilic leaving group X are exposed to water on the same face of the molecule (see projection formulae).

Scheme 12.45 General cinchona alkaloid cage expansion and solvolysis scheme – educt-product correlations; (a) prototype (1-azabicyclo[2.2.2]oct-2-yl)methyl-X rearrangement (X = leaving group); (b) cinchona rearrangements illustrated for Q (6'-R = OMe) and Q' (6'-R = H).

complexation (cf. hydrophilic domain in S_N1 solvolysis, Scheme 12.31). While the 6′-R = OMe-substituent is remote from the C9-C8 region, the so-called *cinch* bases (6′-R = H instead of 6′-R = OMe) show some special properties. Apart from the obvious absence of fluorescence that makes them "invisible" to UV detection at 366 nm, salts of *cinch* bases (e.g., the protonated species) show marked differences from those of quinine and quinidine salts with respect to solubility, HPLC and chromatographic behavior, and enhanced crystallizability. They are thus often preferred for producing crystalline diastereomeric salts for classical resolution. Addition of tartaric acid to water helps dissolve quinine and quinidine and also suppresses the first and second cinchona rearrangements. Commercially available cinchona alkaloids are mainly provided in technical quality only (i.e., 85–90% purity) accompanied by their corresponding C10-C11-dihydro derivatives. This fact may complicate purification by column chromatography in some cases. On the other hand, careful column chromatography or the use of modern separation techniques allows to separate diastereomers quite easily. The fluorescence-based assay is a sensitive method for determining quinine content in tonic water/bitter lemon.

12.9
Conclusions

So far, cinchona chemistry has focused mainly on applications as chiral ligands and chiral organocatalysts. Chemical alterations in alkaloids have usually been a minor issue and also unfashionable in this context. In fact, the first 10 chapters of this handbook provide many impressive examples for these applications, but there is more to the chemistry of cinchona alkaloids than that. New reactions and new structures have come to light only recently in both fundamental and applied works. Landmark developments are

- The discovery, preparation, and industrial manufacture of QCI and QCD, and the emerging chemistry of these building blocks. Bridgehead chirality ("N-chirality") and the proximity rule maximize stereodifferentiation of alternative diastereomeric transition states.

- Cinchona alkynes. They deserve further attention as versatile building blocks, for example, for the construction of 5-membered heterocycles and for all the reactions of the carbon–carbon triple bond.

- (Oxaza)twistanes with phenolic OH group and nonnatural configuration at C3 proved to be useful chiral Baylis–Hillman catalysts.

- The preparation and occurrence of N,O-acetals, which are potentially sensitive in a conventional setting, seem inevitable in alkaloid (iminium ion) chemistry when a proximate hydroxy group allows 5-ring closure. The resulting N,O-acetals are stabilized by concatenation within a polycyclic framework, especially as a [3.3.0], [4.3.0], and a [3.2.1] unit. There is no N-protecting group.

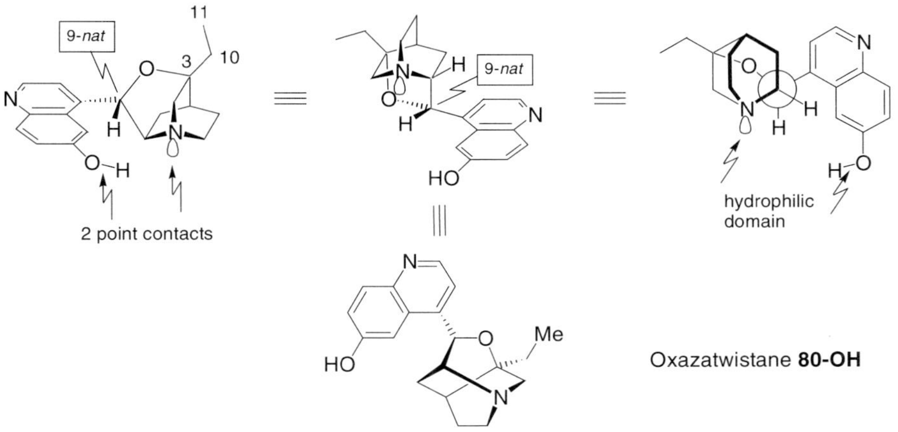

12.9 Conclusions

Scheme 12.46

- Selective nucleophilic addition at carbon C4′ of quinoline is feasible via directed metal chelation, as in the preparation of QCI and QCD, and also [3.2.1]-bicyclic N, O-acetals.

- A series of bridgehead lactam bicyclics has been prepared from rubanone. The sensitive parent lactam has also been generated.

- The stereochemistry of nucleophilic attack at 3-quinuclidinones is controlled by twist sense of the azabicyclic core.

- Simple C9-mesylates (rather than tosylates) are particularly useful for C9 nucleophilic displacements.

- Preparation of nonnatural 9-*epi*-diastereomers and S_N1 reactions with retention of configuration are feasible. A helpful conformational model mapping the hydrophilic domain of molecules has been developed.

- The [2.2.2]azabicyclic cage of the alkaloids is not immutable. The first and second cage expansions have been discovered only during the past decade. A general cinchona alkaloid solvolysis and rearrangement scheme and a unified theory have been presented. Classical solvolysis chemistry with conventional hydrocarbon substrates is complemented. The resulting azabicyclo[3.2.2]nonanes have hardly been studied.

- Conformational model and projection formulae (e.g., Scheme 12.46) have been useful and stimulating for the design of organocatalysts.

The chemistry of cinchona alkaloids is more than just a collection of exotic organic compounds. It is also about stereochemistry of a privileged class of natural products and, in its widest sense, about ideas, reaction mechanism, supramolecular chemistry, biological and catalytic activity, and beyond. The fascination and challenge of the organic chemistry of cinchona alkaloids need never stop.

Acknowledgments

The senior author thanks all his coworkers and associates who have accompanied him on this adventurous journey into the cinchona area. We thank Cornelius von Riesen, Heiko Franz, Ion Neda, Guido Bojack, Elisabeth Peschel, Nicole Blaha, and Rudolf Wartchow for helpful discussions. This work was supported by Buchler GmbH Braunschweig, InnoChemTech GmbH Braunschweig, the Fonds der Chemischen Industrie, and the Volkswagen Foundation.

Appendix 1

Selected examples of novel commercially available Cinchona alkaloid products.

epi-Quinidine	epi-Quinine	(1S,2R,3S,5S,6R)-2-Quinolinyl-8-yl-6-vinyl-1-azabicyclo[3.2.2]nonan-3-ol	(1S,2S,3R,5S,6R)-
3-Hydroxyquinidine	3-Hydroxyquinine	Quincoridine-C10-ester	Quincorine-C9-cyanide
Quincoridine	Quincorine	Didehydro-quincoridine	Didehydro-quincorine
Quincoridine-amine	Quincorine-amine	Dihydro-quincoridinamine	Dihydro-quincorinamine

References

1. (a) Pasteur, L. (1853) *Compt. Rend. Acad. Sci.*, **37**, 113; (b) Miller, W.v. and Rohde, G. (1894) *Ber. Dtsch. Chem. Ges.*, **27**, 1187; (c) Miller, W.v. and Rohde, G. (1894) *Ber. Dtsch. Chem. Ges.*, **27**, 1279; (d) Miller, W.v. and Rohde, G. (1895) *Ber. Dtsch. Chem. Ges.*, **28**, 1056; (e) Miller, W.v. and Rohde, G. (1901) *Ber. Dtsch. Chem. Ges.*, **33**, 3214; (f) Rohde, G. and Schwab, G. (1905) *Ber. Dtsch. Chem. Ges.*, **34**, 306; (g) Rabe, P. and Schuler, W. (1948) *Ber. Dtsch. Chem. Ges.*, **81**, 139; (h) Biddle, C.H. (1912) *J. Am. Chem. Soc.*, **34**, 500.

2. (a) Schmidt, E.A. and Hoffmann, H.M.R. (1972) *J. Am. Chem. Soc.*, **94**, 7832; (b) Hoffmann, H.M.R. and Schmidt, E.A. (1973) *Angew. Chem.*, **85**, 227; (c) Chenoweth, D.M., Chenoweth, K., and Godderd III, W.A. (2008) *J. Org. Chem.*, **73**, 6853.

3. (a) Hoffmann, H.M.R. (1984) *Angew. Chem.*, **96**, 29; (b) Hartung, I.V. and Hoffmann, H.M.R. (2004) *Angew. Chem.*, **116**, 1968.

4. Doering, W.E. and Chanley, J.D. (1946) *J. Am. Chem. Soc.*, **68**, 586.

5. (a) Kolb, H.C., VanNieuwenhze, M.S., and Sharpless, K.B. (1994) *Chem. Rev.*, **94**, 2483; (b) Johnson, R.A. and Sharpless, K.B. (1993) Catalytic asymmetric dihydroxylation, in *Catalytic Asymmetric Synthesis* (ed. I. Ojima), Wiley-VCH Verlag GmbH, Weinheim; (c) Nicolaou, K.C. and Sorensen, E. (1996) *Classics in Total Synthesis*, Wiley-VCH Verlag GmbH, Weinheim; (d) Corey, E.J., Noe, M.C., and Grogan, M.J. (1994) *Tetrahedron Lett.*, **35**, 6427.

6. (a) O'Donnell, M.J., Fang, Z., Ma, X., and Huffman, J.C. (1997) *Heterocycles*, **46**, 617; (b) O'Donnell, M.J., Bennett, W.D., and Wu, S. (1989) *J. Am. Chem. Soc.*, **111**, 2353; (c) Lygo, B. and Wainwright, P.G. (1997) *Tetrahedron Lett.*, **38**, 8595; (d) Corey, E.J., Xu, F., and Noe, M.C. (1997) *J. Am. Chem. Soc.*, **119**, 12414; (e) Corey, E.J., Noe, M.C., and Xu, F. (1998) *Tetrahedron Lett.*, **39**, 5347; (f) Corey, E.J. and Zhang, F.-Y. (2000) *Org. Lett.*, **2**, 1097; (g) Horikawa, M., Busch-Petersen, J., and Corey, E.J. (1999) *Tetrahedron Lett.*, **40**, 3843; (h) Corey, E.J. and Zhang, F.-Y. (1999) *Org. Lett.*, **2**, 2057; (i) Corey, E.J. and Zhang, F.-Y. (1999) *Angew. Chem.*, **111**, 2057.

7. Crosby, J. (1992) *Chirality in Industry* (eds A.N. Collins, G.N. Sheldrake, and J. Crosby), John Wiley & Sons, Ltd, Chichester, pp. 19–20.

8. (a) Hoffmann, H.M.R., Plessner, T., and von Riesen, C. (1996) *Synlett*, 690; (b) quincorine and quincoridine represent the core of quinine and quinidine. In German "cor" can also stand for *C ohne Rest* (C without residue) as in "nor" *N ohne Rest* (N without residue, e.g., adrenaline and noradrenaline).

9. (a) Direct addition of Ph-Li to quinine: Mead, J.F., Rapport, M.M., and Koepfli, J.B., (1946) *J. Am. Chem. Soc.*, **68**, 2704; (b) addition to O-alkylated alkaloids: Johansson, C.C.C., Bremeyer, N., Ley, S.V., Owen, D.R., Smith, S.C., and Gaunt, M.J., (2006) *Angew. Chem.*, **118**, 6170.

10. Hintermann, L., Schmitz, M., and Englert, U. (2007) *Angew. Chem. Int. Ed.*, **46**, 5164.

11. Berkessel, A., Guixa, M., Schmidt, F., Neudörfl, J.M., and Lex, J. (2007) *Chem. Eur. J.*, **13**, 4483.

12. Marcelli, T., van der Haas, R.N.S., van Maarseveen, J.H., and Hiemstra, H. (2006) *Angew. Chem. Int. Ed.*, **45**, 929.

13. Singh, R.P., Bartelson, K., Wang, Y., Su, H., Lu, X., and Deng, L. (2008) *J. Am. Chem. Soc.*, **130**, 2422.

14. Brandes, S., Niess, B., Bella, M., Prieto, A., Overgaard, J., and Jørgensen, K.A. (2006) *Chem. Eur. J.*, **12**, 6039.

15. Marcelli, T., van Maarseveen, J.H., and Hiemstra, H. (2006) *Angew. Chem.*, **118**, 7658.

16. Christensen, B. (1904) *J. Prakt. Chem.*, 217.

17 (a) Braje, W.M., Frackenpohl, J., Schrake, O., Wartchow, R., Beil, W., and Hoffmann, H.M.R. (2000) *Helv. Chim. Acta*, **83**, 777; (b) Schrake, O., Braje, W.M., Hoffmann, H.M.R., and Wartchow, R. (1998) *Tetrahedron: Asymmetry*, **9**, 3717; (c) for further recent investigations into the conditions of the alkyne synthesis, see Kacprzak, K.M., Lindner, W., and Maier, N.M., (2008) *Chirality*, **20**, 441.

18 Frackenpohl, J., Braje, W., and Hoffmann, H.M.R. (2001) *J. Chem. Soc., Perkin Trans. I*, 47.

19 (a) Erben, F.X., Philippi, E., and Schniderschitz, N. (1925) *Chem. Ber.*, **58**, 2854; (b) Kretschmar, H.C. and Erman, W.F. (1970) *Tetrahedron Lett.*, **11**, 41; (c) Schriesheim, A. and Rowe, C.A. (1962) *J. Am. Chem. Soc.*, **84**, 3160.

20 Dehmlow, E.V. and Lissel, M. (1981) *Tetrahedron*, **37**, 1653.

21 Smaardijk, A. and Wynberg, H. (1987) *J. Org. Chem.*, **52**, 135.

22 Thorand, S. and Krause, N. (1998) *J. Org. Chem.*, **63**, 8551; Li, J., Mau, A.W.-H., and Strauss, C.R. (1997) *Chem. Commun.*, 1275.

23 Adam, W.M. (2003) Ph.D. thesis. University of Hannover.

24 Wartchow, R., Frackenpohl, J., and Hoffmann, H.M.R. (2001) *Z. Kristallogr. NCS*, **216**, 235.

25 Seidel, M.C., Smits, R., Stark, C.B.W., Gaertzen, O., Frackenpohl, J., and Hoffmann, H.M.R. (2004) *Synthesis*, 1391.

26 Bing, M., Parkinson, J.L., and Castle, S.L. (2007) *Tetrahedron Lett.*, **48**, 2083.

27 Eliel, E.L., Wilen, S.H., and Doyle, M.P. (2001) *Basic Organic Stereochemistry*, Wiley-Interscience, New York.

28 For further discussion of this multifactorial problem see Rademacher, P., Kowski, K., Hoffmann, H.M.R., Haustedt, L.O., and Holzgrefe, J., (2002) *ChemPhysChem*, **3**, 957.

29 (a) Jeschke, P. (2004) *ChemBioChem*, **5**, 571; (b) Maienfisch, P. and Hall, R.G. (2004) *Chimia*, **58**, 93.

30 Debarge, S., Thibaudeau, S., Violeau, B., Martin-Mingot, A., Jouannetaud, M.-P., Jacquesy, J.-C., and Cousson, A. (2005) *Tetrahedron*, **61**, 2065.

31 Chagnault, V., Jouannetaud, M.-P., and Jacquesy, J.-C. (2006) *Tetrahedron Lett.*, **47**, 5723.

32 (a) Debarge, S., Violeau, B., Jouannetaud, M.-P., Jacquesy, J.-C., and Cousson, A. (2006) *Tetrahedron*, **62**, 602; (b) Chagnault, V., Jouannetaud, M.-P., Jacquesy, J.-C., and Marrot, J. (2006) *Tetrahedron*, **62**, 10248.

33 Cantet, A.-C., Carreyre, H., Gesson, J.-P., Jouannetaud, M.-P., and Renoux, B. (2008) *J. Org. Chem.*, **73**, 2875.

34 Waddell, T.G., Rambalakos, T., and Christie, K.R. (1990) *J. Org. Chem.*, **55**, 4765.

35 (a) Braje, W., Frackenpohl, J., Langer, P., and Hoffmann, H.M.R. (1998) *Tetrahedron*, **54**, 3495; (b) for a general methodology to prevent epimerization, see Daumas, M., Vo-Quang, Y., Vo-Quang, L., and Le Goffic, F., (1989) *Synthesis*, 64.

36 Epperson, M.T., Hadden, C.E., and Waddell, T.G. (1995) *J. Org. Chem.*, **60**, 8113.

37 (a) Braje, W. (1999) Ph.D. thesis. University of Hannover; (b) Frackenpohl, J. (2000) Ph.D. thesis. University of Hannover.

38 Rowan, S.J. and Sanders, J.K.M. (1998) *J. Org. Chem.*, **63**, 1536.

39 Reeder, J., Castro, P.P., Knobler, C.B., Martinborough, E., Owens, L., and Diederich, F. (1994) *J. Org. Chem.*, **59**, 3151.

40 (a) Rowan, S.J., Brady, P.A., and Sanders, J.K.M. (1996) *Angew. Chem.*, **108**, 2283; (b) Rowan, S.J., Hamilton, D.G., Brady, P.A., and Sanders, J.K.M. (1997) *J. Am. Chem. Soc.*, **119**, 2578.

41 (a) Rowan, S.J., Reynolds, D.J., and Sanders, J.K.M. (1999) *J. Org. Chem.*, **64**, 5804; (b) Rowan, S.J., Lukeman, P.S., Reynolds, D.J., and Sanders, J.K.M. (1998) *New J. Chem.*, 1015.

42 Prelog, V. and Cerkovnikov, E. (1940) *Liebigs Ann.*, **545**, 229; 262.

43 Schrake, O., Rahn, V.S., Frackenpohl, J., Braje, W.M., and Hoffmann, H.M.R. (1999) *Org. Lett.*, **1**, 1607.

44 (a) Prelog, V. and Cherkovnikov, E. (1937) *Ann. Chem.*, **532**, 83; (b) Renk, E. and Grob, C.A. (1954) *Helv. Chim. Acta*, **37**, 2119; (c) von Pracejus, H. and Kohl, G. (1969) *Ann. Chem.*, **722**, 1; (d) Bulacinski, A.B. (1978) *Pol. J. Chem.*, **52**, 2181.

45 Mi, Y. and Corey, E.J. (2006) *Tetrahedron Lett.*, **47**, 2515.

46 Cook, J. (2001) *J. Org. Chem.*, **66**, 1509.

47 von Riesen, C., Jones, P.G., and Hoffmann, H.M.R. (1996) *Chem. Eur. J.*, **2**, 673; von Riesen, C. and Hoffmann, H.M.R. (1996) *Chem. Eur. J.*, **2**, 680.

48 (a) Carroll, F.I., Abraham, P., Gaetano, K., Mascarella, S.W., Wohl, R.A., Lind, J., and Petzoldt, K. (1991) *J. Chem. Soc., Perkin Trans. I*, 3017; (b) Henry, T.A., Solomon, W., and Gibbs, E.M. (1935) *J. Chem. Soc.*, 966.

49 Dega-Szafran, Z. and Mitura, W. (1965) *Bull. Acad. Polon. Sci. Ser. Sci. Chim.*, **13**, 591; Suszko, J. and Thiel, J. (1974) *Rocz. Chem. Ann. Soc. Chim. Polon.*, **48**, 1281. The NMR data given by these authors are subject to an offset error (about 0.8 ppm).

50 Portlock, D.E., Naskar, D., West, L., Seibel, W.L., Gu, T., Krauss, H.J., Peng, X.S., Dybas, P.M., Soyke, E.G., Ashton, S.B., and Burton, J. (2003) *Tetrahedron Lett.*, **44**, 5365.

51 Frackenpohl, J. and Hoffmann, H.M.R. (2000) *J. Org. Chem.*, **65**, 3982.

52 (a) Iwabuchi, Y., Nakatani, M., Yokoyama, N., and Hatakeyama, S. (1999) *J. Am. Chem. Soc.*, **121**, 10219; (b) Langer, P. (2000) *Angew. Chem.*, **112**, (2000,) 4503; (2000) *Angew. Chem. Int. Ed.*, **39**, 3049.

53 (a) Iwabuchi, Y., Sugihara, T., Esumi, T., and Hatakeyama, S. (2001) *Tetrahedron Lett.*, **42**, 7867; (b) Iwabuchi, Y., Furukawa, M., Esumi, T., and Hatakeyama, S. (2001) *J. Chem. Soc., Chem. Commun.*, 2030; (c) Shi, M. and Jiang, J.-K. (2002) *Tetrahedron: Asymmetry*, **13**, 1941; (d) Balan, D. and Adolfsson, H. (2003) *Tetrahedron Lett.*, **44**, 2521; (e) for early work on the Baylis–Hillman reaction, see Hoffmann, H.M.R. and Rabe, J. (1983) *Angew. Chem.*, **95**, 795; (1983) *Angew. Chem. Int. Ed.*, **22**, 795; Rabe, J. and Hoffmann, H.M.R. (1983) *Angew. Chem.*, **95**, 796; (1983) *Angew. Chem. Int. Ed.*, **22**, 796; Hoffmann, H.M.R. and Rabe, J. (1985) *J. Org. Chem.*, **50**, 3849.

54 Baur, M.A., Riahi, A., Hénin, F., and Muzart, J. (2003) *Tetrahedron: Asymmetry*, **14**, 2755.

55 (a) Thiel, J. and Katrusiak, A. (2002) *Tetrahedron: Asymmetry*, **13**, 47; (b) Thiel, J. and Fiedorow, P. (1998) *J. Mol. Struct.*, **440**, 203.

56 (a) Nordvall, G., Sundquist, S., Nilvebrant, L., and Hacksell, U. (1994) *Bioorg. Med. Chem. Lett.*, **4**, 2837; (b) Johansson, G., Sundquist, S., Nordvall, G., Nilsson, B.M., Brisander, M., Nilvebrant, L., and Hacksell, U. (1997) *J. Med. Chem.*, **40**, 3804.

57 Wartchow, R., Frackenpohl, J., and Hoffmann, H.M.R. (2000) *Z. Kristallogr. NCS*, **215**, 435.

58 (a) Langer, P. and Hoffmann, H.M.R. (1997) *Tetrahedron*, **53**, 9145; (b) for the influence of helicity on stereoselectivity, see Rouhi, A.M., (2003) *Chem. Eng. News*, **29**, 34.

59 (a) Snider, R.M., Constantine, J.W., Lowe, J.A., III, Longo, K.P., Lebel, W.S., Woody, H.A., Drozda, S.E., Desai, M.C., Vinick, F.J., and Spencer, R.W. (1991) *Science*, **251**, 435; (b) McLean, S., Ganong, A.H., Seeger, T.F., Bryce, D.K., Pratt, K.G., Reynolds, L.S., Siok, C.J., Lowe, J.A., III, and Heym, J. (1991) *Science*, **251**, 437; (c) Swain, C.J., Fong, T.M., Haworth, K., Owen, S.N., Seeward, E.M., and Strader, C.D. (1995) *Bioorg. Med. Chem. Lett.*, **5**, 1261; (d) Snider, R.M. and Lowe, J.A. III (1991) *Chem. Ind.*, 793; (e) Fong, T.M., Cascieri, M.A., Yu, H., Bansai, A., Swain, C., and Strader, C.D. (1993) *Nature*, **362**, 350.

60 (a) Warawa, E.J. and Campbell, J.R. (1974) *J. Org. Chem.*, **39**, 3511; (b) Bender, D.R. and Coffen, D.L. (1967) *J. Org. Chem.*, **33**, 2504.

61 Langer, P. (1998) Ph.D. thesis. University of Hannover.

62 Langer, P., Frackenpohl, J., and Hoffmann, H.M.R. (1998) *J. Chem. Soc., Perkin Trans. I*, 801.

63 Frackenpohl, J., Langer, P., and Hoffmann, H.M.R. (1998) *Helv. Chim. Acta*, **81**, 1429.

64 Ly, T., Krout, M., Pham, D.K., Tani, K., Stoltz, B.M., and Julian, R.R. (2007) *J. Am. Chem. Soc.*, **129**, 1864; Tani, K. and Stoltz, B.M. (2006) *Nature*, **441**, 731.

65 Thibaudeau, S., Violeau, B., Martin-Mingot, A., Jouannetaud, M.-P., and Jacquesy, J.-C. (2002) *Tetrahedron Lett.*, **43**, 8773.

66 Grethe, G., Lee, H.L., Mitt, T., and Uskokovic, M.R. (1978) *J. Am. Chem. Soc.*, **100**, 581.

67 (a) Stork, G., Niu, D., Fujimoto, A., Koft, E.R., Balkovec, J.M., Tata, J.R., and Dake, G.R. (2001) *J. Am. Chem. Soc.*, **123**, 3239; (b) Raheem, I.T., Goodman, S.N., and Jacobsen, E.N. (2004) *J. Am. Chem. Soc.*, **126**, 3706.

68 Johns, D.M., Mori, M., and Williams, R.M. (2006) *Org. Lett.*, **8**, 4051.

69 Braje, W., Holzgrefe, J., Wartchow, R., and Hoffmann, H.M.R. (2000) *Angew. Chem.*, **112**, 2165; (2000) *Angew. Chem. Int. Ed.*, **39**, 2085.

70 (a) Zielinska-Blajet, M., Kucharska, M., and Skarzewski, J. (2006) *Synthesis*, 1176; (b) for thiolate ion as a nucleophile in Walden cycles of secondary mesylates, see Hoffmann, H.M.R. (1964) *J. Chem. Soc.*, 1249.

71 Zielinska-Blajet, M., Siedlecka, R., and Skarzewski, J. (2007) *Tetrahedron: Asymmetry*, **18**, 131.

72 Boratynski, P.J., Turowska-Turk, I., and Skarzewski, J. (2008) *Org. Lett.*, **10**, 385.

73 Connon, S.J. (2008) *J. Chem. Soc., Chem. Commun.*, 2499.

74 (a) Brunner, H., Bügler, J., and Nuber, R. (1995) *Tetrahedron: Asymmetry*, **6**, 1699; (b) Brunner, H. and Schmidt, P. (2000) *Eur. J. Org. Chem.*, 2119.

75 Vakulya, B., Varga, S., Csampai, A., and Soos, T. (2005) *Org. Lett.*, **7**, 1967.

76 McCooey, S.H. and Connon, S.J. (2005) *Angew. Chem. Int. Ed.*, **44**, 6367.

77 Schrake, O., Franz, M.H., and Hoffmann, H.M.R. (2000) *Tetrahedron*, **56**, 4453.

78 (a) Röper, S., Frackenpohl, J., Schrake, O., Wartchow, R., and Hoffmann, H.M.R. (2000) *Org. Lett.*, **2**, 1661; (b) Röper, S., Wartchow, R., and Hoffmann, H.M.R. (2002) *Org. Lett.*, **4**, 3179.

79 Neda, I., Kaukorat, T., and Hrib, C. (2002) *Tetrahedron: Asymmetry*, **13**, 1327.

80 von Riesen, C., Wartchow, R., and Hoffmann, H.M.R. (1998) *Z. Kristallogr. NCS*, **213**, 483.

81 Neda, I., Kaukorat, T., and Fischer, A. (2003) *Eur. J. Org. Chem.*, 3784.

82 (a) Saluzzo, C., Breuzard, J., Pellet-Rostaing, S., Vallet, M., Le Guyader, F., and Lemaire, M. (2002) *J. Organomet. Chem.*, **643–644**, 98–104; (b) Lemaire, M., Pellet-Rostaing, S., Breuzard, J., Halle, R.T., Saluzzo, C., and Vallet, M. (2001) Fr. Demande, 2810666; (c) Pellet-Rostaing, S., Saluzzo, C., Halle, R.T., Breuzard, J., Vial, L., Guyader, F.L., and Lemaire, M. (2001) *Tetrahedron: Asymmetry*, **12**, 1983.

83 (a) Hartikka, A., Modin, S.A., Andersson, P.G., and Arvidsson, P.I. (2003) *Org. Biomol. Chem.*, **1**, 2522; (b) Hedberg, C., Källström, K., Arvidsson, P.I., Brandt, P., and Andersson, P.G. (2005) *J. Am. Chem. Soc.*, **127**, 15083.

84 Kuendig, E.P., Garcia, A.E., Lomberget, T., Perez Garcia, P., and Romanens, P. (2008) *Chem. Commun.*, 3519.

85 Braje, W.M., Wartchow, R., and Hoffmann, H.M.R. (1999) *Angew. Chem.*, **111**, 2698; (1999) *Angew. Chem. Int. Ed.*, **38**, 2540.

86 Wartchow, R., Frackenpohl, J., Röper, S., and Hoffmann, H.M.R. (2000) *Z. Kristallogr. NCS*, **215**, 437.

87 Röper, S., Franz, M.H., Wartchow, R., and Hoffmann, H.M.R. (2003) *J. Org. Chem.*, **68**, 4944.

88 (a) Olah, G.A., Prakash, G.K.S., and Saunders, M. (1983) *Acc. Chem. Res.*, **16**, 440; (b) Olah, G.A. (1995) *Angew. Chem.*, **107**, 1519; (1995) *Angew. Chem. Int. Ed.*, **34**, 1393; (c) X-ray crystal structures: Laube, T. (1995) *Acc. Chem. Res.*, **28**, 339.

89 Röper, S., Franz, M.H., Wartchow, R., and Hoffmann, H.M.R. (2003) *Org. Lett.*, **5**, 2773.

90 Franz, M.H., Röper, S., Wartchow, R., and Hoffmann, H.M.R. (2004) *J. Org. Chem.*, **69**, 2983.

Part Four
Cinchona Alkaloid and Their Derivatives in Analytics

13
Resolution of Racemates and Enantioselective Analytics by Cinchona Alkaloids and Their Derivatives

Karol Kacprzak and Jacek Gawronski

13.1
Introduction

Cinchona alkaloids constitute an exceptional class of natural products making a major impact on human civilization [1]. They were introduced in Europe as early as in eighteenth century, with the discovery of antimalarial activity of the cinchona bark and subsequent isolation of its active compound – quinine (QN) in 1820. An enormous demand for cinchona alkaloids later on has been a subject of economic and political manipulations, counterbalanced by an intense research on their synthesis [2]. Even today, quinine is an important antimalarial drug and the only bittering agent used in food and beverage industry. It is also well known that cinchona alkaloids paved the way for the development of stereochemistry and consequently the ascent of modern asymmetric synthesis. The resolution of tartaric acid with quinicine (**1**), a simple derivative of quinine or quinidine (QD) obtained by treating these alkaloids with diluted mineral acid by Pasteur in 1853, was the milestone that ushered in an era of racemate resolutions by crystallization of diastereomeric salts (Figure 13.1) [3].

The first asymmetric reaction carried out by using cinchona base was published as early as in 1912 by Bredig and Fischke [4]. In the 1980s, Wynberg summarized the use of cinchona alkaloids in asymmetric synthesis concluding that they constituted a class of "miracle catalysts" [5]. Even now, despite enormous activity in stereoselective synthesis and introduction of thousands of chiral catalysts, cinchona alkaloids and their derivatives fall into a class of "privileged ligands" [6], efficiently catalyzing more than 50 types of asymmetric reactions in a highly stereoselective fashion [7].

Interestingly, besides the well-known classical resolutions, a significant progress in the field of cinchona-based enantioseparation and their use as enantioselective analytical tools has been made in the last two decades only. This chapter reviews and summarizes analytical applications of the cinchona alkaloids and their derivatives, with the emphasis on modern enantioselective chromatographic techniques.

What makes cinchona alkaloids so unique in the molecular world? The complex structure and multifunctional character of the four principal members of the cinchona alkaloid family – quinine, quinidine, cinchonine (CN), and cinchonidine

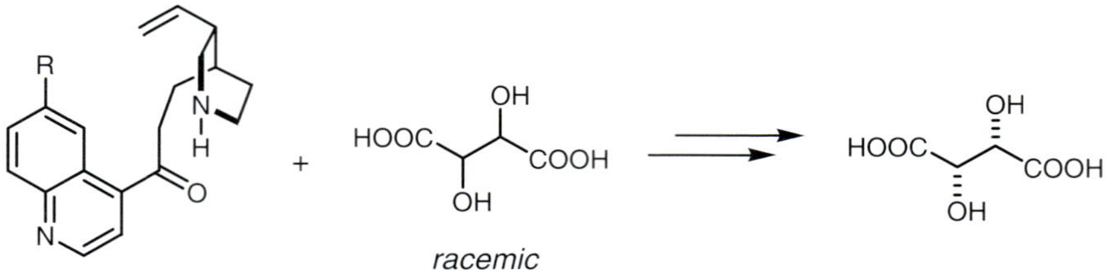

422 | *13 Resolution of Racemates and Enantioselective Analytics by Cinchona Alkaloids and Their Derivatives*

Quinicine (1), R = OMe
Cinchonicine (2) R = H

Figure 13.1 Pasteur resolution of tartaric acid by quinicine (1).

(CD) – nearly perfectly suit their catalytic and chiral discrimination activity (Figure 13.2). The 1,2-aminoalcohol structural motif, including the highly basic and bulky quinuclidine, is primarily responsible for the complexation and catalytic activity. The quinuclidine nitrogen atom readily undergoes protonation

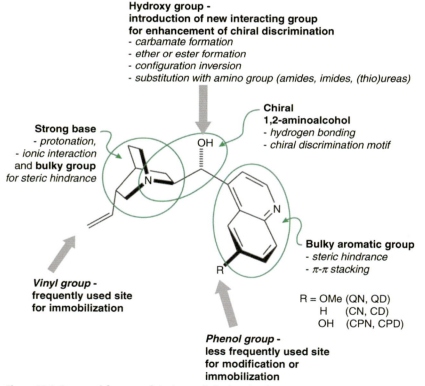

Figure 13.2 Structural features of cinchona alkaloid molecules (QN, quinine; QD, quinidine; CN, cinchonine; CD, cinchonidine; CPN, cupreine; CPD, cupreidine).

and quaternization. The secondary 9-hydroxy group can be easily substituted or transformed into various derivatives, with either retention or inversion of configuration. The peripheral 10,11-vinyl group is an ideal immobilization site using a radical addition. The aromatic 6'-methoxyquinoline ring present in QN and QD or quinoline ring in CD and CN form a steric barrier and also act as reporter groups with a strong UV absorption (λ_{max} about 224 nm, ε about 25 000). This minor structural difference is responsible for a strong fluorescence of quinine and quinidine at 366 nm. The lack of 6'-methoxy group in the so-called cinch bases (CN and CD) results in their lower solubility, increasing their tendency for crystallization compared to QN and QD.

13.2
Resolution of Racemates by Crystallization and Extraction of Diastereoisomers

13.2.1
Resolution of Racemates by Crystallization

Resolution of racemic mixtures using a chiral resolving agent is based on the formation of two diastereoisomeric entities, usually salts differing in their solubility, leading to preferential crystallization of one of them (Figure 13.3). The chemical nature of the racemate usually defines the compatible chemical character of the

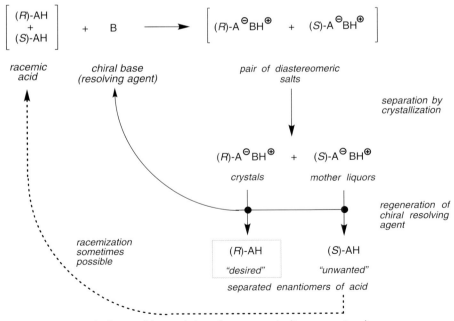

Figure 13.3 General scheme of the resolution of racemates via diastereomeric salts.

resolving agent and most resolutions are carried out using a simple acid–base chemistry. Occasionally, the acidic or basic group lacking in the racemate molecules can be introduced by using an appropriate derivatization step. For example, cinchona bases can be employed for the resolution of alcohols if these were transformed into hemiesters with free carboxylic group in a reaction with succinic, phthalic, or maleic anhydrides [8].

Obviously, the highly basic nature of cinchona alkaloids permits the resolution of racemic acids and other acidic compounds. Pasteur experiment with serendipitous resolution of tartaric acid using quinicine (**1**) and cinchonicine (**2**) [3] has ushered in an era of resolution of racemates by diastereoisomeric salt formation. Since early times, the naturally occurring alkaloids, including those of the cinchona family, have been most frequently used for resolutions both in the industry and in the academia. The availability of two pairs of pseudoenantiomers, that is, QN–QD and CN–CD, significantly facilitated the screening process and acquisition of both enantiomers of the target compound. Moreover, cinchona alkaloids also fulfill most of the criteria for a good resolving agent [9] as they are nontoxic, widely available (annual production about 700 tons per year), moderately expensive, fully recyclable, and soluble in a number of organic solvents. Due to these features, about 25% of all resolutions carried out in the 1990s used the four principal members (QN, QD, CD, and CN) of the cinchona alkaloid group [5]. The application of cinchona bases as resolving agents until 1990s has been reviewed in monographs [10].

At present, the resolution of racemates via classical diastereomer crystallization as a method of chiral target production is somewhat hampered by a rapid development of other methods, mainly asymmetric synthesis, including biocatalysis [9] and enantioselective chromatography [11]. Diastereomer crystallization remains, however, an important technique because of its two fundamental advantages, especially attractive for industry. First, process development (practical know-how and accessibility to wide libraries of the resolving agents) is usually fast and easy. Second, the cost is often low compared to other methods.

Perhaps, one of the most important industrial processes using cinchonidine (or quinidine) as the resolving agent was the production of (*S*)-naproxen (**3**) by resolution of racemic naproxen. It is not clear whether it is still in use (Figure 13.4) [12].

Parent, unmodified cinchona alkaloids remain important resolving agents of acidic compounds, both in the research and in the applied chemistry, for example, in pharmaceutical and agrochemical industries. Selected, recent applications of unmodified cinchona alkaloids as resolving agents include pharmacologically active ingredients or their precursors, nonproteinogenous amino acids, or chiral intermediates, as shown in Table 13.1. Under optimized conditions, highly enriched diastereoisomeric salts can be obtained directly or after few recrystallizations with good to reasonable yields.

The alkaloids have also been used for the resolution of various phosphoric and carboxylic acids **4–10**, including axially chiral compounds, such as biaryl derivatives **4**, **5** and **7** [13], allenic acids **9**, and planar chiral naphthalenophane carboxylic acids **10** (Figure 13.5).

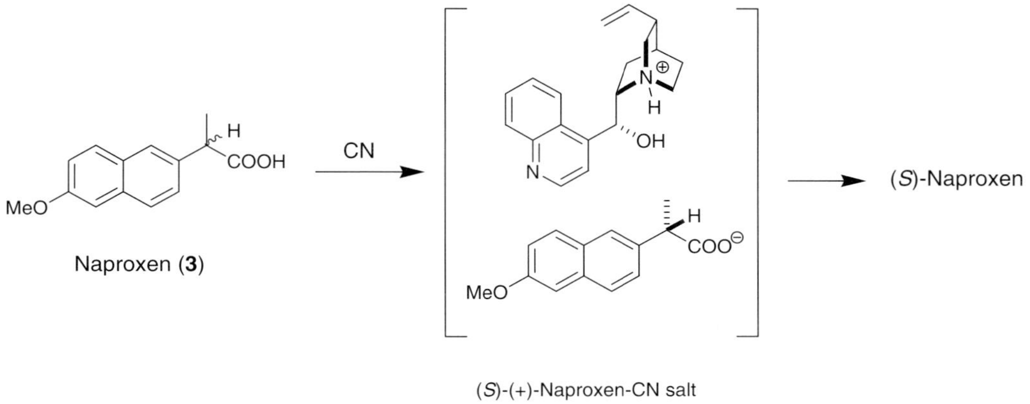

Figure 13.4 Resolution of naproxen (**3**) with the use of CD.

Among other classes of compounds resolved with the aid of cinchona alkaloids are those having stereogenic phosphor and sulfur atoms. Some recent examples (**11–14**) of such resolutions are shown in Figure 13.6 [16–19].

In summary, high probability of success in the resolution of the diverse acids by cinchona alkaloids make them still important and competitive to many existing resolving agents. The four main members (QN, QD, CD, and CN) are often the first choice for screening the resolutions. These alkaloids are included in the resolving set for acids commercially available from Aldrich [20]. Additional benefits that are associated with the use of cinchona alkaloids as resolving agents are high diastereomer enrichment after just one or two crystallization steps (Table 13.1) and possible resolution of the more soluble diasteromeric salt by the substitution of the base with pseudoenantiomeric alkaloid.

Surprisingly, derivatives of cinchona alkaloids have been only rarely used as resolving agents. Fundamental studies on the formation of molecular complexes due to the hydrogen bonding and other molecular interaction were initiated by Toda [37] who has shown that N-benzylcinchonidinium chloride (**15**) can be used for efficient resolution of racemic 1,1′-binaphthalene-2,2′-diol (BINOL, **16**) – an important chiral scaffold widely explored in asymmetric catalysis and supramolecular chemistry [38]. This process provided the (R)-**16** with 99% ee after one crystallization and was later adapted by Cai [39] for a practical, multigram resolution of both enantiomers of BINOL. The use of N-benzylcinchonidinium chloride (**15**) is not restricted to the resolution of **16** only. Many other weak acids, such as **17–21** (Figure 13.7), have been successfully resolved in this way with high enantiomeric excess [37, 40].

The X-ray-determined structure of the complex of **16** and **19** with quaternary salt **15** revealed that the primary discriminative forces leading to an efficient resolution are the formation of directional hydrogen bonds of hydroxy groups of cinchonidine and BINOL with the halide anion as well as aryl–aryl interaction between the naphthyl and the quinoline rings [40].

Table 13.1 Selected examples of racemate resolution using cinchona alkaloids.

Racemate	Target/function	Resolving agent (number of recrystallization steps)	Yield of crystalline diastereomeric salt (%)	Configuration or optical rotation sign of resolved enantiomer	ee or optical purity (op) (%)	Reference
	Ketoprofen (nonsteroidal anti-inflammatory drug)	CD (4)	39	(S)-(+)	99 ee	[23]
	Bicalutamide intermediate (nonsteroidal anti-androgen)	CD (1)	n.a.	S	56 ee	[24]
				R		
	Precursor of PGI$_2$ antiplatelet drug – Beraprost	QN (10)	13	S	99 ee	[25]
		CD (9)	7	R	99 ee	
	Precursor of (R)-aminoglutethimide	CD (2)	30	R	>90 ee	[26]
		QN (1)	n.a.	S	n.a.	

13.2 Resolution of Racemates by Crystallization and Extraction of Diastereoisomers

Structure	Application	Resolving agent	Yield	Config.	ee/op	Ref
(structure)	Inhibitor of 5-lipooxygenase (antagonist of leukotriene receptor)	QN (1)	74	R, R	n.a.	[27]
(structure)	Verapamil precursor	QN (2)	n.a.	S	92 ee	[28]
(structure)	Growth hormone secretagogue intermediate	QD (0)	33	R	96	[29]
		CN (0)	35	R	91	
		QN (0)	33	S	81	
(structure)	(−)-Tetrahydrolipstatin intermediate	QN (1)	30	R	100 op	[30]

(Continued)

Table 13.1 (Continued)

Racemate	Target/function	Resolving agent (number of recrystallization steps)	Yield of crystalline diastereomeric salt (%)	Configuration or optical rotation sign of resolved enantiomer	ee or optical purity (op) (%)	Reference
BocHN—[structure]—OH	GABA-agonist precursor	CD (2)	46	(S)-(+)	46 ee	[31]
BzHN, R / HO—[structure]—COOH (R = Ph) and alanine (R = Bn)	α-Hydroxymethylphenylglycine	CD (R = Ph) CD (R = Bn)	71 72	(S)-(−) (−)	100 op 100 op	[32]
[structure] COOH / HN—CHO	Plant ethylene biosynthesis precursor	QN (3)	32	(R)-(+) (S)-(−)	99 ee[a] 99 ee[a]	[33]
[structure with NH$_2$, P, COOH]	Herbicide	QN (1)	79	L	97 ee[b]	[34]
BocHN [structure] COOH	N-Methyl-D-aspartic acid receptor agonist	CD (3)	18	(R)-(−)	95 ee	[35]
OMe [structure] COOH	Chiral intermediate	QN (2)	90	(S)-(+)	99% ee	[36]

[a] Determined for the corresponding ethyl ester.
[b] Determined for the diastereomeric salt.

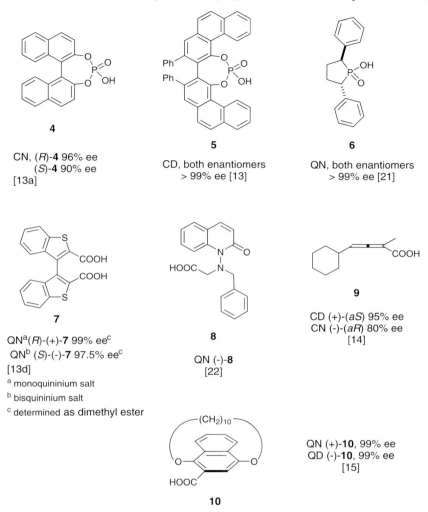

Figure 13.5 Carboxylic and phosphoric acids 4–10 resolved by cinchona alkaloids.

Molecular complexation with cinchonium quaternary salts was later applied for the successful resolution of other, nonsymmetrical biaryl derivatives – 2-amino-2′-hydroxybinaphthyl 20 [41] and isoquinoline 21 [42].

N-Benzylcinchoninium chloride, pseudoenantiomeric to N-benzylcinchonidinium chloride (15), used for the resolution of the racemic binol 16, gave a crystalline complex of the same (R)-configuration as 16. X-ray structure showed a different packing of the components in the crystals, the main attractive force in this case being the electrostatic and C−H−π interactions [43].

Surprisingly, other cinchona derivatives have not been extensively studied as the resolving agents with a few notable exceptions such as cinchona 9-O-carbamates

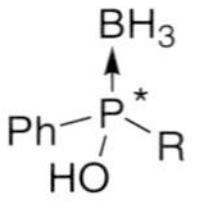

11

R = Alk, Ar, Bn
CN 96-100% op
(determined as methyl ester)
[16]

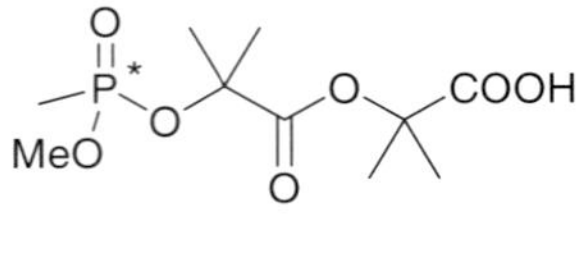

12

QN, CN - LC separation
ee n. a. [17]

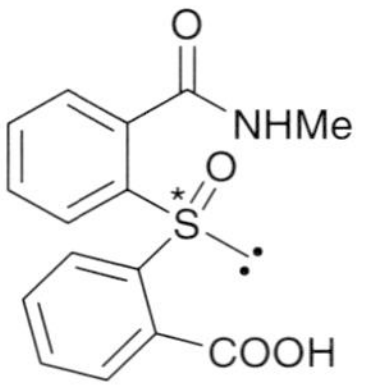

13

CN (R)-(+) 99% ee
(S)-(-) >99% ee
[18]

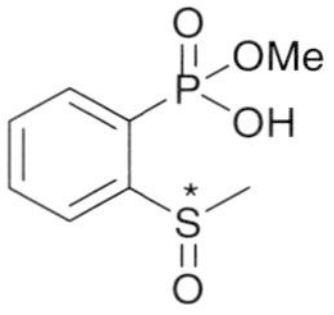

14

CN 100% de
[19]

Figure 13.6 Resolution of compounds with stereogenic phosphor **11–12** and sulfur **13–14** atoms by cinchona alkaloids.

that were used by the Lindner group for the resolution of 3,5-dinitrobenzoyl (DNB)-protected amino acids [44] or for proving the mechanism of chiral recognition by cinchona carbamate-derived chiral stationary phases (CSPs) (see below) [45].

13.2.2
Resolution of Racemates by Enantioselective Extraction

Resolution of racemates by the selective liquid–liquid extraction (Figure 13.8) of diastereomers due to their different solubility in two-phase system is by far a less common technique than crystallization of diastereomers. In this process, a highly discriminative chiral ionic reagent transfers (preferably in one extraction step) a highly enriched enantiomer as an ionic pair from one phase to another, which dissolves selectively the formed diastereomeric associate but not the starting enantiomers. The following back-extraction with an appropriate acid or base produces the desired enantiomer and recovered chiral selector (SO).

Easy protonation of cinchona alkaloids (ionic character) together with the possibility of modification of their lipophilicity makes this class of compounds interesting candidates for the enantioselective extraction. A pioneering work in this

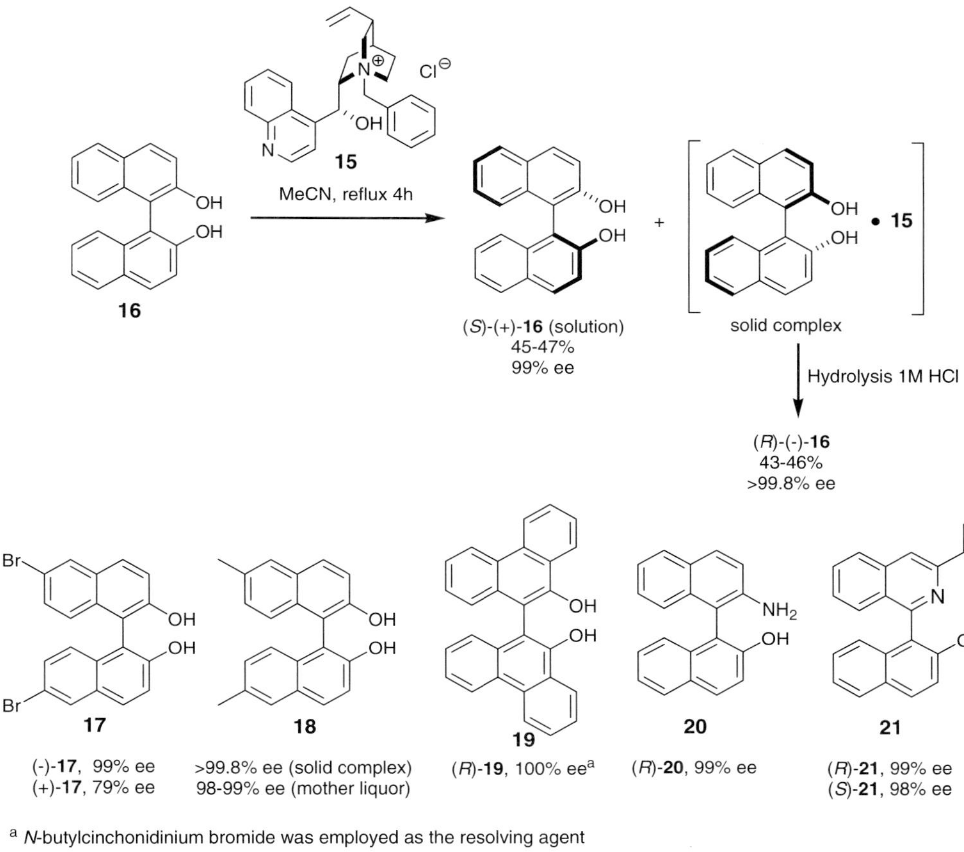

Figure 13.7 Resolution of biaryl diols **16–21** by molecular complexation with cinchona quaternary salts.

[a] N-butylcinchonidinium bromide was employed as the resolving agent

line has been patented by Dow Chemical Company. In these studies, quinine was employed as a chiral phase-transfer carrier for the resolution of several herbicides having the structure of 2-(4-aryloxyphenoxy)propionic acid **22** [46]. In this application, quinine was usually R-enantiomer selective and the recommended solvent system consisted of chlorinated hydrocarbons (mainly CCl_4) and ethylene glycol. Single, short (1 min) extraction gave the enrichment ratio R/S in the range from 51/49 to 2/1.

Very efficient enantioseparation of racemic DNB-leucine **23** exploiting an extraction with a mechanistically designed quinine 1-adamantylcarbamate **24** bearing a highly hydrophobic side arm has been reported by Lindner and coworkers. The most efficient extraction was achieved with aqueous ammonium acetate buffer/dodecane solvent system. After back-extraction of the organic phase with 1 M H_3PO_4 solution, the preferentially complexed S-enantiomer of **23**

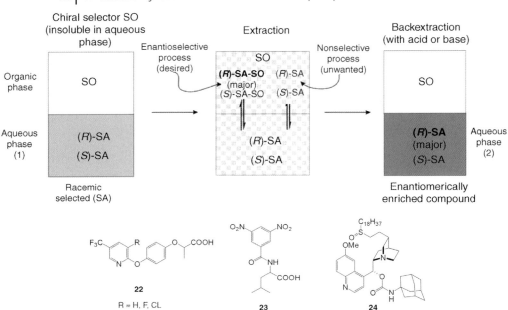

Figure 13.8 Schematic representation of enantioselective extraction.

was separated with 95% ee and 70% yield [47]. Interestingly, other N-protected leucines bearing benzyloxycarbonyl (Z), t-butoxycarbonyl (Boc), and Ac protecting groups showed no enantioseparation, while 9-fluorenylmethoxycarbonyl (Fmoc) derivative gave only 20% ee under similar conditions. This study shows the importance of π–π interaction between π-acidic DNB moiety and π-basic quinoline ring in chiral discrimination. Similar separation has been recently adapted for a continuous process using centrifugal contactor separator. Unfortunately, comparing the batch process much lower enantioselectivity (34% ee) was obtained [48].

The enantioselective extraction was later used by Lindner et al. for the development of a rapid and cost-effective screening protocol for structurally related cinchona-derived CSPs [49]. The protocol is based on solid–liquid extraction of the dissolved racemate (liquid phase) by immobilized chiral selectors (solid phase). After equilibration, the enantiomer excess of the less strongly bound enantiomer is determined in the liquid phase by an appropriate HPLC method. The selectivity (α) data obtained by the extraction procedure for a number of cinchona selectors correlated well with chromatographic selectivities obtained from HPLC experiments with the use of corresponding cinchona CSPs (the same selector but immobilized on the silica support). This allowed a reliable prediction of the corresponding chromatographic selectivity of the new cinchona selectors with a small amount of immobilized sample and without the need of packing the chromatographic column.

13.3
Enantioselective Chromatography and Related Techniques

Enantioselective liquid chromatography (HPLC) and related techniques such as supercritical fluid chromatography (SFC), simulated moving bed (SMB), countercurrent chromatography (CCC), or capillary electrophoresis (CE) and its hybrid technique capillary electrochromatography (CEC) are currently among the most important analytical (HPLC, LC–MS, CE, and CEC) and preparative (LC, SMB, and CCC) separation techniques in chirotechnology [11, 50]. Enantioselective analytics has been revolutionized with the implementation of HPLC, GC, and CE methods, offering reliable and sensitive tools for quality control purposes and for research. In addition, some of the chromatographic techniques are well suited for the production of chiral target compounds in the laboratory (mg-g, HPLC) or on an industrial scale (kg-tons, SMB, LC) [50].

Enantioselective chromatography and related techniques are based principally on the reversible formation of diastereomeric associates between both enantiomers of the chiral analyte (selectand, SA) and the chiral selector (SO) that is usually covalently immobilized or coated on a solid support (Figure 13.9).

A different but experimentally simpler solution is to use a chiral selector (called often in such a case chiral additive) dissolved directly in the mobile phase. The diastereomeric associates thus formed differ in their mobility or adsorption, and in favorable cases they can be separated on an achiral phase. This approach was applied in the 1980s by Pettersson et al. who used native cinchona alkaloids in ion-pairing chromatography [51]. Nowadays, ion-pairing technique is of considerable interest in a number of electrophoretic techniques (see below).

Chiral separation can be observed when there is a suitable difference between free energies (ΔG) of diastereomeric associates formed by R- and S-enantiomers of selectand and a chiral selector (SO). The energy differences can be directly attributed to the chiral recognition phenomena, involving complementarity of size, shape, and molecular interaction of SO and SA molecules. These factors are also controlled by experimental conditions such as temperature, mobile phase

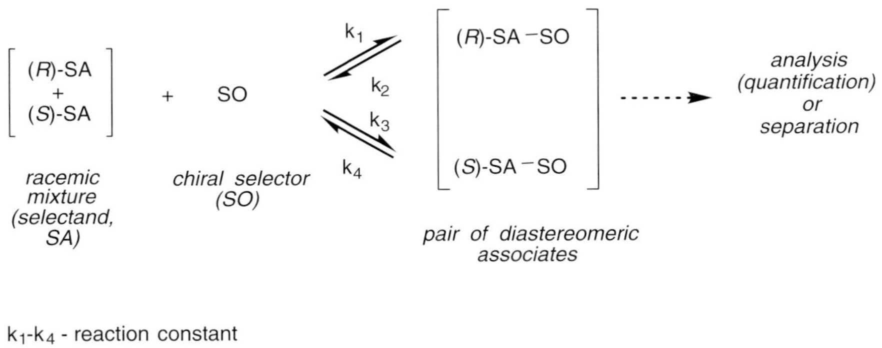

k_1–k_4 - reaction constant

Figure 13.9 General mechanism of the enantioseparation by chromatographic methods.

composition, pH, and ionic strength. Thus, efficient enantioselective separation procedure requires, besides an appropriate chiral selector, considerable experimentation to identify optimal chromatographic conditions.

The multifunctional character of cinchona alkaloid molecules, their accessibility, and easy functionalization/immobilization make this class of compounds an attractive platform for the construction of chiral selectors and chiral stationary phases. However, it must be pointed out that high basicity of the quinuclidine nitrogen (pK_a 9.8) leads to its complete protonation below pH 7.5 of the mobile phase, either under polar organic or reversed phase conditions. Under such a condition, the positively charged quaternary chiral nitrogen ion may interact with the oppositely charged analytes (e.g., carboxylates); therefore, ionic interactions are the principal driving force behind the retention of negatively charged analytes. Cinchona-based CSPs should be then recognized as weak chiral anion exchangers (WAX), and such type of chromatography is classified as ion (anion) exchange. Of course, sole ionic interactions with the chiral anion exchanger may not warrant efficient chiral recognition. Usually, they are complemented by hydrogen bonding and/or π–π and steric interactions. The combination of the structural features of cinchona-based selector with the buffer ionic strength and pH of the mobile phase allows to achieve optimal enantioseparation and a difference in retention. These two parameters can be further adjusted without compromising the enantioselectivity by appropriate selection of counter ions and acidic or basic modifiers.

The protonation of cinchona selectors by the acidic analytes may also occur under normal conditions (hydrocarbons + polar nonaqueous modifier) and can be responsible for the strong retention of the analyzed molecules, making such mobile-phase regime less practical for most of the chromatographic applications.

13.3.1
Early Attempts and Current Status

Pioneering attempts at using cinchona alkaloids as a platform for chiral stationary phase preparation have been reported as early as in the mid-1950s by Grubhofer and Schleith [52]. The chiral anion exchange polymeric materials were prepared by immobilization of quinine (and other cinchona alkaloids) via the 9-hydroxyl group or quinuclidine nitrogen to a polymer support. However, this resulted in very low selectivities of these phases toward racemic mandelic acid as a test analyte. Results of the early studies have been reviewed in detail by Davankov [53].

Radical *anti*-Markovnikov thiol addition to the vinyl group of cinchona alkaloids has been proposed by Rosini *et al.* as a method of immobilization, leaving intact the free 2-aminoalcohol interaction sites [54]. CSPs **25–27** (Figure 13.10) were screened against a large set of neutral analytes **28–32**, initially under normal conditions (solvents of low polarity and polar modifier). Unfortunately, the enantioseparation factors (α) were usually low, rarely exceeding 1.1, resulting in none or partial separation of the analytes. The substrates fully resolved under these conditions were the few sterically congested carbinols (**28**), binaphthol derivatives (**29**), and some other polar compounds (**30–32**) of pharmaceutical interest [54–56].

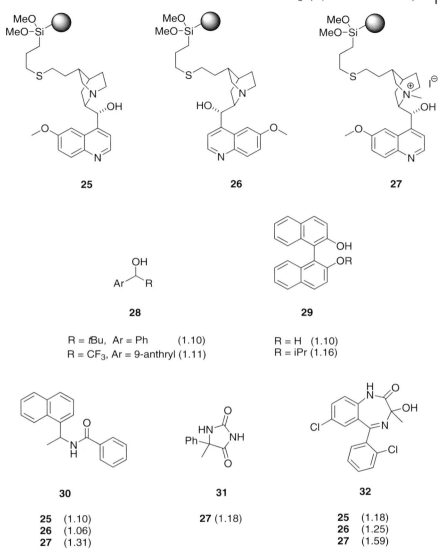

Figure 13.10 First generation of cinchona-based CSPs (**25–27**) and resolved racemates **28–32** (in parentheses are enantioseparation factors α).

The chiral discrimination process studied for quinine and binaphthol **29** (R = i-Pr) by using ^{1}H NMR spectroscopy led to the conclusion that the primary forces responsible for the enantiodiscrimination are the hydrogen bonds and steric interactions [57, 58]. Optimization studies on quinine for the enantioseparation of carbinol **28** (Ar = 9-anthryl, R = CF$_3$) on quinine CSP **25** carried out by Nesterenko et al. showed that under normal conditions the best selectivity (α 1.53) could be achieved by using CCl_4 as the solvent [59]. Investigation of these first-generation quinine and quinidine CSPs under reversed phase and polar organic modes was later initiated by Lindner and coworkers. It has been shown that the overall efficiency of these selectors was higher, leading eventually to higher separation factors for a broad class of racemic analytes, mainly of acidic character [44, 60].

13.3.2
Cinchona 9-O-Carbamates as CSPs in HPLC

A significant breakthrough in cinchona-based enantioselective chromatography came with the demonstration by Lindner *et al.* in 1996 that immobilized quinine or quinidine 9-O-carbamates under polar organic or reversed phase condition efficiently resolved a number of acidic racemates, with opposite elution orders compared to unmodified QN or QD CSPs [44, 61]. Cinchona 9-O-carbamates contain a new functionality that can serve as a binding site of double character, that is, an H-bond donor–acceptor and, depending on the N-substituent, it may also provide a steric barrier or possibly a source of π–π interaction (Figure 13.11).

The conformational preferences of 9-O-carbamoyl cinchona free bases reflect in general those of the native cinchona alkaloids. 6′-Neopentoxy-9-O-*tert*-butylcarbamoylcinchonidine exists as a mixture of two major *anti-closed* and *anti-open* conformers in a 65 : 35 ratio, whereas upon protonation *anti-open* conformation has been observed exclusively [62].

From the practical point of view, 9-O-cinchona carbamates also offer additional benefits: they are stable in a broad range of pH (3–10) and can be readily obtained in a one-step reaction of a cinchona alkaloid with an appropriate isocyanate.

Among the many carbamates screened, quinine and quinidine *tert*-butyl-9-O-carbamates (*t*-BuCQN and *t*-BuCQD) (Figure 13.12) were found as the most versatile and easy-to-get selectors. They were commercialized in the late 1990s, initially by Bischoff GmbH and then by Chiral Technologies [63].

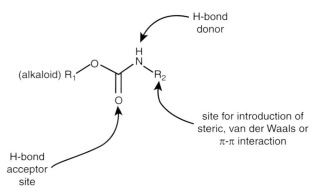

Figure 13.11 Structural features of the carbamate group.

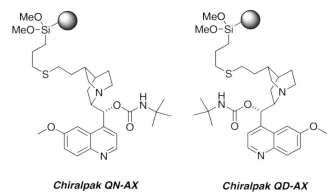

Figure 13.12 The commercialized 9-O-tert-butylcarbamoylquinine (Chiralpak QN-AX) and 9-O-tert-butylcarbamoylquinidine (Chiralpak QD-AX) CSPs.

13.3.2.1 Applications

The major area of applications of cinchona 9-O-carbamate type CSPs embraces enantioseparation of diverse N-protected amino acids and peptides, carboxylic, phosphonic, and phosphoric acids as well as other acidic compounds.

N-Protected amino acids are important building blocks for the chemical and enzymatic peptide and protein synthesis. The 16 proteinogenous racemic amino acids as various N-protected derivatives (Table 13.2) have been baseline separated using cinchona 9-O-carbamate type CSPs under polar organic or reversed phase conditions. Many synthetically important protecting groups, such as Ac, Fmoc, Bz, Z are tolerated in the process of chromatography; however, employment of the π-acidic 3,5-dinitrobenzoyl group (DNB) gave extraordinary high enantioselectivities ($\alpha > 10$) [44, 60, 63–68]. Selected examples of enantioseparation of N-protected amino acids are provided in Table 13.2.

Successful enantioseparation of individual N-protected amino acids stimulated the development of a rapid method of their simultaneous enantioseparation and quantification in a mixture. A feasibility study on this topic has been recently published by Welsch et al. [69]. The two-dimensional HPLC method involves online coupling of a narrow-bore C18 reverse phase (RP) column in the first dimension (separation of racemic amino acids) to a short enantioselective column based on nonporous 1.5 μm particles modified with t-BuCQD in the second dimension (determination of enantiomer composition). Using narrow-bore column resulted in fast analysis time; for example, the mixture of nine racemic N-DNB-protected amino acids was completely analyzed within 16 min.

9-O-tert-Butylcarbamoylquinine (t-BuCQN) and 2,6-diisopropylphenylcarbamoyl-quinine (DIPPCQN) type CSPs have been evaluated for the enantiomer separation of a set of isoindolin-1-one fluorescent derivatives of α-amino acids obtained by their reaction with the ortho-phthalaldehyde, naphthalene-2,3-dicarboxaldehyde, or anthracene-2,3-dicarboxaldehyde. The latter reagent afforded the derivatives of the

Table 13.2 Enantioseparation of N-protected proteinous amino acids on carbamate-type cinchona CSPs[a].

Amino acid	Protecting group	CSP	α	References
Ala	Bz	t-BuCQN	1.77	[64]
	DNB	t-BuCQN	7.93	[64]
Asp	DNB	t-BuCQN	1.45	[66]
Arg	Fmoc	t-BuCQN	1.68	[66]
Cys	Fmoc	t-BuCQN	1.62	[44]
Glu	DNB	t-BuCQN	5.33	[66]
	DNZ	t-BuCQN	1.36	[65, 66]
Ile	Boc	t-BuCQN	1.55	[44, 67]
	DNB	t-BuCQN	13.22	[44, 67]
Leu	Ac	t-BuCQN	1.40	[64–66]
	Bz	t-BuCQN	2.58	[64–66]
	DNB	t-BuCQN	14.70	[64–66]
	DNP	DIPPCQN	1.74	[64–66]
	Z	t-BuCQN	1.31	[64–66]
Lys	Bis-DNZ	t-BuCQN	1.50	[66]
Met	Fmoc	t-BuCQN	1.67	[68]
Phe	Bz	t-BuCQN	2.45	[63, 64, 66]
	Bz	t-BuCQD	2.44	[63, 64, 66]
	DNB	t-BuCQN	10.10	[63, 64, 66]
	Z	t-BuCQN	1.24	[63, 64, 66]
Pro	DNZ	t-BuCQN	1.28	[44, 75]
Ser	DNB	t-BuCQN	5.78	[44, 66]
	Z	t-BuCQN	1.21	[44, 66]
Thr	Boc	t-BuCQN	1.50	[44, 67]
	DNB	t-BuCQN	5.81	[44, 67]
Trp	Ac	t-BuCQN	1.80	[63, 64, 66]
	Boc	t-BuCQN	1.45	[63, 64, 66]
Tyr	Boc	TrCQN	1.44	[66]
Val	Ac	t-BuCQN	1.67	[44]
	DNB	t-BuCQN	11.69	[44]

[a]For experimental conditions, refer to the references.

majority of the analytes with highest retention factors (k) as well as resolution (R_S) and enantioselectivity (α) values [70].

The applicability of the 9-O-cinchona carbamate-type CSPs is not restricted to the natural α-amino acids. Many synthetic and nonproteinogenous α- and β-amino acids 33–37 and their derivatives, including phenylglycines 33 [71], phenylalanines 34 [72, 73], β-amino acids 35 [44, 74, 75] as well as prolines [76], 2-methyltaurine, GABA, and many β-methyl amino acids [77] were also successfully separated (Figure 13.13).

Both enantiomers of hormones thyroxine (D- and L-T4, 38) and triiodothyronine (D- and L-T3, 39) can also be baseline separated and quantitatively determined in pharmaceutical formulations of *levo*-thyroxine with the use of commercial t-BuCQN

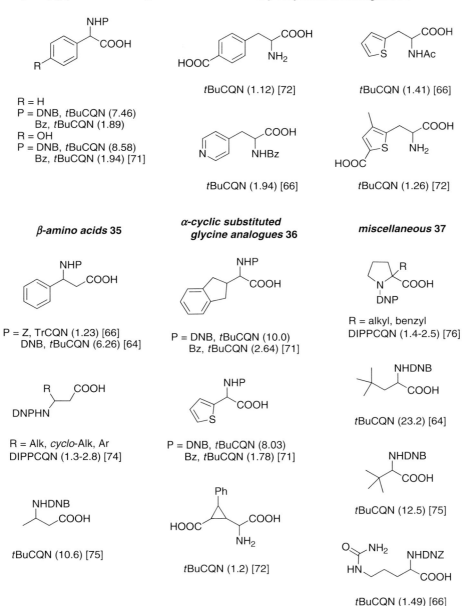

Figure 13.13 Enantioseparation of nonproteinogenous and synthetic amino acids **33–37** (enantioseparation factors α are given in parentheses).

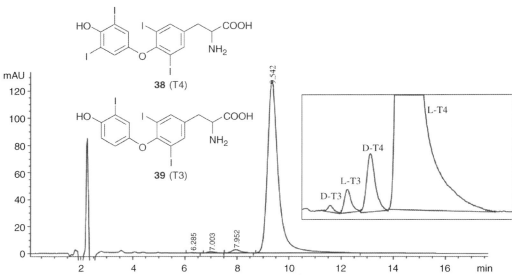

Figure 13.14 Identification of chiral impurities in pharmaceutical *levo*-thyroxine sodium salt [78].

column under reversed phase condition. The published protocol allows a direct (no need for sample derivatization), rapid (15 min), and sensitive determination of as low as 0.1% impurities of unwanted D-enantiomers of both T4 and T3 (Figure 13.14) [78]. Recently, this method has been optimized to overcome the problem of a limited chemoselectivity for the adjacent peak pair of L-triiodothyronine (L-T3) and D-thyroxine (D-T4). Coupling of a reversed phase and a chiral anion exchanger *t*-BuCQN columns resulted in a significant improvement of the separation of L-T3 and D-T4 pair for which resolution $R = 3.7$ was obtained [79].

D-Enantiomers of amino acids in living organisms are attributed to many important bioprocesses or can be markers of certain disorders. Therefore, accurate quantification of the low levels of the D-form in the presence of a large amount of L-form is of considerable interest. Branched D-amino acids in mammalian tissues and body fluids were quantified recently using a sophisticated two-dimensional HPLC system containing a narrow-bore reversed phase and *t*-BuCQN column for the enantioseparation of L- and D-enantiomers. Target analytes were determined as their fluorescent derivatives, precolumn labeled with 4-fluoro-7-nitro-2,1,3-benzoxadiazole. D-Val, allo-Ile, Ile, and Leu were determined at nmol/ml level, making this a valuable method for quantification of D-enantiomers of amino acids [80].

Similar method has been applied for the simultaneous determination of proline enantiomers, *trans*-4-hydroxyproline and *cis*-4-hydroxyproline, in serum and collagen-rich skin tissue of mammals. As the first dimension, a monolithic ODS column of 0.53 mm diameter was used whereas DIPPCQN column was employed in the second dimension. Such a system gave the separation factor in the range

1.44–1.83 and allowed the accurate quantitation of D-enantiomers with the lowest limit of one fmol [81].

The complex stereoisomerism of peptides (large number of diastereomers) and their biological activity associated with one particular stereoisomer make peptide analysis a challenging task [82]. Comprehensive studies in this direction, using a number of dedicated cinchona-based CSPs, have been carried out by Cervenka and Lindner [82–86]. 6′-Neopentoxy-9-O-tert-butylcarbamoylcinchonidine CSP was employed for an efficient enantioseparation of racemic (all-R/S)-di- and trialanine peptides [83]. This study was later extended to include the enantioseparation of all-S/R-oligoalanines up to deca-alanine with the use of "receptor-like" cinchona CSPs (see the following discussion) [84]. Later on, the authors solved yet a more complex problem of separation of all diastereomers and enantiomers of tri- and tetraalanines. Thus, applying a dedicated two-dimensional LC-MS system comprising two columns, achiral reversed phase for the diastereomers separation and enantioselective cinchona-based CSPs for quantifying the enantiomers led to a successful separation of all eight stereoisomers of trialanine and nine out of ten stereoisomers of tetraalanine [85]. The extension of this study to the enantioseparation of all-S/R alanine-glycine di- and tripeptides using t-BuCQN has also been published [86].

A large number of racemic carboxylic acids **40–53** (Figure 13.15) were also successfully resolved on 9-O-cinchona carbamate-type CSPs (Figure 13.15). Besides synthetically important intermediates or auxiliaries, such as Mosher acid **40** [44, 66], its analogues **41** [87] and mandelic acids **42**, these include a number of compounds of industrial importance. Herbicides based on 2-aryloxypropionic acid, for example, dichloroprop **43** and many other analogues, can be separated by using t-BuCQN CSP [44, 63]. Other acids from the agrochemical pool – cis- and trans-permethrinic (**44**) and chrysanthemic acids (**45**), the metabolites of synthetic pyrethroidal insecticides – can also be resolved. The optimization gave a baseline separation of all four stereoisomers of **44** and **45** in a single run [88]. Several pharmacologically active acids have also been successfully separated, including nonsteroidal anti-inflammatory profens such as naproxen **46**, ibuprofen **47**, flurbiprofen **48** [58, 66], planar chiral acid **49** [89], quinazolone derivatives **50** [90], and monomethyl ester of 4-(2′,3′-dichlorophenyl)-2,6-dimethyl-1,4-dihydropyridine-3,5-dicarboxylic acid **51** – an active metabolite of the short-acting calcium antagonist clevidipine [91].

t-BuCQN CSP has also been employed for the resolution and racemization studies of an atropisomeric biphenyl carboxylic acid **52** [93]. Researchers from Eli Lilly developed an HPLC–tandem mass spectrometric method using two parallel t-BuCQN columns for the quantification of chiral metabolites of antiplatelet prasugrel **53** in humans [92].

t-BuCQD CSP has been employed for preparative enantioseparation of all stereoisomers of the N-protected phosphinic pseudodipeptide **54** (leucine aminopeptidase inhibitor) as well as for enantiomeric N-protected-α-aminophosphinic acid **55** and its other structural analogues [95]. Under reversed-phase condition, single enantiomers of analogues of **54** have been isolated in 100 mg quantities [94].

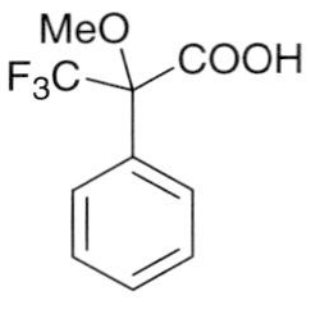

40
TrCQN [66]

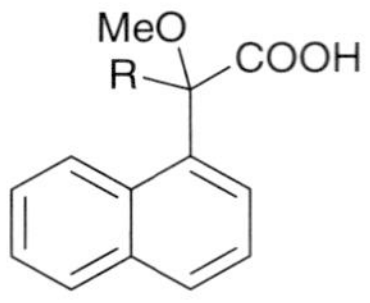

41
R = H, Me, DIPPCQN [87]

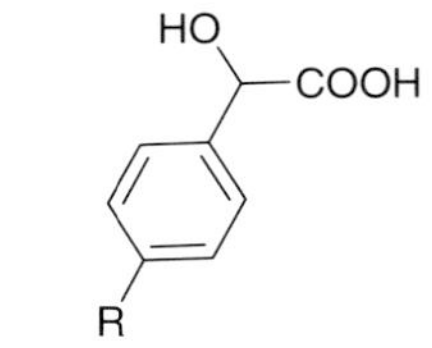

42
R = H, CF_3, *t*BuCQN [102]

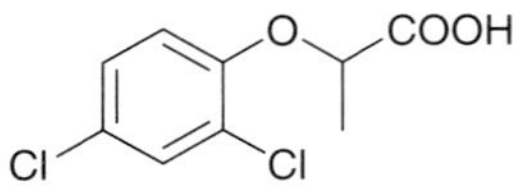

43
*t*BuCQN [63, 66]

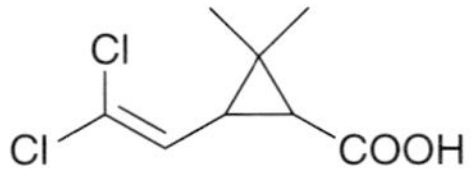

R = Cl, **44**
R = Me, **45**
*t*BuCQN [45]

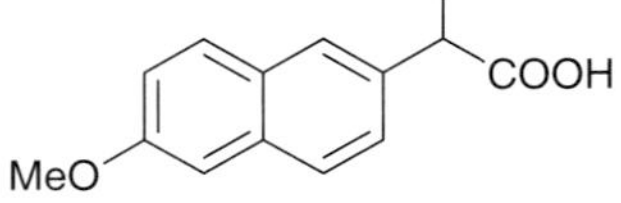

46
DIPPCQN [66]

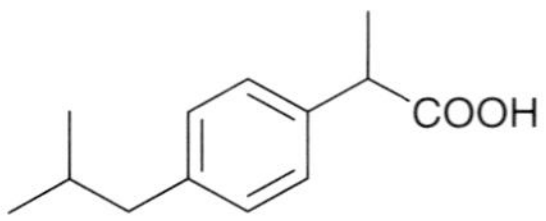

47
DIPPCQN [66]

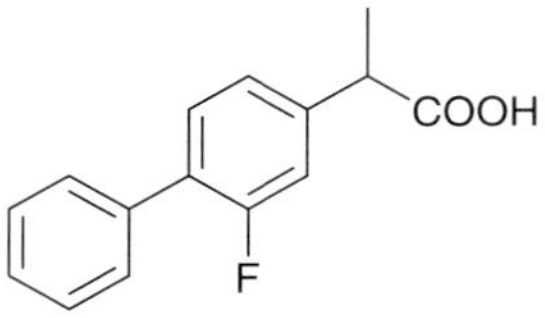

48
DIPPCQN [66]

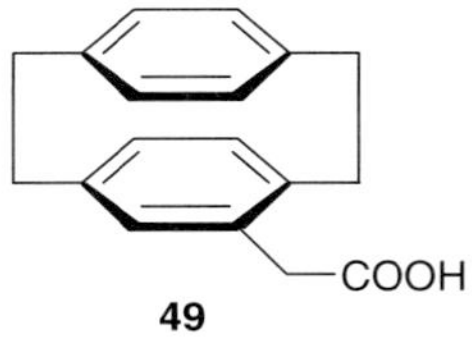

49
*t*BuCQN [89]

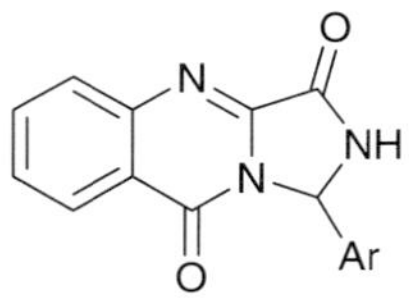

50
*t*BuCQN [90]

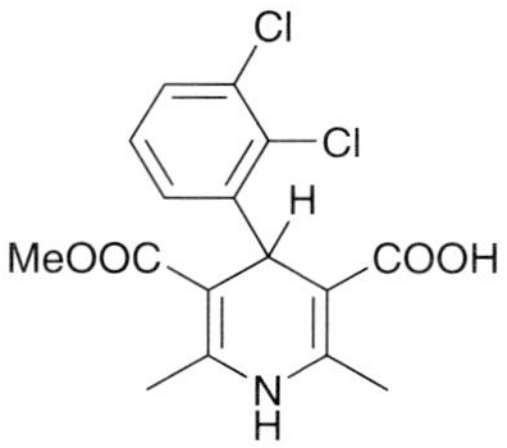

51
*t*BuCQN [91]

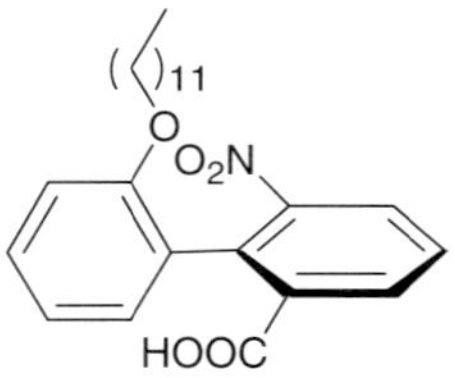

52
*t*BuCQN [93]

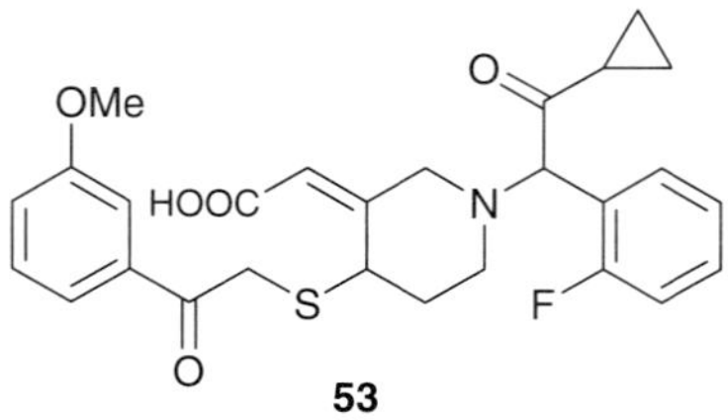

53
*t*BuCQN [92]

Figure 13.15 Enantioseparation of selected carboxylic acids and other acidic compounds **40–53**.

54 (structure with NHZ, COOH, P(=O)(OH)Ph)

55 (structure with NHP, R, P(=O)(OH)H)

P = DNP, DNZ
R = alkyl, aryl, heteroaryl

13.3.2.2 Mechanistic Studies on Chiral Discrimination

The extraordinary selectivities observed for enantioseparation of DNB-leucine **23** on t-BuCQN CSP ($\alpha = 15.8$), or even higher with the use of 6′-neopentoxy-9-O tert-butylcarbamoylcinchonidine selector **56** ($\alpha = 32.6$), make these selectand–selector pairs ideal for the mechanistic studies on chiral discrimination. Extensive investigation of this model has been carried out by the Lindner group with the aid of numerous techniques, including 2D ^{1}H NMR in solution [60] and ^{1}H and ^{29}Si NMR for the immobilized CSP [96], X-ray diffraction [60], microcalorimetry, and CD spectroscopy [97], as well as by molecular modeling [60, 98] and a number of chromatographic methods [64]. All these studies provided a set of data that could be used to explain the chiral discrimination process in a consistent way.

56 (immobilized CSP structure)

57 (soluble analogue structure)

The stereodiscrimination performance of CSPs is usually expressed as a differential free binding energy $\Delta_{R,S}(\Delta G)$ and can be calculated from chromatographic α-values using the equation $\Delta_{R,S}(\Delta G) = -RT \ln \alpha$. For selectors t-BuCQN and **56** and (R,S)-DNB-leucine as the analyte, the corresponding $\Delta_{R,S}(\Delta G)$ values are -6.73 and -8.63 kJ/mol at 298 K, respectively. (S)-DNB-leucine is a more strongly retained enantiomer for all carbamate-type CSPs studied. NMR study of soluble analogue **57** of CSP **56** and DNB-leucine as the selectand shows that both enantiomers form 1 : 1 complexes with the selector, whereas only (S)-DNB-leucine–**57** associate give a set of intermolecular NOEs in the ^{1}H NMR spectra. An analysis of the 2D ^{1}H NMR spectra and X-ray-determined structures revealed that selector **56** in a complex adopts

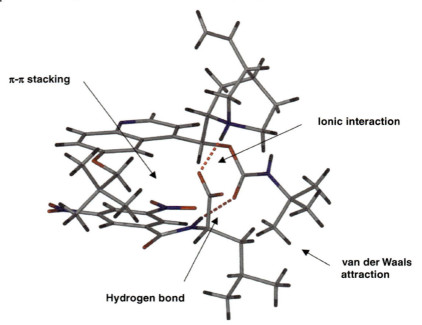

Figure 13.16 X-ray structure of the complex of **57** and (S)-DNB-Leu (**23**). Dotted lines denote the principal interaction in selector–selectand pair. (Data taken from Cambridge Crystallographic Data Centre.)

anti-open conformation with the bulky 6′-O-alkyl and 9-O-carbamoyl groups occupying remote positions and forming a shallow cleft for the selectand. Bulky quinuclidine moiety provides a basic acceptor site for the acid–base interaction. In such a spatial arrangement, binding of (S)-DNB-Leu results in intercalation between 6′-O and 9-O substituents (van der Waals interaction of the isobutyl chain of DNB-Leu with the *tert*-butylcarbamate group). The docking mode orients the aromatic moiety of DNB-Leu parallel to the lower face of the quinoline ring making it an ideal arrangement for intermolecular face-to-face π–π stacking, as evident from both the NMR and the X-ray data. In addition, from the X-ray structure, the presence of the hydrogen bond between the carbonyl group of the carbamate and the proton of the amide group can be deduced (Figure 13.16).

(R)-DNB-Leu molecule also binds to the selector via an ion-pairing mechanism, but it cannot be stabilized by π–π stacking interaction that is apparently of crucial importance for the chiral discrimination. In summary, the enantiodiscrimination is achieved by cooperative multiple contact interactions, such as ion pairing, hydrogen bonding, and π–π stacking [62]. A mechanistically similar enantiodiscrimination process of peptide enantiomers on CSP **56** has been proposed [83].

Extensive studies on the resolving power of cinchona 9-O-carbamate CSPs allowed to identify a set of critical structural parameters influencing the enantioseparation

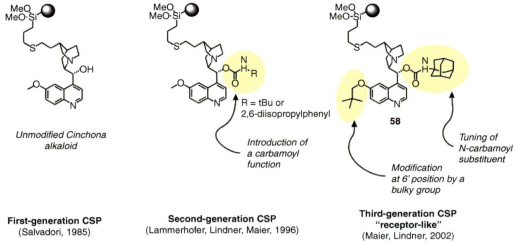

Figure 13.17 Evolution of cinchona-9-O-carbamate-based CSPs.

efficiency and to introduce a new generation of carbamate-type CSPs. It has been shown that besides the introduction of an appropriate 9-O-carbamoyl residue in second-generation CSPs, the replacement of 6′-methoxy group by a more bulky neopentoxy moiety led to a significant enhancement of discrimination power in third-generation CSPs. The evolution of 9-O-carbamate-type cinchona-based CSPs is illustrated in Figure 13.17 [99, 100, 106].

The major factors controlling the performance of cinchona carbamate CSPs are listed as follows:

i. **Configuration at 9-carbon of cinchona alkaloid**: Only naturally configured quinine (8S,9R) and quinidine (8R,9S) derivatives are active, their 9-epimers exhibit much reduced selectivity [150]. In specific cases, however, *epi*-stereoisomers perform better than the naturally configured congeners, as is the case with oligoalanine enantiomers [84].

ii. **9-O-Carbamoyl group**: The substituent located at the nitrogen atom of the carbamoyl group plays a crucial role in determining efficiency of chiral discrimination. Screening a large set of carbamates revealed that *tert*-butyl group is the most convenient in terms of selectivity and applicability. Bulky aromatic 2,6-diisopropylphenyl group has been found to give a better enantiodiscrimination of certain analytes (see Figure 13.15). High selectivity has also been observed for 1-adamantylcarbamates, especially in conjunction with 6′-neopentyloxy group. In contrast to the N-monosubstituted (-NHR) carbamates, their N-methylated (-NMeR) counterparts exhibit very poor performance [60]. Similarly, the replacement of the carbamoyl group with N-substituted hydrazide resulted in a significant drop of chiral discrimination toward typical analytes [101] but resulted in a better separations of the DNP-derivatized amino acids and certain acidic drugs, such as the profens.

iii. **Immobilization type and site:** The thioether linkage is the linker of choice for the immobilization of cinchona alkaloids as it is easy to install, is stable, and exhibits a negligible level of nonspecific absorption. Recently, click chemistry (Huisgen 1,3-dipolar cycloaddition) was employed for a convenient and controlled immobilization of 10,11-didehydro-9-O-*tert*-butylcarbamoylquinine to 3-azidopropylsilica giving fully active CSPs of comparable resolving power with the thioether-linked analogues. The rigid 1,2,3-triazole linker in this case modulates the chiral discrimination processes, and the click-CSPs performed better than the corresponding thioether-linked CSPs for a set of mandelic and aryloxypropionic acids and profens (Figure 13.18) [102].

Although the remote vinyl group of cinchona bases is the optimal site of immobilization in terms of efficiency and accessibility, linking via 9-O-carbamate with a flexible chain also results in active CSPs (Figure 13.18) [61, 103].

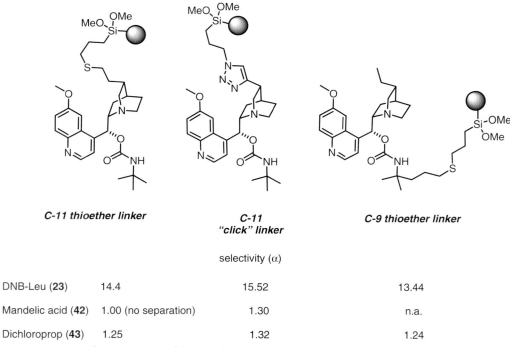

	C-11 thioether linker	C-11 "click" linker	C-9 thioether linker
		selectivity (α)	
DNB-Leu (**23**)	14.4	15.52	13.44
Mandelic acid (**42**)	1.00 (no separation)	1.30	n.a.
Dichloroprop (**43**)	1.25	1.32	1.24

Figure 13.18 Immobilization chemistry and its impact on enantioseparation selectivity of DNB-Leu, mandelic acid, and dichloroprop on cinchona 9-O-carbamate-based CSPs. Chromatographic conditions: mobile phase MeOH : 0.1 M NH$_4$OAc : AcOH 98 : 2 : 0.5 v/v/v, 1 ml/min, l = 10 cm for C11 linkers [102], MeOH : 0.1 M NH$_4$OAc 80 : 20 v/v pH$_a$ 6, 1 ml/min, l = 15 cm for C9 linker [103].

iv. **Quinoline ring**: The presence of 6′-methoxy group is of crucial importance for the performance of CSPs (the corresponding CD or CN CSPs are much less selective). Substituting the methoxy group for bulkier branched substituents, for example, neopentyl or 1-adamantyl unit, usually results in a significantly higher enantiodiscrimination.

v. **Support**: Most frequently non-porous silica gel is used as a support of good mechanical properties and reasonable stability. End-capping procedure offers a decrease in nonspecific absorption of analytes. Silica-based monolithic materials were also successfully used and they are discussed in the last section of this chapter. Carbon-clad zirconia support has been used for noncovalent immobilization of quinine 9-O-phenylcarbamate giving CSPs with the performance characteristics comparable to that of silica-immobilized CSPs [104].

13.3.3
Other Cinchona-Based Selectors: Toward "Receptor-Like" CSPs

Development of highly (enantio)selective synthetic receptors resembling those occurring in the nature is one of the primary goals of the target-specific application in life sciences, for example, as diagnostic tools or in separation technology (membrane enantioseparation) [105].

Detailed studies on cinchona 9-O-carbamates by Lindner *et al.* allowed to identify structural features of cinchona-based chiral selectors that are responsible for the increased selectivity toward some analytes. Usually, a combination of 6′-neopentoxy group with *tert*-butyl or 1-adamantyl group in 9-O-carbamate function led to the highest discrimination level (Figure 13.17). For example, the separation of DNB-neopentylglycine **59** on the CSP containing 6′-neopentyloxy-9-O-(1-adamantyl) carbamoylcinchonidine (**58**, Figure 13.17) as a chiral selector is remarkable for its exceptionally high selectivity, $\alpha > 70$ [106].

Dimerization of cinchona alkaloid 9-O-carbamates via a bifunctional aliphatic or cyclic linker was another attempt to prepare novel, highly effective selectors. Examination of the corresponding CSPs revealed that both the bulkiness and the length of the spacer affected the chiral recognition abilities of dimeric selectors. Shorter spacers reduced the chiral discrimination power whereas the 1,6-hexylene bridge allowed a nearly independent interaction of the two QN subunits with the racemic analytes. Unfortunately, a comparison with analogous monomeric CSP showed that usually the dimeric selectors retained stronger the analytes but without any substantial improvement in the enantioseparation. Nevertheless, the dimeric selector **60** with a *trans*-1,2-diaminocyclohexane linker exhibited good resolution abilities in the separation of DNP-protected amino acids and certain acidic drugs of therapeutical interest (e.g., profens) [107].

Although carbamate-type derivatives remain the most important group of cinchona-based selectors, Maier and Lindner have shown that bis-quinidine phthalazine ((DHQD)$_2$PHAL) type ligands **61**, well known for their applications in

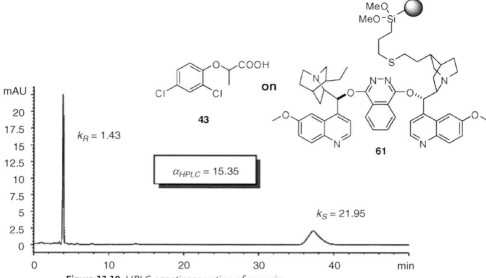

Figure 13.19 HPLC enantioseparation of racemic dichloroprop 43 on the (DHQD)$_2$PHAL-type CSP (61). Column (150 mm × 4 mm i.d.); mobile phase methanol/acetic acid/ ammonium acetate (92 : 2 : 0.5, v/v/w); flow rate 1 ml/min; UV detection 254 nm; column temperature 25 °C [108].

stereoselective synthesis, after immobilization on silica were able to effectively separate racemic dichloroprop with an extraordinarily high selectivity, α = 15.3 (Figure 13.19) [108]. Standard t-BuCQN-based CSP gave an unimpressive value, α = 1.18 (note the opposite elution order due to the change of alkaloid configuration). The phthalazine CSP 61 was also found to efficiently separate enantiomers of DNB-oligoalanines having up to six alanine residues in a molecule. Very high selectivity was observed for DNB-alanine (α = 18.9) and even for DNB-(alanine)$_6$ the selectivity was still at the level of α = 2.1 [84].

The corresponding support-free 1,4-(dihydroquinidinyl)phthalazine derivative 62, bearing a lipophilic aliphatic chain has been applied in centrifugal partition chromatography (CPC) for preparative separation of dichloroprop 43. In both preparative HPLC and CPC methods, as high as equimolar ratio of racemate to selector could be used, resulting in complete utilization of chiral selector in a single separation process – a favorable situation, not very common in enantioseparation. The upscaled data from both methods compared in terms of productivity gave 2.1 and 7.8 kg of resolved racemate per day and mole of chiral selector for CPC and HPLC, respectively [108].

Further studies by Lindner and Maier have shown that substitution of one of the cinchona alkaloid moieties in dimeric cinchona phthalazine-type CSPs with a bulky naphthalene group still allows to maintain the high enantioselectivity of separation. For example, enantioseparation of DNZ-protected β-neopentylglycine on mixed CSP 63 occurs with excellent selectivity, α = 34.2 [106].

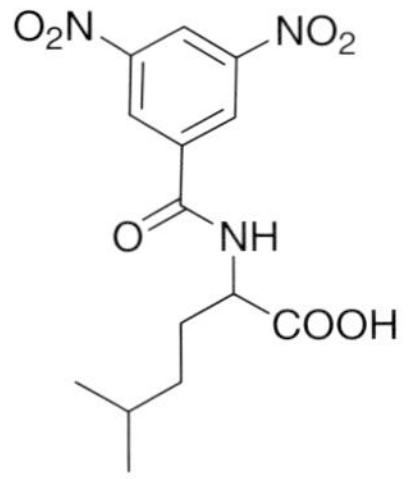

59

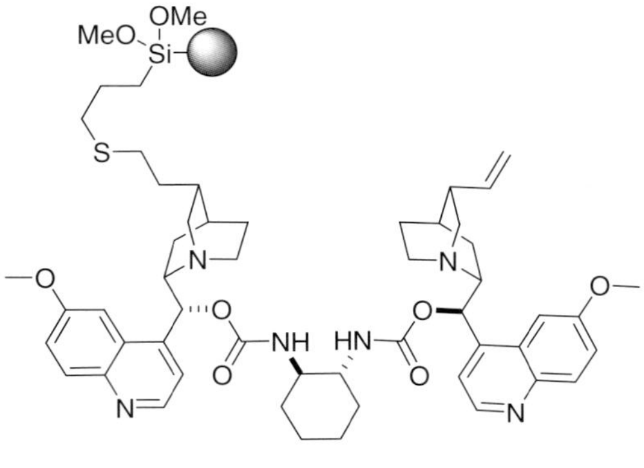

60

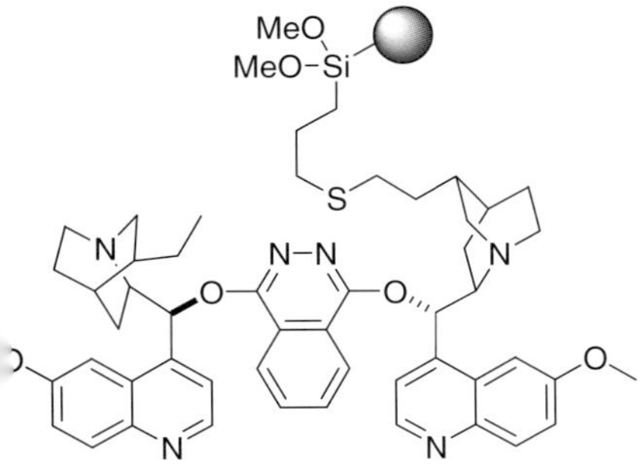

61

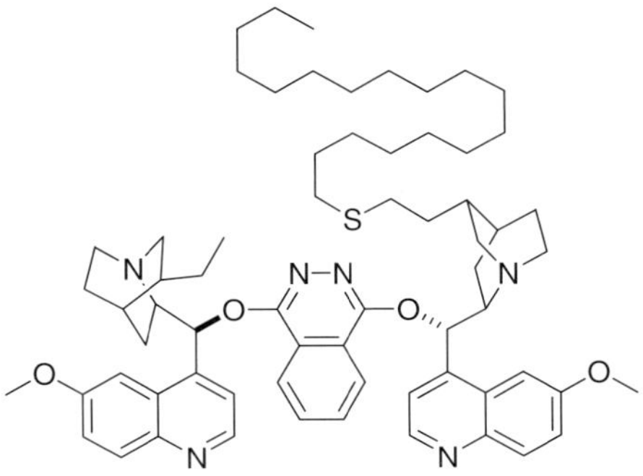

62

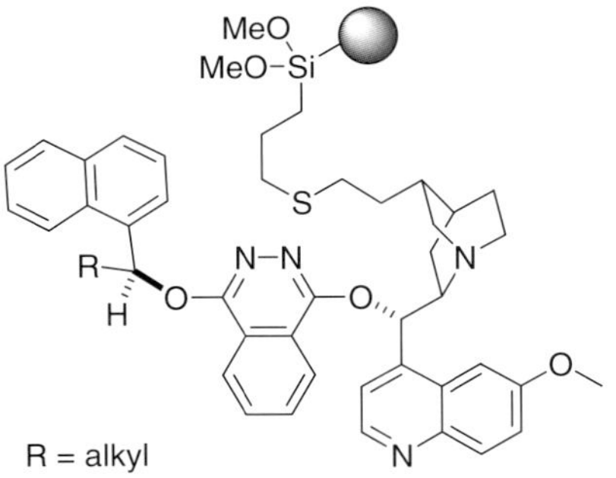

R = alkyl

63

Cinchona-calix[4]arene hybrids **64** and **65** have also been evaluated as CSPs by Maier and Lindner. Interestingly, **64** showed a lower degree of selectivity toward a set of N-protected amino acids, compared to t-BuCQN CSP independently of the 9-configuration of cinchona alkaloid (QN or epi-QN), and no significant level of cooperativity between the cinchona and calix[4]arene units [109]. However, minor modification of **64** by changing the carbamoyl for the urea linkage (CSP **65**, with epiquinine) resulted in a significant enhancement of enantiodiscrimination profile toward Boc-, Z-, and Fmoc-protected cyclic amino acids, such as proline, azetidine-2-carboxylic acid, and pipecolinic acid. These analytes are rather poorly resolved on *t*-BuCQN phase. When chloroform was used as a mobile phase, Boc-proline could be separated with selectivity, $\alpha = 5.4$, in contrast to the standard resolution with *t*-BuCQN CSP, $\alpha = 1.7$, under similar conditions [110]. A comparison of the performance characteristics of selector **65** with structurally closely related analogues uncovers an active involvement of the urea and calixarene units in the chiral recognition process. The urea linker motif was shown to contribute to the analyte binding via multiple hydrogen-bonding interactions, while the calixarene unit is believed to support stereodiscrimination by enhancing the shape complementarity of the selector binding site.

64

(8S,9S)-epi-65
(8R,9S)-65

13.3.4
Cinchona-Based Chiral Modifiers and Phases in Capillary Electrophoresis and Capillary Electrochromatography

Capillary electrophoresis and its most popular hybrid technique – capillary electrochromatography – are complementary to HPLC, offering rapid analysis, low consumption of sample and solvents, and usually a higher efficiency of separation (due to a larger number of theoretical plates). Similar to HPLC, enantioseparation with the use of electrophoretic methods can be conducted by direct (chiral phase or additive) or indirect modes (separation of diastereomers on achiral phase). Most of the 9-O-cinchona carbamate selectors that have been successfully used in HPLC have also been adapted for CE and CEC and showed similar analyte selectivity [111].

Besides the routine stereoselective analysis, low cost and simplicity of CE and CEC methods make them popular tools for fast screening of chiral selectors [112], for studying mechanisms of discrimination, or for determining the binding constant, as it has been shown for t-BuCQN and (S)-DNB-leucine complex [113].

Stalcup and Gham were the first to use quinine as chiral ion-pairing additive in CE [114]. As in HPLC, t-BuCQN in comparison to native QN showed higher selectivity as chiral ion-pairing additive for enantioseparation of N-protected amino acids by nonaqueous CE. Under optimized conditions, with the use of 10 mM of selector, 100 mM octanoic acid, and 12.5 mM of ammonia in methanol–ethanol mixture (40 : 60), racemic DNB-leucine has been resolved with selectivity $\alpha = 1.6$, resolution $R = 64.3$, and plate number $(N) = 127\,000$ [115]. These studies were expanded for the screening of efficiency of four major cinchona alkaloids and a number of quinine and quinidine carbamates bearing 3,5-dinitrophenyl, cyclohexyl, 1-adamantyl, 3,4-dichlorophenyl, or allyl groups at carbamate nitrogen atom as well as of six dimeric quinine carbamates as selectors toward a set of N-protected amino acids as analytes [116]. In a separate study, a good correlation between the data obtained by HPLC and CE has been found, thus confirming the suitability of CE as a reliable screening tool for chiral selectors [112].

In conventional CE, the problem associated with high absorption background of the electrolyte due to the dissolved cinchona selector has been solved by the use of partial filling mode of CE, where the inlet and the outlet are filled with a background electrolyte whereas the absorbing selector occupies only a defined space in the capillary [117]. This modification has been employed for the separation of DNB-oligoalanines with t-BuCQN as chiral selector [118] and for simultaneous separation of the enantiomers and diastereomers of 1-amino-2-hydroxypropane and 2-amino-1-hydroxypropane phosphonic acid derivatives [119]. In the latter case, an optimized protocol allowed a baseline separation of all eight stereoisomers of N-DNP-protected α- and β-aminophosphonic acids in a single run [119].

The enantioseparation by CE using cinchona derivatives as chiral ion-pairing reagents has been the subject of in-depth reviews [120].

The increased popularity of CEC is the result of its advantages that include the vastly increased column efficiency compared to the conventional pressure-driven chromatography. CEC was also adopted for the separations of enantiomers, employing two major modes similar to those used in CE. In the simplest mode, a chiral selector is added to the mobile phase. At present, a less erroneous and more elegant technique uses an open tubular conventional packing or preferably a monolithic column containing an immobilized chiral selector that discriminates the enantiomers.

A nonaqueous CEC method for an efficient, simultaneous separation of the four stereoisomers of N-Z-phosphinic pseudodipeptides **54** has been developed by Lindner et al. As a chiral stationary phase, a monolithic silica capillary column modified with t-BuCQD was used. Under optimized conditions, a baseline separation of all four stereoisomers of **54** has been achieved. The results of this CEC application proved to be superior to the corresponding results of HPLC separations due to significantly higher plate numbers (between 200 000 and 600 000 m^{-1} in CEC) [121]. Similar favorable results have been obtained in CEC separation of a number of N-protected amino acids, with the use of t-BuCQN selector in a capillary column [122] or when this selector served as a chiral ion-pairing reagent [123].

13.3.5
Other Chromatographic Techniques

A single report on the resolution of racemic amino acids by TLC has been recently published. The use of silica gel TLC plates impregnated with QN (0.1%), in combination with appropriate mobile phase system, gave a successful enantioresolution of methionine, alanine, threonine, valine, leucine, and isoleucine. This method combines the simplicity of TLC technique with a good sensitivity (0.9–3.7 µg) and enantioselectivity (e.g., for methionine, the corresponding R_f values for D- and L-enantiomers were 25 and 50, respectively) [124].

Countercurrent chromatography has been adapted for the preparative enantioseparation of some model analytes such as DNB- and DNZ-protected amino acids and 2-aryloxypropionic acids, using rationally designed quinine- and quinidine-derived chiral selectors **66** and **67** bearing 1-adamantylcarbamoyl function for the enhancement of selectivity. Highly lipophilic octadecyl chain was added to suppress the solubility in the mobile phase. The optimized conditions involving a binary solvent system (ammonium acetate buffer/methyl isobutyl ketone or diisopropyl ether) allowed to resolve in a single run up to 300 mg of DNB-leucine with the use of 10 mM selector and 122 ml stationary phase. When the pH zone-refining mode was applied, the amount of racemate could be further increased up to 900 mg [125].

66

67

13.4
Cinchona Alkaloids as Chiral Solvating (Shift) Agents in NMR Spectroscopy

A method for ee determination by NMR spectroscopy uses nonequivalence of shielding of selected atoms in a racemic or enriched sample induced by an appropriate chiral solvating agent (CSA) [126]. In such a case, the originally enantiotopic atoms become distinguishable diastereotopic ones. If at least one pair of signals coming from the diastereomeric complexes is baseline separated, the quantification can be simply achieved by integration of the signal area. The CSA method offers many advantages – it is simple and quick (just acquisition of the NMR spectra), is cheap, and usually does not require special preparation (derivatization) of the sample. Different nuclei can be used as probes. Besides the most popular ^{1}H and ^{13}C spectra, experiments involving ^{19}F and ^{31}P expand the applicability of this method giving often a clearer picture than complex, overlapping proton and carbon spectra.

Ready availability, low cost, and multifunctional character of cinchona alkaloids, as well as their good solubility in many of the deuterated solvents, make them attractive CSA candidates. The early studies in this field were initiated in the mid-1980s by the Salvadori group. It has been shown that quinine discriminated many classes of the racemates, such as binaphthol, alkylaryl (or heteroaryl)carbinols, N-acylphenethylamines, diuretic penflutizide [127], and β-hydroxyesters [128]. Later on, other groups used quinine as a CSA for ee determination of several classes of fluoro- (e.g., **68**) and phosphorganic compounds employing ^{19}F and ^{31}P spectra, respectively. Simple NMR spectrum, not overlapped by the proton or carbon signals of the cinchona residue, is the major advantage of using one of these two nuclei (Figure 13.20). A summary of quinine and quinidine applications as CSAs is given in Table 13.3.

Extensive studies on the optimization of the cinchona-based CSA have been conducted by the Salvadori group. A number of diverse quinine derivatives, including 9-O-acetyl ester, N-benzylquininium chloride, and 9-O-carbamates **69**, were screened [134]. Among these, rigid 9-O-carbamoyl **69** [135] and more flexible 11-O-carbamoyl quinine **70** [136] bearing (S)-configured 1-(1-naphthyl)ethyl group located in the carbamoyl function were identified as efficient CSAs of broader applicability than the parent quinine. The most efficient enantiodiscrimination takes place when DNB- or DNP-derivatized amino acids, amines, alcohols, and acids, including anti-inflammatory drugs ibuprofen and flurbiprofen, are used. This suggests an additional π–π interaction between π-acidic DNB of the selectand and 1-naphthyl group of the selector [135].

Mechanistic studies on the interaction of quinine or quininyl-9-O-carbamates **69** and **70** with selected enantiomer pairs, employing sophisticated NMR experiments and molecular modeling, have been carried out by the Salvadori group. These studies revealed that due to the multifunctional character of the CSA molecules and their specific conformational preferences, each type of analyte induced its own specific interaction pattern [134, 137]. The main driving forces leading to enantiodiscrimination are due to hydrogen bonds between hydroxy group in QN or

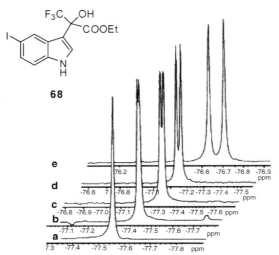

Figure 13.20 ^{19}F NMR resonances of CF$_3$ group in racemic **68** (a) no alkaloid added, (b) CN, (c) CD, (d) QN, (e) QD added (376 MHz, CDCl$_3$, 25 °C) [130].

carbamoyl group in carbamates together with the π–π interaction of the quinoline ring and the aromatic moiety of analyte rather than the interaction with the quinuclidine nitrogen.

At the end of this section, we would like to mention that there are some limitations associated with the use of cinchona alkaloids as CSA. These include the possibility of self-association of alkaloids that has been observed, for example, with dihydroquinidine as early as in 1969 [138]. The rich and complex ^{1}H NMR spectra of cinchona alkaloids can overlap with the sample absorption in the region of interest. Nevertheless, the simplicity of the CSA approach, low price of parent cinchona alkaloids, and numerous cinchona derivatives synthesized so far should stimulate further development of new target-specific applications in this field.

Table 13.3 Quinine (entries 1-2,4,6–8) or quinidine (entries 3,5) as CSAs for ee determination of diverse racemates.

Entry	Compound		Nucleus	$\Delta\delta$ (ppm)[a]	Reference
1	BINOL derivative with R_1, R_2, OH	R_1 = H, OMe; R_2 = COOR, OR	^{1}H	0.02–0.2	[127]
2	Ar–CH(OH)–R	Ar = Ph, 9-anthryl, 2-thienyl, 2-furyl; R = alkyl	^{1}H	About 0.02	[127, 129]
3	Ar–CH(OH)–CF$_3$	Ar = Ph, heteroaryl	^{19}F	0.02–0.07	[130]
4	R_1–CH(OH)–COOR$_2$	R_1 = Ph, cHex; R_2 = Me, Et, t-Bu	^{1}H	n.a.	[128]
5	F$_3$C–C(Ar)(OH)–(CH$_2$)$_n$–COOEt	Ar = Ph, heteroaryl; n = 0, 1	^{19}F	0.01–0.12	[130]
6	R–CH(OH)–P(OEt)$_2$=O	R = Ph, alkyl	^{31}P	0.08–0.21	[131]
7	Ph(R)–CH(OH)–CH(OH)–P(=O)(OEt)$_2$	R = H, Me, halogen	^{31}P	0.04–0.18 (threo) 0.03–0.12 (erythro)	[132]
8	F$_3$C–CH(OH)–(CH$_2$)$_n$–P(=O)(OEt)$_2$	n = 0, 1	^{19}F	n.a.	[133]

[a] $\Delta\delta$ is the difference of chemical shifts of signals due to diastereotopic nuclei.

13.5 Cinchona-Based Sensors, Receptors, and Materials for Separation and Analytics

An emerging area of modern enantioselective analytics is the development of dedicated, target-specific molecular tools that meet the criteria of being robust, high throughput, and specific [105]. Important groups of such tools constitute chiral artificial sensors and receptors as well as polymer-based chiral materials, such as molecularly imprinted polymers (MIPs) [139], monolithic phases, and membranes. Cinchona alkaloid derivatives are apparently well suited to act as chiral scaffold for the above-mentioned analytical applications.

Aromatic diimides **71** and **72** derived from 9-*epi*-9-amino cinchona alkaloids and pyromellitic or 1,4,5,8-naphthalenetetracarboxylic anhydride have been designed by Gawroński and Kacprzak as novel chiral receptors. These triads are present at room temperature as equimolar mixture of *syn* and *anti* conformers, resulting from restricted rotation around the imide C−N bonds. These triads show high affinity toward carboxylic acids. The *syn* conformer binds preferentially 1,2-dicarboxylates in a 1:1 molecular ratio whereas the *anti* conformer is selective toward monocarboxylates forming a complex in a ratio 1:2 (triad:acid). The response is sensitive and could be observed by either ^{1}H NMR or by CD spectroscopy even when equimolar amounts of the acids are present in the solution. The competition experiments have shown a higher selectivity of triads toward 1,2-dicarboxylates [140].

Cinchona diimides have also been used to develop a simple colorimetric indicator displacement assay for the determination and chiral discrimination of tartaric acid. Thus, conjugation of **72** with bromophenol blue **73** resulted in the formation of a dark blue complex (λ_{max} at 597 nm) that after treatment by an appropriate acid released the protonated form of the indicator with a well-separated absorption maximum at 426 nm. This assay allowed to determine (*R,R*)-tartaric acid with the detection limit as low as 0.02 mg/ml and efficiently discriminated the enantiomers of tartaric acids,

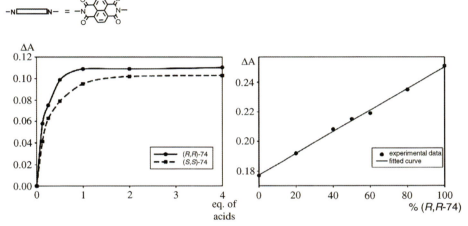

Figure 13.21 Indicator displacement assay for tartrate sensing (upper panel). Displacement isotherms at 597 nm for the addition of (R,R)- and (S,S)-tartaric acid **74** to the complex of **72** and bromophenol blue (**73**) (c**72** and **73** = 1 × 10^{-4} mol/dm^3) (left bottom panel). Change of the absorbance (ΔA) of **72–73** complex at 599 nm as a function of the enantiomeric ratio of tartaric acid **74** added (c = 1.17 × 10^{-4} mol/dm^3) (right bottom panel). All measurements in methanol solution [141].

giving the association constant of 8300 and 5500 mol^{-1} for (R,R)- and (S,S)-enantiomers, respectively (Figure 13.21) [141].

The fluorescence of quinine and the possibility of its quenching or modulation in the presence of external molecules could provide a method for sensing and assaying these molecules. For example, diastereomeric complexes of quinine and quinidine with (+)-10-camphorsulfonic acid can be discriminated on the basis of their phase modulation-resolved fluorescence (different fluorescence lifetimes for QN and QD). Thus fluorescence lifetimes of 21.79 and 22.89 ns for QN and QD complex, respectively, have been measured, allowing a quantitative determination of QN and QD with a detection limit of 1.8 and 0.97 µM, respectively [142]. Similarly, room-temperature phosphorescence lifetimes were also shown to differ for diastereoisomeric complexes of QN and QD [143].

Molecular imprinting technology offers a unique opportunity to provide chiral stationary phases or sensing devices with predefined chiral recognition properties by employing the enantiomers of a molecule of interest as binding site-forming templates. The advantages of MIPs, such as the ease of preparation, low-cost production, and a broad possibility of shaping in various self-supporting formats, make them attractive materials for a broad range of chiral recognition applications [139].

A simple strategy for preserving and enhancing the chiral recognition capacity of polymer-embedded chiral selectors has been demonstrated by Lindner and Maier. The novelty of this approach lies in temporary blockage of the receptor binding site with tightly binding analytes during the polymerization process (Figure 13.22).

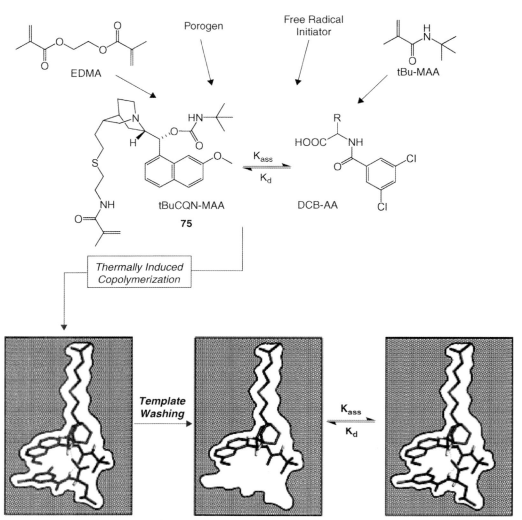

Figure 13.22 Schematic representation of the polymerization protocol employed for the preparation of the analyte-templated polymers [144].

13.5 Cinchona-Based Sensors, Receptors, and Materials for Separation and Analytics

The copolymerization of the quinine *tert*-butylcarbamate selector monomer, *t*-BuCQN-MAA (**75**), with chiral 3,5-dichlorobenzoyl amino acids allowed to control to a certain extent the binding characteristics of the resultant polymeric chiral stationary phases. The comparison of four different templated polymeric phases with analogous silica gel-based CSP and similarly prepared but nontemplated polymers has shown that all templated polymers retained, and in one case even exceeded, the selectivity of silica-based CSP, whereas the nontemplated polymers did neither retained nor recognized the analytes at all. Further studies indicated the crucial role of high affinity of selector and template for the successful shaping of the recognition materials. The analysis of adsorption behavior revealed that the enhancement of selectivity arose from the higher population of template-specific binding sites and reduction of the binding affinity of nontemplating enantiomer [144].

An attractive approach to develop faster and high-throughput HPLC and CE methods is to use the monolithic materials as chromatographic packings. Monolithic columns are composed of a continuous bed of inorganic or organic (polymeric) matrix with through-pores allowing mass transfer [145]. The monolithic materials can also be used in an enantioselective fashion when they are prepared or modified (usually *in situ*) with an appropriate chiral selector (Figure 13.23). This can be achieved by direct copolymerization of a mixture of monomers including the chiral monomer or by generation of reactive groups within the monolite pores followed by *in situ* covalent immobilization of the desired chiral selector. Both strategies have been successfully used for the preparation of chiral monolithic materials, using QN and QD derivatives **76–78**, structurally related to those previously employed in the classical HPLC and CE formats.

76 (from QD) **77** (from QN)

78 (from QN)

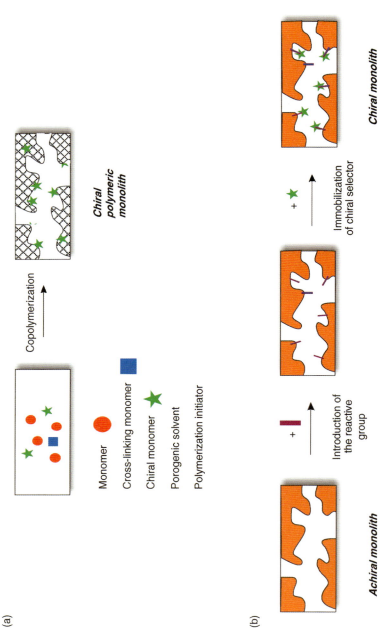

Figure 13.23 Schematic representation of chiral monoliths preparation: (a) direct preparation of chiral monolith; (b) indirect preparation involving immobilization of chiral selector within the preformed monolith.

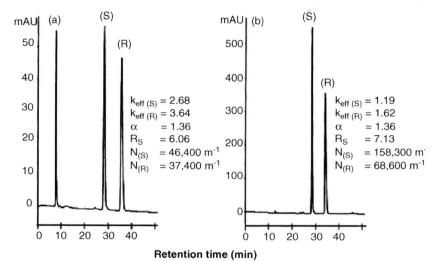

Figure 13.24 Separation of racemic DNZ-leucine on polymeric monolithic material. Conditions: polymerization mixture, chiral monomer **76** 8 wt%, 2-hydroxyethyl methacrylate 24 wt%; ethylene dimethacrylate 8 wt%, 1-dodecanol 45 wt%, and cyclohexanol 15 wt%; UV-initiated polymerization for 16 h at room temperature; pore diameter, 1265 nm' capillary column, 335 mm (250-mm active length) × 0.1 mm i.d.; EOF marker, acetone; mobile phase, 0.4 mol/l acetic acid and 4 mmol/l triethylamine in 80:20 mixtures of acetonitrile/methanol (a) and acetonitrile/water (b); separation temperature, 50 °C; voltage, −15 kV [146].

One-step method for the preparation of highly enantioselective monolithic columns for CEC has been developed by Frechet et al. The chiral polymer bed of defined pore distribution and chiral ligand concentration has been synthesized within the confines of untreated fused silica capillaries using a mixture of O-[2-(methacryloyloxy)ethylcarbamoyl]-10,11-dihydroquinidine **76**, ethylene dimethacrylate (EDMA), and glycidyl methacrylate or 2-hydroxyethyl methacrylate (HEMA) in the mixture of cyclohexanol and 1-dodecanol as porogenic solvents. Under optimized synthetic and chromatographic conditions, these materials with the desired characteristics were demonstrated to efficiently separate a model racemic DNZ-Leu, Figure 13.24 [146].

Further continuation of this research led to the development of more beneficial monolithic materials composed of 9-O-(*tert*-butylcarbamoyl)-11-[2-(methacryloyloxyethylthio]-10,11-dihydroquinine **77** as a chiral monomer. The copolymerization of **77** with EDMA and HEMA in the presence of cyclohexanol and 1-dodecanol as porogens gave a new poly(1-co-HEMA-co-EDMA) monolithic copolymer showing enhanced enantioselectivities and reverse elution order due to the change of alkaloid configuration, as well as faster separation time compared to the poly(4-co-HEMA-co-EDMA) analogue. Fluorescence-labeled 9-fluorenylmethoxycarbonyl, dansyl (DNS), 7-dimethylaminosulfonyl-1,3,2-benzoxadiazol-4-yl

(DBD), and carbazole-9-carbonyl (CC) amino acids have been separated with resolution values in the range of 2–4 and efficiencies between 100 000 and 200 000 plates/m [146].

Another strategy for the preparation of chiral monolithic capillary columns for CEC has been developed by Lindner and coworkers. In the first approach, the pendent 2,3-epoxypropyl groups of poly(glycidyl methacrylate-co-EDMA) of preformed monoliths were transformed into the corresponding 3-mercapto-2-hydroxypropyl residues. The reactive thiol groups were then used for radical immobilization of t-BuCQN as chiral selector [147]. The monolithic silica based column for HPLC enantioseparation has been prepared by the same group. In this case, however, the commercially available silica monolithic material was *in situ* modified by 3-mercaptopropyltrimethoxysilane, followed by known radical addition of t-BuCQN. Comparison of this packing with particulate 5 μm column containing the same chiral selector revealed that monolithic columns exhibited similar selectivity but significantly higher efficiency (theoretical plates number), especially in polar organic mode, and reduced analysis time [148].

An interesting application of the supported liquid membrane containing 9-O-(1-adamantyl)-carbamoylquinine or quinidine bearing octadecyl chain as chiral carriers in the separation of enantiomeric N-protected amino acid derivatives has been reported by Chmiel *et al.* The prototypic preparative plant device consisted of two hollow fiber membrane modules with 250 individual polysulfone hollow fibers with a total membrane surface of $0.1\,m^2$. The liquid membrane was installed in the pores of the membrane and contained chiral carriers dissolved in 1-decanole/pentadecane. The separation process was carried out in a continuous mode and allowed the highly enantioselective separation (selectivity 2–4) of N-protected amino acids with the purity of 99% on a gram scale. For example, racemic DNB-leucine (**23**) after five separation steps gave 99% pure enantiomers with a transmembrane flux of more than $20\,mmol/(m^2\,h)$ [151]. A less successful example of enantioselective separation of mandelic acid and phenylglycine using a membrane modified with cinchonidine as a chiral carrier has also been published [152].

List of Less Common Abbreviations

α	chromatographic selectivity (see Appendix 13.A)
t-BuCQN	9-O-*tert*-butylcarbamoylquinine
t-BuCQD	9-O-*tert*-butylcarbamyolquinidine
CD	cinchonidine
CE	capillary electrophoresis
CEC	capillary electrochromatography
CN	cinchonine
CSA	chiral solvating agent
CSP	chiral stationary phase
CPD	cupreidine

CPN	cupreine
DIPPCQN	2,6-diisopropylphenylcarbamoylquinine
DNB	3,5-dinitrobenzoyl
DNP	3,5-dinitrophenyl
DNZ	3,5-dinitrobenzyloxycarbonyl
ee	enantiomeric excess
EDMA	ethylene dimethacrylate
HEMA	2-hydroxyethylmethacrylate
HPLC	high-performance liquid chromatography
LC	liquid chromatography
N	efficiency (number of theoretical plates)
QN	quinine
QD	quinidine
R_s	chromatographic resolution (see Appendix 13.A)
RP	reversed-phase (chromatography)
SA	selectand (discriminated enantiomer)
SO	selector (chiral discriminating agent)
TrCQN	9-O-Tritylcarbamoylquimine

Appendix 13.A

- **Efficiency (or number of theoretical plates) (N)**: A measure of peak band spreading determined by various methods, some of which are sensitive to peak asymmetry. In practice, a higher N gives more efficient chromatographic conditions.

- **Resolution (R_s)**: Ability of a column to separate chromatographic peaks. Resolution can be improved by increasing column length, decreasing particle size, or changing temperature, eluent, or stationary phase. It can also be expressed in terms of the separation of the apex of two peaks divided by the tangential width average of the peaks.

- **Selectivity (or separation factor) (α)**: A thermodynamic factor that is a measure of relative retention of two substances, in the case of enantioselective separation enantiomers, fixed by a certain stationary phase and mobile phase composition. Higher selectivity value indicates highly discriminative separation conditions due to CSP and other chromatographic parameters, such as mobile phase composition, pH, temperature, etc.).

$$a = \frac{k_2}{k_1}$$

k_1 and k_2 are the respective capacity (or retention) factors.

where $k = \frac{t_R - t_0}{t_0}$

t_R denotes retention time and t_0 breakthrough time (or void time, the time for elution of a nonretained marker).

References

1. Garfield, S. (2000) *Mauve*, Faber & Faber, London, p. 224.
2. Excellent reviews about quinine total synthesis with historical background: (a) Kaufmann, T.S. and Ruveda, E.A. (2005) *Angew. Chem. Int. Ed.*, **44**, 854–885; (b) Nicolaou, K.C. and Sorensen, E.J. (1996) in *Classics in Total Synthesis: Targets, Strategies, Methods*, Wiley-VCH Verlag GmbH, Weinheim; Chapter 15.
3. (a) Pasteur, L. (1853) *C. R. Acad. Sci.*, **37**, 162; (b) Pasteur, L. (1853) *Liebigs Ann. Chem.*, **88**, 209.
4. (a) Bredig, G. and Fiske, P.S. (1912) *Biochem. Z.*, **46**, 7; (b) Bredig, G. and Minaeff, M. (1932) *Biochem. Z.*, **249**, 241.
5. Wynberg, H. (1986) *Top. Stereochem.*, **16**, 87–129.
6. Yoon, T.P. and Jacobsen, E.N. (2003) *Nature*, **299**, 1691–1693.
7. This monograph and (a) Kacprzak, K. and Gawroński, J., (2001) *Synthesis*, 961–999; (b) Tian, S.-K., Chen, Y., Hang, J., Tang, L., Mcdaid, P., and Deng, L. (2004) *Acc. Chem. Res.*, **8**, 518.
8. Kiss, V., Egri, G., Balint, J., and Fogassy, E. (2006) *Chirality*, **18**, 116–120.
9. Collins, A.N., Sheldrake, G.N., and Crosby, J.(eds) (1992/1997) *Chirality in Industry*, vols **1** and **2**, John Wiley & Sons, Ltd, Chichester.
10. (a) Newman, P. (1981) *Optical Resolution Procedures for Chemical Compounds: Acids*, vol. **2**, Optical Resolution Information Center; Manhattan College, Riverdale, New York, pp. 7–22; (b) Jacques, J., Collet, A., and Wilen, S.H. (1981) *Enantiomers, Racemates and Resolutions*, John Wiley & Sons, Inc., New York, pp. 254–257; (c) Sheldon, R.A. (1993) *Chirotechnology*, Marcel Dekker, New York; Chapter 6; (d) Kozma, D. (2002) *CRC Handbook of Optical Resolutions via Diastereomeric Salts*, CRC Press, Boca Raton.
11. (a) Lindner, W. and Francotte, E.(eds) (2006) *Chirality in Drug Research*, Wiley-VCH Verlag GmbH, Weinheim; (b) Subramanian, G.(ed.) (2001) *Chiral Separation Techniques: A Practical Approach*, Wiley-VCH Verlag GmbH, Weinheim.
12. Dagni, R. (1995) *Chem. Eng. News*, **73**, 33; Stinson, S.C. (1995) *Chem. Eng. News*, **73**, 44.
13. (a) Jacques, J. and Fouquey, C. (1988) *Org. Synth.*, **67**, 1; (b) Jacques, J. and Fouquey, C. (1971) *Tetrahedron Lett.*, **12**, 4617; (c) Truesdale, L.K. (1989) *Org. Synth.*, **67**, 13; (d) Mezlova, M., Petrickova, H., Malon, P., Kozmik, V., and Svoboda, J. (2003) *Collect. Czech. Chem. Commun.*, **68**, 1020–1038; (e) Bao, J., Wulff, W.D., Dominy, J.B., Fumo, M.J., Grant, E.G., Rob, A.C., Whitcomb, M.C., Yeung, S.-M., Ostrander, R.L., and Rheingold, A. (1996) *J. Am. Chem. Soc.*, **118**, 3392–3405.
14. Nyhlen, J., Eriksson, L., and Backvall, J.-E. (2008) *Chirality*, **20**, 47–50.
15. Hattori, T., Harada, N., Oi, S., Abe, H., and Miyano, S. (1995) *Tetrahedron: Asymmetry*, **6**, 1043–1046.
16. Stankievic, M. and Pietrusiewicz, K.M. (2007) *Tetrahedron: Asymmetry*, **18**, 552.
17. Białczewski, P., Szadkowiak, A., Białas, T., Wieczorek, W.M., and Balińska, A. (2006) *Tetrahedron: Asymmetry*, **17**, 1209–1216.
18. Varga, J., Szabo, D., Sar, C.P., and Kapovits, I. (2001) *Tetrahedron: Asymmetry*, **12**, 745–753.
19. Mauger, C., Masson, S., Vazeux, M., Saint-Clair, J.-F., Midura, W.H., Drabowicz, J., and Mikolajczyk, M. (2001) *Tetrahedron: Asymmetry*, **12**, 167–174.
20. ChiroSolv® (from Aldrich), see www.aldrich.com.
21. Guillen, F., Rivard, M., Toffano, M., Legros, J.-Y., Daran, J.-C., and Fiaud, J.-C. (2002) *Tetrahedron*, **58**, 5895–5904.
22. Atkinson, R.S. and Judkins, B.D. (1981) *J. Chem. Soc., Perkin Trans. I*, 1309–1311.
23. Manimaran, T. and Potter, A. (1992) WO 018455.

24 Ekwuribe, N. N. and James, K. D. (2003) US 6,593,492.
25 Wakita, H., Yoshiwara, H., Kitano, Y., Nishiyama, H., and Nagase, H. (2000) *Tetrahedron: Asymmetry*, **11**, 2981–2989.
26 Bunegar, M.J., Dyer, U.C., Evans, G.R., Hewitt, R.P., Jones, S.W., Henderson, N., Richards, C.J., Sivaprasad, S., Skead, B.M., Stark, M.A., and Teale, E. (1999) *Org. Proc. Res. Dev.*, **3**, 442–450.
27 Friedmann, R.C. and Quallich, G.J. (1990) EP 0 411 813 A1.
28 Evans, R. M. and Skead, G. R. (1997) WO 029081.
29 Kress, T.J., Robey, R.L., Wepsiec, J.P., Alt, Ch.A., and Rhodes, G.A. (2002) WO 032878.
30 Yin, J., Yang, X.B., Chen, Z.X., and Zhang, Y.H. (2005) *Chin. Chem. Lett.*, **16**, 1448–1450.
31 Krogsgaard-Larsen, P., Nielsen, L., Falch, E., and Curtis, D.R. (1985) *J. Med. Chem.*, **28**, 1612–1617.
32 (a) Shiraiwa, T., Suzuki, M., Sakai, Y., Nagasawa, H., Takatani, K., Noshi, D., and Yamanashi, K. (2002) *Chem. Pharm. Bull.*, **50**, 1362–1366; (b) Olma, A. (1996) *Pol. J. Chem.*, **70**, 1442–1447.
33 Kirihata, M., Sakamoto, A., Ichimoto, I., Ueda, H., and Honma, M. (1990) *Agric. Biol. Chem.*, **54**, 1845.
34 Knorr, H., Schlegel, G., and Stark, H. (1995) WO 95/23805.
35 Madsen, U., Frydenvang, K., Ebert, B., Johansen, T. N., Brehm, L., and Krogsgaard-Larsen, P. (1996) *J. Med. Chem.*, **39**, 183–190.
36 Mutti, S., Daubié, C., Decalogne, F., Fournier, R., Montuori, O., and Rossi, P. (1996) *Synth. Commun.*, **26**, 2349–2354.
37 Tanaka, K., Okada, T., and Toda, F. (1993) *Angew. Chem. Int. Ed.*, **32**, 1147–1148.
38 Brunel, J.M. (2005) *Chem. Rev.*, **105**, 857–897.
39 (a) Cai, D., Hughes, D.L., Verhoeven, T.R., and Reider, P.J. (1999) *Org. Synth.*, **76**, 1–5; (b) Cai, D., Hughes, D.L., Verhoeven, T.R., and Reider, P.J. (1995) *Tetrahedron Lett.*, **36**, 7991–7994.
40 Toda, F., Tanaka, K., Stein, Z., and Goldberg, I. (1994) *J. Org. Chem.*, **59**, 5748–5751.
41 Ding, K., Wang, Y., Yun, H., Liu, J., Wu, Y., Terada, M., Okubo, Y., and Mikami, K. (1999) *Chem. Eur. J.*, **5**, 1734–1737.
42 Sweetman, B.A., Muller-Bunz, H., and Guiry, P.J. (2005) *Tetrahedron Lett.*, **46**, 4643–4646.
43 Wang, Y., Sun, J., and Ding, K. (2000) *Tetrahedron*, **56**, 4447–4451.
44 Lammerhofer, M., Maier, N. M., and Lindner, W. (1997) WO 97/46557.
45 Bicker, W., Chiorescu, I., Arion, V.B., Lammerhofer, M., and Lindner, W. (2008) *Tetrahedron: Asymmetry*, **19**, 97–110.
46 Russell, J. W. (1989) US 4,831,147.
47 Kellner, K.-H., Blasch, A., Chmiel, H., Lammerhofer, M., and Lindner, W. (1997) *Chirality*, **9**, 268–273.
48 Schuur, B., Floure, J., Hallett, A.J., Winkelman, J.G.M., deVries, J.G., and Heeres, H.J. (2008) *Org. Proc. Res. Dev.*, **12**, 950–955.
49 Tobler, E., Lammerhofer, M., Oberleitner, W.R., Maier, N.M., and Lindner, W. (2000) *Chromatographia*, **51**, 65–70.
50 Cox, G.B.(ed.) (2005) *Preparative Enantioselective Chromatography*, Blackwell Publishing.
51 Pettersson, C. and No, K. (1983) *J. Chromatogr. A*, **282**, 671–684.
52 Grubhofer, N. and Schleith, L. (1953) *Naturwissenschaften*, **40**, 508.
53 Rogozhin, S.V. and Davankov, V.A. (1968) *Russ. Chem. Rev.*, **37**, 565–575.
54 Rosini, C., Bertucci, C., Pini, D., Altemura, P., and Salvadori, P. (1985) *Tetrahedron Lett.*, **26**, 3361–3364.
55 Rosini, C., Altemura, P., Pini, D., Bertucci, C., Zullino, G., and Salvadori, P. (1985) *J. Chromatogr. A*, **348**, 79–87.
56 Salvadori, P., Rosini, C., Pini, D., Bertucci, C., Altemura, P., Uccello-Barretta, G., and Raffaelli, A. (1987) *Tetrahedron*, **43**, 4969–4978.
57 Rosini, C., Bertucci, C., Pini, D., Altemura, P., and Salvadori, P. (1987) *Chromatographia*, **24**, 671–676.

58 Salvadori, P., Pini, D., Rosini, C., Bertucci, C., and Uccello-Barretta, G. (1992) *Chirality*, **4**, 43–49.

59 Nesterenko, P.N., Krotov, V.V., and Staroverov, S.M. (1994) *J. Chromatogr. A*, **667**, 19–28.

60 Mandl, A., Nicoletti, L., Lammerhofer, M., and Lindner, W. (1999) *J. Chromatogr. A*, **858**, 1–11.

61 Lämmerhofer, M. and Lindner, W. (1996) *J. Chromatogr. A*, **741**, 33–48.

62 Maier, N.M., Schefzick, S., Lombardo, G.M., Feliz, M., Rissanen, K., Lindner, W., and Lipkowitz, K.B. (2002) *J. Am. Chem. Soc.*, **124**, 8611–8629.

63 www.chiraltech.com, Chiralpak QN-AX and QD-AX application notes.

64 Oberleitner, W.R., Maier, N.M., and Lindner, W. (2002) *J. Chromatogr. A*, **960**, 97–108.

65 Piette, V., Lammerhofer, M., Bischoff, K., and Lindner, W. (1997) *Chirality*, **9**, 157–161.

66 Lammerhofer, M., Maier, N.M., and Lindner, W. (1998) *Am. Lab.*, **30**, 71–78.

67 Lee, M., Jang, M.D., Lee, W., and Park, J.H. (2008) *Bull. Korean Chem. Soc.*, **29**, 257–260.

68 Xiong, X., Baeyens, W.R.G., Aboul-Enein, H.Y., Delanghe, J.R., Tu, T., and Ouyang, J. (2007) *Talanta*, **71**, 573–581.

69 Welsch, T., Schmidtkunz, C., Müller, B., Meier, F., Chlup, M., Köhne, A., Lämmerhofer, M., and Lindner, W. (2007) *Anal. Bioanal. Chem.*, **388**, 1717–1724.

70 Gyimesi-Forras, K., Leitner, A., Akasaka, K., and Lindner, W. (2005) *J. Chromatogr. A*, **1083**, 80–88.

71 Torok, R., Berkecz, R., and Péter, A. (2006) *J. Chromatogr. A*, **1120**, 61–68.

72 Sardella, R., Lammerhofer, M., Natalini, B., and Lindner, W. (2008) *Chirality*, **20**, 571–576.

73 Torok, R., Berkecz, R., and Péter, A. (2006) *J. Sep. Sci.*, **29**, 2523–2532.

74 Peter, A. (2002) *J. Chromatogr. A*, **955**, 141–150.

75 Piette, V., Lammerhofer, M., Lindner, W., and Crommen, J. (2003) *J. Chromatogr. A*, **987**, 421–427.

76 Peter, A., Vekes, E., Arki, A., Tourvé, D., and Lindner, W. (2003) *J. Sep. Sci.*, **26**, 1125–1132.

77 Peter, A., Torok, G., Tóth, G., and Lindner, W. (2000) *J. High Resolut. Chromatogr.*, **23**, 628–636.

78 Gika, H., Lammerhofer, M., Papadoyannis, I., and Lindner, W. (2004) *J. Chromatogr. B*, **800**, 193–201.

79 Sardella, R., Lammerhofer, M., Natalini, B., and Lindner, W. (2008) *J. Sep. Sci.*, **31**, 1702–1711.

80 Hamase, K., Morikawa, A., Ohgusu, T., Lindner, W., and Zaitsu, K. (2007) *J. Chromatogr. A*, **1143**, 105–111.

81 Tojo, Y., Hamase, K., Nakata, M., Morikawa, A., Mita, M., Ashida, Y., Lindner, M., and Zaitsu, K., (2008) *J. Chromatogr. B*, **875**, 174–179.

82 Czerwenka, C. and Lindner, W. (2005) *Anal. Bioanal. Chem.*, **382**, 599–638.

83 Czerwenka, C., Zhang, M.M., Kahlig, H., Maier, N.M., Lipkowitz, K.B., and Lindner, W. (2003) *J. Org. Chem.*, **68**, 8315–8327.

84 Czerwenka, C., Lammerhofer, M., Maier, N.M., Rissanen, K., and Lindner, W. (2002) *Anal. Chem.*, **74**, 5658–5666.

85 Czerwenka, C., Maier, N.M., and Lindner, W. (2004) *J. Chromatogr. A*, **1038**, 85–95.

86 Czerwenka, C., Polásková, P., and Lindner, W. (2005) *J. Chromatogr. A*, **1093**, 81–88.

87 Gyimesi-Forrás, K., Akasaka, K., Lammerhofer, M., Maier, N.M., Fujita, T., Watanabe, M., Harada, N., and Lindner, W. (2005) *Chirality*, **17** (Suppl.), S134–S142.

88 Bicker, W., Lammerhofer, M., and Lindner, W. (2004) *J. Chromatogr. A*, **1035**, 37–46.

89 Lammerhofer, M., Imming, P., and Lindner, W. (2004) *Chromatographia*, **60** (Suppl.), S13–S17.

90 Gyimesi-Forrás, K., Kokosi, J., Szász, G., Gergely, A., and Lindner, W. (2004) *J. Chromatogr. A*, **1047**, 59–67.

91 Lammerhofer, M., Gyllenhaal, O., and Lindner, W. (2004) *J. Pharm. Biomed. Anal.*, **35**, 259–266.

92 Wickremsinhe, E.R., Tian, Y., Ruterbories, K.J., Verburg, E.M., Weerakkody, G.J., Kurihara, A., and Farid, N.A. (2007) *Drug Metab. Dispos.*, **35**, 917–921.

93 Tobler, E., Lammerhofer, M., Mancini, G., and Lindner, W. (2001) *Chirality*, **13**, 641–647.

94 (a) Mucha, A., Lammerhofer, M., Lindner, W., Pawełczak, M., and Kafarski, P. (2008) *Bioorg. Med. Chem. Lett.*, **18**, 1550–1554; (b) Lammerhofer, M., Hebenstreit, D., Gavioli, E., Lindner, W., Mucha, A., Kafarski, P., and Wieczorek, P. (2003) *Tetrahedron: Asymmetry*, **14**, 2557–2565.

95 Zarbl, E., Lammerhofer, M., Hammerschmidt, F., Wuggenig, F., Hanbauer, M., Maier, N.M., Sajovic, L., and Lindner, W. (2000) *Anal. Chim. Acta*, **404**, 169–177.

96 Hellriegel, C., Skogsberg, U., Albert, K., Lammerhofer, M., Maier, N.M., and Lindner, W. (2004) *J. Am. Chem. Soc.*, **126**, 3809–3816.

97 (a) Julínek, O., Urbanová, M., and Lindner, W. (2009) *Anal. Bioanal. Chem.*, **393**, 303–312; (b) Lan, J., Maier, N.M., Lindner, W., and Vesnaver, G. (2001) *J. Phys. Chem. B*, **105**, 1670–1678.

98 Schefzick, S., Lindner, W., Lipkowitz, K.B., and Jalaie, M. (2000) *Chirality*, **12**, 7–15.

99 Schefzick, S., Lammerhofer, M., Lindner, W., Lipkowitz, K.B., and Jalaie, M. (2000) *Chirality*, **12**, 742–750.

100 Lammerhofer, M., Franco, P., and Lindner, W. (2006) *J. Sep. Sci.*, **29**, 1486–1496.

101 Franco, P., Lammerhofer, M., Klaus, P.M., and Lindner, W. (2000) *Chromatographia*, **51**, 139–146.

102 Kacprzak, K.M., Maier, N.M., and Lindner, W. (2006) *Tetrahedron Lett.*, **47**, 8721–8726.

103 Krawinkler, K.H., Gavioli, E., Maier, N.M., and Lindner, W. (2003) *Chromatographia*, **58**, 555–564.

104 (a) Park, J.H., Lee, J.W., Kwon, S.H., Cha, J.S., Carr, P.W., and McNeff, C.V. (2004) *J. Chromatogr. A*, **1050**, 151–157; (b) Park, J.H., Lee, J.W., Song, Y.T., Ra, C.S., Cha, J.S., Ryoo, J.J., Lee, W., Kim, I.W., and Jang, M.D. (2004) *J. Sep. Sci.*, **27**, 977–982.

105 Maier, N.M., Franco, P., and Lindner, W. (2001) *J. Chromatogr. A*, **906**, 3–33.

106 Lindner, W. and Maier, N. (2006) International Symposium on HPLC, San Francisco.

107 Franco, P., Lammerhofer, M., Klaus, P.M., and Lindner, W. (2000) *J. Chromatogr. A*, **869**, 111–127.

108 Gavioli, B., Maier, N.M., Minguillón, C., and Lindner, W. (2004) *Anal. Chem.*, **76**, 5837–5848.

109 Krawinkler, K.H., Maier, N.M., Ungaro, R., Sansone, F., Casnati, A., and Lindner, W. (2003) *Chirality*, **15** (Suppl.), S17–S29.

110 Krawinkler, K.H., Maier, N.M., Sajovic, E., and Lindner, W. (2004) *J. Chromatogr. A*, **1053**, 119–131.

111 Lammerhofer, M., Svec, F., Frechet, J.M.J., and Lindner, W. (2000) *Trends Anal. Chem.*, **19**, 676–698.

112 Piette, V., Fillet, M., Lindner, W., and Crommen, J. (2000) *J. Chromatogr. A*, **875**, 353–360.

113 Bartak, P., Bednar, P., Kubacek, L., Lammerhofer, M., Lindner, W., and Stránský, Z. (2004) *Anal. Chim. Acta*, **506**, 105–113.

114 Stalcup, A.M. and Gahm, K.H. (1996) *J. Microcolumn Sep.*, **8**, 145–150.

115 Piette, V., Lammerhofer, M., Lindner, W., and Crommen, J. (1999) *Chirality*, **11**, 622–630.

116 (a) Piette, V., Lindner, W., and Crommen, J. (2002) *J. Chromatogr. A*, **948**, 295–302; (b) Piette, V., Lindner, W., and Crommen, J. (2000) *J. Chromatogr. A*, **894**, 63–71.

117 Lammerhofer, M., Zarbl, E., and Lindner, W. (2000) *J. Chromatogr. A*, **892**, 509–521.

118 Czervenka, C., Lammerhofer, M., and Lindner, W. (2002) *Electrophoresis*, **23**, 1887–1899.

119 Lammerhofer, M., Zarbl, E., Lindner, W., Simov, B.P., and Hammerschmidt, F. (2001) *Electrophoresis*, **22**, 1182–1187.

120 (a) Lammerhofer, M. and Lindner, W. (2004) *Methods Mol. Biol.*, **243**, 323–342; (b) Lammerhofer, M., Lindner, W. (2008) *Adv. Chromatogr.*, **46**, 1–107.

121 Preinerstorfer, B., Lubda, D., Mucha, A., Kafarski, P., Lindner, W., and Lammerhofer, M. (2006) *Electrophoresis*, **27**, 4312–4320.

122 (a) Lammerhofer, M., Tobler, E., and Lindner, W. (2000) *J. Chromatogr. A*, **887**, 421–437; (b) Lammerhofer, M. and Lindner, W. (1998) *J. Chromatogr. A*, **829**, 115–125.

123 Lammerhofer, M. and Lindner, W. (1999) *J. Chromatogr. A*, **839**, 167–182.

124 Bhushan, R. and Arora, M. (2001) *Biomed. Chromatogr.*, **15**, 433–436.

125 Franco, P., Blanc, J., Oberleitner, W.R., Maier, N.M., Lindner, W., and Minguillón, C. (2002) *Anal. Chem.*, **74**, 4175–4183.

126 (a) Uccello-Barretta, G., Balzano, F., and Salvadori, P. (2006) *Curr. Pharm. Des.*, **12**, 4023–4045; (b) Parker, D. (1991) *Chem. Rev.*, **91**, 1441–1457.

127 Rosini, C., Uccello-Barretta, G., Pini, D., Abete, C., and Salvadori, P. (1988) *J. Org. Chem.*, **53**, 4579–4581.

128 Uccello-Barretta, G., Pini, D., Mastantuono, A., and Salvadori, P. (1995) *Tetrahedron: Asymmetry*, **6**, 1965–1972.

129 Van Oeveren, A., Menge, W., and Feringa, B.L. (1989) *Tetrahedron Lett.*, **30**, 6427–6430.

130 Abid, M. and Torok, B. (2005) *Tetrahedron: Asymmetry*, **16**, 1547–1555.

131 Zymanczyk-Duda, E., Skwarczynski, M., Lejczak, B., and Kafarski, P. (1996) *Tetrahedron: Asymmetry*, **7**, 1277–1280.

132 Maly, A., Lejczak, B., and Kafarski, P. (2003) *Tetrahedron: Asymmetry*, **14**, 1019–1024.

133 Zhang, Y., Li, J.-F., and Yuan, C.-Y. (2003) *Tetrahedron*, **59**, 473–479.

134 Uccello-Barretta, G., Balzano, F., Quintavalli, C., and Salvadori, P. (2000) *J. Org. Chem.*, **65**, 3596–3602.

135 Uccello-Barretta, G., Bardoni, S., Balzano, F., and Salvadori, P. (2001) *Tetrahedron: Asymmetry*, **12**, 2019.

136 Uccello-Barretta, G., Mirabella, F., Balzano, F., and Salvadori, P. (2003) *Tetrahedron: Asymmetry*, **14**, 1511–1516.

137 (a) Uccello-Barretta, G., Balzano, F., and Salvadori, P. (2005) *Chirality*, **17** (Suppl.), S243–S248; (b) Uccello-Barretta, G., Balzano, F., Bardoni, S., Vanni, L., Giurato, L., and Guccione, S. (2008) *Tetrahedron: Asymmetry*, **19**, 1084–1093.

138 (a) Williams, T., Pitcher, R.G., Bommer, P., Gutzwiller, J., and Uskokovic, M.J. (1969) *J. Am. Chem. Soc.*, **91**, 1871–1872; (b) Uccello-Barretta, G., Bari, L.D., and Salvadori, P. (1992) *Magn. Reson. Chem.*, **30**, 1054–1063; (c) Casabianca, L.B. and De Dios, A.C. (2004) *J. Phys. Chem. A*, **108**, 8505–8513.

139 Maier, N.M. and Lindner, W. (2007) *Anal. Bioanal. Chem.*, **389**, 377–397.

140 Kacprzak, K. and Gawroński, J. (2003) *Chem. Commun.*, 1532–1533.

141 Kacprzak, K., Grajewski, J., and Gawroński, J. (2006) *Tetrahedron: Asymmetry*, **17**, 1332–1336.

142 (a) Garcia Sanchez, F., Navas Diaz, A., and Torijas, C. (1999) *Biomed. Chromatogr.*, **13**, 179–180; (b) Navas Diaz, A., Garcia Sanchez, F., and Torijas, M.C. (1999) *Anal. Chim. Acta*, **381**, 11–16.

143 Wei, Y., Chan, W.-H., Lee, A.W.M., and Huie, C.W. (2004) *Chem. Commun.*, 288–289.

144 Gavioli, E., Maier, N.M., Haupt, K., Mosbach, K., and Lindner, W. (2005) *Anal. Chem.*, **77**, 5009–5018.

145 Guiochon, G. (2007) *J. Chromatogr. A.*, **1168**, 101–168.

146 (a) Lammerhofer, M., Peters, E.C., Yu, C., Svec, F., Fréchet, J.M.J., and Lindner, W. (2000) *Anal. Chem.*, **72**, 4614–4622; (b) Lammerhofer, M., Svec, F., Fréchet, J.M.J., and Lindner, W. (2000) *Anal. Chem.*, **72**, 4623–4628.

147 Lammerhofer, M., Tobler, E., Zarbl, E., Lindner, W., Svec, F., and Fréchet, J.M.J. (2003) *Electrophoresis*, **24**, 2986–2999.

148 Preinerstorfer, B., Bicker, W., Lindner, W., and Lammerhofer, M. (2004) *J. Chromatogr. A*, **1044**, 187–199.
149 Lubda, D. and Lindner, W. (2004) *J. Chromatogr. A*, **1036**, 135–143.
150 Maier, N.M., Nicoletti, L., Lammerhofer, M., and Lindner, W. (1999) *Chirality*, **11**, 522–528.
151 Maximini, A., Chmiel, H., Holdik, H., and Maier, N.W. (2006) *J. Membr. Sci.*, **276**, 221–231.
152 Stella, D., Calzado, J.A., Girelli, A.M., Canepari, S., Bucci, R., Palet, C., and Valiente, M. (2002) *J. Sep. Sci.*, **25**, 229–238.

Appendix: Tabular Survey of Selected Cinchona-Promoted Asymmetric Reactions

Ji Woong Lee and Choong Eui Song

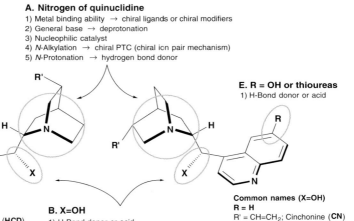

The appendix lists selected examples of cinchona-promoted asymmetric reactions. The table is organized according to the reaction types. Detailed information on the reaction procedures and the availability of cinchona-based chiral inducers is found in the corresponding chapter(s) and reference(s).

Cinchona Alkaloids in Synthesis and Catalysis, Ligands, Immobilization and Organocatalysis
Edited by Choong Eui Song
Copyright © 2009 WILEY-VCH Verlag GmbH & Co. KGaA, Weinheim
ISBN: 978-3-527-32416-3

Appendix: Tabular Survey of Selected Cinchona-Promoted Asymmetric Reactions

Structure	Active sites (mode of action)	Reaction	Typical substrate	Selected references (Chapter No., Reference(s))
Metal-catalyzed asymmetric reductions				
(Hydrocinchonidine-like structure with NH$_2$)	A (ligand)	Homogeneous Rh- and Ir-catalyzed transfer hydrogenation	Aryl ketone → alcohol; [M(cod)Cl]$_2$ (1 mol%) (M = Ir, Rh), Ligand, i-PrOH / KOH, −20°C, 24–48 h; R = H, Me, OMe, Cl, CF$_3$; R' = Me, Et, n-Pr, i-Pr; 70–90%, 72–97% ee	Chapter 2, Ref. [8]
Hydrocinchonidine	A (surface modifier)	Platinum-catalyzed heterogeneous hydrogenation of β-keto esters	β-ketoester → COOEt; Pt/Al$_2$O$_3$, HCD; 92% ee	Chapter 2, Ref. [19]
Metal-catalyzed asymmetric oxidations				
(DHQD)$_2$PHAL or (DHQ)$_2$PHAL	A (ligand)	Osmium-catalyzed asymmetric dihydroxylation	Ph-CH=CH$_2$ → Ph-CH(OH)-CH$_2$OH; OsO$_4$, Ligand, K$_3$Fe(CN)$_6$, t-BuOH/H$_2$O; (DHQD)$_2$PHAL: 97% ee (R); (DHQ)$_2$PHAL: 97% ee (S)	Chapter 3, Ref. [2]
		Osmium-catalyzed asymmetric hydroxylation	CO$_2$i-Pr cinnamate → NHAc/OH product; AcNHBr, LiOH, K$_2$OsO$_2$(OH)$_4$, (DHQ)$_2$PHAL, t-BuOH/H$_2$O, 4 °C; 99% ee	Chapter 3, Ref. [3]

Appendix: Tabular Survey of Selected Cinchona-Promoted Asymmetric Reactions | 473

Ligand	Reaction	Conditions/Results	Reference
A (ligand) (DHQD)₂PYR	WO₃-catalyzed asymmetric oxidation of sulfides	(DHQD)₂PYR, WO₃, H₂O₂, THF; 70–90%, 35–65% ee	Chapter 3, Refs [375]
	WO₃-catalyzed kinetic resolution of racemic sulfoxides	(DHQD)₂PYR, WO₃, H₂O₂, THF; 24–44%, 44–90% ee	Chapter 3, Ref. [376]

Metal-promoted enantioselective C–C bond forming reactions

Ligand	Reaction	Conditions/Results	Reference
A, C (ligand)	Diethylzinc addition to aromatic aldehydes	Et₂Zn / Cat. (10 mol%), Toluene, rt, 3 h; 92% ee	Chapter 4, Ref. [6]
A, C (ligand)	Diethylzinc addition to imines	CD, Et₂Zn, toluene; 93% ee	Chapter 4, Ref. [8]
Cinchonine or cinchonidine	Alkynylzinc addition to aldehydes	Zn(OTf)₂ (20 mol%), CD (22 mol%), NEt₃, toluene; 61% yield, 89% ee	Chapter 4, Ref. [15]

(Continued)

Appendix: Tabular Survey of Selected Cinchona-Promoted Asymmetric Reactions

Structure	Active sites (mode of action)	Reaction	Typical substrate	Selected references (Chapter No., Reference(s))
Cinchonine or cinchonidine	A, C (ligand)	Reformatsky reactions	BrZn-CH$_2$-CO-Ot-Bu (3.5 equiv); CN (1.5 equiv), pyridine (4.0 equiv), -40 °C, 4 h; substrate: 4-benzoyl-N-trityl imidazole → HO-C(Ph)(CO$_2t$-Bu)(imidazole-NTr), 99%, 97% ee	Chapter 4, Ref. [19]
		Indium-mediated allylation of aldehydes	R-CHO (R = aryl, alkyl) + allyl bromide R'CH=CHCH$_2$Br (R' = H, CH$_3$) (6 equiv), In (2 equiv), CD or CN (2 equiv), solvent, 5 h, rt → R-CH(OH)-CH(R')-CH=CH$_2$; R = H, up to 75% ee; R = CH$_3$, up to 90% ee	Chapter 4, Ref. [21]
		Indium-mediated propargylation of aldehydes	R-CHO (R = alkyl, aryl) + HC≡C-CH$_2$Br (3 equiv), In (1 equiv), CD (1 equiv), THF : hexane (3:1), rt, 15 h → R-CH(OH)-CH$_2$-C≡CH; R^1 = H, up to 85% ee	Chapter 4, Ref. [23]
		Strecker synthesis	Ph-C(=NTs)-Me, biphenol, Ti(OiPr)$_4$, CN, TMSCN, toluene, -20 °C → Ph-C(NHTs)(Me)(CN), 99%, 99% ee	Chapter 4, Ref. [34]

Cinchonine				
1.A (nucleophilic cocatalyst)	Cyanohydrin synthesis	R-CHO + EtOC(O)CN →[CN (10 mol%), (S)-AlLi(binaphthoxide) (10 mol%)][CH$_2$Cl$_2$, -20 °C] Ph-CH(OC(O)OEt)-CN; R = alkyl, aryl; up to 99%, 74–95% ee		Chapter 4, Ref. [29]
2.Al (electrophile activator)				
1.A (nucleophilic cocatalyst)	Nucleophilic addition of ketene to aldehyde for β-Lactone synthesis	Cl-CH$_2$-C(O)-Me + PhCHO →[QN][LiClO$_4$, CH$_2$Cl$_2$/Et$_2$O, i-Pr$_2$NEt, -25 °C] β-lactone (Me, Ph); 68–85%, 99% ee		Chapter 4, Ref. [37]
2.Li (electrophile activator)				
1.A (nucleophilic cocatalyst)	Nucleophilic addition of ketene to aldehyde for β-Lactone synthesis	R-CHO + ketene →[Cat. (1 mol%)][CH$_2$Cl$_2$, -78 °C] β-lactone (R); R = BnOCH$_2$- or Ar; up to >99% ee		Chapter 4, Refs [43, 44]
2.Co (electrophile activator)				

(Continued)

Structure	Active sites (mode of action)	Reaction	Typical substrate	Selected references (Chapter No., Reference(s))
(cinchona alkaloid with OMe, OBz)	1.A (nucleophilic cocatalyst) (electrophile activator)	[4+2] Cycloaddition of ketene enolates and o-benzoquinone diimides	Cat. (10 mol%), Zn(OTf)₂, THF, −78 °C; 82%, >99% ee	Chapter 10, Ref. [10]
(cinchona alkaloid with OMe, OCOPh)	1.A (nucleophilic cocatalyst) 2.In (electrophile activator)	Nucleophilic addition of ketene to aldehyde for β-lactam synthesis	Cat. (10 mol%), toluene, −78 °C; with In(OTf)₃ (cocatalyst); 92-98%, 96-98% ee	Chapter 4, Ref. [46]

(DHQ)₂PYR	1.A (nucleophilic cocatalyst)	[4+2] Cycloaddition of ketene enolates and to o-quinone		Chapter 10, Ref. [11]
	2.Pd (electrophile activator)			
	1.A (nucleophilic cocatalyst)	Synthesis of β-sultones	up to 87% up to 99% ee dr up to >100:1	Chapter 4, Ref. [40]
	2.In, Bi, etc. (electrophile activator)			
Quinine	1.A (cobase catalyst)	Aza-Henry reaction	98% ee (14:1 dr)	Chapter 4, Ref. [50]
	2.Cu (electrophile activator)			

(Continued)

Appendix: Tabular Survey of Selected Cinchona-Promoted Asymmetric Reactions

Structure	Active sites (mode of action)	Reaction	Typical substrate	Selected references (Chapter No., Reference(s))
Cinchonidine	1. A (cobase catalyst)	Hydrophosphonylation	Ph-CHO + HPO(OMe)₂ →[Binol derivative **CD** (2.5 mol%) / Ti(OiPr)₄ (2.5 mol%) / m-xylene, -20 °C, 6-24 h] Ph-CH(OH)-P(O)(OMe)₂, 99% ee	Chapter 4, Ref. [52]
	2. B (ligand)			
	3. Ti (electrophile activator)			
Quinine	A, B-metal coordination	Claisen rearrangement	R¹,R²-allyl vinyl ether (X = Tfa (trifluoroacetyl)) →[LHMDS (5.5 equiv), Al(OiPr)₃ (1.2 equiv), **QN** (2.5 equiv), -78 °C → rt] 1. H₃O⁺ 2. CH₂N₂ → product with R¹, R² groups, OMe. For R¹ = t-Bu, R² = Me, 98% ds and 93% ee	Chapter 4, Refs [54, 55]
	A, B (ligand)	Pd-catalyzed allylic substitution	Ph-CH(OAc)-CH=CH-Ph + MeO-C(O)-CH₂-C(O)-OMe →[**Ligand** (6 mol%), [Pd(C₃H₅)Cl]₂ (2 mol%), KOAc (3 mmol%), BSA (3 equiv), CH₂Cl₂] product, 72%, 94% ee	Chapter 4, Ref. [58]

Cinchona-catalyzed asymmetric oxidation

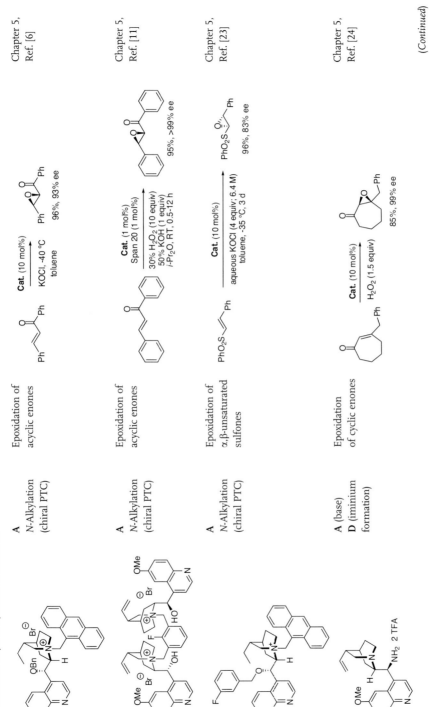

Appendix: Tabular Survey of Selected Cinchona-Promoted Asymmetric Reactions

Structure	Active sites (mode of action)	Reaction	Typical substrate	Selected references (Chapter No., Reference(s))
	A (base) D (iminium formation)	Epoxidation of acyclic enones		Chapter 5, Ref. [25]
	A (base) D (iminium formation)	Synthesis of cyclic peroxyhemiketals		Chapter 5, Ref. [26]
	A (base) D (iminium formation)	Aziridination of enones		Chapter 5, Ref. [33]

Cinchona-catalyzed reduction of carbonyl compounds

Structure	Active sites (mode of action)	Reaction	Typical substrate	Selected references (Chapter No., Reference(s))
	A (N-Alkylation) F⁻ (silane activator)	Reduction of ketones		Chapter 5, Ref. [36]

Cinchona-catalyzed nucleophilic α-substitution of carbonyl derivatives

Reaction type	Catalyst / Conditions	Product (yield, ee)	Reference
α-Alkylation of glycine derivatives	**A** N-Alkylation (chiral PTC); PhCH$_2$Br, Cat. (10 mol%), CsOH·H$_2$O, CH$_2$Cl$_2$, −78 °C	84%, 94% ee	Chapter 6, Ref. [14]
α-Alkylation of β-keto esters with aziridines	**A** N-Alkylation (chiral PTC); Ar = o-CF$_3$C$_6$H$_4$, Cat. (10 mol%), 50% K$_2$HPO$_4$ aq, −20 °C, 48–72 h	100%, 97% ee	Chapter 6, Ref. [48]
α-Arylation of β-ketoester	**A** (base), **B** (H-bond donor); QN (20 mol%), CH$_2$Cl$_2$, −20 °C, O$_2$	59%, 96% ee	Chapter 6, Ref. [58]
Nucleophilic aromatic substitution	**A** N-Alkylation (chiral PTC); Cat. (15 mol%), CsOH·H$_2$O, PhMe, −20 °C	84%, 80% ee	Chapter 6, Ref. [49]

(Continued)

Appendix: Tabular Survey of Selected Cinchona-Promoted Asymmetric Reactions

Structure	Active sites (mode of action)	Reaction	Typical substrate	Selected references (Chapter No, Reference(s))
	A N-Alkylation (chiral PTC)	Vinylic substitution	Cat. (3 mol%), 33% K_2CO_3 aq, o-Xylene-CHCl$_3$ (7:1), -20 °C; R = C_6H_5, 85%, 94% ee	Chapter 6, Ref. [50]
	A (nucleophilic catalyst)	α-Halogenation	Cat. (10 mol%), BEMP resin, halogenating reagent; R^1 = aryl, alkyl, oxo, azo, alkenyl; X = Cl, Br; $R^2 = C_6Cl_5, C_6H_2Br_3$; 43–81%, 88–99% ee	Chapter 6, Refs [64, 65]
	A N-Alkylation (chiral reagent)	α-Fluorination of carbonyl derivatives	1) LiHMDS / THF; 2) PTC (1.1–1.2 eq), -78 °C; 65%, 92% ee	Chapter 6, Ref. [56]
(DHQ)$_2$PYR	A-Base, Activation of NFSI	Fluorodesilylation	Cat. (10 mol%), NFSI, K_2CO_3, MeCN, -20 or -40 °C; 75%, 94% ee	Chapter 6, Ref. [62]

Appendix: Tabular Survey of Selected Cinchona-Promoted Asymmetric Reactions

Enantioselective protonation

Catalyst	Reaction type	Scheme / Conditions	Results	Reference
Quinine **A** (base) **D** (H-bond donor)	α-Amination of carbonyl derivatives	NC–CH(Ar)–CO₂t-Bu + Boc-N=N-Boc → NC–C(Ar)(N(Boc)NHBoc)–CO₂t-Bu **Cat.** (5 mol%), PhMe, –78 °C Ar = 2-Naph	99%, 98% ee	Chapter 6, Ref. [69]
(DHQ)₂AQN **A** (chiral proton source)	Enantioselective protonation of silyl enol ethers	2-Bn-1-OTMS-5-OMe-dihydronaphthalene → 2-Bn-1-oxo-5-OMe-tetralone **(DHQ)₂AQN** (10 mol%), PhCOF/EtOH, DMF, rt, 12 h	98%, 92% ee	Chapter 7, Ref. [5]
QN **A-H⁺** (chiral proton source)	Palladium catalyzed decarboxylative protonation	2-F-2-(CO₂Bn)-1-oxo-tetralone → 2-F-1-oxo-tetralone Pd/C, H₂, **QN** (30 mol%), CH₃CN, rt, 1 h	92%, 65% ee	Chapter 7, Ref. [30]
Thiourea (bis-CF₃-phenyl, quinine-derived) **A** (base) **E** (H-bond donor)	Enantioselective decarboxylative protonation	2-CO₂H-2-CO₂Et-N-acetylpiperidine → 2-CO₂Et-N-acetylpiperidine **Cat.** (1 equiv), Acetone, 0 °C, 7 d	90%, 93% ee	Chapter 7, Ref. [41]

(Continued)

Appendix: Tabular Survey of Selected Cinchona-Promoted Asymmetric Reactions

Structure	Active sites (mode of action)	Reaction	Typical substrate	Selected references (Chapter No, Reference(s))

Cinchona-catalyzed enantioselective nucleophilic 1,2-addition reactions

Reaction with ammonium ketene enolate

Structure	Active sites (mode of action)	Reaction	Typical substrate	Selected references
(quinidine acetate, OMe)	A (nucleophilic catalyst)	Nucleophilic addition to ketene	Ph–C(=O)–Me + MeOH, Cat. (1 mol%), toluene, −110 °C → Ph–C(OMe)(Me) 74% ee	Chapter 7, Ref. [8]
(cinchona benzoate, OMe)	A (nucleophilic catalyst)	Nucleophilic addition to ketenes	Me₃Si–C(=O)–CH₂R + PhSH, Cat. (10 mol%), toluene, −78 °C → Me₃Si–C(=O)–CH(SPh)(2-Naphthyl); R = 2-Naphthyl; 99%, 93% ee	Chapter 7, Ref. [10]
(OBn cinchona, OMe)	A (nucleophilic catalyst)	[4+2] Cycloaddition of ketene enolates and o-quinones	Et–CH₂–C(=O)–Cl + tetrachloro-o-benzoquinone, Cat. (10 mol%), DIPEA, THF, −78 °C → benzodioxinone product; 91%, 99% ee	Chapter 10, Ref. [8]
(OBn cinchona, OMe)	A (nucleophilic catalyst)	[4+2] Cycloaddition of ketene enolates and o-benzoquinone imides	i-Bu–CH₂–C(=O)–Cl + tetrachloro-o-benzoquinone N-CO-p-NO₂C₆H₄ imide, Cat. (10 mol%), DIPEA, THF, −78 °C → benzoxazinone product (i-Bu, N-CO-p-NO₂C₆H₄); 65%, 99% ee	Chapter 10, Ref. [9]

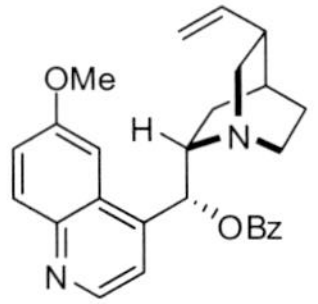

A (nucleophilic catalyst)	[4+2] Cycloaddition of ketene enolates and o-benzoquinone diimides		82%, >99% ee	Chapter 10, Ref. [10]
A (nucleophilic catalyst)	[4+2] Cycloaddition of ketene enolates and N-thioacyl imines		76%, dr 95:5 98% ee (syn)	Chapter 10, Ref. [12]
A (nucleophilic catalyst)	[4+2] Cycloaddition of vinyl ketene enolates to aldehydes		80%, 95% ee	Chapter 10, Ref. [13]
A N-Alkylation (chiral PTC)	Mukaiyama-type aldol reactions		48–81% dr 1:1–13:1 (syn:anti) 72–95% ee for syn	Chapter 8, Ref. [5]

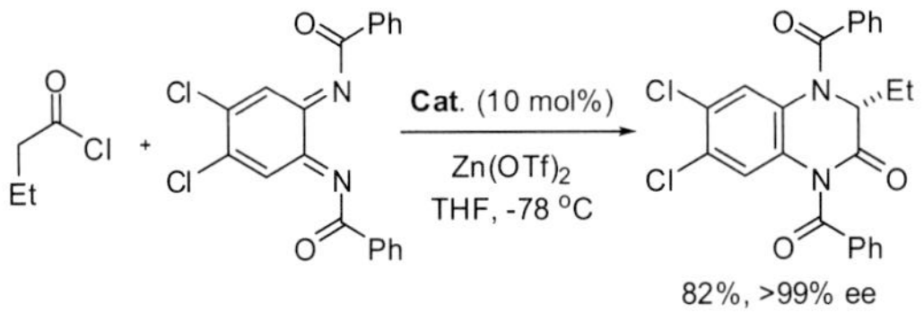

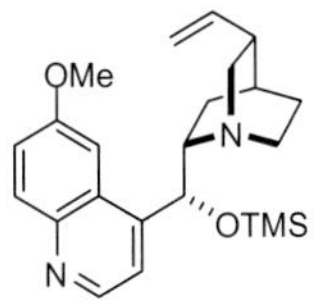

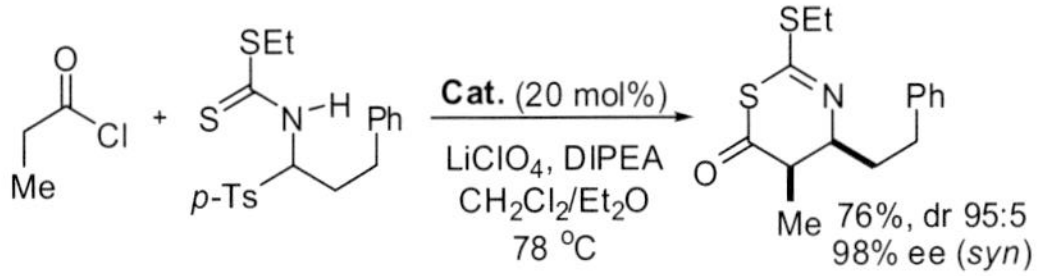

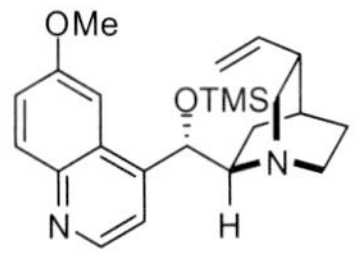

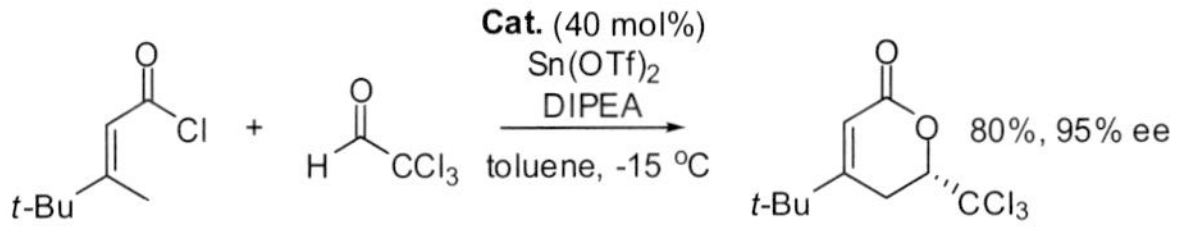

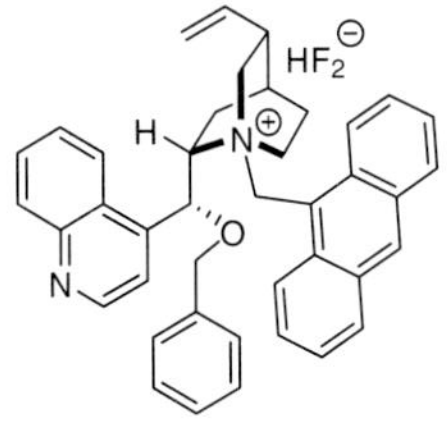

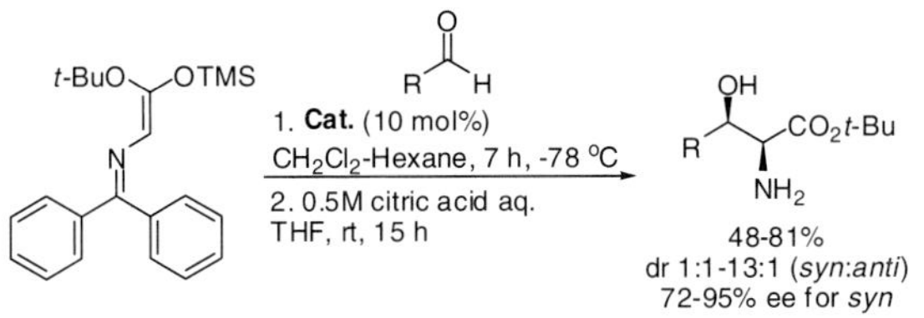

Aldol reactions

(Continued)

Structure	Active sites (mode of action)	Reaction	Typical substrate	Selected references (Chapter No., Reference(s))
Cinchona-derived catalyst with anthracenyl-N+ group, CF₃-substituted aryl, ⊖OPh·HOPh counterion	A N-Alkylation (chiral PTC)	Mukaiyama-type aldol reactions	Ph–CH₂CH₂–CHO + 2-(OSiMe₃)-5-methylfuran, Cat. (10 mol%), CH₂Cl₂, –78 °C, 1 h, then 1M HCl, THF; 75%, 78:22 (syn:anti), 97% ee (syn)	Chapter 8, Ref. [7]
9-amino-9-deoxyepi-cinchona alkaloid (NH₂ on C9)	A·H⁺ (H-bond donor) D (enamine formation)	Direct aldol reactions	Ar–CHO + cyclohexanone, Cat. (10 mol%), TfOH (15 mol%); Ar = o,m,p-NO₂C₆H₄ 74–99%, 3.2:1–9.2:1 (dr for anti), 97–99% ee; Ar = Ph, naphthyl, p-OMeC₆H₄ 19–31%, 1:1–4.9:1 (dr for anti), 86–93% ee	Chapter 8, Ref. [13]
Cinchona alkaloid with prolinamide at C9	A·H⁺ (H-bond donor) D (enamine formation)	Direct aldol reaction	R–CHO + acetone, Cat. (10 mol%), HOAc (20 mol%), neat, –30 °C, 23–96 h; up to 97% ee	Chapter 8, Ref. [17]

Appendix: Tabular Survey of Selected Cinchona-Promoted Asymmetric Reactions

(DHQ)$_2$PHAL or (DHQD)$_2$PHAL

Catalyst role	Reaction	Conditions / Yield / ee	Reference
A (base) **A-H**$^+$ (H-bond donor)	Direct aldol-type reaction of oxindoles with trifluoropyruvate	(DHQD)$_2$PHAL (10 mol%), 0 °C then RT, Et$_2$O, 2 equiv R = Me, Et, benzyl, p-substituted benzyl 89–99%, 86:14–94:6 dr, 92–99% ee	Chapter 8, Ref. [18]
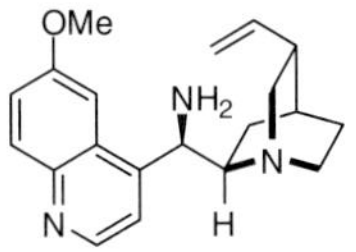 **A-H**$^+$ (H-bond donor) **D** (enamine formation)	Intramolecular aldol reaction	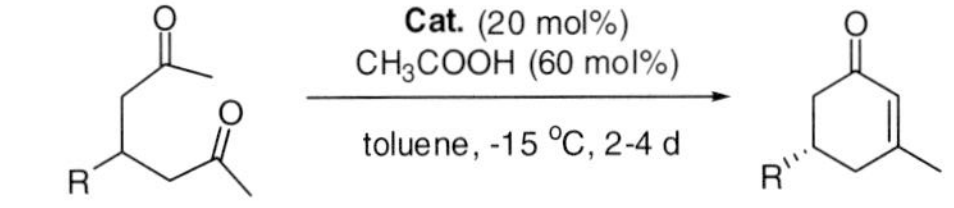 Cat. (20 mol%), CH$_3$COOH (60 mol%), toluene, −15 °C, 2–4 d R = alkyl, aryl, 2-thienyl 80–97%, 86–93% ee	Chapter 8, Ref. [14]
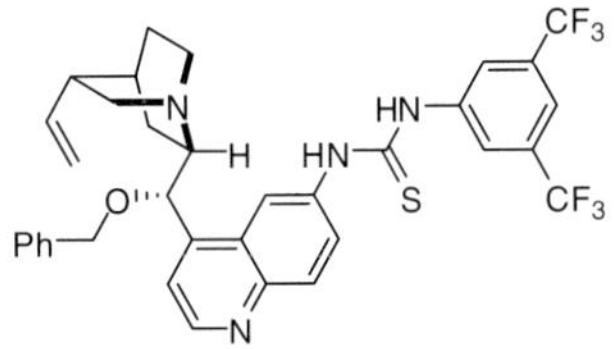 **A** (base) **D** (H-bond donor)	Henry reaction	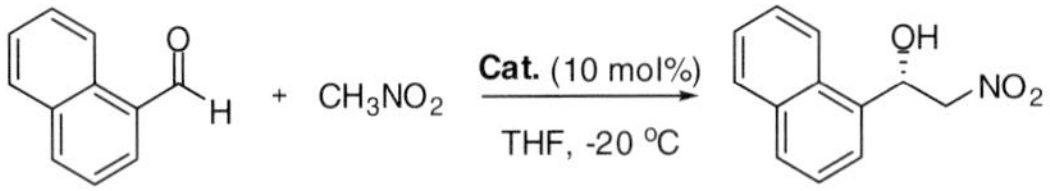 Cat. (10 mol%), THF, −20 °C 48 h, 99%, 92% ee	Chapter 8, Ref. [22]
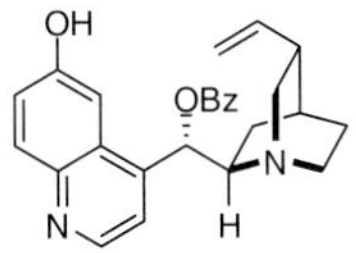 **A** (base) **E** (H-bond donor)	Henry reaction	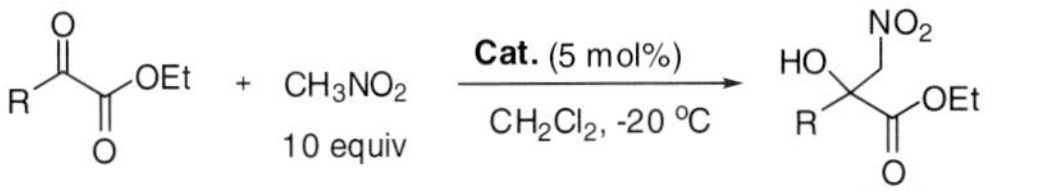 Cat. (5 mol%), CH$_2$Cl$_2$, −20 °C, CH$_3$NO$_2$ 10 equiv up to 97% ee	Chapter 8, Ref. [23]

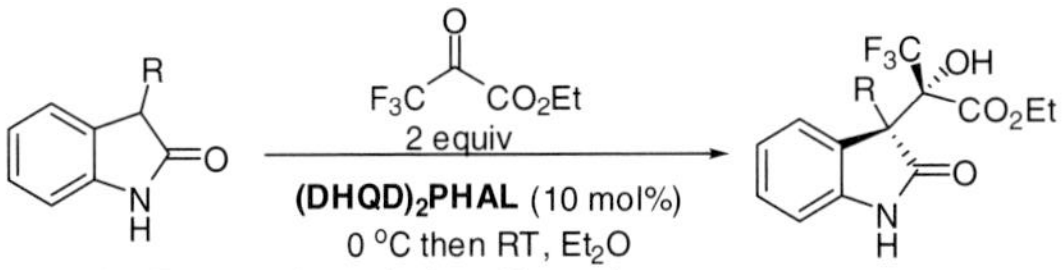

Henry reaction

(Continued)

Appendix: Tabular Survey of Selected Cinchona-Promoted Asymmetric Reactions

Mannich reaction

Structure	Active sites (mode of action)	Reaction	Typical substrate	Selected references (Chapter No., Reference(s))
Cinchonine	A (base) B (H-bond donor)	Mannich reaction	H₃C–C(O)–CH₂–C(O)–OR¹ + MeO–C(O)–N=CH–Ar; CN (10 mol%), CH₂Cl₂, −35 °C, 16 h, under Ar; product: HN(C(O)OMe)–CH(Ar)–CH(C(O)OR¹)(CH₃); 82–99%, 1:1–20:1 dr, 80–96% ee	Chapter 8, Ref. [25]
(Cinchona thiourea with 3,5-bis(CF₃)phenyl group)	A (base) D (H-bond donor)	Mannich reaction	PG–NH–CH(R)(SO₂Ar); Cat. (20 mol%), CH₂(COOBn)₂, CsOH (0.1 M), CH₂Cl₂; product: PG–NH–CHR–CH(COOBn)₂; R = alkyl, aryl; PG = Boc or Cbz; 45–99%, 85–96% ee	Chapter 8, Ref. [34]
(N-allyl cinchoninium bromide with 2-cyanobenzyl group)	A N-Alkylation (chiral PTC)	Mannich reaction	PG–NH–CH(R¹)(SO₂-p-tol) + R²O₂C–CH₂–CO₂R²; Cat. (1 mol%), toluene–50% K₂CO₃ aq, −20 °C, 48 h; product: PG–NH–CH(R¹)–CH(CO₂R²)₂; PG = Boc or Cbz; R¹ = aryl, alkyl; R² = 4-MeO-C₆H₄; R¹ = Ph: 81%, 98% ee; i-Pr: 80%, 86% ee	Chapter 8, Ref. [35]

Appendix: Tabular Survey of Selected Cinchona-Promoted Asymmetric Reactions | 489

A N-Alkylation (chiral PTC)	Fluorobis-(phenylsulfonyl)methylation of imines	NHBoc + PhO₂S⟶SO₂Ph / F **Cat.** (5 mol%), CsOH·H₂O (1.2 equiv), CH₂Cl₂, −80 °C, 1–2 d → NHBoc, R—C(SO₂Ph)(F)(SO₂Ph); up to 93%, 99% ee	Chapter 8, Ref. [37]
A N-Alkylation (chiral PTC)	Nitro-Mannich reaction	HN-Boc / R-SO₂Ar + MeNO₂; R = aryl, alkyl **Cat.** (10 mol%), KOH, toluene, −45 °C, 40–44 h → HN-Boc, R—CH(NO₂); R = alkyl; Ar = o-tolyl: 93–98% ee; R = aryl; Ar = Ph: 75–88% ee	Chapter 8, Ref. [38]
A (base) D (H-bond donor)	Nitro-Mannich reaction	N-PG / Ar-H + CH₃NO₂; Ar = aryl, 2-furyl, 2-thienyl; PG = Cbz, Boc, Fmoc **Cat.** (20 mol%), toluene, −24 or −40 °C → PG-NH, Ar—CH(NO₂); 50–95%, 52–95% ee	Chapter 8, Ref. [40]
A N-Alkylation (chiral PTC)	Darzens reaction	(tetralone with Cl) + R-CHO; R = alkyl; X = H, OMe **Cat.** (10 mol%), LiOH, n-Bu₂O, rt → epoxide product; 65–99%, 50–86% ee	Chapter 8, Ref. [42]

Nitro-mannich (aza-Henry) reactions

Darzens reaction

(Continued)

Appendix: Tabular Survey of Selected Cinchona-Promoted Asymmetric Reactions

Structure	Active sites (mode of action)	Reaction	Typical substrate	Selected references (Chapter No., Reference(s))
(structure with OMe, Br, N, OH, fluorinated benzyl)	A N-Alkylation (chiral PTC)	Darzens reaction	ArCHO + Cl–CH₂SO₂Ph, Cat. (10 mol%), toluene, RbOH, rt → Ar–(epoxide)–SO₂Ph, 80–95%, 74–97% ee	Chapter 8, Ref. [44]
Morita–Baylis–Hillman Reaction				
(cinchona-derived structure with OH)	A (nucleophilic catalyst) E (H-bond donor)	Morita–Baylis–Hillman Reactions	Aldehyde + acrylate with CF₃/CF₃ ester, Cat. (10 mol%), HFIPA, DMF, –55 °C; R = Ph, p-NO₂Ph, (E)-PhCH=CH, Et, i-Bu, i-Pr, c-Hex → OH/CF₃/CF₃ product (R), 31–58%, 91–99% ee	Chapter 8, Ref. [45]
(cinchona-derived structure with OH)	A (nucleophilic catalyst) E (H-bond donor)	Aza-Morita–Baylis–Hillman Reactions	N-Ts imine + vinyl ketone, R = Me, Et, Cat. (10 mol%), MeCN/DMF (1:1), –30 °C, 24 h → Ts-NH/Ar/R product, R = Me, 55–80%, 73–99% ee; R = Et, 46–54%, 82–94% ee	Chapter 8, Ref. [51]
Cyanohydrin synthesis				
(DHQD)₂AQN	A (nucleophilic catalyst)	Cyanohydrin synthesis	2,2-dimethylcyclopentanone, EtOCOCN, Cat. (15 mol%), chloroform, –24 or –12 °C, 12h–7d → NC/OCO₂Et product, 66%, 97% ee	Chapter 8, Refs [58, 59]

Appendix: Tabular Survey of Selected Cinchona-Promoted Asymmetric Reactions | 491

Strecker synthesis

(DHQ)₂AQN or (DHQD)₂PHAL

| | A (nucleophilic catalyst) | Cyanohydrin synthesis | R¹–C(=O)–OR² / OR² → TMSCN (2-4 equiv), Cat. (2-20 mol%), −30 to −50 °C → NC–C*(OTMS)(R¹)(OR²/OR²) 81-98%, 90-98% ee | Chapter 8, Ref. [61] |
| 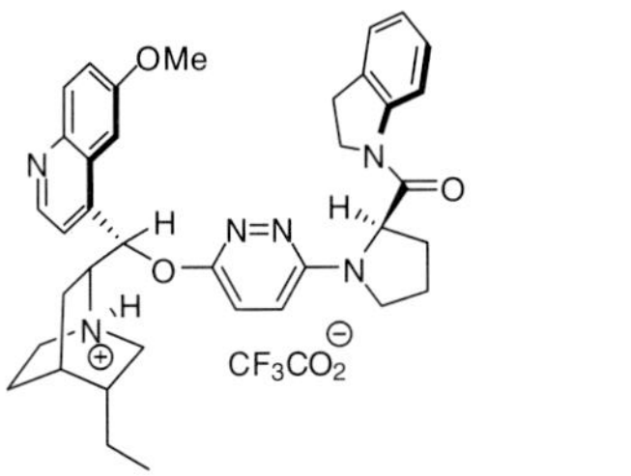 | A·H⁺ (H-bond donor) | Strecker synthesis | 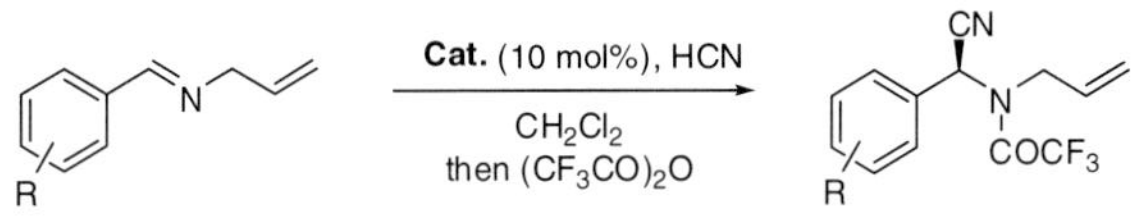 Ar–CH=N–allyl → Cat. (10 mol%), HCN, CH₂Cl₂, then (CF₃CO)₂O → Ar–CH(CN)(N(allyl)COCF₃) 86–98%, 79– >99% ee | Chapter 8, Ref. [63] |

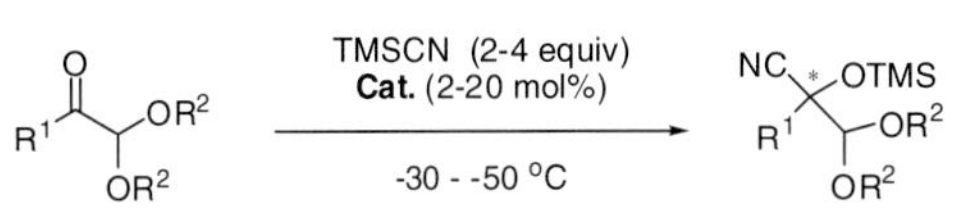

Trifluoromethylation

| 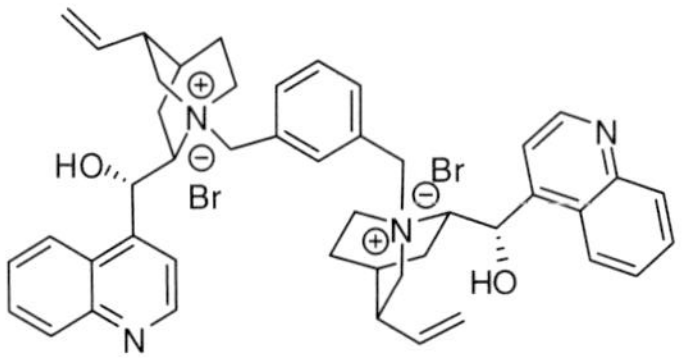 | A N-Alkylation (chiral PTC) | Trifluoro-methylation | 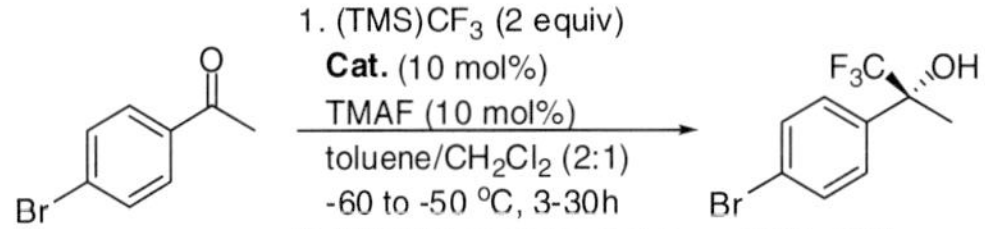 4-Br-C₆H₄–C(=O)–CH₃ → 1. (TMS)CF₃ (2 equiv), Cat. (10 mol%), TMAF (10 mol%), toluene/CH₂Cl₂ (2:1), −60 to −50 °C, 3-30 h; 2. TBAF/H₂O, THF, rt, 1 h → 4-Br-C₆H₄–C(CF₃)(OH)–CH₃ 81%, 86% ee | Chapter 8, Ref. [68] |

Friedel–Crafts type alkylation

Cinchonidine or cinchonine

| | A (H-bond acceptor) B (H-bond donor) | Friedel–Crafts type alkylation | indole + F₃C–C(=O)–COOEt → CD (5 mol%), ether, −8 °C, 2h → indol-3-yl–C(CF₃)(OH)–COOEt up to 95% ee | Chapter 8, Ref. [71] |

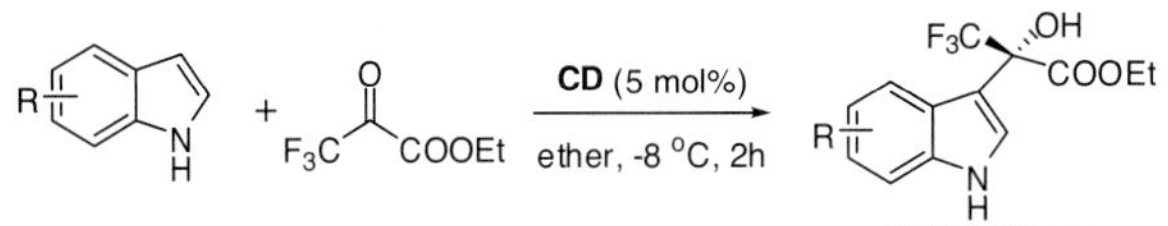

(Continued)

Appendix: Tabular Survey of Selected Cinchona-Promoted Asymmetric Reactions

Structure	Active sites (mode of action)	Reaction	Typical substrate	Selected references (Chapter No., Reference(s))
Hydrophosphonylation				
Hydroquinine or hydroquinidine	**A** (base) **B** (H-bond donor)	Hydrophosphonylation	HQN (10 mol%), toluene, −40 °C, 5–6h; 98%, 97%	Chapter 8, Ref. [82]
Cinchona-catalyzed enantioselective conjugate addition reactions				
α,β-Unsaturated ketones, amides, and nitriles as acceptors				
(structure with Br⁻, anthracenyl, allyl ether, quinoline, OMe)	**A** N-Alkylation (chiral PTC)	Michael addition of glycinate derivatives	Cat. (10 mol%), CsOH·H₂O, CH₂Cl₂, −78 °C; 85%, 95% ee	Chapter 9, Ref. [7]
(structure with Cl⁻, anthracenyl, OH, quinoline, OMe)	**A** N-Alkylation (chiral PTC)	Michael addition	Cat. (0.11 equiv), K₂CO₃ (0.14 equiv), −20 °C, 2 d; 91%, 90% ee	Chapter 9, Ref. [11]

A N-Alkylation (chiral PTC)	Michael addition of silyl enolates	Ph–CH=CH–C(O)Ph + OSiMe$_3$/OPh/i-Pr → δ-lactone (Ph, i-Pr, Ph substituents) **Cat.** (5 mol%) THF, −78 °C, 0.5 h 99% *anti/syn*: >99/1 96% ee	Chapter 9, Ref. [12]
A N-Alkylation (chiral PTC)	1,6-Conjugate addition of β-keto esters	indanone-CO$_2$t-Bu + MeO-dienone → 96%, 97% ee **Cat.** (3 mol%) o-xylene/CHCl$_3$ 7:1 40% Cs$_2$CO$_3$, 4 °C	Chapter 9, Ref. [13]
A N-Alkylation (chiral PTC)	1,6-Conjugate addition to δ-unsubstituted dienes	Ph$_2$C=N–CH$_2$–CO$_2$t-Bu + EWG-diene → Ph$_2$C=N–C(CO$_2$t-Bu)(CH$_2$CH=CHCH$_2$EWG) **Cat.** (3 mol%) o-xylene/CHCl$_3$ 7:1 50% KOH, −40 °C, 1 h 6 examples 49–98% yield 60–98% ee	Chapter 9, Ref. [13]

(Continued)

Appendix: Tabular Survey of Selected Cinchona-Promoted Asymmetric Reactions

Structure	Active sites (mode of action)	Reaction	Typical substrate	Selected references (Chapter No., Reference(s))
	A N-Alkylation (chiral PTC)	Conjugate addition to allenic esters	cyclohexanone-CO_2t-Bu + allenic acrylate (EtO), Cat. (3 mol%), o-xylene/CHCl$_3$ (7:1), 50% K$_3$PO$_4$ aq. 4 °C	Chapter 9, Ref. [14]
	A N-Alkylation (chiral PTC)	Anti-Michael reaction	indanone-CO_2t-Bu + PhO$_2$S-CN, Cat. (6 mol%), Cs$_2$CO$_3$ aq. −20 °C; 90%, 86% ee	Chapter 9, Ref. [15]
	A N-Alkylation (chiral PTC)	Monofluoromethylation via conjugate addition	Ph-CH=CH-C(O)-Ph + PhO$_2$S-CHF-SO$_2$Ph (FBSM), Cat. (5 mol%), Cs$_2$CO$_3$ (3 equiv), CH$_2$Cl$_2$, −40 °C, 1–2 d; 80%, 97% ee	Chapter 9, Ref. [16]

Catalyst	Reaction type	Scheme	Reference
(DHQD)₂PHAL; A (base)	Conjugate addition to alkynones	96%, 96% ee; dr up to 1.8:1, up to 99% yield, 95% ee	Chapter 9, Ref. [17]
A (base)	Conjugate addition to α,β-unsaturated ketones	96%, 96% ee	Chapter 9, Ref. [18]
E (H-bond donor)	Tandem conjugate addition–protonation reactions	89%, dr 5:1, 85% ee	Chapter 9, Ref. [21]
A (base); D (H-bond donor)	Conjugate addition of nitromethane to chalcone	99h, 93%, 96% ee	Chapter 9, Ref. [23]

(Continued)

Appendix: Tabular Survey of Selected Cinchona-Promoted Asymmetric Reactions

Structure	Active sites (mode of action)	Reaction	Typical substrate	Selected references (Chapter No., Reference(s))
(cinchona-derived thiourea with OMe quinoline and 3,5-bis(CF$_3$)phenyl thiourea)	A (base) D (H-bond donor)	Conjugate addition to chalcone	X–CHR–Y + Ph–CH=CH–C(O)–Ph → Cat. (10 mol%), xylene, rt → product, up to 98% ee; X,Y = EWG	Chapter 9, Ref. [24]
(cinchona-derived thiourea with OBn and 3,5-bis(CF$_3$)phenyl thiourea)	A (base) E (H-bond donor)	Intramolecular oxo-Michael addition	2'-OH-chalcone with CO$_2$t-Bu → Cat. (10 mol%), toluene, −25 °C, d.r. > 20:1 → chromanone, 85%, 92% ee	Chapter 9, Ref. [28]
(cinchona primary amine, 9-NH$_2$)	A (base) D (iminium formation)	Aza-Michael addition	Ph–CH=CH–C(O)–Me + Boc-N(H)-OBn → Cat. (20 mol%), TFA (20–40 mol%), CH$_2$Cl$_2$, rt → Boc-N(OBn)-CHPh-CH$_2$-C(O)-Me, 24 h, 13%, 96% ee	Chapter 9, Ref. [30]

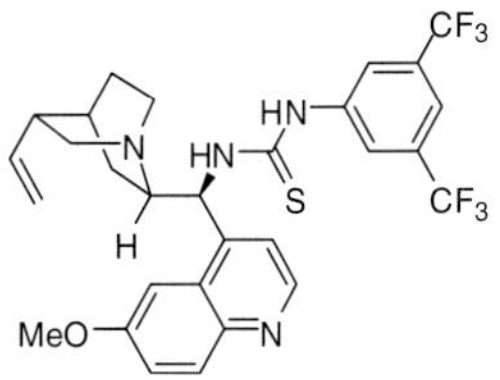

Appendix: Tabular Survey of Selected Cinchona-Promoted Asymmetric Reactions | 497

A (base) D (H-bond donor)	Tandem thio-Michael-aldol reaction		75-97% dr > 20:1 91-99% ee	Chapter 9, Ref. [29a]
A N-Alkylation (chiral PTC)	Conjugate addition/ cyclization for γ-lactam synthesis		up to 99% ee	Chapter 10, Ref. [36]
A (base) D (H-bond donor)	Michael addition of malonate to nitroalkenes		91-95% 87-99% ee	Chapter 9, Ref. [38, 39]

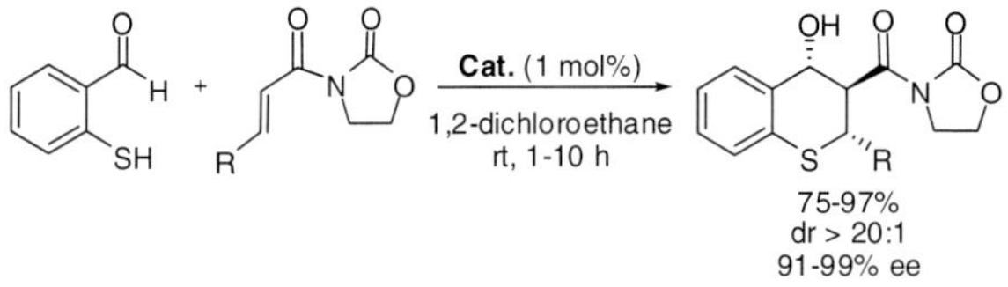

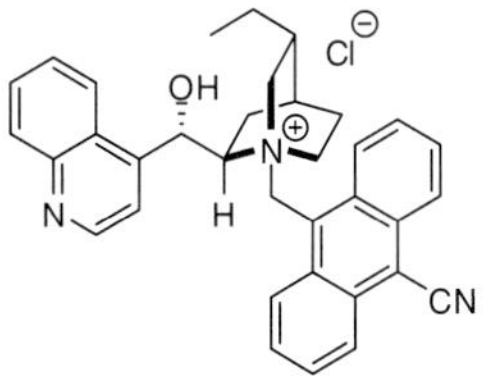

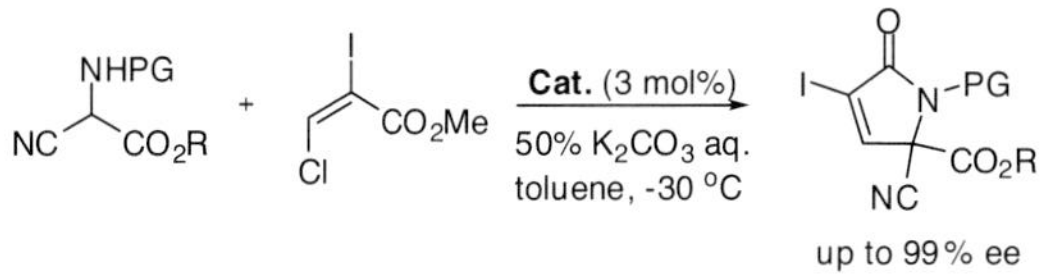

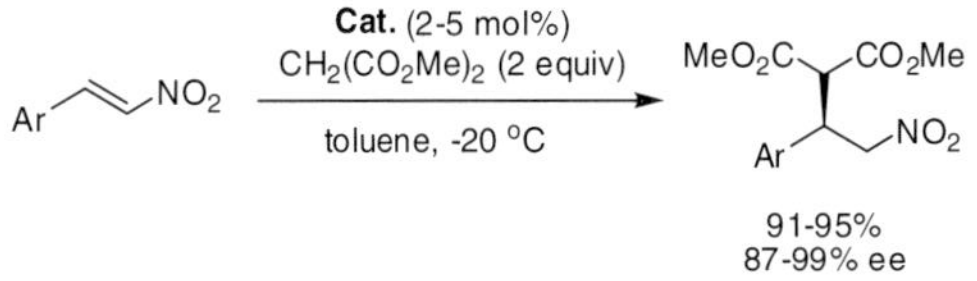

Nitro alkenes as acceptors

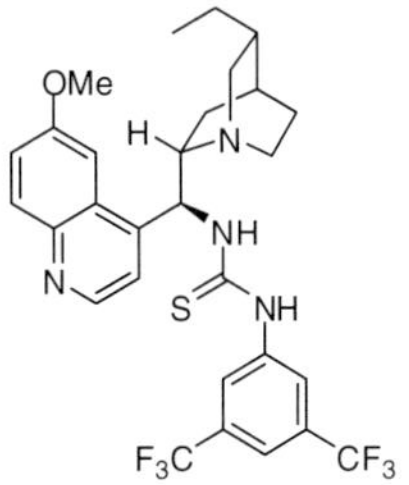

(Continued)

498 | Appendix: Tabular Survey of Selected Cinchona-Promoted Asymmetric Reactions

Structure	Active sites (mode of action)	Reaction	Typical substrate	Selected references (Chapter No., Reference(s))
(squaramide-cinchona structure with CF$_3$ groups)	A (base) D (H-bond donor)	Conjugate addition of carbon nucleophiles	R^{1}COCHR^{2}COR^3 + Ar-CH=CH-NO$_2$ → product, Cat. (0.5 mol%), CH$_2$Cl$_2$, rt, up to 99% ee	Chapter 9, Ref. [47]
(cinchona-OCOPh structure with OH)	A (base) E (H-bond donor)	Michael addition of anthron	anthrone + R-CH=CH-NO$_2$ → product, Cat. (5 mol%), CH$_2$Cl$_2$, −40 °C, 12 h, 91–99%, up to 99% ee	Chapter 9, Ref. [36]
(thiourea-cinchona structure with 3,5-(CF$_3$)$_2$-C$_6$H$_3$)	A (base) D (H-bond donor)	Michael addition of dioxolans	dioxolanone (Ar, CF$_3$, CF$_3$) + R-CH=CH-NO$_2$ → product, Cat. (0.05 equiv), 0 °C, 58–88%, dr < 99:1, 60–89% ee	Chapter 9, Ref. [40]

Appendix: Tabular Survey of Selected Cinchona-Promoted Asymmetric Reactions | 499

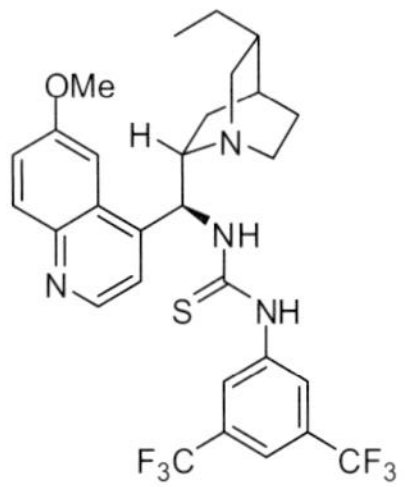 A (base) D (H-bond donor)	Michael addition of oxazolones	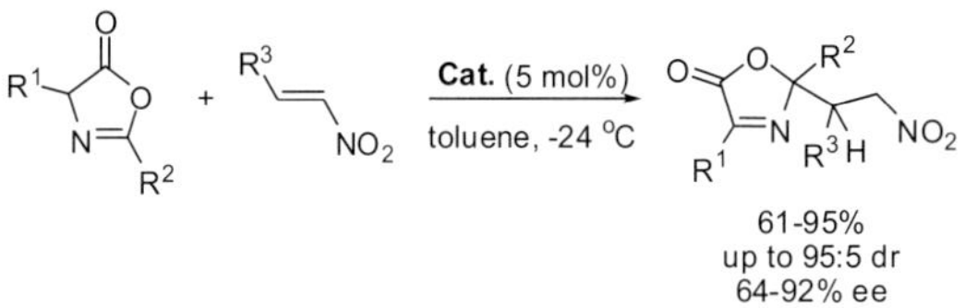 61–95% up to 95:5 dr 64–92% ee	Chapter 9, Ref. [41]
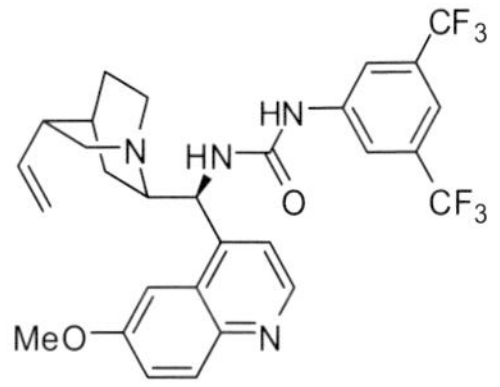 A (base) D (H-bond donor)	Decarboxylative Michael addition	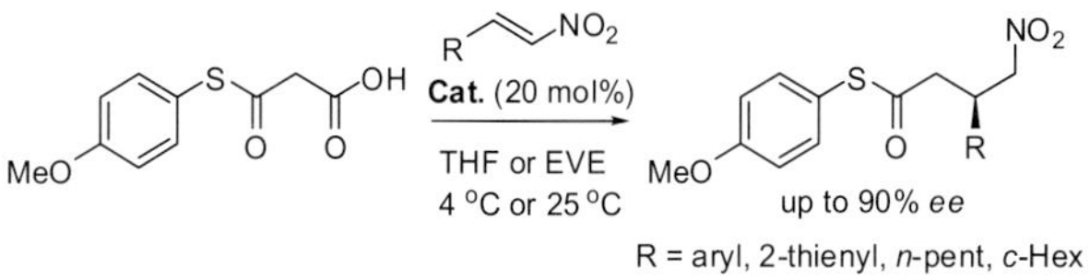 up to 90% ee R = aryl, 2-thienyl, n-pent, c-Hex	Chapter 9, Ref. [42]
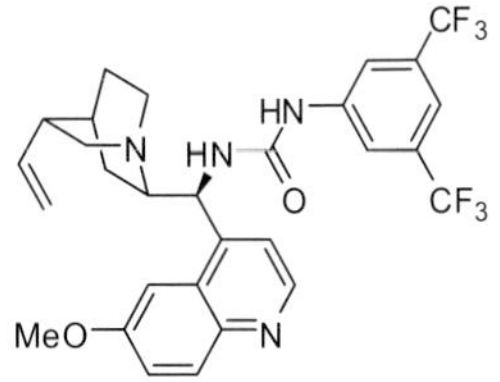 A (base) D (H-bond donor)	Nucleophilic addition of the silylated thiazolium salt	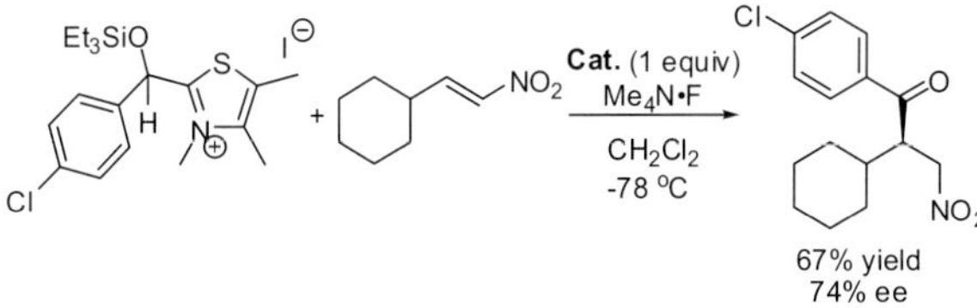 67% yield 74% ee	Chapter 9, Ref. [44]
A (base)	Vinylogous Michael addition of vinyl malonitriles	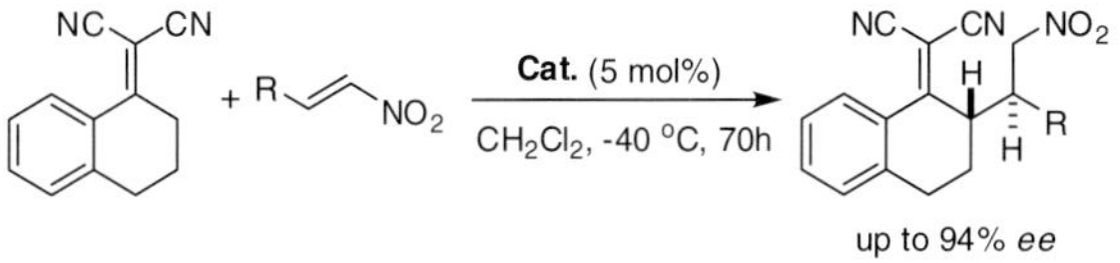 up to 94% ee	Chapter 9, Ref. [50]

(DHQD)₂PYR

(*Continued*)

Appendix: Tabular Survey of Selected Cinchona-Promoted Asymmetric Reactions

Structure	Active sites (mode of action)	Reaction	Typical substrate	Selected references (Chapter No., Reference(s))
(cinchona alkaloid with OH)	A (base) E (H-bond donor)	Michael addition of malonates to nitroalkenes	R–CH=CH–NO$_2$ + CH$_2$(CO$_2$Me)$_2$ (3.0 equiv), Cat. (10 mol%), THF, −20 °C → MeOOC–CH(R)–CH$_2$–NO$_2$ (COOMe); R = aryl, heteroaryl, alkyl; 71–99%, 91–98% ee	Chapter 9, Ref. [33]
(cinchona alkaloid with OBz)	A (base) E (H-bond donor)	Conjugate addition to nitroolefine and cyclization	MeOOC–CH(Ph)–N≡C + p-CF$_3$-C$_6$H$_4$–CH=CH–NO$_2$, Cat. (20 mol%), CH$_2$Cl$_2$, 35 °C, 24 h → pyrroline product; 68%, 10:1 dr, 98% ee	Chapter 10, Ref. [23]
(cinchona alkaloid with NH$_2$, OMe)	A (base) D (H-bond donor)	Domino Michael-Henry reaction	diketoester + R^2–CH=CH–NO$_2$, Cat. (10–15 mol%), toluene or Et$_2$O, 4 °C to rt, n = 0 or 1 → cyclohexanol/cyclopentanol products; up to 99% ee, dr 93:1	Chapter 10, Refs [31, 32]

Vinyl sulfones and vinyl phosphates as acceptors

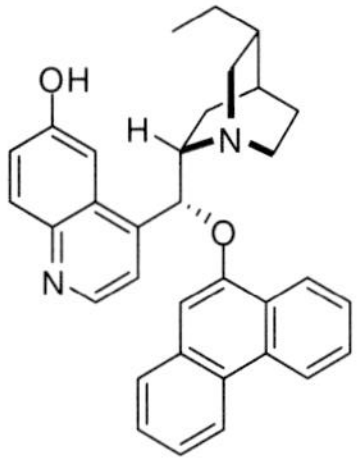

	Conjugate addition to vinylsulfone	**A** (base) **E** (H-bond donor)	EtO₂C CN + SO₂Ph → (Cat. 20 mol%, toluene) → EtO₂C, CN, SO₂Ph; R = aryl, heteroaryl; 80–96%, 88–97% ee	Chapter 9, Ref. [54]
	Michael addition of cyclic ketones to vinyl sulfone	**A** (base) **D** (enamine formation)	cyclic ketone + CH₂=C(SO₂Ph)₂ → (Cat. 20 mol%, CHCl₃, 0 °C) → product; 78–93%, 88–97% ee	Chapter 9, Ref. [56]
	Conjugate addition to vinylphosphate	**A** (base) **B** (H-bond donor)	(EtO)₂OP–PO(OEt)₂ vinyl + cyclopentenone-CO₂t-Bu → (HQN 20 mol%, toluene 0.1 M, −60 °C) → product; 98%, 99% ee	Chapter 9, Ref. [57]

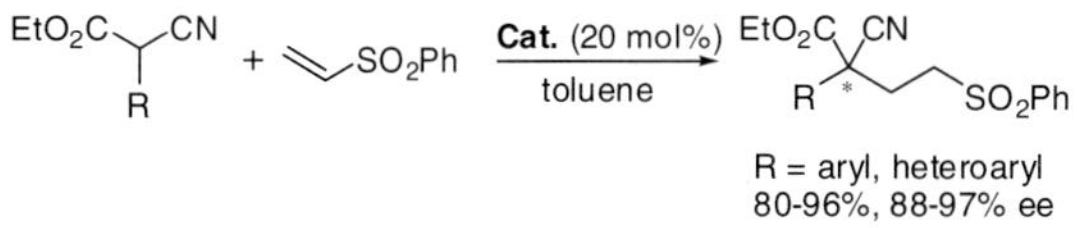

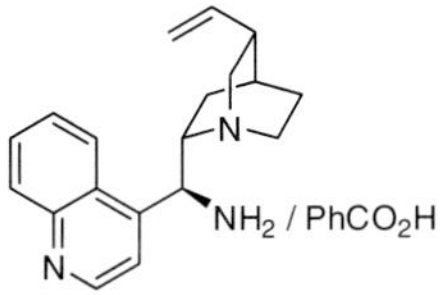

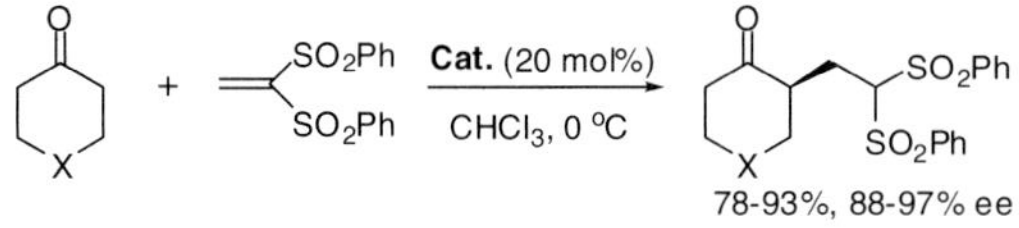

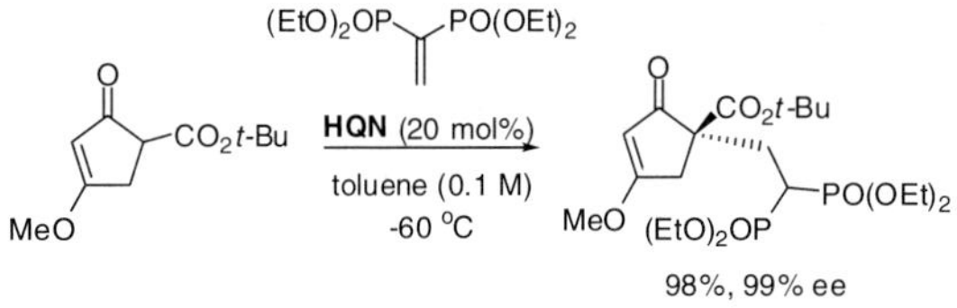

Cyclopropanation

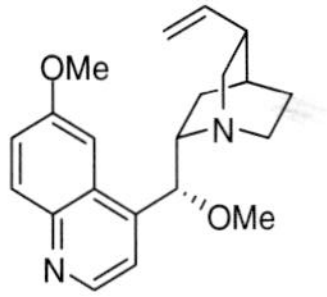

| | Cyclopropanation | **A** (nucleophilic catalyst) | 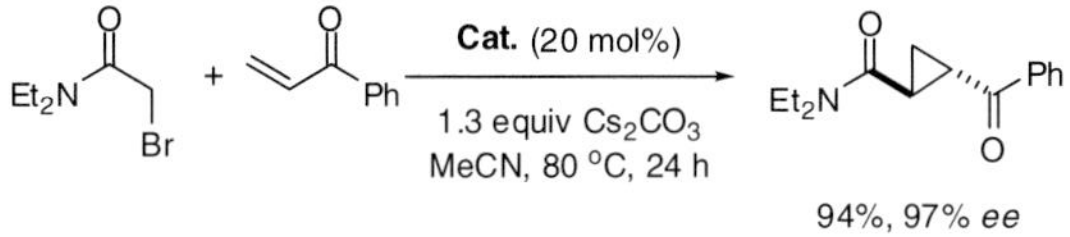 Et₂N-CO-CH=CH-Ph + Et₂N-CO-CH₂Br → (Cat. 20 mol%, 1.3 equiv Cs₂CO₃, MeCN, 80 °C, 24 h) → cyclopropane product; 94%, 97% ee | Chapter 9, Ref. [60] |

Hydroquinine or hydroquinidine catalyst (structure with NH₂ / PhCO₂H)

Cinchona alkaloid catalyst with OMe/OMe substituents

(Continued)

Appendix: Tabular Survey of Selected Cinchona-Promoted Asymmetric Reactions

Structure	Active sites (mode of action)	Reaction	Typical substrate	Selected references (Chapter No., Reference(s))

Epoxidation and Aziridination

See Chapter 5, Refs [6, 11, 23, 24, 25, 26, 33]

Cinchona-catalyzed enantioselective cycloaddition reactions

Structure	Active sites (mode of action)	Reaction	Typical substrate	Selected references
	A (base) E (hydrogen bonding donor)	Diels–Alder reaction	Cat. (5 mol%), Et₂O, rt; exo:endo (93:7), 87%, 94% ee (exo)	Chapter 10, Ref. [20]
	A (base) D (iminium catalysis)	Diels–Alder reaction	Cat. (5 mol%), Et₂O, rt; exo:endo (82:18), 91%, 96% ee (exo)	Chapter 10, Ref. [30]
	A (base) D (iminium catalysis) E (hydrogen bonding donor)	1,3-Dipolar cycloaddition	Cat. 10 mol%, TIPBA (20 mol%), THF, 40 °C; 89%, 90% ee	Chapter 10, Ref. [34]

Appendix: Tabular Survey of Selected Cinchona-Promoted Asymmetric Reactions

Cinchona-catalyzed desymmetrization of meso-compounds

Catalyst	Mode	Reaction	Conditions / Results	Reference
Quinidine-NH₂ derivative (OMe)	A (base) / D (iminium/enamine catalysis)	Desymmetrization/domino Michael–Michael addition	Cat. (20 mol%), TFA (40 mol%), DIPEA, THF, −20 °C; 81%, 99% ee	Chapter 10, Ref. [29]
Quinidine	A (nucleophilic catalyst) / B (hydrogen bonding donor)	Desymmetrization of meso-anhydride	QD (10 mol%), pempidine (1 equiv), MeOH (3 equiv), toluene/CCl₄, −55 °C, 6d; 74%, 98% ee	Chapter 11, Ref. [6]
(DHQD)₂AQN thiourea derivative	A (base) / D (hydrogen bonding donor)	1,3-Dipolar cycloaddition	Cat. (10 mol%), MeOH (10 equiv), MTBE, 5h; 98% ee	Chapter 11, Ref. [15]
(DHQD)₂AQN	A (base)	Kinetic resolution of racemic cyclic anhydrides	Cat. (10 mol%), R'OH (1.5 equiv), Et₂O, −20 to 78 °C; 46–48%, 90–96% ee; 32–40%, 85–95% ee	Chapter 11, Ref. [41]
(DHQD)₂AQN	A (base)	(Dynamic) kinetic resolution of racemic cyclic anhydrides	Cat. (10–20 mol%), EtOH or Allylalcohol, via DKR; up to 85%, 95% ee; up to 98%, 92% ee	Chapter 11, Refs [42, 43]

(Continued)

504 | Appendix: Tabular Survey of Selected Cinchona-Promoted Asymmetric Reactions

Structure	Active sites (mode of action)	Reaction	Typical substrate	Selected references (Chapter No., Reference(s))
	A (base) D (H-bond donor)	Dynamic kinetic resolution of azlactones		Chapter 11, Ref. [12]
contains 2% of phosphinate	A, C (base/nucleophilic catalyst)	Desymmetrization of *meso*-diols		Chapter 11, Ref. [30]
	A, NMe$_2$ (nucleophilic catalyst)	Desymmetrization of *meso*-diols		Chapter 11, Ref. [31]
	A (base) E (H-bond donor)	Desymmetrization of *meso*-endoperoxides		Chapter 11, Ref. [35]

Catalyst	Reaction	Conditions & Product	Reference
A N-Alkylation (chiral PTC)			Chapter 11, Ref. [40]
A (nucleophilic reagent)	Desymmetrization of prochiral ketone via Horner–Wadsworth–Emmons reaction	$\underset{t\text{-Bu}}{\text{cyclohexanone}}$ + (EtO)$_2$P(O)CH$_2$CO$_2$Et → 1. **Cat.** (20 mol%), benzene, RbOH, rt; 2. HCl/EtOH, 60 °C → product, 69%, 57% ee	
A (nucleophilic reagent)	Sulfinyl transfer reaction via DKR	Ar–S(O)–Cl (±) + t-BuOH, **QD(C9-OAc)**, −78 °C / 30 min, *via DKR* → sulfinate (MeO, Ot-Bu), 74%, 96% ee	Chapter 11, Ref. [50]
A (nucleophilic catalyst)	Sulfinyl transfer reaction via DKR	t-Bu–S(O)–Cl, **Cat.** (10 mol%), Proton sponge (2.5 equiv), 9-fluorenol, toluene, −40 °C, 36–48 h → R–S*(=O)–O-fluorenyl, 90%, 94% ee	Chapter 11, Refs [51, 52]

Index

a
α-acetal ketones 232
N,O-acetals, formation 391
N-acetophenone moiety 146
– electron-withdrawing 146
acetylenic quinidine acetate 375, 376
O-acetyl quinidine 342
O-acetyl quinine 240
Achilles' heel 363
achiral osmium oxidation 34
acid cocatalyst 203
acid halide, α-chlorination 162
π-acidic DNB moiety 432
– π-π interaction 432
acidic drugs, separations 445
acrylates, asymmetric aziridination 121
N-acyl-N-arylhydroxylamines 121
– nitrogen transfer reagents 121
acyclic aldol donors, acetone 202
acyclic benzyl β-oxo-esters 186
– hydrogenolysis/decarboxylation/asymmetric protonation reaction cascade 186
acyclic enones 108, 293
– asymmetric epoxidation 108–113
– enantioselective epoxidation, cinchona-based PTCs 293
acyclic β-ketoesters 187
acyl enol esters 160
alanine acetamide 190
aldehydes 74, 80, 83, 85
– azomethines 216
– dialkylzinc addition 74, 75
– enantioselective cyanation 83, 85
  – Lewis acid-Lewis base activation 85
– enantioselective indium-mediated allylation reaction 80
aldol reactions 198–206, 394
– 2-butanone 204

– diazoacetates 201
– direct reactions 200–206
– Mannich-related reactions 218–229
  – ammonium ketene enolate, nucleophilic addition 228, 229
  – aza-Morita–Baylis–Hillman reactions 225–228
  – Darzens reactions 218–221
  – Morita–Baylis–Hillman reactions 221–225
– Mukaiyama-type aldol reactions 198–200
aliphatic aldehydes, isobutyraldehyde 204
alkaloid derivatives 230
– (DHQD)$_2$AQN 230
– (DHQD)PHN 230
alkaloid esters 30
– pseudoenantiomeric ligands 30, 31
alkene(s) 30, 38, 50
– achiral osmium-catalyzed oxidations 35
– cis-alkene 43
– asymmetric dihydroxylation 30, 50
  – bisalkaloid ligands 33–35
  – catalytic cycles, hypothesis 30, 31
  – 1,2-diols 51–56
  – double asymmetric induction 44–50
  – mechanism 35, 36
  – reactions 30–33
  – resolutions 50, 51
  – substrates/selectivity 38–51
  – variations 36, 37
– asymmetric osmium-catalyzed oxidations 35
– ligand accelerating effects 38
– oxazolidinone 58
– substrates, asymmetric dihydroxylation reaction 39
alkyl α-amidosulfones 214
α-alkylated products 151, 154

Cinchona Alkaloids in Synthesis and Catalysis, Ligands, Immobilization and Organocatalysis
Edited by Choong Eui Song
Copyright © 2009 WILEY-VCH Verlag GmbH & Co. KGaA, Weinheim
ISBN: 978-3-527-32416-3

N-alkylated cinchona derivative 14
α-alkylcysteine derivatives 150
5-alkyl dioxolanes 347
α-alkylglycolates 152
α-alkyl-α-hydroxycarbonyl compounds 151
alkyl imines, N-protected 212
α-alkylserine derivatives 150
alkyne cinchona alkaloids 368, 371
– AD reactions 371
– bromination-double-dehydrobromination sequence 368
– derivatives 368
– efficiency 371
– preparation 371
– transformations 368
alkyne quinuclidines 374
alkynylated products 156
allyl alcohol 348
N-allyl aldimines 233
O-allyl- N-anthracenylmethylcinchonidinium p-nitrophenoxide 137
– X-ray analysis 137
O-allyl derivative 144
allylindium-alkaloid ligand complex 80
α-amido sulfones 244
amino acids 440
– carbazole-9-carbonyl, separation 462
– D-enantiomers 440
– 3,5-dinitrobenzoyl (DNB) 453
– 3,5-dinitrophenyl (DNP) 453
– N-protected 437
α-amino acid 136, 150, 232
– α-alkylation 150
– asymmetric synthesis 232
– derivatives, α-monosubstituted alkylation 148
– enantioselective alkylation 136
β-amino acid synthesis 337
γ-amino acid synthesis 337
1,2-amino alcohols 53
β-amino alcohols, 3-exo-dimethylaminoisoborneol (DAIB) 74
aminocatalysis 286
9-amino-9-deoxyepicinchona alkaloids catalyst, role 319
9-amino-9-deoxyepiquinine 118, 202
amino esters, (R)-enantiomers 191
α-amino esters, asymmetric synthesis 191
β-amino esters, Soloshonok's biomimetic asymmetric synthesis 193
aminohydroxylation 56–61
α-amino sulfones 215, 216
aminoxylation 164
ammonium carbamate 401

ammonium ketene enolate
– nucleophilic addition 228, 229
– reactions 86, 484
analgesic agent (–)-Wy-16 225 synthesis 151
angiotensin-converting enzyme (ACE) 18
anionic repulsion 143
anthrone 275, 276
anti conformer binding 456
aqueous tert-butanol 36
aromatic α-amido p-tolylsulfones 213
aromatic cyclic ketones 156
aromatic diimides 456
aromatic electrophiles 154
5-aryl N-carboxyanhydrides 348
α-aryl-α-cyanoacetates 163, 164
– enantioselective organocatalytic α-amination 163, 164
5-aryl 1,3-dioxolanes 347
aryl Grignard reagents 367
– electron-donating/electron-accepting groups 367
β-aryl α-hydroxy esters 54
aryl ketones 163
– α,α-dicyanoalkenes 313
– enantioselective α-amination 163, 164
aryl oxide anions 235
– binaphthoxide 235
– phenoxide 235
asymmetric aldol reaction, see aldol reaction
asymmetric allylic alkylation reaction, P,N-bidentate ligands 97
asymmetric aminohydroxylation (AA) reactions 29, 55, 57, 61
– α-arylglycinols 59
– nitrogen nucleophile 59
– osmium catalyst 61
– silyl enol ethers 60
asymmetric annulation reaction 305
asymmetric Baylis–Hillman reaction 385
asymmetric C−C bond formation, synthetic methods 95
asymmetric conjugate addition reactions 249
asymmetric cyanation reaction 81–86
asymmetric cycloaddition reaction(s) 297, 308, 312, 320
– bifunctional cinchona alkaloids catalyzed 308–312
– cinchona-based phase-transfer catalysts 320–323
– cinchona-based primary amines catalyzed 312–320
– quinuclidine tertiary amine catalyzed 297–308

asymmetric dihydroxylation (AD)   29, 35, 38, 40, 45, 373
– alkene substitution patterns   38
– γ-amino-α,β-unsaturated esters   49
– asymmetric dihydroxylation reaction   45, 46
– conjugated polyalkene substrates   47, 48
– enol ether substrate   53
– functionalized alkene substrates   40–42
– $K_2OsO_2(OH)_4$ 34
– ligand acceleration effects   38
– nitrogen nucleophile   60
– nonconjugated polyenes   44
– steric requirements   35
asymmetric Michael addition reaction, phase-transfer conditions   252
asymmetric organocatalysis   131
asymmetric pericyclic reaction   297
asymmetric phase-transfer catalysis   136
– alkylation   151
asymmetric reductions, cinchona-based organocatalysts   125–127
asymmetric Reformatsky reaction   78, 79
asymmetric sulfinyl transfer reactions   351
– chiral sulfur-containing compounds, synthesis   351
– pharmaceuticals   351
axinellamines   338
azaallylic radical   365
azabicyclic analogues   375, 376
azabicyclic moiety rearrangements   403–409
1-azabicyclo[2.2.2]octane   372, 382, 384, 397, 403
– cinchona alkaloids   403
– H(C9) hydrogen   397
– helicity   384
aza-Henry reaction,   212. see also nitro-Mannich reactions
aza-Michael addition reaction   251, 252, 496
aza-Morita–Baylis–Hillman reactions   225–228
azide ion   410
azido ketone, Schmidt–Aubé reaction   390
aziridines   120, 153
azlactones, dynamic kinetic resolution   350
azodicarboxylates, electrophilic α-amination methods   163
azomethine imines, asymmetric 1,3-dipolar cycloaddition   319
azomethine ylides   308, 309
– tert-butyl acrylate   309
– 1,3-dipolar cycloaddition   308

b
Baeyer–Villiger oxidation   152
basic additives, pyridine   79

Baylis–Hillman reactions   382
benzaldehyde, enantioselective phenylacetylene addition   77
benzophenone   253, 258
– glycinate Schiff base   253, 258
  – Michael addition reaction   253
  – PTC conditions   253
O-benzoyl derivative, uses   155
N-benzoyl piperidino hemimalonate   190
9-O-benzoylquinidine   176
O-benzoylquinidine   336
O-benzoylquinine   161, 162
– moiety   161
O-benzylated N-anthracenylmethyl dihydrocinchondinium hydroxide   109
N-benzylcinchonidinium chloride   116, 425
N-benzylcinchoninium chloride   116, 136, 429
benzyl quininium chloride   113, 219
– chiral catalyst   219
N-benzylquinidinium fluoride   126
N-benzylquininium fluoride   127
– pseudoenantiomeric catalyst   127
Bergman cycloaromatization   371
betaine intermediates   222
biaryl diols resolution   431
bicyclicaminoketones, preparation   409
bicyliclactams, formation   391
bifunctional catalysis   310
– Diels–Alder reaction   310
– ligand   76
bifunctional cycloaddition reaction   300
– of o-benzoquinone diimides   300
– ketene enolates   300
bimetallic chelate-bridged ester enolate complex   95
binaphthol, chiral discrimination process   435
binding pockets, U/L-shape   36
Biotin synthesis   337
bipol ligands   85
bisalkaloid ligands   32–35
2,3-bisamino epoxide   391
bis-cinchona alkaloid(s)   204, 329, 350
– $(DHQD)_2PHAL$ 204
– $(DHQD)_2PYR$ 282, 283
  – enantioselectivity   283
– dual catalytic motion   350
bisligand systems, uses   32
N,O-bis-(trimethylsilyl)acetamide (BSA)-KOAc   95
Boc-proline   450
Bolm's protocol   329, 330

bridgehead bicyclic lactams, formation 389–392
bridgehead tertiary amine 133
bromide reaction 54
α-bromination 161
– acylated rubanones 389
bromodihydroquinine, Grob fragmentation 385
α-bromo hemiaminal 391
Bryostatin-C, synthesis 39
Buchler Chininfabrik GmbH, OCI/QCD/quinolines 366
butadiene, dimerization 98, 99
tert-butanesulfinyl chloride 353
– sulfinyl transfer reaction 353
tert-butyl-9-*O*-carbamates quinidine (*t*-BuCQD) 436
9-*O*-*tert*-butylcarbamoylquinidine 437
9-*O*-*tert*-butylcarbamoylquinine (*t*-BuCQN) 437
tert-butyl 2-(diphenylmethyleneamino)acetate/nitroalkenes 311
– asymmetric 1,3-dipolar cycloaddition 311
tert-butyl 2-(6-methoxynaphthalen-2-yl) acetate 153
– asymmetric methylation 153
t-butyl sulfinyl chloride, dynamic kinetic resolution 352

c

C=C bond 21
– hydrogenation 22
– Pd/cinchona-catalyzed hydrogenation 21
C9-*epi*-quinine 395
C=F bond forming reactions, cinchona-based chiral ligands 99
C6′-modified phase-transfer catalysts 369
– preparation 369
– reagents/conditions 369
C9-OH analogue, dimeric cinchona structure 141
C9-protected quinine 242
camphor 192
capillary electrophoresis 433
– chiral ion-pairing additive 451
– enantioseparation 451
carbamate 58
9-*O*-carbamoyl 453, 454
– cinchona alkaloids 436
11-*O*-carbamoyl quinine 453, 454
carbazole-9-carbonyl (CC) 462
2-(carboethoxy)cyclopentanone 154, 155
carbon C6, functionalization 389–392
carbon C9 394, 398, 399, 407

– C9-*epi*-configured stereoisomers 394
– cyanide ion, bond formation 399
– Grignard reactions 398
– inversion/retention 398
– nucleophilic attack 407
– nucleophilic substitution 394
　– C9-activated alkaloids 394–396
　– C9-hydroxy group replacement 396–399
– transformations 399
　– QCI/QCD 399–403
carbon-carbon bond formation 399
carbon-clad zirconia 447
carbonyl compounds 77, 79, 133
– alkynylzincs addition 77, 78
– indium-promoted allylation 79
– α-monosubstitution, mechanistic scheme 133
carbonyl derivatives 131
– cinchona-catalyzed nucleophilic α-substitution 131
– nucleophilic α-substitution 157
　– α-amination 162–165
　– α-arylation 158, 159
　– α-halogenation 160–162
　– α-hydroxylation 159, 160
　– *via* non-PTC 157
　– α-sulfenylation 165
– organocatalytic nucleophilic α-substitution pathways 131–133, 150, 157
　– β-keto carbonyl compounds 153–156
　– monocarbonyl compounds 150–153
carboxylic acids, enantioseparation 442
catalytic process 265, 288, 301, 302, 312
– dual-function catalyst 265
– enantioselective Michael addition reaction 265
– transient enol protonation 265
catalytic selenium, photochemical oxidation 34
catalytic sulfinyl transfer reaction 351
CatASium F214 catalyst 16
centrifugal partition chromatography (CPC) 448
chalcone, π-π stacking interaction 112
chalcone-type enones 111
– phase-transfer-catalyzed epoxidation methods 111
chelate-bridged enolates 94
chiral α-amino acids, synthesis method 84
chiral β-aminoalcohols 77
chiral ammonium ketene enolates reactions 86–92
chiral ammonium salt 232

chiral anion exchanger, polymeric
  materials 434
chiral artificial receptors 455
chiral artificial sensors 455
chiral Brønsted bases 308
chiral building blocks 206, 215
– 1,2-amino alcohols 206
– vicinal diamines 215
chiral catalyst 133
chiral cyanide/acylammonium ion pair 230
chiral α,α-dialkyl-α-amino acids (ααAAs)
  148, 149
– enzymatic inhibitors 149
chiral 2,3-dihydropyrroles synthesis 313
chiral discrimination 432
– activity 422
– π-basic quinoline ring 432
– binaphthol 435
– π-π interaction 432
– mechanistic studies 443–447
– process 443
– quinine 435
– tartaric acid 456
chiral fluorinating reagent 100
– N-fluorobenzenesulfonamide (NFSI) 100
chiral α-halocarbonyl products 161, 162
chiral hemiester 334
chiral α-hydroxyl carboxylic acids synthesis
  299
chiral inductor 186
chiral ion-pairing reagents 451
chiral ketene enolates 92, 302
– applications 92
– catalytic application 302
chiral ligands, uses 30
chiral lithium base-promoted rearrangement
  reaction 345
chiral monolithic materials 459
– capillary columns 462
– preparation 459, 460
    – direct/indirect, schematic
      representation 460
chiral organofluorine compounds 157
chiral phase-transfer catalyst 320
chiral phosphonates 242
chiral polymer bed, definition 461
chiral recognition phenomenon 433, 434
– applications 458
– capacity 458
chiral Rh-complex 88
chiral separation 433
chiral solvating agent (CSA) 453
– advantages 453
– multifunctional character 453

– quinine/quinidine 455
– ee determination 453, 455
chiral stationary phases (CSPs) 430
– analogous silica gel 459
– t-BuCQD 441
– t-BuCQN 441, 449
– cinchona carbamate controlling factors 445
    – 9-O-carbamoyl group 445
    – cinchona alkaloid, 9-carbon
      configuration 445
    – 9-O-cinchona carbamate 441
– 6′-neopentoxy-9-O-tert-
  butylcarbamoylcinchonidine 441
– immobilization chemistry 446
– quinoline ring 447
– support 447
– stereodiscrimination performance 443
– urea linkage 450
chiral tertiary amine 175, 177
– uses 175
chiralpak QD-AX, see 9-O-tert-
  butylcarbamoylquinidine
α-chlorination 161
α-chloroacrylonitrile, model reaction 182
chloroform, mobile phase 450
chloromethyl phenyl sulfone reaction 220
β-chloropropenones 155
9-chloro-substituted quinine, Grignard
  reagents 397
chromane, asymmetric synthesis 251
chromatographic packings 459
cinchona alkaloid-based thiouera catalyst
– dimethylchloromalonate catalysis 290
cinchona alkaloid 9-O-carbamates, 429, 447
– dimerization 429, 447
cinchona alkaloid derivatives 5, 82, 86, 425
– chemical structural 5
– monofunctional/bifunctional 98
– quaternary ammonium salts 215
– O-trimethylsilylquinidine 86
– O-trimethylsilylquinine 86
cinchona alkaloid esters,
  macrolactonization 379
cinchona alkaloid family
– cinchonidine 421
– cinchonine 421
– quinidine 421
– quinine 421
cinchona alkaloid ketones 381, 389
– synthesis 381
cinchona alkaloid ligands, side chain
  modification impact 373
cinchona alkaloids 1, 73, 74, 75, 77, 86, 92, 93,
  134, 171, 197, 198, 209, 237, 228, 374, 424

- active sites 3
- derivatives 3, 4
- alkynes derived, X-ray crystal structure 374
- α-amination 162
- β-amino alcohols 192
- ammonium salts 4
- *anti*-Markovnikov thiol addition reaction 434
- aryl alkyl ketones 13
- asymmetric organocatalysis 133, 134
- asymmetric oxidations 33
- bifunctional chiral catalysts 177
- BQD, bifunctional catalysis 300
- cage expansion 411
- C9-amino analogues 398
- C4′−C bonds 366
- chiral auxiliaries/chirality transmitters 13
- chiral inducting effect 188
- chirality inducers 2
- chiral ligands 32, 56
  - alkenes, asymmetric dihydroxylation 30
  - aminohydroxylation 56–61
  - sulfur oxidations 61
- chiral solvating agent 453–455
- chiral stationary phase (CSPs) 434, 435
  - first generation 435
  - preparation 434
- cinchona tree, bark 2
- cinchonidine 209, 237
- cinchonine 209, 237
- combinations 93
- conformational behavior 6
  - *anti*-closed/open 7
  - pivotal role 6
- conformational investigations 4
  - *anti/syn*-closed/open 6
  - computational/spectroscopic methods 4
- derived catalyst 274
  - fluorinating agent 160
- direct protonating agent, enolates 171
- feature 7
- functional groups 133
- heterogeneous Pd catalysts modification 15–22
  - catalysts 16
  - C=C bonds, hydrogenation 16
  - substrate scope 21, 22
- heterogeneous Pt catalysts modification 15–22
  - catalysts 16
  - solvents 16, 17
  - substrate scope 17–21
- ^{1}H NMR spectra 454
- homogeneous catalysts 15
- Rh/Ir-catalyzed transfer hydrogenation 15
- α-hydroxyamination 164
- hydroxylated amines 382
- 7-hydroxyquinine 393
- immobilization, thioether linkage 446
- industrial applications, HPB ester 22
- isocyanate 436
- macrocycles 379
- Michael donor/acceptor 181
- modes of action 471
- moieties 448
- nonnatural *epi*-configuration 395
- organocatalysts 197
- organic chemistry 361
  - C9-H proton 361
  - conventional β-elimination 361
  - deprotonation 362
  - 1,2-hydride shift 361
  - tautomerization 361
- oxophilic Lewis acid complexes 100
- polymer-bound 176
- protonation 6, 430
- QN/QD/CN/CD/CPN/CPD, structural features 422, 424
- quaternary ammonium salts 134, 138
- quincoridine 364
- quincorine 364
- quinoline moiety 368
- quinuclidine
  - nitrogen 471
  - novel transformations 382–394
- racemate resolution 426–428
- racemic acids resolution 424
- resolving agents 424
  - benefits 425
  - *N*-butylcinchonidinium bromide 431
  - carboxylic/phosphoric acids 429
- roles 1
- stereogenic phosphor/sulfur atoms resolution 430
- structure 74, 86, 198
  - abbreviations 14
  - information 4–8
- thiourea catalyst, dimethylchloromalonate catalysis 290
- thiourea derivatives 191
- uses 2
  - chiral Lewis base 2
  - nucleophilic catalysts 2
cinchona alkyne. Sonogashira coupling 370
cinchona-based H-bond donor catalysts 281
- cinchonine-derived squaramide catalyst 281

– squaramide moiety 281
cinchona-based *P*, *N*-bidentate ligands 96
cinchona-based quaternary ammonium salts 200, 201, 202, 235
cinchona-based sensors 455
– separation and analytics 455–462
cinchona-based thioureas 211, 217
– catalyst 286, 333
cinchona-calix[4]arene hybrids 449
cinchona-catalyzed Mannich reactions 213
– imine precursors, α-amidosulfones 213
cinchona-catalyzed sulfinyl transfer reactions 353
cinchona-derived anthracenylmethylated ammonium salts 123
cinchona-derived catalyst 119, 198, 206
– modification 18
  – Pt catalyst 22
  – substrates structures 18
– PTCs 109, 118, 126
– salt 124
cinchona-promoted asymmetric reactions 471
– tabular survey 471–505
  – aldol reactions 485
  – ammonium ketene–enolate 484
  – aziridination 502
  – cinchona-catalyzed asymmetric oxidation 479
  – cinchona-catalyzed desymmetrization 503
  – cinchona-catalyzed enantioselective conjugate addition reactions 492, 502
  – cinchona-catalyzed enantioselective nucleophilic 1,2-addition reactions 484
  – cinchona-catalyzed nucleophilic α-substitution of carbonyl derivatives 481
  – cinchona-catalyzed reduction of carbonyl compounds 480
  – cinchonine/cinchonidine 473, 474, 475, 478, 488, 491
  – cyanohydrin synthesis 490
  – cyclopropanation 501
  – Darzens reaction 489
  – (DHQ)$_2$AQN/(DHQD)$_2$PHAL 490, 491, 503
  – (DHQD)$_2$PYR 499
  – (DHQ)$_2$PHAL/(DHQD)$_2$PHAL 487, 495
  – (DHQ)$_2$PYR 477, 482
  – enantioselective protonation 483
  – epoxidation 502
  – Friedel–Crafts type alkylation 491
  – Henry reaction 487
  – hydrocinchonidine 472

– hydrophosphonylation 492
– hydroquinine/hydroquinidine 492, 501
– Mannich reaction 488
– metal-catalyzed asymmetric oxidations/reductions 472
– metal-promoted enantioselective C–C bond forming reactions 473
– Morita–Baylis–Hillman reaction 490
– nitro alkenes as acceptors 497
– nitro-Mannich (aza-Henry) reactions 489
– phosphinate 504
– quinidine 503
– quinine 477, 478, 481, 483
– Strecker synthesis 491
– trifluoromethylation 491
– α,β-unsaturated ketones, amides, and nitriles as acceptors 492
– vinyl phosphates 501
– vinyl sulfones 501
cinchona bark, antimalarial property 1
cinchona bases 446
– remote vinyl group 446
cinchona-catalyzed ketene enolates applications 302
cinchona diimides 456
Cinchona ledgeriana, 1. *see also* cinchona trees
cinchona modifiers 17
cinchona phase transfer catalysts 139, 143
– α-alkylalanine derivatives 149
– electronic factor 143
– stereoselectivity, origin 139
cinchona quaternary salts, molecular complexation 431
cinchona thiourea 191
– α-amino esters, asymmetric synthesis 191
cinchona trees 1
– cinchona alkaloids 2
– quinine 2
cinchonidine 22, 25, 203, 424
– based catalyst 203, 273
  – multifunctional 273
– derived thioureas 211
– hydrogen bonds 425
– naproxen resolution 425
cinchonidinium salts 144, 151
– 3,5-dialkoxybenzyl group 144
cinchonine 13, 421
– derived thioureas 211
Claisen condensation 23, 111
– epoxidation sequence 111
Claisen rearrangements 94, 95
cobalt alkyne complex 98
cobalt-mediated Pauson–Khand reaction 97, 98

Co(II)salen complex 89, 90
compactin lactone synthesis 54
conventional β-elimination, deprotonation at C9 362
cooxidant uses, osmium 29
copper(II) bisoxazoline complex 92
Corey model 36
– bifluoride catalyst 199
– phase-transfer catalyst 258
– sulfur ylide reagent 388
countercurrent chromatography 452
cumyl hydroperoxide (CHP) 117, 159
cupreidine 422
cupreine 422
cyanation reactions 229–234
– aldehydes/ketones 229
– cyanohydrin synthesis 229–232
– Strecker synthesis 232–234
cyanide source 81, 229, 233
– acetone cyanohydrin 233
– acetyl cyanide (CH$_3$C(O)CN) 229
– cyanoformic esters (ROC(O)CN) 229
– trimethyl-silyl cyanide (TMSCN) 229
α-cyanocyclopentanone, model reaction 182
α-cyano ester, fluorination 157
cyanoformic esters (EtOCOCN) 82
cyanohydrin synthesis 81–84, 229–232
α-cyanoketones, addition-protonation reactions 264
cyclic enone systems 113, 125, 254, 293, 319
– asymmetric 1,3-dipolar cycloaddition 319
– cyclohexenone 125
– epoxidation 113–115, 293
  – cinchona-based bifunctional primary amine catalysts 293
– self-dimerization, PTC conditions 254
cyclic β-ketoester, 2-(tert-butoxycarbonyl) cyclopentanone
  – asymmetric benzylation, 144, 153
cyclic peroxyhemiketals 119, 120
cyclic sulfate chemistry 54, 55
– epoxide 54, 55
– comparison 55
cycloaddition reaction 297, 298
– o-benzoquinone imides 298, 300
  – with ketene enolates 299
– cinchona-catalyzed 297
– of ketene enolates/o-quinones 298, 299
– ketene-N-thioacyl imine [4+2] cycloadditions 303
  – cinchona alkaloid catalyzed 303
– nonnucleophilic Hunig base 298
– Pd(II) cocatalyzed, bifunctional mechanism 301

– ring opening reaction 301
[2+2]-cycloadducts 228
– chiral β-lactones 228
– β-lactams 228
cyclohexanes 306, 317
– construction 317
– [3+3]-type reaction 306
cyclopentanes, construction 317
cyclopropanation reactions 288–293
– aziridination 292
– epoxidation 292

d
DBU-promoted double-bond isomerization/ cyclization 259
deazacinchonidine derivative catalyst 114
deconjugative Michael reaction 252
Dess–Martin oxidation 379
Dess–Martin periodinane 379, 392
Desvinyl-quincorine ester 381
(DHQ)$_2$PYR 161
(DHQD)$_2$AQN 346, 347, 348, 349
(DHQD)$_2$AQN-OMs 410
– azide ion 410
– cascade reactions 410
diamines, catalytic efficiency 339
diaminopropanoic acid linker, configuration 90
diastereomer crystallization 424
diastereomeric quinidine-based diols 375, 377
diastereomeric sulfinylammonium salts 352
diastereoselective hydrogenation 344
α-diazo-β-hydroxyesters 201
dibenzyl azodicarboxylate 162, 163
dibromoisocyanuric acid (DBI) 57
1,5-dibromopentane 150
β-dicarbonyl compounds 165
– C–H bonds 165
  – α-sulfenylation 165
1,3-dicarbonyl systems 159
– α-hydroxylation 159
– substrate systems 162
dichloroisocyanuric acid 58
dichloromethane 401
– QCI 401
  – quaternization 401
dichloromethane-hexane 116
α,α-dicyanoalkene 313
– domino Michael–Michael addition reaction 313
Diels–Alder reaction 310, 311, 315
– bifunctional thiourea catalysts 311

difluoroketones 208
dihydro analogues 370
dihydropyrimidones, synthesis 213
2,3-dihydropyrroles, application 311
dihydroquinidine (DHQD)$_2$AQN 330
– anthraquinone-bridged dimers 330
– ee values 331
dihydroxylation procedure 37, 50
– uses 37
2,6-diisopropylphenylcarbamoyl-quinine (DIPPCQN) 437
α,γ-diketo acid derivatives, hydrogenation 19
α,γ-diketo esters 19
dimeric cinchona alkaloids 225, 230, 347, 349
– (DHQD)$_2$AQN 349
– kinetic resolution 347
– monocarboxylic acid salts 225
7-dimethylaminosulfonyl-1,3,2-benzoxadiazol-4-yl (DBD) 461
9,10-dimethylanthracenyl moiety 141
cis-2,4-dimethyl glutaric anhydride, desymmetrization 328
2,7-dimethylnaphthalene moiety 140
dinitrobenzoyl-alanine 448
– enantioseparation 446
– hydrogen bond 444
– immobilization chemistry 446
– ionic interaction 444
– leucine 431, 443
– π-π stacking 444
– van der Waals attraction 444
– X-ray structure 444
(R,S)-dinitrobenzoyl-leucine 443
dinitrobenzoyl-oligoalanines 448, 451
1,2-diols reactions 51
– asymmetric oxidation 51
– cyclic sulfates 54–56
– Jacobsen asymmetric epoxidation 52
– sulfites 55
dinitrophenyl-derivatized amino acids, separations 445
dioxanone derivatives 222
1,3-diphenyl-3-acetoxyprop-1-ene 96
diphenylmethyloxy-2,5-dimethoxyacetophenone 152
diphenylphosphinoyl aryl imines 225
N-diphenylphosphinoylimines 75
dirhodium(II) complexes 88
– Rh$_2$(4S-MEAZ)$_4$ 88
– rhodium(II) acetate 88
α-disubstituted indanone, photodeconjugation 193
di-tert-butyl azodicarboxylate (Dt-BuAD) 163, 164, 368

domino Michael–Michael addition reaction 314, 317, 318
– iminium-enamine catalysis 314
domino process 305
double activation model 203
dynamic kinetic resolution (DKR) 325

e
electron deficiency 146
– alkene
 – Morita–Baylis–Hillman reactions 221–228
– allenic esters/ketones, uses 259
– aromatic aldehydes 202
– olefins 108, 123
 – asymmetric epoxidation 108
electron-deficient
– C=C double bonds 249
electron delocalization 407
electron-donating groups, OMe 201
electron-rich aromatic aldehydes 204
electrophile 350
electrophilic aldehyde 365
electrophilic sulfur reagent 165
enamine catalysis approach 202
enamine formation 164
– 4 Å molecular sieve 164
– enol ethers 172
enantiodiscrimination 435
enantioenriched β-hydroxyisobutyric acid 189
enantioenriched β-trifluoromethyl-β-amino acid 193
enantioselective chromatography 433
– chiral stationary phase preparation 434
 – first generation 435
 – cinchona-based chiral modifiers 450
  – capillary electrochromatography 450
  – capillary electrophoresis 450
  – nonaqueous method 451
– cinchona-based chiral selectors 447–450
– cinchona 9-O-carbamates 436
 – applications 437–443
 – chiral discrimination, mechanistic studies 443–447
– related techniques 433
– silica gel TLC plates 452
enantioselective cyclopropanation process 288
– cinchona alkaloid derivatives 288
enantioselective decarboxylative protonation (EDP) 184, 191
– β-amino alcohols 187
– comparative performances 187

- α-amino esters, Brunner's asymmetric synthesis 190
- asymmetric synthesis 187
  - fluorotetralone *via* debenzylative 187
- chiral inductor 184
- 2-cyanopropanoic derivative 189, 190
- copper-catalyzed 184, 185
  - prochiral malonic acids 185
  - prochiral phenylalkylmalonic acids 184
  - racemic hemimalonates 185
  - substituted malonic acid 184
- cyclic β-ketoesters 187
  - palladium-catalyzed tandem debenzylation 187
- enantioenriched proteinogenic amino acid precursors, synthesis 189
- β-hydroxyisobutyric acid synthesis 189
- malonic acid synthesis 184
- mechanisms 184
- optically enriched pipecolate 190
  - preparation 190
- organocatalysts 188–192
  - Brunner's 188
  - Muzart/Henin's 189
- palladium-catalysts 185–188
  - allyl-/benzyl-carboxylated compounds 185
  - chiral proton 185
  - cyclic β-keto esters 187
  - debenzylation 186
  - deprotection 186
  - prochiral enol carbonates 185
  - racemic β-keto esters 185
enantioselective extraction 432
- schematic representation 432
enantioselective hydrophosphonylation 93
enantioselective liquid chromatography 433, 436
- capillary electrochromatography 433
- capillary electrophoresis 433
- cinchona 9-*O*-carbamates 436
  - applications 437–443
  - chiral discrimination, mechanistic studies 443–447
- counter-current chromatography (CCC) 433
- hormones thyroxine 438
- mechanism 433
- simulated moving bed (SMB) 433
- supercritical fluid chromatography (SFC) 433
enantioselective Michael addition reactions 180, 252, 261
- cyclopentenone 180

- non-PTC-catalyzed 261–274
  - amino-derivatized cinchona alkaloids 272–274
- PTC-catalyzed 252–261
enantioselective protonation 171
- camphor, chiral aminoalcohols 192
- chiral inductor 183
- enols/enolates/enamines 171, 173
- equivalents 172–175
- groups 171
- indanone 188
- ketenes 175
  - nucleophilic addition 175–178
- Levacher's organocatalytic 174
- lithium enolate 191
- metal enolates 172
- metallic nucleophiles 175
- Michael addition reaction 178–183
  - acyclic cyanoacetates 182, 183
  - optimization 183
- preformed enolates 172–175
- proton Migrationr 192–194
- quinidine 173
- silyl enol ether 174
enantioselectivity 149, 278
- producing reaction system 149
end-capping procedure 447
eniminium ion intermediate 119
enolate 137, 141, 173
- alkylation 141
- asymmetric protonation 173
- generation reactions 172
enol, enantioselective tautomerization 180
enol ether
- dihydroxylation 53
- substrate, uses 53
enolizable β,γ-unsaturated ester, alkylation 151
enones 110, 115, 118, 120, 318
- asymmetric epoxidation, synthetic applications 115–117
- aziridination 120–125
- organocatalytic asymmetric epoxidation 118–120
- organocatalytic 1,3-dipolar cycloaddition 318
enzymatic hydrolysis 149
enzyme system 279
ephedra alkaloids 78
9-*epi*-amino-9-deoxycinchona alkaloids 273
9-*epi*-aminoquinine 399
- novel bifunctional organocatalysts, synthesis 399
9-*epi*-cinchonine benzamide derivative 190

9-*epi*-configured bromoquinine   404
– *epi*-1a-Br/*epi*-2a-Cl, silver ion   404
– first cinchona cage expansion   404
9-*epi*-quinidine *epi*-2a-OH   395
9-*epi*-quinine *epi*-1a-OH   394
9-*epi*-quinine mesylate *epi*-1a-OMs   394
9-epiquinine thiourea, dimeric structure   8
α,β-epoxysulfonyl compounds   218
ester product   152
ethylene dimethacrylate (EDMA)   461
ethylene glycol   431
ethyl nosyloxycarbamate   122
– asymmetric aza-Michael-initiated ring closure additions   122
ethyl pipecolate, *N*-protected synthesis   190
ethynyl (vinyl) acceptor   409

f

face selectivity, investigation   388
facile ring   55
Feist–Bénary reaction   306, 307
– catalytic mechanism   307
first cinchona cage expansion   403, 404
– bromoquinine   404
– *O*-mesylated quinine   406
– quincoridine   405
– quincorine   405
9-fluorenylmethoxycarbonyl (Fmoc)   432
– protected cyclic amino acids   450
fluorinated ketones, hydrogenation   20
α-fluorination   157, 161
fluorine-containing peptides   157
– H···F bonding   157
N-fluoro-*O* (9)-acyl hydroquininium salt   157
N-fluorobenzenesulfonimide (NFSI)   157, 160
2-fluoro benzyl β-keto ester   186
– catalytic enantioselective decarboxylative protonation   186
fluorobis(phenylsulfonyl)methane (FBSM)   214, 261
– Mannich-type reaction   214
N-fluoro-cinchona reagents   157
α-fluoro-keto esters   157
3-fluorophenyl methyl ether   118
2-fluoro-1-tetralone   186
α-fluorotetralone, uses   151
^{19}F NMR resonances   454
– CF$_3$ group   454
Friedel–Crafts reaction
– alkylation   237–240
– Michael addition reactions   279
  – naphthols, use   279

g

glycidyl methacrylate   461
glycine allylic esters, *N*-protected   94
glycine-derived enol silyl ethers   200
glycine esters   136
– benzophenone imines   136
  – asymmetric alkylation   136
Grignard reagents   54, 366, 367
– nucleophilic addition   367
Grob fragmentation   385, 386

h

α-halogenation   156
– ketones   390, 391
3-halo-3-pyrrolin-2-ones preparation
– synthetic scope for   322
hemibenzyl esters   329
hemimalonate reaction   188
Henry reactions   206–209
heterobimetallic (*S*)-AlLi(binaphthoxide) ((*S*)-ALB) complex   82
heterobimetallic multifunctional catalyst   93
N-heterocycle, Michael addition reaction   276
hexafluoroisopropyl acrylate (HFIPA)   222
hexafluorophosphate, counteranion effect   142
hormone receptor antagonist, melanin concentrating   211
– SNAP-7941   211
Horner–Wadsworth–Emmons reaction   346
– olefination   287
hydrazones, aza-Michael addition reaction   251
hydrocinchoninium catalyst   155
hydrogen bonding   36, 143, 318, 435
– interaction   219, 306
hydrophosphonylation   240–244
hydroquinine   159
– thiourea catalyst   217, 281
hydrosilane, polymethylhydrosiloxane (PMHS)   127
hydroxyamination   164
γ-hydroxyenone synthesis   344
α-hydroxy esters synthesis   302
2-hydroxyethyl methacrylate (HEMA)   461
α-hydroxyindanone   156
α-hydroxy-β-keto esters   159, 160
α-hydroxy ketones   156
α-hydroxylation   156
2-hydroxymethyl quinuclidin-5-ols, synthesis   387
2-hydroxy-4-phenyl butyrate (HPB) ester   22, 23
– catalyst   23

- Fluka, related compounds 24
- industrial manufacturing 24
- modifier 23
- reaction conditions 23
- second-generation synthesis 23
- solvent 23
- substrate quality 23
3-hydroxy-2-pyrone 309, 310
- Diels–Alder reaction 309
  - bifunctional catalysis 310
  - with N-methylmaleimide 309
- HOMO energy 310
- rigid complex formation 309
α-hydroxytetralone 156
hypochlorite salt 110

i

4-imidazolyl 78
imines, dialkylzinc addition 75–77
iminium catalysis 293
- enamine catalysis 313
α-imino ester-metal complex 308
- dipol-chiral base ion pair 308
immunosuppressor, mycestericin E 223
indium-mediated addition 79–81
- allylation 80
indole-cinchona alkaloids 371
iodo ketones 406
2-iodomethylquinuclidin-5-ones, ring expansion 406
3-iodo-3-pyrrolin-2-one, coupling reactions 321
ionic liquid-water-$tert$-butanol solvent system 37
ion-pairing mechanism 293, 444
β-isocupreidine (β-ICD) 212, 221, 225
isocyanate 398
isocyanoesters, cycloaddition to nitroalkenes 312
isoflavone epoxide 117

j

Jacobsen asymmetric epoxidation 52
Jones oxidation 379
Jones reagent 379

k

ketene addition, nucleophilic additions 172
ketene intermediate mechanism 161
ketenes 86, 90, 175
- nucleophilic addition 175–178
  - Lewis acid 86–92
α-keto acetals, hydrogenation 20
α-keto acid derivatives 18
- ee values 19
- structures 18
β-keto carbonyl derivatives 153
- asymmetric alkylation 153
- catalytic enantioselective α-alkylations 153
β-keto carbonyl substrates 153
- N-protected aziridine 153, 154
α-keto esters 22, 207
- Henry reaction 207
- Pt-cinchona-catalyzed hydrogenation 22
β-ketoester(s) 155, 156, 158, 159, 162, 163
- α-alkynylation 156
- anti-Michael reaction 260
- asymmetric α-arylation 158
- asymmetric 1,6-addition 256
- 2-($tert$-butoxycarbonyl)indanone 156
- conjugate addition reactions 258
- enantioselective α-amination 162
- α-hydroxylation 159
- β-keto nitrile, nucleophilic addition 180
- Michael addition reaction 263
- α-position 155
- α-substituted, Michael addition reaction 262
α-keto ethers, kinetic resolution 21
α-keto ketals, hydrogenation 20
α-ketolactone hydrogenation 25
ketone reductions, homogeneous systems 14
ketones 16
- enol carbonates 186
- heteroatomic nucleophiles 175
  - chiral tertiary amine catalyzed addition 175
- hydrogenation 16
- keto esters 186
- Michael–Henry reaction 282
ketopantolactone, hydrogenation 25
kinetic isotope effect 328
kinetic resolution 50
Kornblum–DeLaMare rearrangement 342
Krapcho procedure 180
Kumada–Corriu coupling reaction 401, 402

l

β-lactam 56, 90, 91
- adducts formation 302
- substituted alkenes 50
- synthesis 56, 90, 91
γ-lactams 320, 321
- optically active 320, 321
- preparation 320, 321
lactones 165
- trichloromethyl moiety 304

layered double hydroxide (LDH)   61
levo-thyroxine sodium salt   440
– chiral impurities, identification   440
Lewis acid(s)   87, 363, 403
– additive   220
– cocatalyst   91, 303
– dimeric salen-Ti complex   82
– hydroxyl function   3
– LiClO$_4$ 87
Lewis base
– catalysts   306
– trivalent phosphinite group   339
ligand-accelerated catalysis (LAC)   4
ligand acceleration effect (LAE)   29
lipophilic octadecyl chain   452

m

malonic acid half-thioesters, CoA-bound   279
mandelic acids   434, 441, 446, 462
Mannich α-aminomethylation, carbonyl system   382
Mannich reactions   209–215
– catalysts   211
– *N*-protected aromatic imines substrates   212
masochistic nucleophilic attack   388
meroquinene ester, synthesis   363
meso-anhydrides   333
– alcoholytic desymmetrization, applications   337
– cinchona-catalyzed alcoholysis   336
– heterogeneous methanolytic desymmetrization   334
– methanolysis   333
meso-aziridine anhydride   329
meso-compounds   325, 326
– desymmetrization   325, 326
    – cinchona-catalyzed   325
– types   325
meso-cyclic anhydrides   98, 99, 326–336
– alcoholytic desymmetrization reactions   330
– asymmetric methanolysis   99
– enantiotopic differentiation reaction   98
– methanolytic desymmetrization reaction   331
meso-diols   336–341
– benzoyl chloride   338
– desymmetrization reaction   336, 338, 341
    – cinchona-mediated   336
    – enzymatic/nonenzymatic catalytic   341
– hydroxyl groups   336
meso-endoperoxides   326, 341–344
– desymmetrization   342

meso-epoxy anhydride, methanolytic desymmetrization   329
meso-phospholenes   326, 344, 345
– desymmetrization   344
    – alkene isomerization   344
    – allyl alcohols   345
meso-phospholenamide   326
meta-di(1-naphthyl)-substituted catalyst   145, 146
metal binding ability   3
– chiral ligands/chiral modifiers   3
metal-catalyzed asymmetric reductions   13
– cinchona alkaloids   13
    – chirality transmitters   13–25
metal salts   94, 308
methanol, achiral additives   75
methanolysis   333, 395
– enantioselectivity   334
    – concentration effect   333, 334
    – temperature effect   333, 334
methanolytic desymmetrization, with catalysts   335
β-methoxy amines   411
methyl dihydrojasmonate, synthesis   181, 255
methyl (*R*)-2-hydroxy-4-phenyl butyrate, synthesis   22
N-methyl-morpholine-*N*-oxide (NMMO)   29
– potassium ferricyanide   30
– uses   30
2-methylnaphthoquinone, asymmetric epoxidation   114
2-methylpropionaldehyde, phenylacetylene addition   78
N-(6-methyl-2-pyriylsulfonyl)-imines   241
methyltrioxorhenium   34
methyl vinyl ketone (MVK)   223, 227, 270
Michael acceptors   253, 267
– broad spectrum   267
Michael addition reactions   172, 273
– dimethyl malonate   179
– double bond transposition sequence   254
– enantioselective protonation   178
– β-keto ester/nitrile   182
– nucleophiles   178, 183
– thiobenzoic acid   179
– thiophenol   178
Michael donors   221, 253, 256, 267, 286
– amines/phosphines   221
– broad spectrum   267
Michael–Henry reaction   315
– C–C bond forming processes   315
Michael–Michael cascade process   271
– DKR-mediated   271
– quinine-based thiourea catalyzed   271

Mitsunobu reaction, hydrazoic acid   398
molecularly imprinted polymers (MIPs)   456, 458
– advantages   458
molecular oxygen, electrophile   156
monobenzoylated product   340
– ee values   340
– kinetic resolution   340
monocarboxylic acid salts   225
monolithic materials   459, 461, 462
monomeric cinchona phase-transfer catalysts   144
– catalytic asymmetric organic reactions   144
Morita–Baylis–Hillman carbonates   305
– intramolecular Michael addition reaction   305
Morita–Baylis–Hillman reactions   221–228
Mukaiyama-type aldol reactions   198–200
Muzart/Hénin's procedure   188
– Baiker's reinvestigation   188

n

naproxen   153
– asymmetric synthesis   180
– resolution   425
naphthyl aldimine   149
N-(α-naphthylmethyl) quinidinum chloride   114
9-*nat*-chlorides   398
– Grignard reaction   398
9-*nat*-cinchonidine/cinchonine   411
natural cinchona alkaloids   249–251, 266, 283
– as catalysts   249–252
– HCD catalysts   251
– stereochemistry   266
– structures   250
– use   250
natural cinchonidine, functions   289
natural quinidine catalyst, racemic product   268
N–F fluorinating reagents   99
nitroalkenes   274, 275
– conjugate addition   274–284
p-nitrobenzyl enaminoester   193
α-nitro esters, stereoselective α-fluorination   160
nitrogen nucleophile   51, 58
nitro-Mannich reactions   212, 215–218
– catalysts   217
nitromethane acid   207
nitrosobenzene   164
NMR spectroscopy, ee determination   453
nonnatural 9-*epi*-quinine   403
– silver benzoate-mediated reaction   403

nonproteinogenous amino acids   424
nonsteroidal anti-inflammatory drugs (NSAIDs)   52, 153
novel cinchona alkaloid organocatalysts, synthesis   370
nucleophile(s)   348, 350, 364
nucleophilic addition   175
– tertiary amine, catalytic role   175
nucleophilic catalysis mechanism   305
nucleophilic catalysts   263
– DABCO   263
– β-isocupreidine (β-ICD)   263
– quinuclidine   263
nucleophilic displacements   397
– configuration, inversion   397
nucleophilic monoanions   399

o

Oda catalysis mechanism, schematic presentation   328
6'-OH cinchona alkaloid catalyst   264
olefins   2, 122
– asymmetric epoxidation   292
– electron-deficient   292
– osmium-catalyzed asymmetric dihydroxylation   2, 3
– π-bond   122
one pot cyclopropanation processes   297
optically active bicyclic hydroxyketone, synthesis   81
optically active chiral amines, synthesis   209
organolithium reagents   366
organophosphorous materials   240
organozinc addition reaction   74–78
osmate esters, hydrolysis   32
osteoporosis medicine   338
oxazatwistane   368, 382, 385
– OH group   368
oxazoline moieties   150
– function   150
– α-proton activation   150
– side chain hydroxy groups, protection   150
oxindoles   160, 161
2-*oxo*-cyclopentanecarboxylic acid *tert*-butyl ester   100
– enantioselective fluorination   100
oxygen-free phase-transfer conditions   151
N-oxypyridine-based series   146

p

palladium
– catalyzed asymmetric allylic substitutions   95–97
– mediated cascade reaction   186

Pauson–Khand reaction 97, 98, 371, 372
pentamethylphenyl moiety 151
peptides 149, 441
– αα asymmetric aminohydroxylation 149
– biological activities 149
– complex stereoisomerism 441
phase-transfer catalyst (PTC) reaction 107, 132, 252, 256, 261
– α-amino acid derivatives 135
  – α-substitution 135–150
– 9-anthracene 149
– bridgehead nitrogen, Merrifield resin 146
– carbonyl derivatives 150
  – α-substitution 150–156
– catalytic conditions 154
– chiral onium 132
– cinchona-based 107, 139, 142
– cinchonidine-derived 256
– cinchonidinium model 137, 139
– complex ion pair 135
– diacetamido-PEG2000 chloride 148
– dimeric cinchona 140, 142
– 2,4-dinitrofluorobenzene 155
– α-heteroatom substitution 156
  – carbonyl derivatives 156, 157
– hydrogen bonding 144
– Lygo–Corey 147
– Park–Jew 152
– PEG-bound 147
– plausible model, stereoview 141
– polymerization, advantage 139
– quinine-derived 216
– quinidine-derived 261
– quinoline ring 261
  – aromatic π-π interactions 261
– rubidium hydroxide 150
– stereopair representation 138
phase-transfer catalytic α-substitution 134, 135
– α-amino acid derivatives 135
  – benzophenone imines, monoalkylation 135–148
  – glycine esters 135
– α-monosubstituted α-amino acid derivatives 148
  – alkylation 148–150
phase-transfer glycolate alkylation 152
phenanthryl 342, 343
phenylacetylene 77
– asymmetric addition 77
– Et_2Zn-mediated addition 77
α-phenylacrylates 179

– tandem thiophenol addition/enantioselective protonation 179
α-phenyl cyanoacetate, Michael addition reaction 286
phenylmethylketene 176
– asymmetric protonation 176
– methanol addition, Nakamura's supported catalysis 176
1-phenylpropargylic bromide 80
phenyl selenides, cinchona alkaloid derived 96
phosphazene base 200
– t-butyliminotri(pyrrolidino)phosphorane (BTTP) 200
N-Z-phosphinic pseudodipeptides, stereoisomers 451
phospholane ligand, synthetic precursor 344
phospholenes, catalytic asymmetric synthesis 344
phthaloylphenylglycinonitrile 157
– α-fluorination 157, 158
(−)-pironetin, catalytic asymmetric total synthesis 88
polyenes 43
– conjugated polyenes 43, 44
– nonconjugate olefins 43
– systems 44
poly(ethylene glycol) (PEG) 113, 147
– acetamido-moiety-connected cinchona 148
– cinchona PTCs 147
polyketide synthesis 280
– active site, MAHT activation 280
polymer-bound catalysts, advantages 146
polymeric monolithic material 461
– racemic DNZ-leucine separation 461
polymerization process, receptor binding site 458
polymer-supported cinchona catalyst 162
polystyrene-bound cinchona alkaloids (PS-CA) 160, 161
potassium ferricyanide 34
P,P-diphenyl-N-(phenylmethylene)phosphinic amide 75, 76
– diethylzinc addition 76
prochiral C=O/C=N moieties 197
– nucleophilic addition 197
prochiral enolate 175
prochiral ketone(s) 177, 326, 346
– desymmetrization 346
prochiral silylketenes 176
propargyl bromide 81
– enantioselective indium-mediated addition 81
propargylic bromide 80

propiophenones 192
O-protonation 171
proteinous amino acids, N-protected 438
– enantioseparation 438
proton shift 193
pseudoenantiomeric N-benzylcinchonidinium chloride 137
pseudoenantiomeric cinchonidine 344
pseudoenantiomeric quinidine 240
– analogue 202
pseudoenantiomeric quinine-derived catalyst 207
pseudoenantiomeric-substituted 3-quinuclidinones synthesis 383
pseudolanthanide triflates 87
Pt catalysts 17
– cyclic imidoketone 21
– α,γ-diketo esters 18
– fluorinated ketones 18–20
– α-keto acetals 20
– α-keto acid derivatives 17, 18
– α-keto ethers 20, 21
– ketopantolactone 21
– substrate scope 17
Pt-cinchona system 16
1-pyrenylmethyl group 145, 146
2-pyridyl ketones 78
2-pyrones 310, 314, 316
– Diels–Alder reaction 310, 314, 316
  – bifunctional catalysis 310

q

quincoridine (QCD) 364, 403, 405
– diamine ligands 401
– En route 365
– π-face selectivity 386
– first cinchona cage expansion 403, 405
– 2-iodomethyl ketones 406
– Jones oxidation 380
– oxidative functionalization 380, 381
– phosphine/amino ligands 402
  – asymmetric catalysis 402
– preparation 364–366
– quinidine 364
– S_N2 displacements 400
– vinyl side chain 386
quincorine (QCI) 364, 403, 405
– diamine ligands 401
– first cinchona cage expansion 403, 405
– 2-iodomethyl ketones 406
– Jones oxidation 380
– oxidative functionalization 380, 381
– phosphine/amino ligands 402
  – asymmetric catalysis 402
– preparation 364–366
– quinine 364
– S_N2 displacements 400
– vinyl side chain 386
quincorine (QCI)-alkyne 372
– synthesis 372
– transformations 372
quinidine (QD) 2, 13, 241, 421, 424
– anti/syn conformers 6
– tert-BuMgCl 352
– 9-O-carbamates 436
– derived thiourea derivative 206
– Grignard reagents 397
– hydrobromination 381
– hydrolysis 396
– metabolism 381
– QCD formation 365
– thionyl chloride 352
– thiourea derivatives 182
quinidine-derived catalysts 206, 238, 262, 275
– diastereoselectivity 275
– free OH group 262
– enantioselectivity 275
– Gauche-open conformer 276
– Michael donors, application 275
quinine (QN) 1, 13, 134, 362
– antimalarial drug 421
– aromatic 6'-methoxyquinoline ring 423
– tert-butylcarbamate selector monomer, copolymerization 459
– chiral ion-pairing additive 451
– chiral phase-transfer carrier 431
– chiral sulfinyl transfer agents 352
– cinchona tree, bark 2, 421
– dimer, conformation 7
– R-enantiomer selective 431
– fluorescence-based assay 412, 457
– hydrolysis 396
– immobilization, chiral anion exchange polymeric materials 434
– molecular weight measurement 7
– 1-naphthoate 353
– oxazatwistanes 384
  – synthesis 385
– Pd complexes 14
– 9-O-phenylcarbamate 447
– quinotoxine 362
– thiourea catalyst 8
– thiourea derivatives 182
quinine-derived catalyst 206, 262, 275
– diastereoselectivity 275
– enantioselectivity 275
– free OH group 262
– Michael donors, application 275

quininone, oxidative cleavage 363
o-quinone methide (o-QM) 320
– enantioselective cycloaddition 320
  – with silyl ketene acetal 320
– Lewis acid (LA) complexation 300
quinoline moiety 328, 366
– C2'-position substitution 366
– transformations 366–368
quinoline ring 7, 344
–π-π interaction 454
quinotoxine 361, 363
– cleavage, reagents/conditions 362
quinuclidine 3, 328, 384
– carbons, challenging sites 393
– 2-carboxylic acid ester, synthesis 381
– functionalization 393, 394
– nitrogen 3, 344, 353, 422, 454
3-quinuclidinols 386
quinuclidinone 382, 386, 387
– helix, masochistic nucleophilic 387
– nucleophilic attack 382–388
– synthesis 392

r
racemates resolution 423, 424
– crystallization 423–430
– diastereomeric salts 423
– naproxen resolution 424
racemic amino acids resolution, TLC
  technique 452
racemic carboxylic acids 441
racemic compounds 327, 346
– dynamic kinetic resolution 325, 327, 346
  – catalytic sulfinyl transfer reaction 351–354
  – racemic azlactones 350
  – racemic cyclic anhydrides 346–348
  – racemic N-cyclic anhydrides 348–350
racemic cyclic anhydrides 346–348
– (DHQD)₂AQN 346
– dynamic kinetic resolution 346
racemic dichloroprop, HPLC
  enantioseparation 448
racemic sulfinyl chloride 352
racemic tartaric acid 363
racemization reaction, DKR process 347
recrystallization process 152
rubanones transformations 389
rubidium hydroxide 150
Ruppert–Prakash reagent, TMSCF₃ 199, 234
ruthenium-catalyzed oxidation 54

s
Schiff base 136
second cinchona cage expansion 407
second cinchona rearrangement 408
– change effect 408
– cinchonidine motif 408
– cinchonine motif 408
– O-mesyl-cinchonidine 408
selenoxide 34
serines, oxazoline-based substrate 150
Sharpless ligands, see bis-cinchona alkaloids;
  monocarboxylic acid salts
Sharpless model 36
silica, immobilization 448
silicium atom 176
silyl enol ethers 160, 161, 174, 175, 198, 199
– activator 199
– enantioselective protonation 174
– nucleophilic activator 198
– organocatalytic protonation 175
silylated quinidine-based ketone 389
– aldol reactions 389
silylated thiazolium salt, desilylation 280
Simpkins's thiophenol addition reaction,
  silylalkylketenes 177
Soloshonosk's biomimetic procedure 194
– Plaquevent's optimization 194
solvolytic displacements 394
Sonogashira coupling reaction 371, 372
Staudinger protocol 400
stereoselective alcoholysis 326
stereoselective analysis 261, 451
stereoselective reaction systems 131
stereoselective synthesis 448
Strecker synthesis 84–86
styrene-divinylbenzene copolymeric
  catalyst 334
styrenes 57, 59
– α-arylglycinols preparation 59
– tert-butyl carbamates, use 57
α-substituted β-aminophosphonates
  synthesis 284
– chiral precursors 284
α-substituted azlactones, catalytic dynamic
  kinetic resolution 350
α-substituted enones 118
α-substituted Michael acceptor 178
– Pracejus's benzyl thiol addition 178
α-sulfenylated β-keto esters, diastereoselective
  reduction 165
sulfinate ester 353
sulfonamide 32, 334
– catalyst 334
– uses 32
sulfur oxidations 61
surfactant 112
– Span 20 112

- Triton X-100 112
- Tween 20 112
syn-aldol product 200
syn conformer binding 456
syn-β-proton 345
synthetic amino acids, enantioseparation 439

t

Takemoto's catalyst 212, 270, 350
- Brønsted basic tertiary amine group 350
- Lewis acidic (thio)urea moiety 350
Takeuchi's benzil-to-benzoin reductive asymmetric protonation 173
tartaric acid 412, 456
- chiral discrimination 456
- resolution, quinicine 422
- water, addition 412
tartrate sensing 457
- indicator displacement assay 457
tautomerization 362
tetrabutylammonium bromide (TBAB) 308
- S_N2 mechanism 308
tetrafluoroborate, counteranion effect 142
tetramethylammonium fluoride (TMAF) 235, 280
thermodynamic trans-diastereoisomer 180
thiazoline-based substrate 150
thiazoline moieties 150
thioisocyanate 398
thionyl chloride 54
thiourea catalysts 265, 268, 270, 277, 334
- activity 277
- epi-cinchonidine/cinchonine-derived 265
- quinine/quinidine-derived 270
  - catalytic efficiency 270
- synergistic cocatalytic effect 277
thiourea-cinchona alkaloid catalysts 280
thiourea derivatives 238, 266, 368, 370
- epi-quinine/quinidine-derived 266
thiourea organocatalysts, cinchona-based 278
titanium silicate-silica gel system 37
- uses 37
(TMS)CF$_3$ 235
TMS-ketene 89
- enantioselective [2 + 2] cycloaddition 89
- dual Lewis acid–Lewis base activation 89
p-toluenesulfonic acid 163
trans-alkene 43
transitional metal-based catalysts 197
transition-state model 93, 112, 212, 243, 251, 287, 288, 332
- DFT-computed structure 332

- quinine-catalyzed Michael addition of thiol 251
triazole-based cinchoninium PTC 146
trichloroisocyanuric acid (TCCA) 110
- stoichiometric oxidant 110
trichloromethyl group, synthetic modifications 304
tricyclic N,O-acetals, formation 390
(N)-trifluoroacetyl-glycine allylic esters 95
- asymmetric ester enolate Claisen rearrangement 95
2,3,4-trifluorobenzyl-incorporated quinidinium bromide 221
trifluoromethylated oxindoles 205
trifluoromethylation 234–237
trifluoromethyl ketones, Henry reaction 208
trifluoropyruvate 237
triisopropylbenzenesulfonic acid (TIPBA) 319
trimethylsilyl cyanide (TMSCN) 83
- asymmetric addition reaction 83
trimethylsilylenol ethers, fluorination 157
tris(trimethoxyphenyl) phosphine (TTMPP) 406
N-Ts aldimines/ketoimines 86
- asymmetric Strecker reaction 86
27-OMs, cascade reactions 410

u

unactivated ketones, acetophenone 208
α,β-unsaturated acid chloride/chloral 303
- cycloaddition 303
α,β-unsaturated amides 249
- conjugate reaction 249–274
α,β-unsaturated electron withdrawing groups 178
α,β-unsaturated esters 192
- Pete's photodeconjugation 192
α,β-unsaturated ketones 249, 316
- asymmetric Diels–Alder reaction 316
- conjugate reaction 249–274
α,β-unsaturated δ-lactones, preparation 304
α,β-unsaturated nitriles, conjugate reaction 249–274
α,β-unsaturated sulfones, epoxidation 117
urea linker motif 450
N-urethane-protected N-carboxyanhydrides (UNCAs) 348, 349
- kinetic resolution 348
- racemization 349
  - bimolecular mechanism 350
- termolecular mechanism, alcoholysis 350
UV absorption 423

v

vinyl ether   39
vinyl Grignard reagents   367
vinyl magnesium bromide   388
vinylogous Michael addition reaction   314
vinyl phosphates   284, 286
– conjugate addition   284–287
5-vinyl-2-quinuclidinone   390
vinyl side chain   368, 376
– basic transformations   368
  – alkyne cinchona alkaloids   368–374
  – degradation   381–382
  – fluorination   374, 375
  – oxidation/oxidative cleavage   375–381
– carbon C11, oxidation   379
– degradation   383
– *en* route   377
– fluorination   376
– oxidative cleavage   377, 378
    – C10-aldehydes   377
    – C10-esters   378

vinyl sulfones   39, 284
– carbon nucleophiles, Michael addition reaction   284
– conjugate addition   284–287

w

weak chiral anion exchangers (WAX)   434
Weinreb amide   375
Wynberg's catalysts   109, 113, 221

x

X-ray crystallographic   141, 143
X-ray diffraction   443

z

Zimmerman–Traxler transition state   87
zinc complex   74
Z-vinyl iodide   371
zwitterionic acyl ammonium enolate   86
zwitterionic enolates   91